NANOHYBRID FUNGICIDES

Nanobiotechnology for Plant Protection

NANOHYBRID FUNGICIDES

New Frontiers in Plant Pathology

Edited by

KAMEL A. ABD-ELSALAM

Plant Pathology Research Institute, Agricultural Research Center, Giza, Egypt

ELSEVIER

Elsevier
Radarweg 29, PO Box 211, 1000 AE Amsterdam, Netherlands
The Boulevard, Langford Lane, Kidlington, Oxford OX5 1GB, United Kingdom
50 Hampshire Street, 5th Floor, Cambridge, MA 02139, United States

ISBN: 978-0-443-23950-2

For information on all Elsevier publications
visit our website at https://www.elsevier.com/books-and-journals

Publisher: Jonathan Simpson
Acquisitions Editor: Nina Bandeira
Editorial Project Manager: Shivangi Mishra
Production Project Manager: Swapna Srinivasan
Cover Designer: Vicky Pearson Esser

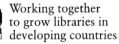

Typeset by STRAIVE, India

Contents

5. Nanoencapsulation of fungicides: New trend in plant disease control

Pallavi Nayak

6. Antifungal potential of nano- and microencapsulated phytochemical compounds and their impact on plant heath

Nasreen Musheer, Anam Choudhary, Arshi Jamil, and Sabiha Saeed

7. The antagonistic yeasts: Novel nano/biofungicides for controlling plant pathogens

Parissa Taheri, Saeed Tarighi, and Farah K. Ahmed

8. Antifungal activity of microbial secondary metabolites

Ragini Bodade and Krutika Lonkar

9. Fungal metabolites as novel plant pathogen antagonists

Jagriti Singh, Shweta Mishra, and Vineeta Singh

10. Trichoderma secondary metabolites for effective plant pathogen control

A.D. Lopes, W.R. Rivadavea, and G.J. Silva

11. Exploring biological control strategies for managing *Fusarium* mycotoxins

Mirza Abid Mehmood, Areeba Rauf, Muhammad Ashfaq, and Furqan Ahmad

2

General applications, commercialization, and remediation

12. Applications of nanofungicides in plant diseases control

Rajkuberan Chandrasekaran, P. Rajiv, Farah K. Ahmed, and Karungan Selvaraj Vijai Selvaraj

13. Nanostructures for fungal disease management in the agri-food industry

R. Britto Hurtado, S. Horta-Piñeres, J.M. Gutierrez Villarreal, M. Cortez-Valadez, and M. Flores-Acosta

14. Nano-biofungicides for the reduction of mycotoxin contamination in food and feed

Mohamed Amine Gacem, Badreddine Boudjemaa, Valeria Terzi, Aminata Ould El Hadj-Khelil, and Kamel A. Abd-Elsalam

15. Nanobiopesticides: Significance, preparation technologies, safety aspects, and commercialization for sustainable agriculture

P. Karthik, A. Saravanaraj, V. Vijayalakshmi, K.V. Ragavan, and Vinoth Kumar Vaidyanathan

Contributors

Kamel A. Abd-Elsalam Plant Pathology Research Institute, Agricultural Research Center, Giza, Egypt

Cristóbal Noé Aguilar-González Department of Food Research, Chemical Sciences School of the Autonomous University of Coahuila, Saltillo, Coahuila, Mexico

Furqan Ahmad Institute of Plant Breeding and Biotechnology, Muhammad Nawaz Shareef University of Agriculture, Multan, Pakistan

Farah K. Ahmed Biotechnology English Program, Faculty of Agriculture, Cairo University, Giza, Egypt

Mousa A. Alghuthaymi Biology Department, Science and Humanities College, Shaqra University, Alquwayiyah, Saudi Arabia

Roberto Arredondo-Valdés Nanobioscience Group, Chemical Sciences School of the Autonomous University of Coahuila, Saltillo, Coahuila, Mexico

Muhammad Ashfaq Plant Pathology, Institute of Plant Protection, Muhammad Nawaz Shareef University of Agriculture, Multan, Pakistan

Ragini Bodade Department of Microbiology, Savitribai Phule Pune University, Pune, Maharashtra, India

Badreddine Boudjemaa Department of Biology, Faculty of Science, University of Amar Tlidji, Laghouat, Algeria

R. Britto Hurtado Department of Physics Research, University of Sonora, Hermosillo, Sonora, Mexico

Rajkuberan Chandrasekaran Department of Biotechnology, Karpagam Academy of Higher Education, Coimbatore, Tamil Nadu, India

Anam Choudhary Aligarh Muslim University, Aligarh, Uttar Pradesh, India

M. Cortez-Valadez CONACYT—Department of Physics Research, University of Sonora, Hermosillo, Sonora, Mexico

Jéssica de Matos Fonseca Department of Chemical and Food Engineering, Federal University of Santa Catarina, Florianópolis, Brazil

Aminata Ould El Hadj-Khelil Laboratory of Ecosystems Protection in Arid and Semi-Arid Area, University of Kasdi Merbah, Ouargla, Algeria

Alshaimaa M. Elsayed Molecular Biology Department, Biotechnology Research Institute, National Research Centre, Giza, Egypt

Liliana Sofía Farías-Vázquez Nanobioscience Group, Chemical Sciences School of the Autonomous University of Coahuila, Saltillo, Coahuila, Mexico

M. Flores-Acosta Department of Physics Research, University of Sonora, Hermosillo, Sonora, Mexico

Mohamed Amine Gacem Department of Biology, Faculty of Science, University of Amar Tlidji, Laghouat, Algeria

Prabuddha Ganguli Vision-IPR, Mumbai, India and Adjunct Faculty, Indian Institute of Technology, Jodhpur, Rajasthan, India

J.M. Gutierrez Villarreal Technological University of South Sonora, Obregon, Sonora, Mexico

Ayat F. Hashim Fats and Oils Department, Food Industries and Nutrition Research Institute, National Research Centre, Giza, Egypt

S. Horta-Piñeres Physics Department, Popular University of Cesar, Valledupar, Colombia

Anna Iliná Nanobioscience Group, Chemical Sciences School of the Autonomous University of Coahuila, Saltillo, Coahuila, Mexico

Avinash P. Ingle Biotechnology Centre, Department of Agricultural Botany, Dr. Panjabrao Deshmukh Agricultural University, Akola, Maharashtra, India

Arshi Jamil Aligarh Muslim University, Aligarh, Uttar Pradesh, India

P. Karthik Centre for Food Nanotechnology, Department of Food Technology, Faculty of Engineering, Karpagam Academy of Higher Education (Deemed to be University), Coimbatore, India

Bipin D. Lade Vegetable and Fruit Improvement Center, USDA National Center of Excellence Department of Horticultural Sciences, Texas A&M University, College Station, TX, United States

Krutika Lonkar Department of Microbiology, Savitribai Phule Pune University, Pune, Maharashtra, India

A.D. Lopes Biotechnology Department, Post-Graduate Program in Biotechnology Applied to Agriculture, Paranaense University, Umuarama, Paraná, Brazil

Amanda Galvão Maciel Department of Chemical and Food Engineering, Federal University of Santa Catarina, Florianópolis, Brazil

José Luis Martínez-Hernández Nanobioscience Group, Chemical Sciences School of the Autonomous University of Coahuila, Saltillo, Coahuila, Mexico

Mirza Abid Mehmood Plant Pathology, Institute of Plant Protection, Muhammad Nawaz Shareef University of Agriculture, Multan, Pakistan

Shweta Mishra Department of Biotechnology, Institute of Engineering and Technology, Dr. A.P.J. Abdul Kalam Technical University, Lucknow, Uttar Pradesh, India

Mangesh Moharil Biotechnology Centre, Department of Agricultural Botany, Dr. Panjabrao Deshmukh Agricultural University, Akola, Maharashtra, India

Alcilene Rodrigues Monteiro Department of Chemical and Food Engineering, Federal University of Santa Catarina, Florianópolis, Brazil

Nasreen Musheer Glocal University, Saharanpur, Uttar Pradesh, India

Pallavi Nayak Nuclear Medicine Unit, University Hospital Sant'Andrea; Department of Medical-Surgical Sciences and of Translational Medicine, Faculty of Medicine and Psychology, "Sapienza" University of Rome, Rome, Italy

Abdelmageed M. Othman Microbial Chemistry Department, Biotechnology Research Institute, National Research Centre, Giza, Egypt

Sandra Pacios-Michelena Nanobioscience Group, Chemical Sciences School of the Autonomous University of Coahuila, Saltillo, Coahuila, Mexico

Bhimanagouda S. Patil Vegetable and Fruit Improvement Center, USDA National Center of Excellence Department of Horticultural Sciences, Texas A&M University, College Station, TX, United States

K.V. Ragavan Agro-Processing and Technology Division, CSIR-National Institute for Interdisciplinary Science and Technology, Thiruvananthapuram; Academy of Scientific and Innovative Research (AcSIR), Ghaziabad, India

P. Rajiv Department of Biotechnology, PSG College of Arts & Science, Coimbatore, Tamil Nadu, India

Rodolfo Ramos-González CONAHCYT-Autonomous University of Coahuila, Saltillo, Coahuila, Mexico

Areeba Rauf Plant Pathology, Institute of Plant Protection, Muhammad Nawaz Shareef University of Agriculture, Multan, Pakistan

W.R. Rivadavea Agronomy Department, Graduate program in Agronomy Engineering,

Paranaense University, Umuarama, Paraná, Brazil

Raúl Rodríguez-Herrera Department of Food Research, Chemical Sciences School of the Autonomous University of Coahuila, Saltillo, Coahuila, Mexico

Sabiha Saeed Aligarh Muslim University, Aligarh, Uttar Pradesh, India

A. Saravanaraj Department of Chemical Engineering, Vel Tech High Tech Dr. Rangarajan Dr. Sakunthala Engineering College, Chennai, India

Elda Patricia Segura-Ceniceros Nanobioscience Group, Chemical Sciences School of the Autonomous University of Coahuila, Saltillo, Coahuila, Mexico

G.J. Silva Biotechnology Department, Post-Graduate Program in Biotechnology Applied to Agriculture, Paranaense University, Toledo, Paraná, Brazil

Jagriti Singh Department of Biotechnology, Institute of Engineering and Technology, Dr. A.P.J. Abdul Kalam Technical University, Lucknow, Uttar Pradesh, India

Vineeta Singh Department of Biotechnology, Institute of Engineering and Technology, Dr. A.P.J. Abdul Kalam Technical University, Lucknow, Uttar Pradesh, India

Parissa Taheri Department of Plant Protection, Faculty of Agriculture, Ferdowsi University of Mashhad, Mashhad, Iran

Saeed Tarighi Department of Plant Protection, Faculty of Agriculture, Ferdowsi University of Mashhad, Mashhad, Iran

Valeria Terzi Council for Research in Agriculture and Economics, Research Centre for Genomics and Bioinformatics (CREA-GB), Fiorenzuola d'Arda, Italy

Vinoth Kumar Vaidyanathan Integrated Bioprocessing Laboratory, Department of Biotechnology, School of Bioengineering, Faculty of Engineering and Technology, SRM Institute of Science and Technology (SRM IST), Kattankulathur, Chengalpattu, Tamil Nadu, India

Karungan Selvaraj Vijai Selvaraj Vegetable Research Station, Tamil Nadu Agricultural University, Palur, Cuddalore, Tamil Nadu, India

V. Vijayalakshmi Department of Biotechnology, Vel Tech High Tech Dr. Rangarajan Dr. Sakunthala Engineering College, Chennai, India

Khamis Youssef Plant Pathology Research Institute, Agricultural Research Center (ARC), Giza; Agricultural and Food Research Council, Academy of Scientific Research and Technology, Cairo, Egypt

Preface

Powerful fungicides are required to combat the myriad economic and environmental problems that pathogenic fungi cause around the world. Research and development on new fungicides are crucial in the fight against hazardous fungal strains that are quickly evolving resistance. Recent studies have investigated several nanomaterials, including polymers, lipids, essential oils, and metallic NPs such as Cu, S, Se, Si, Ag, and Zn, with encouraging results for reducing the use of synthetic fungicides in the treatment of plant pathogenic fungi. Because of their solubility, permeability, low dose-dependent toxicity, low dose, increased bioavailability, targeted delivery, increased bioavailability, and controlled release, nanofungicides are effective. Therefore, it is anticipated that nano- and hybrid nanofungicides will meet the needs of growers, consumers, and environmental activists by providing quick, effective, and comparably improved ecosafety attributes for reducing the potential of fungal phytopathogens to reduce the quality and yield of produce. The book "**Nanohybrid Fungicides: New Frontiers in Plant Pathology**" discusses the knowledge, discoveries, and beneficial results of nano- and hybrid nanofungicides and their applications in agro-ecosystems. Nanohybrid fungicides represent a new frontier in plant pathogen control. These compounds combine the benefits of standard fungicides with the unique features of nanomaterials, resulting in more effective, targeted treatments that are less likely to cause environmental damage. This collection includes 18 chapters written by distinguished scholars from Algeria, Brazil, Bangladesh, Chile, Iran, India, Italy, Egypt, Mexico, Turkey, Malaysia, Nigeria, Oman, Pakistan, Poland, Slovakia, South Africa, South Korea, and the United States.

Part 1 of the proposed research project focuses on exploring various types of nanohybrid fungicides, including: A case study on nano-based fungicides: This section will present a detailed analysis of fungicides that incorporate nanotechnology, examining their effectiveness in managing fungal plant diseases. Chitosan-based agro-nanofungicides: A sustainable alternative in fungal plant disease management: This case study will investigate the use of chitosan-based nanofungicides as an environmentally friendly and sustainable approach to combat fungal plant diseases. Gum nanocomposites for postharvest fungal disease control in fruits: This part will explore the application of gum nanocomposites in controlling postharvest fungal diseases in fruits, assessing their efficacy and potential for commercial use. Nanoencapsulation applications in fungicides: This section will delve into the applications of nanoencapsulation in fungicides, studying their ability to enhance the targeted delivery of active ingredients and improve the efficacy of fungal disease management. Antifungal potential of nano- and microencapsulated phytochemical compounds and their impact on plant health: This case study will examine the antifungal properties of nano- and microencapsulated phytochemical compounds, evaluating their impact on plant health and disease prevention. The antagonistic yeasts: As novel and safe biological agents: This part of the research will focus on the exploration of antagonistic yeasts as promising and safe biological agents for controlling fungal plant pathogens, highlighting their potential in integrated disease management strategies. Nano/biofungicides for controlling fungal plant pathogens: This section will discuss the development and utilization of nano/biofungicides, which combine nanotechnology and biological agents, as effective solutions for managing fungal plant pathogens. By examining these various types of nanohybrid fungicides, the research aims to contribute to the understanding and development of innovative and sustainable approaches for fungal plant disease management.

Part 2 of the proposed research project focuses on the general applications, commercialization, and remediation aspects related to nanohybrid fungicides, including: Applications of nanofungicides in plant disease control: This section examines the various applications of nanofungicides and their effectiveness in managing plant diseases, highlighting their potential impact on agricultural practices. Nanoarchitectures for fungal disease management in the agri-food sector: The research will explore nanoarchitectures designed specifically for managing fungal diseases in the agri-food sector, discussing their potential benefits and applications. Nanobiofungicides for the reduction of mycotoxin contamination in food and feed: This part focuses on the use of nanobiofungicides to mitigate mycotoxin contamination in food and feed, evaluating their efficacy and potential for commercial use. Nanobiopesticides: Some bioactive materials used in the management of plant pathogens: This section investigates the use of bioactive materials in nanobiopesticides for effectively managing plant pathogens, highlighting their significance and potential applications. Fungal metabolites as novel plant pathogen antagonists: The research explores the potential of fungal metabolites as innovative plant pathogen antagonists, highlighting their effectiveness and applications in disease management. Secondary metabolites from *Trichoderma* for effective plant pathogen control: This part focuses on secondary metabolites derived from *Trichoderma* species and their applications in controlling plant pathogens, discussing their efficacy and potential for commercialization. Use of biocontrol agents for Fusarium mycotoxins: use potential as biocontrol agents, significance, preparation technologies, safety aspects, and commercialization for sustainable agriculture. A nanoagrochemicals start-up providing sustainable crop protection options: This case study examines a nanoagrochemicals start-up that offers sustainable crop protection options, highlighting their innovative approaches and contributions to the agricultural sector. Patent landscape in biofungicides, nanofungicides, and nanobiofungicides: This research will analyze the patent landscape in the field of biofungicides, nanofungicides, and nanobiofungicides, providing insights into the intellectual property landscape and technological trends. Microbial bioremediation of fungicides: This section focuses on microbial bioremediation as a potential remediation approach for fungicides, exploring its effectiveness and applications in reducing environmental impacts. By exploring these topics, Part 2 aims to provide a comprehensive understanding of the general applications, commercialization potential, and remediation aspects associated with nanohybrid fungicides.

The nanohybrid fungicide volume is of an interdisciplinary nature and will be very useful for students, teachers, and researchers, agri-food environmental scientists, and agrochemical companies working in nanotechnology, materials science, biology, chemistry, physics, plant pathology, chemical technology, microbiology, plant physiology, biotechnology, and further groups of interests such as the agrochemicals industry. Additionally, this will be a useful tool for industrial scientists investigating technology to update their understanding of antifungal activity, mechanisms, nanotoxicology, and nanosafety. Several new topics will be covered, including the application of nanotechnology in plant disease management, antifungal nanotherapy, mycotoxin reduction, and veterinary applications. We really hope that we have provided a balanced, fascinating, and innovation-based perspective not just for expert readers but also for industrial decision-makers and others with limited knowledge.

The expected audience comprises researchers, graduates, postgraduate students, and other agrochemical businesses from many disciplines of science and technology. Our objective is to provide a reliable and informed perspective that appeals to both experts and nonexperts, including technical decision-makers and individuals with limited knowledge of the subject matter. The target audience for this book is undergraduate and postgraduate students from various scientific and technical sectors.

We are thankful to the writers for allowing us to exhibit this high-quality collection of papers. We thank all the writers who contributed recommendations and relevant experiences to the book's chapters, and this edited book. This book would not have been possible without their cooperation and devotion. Elsevier is likewise well regarded, owing to the project's extensive expertise, dependability, and tolerance. First and foremost, we would like to express our gratitude to all Elsevier employees who collaborated with us and made an effort to help us, especially Nancy Maragioglio (Senior Acquisitions Editor), Moises Carlo Catain, and Howell Angelo M. De Ramos (Editorial Project Manager). We appreciate everyone who took the time to discuss each chapter in the reviews. We also acknowledge the family members for their ongoing support and attention.

Kamel A. Abd-Elsalam
Plant Pathology Research Institute, Agricultural Research Center, Giza, Egypt

Series Preface

The field application of engineered nanomaterials (ENMs) has not been well investigated in plant promotion and protection in the agro-environment yet. Many components have only been taken into consideration theoretically or with prototypes, which makes it hard to evaluate the utility of ENMs for plant promotion and protection. Examples of nanotechnology applications in the food industry include encapsulation and delivery of materials to targeted sites; flavor development; introducing nanoantimicrobial agents into food; improving shelf life; sensing contamination; improving food preservatives; and monitoring, tracing, and logo protection. The list of environmental problems that the world faces may be huge, but the few strategies for fixing them are not. Scientists worldwide are developing nanomaterials that could use selected nanomaterials to capture poisonous pollutants from water and degrade solid waste into useful products. The market intake of nanomaterials is booming, and the Freedonia Group predicts that nanostructures will grow to $100 billion through 2025. Nanotechnology research and development have been growing on a steep slope across all scientific disciplines and industries. Based on this background, the scientific series "Nanobiotechnology for Plant Protection" was inspired by the desire of the editor, Kamel A. Abd-Elsalam, to put together detailed, up-to-date, and applicable studies on the field of nanobiotechnology applications in agroecosystems to foster awareness and extend our view of future perspectives.

The main appeal of this book series is its specific focus on plant protection in agri-food and the environment, which is one of the most topical nexus areas in the many challenges faced by humanity today. The discovery and highlighting of new crop inputs based on nanobiotechnology that can be used at lower application rates will be critical to eco-agriculture sustainability. The research in the relevant fields is dispersed and not concentrated in a single location. This book series will cover the applications in the agri-food and environmental sectors, which are the new topics of research in the field of nanobiotechnology. This series will be a comprehensive account of the literature on specific nanomaterials and their applications in agriculture, food, and the environment. The audience will be able to gather information from a single series. Students, teachers, and researchers from colleges, universities, and research institutes as well as industry will benefit from this series. Four specific features make the current series unique. First, the series has a very specific editorial focus, so researchers can locate nanotechnology information precisely without looking into various full-text sources. Second, and more importantly, the series offers a crucial evaluation of the content material along with nanomaterials, technologies, applications, methods, and equipment as well as safety and regulatory aspects in agri-food and environmental sciences. Third, the series provides the reader with content precision; it will provide nanoscientists with clarity and deep information. Finally, presenting researchers with insights on discoveries, this series gives researchers a sense of what to do, what they need to do, and how to do it properly by searching for others who have done it. The book "**Nanohybrid Fungicides: New Frontiers in Plant Pathology**" discusses the knowledge, discoveries, and beneficial results of nano- and hybrid nanofungicides and their applications in agroecosystems. Nanohybrid fungicides are a cutting-edge development in plant pathology with great promise for combating plant diseases. These cutting-edge fungicides use hybrid materials and nanotechnology to prevent disease more effectively and efficiently in crops. The expected readership for the current series is researchers in the fields of environmental science, food science, and agriculture science. Some readers may also come from chemists, those in the green chemistry industry, material scientists, government regulatory agencies, agro and food industry players, and academicians. A few readers from the industrial personnel field would also be interested. This series is useful to a wide audience of food, agriculture, and environmental sciences researchers, including undergraduate and graduate students, postgraduates, etc. In addition, agricultural producers could benefit from the applied knowledge that will be highlighted in the book, which, otherwise, would be buried in different journals. Both primary and secondary audiences are seeking up-to-date knowledge of the applications of nanotechnology in environmental science, agriculture, and food science. It is a trending area, and lots of new studies get published every week. Readers need some good summaries to help them learn the latest key findings, which could be review articles and/or books. This series will help to put these pockets of knowledge together and make them more easily accessible globally.

Kamel A. Abd-Elsalam
Plant Pathology Research Institute, Agricultural Research Center, Giza, Egypt

PART 1

Types of nanohybrid fungicides

CHAPTER

1

Next-generation fungicides based on nanohybrids: A preliminary review

Kamel A. Abd-Elsalam

Plant Pathology Research Institute, Agricultural Research Center, Giza, Egypt

1 Introduction

Every year, fungal diseases can cause significant reductions in crop yields around the world (Worrall et al., 2018). Currently, disease control relies on the use of agrochemicals such as fungicides. As a result, the use of synthetic pesticides has become an essential aspect of agriculture. Since the first synthetic fungicide, phenylmercury acetate, was discovered in 1913, over 110 unique fungicides have been created, resulting in an increase in food output worth USD 12.8 billion annually in the United States (Carvalho, 2017). Nanotechnology can help by lowering the harmful effects of fungicides in a sustainable and environmentally benign way, for example, by making low water-soluble fungicides more soluble, extending their shelf life, and reducing their toxicity (Kutawa et al., 2021). Many researchers have created nanoparticles with various properties such as pore size, surface properties, and shape that can be used as protectants or for precise administration via encapsulation and adsorption of an active substance (Maluin et al., 2020). The nanohybrid antifungal is expected to meet the needs of producers, consumers, and environmental activists by providing quick, effective, and comparatively improved ecosafety features for managing yield and producing quality by repelling the fungal plant diseases (Alghuthaymi et al., 2021). Finally, when fungicide-compatible biocontrol agents are integrated with some nanosystems, fungal disease control can be improved. Because the risks of resistance development are lower and the fungicide dose may be reduced compared to standard treatment with single fungicides, these treatments may offer the potential to develop novel antifungal techniques for integrated pest control. Secondary metabolites derived from biological materials, for example, are essential in the production of nanofungicides capable of killing fungi or reducing the proliferation of fungal diseases (Periakaruppan et al., 2023).

https://doi.org/10.1016/B978-0-443-23950-2.00017-5

The title of Volume 2 is *Nanohybrid Fungicides: New Frontiers in Plant Pathology*, and it suggests that the information, discoveries, and beneficial results of nano- and hybrid nanofungicides and their applications in agroecosystems have been compiled in 18 chapters. Also included in this volume is a discussion of the implications of nanofungicides for plant pathology. First, we discuss many forms of nanohybrid fungicides, such as "A Case Study on Nano-based Fungicides," "Chitosan-Based Agronanofungicides: A Sustainable Alternative in Fungal Plant Diseases Management," and "Nanohybrid Fungicides: An Emerging Class of Fungicides." Nanocomposites were employed for the prevention of postharvest fungal diseases in fruits. For instance, nano- and microencapsulated phytochemical substances were used to develop new fungicides and investigate the effect these compounds have on the health of plants. The antagonistic yeasts have been proposed as potential innovative biological agents, nano/biofungicides, for the control of fungal plant diseases. Part 2 outlines general uses, commercialization, and remediation, such as applications of nanofungicides in the control of plant diseases. Use nanoarchitectures for fungal disease management in the agrifood sector. Employ nanobiofungicides for the reduction of mycotoxin contamination in food and feed. Develop nanobiopesticides: some bioactive materials used in the management of plant pathogens, such as fungal metabolites, as novel plant pathogen antagonists. Investigate the secondary metabolites from Trichoderma for effective plant pathogen conquest, in addition to managing Fusarium mycotoxins using some biocontrol agents. Also, the significance, preparation technologies, safety aspects, and commercialization of nanofungicides for sustainable agriculture were demonstrated. This volume covers various topics, including sustainable crop protection options offered by a nanoagrochemicals start-up, an analysis of the patent landscape in biofungicides, nanofungicides, and nanobiofungicides, as well as the application of microbial bioremediation in fungicide management. The current volume focuses on the use of various nanodelivery methods loaded with fungicide active ingredients in nanoformulation, nanocapsulations, and nanostructures in agricultural applications. These nanoparticles are more reactive and can create covalent connections with fungicides or biofungicides. One advantage of nanohybrid fungicides is that they can be more effective at lower doses than traditional fungicides, reducing the amount of chemicals needed to control fungal infections. This can help to reduce the environmental impact of fungicide use and minimize the risk of developing fungicide-resistant strains of fungi.

2 Types of nanopesticides

There are two primary types of nanopesticides: Type 1 contains metallic nanoparticles such as silver, copper, and titanium, which do not have additional carriers, whereas Type 2 includes active substances in nanocarriers such as polymers and clays, or in the form of an emulsion or liposome. Additionally, Type 3 combines application methods from Types 1 and 2. Type 1 nanopesticides seem to have the most significant influence on antimicrobial activity due to their adhesion, oxidative stress, and genotoxicity-induced cell death (Vasseghian et al., 2022). However, Type 2 focuses on the delivery platform and aims to reach sustainable agriculture goals through the encapsulation and release of active substances.

3 Nanohybrid fungicides

A substance that combines two or more different ingredients is called a hybrid (Kickelbick, 2006). Although different arrangements have been recorded, typically one of the ingredients is organic and the other is inorganic (Kaur et al., 2015). When a polymer and a nanomaterial are present in a material, it is referred to as a nanocomposite or nanostructured polymeric hybrid material (Kaur et al., 2015; Baigorria et al., 2021; Baigorria and Fraceto, 2022). The term "hybrid material," according to Kickelbick (2006), refers to a broad spectrum of systems that include substances like organized crystalline coordination polymers, amorphous sol-gel composites, and substances with or without interactions between inorganic and organic units. Because the blend of these materials' components is intrinsic to their organic and/or inorganic phases rather than just a physical combination, it greatly influences their characteristics and attributes (Kaur et al., 2015).

Because they combine qualities like the flexibility of the organic portions and stiffness/stiffness of the inorganic halves, organic/inorganic hybrid materials have attracted the attention of the scientific community (Kaur et al., 2015). They have been divided into groups depending on the interactions that may occur between the organic and inorganic parts of the material as well as interactions between its phases. Van der Waals interactions, electrostatic connections, or hydrogen bonds are weak contacts between the phases of class I hybrid materials, whereas chemical interactions are stronger in class II hybrid materials. By considering their structural characteristics, hybrid materials can be divided into other subcategories (Kaur et al., 2015). An organic phase with functional groups that can bond to an inorganic network, for instance, or mixtures of organic and inorganic materials with unconnected organic and inorganic building blocks are examples of hybrid materials. An example of the latter is a polymeric hybrid material produced by an organic polymer that has discrete inorganic components trapped inside of it. Inorganic particles that are trapped in the polymeric matrix or by physical interactions in these materials exhibit weak crosslinking (Kaur et al., 2015). Materials with nanostructured polymeric hybrids can be produced synthetically in various methods. Some authors refer to two different approaches. The first method is known as "building block," in which the elements are applied in a preset way to interact with one another and produce the finished hybrid material. They thus either generate new structural units with fresh properties or retain part of the original integrity of their forebears. The second tactic, known as in situ component production, involves a chemical transformation of the precursors while the material is being manufactured (Kaur et al., 2015).

Nanostructured hybrid materials are among the most significant advances in chemistry in recent decades. The numerous possibilities for integrating the features of isolated materials into a single hybrid material have sparked a tremendous burst of applications. One of these applications is the utilization of nanostructured polymeric hybrid materials for pesticide removal from water basins.

The majority of nanocarriers are polymer- and clay-based nanocarriers that are biocompatible, cheap, and stimulus-responsive. While clay-based mesoporous silica and montmorillonite have demonstrated high active ingredient encapsulation capability, natural polymers such as chitosan, cellulose, and polylactide are frequently used to create nanocapsules, nanospheres, nanohydrogels, and nanomicelles containing active ingredients (Wang et al., 2022).

Pathogens struggle to develop true resistance to biofungicides because they usually employ two or more nonspecific mechanisms of action. As a result, they are good rotating partners for many synthetic fungicides because they can help prevent or delay the development of fungicide resistance. They work best against certain crop diseases when supplied alone at low to moderate pathogen levels under normal disease development conditions. They are unlikely to be successful in instances where disease pressures are severe, such as when pathogen levels are high and conditions become suddenly highly favorable to disease development.

The terms "nanohybrid fungicides" and "hybrid nanofungicides" are vague and might be used interchangeably, leading to misunderstanding. However, certain distinctions may be drawn from the terminology. A nanohybrid fungicide combines two or more types of nanoparticles or nanomaterials to produce a hybrid material with increased antifungal activity. A nanohybrid fungicide, for example, might mix silver nanoparticles and zinc oxide nanoparticles to produce a substance with broad-spectrum antifungal activity. A hybrid nanofungicide, on the other hand, is a fungicide that mixes a nanoparticle or nanomaterial with a conventional fungicide to produce a hybrid material with higher efficacy. A hybrid nanofungicide, for example, may mix a copper-based fungicide with copper nanoparticles to produce a substance with increased antifungal action and lower environmental effects. In conclusion, while the phrases "nanohybrid fungicides" and "hybrid nanofungicides" can be used interchangeably, they may refer to slightly distinct approaches to developing fungicides with improved efficacy and lower environmental impact. Fungicides are classified into different types based on their structures and resources. Table 1 displays some meanings for the terms they employ. However, because nanoparticles might have unique features that may pose dangers to human health and the environment, it is critical to carefully analyze the safety and potential risks of utilizing nanohybrid fungicides. Before nanohybrid fungicides and other nanotechnology-based treatments may be certified for use, regulatory organizations such as the Environmental Protection Agency (EPA) must assess their safety.

As a result, they are often used to treat infections in conjunction with appropriate synthetic and biological fungicides. The structure and resources of nanohybrid fungicides are split into three classes, as shown in Fig. 1. A hybrid solution differs from typical fungicides in that it creates a long-lasting and very effective solution for managing plant pathogenic fungi.

4 Bioactive materials nanoparticles

4.1 Plant or microbe-derived compounds

Secondary metabolites, such as poisons and enzymes, are produced by some plants and microorganisms and contribute to their virulence and pathogenicity. These substances can also harm human and animal health, as well as the environment. Understanding the interactions between plants and their pathogens, as well as the involvement of secondary metabolites in these interactions, is critical for designing efficient disease control and management techniques. Terpenoids, phenolics, and alkaloids are the three biosynthetic groups of secondary metabolites found in plants (Eljounaidi and Lichman, 2020). Terpenes are a large and

TABLE 1 Definition of terms used for different types of fungicides based on structures and resources.

Chemical fungicides	Chemical fungicides are synthetic compounds that are designed to kill or inhibit the growth of pathogenic fungi. They are typically effective against a wide range of fungal pathogens, but they can also have negative impacts on the environment and human health.
Biofungicides	Biofungicides, are derived from natural sources such as bacteria, fungi, and plants. They work by either directly killing the fungal pathogen or by stimulating the plant's natural defense mechanisms. Biofungicides are generally considered to be safer and more environmentally friendly than chemical fungicides, but they may not be as effective against all types of fungal pathogens.
Nanofungicides	Nanofungicides are fungicides that incorporate nanoparticles into their formulation. These nanoparticles can enhance the fungicidal activity of the compound, improve its stability, and reduce its environmental impact. However, there are concerns about the potential toxicity of nanoparticles and their impact on the environment.
Nano-biofungicides	Nano-biofungicides are a combination of biofungicides and nanofungicides. They are designed to provide the benefits of both types of fungicides, including enhanced efficacy and reduced environmental impact. However, more research is needed to fully understand the potential risks and benefits of nano biofungicides.
Nanohybrid fungicides	A nanohybrid fungicides, typically refers to a fungicide that combines two or more different types of nanoparticles or nanomaterials to create a hybrid material with enhanced antifungal activity. For example, a nanohybrid fungicide may combine silver nanoparticles and zinc oxide nanoparticles to create a material with broad-spectrum antifungal activity.
Hybrid nanofungicides	A hybrid nanofungicides typically refers to a fungicide that combines a nanoparticle or nanomaterial with a conventional fungicide to create a hybrid material with improved efficacy. For example, a hybrid nanofungicide may combine a copper-based fungicide with copper nanoparticles to create a material with improved antifungal activity and reduced environmental impact.

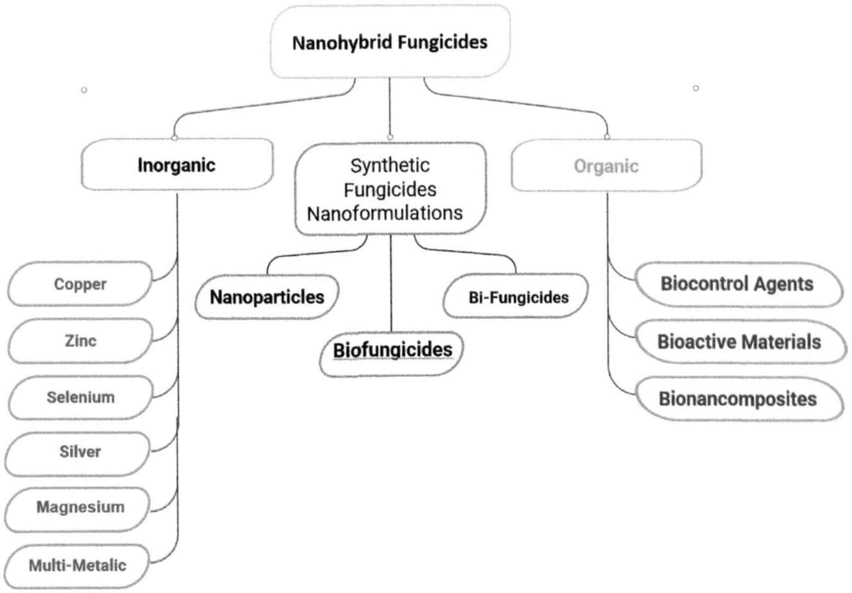

FIG. 1 Classifications of nanohybrid fungicides based on organic, inorganic, and synthetic structure and resources.

diversified class of naturally occurring, highly enriched secondary plant metabolites. These chemicals' antifungal effects have been examined for two decades, and various investigations have found that thymol, carvacrol, eugenol, and menthol have significant activity against several infections (Kusakizako et al., 2020). Curcumin-loaded electrospun zein nanofibers (CLZN), for example, were utilized to coat apple fruits contaminated with *Penicillium expansum* and *Botrytis cinerea*. CLZN mats pave the way for novel uses of edible and biodegradable antifungal protective materials capable of reducing fungus development in covered apples during storage (Yilmaz et al., 2016).

Secondary metabolites combined with nanotechnology have the potential to improve the properties and applications of these natural substances. Nanotechnology is the modification of materials at the nanoscale, which can improve secondary metabolite solubility, stability, and bioavailability, as well as their tailored transport to specific cells or tissues. Nanoparticles, for example, can be utilized to encapsulate secondary metabolites from plants or microbes, preventing degradation and enhancing absorption and distribution in the body. Nanoparticles can also be functionalized with ligands or antibodies that target specific cells or tissues, making secondary metabolite distribution more efficient. Furthermore, combining secondary metabolites with nanotechnology can result in the development of novel materials with distinct features and applications. Secondary metabolites, for example, can be employed to construct nanomaterials with specialized functionalities, such as antibacterial or antioxidant activity. Overall, the combination of secondary metabolites and nanotechnology has the potential to revolutionize the natural product area and lead to the development of new and novel products with superior features and applications. Combinations of different plant species, as well as mixes of different phytochemicals and microbial secondary metabolites, have been found to exhibit antifungal efficacy against a variety of plant diseases with varying modes of action.

5 Biocontrol agents combined with NPs

Recent worries about the use of chemical pesticides, as well as the growing need for low-input sustainable agriculture, have pushed the use of microbial biocontrol agents and biopesticides against plant pathogenic fungi to the forefront. Biocontrol agents are microorganisms that can be employed to control plant diseases by inhibiting plant pathogen growth and activity (Li et al., 2019). Biocontrol drugs can be more successful in managing plant diseases when coupled with nanoparticles (NPs). NPs can improve the delivery and uptake of biocontrol chemicals by plants, hence increasing their efficacy. Furthermore, NPs can shield biocontrol agents from environmental challenges such as UV radiation and temperature variations, enhancing their survival and activity in the field. Metal NPs, metal oxide NPs, and carbon-based NPs are among the NPs that can be utilized in conjunction with biocontrol agents. Silver nanoparticles, for example, have been found to improve the efficacy of biocontrol agents against plant diseases by boosting their adherence to plant surfaces and penetration into plant tissues. Similarly, zinc oxide nanoparticles have been found to enhance the antifungal action of biocontrol agents against fungal infections.

Biocontrol agents have enormous potential as a novel method, either alone or in combination (Patibanda and Ranganathswamy, 2018). For example, using *Trichoderma* genus for the synthesis of metallic nanoparticles is environmentally friendly, time-saving, and cost-effective. Additionally, possible agrochemicals can be created by combining NPs with *Trichoderma* strains to create more sustainable goods (Alghuthaymi et al., 2022). The effect of metal-chitosan nanocomposites at $100\,g\,mL^{-1}$ paired with Cu-tolerant *Trichoderma longibrachiatum* strains on cotton seedling damping-off under greenhouse conditions was also investigated. Cotton seedling disease induced by *Rhizoctonia solani* may be suppressed in vivo by Cu-chitosan nanocomposite and bimetallic blends mixed with *Trichoderma* (Abd-Elsalam et al., 2018). A sort of agricultural technology that combines the use of helpful microorganisms (biocontrol agents) with nanoparticles to improve their effectiveness in managing plant diseases is biocontrol agents mixed with nanoparticles. Biocontrol agents are naturally occurring microorganisms, such as bacteria and fungi, that have the ability to suppress plant diseases while promoting plant development. However, the use of NPs in conjunction with biocontrol agents raises issues concerning nontarget organisms and environmental toxicity. As a result, before NPs are widely used in agriculture, it is critical to carefully assess the benefits and hazards of employing them in conjunction with biocontrol agents.

6 Synthetic fungicides combined with metal NPs

Metal NPs with antibacterial capabilities, such as silver and copper NPs, have been demonstrated to increase the efficacy of fungicides. When fungicides are mixed with metal nanoparticles, the combined activity is larger than the sum of the individual activities. As a result, smaller fungicide doses may be necessary to achieve the same degree of disease control, reducing the environmental impact of fungicide use. However, there are worries regarding metal NPs' potential toxicity to nontarget creatures and the environment, which must be thoroughly investigated. Metal nanoparticles have been demonstrated to have a considerable toxic effect on fungal diseases that are resistant to traditional fungicides whether used alone or in combination with fungicides such as thiram, tebuconazole, propineb, carbendazim, fludioxonil, and mancozeb (Malandrakis et al., 2020, 2021). Despite the resistant phenotype, a synergistic pattern between Cu-NPs and the oxidative phosphorylation inhibitor fluazinam (FM) was found, suggesting that Cu-NPs' method of action may include ATP metabolism. Further evidence for the significance of copper ions in the fungitoxic action of Cu-NPs was provided by the observed cross-sensitivity and antagonistic activity of Cu-NPs against various plant pathogens. The finding suggests that by reducing fungicide use and addressing *B. cinerea* resistance, using Cu-NPs in addition to conventional fungicides can provide a means for an environmentally friendly, long-term resistance management strategy (Malandrakis et al., 2020). Cu-NPs' capacity to modify *M. fructicola* strains. Investigations were made into the sensitivity or resistance of *Monilinia fructicola* isolates to benzimidazoles used alone or in conjunction with fungicides. Both in vitro and when applied to plum fruit, Cu-NPs showed a significant synergy with thiophanate methyl (TM) against *M. fructicola* isolates, suggesting increased availability or carbendazim transformation caused by nanoparticles. The interaction between Cu-NPs and the oxidative phosphorylationuncoupler fluazinam was found to be synergistic, indicating that ATP-dependent metabolism is probably involved in the way Cu-NPs cause fungitoxicity (FM). Copper ions were released, which facilitated the toxic activity of Cu-NPs against *M. fructicola* strains. The absence of a correlation between the nano and bulk/ionic copper forms, however, points to a different nanoproperty-mediated mechanism of fungitoxic action (Malandrakis et al., 2021). Combined NPs have been used to generate pesticides with greater biological potential and environmental safety (Alghuthaymi et al., 2021). Overall, combining fungicides with metal NPs has the potential to improve fungicide efficacy while reducing environmental impact, but more study is needed to fully understand the benefits and hazards of this method.

7 Nanoencapsulation of fungicides

Fungicide nanoencapsulation is a technology that involves encapsulating fungicides into nanoparticles to boost efficacy and reduce environmental impact. The encapsulation technique can protect fungicides from degradation and increase their solubility, improving bioavailability and plant uptake. Furthermore, nanoparticles can enable the regulated release of fungicides, extending their action and reducing the frequency of administration. The use of nanoencapsulation can also reduce the number of fungicides necessary to achieve the same degree of disease control, lowering the environmental effect of fungicide use. However, there are worries concerning nanoparticle toxicity to nontarget creatures and the environment, which must be thoroughly investigated. Overall, fungicide nanoencapsulation has the potential to improve fungicide effectiveness and sustainability, but more research is needed to fully understand the benefits and dangers of this technique. Fungicide nanoencapsulation includes the creation of fungicide-loaded or -entrapped particles with diameters in the nanorange. This size range should be 1100nm in at least one dimension, according to the definition of nanoparticle (Auffan et al., 2009). The nanoencapsulation of fertilizers, insecticides, fungicides, and herbicides ensures the controlled release and targeted distribution of these agrochemicals, which are necessary for effective nutrient uptake, disease control, and improved crop growth (Wani et al., 2019). In an emerging field of study, fungicides are being "nanoencapsulated," or packaged, to lessen their adverse effects and increase their efficiency. Low water-soluble fungicides' solubility can be increased through the application of nanotechnology, which can increase their bioavailability and effectiveness (Kutawa et al., 2021). One type of nanoscale delayed delivery technology used to replace traditional fungicides is nanofungicides. These nanofungicides are made up of fungicides that have been encapsulated, and they work to more effectively and precisely control fungal diseases (Bhattacharyya et al., 2016). Recent research has demonstrated that nanoencapsulation can also be a successful method for administering herbicides. The potential of nanoencapsulation to increase the efficacy of herbicides, as well as fungicides, has been demonstrated, for instance, by the discovery that atrazine-containing PCL nanocapsules have very effective postemergence herbicidal action (Oliveira et al., 2015). Overall, the nanoencapsulation of fungicides and herbicides is a fascinating field of study with enormous potential for enhancing the efficacy of these crucial agricultural chemicals. The fungicidal active ingredients (hexaconazole) were encapsulated into chitosan nanoparticles to create a fungicide nanodelivery system that can transport them more effectively to the target a virulent pathogenic fungus, *Ganoderma*

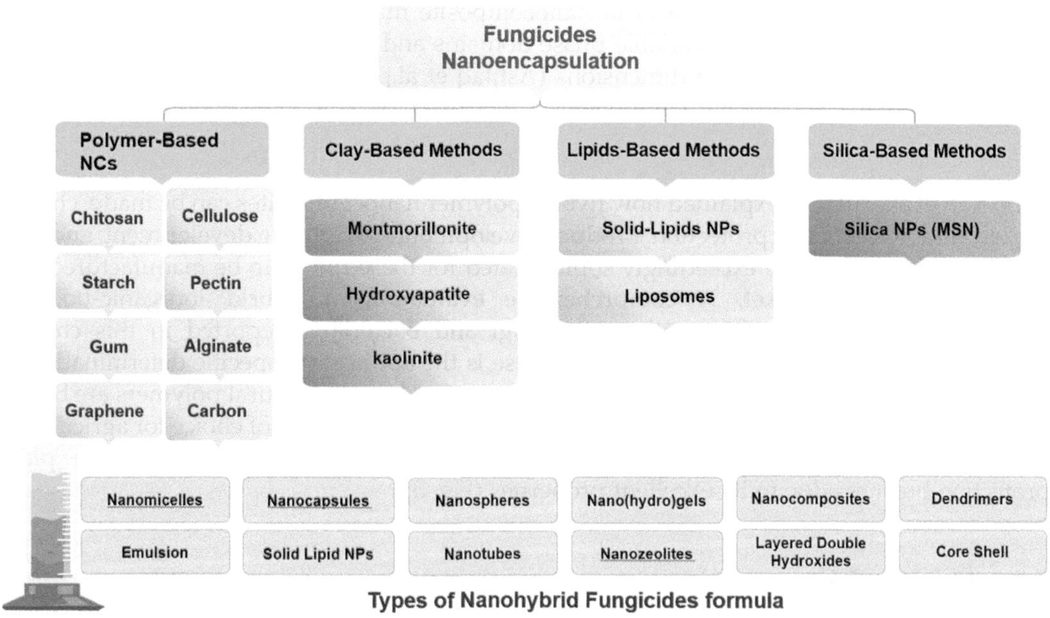

FIG. 2 Diagram illustrating types of promising nanoencapsulation materials for effective fungicide delivery.

boninense, which causes significant loss to Asian oil palm plantations. The in vitro antifungal assay demonstrated that chitosan-hexaconazole nanoparticles with smaller particle sizes inhibited *G. boninense* more effectively, with lower EC50 values (Maluin et al., 2019). Based on the nanocolloidal antifungal ingredient, Shepros created a new product called Ganoderma eliminator. Its main constituents are Nano Alpha 10, nanosilver, plant derivatives, and food additives. This product was discovered to infiltrate through inaccessible places, such as the septa of fungus, and to suppress the production of ergosterol in the mushroom's cell membrane (Maluin et al., 2020). Fig. 2 shows how nanoencapsulation materials deliver fungicides effectively.

7.1 Polymer nanoparticles and nanocomposites as nanocarriers

Polymer NPs such as polymeric nanocapsules and nanospheres are used as polymer nanocarriers because of their adaptability in building advanced pesticide delivery systems with various modes of action, biocompatibility, scalable manufacture, and biodegradation. Polymer nanocapsules are more effective than larger capsules in terms of absorption, spraying solution stability, spraying surface, and homogenous dispersion, resulting in improved insect suppression. Polymer nanocarriers can be cost-effective by maintaining the bare minimum effective dosage, protecting essential oils (EOs) and secondary metabolites from instability during application due to high temperature, moisture, and air (Ghormade et al., 2011). Polysaccharides include chitosan, starch, and alginates, while poly-ε-caprolactone (PEC), polyethylene glycol (PEG), beeswax, cashew gum, lecithin, and maize oil are biologically derived biodegradable compounds (Abreu et al., 2012).

Multiphase components are used in nanocomposite materials. These materials could be made up of components with variable phase domains and at least one continuous phase, as well as another with nanoscale dimensions (Ashfaq et al., 2020). Cosynthesis/impregnation of several inorganic and organic components can be used to create these hybrid nanomaterials (Winter et al., 2020). Plant disease management with hybrid polymer nanocomposites focuses on creating weed-controlling mulch films, nanoinsecticides, and biostatic agents (Adisa et al., 2019; Kah et al., 2016). It explained how hybrid polymer nanocomposites can be made, characterized, and used in plant protection. Product development, advertising development, and consumer education must be exceedingly sophisticated for these things to be manufactured and marketed on the market. A comprehensive evaluation of hybrid inorganic-polymer nanocomposites against plant pathogenic fungi and bacteria is reported in this chapter (Hashim et al., 2020). A crucial and critical phase is the application-specific determination of how polymer nanocomposites can be used. Furthermore, because natural polymers are biodegradable and exhibit regulated release behavior, they are the material of choice for agricultural applications. Chitosan polymer is increasingly becoming a feasible choice for usage in the plant protection business due to its excellent properties (Fig. 3).

7.1.1 Chitosan NCs

It is necessary to comprehend chitosan's mode of action at the molecular level in order to get a deeper grasp of its mechanisms, given the present understanding of chitosan and its various advantages on biocontrol agents, including maintenance and contamination. This can be achieved by looking at the chitosan's structure. This polymer's features in regulating the release of encapsulated components and their chemical stability may also be improved by combining it with other biopolymers. Because they are simple to make and can include both hydrophilic and hydrophobic fungicides, chitosan nanoparticles are a good option for the role of the nanocarrier (Agarwal et al., 2015). The impact of Cu@Chit NCs on the development of two tested fungi shows significant inhibitory action. When tested against *R. solani* and *Sclerotium rolfsii*, high concentrations of Cu@Chit NC made with acetone as the solvent exhibited the maximum amount of inhibition. These findings show that *R. solani* and *S. rolfsii* sclerotia development was significantly inhibited by Cu@Chit nanocomposites in an in vitro setting. Because the growth of the fungus transferred from NCs-challenged plates to nonchallenged conditions was the same as the growth of the fungus initiated from untreated plates, the antifungal activity of the Cu@Chit nanocomposites against *S. rolfsii* appears to be fungistatic rather than fungicidal (Rubina et al., 2017). A green chemistry approach was used to create chitosan-carrageenan nanocomposites with a size range of 66.6–231.82 nm that contain the chemical fungicide mancozeb (nano CSCRG-M). Mancozeb-loaded nanoparticles were found to have a relatively greater in vivo disease control efficacy (79.4 ± 1.7%) against *R. solani* in plants than commercial fungicides (76 ± 1.1%) in pot conditions. This was because mancozeb-loaded nanoparticles carried a sizable quantity of the drug. For plant growth indicators such as germination rate, root-shoot ratio, and dry biomass, nanomancozeb showed enhanced efficacy (Kumar et al., 2022a,b). The antifungal potential of chitosan and CuO nanocomposite was evaluated against *B. cinerea*, the fungus responsible for tomato gray mold, using a range of different concentrations of the nanocomposite. In vitro and in vivo studies on either detached tomato leaves, complete plants, or their fruits artificially infected by gray mold disease indicated that CH@CuO NPs improved a powerful antifungal efficacy at concentrations of 100 and $250\,mg\,L^{-1}$ (Ismail et al., 2023).

7.1.2 Cellulose NCs

The remaining edible coatings that are used in the treatment of postharvest diseases in berries are based on being carriers for various antimicrobial chemical compounds or biological control agents (Romero et al., 2022). This allows for diseases to be prevented after harvesting the berries. For instance, carboxymethylcellulose (CMC) is a polysaccharide that is formed after the carboxymethylation of cellulose. Its use as an edible coating has increased in recent years due to its biodegradability, absence of toxicity, and solubility in water (Salama et al., 2019). CMC also has a good ability to form transparent films. Along with the EOs of *Lippia sidoides*, CMC has been applied to the outside of strawberries to create an edible covering. According to research carried out by Oliveira et al. (2019), the presence of antifungal phenolic compounds in the EOs of these fruits greatly reduced the infections that were brought on by the pathogens *Rhizopus stolonifer* and *Colletotrichum acutatum*. The hydrophobic fungicides captan and pyraclostrobin were placed in situ onto the NCs while they were undergoing the crosslinking procedure. We were successful in obtaining NCs with an agrochemical load of 20 wt% and average diameters ranging from 200 to 300 nm. Cellulose NCs are an environmentally friendly alternative to the existing method of treating Apple Canker fungus, which involves the indiscriminate spraying of agrochemicals (Machado et al., 2021).

7.1.3 Starch NCs

Many studies have focused on the use of starch to create edible films and coatings since it is one of the most common natural biopolymers and is particularly appealing due to its cheap cost biodegradability, edibility, simplicity of chemical modification, and sustainability (Hu et al., 2019). Investigations were made into the structural, physical, and antifungal properties of starch edible films that had been combined with nanocomposites and Mexican oregano (*Lippia berlandieri* Schauer) to evaluate antifungal activity against *Rhizopus* species, *Fusarium* species, and *Aspergillus niger*. According to Aguilar-Sánchez et al. (2019), adding Mexican oregano EO to edible starch films generally provides a sufficient fungicidal impact. To enhance banana seedling germination and rooting, an antifungal starch nanocomposite (St@CuONPs) based on starch and biosynthesized copper oxide nanoparticles (CuONPs) was created utilizing *A. niger* AH1. St@CuONPs, as a nanocomposite, has the potential to enhance banana seedling germination and protection from fungus-related plant diseases (Hasanin et al., 2021). Starch can also be easily chemically altered to enhance its qualities, such as its tensile strength and water resistance. All things considered, using starch-based films and coatings for food packaging has the potential to lessen the negative effects of packaging waste on the environment and offer a more sustainable option for the food sector.

7.1.4 Pectin NCs

In order to create nanocomposite films and assess the impact of neem oil nanoemulsions on the coating characteristics of soybean seeds, pectin matrices, and neem oil nanoemulsions were combined. The antifungal, morphological, mechanical, and barrier properties of the

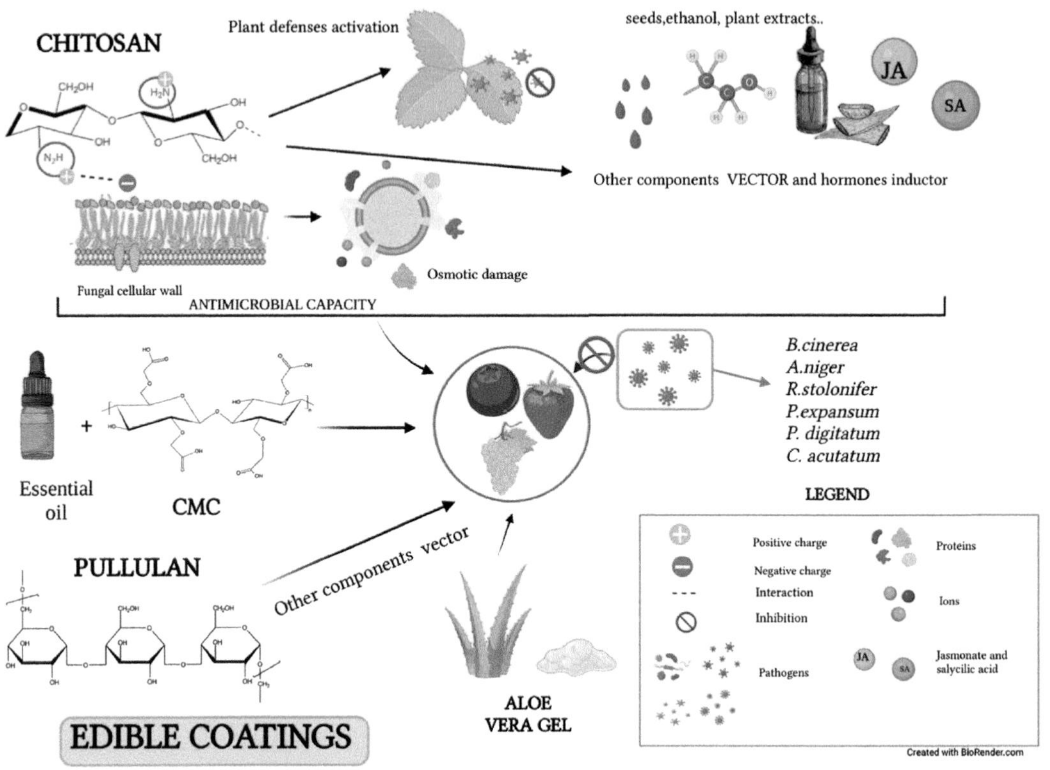

FIG. 3 Infographic summary on the use of edible coatings to suppress postharvest infections in berries. *Data from Romero, J., Albertos, I., Díez-Méndez, A., Poveda, J., 2022. Control of postharvest diseases in berries through edible coatings and bacterial probiotics. Sci. Hortic. 304, 111326. This is an open-access article distributed under the terms of the Creative Commons CC-BY license, which permits unrestricted use, distribution, and reproduction in any medium, provided the original work is properly cited with permission from Elsevier.*

nanocomposite were studied. *Aspergillus flavus* and *Penicillium citrinum* were both resistant to the fungal effects of neem oil. The seed coats encouraged a favorable impact on the germination of soybean seeds. The creation of antifungal nanocomposite films using renewable resources was therefore accomplished. These novel substances are promising to create seed coverings due to the fungicidal inhibition of Neem oil as a nanoemulsion (de Castro e Silva et al., 2019).

7.1.5 Gum NCs

Using chitosan-gum Arabic-coated liposome 5ID nanoparticles, the bipolymeric drug delivery system was created. These NCs were examined and put to the test mechanically against *B. cinerea*, a plant pathogen. Stabilized microtubule polymerization was identified as the mechanism underlying the suppression of *B. cinerea* by chitosan-gum Arabic-coated in vitro and in silico. This research creates a new path for creating polymeric NPs that act as antifungal agents (Raj et al., 2021). Using the ionic gelation and polyelectrolyte complexation methods, the commercial fungicide mancozeb (M) was loaded into chitosan-gum acacia

(CSGA) polymers to create nanocomposite (NC) CSGA-M (mancozeb-laden) measuring 363.6 nm. The antifungal activity of nano-CSGA-M was shown to be comparable to mancozeb in vitro and in vivo, but less harmful to Vero cell lines; as a result, this formulation may be utilized for sustainable agriculture in the future (Kumar et al., 2022a,b). Green science principles were used to create an in vitro highly stable gum acacia-gold nanocomposite (GA-AuNC-NT) made using the food preservative natamycin. By measuring the zone of inhibition, high rate of hyphae fragmentation, and pronounced spore germination inhibition against the tested fungal strain, the synthesized nanocomposite's potential antifungal efficacy was confirmed. The complete lack of fungi development also demonstrates how special the nanocomposite is (Namasivayam et al., 2022).

7.1.6 Lipid-based nanocarriers

Reduced losses due to leaching or degradation, decreased toxicity in both humans and the environment, and altered bioactive component release patterns are just a few benefits that solid lipid nanoparticles and polymeric nanocapsules offer as carrier vehicles. Tebuconazole (TBZ) and carbendazim (MBC) are frequently used in agriculture to prevent and control fungus-related illnesses. In order to use solid lipid nanoparticles and polymeric nanocapsules as carriers for a carbendazim and tebuconazole mixture, they were created. The cytotoxicities of the formulations were assessed, and the encapsulation efficiencies and release profiles of the fungicides were established in vitro. These fungicide systems provide fresh possibilities for the management of plant fungal diseases (Campos et al., 2015).

7.1.7 Lignin-based nano- and microcarriers

Utilizing nano- and microcarriers in agriculture has several advantages, including prolonged and targeted administration, effective fungicide uptake, and minimal environmental impact (Liang et al., 2018). We created stable aqueous dispersions of lignin nanocarriers with a core-shell structure such that fungicides could be loaded there and released across a range of times depending on the degree of crosslinking. This formulation could be employed as a controlled-release technique for agrochemicals for long-term crop protection by timed diffusion by stem injection or spray application, in addition to the previously demonstrated enzymatic degradability of lignin nanocarriers (Beckers et al., 2020). By combining copper(II) salts with two different types of engineered lignins, novel organic-inorganic hybrid materials were created, and their antifungal and antibacterial capabilities were tested against many significant agronomic diseases. Furthermore, preliminary studies on greenhouse crops found that Lignin@Cu was more effective against *R. solani* on tomato plants than commercial copper(II) hydroxide-type insecticides, indicating that these compounds have a lot of potential as crop protection agents (Sinisi et al., 2018). After adding Aza-Michael to the miniemulsion and letting the solvent drain, chemically crosslinked lignin nanocarriers (NCs) were created. Crosslinking lignin with biobased amines (spermine and spermidine). The versatility of the miniemulsion polymerization technique was demonstrated by the ability to encapsulate in place a number of fungicides, including azoxystrobin, pyraclostrobin, tebuconazole, and boscalid. Depending on the drug's solubility, lignin NCs with dynamic light scattering-measured diameters of 200–300 nm show good encapsulation efficiencies (70%–99%). The fungus *Phaeomoniella chlamydospora* and *Phaeoacremonium minimum* did not grow when treated with lignin NCs (Machado et al., 2020).

7.1.8 Clay NCs

The current advances in clay/metallic, clay-polymer, and clay-carbon composites exhibiting their application in plant promotion and protection were given careful examination. The antibacterial characteristics of clay-based nanocomposites, as well as their manufacture and utilization as a delivery vehicle for agrochemicals and nucleic acids, were examined. This evaluation would aid in the identification and resolution of current challenges concerning the delivery of active chemicals and biological macromolecules for crop protection, enhanced growth, and sustainability (Pal et al., 2022). The fungistatic characteristics of epoxy-organo-montmorillonite nanocomposites were investigated in relation to the type of alkylammonium-modified clay used. The developed polymer materials showed remarkable fungal resistance to the most aggressive polymer-destructor microorganisms, such as *A. niger*, *Penicillium chrysogenum*, and *Trichoderma viridescens* (Mykola et al., 2016). Clay-based nanocomposite materials have been shown to be a more cost-effective technique than traditional treatment procedures. To combat the pathogen, a clay-chitosan nanocomposite (CCNC) was created and tested in vitro; at 20 g mL^{-1}, the nanocomposite prevented the growth of *Penicillium digitatum*. In in vivo tests, the nanocomposite reduced lesions by 70% and inhibited disease in orange. The CNCC-coated orange was found to be disease-free and to have higher pH, chroma, peel moisture, and firmness than the control (Youssef and Hashim, 2020).

7.1.9 Hybrid silica nanoparticles

Biomedical applications for amorphous hybrid SiNPs with varying bioactivity and stability include cellular imaging, gene transport, and drug delivery (Li et al., 2021). For example, Zhao et al. (2017) loaded pyrimethanil with mesoporous silica nanoparticles (MSNs) and studied its uptake in cucumber for 7 weeks. The dosage and uptake of fungicide-loaded MSN in cucumber plant leaves had no effect on the rate of dispersion and dissemination in the plants. According to their findings, pyrimethanil-loaded MSN has a low risk of accumulating in the edible section of cucumber plants. Zhao et al. (2017) have advanced our understanding of the spread and mobility of MSN laden with fungicides when sprayed on the leaves. The antifungal activity of synthesized captan&ZnO and captan&SiO_2 was 43%–61% higher than captan, despite the low captan level. The reported fungicidal action of nanofungicides was due to captan components, as neither nanocarrier alone demonstrated antifungal activity. These findings support the hypothesis that nanoformulation improves fungicide efficiency, implying that incorporating nanotechnology may result in a reduction in pesticide usage (Grillo et al., 2021). Si-Ag nanoparticles inhibited fungal growth at 10 ppm, while bacteria were effective at 100 ppm. Furthermore, Si-Ag nanoparticles inhibited downy mildew disease in cucurbits and demonstrated phytotoxicity in cucumber and pansy plants at a dosage of 3200 ppm. Nanosilica kills insects by physically absorbing and killing them in their cuticle lipid (Zhu et al., 2022).

7.1.10 Bimetallic NPs

In vitro and in vivo studies of the antifungal activity of Cu-chitosan, Zn-chitosan, and bimetallic blends (BBs) at concentrations of 30, 60, and 100 g mL^{-1} were conducted against two anastomosis groups of *R. solani* for the control of cotton seedling damping-off. The results showed that Cu-chitosan and Zn-chitosan were more effective than BB. Under greenhouse circumstances, the impact of metal-chitosan nanocomposites at $100\,g\,mL^{-1}$ mixed with Cu-tolerant *T. longibrachiatum* strains was also studied for the management of cotton seedling damping-off. In vitro testing revealed that the BBs and the Cu-chitosan nanocomposite demonstrated the greatest antifungal effectiveness against both anastomosis groups of *R. solani*. Based on these findings, it appears that BBs, Cu-chitosan nanocomposite, and BBs mixed with *Trichoderma* may be able to inhibit the cotton seedling disease caused by *R. solani* in vivo. Synergistic inhibitory effects were observed when *R. solani* was tested in a greenhouse alongside a *Trichoderma* strain (Abd-Elsalam et al., 2018). Diethylene glycol (DEG) and PEG, as well as CuZn@DEG and ZnO@PEG nanofertilizers (NFs), were tested in vitro to see whether they were able to inhibit the growth of *B. cinerea* and *Sclerotinia sclerotiorum*. There was clear evidence of a dose-dependent pattern of significant growth suppression exerted by ZnO@PEG and CuZn@DEG NFs on both fungi. This is the first study that we are aware of that describes NFs being examined for their ability to suppress phytopathogenic fungus. The composition synergetic effect of the two bioessential metals (Cu/Zn) was indicated by the fact that the hydrodynamic sizes were found to be similar in the bimetallic CuZn@DEG NFs and the ZnO@PEG NFs. Bimetallic CuZn@DEG NFs showed higher fungal growth inhibition compared to single metal oxide NFs (ZnO@PEG NFs) (Fig. 4). In addition to this, the capacity of ZnO@PEG and CuZn@DEG NFs to lessen the severity of the disease known as White Rot (Sclerotiniasis) was demonstrated on lettuce plants (Tryfon et al., 2021).

7.1.11 Multimetallic NPs

Due to their novel functions, multimetallic NPs, which combine two or more distinct metals to produce alloys or core-shell nanocomposites, have sparked substantial interest as novel materials. In a chemical transformation, the combined action of several metals and metal oxides improves the catalytic efficacy of multimetallic NPs (Buchwalter et al., 2015). However, only a few studies have looked at the synergistic effects of multimetallic NPs, including bi-, tri-, and quadrametallic nanocomposites. The synergistic effects of these multimetallic NPs have piqued the interest of researchers due to their various and customizable physicochemical features, as well as their superior catalytic properties compared to other multimetallic NPs (Basavegowda and Baek, 2022). Bimetallic NPs and their composites, for example, can be used to create more effective antibacterial agents to prevent the emergence and spread of pathogenic fungi. Bimetallic blends and Zn-Chitosan, as well as Cu-Chitosan, reduced the development of *R. solani* at concentrations of 30, 60, and $90\,g\,mL^{-1}$. Furthermore, it demonstrated efficient damping-off management of cotton seedlings under greenhouse conditions (Abd-Elsalam et al., 2018).

8 Synergism mechanisms

The mechanism of action for these nanoparticle-based antifungals is complex, but it largely involves suppressing mycelial development, which permits fungi cells to increase in length or diameter via branching filaments called hyphae. Inhibiting this mechanism prevents the virus from spreading farther across the body or environment from whence it started. Nanoparticles

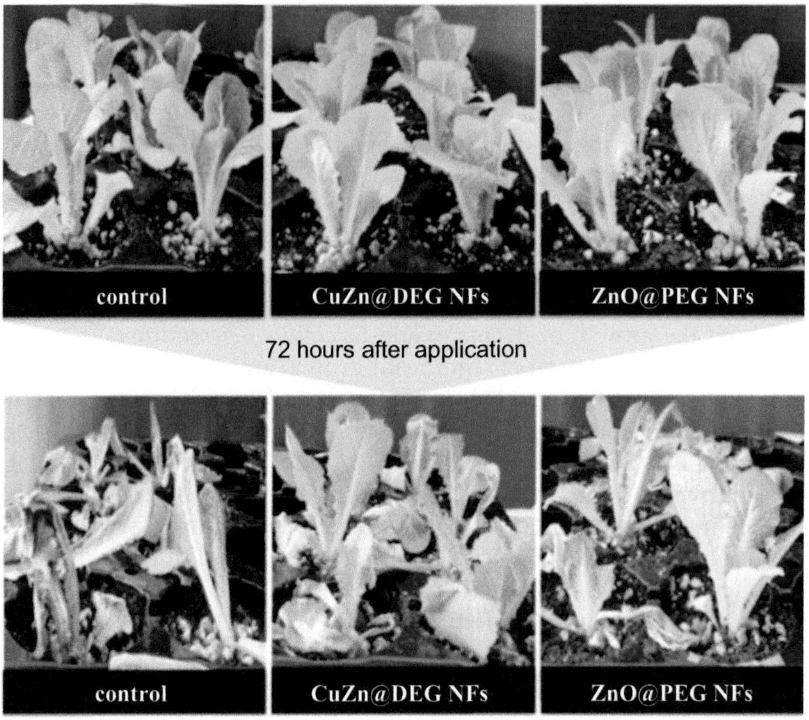

FIG. 4 Lettuce plants inoculated by *S. sclerotiorum* and treated with CuZn@DEG and ZnO@PEG NFs at 0h and 72h after application. *Data from Tryfon, P., Kamou, N.N., Mourdikoudis, S., Karamanoli, K., Menkissoglu-Spiroudi, U., Dendrinou-Samara, C., 2021. CuZn and ZnO nanoflowers as nano-fungicides against* Botrytis cinerea *and* Sclerotinia sclerotiorum: *phytoprotection, translocation, and impact after foliar application. Materials 14, 7600. This is an open-access article distributed under the terms of the Creative Commons CC-BY license, which permits unrestricted use, distribution, and reproduction in any medium, provided the original work is properly cited with permission from MDPI.*

also disrupt cell wall structures by binding to specific points on their surface, preventing new material from being added to them during replication processes like mitosis or meiosis; this results in weakened fungal cells that are more susceptible to other treatments like antibiotics or fungicides. Furthermore, these particles interact with protein molecules found inside fungi, disrupting normal functioning, and eventually leading to death due to the production of ROS (Reactive Oxygen Species)—highly reactive molecules containing oxygen atoms capable of damaging DNA strands, resulting in cellular damage/death if left unchecked over time. If each antibiotic invades a separate target or signaling pathway with a different mechanism of action, the use of antimicrobial medicines in combination can create synergistic effects. The synergy between EOs and antibiotics may be due in part to the EO-induced permeabilization of the cell membrane, which results in the quick transport of antibiotics into the cell's interior (Aleksic et al., 2014). The NPs functioned against plant diseases by disrupting protein, forming, and depriving ROS, impeding membrane permeability, and inhibiting the gene transporter. All the following mechanisms were interdependent and worked in tandem with several plant pathogens (Alkhattaf, 2021).

Several investigations on antifungal mechanisms at the molecular level utilizing diverse animal models are still needed to completely show the mode of action involved in plant pathogen management. This will help us to develop a better understanding of how these substances interact with various fungal diseases, allowing for more effective management approaches that are tailored to particular instances. Furthermore, it may provide information regarding potential negative side effects linked with the long-term use of these pesticides when applied topically to crops.

9 Challenges

There are some concerns about the long-term effects of using nanofungicide formulations on human health and the environment (Zahid et al., 2019). As a result, scientists should analyze the effects of nanofungicides on human health and the environment, as well as improve the methods used to assess and regulate any risks they may pose (Zhang et al., 2019). However, widespread pesticide usage harms a wide range of nonpathogenic species, and there has also been an increase in infections in foods, resulting in additional deterioration of soil and water quality. The research and development of novel nanofungicides, NFs, and nanosorbents derived from biocontrol agents, as well as the evaluation of these nanoformulations for the remediation of heavy metal-contaminated industrial wastewater, are still in their early stages. Several challenges must be overcome to scale up biogenic nanoparticles to meet real-world demands (Alghuthaymi et al., 2022). Many studies are needed to explain the interactions between plants, phylloplane microflora, nanomaterials, soil microorganisms, and endophytes, as well as the pathogenic and positive effects on plant health. Finally, educational and developmental activities should be formed through research strategies in order to bring users and professionals together to meet potential client interests.

10 Future trends

Soon, encapsulation will be utilized to blend various active components together in one compound, such as pairing two microbial bioagents, a chemical pesticide along with a biocontrol agent, or a biocontrol agent with enhancing compounds like nanoparticles. As polymer science advances, newly developed gels will have a direct impact on the structure of capsules, leading to exceptional multicompartmentalized capsules. In light of this method, more effort needs to be put into adapting already established techniques to accommodate biocontrol microorganisms, using a smart arrangement of previously tested strategies to manage issues such as shelf life. In the coming days, this technique, with all of its advantages including slow release of helpful microbes, might lead to a new course in the creation of unfriendly chemicals (Saberi-Riseh et al., 2021).

As a result, integrated pest control approaches that combine systemic or nonsystemic fungicides with resistance antagonists or inducers are advocated (Ons et al., 2020). To regulate harmful fungus, we will soon be able to mix heavy-tolerant biocontrol agents with modest doses of metal nanosystems. The efficacy of using biopolymers to encapsulate heavy-metal tolerant biocontrol agents with NPs warrants further investigation, particularly in terms of the preparation and production of new formulations based on biodegradable polymers, as well as their ability to increase the quantity and quality of agricultural products while controlling pests and plant diseases. As a result, nanoparticle preparation procedures should be improved to obtain dispersed and stable nanoparticles. To attain superior antimicrobial results, more study is needed to investigate the synergistic effects of combinations of plant extracts/EOs, plant extracts/nanomaterials, and nanomaterials/EOs. Furthermore, more research is needed to study the synergistic effects of plant extracts/EOs/nanomaterial combinations.

11 Conclusion

In the current volume, we cover the combined effects of various antifungal agents such as plant extracts, EOs, with nanomaterials. Furthermore, we discuss the synergistic interactions and antibacterial properties of several antifungal drugs in combination, as well as the mechanism of action, toxicity, and future directions. Combinations of three distinct antimicrobial medicines may have more synergistic antimicrobial activity than a combination of only two antifungal agents. These new hybrid nanosystems and innovative antimicrobial techniques may have a key synergistic effect and serve as alternatives to traditional antibiotics in the Agri-Food sector for preventing the spread of plant pathogenic fungus. Nanotechnology-based biofungicides offer environmentally acceptable and effective disease control methods that are not detrimental to the environment. These nanosystems have proven a great ability for controlled-release behavior, ensuring their long-term usefulness and the possibility of tackling eutrophication and pesticide residue accumulation.

To summarize, nanohybrid fungicides are a type of fungicide that combines the benefits of traditional fungicides and nanotechnology. They aim to improve the management of fungal diseases in crops while utilizing smaller amounts of fungicide and causing less harm to the environment. Typically, nanohybrid fungicides blend a conventional fungicide with nanoparticles, like silver or copper particles, to increase the effectiveness of fungicidal action and fungicide delivery and absorption. Nanoparticles also provide benefits like greater stability and longer shelf life. Nanohybrid fungicides are under development as a potential alternative to traditional fungicides, which can pose risks to both human health and the environment.

References

Abd-Elsalam, K.A., Vasil'kov, A.Y., Said-Galiev, E.E., Rubina, M.S., Khokhlov, A.R., Naumkin, A.V., Shtykova, E.V., Alghuthaymi, M.A., 2018. Bimetallic blends and chitosan nanocomposites: novel antifungal agents against cotton seedling damping-off. Eur. J. Plant Pathol. 151, 57–72.

Abreu, F.O.M.S., Oliveira, E.F., Paula, H.C.B., de Paula, R.C.M., 2012. Chitosan/cashew gum nanogels for essential oil encapsulation. Carbohydr. Polym. 89, 1277–1282.

Adisa, I.O., Pullagurala, V.L.R., Peralta-Videa, J.R., Dimkpa, C.O., Elmer, W.H., Gardea-Torresdey, J.L., White, J.C., 2019. Recent advances in nano-enabled fertilizers and pesticides: a critical review of mechanisms of action. Environ. Sci. Nano 6, 2002–2030.

Agarwal, M., Nagar, D., Srivastava, N., Agarwal, M., 2015. Chitosan nanoparticles-based drug delivery: an update. Int. J. Adv. Multidiscip. Res. 2, 1–13.

Aguilar-Sánchez, R., Munguía-Pérez, R., Reyes-Jurado, F., Navarro-Cruz, A.R., Cid-Pérez, T.S., Hernández-Carranza, P., Beristain-Bauza, S.D.C., Ochoa-Velasco, C.E., Avila-Sosa, R., 2019. Structural, physical, and antifungal characterization of starch edible films added with nanocomposites and Mexican oregano (*Lippia berlandieri* Schauer) essential oil. Molecules 24 (12), 2340.

Aleksic, V., Mimica-Dukic, N., Simin, N., Nedeljkovic, N.S., Knezevic, P., 2014. Synergistic effect of *Myrtus communis* L. essential oils and conventional antibiotics against multi-drug resistant *Acinetobacter baumannii* wound isolates. Phytomedicine 21, 1666–1674.

Alghuthaymi, M.A., Kalia, A., Bhardwaj, K., Bhardwaj, P., Abd-Elsalam, K.A., Valis, M., Kuca, K., 2021. Nanohybrid antifungals for control of plant diseases: current status and future perspectives. J. Fungi 7 (1), 48.

Alghuthaymi, M.A., Abd-Elsalam, K.A., AboDalam, H.M., Ahmed, F.K., Ravichandran, M., Kalia, A., Rai, M., 2022. *Trichoderma*: an eco-friendly source of nanomaterials for sustainable agroecosystems. J. Fungi 8 (4), 367.

Alkhattaf, F.S., 2021. Gold and silver nanoparticles: green synthesis, microbes, mechanism, factors, plant disease management and environmental risks. Saudi J. Biol. Sci. 28 (6), 3624–3631.

Ashfaq, M., Talreja, N., Chuahan, D., Srituravanich, W., 2020. Polymeric nanocomposite-based agriculture delivery system: emerging technology for agriculture. Genet. Eng. Glimpse Tech. Appl., 1–16.

Auffan, M., Rose, J., Bottero, J.-Y., Lowry, G.V., Jolivet, J.-P., Wiesner, M.R., 2009. Towards a definition of inorganic nanoparticles from an environmental, health and safety perspective. Nat. Nanotechnol. 4, 634–641.

Baigorria, E., Fraceto, L.F., 2022. Biopolymer-nanocomposite hybrid materials as potential strategy to remove pesticides in water: occurrence and perspectives. Adv. Sustain. Syst. 6 (1), 2100243.

Baigorria, E., Ollier, R.P., Alvarez, V.A., 2021. In: Mallakpour, S., Hussain, C.M. (Eds.), Handbook of Consumer Nanoproducts. Springer Singapore, Singapore, pp. 1–24.

Basavegowda, N., Baek, K.H., 2022. Combination strategies of different antimicrobials: an efficient and alternative tool for pathogen inactivation. Biomedicine 10 (9), 2219.

Beckers, S., Peil, S., Wurm, F.R., 2020. Pesticide-loaded nanocarriers from lignin sulfonates—a promising tool for sustainable plant protection. ACS Sustain. Chem. Eng. 8 (50), 18468–18475.

Bhattacharyya, A., Duraisamy, P., Govindarajan, M., Buhroo, A.A., Prasad, R., 2016. Nano-biofungicides: emerging trend in insect pest control. In: Advances and Applications Through Fungal Nanobiotechnology. Springer, pp. 307–319.

Buchwalter, P., Rosé, J., Braunstein, P., 2015. Multimetallic catalysis based on heterometallic complexes and clusters. Chem. Rev. 115, 28–126.

Campos, E.V., de Oliveira, J.L., da Silva, C.M., Pascoli, M., Pasquoto, T., Lima, R., Abhilash, P.C., Fraceto, L.F., 2015. Polymeric and solid lipid nanoparticles for sustained release of carbendazim and tebuconazole in agricultural applications. Sci. Rep. 5, 13809.

Carvalho, F.P., 2017. Pesticides, environment, and food safety. Food Energy Secur. 6, 48–60.

de Castro e Silva, P., Pereira, L.A.S., Lago, A.M.T., Valquíria, M., de Rezende, É.M., Carvalho, G.R., Oliveira, J.E., Marconcini, J.M., 2019. Physical-mechanical and antifungal properties of pectin nanocomposites/neem oil nanoemulsion for seed coating. Food Biophys. 14, 456–466.

Eljounaidi, K., Lichman, B.R., 2020. Nature's chemists: the discovery and engineering of phytochemical biosynthesis. Front. Chem. 8, 1041.

Ghormade, V., Deshpande, M.V., Paknikar, K.M., 2011. Perspectives for nano-biotechnology enabled protection and nutrition of plants. Biotechnol. Adv. 29, 792–803.

Grillo, R., Fraceto, L.F., Amorim, M.J., Scott-Fordsmand, J.J., Schoonjans, R., Chaudhry, Q., 2021. Ecotoxicological and regulatory aspects of environmental sustainability of nanopesticides. J. Hazard. Mater. 404, 124148.

Hasanin, M., Hashem, A.H., Lashin, I., Hassan, S.A., 2021. In vitro improvement and rooting of banana plantlets using antifungal nanocomposite based on myco-synthesized copper oxide nanoparticles and starch. Biomass Convers. Biorefinery, 1–11.

Hashim, A.F., Youssef, K., Roberto, S.R., Abd-Elsalam, K.A., 2020. Hybrid inorganic-polymer nanocomposites: synthesis, characterization, and plant-protection applications. In: Multifunctional Hybrid Nanomaterials for Sustainable Agri-Food and Ecosystems. Elsevier, pp. 33–49.

Hu, X., Jia, X., Zhi, C., Jin, Z., Miao, M., 2019. Improving the properties of starch-based antimicrobial composite films using ZnO-chitosan nanoparticles. Carbohydr. Polym. 210, 204–209.

Ismail, A.M., Mosa, M.A., El-Ganainy, S.M., 2023. Chitosan-decorated copper oxide nanocomposite: investigation of its antifungal activity against tomato gray mold caused by *Botrytis cinerea*. Polymers 15, 1099.

Kah, M., Weniger, A., Hofmann, T., 2016. Impacts of (nano)formulations on the fate of an insecticide in soil and consequences for environmental exposure assessment. Environ. Sci. Technol. 50, 10960–10967.
Kaur, S., Gallei, M., Ionescu, E., 2015. In: Kalia, S., Haldorai, Y. (Eds.), Advanced Polymer Science. Springer International Publishing, Cham.
Kickelbick, G., 2006. Hybrid Materials: Synthesis, Characterization, and Applications. Wiley, New York.
Kumar, R., Duhan, J.S., Manuja, A., Kaur, P., Kumar, B., Sadh, P.K., 2022a. Toxicity assessment and control of early blight and stem rot of *Solanum tuberosum* L. by mancozeb-loaded chitosan–gum acacia nanocomposites. J. Xenobiotics 12 (2), 74–90.
Kumar, R., Najda, A., Duhan, J.S., Kumar, B., Chawla, P., Klepacka, J., Malawski, S., Kumar Sadh, P., Poonia, A.K., 2022b. Assessment of antifungal efficacy and release behavior of fungicide-loaded chitosan-carrageenan nanoparticles against phytopathogenic fungi. Polymers 14 (1), 41.
Kusakizako, T., Miyauchi, H., Ishitani, R., Nureki, O., 2020. Structural biology of the multidrug and toxic compound extrusion superfamily transporters. Biochim. Biophys. Acta Biomembr. 1862, 183154.
Kutawa, A.B., Ahmad, K., Ali, A., Hussein, M.Z., Abdul Wahab, M.A., Adamu, A., Ismaila, A.A., Gunasena, M.T., Rahman, M.Z., Hossain, M.I., 2021. Trends in nanotechnology and its potentialities to control plant pathogenic fungi: a review. Biology 10 (9), 881.
Li, L., Xu, Z., Kah, M., Lin, D., Filser, J., 2019. Nanopesticides: a comprehensive assessment of environmental risk is needed before widespread agricultural application. Environ. Sci. Technol. 53 (14), 7923–7924.
Li, W., Qamar, S.A., Qamar, M., Basharat, A., Bilal, M., Iqbal, H.M., 2021. Carrageenan-based nano-hybrid materials for the mitigation of hazardous environmental pollutants. Int. J. Biol. Macromol. 190, 700–712.
Liang, Y., Fan, C., Dong, H., Zhang, W., Tang, G., Yang, J., Jiang, N., Cao, Y., 2018. Preparation of MSNs-Chitosan@Prochloraz nanoparticles for reducing toxicity and improving release properties of prochloraz. ACS Sustain. Chem. Eng. 6, 10211–10220.
Machado, T.O., Beckers, S.J., Fischer, J., Müller, B., Sayer, C., de Araújo, P.H., Landfester, K., Wurm, F.R., 2020. Bio-based lignin nanocarriers loaded with fungicides as a versatile platform for drug delivery in plants. Biomacromolecules 21 (7), 2755–2763.
Machado, T.O., Beckers, S.J., Fischer, J., Sayer, C., de Araújo, P.H., Landfester, K., Wurm, F.R., 2021. Cellulose nanocarriers via miniemulsion allow Pathogen-Specific agrochemical delivery. J. Colloid Interface Sci. 601, 678–688.
Malandrakis, A.A., Kavroulakis, N., Chrysikopoulos, C.V., 2020. Synergy between Cu-NPs and fungicides against *Botrytis cinerea*. Sci. Total Environ. 703, 135557.
Malandrakis, A.A., Kavroulakis, N., Chrysikopoulos, C.V., 2021. Copper nanoparticles against benzimidazole-resistant *Monilinia fructicola* field isolates. Pestic. Biochem. Physiol. 173, 104796.
Maluin, F.N., Hussein, M.Z., Yusof, N.A., Fakurazi, S., Idris, A.S., Zainol Hilmi, N.H., Jeffery Daim, L.D., 2019. Preparation of chitosan–hexaconazole nanoparticles as fungicide nanodelivery system for combating Ganoderma disease in oil palm. Molecules 24 (13), 2498.
Maluin, F.N., Hussein, M.Z., Idris, A.S., 2020. An overview of the oil palm industry: challenges and some emerging opportunities for nanotechnology development. Agronomy 10, 356.
Mykola, S., Olga, N., Dmitry, M., 2016. The influence of alkylammonium modified clays on the fungal resistance and biodeterioration of epoxy-clay nanocomposites. Int. Biodeterior. Biodegrad. 110, 136–140.
Namasivayam, S., Raja, K., Manohar, M., Aravind Kumar, J., Samrat, K., Kande, A., Arvind Bharani, R.S., Jayaprakash, C., Lokesh, S., 2022. Green chemistry principles for the synthesis of anti fungal active gum acacia-gold nanocomposite-natamycin (GA-AuNC–NT) against food spoilage fungal strain *Aspergillus ochraceopealiformis* and its marked Congo red dye adsorption efficacy. Environ. Res. 212, 113386.
Oliveira, H.C., Stolf-Moreira, R., Martinez, C.B.R., Grillo, R., de Jesus, M.B., Fraceto, L.F., 2015. Nanoencapsulation enhances the post-emergence herbicidal activity of atrazine against mustard plants. PLoS One 10 (7), e0132971.
Oliveira, J., Gloria, E.M., Parisi, M.C.M., Baggio, J.S., Silva, P.P.M., Ambrosio, C.M.S., Spoto, M.H.F., 2019. Antifungal activity of essential oils associated with carboxymethylcellulose against *Colletotrichum acutatum* in strawberries. Sci. Hortic. 243, 261–267.
Ons, L., Bylemans, D., Thevissen, K., Cammue, B.P., 2020. Combining biocontrol agents with chemical fungicides for integrated plant fungal disease control. Microorganisms 8 (12), 1930.
Pal, A., Kaur, P., Dwivedi, N., Rookes, J., Bohidar, H.B., Yang, W., Cahill, D.M., Manna, P.K., 2022. Clay-nanocomposite based smart delivery systems: a promising tool for sustainable farming. ACS Agric. Sci. Technol. 22. https://doi.org/10.1016/j.mtbio.2023.100759.

Patibanda, A.K., Ranganathswamy, M., 2018. Effect of agrichemicals on biocontrol agents of plant disease control. In: Microorganisms for Green Revolution. Springer, pp. 1–21.

Periakaruppan, R., Palanimuthu, V., Abed, S.A., Danaraj, J., 2023. New perception about the use of nanofungicides in sustainable agriculture practices. Arch. Microbiol. 205 (1), 4.

Raj, V., Raorane, C.J., Lee, J.H., Lee, J., 2021. Appraisal of chitosan-gum arabic-coated bipolymeric nanocarriers for efficient dye removal and eradication of the plant pathogen *Botrytis cinerea*. ACS Appl. Mater. Interfaces 13 (40), 47354–47370.

Romero, J., Albertos, I., Díez-Méndez, A., Poveda, J., 2022. Control of postharvest diseases in berries through edible coatings and bacterial probiotics. Sci. Hortic. 304, 111326.

Rubina, M.S., Vasil'kov, A.Y., Naumkin, A.V., Shtykova, E.V., Abramchuk, S.S., Alghuthaymi, M.A., Abd-Elsalam, K.A., 2017. Synthesis and characterization of chitosan–copper nanocomposites and their fungicidal activity against two sclerotia-forming plant pathogenic fungi. J. Nanostruct. Chem. 7, 249–258.

Saberi-Riseh, R., Moradi-Pour, M., Mohammadinejad, R., Thakur, V.K., 2021. Biopolymers for biological control of plant pathogens: advances in microencapsulation of beneficial microorganisms. Polymers 13 (12), 1938.

Salama, H.E., Aziz, M.S.A., Alsehli, M., 2019. Carboxymethyl cellulose/sodium alginate/chitosan biguanidine hydrochloride ternary system for edible coatings. Int. J. Biol. Macromol. 139, 614–620.

Sinisi, V., Pelagatti, P., Carcelli, M., Migliori, A., Mantovani, L., Righi, L., Leonardi, G., Pietarinen, S., Hubsch, C., Rogolino, D., 2018. A green approach to copper-containing pesticides: antimicrobial and antifungal activity of brochantite supported on lignin for the development of biobased plant protection products. ACS Sustain. Chem. Eng. 7 (3), 3213–3221.

Tryfon, P., Kamou, N.N., Mourdikoudis, S., Karamanoli, K., Menkissoglu-Spiroudi, U., Dendrinou-Samara, C., 2021. CuZn and ZnO nanoflowers as nano-fungicides against *Botrytis cinerea* and *Sclerotinia sclerotiorum*: phytoprotection, translocation, and impact after foliar application. Materials 14 (24), 7600.

Vasseghian, Y., Arunkumar, P., Joo, S.W., Gnanasekaran, L., Kamyab, H., Rajendran, S., Balakrishnan, D., Chelliapan, S., Klemeš, J.J., 2022. Metal-organic framework-enabled pesticides are an emerging tool for sustainable cleaner production and environmental hazard reduction. J. Clean. Prod., 133966.

Wang, D., Saleh, N.B., Byro, A., Zepp, R., Sahle-Demessie, E., Luxton, T., Ho, K.T., Burgess, R.M., Flury, M., White, J.-C., Su, C., 2022. Nano-enabled pesticides for sustainable agriculture and global food security. Nat. Nanotechnol. 17 (4), 347–360.

Wani, T.A., Masoodi, F.A., Baba, W.N., Ahmad, M., Rahmanian, N., Jafari, S.M., 2019. Nanoencapsulation of agrochemicals, fertilizers, and pesticides for improved plant production. In: Advances in Phytonanotechnology. Academic Press, pp. 279–298.

Winter, J., Nicolas, J., Ruan, G., 2020. Hybrid nanoparticle composites. J. Mater. Chem. B 8, 4713–4714.

Worrall, E.A., Hamid, A., Mody, K.T., Mitter, N., Pappu, H.R., 2018. Nanotechnology for plant disease management. Agronomy 8, 285.

Yilmaz, A., Bozkurt, F., Cicek, P., Dertli, E., Durak, M.Z., Yilmaz, M.T., 2016. A novel antifungal surface–coating application to limit postharvest decay on coated apples: molecular, thermal and morphological properties of electrospun zein–nanofiber mats loaded with curcumin. Innov. Food Sci. Emerg. Technol., 74–83.

Youssef, K., Hashim, A.F., 2020. Inhibitory effect of clay/chitosan nanocomposite against *Penicillium digitatum* on citrus and its possible mode of action. Jordan J. Biol. Sci. 13 (3), 349–355.

Zahid, N., Maqbool, M., Ali, A., Siddiqui, Y., Abbas Bhatti, Q., 2019. Inhibition in production of cellulolytic and pectinolytic enzymes of *Colletotrichum gloeosporioides* isolated from dragon fruit plants in response to submicron chitosan dispersions. Sci. Hortic. 243, 314–319.

Zhang, X., Xu, Z., Wu, M., Qian, X., Lin, D., Zhang, H., Tang, J., Zeng, T., Yao, W., Filser, J., Li, L., Sharma, V.K., 2019. Potential environmental risks of nanopesticides: application of $Cu(OH)_2$ nanopesticides to soil mitigates the degradation of neonicotinoid thiacloprid. Environ. Int. 129, 42–50.

Zhao, P., Cao, L., Ma, D., Zhou, Z., Huang, Q., Pan, C., 2017. Synthesis of pyrimethanil-loaded mesoporous silica nanoparticles and its distribution and dissipation in cucumber plants. Molecules 22, 817.

Zhu, L., Chen, L., Gu, J., Ma, H., Wu, H., 2022. Carbon-based nanomaterials for sustainable agriculture: their application as light converters, nanosensors, and delivery tools. Plants 11 (4), 511.

CHAPTER

2

Current topics of nanotechnological approach in agriculture: A case study on nano-based fungicides

Liliana Sofía Farías-Vázquez[a], *Rodolfo Ramos-González*[b], *Sandra Pacios-Michelena*[a], *Cristóbal Noé Aguilar-González*[c], *Roberto Arredondo-Valdés*[a], *Raúl Rodríguez-Herrera*[c], *José Luis Martínez-Hernández*[a], *Elda Patricia Segura-Ceniceros*[a], *and Anna Ilíná*[a]

[a]Nanobioscience Group, Chemical Sciences School of the Autonomous University of Coahuila, Saltillo, Coahuila, Mexico [b]CONAHCYT-Autonomous University of Coahuila, Saltillo, Coahuila, Mexico [c]Department of Food Research, Chemical Sciences School of the Autonomous University of Coahuila, Saltillo, Coahuila, Mexico

HIGHLIGHTS

- The use of nanotechnology has helped resolve problems related to food production.
- Nanoparticles (NPs) could enhance or inhibit microorganisms.
- The nanoscale formulation can potentialize the effect of pesticides and fertilizers.
- The type of NP and the type of organism determines their interactions.

1 Introduction

According to previous reports, the world's population will reach nine billion by 2050. This increase will cause an expansion in resource demand, including food, which will go from 59% to 98% (Neme et al., 2021). Currently, there are approximately 815 million starving people,

Nanohybrid Fungicides
https://doi.org/10.1016/B978-0-443-23950-2.00006-0

and it is expected to increase to 2 billion by 2050 (Usman et al., 2020). Farmers worldwide are focusing on using recent technologies to enhance food production (Neme et al., 2021). Agriculture is a science that provides food for the world through food crops, which are vital for the human population (Hoang et al., 2022; Keswani et al., 2019). One of the sustainable development goals set by the United Nations (UN) is to achieve food security and engender agricultural sustainability by 2030. Therefore, it is crucial to take proactive steps to improve food yield (Falade et al., 2021). Sustainable agriculture is vital for achieving this objective. Global annual crop production exceeds three billion tonnes. To obtain this amount of products, agrochemicals are required. One hundred eighty-seven million tonnes of fertilizer and four million tonnes of pesticides are used worldwide (Usman et al., 2020). However, improper and extensive use of agrochemicals can be dangerous to the environment and causes damage to human physical and mental health (Hoque et al., 2022).

Nanotechnology has proved to be a potential tool in different sciences. One of them is agriculture, where nanotechnology has been shown to have multiple direct or indirect uses (Zobir et al., 2021). Nanotechnology helps suppress phytopathogens, improve plant nutrition, detect diseases, and control the release of functional molecules, biosensors, and biopesticides (Ashraf et al., 2022; Baniamerian et al., 2021). Different studies have demonstrated that nanomaterials can be a solution to plant pathogens (Nisha Raj et al., 2021). That is why researchers are focusing on tracking elements in nanometer-sized materials. The positive effects of using nanoparticles (NPs) have been attributed to their high specific area, reactivity, specificity, and dispersibility (Baniamerian et al., 2021). Therefore, nanotechnology applications to resolve crop problems have gained attention in the last years (Usman et al., 2020).

Nanotechnology can have many applications in agriculture, which can be classified into five main categories:

(a) Nanopesticides: Nanomaterials can have antifungal activities and are applied to control fungal destruction.
(b) Stabilization of biopesticides using nanotechnology: Nanotechnology can improve the efficacy of existing pesticides (Acharya and Pal, 2020).
(c) Nanocarriers: Plant nutritional deficits can be ameliorated by designing a smart delivery system for controlled nutrient release (Acharya and Pal, 2020; He et al., 2019).
(d) The transport of genetic material by nanomaterials: Nanotechnology can improve crops, making them resistant to pathogens and stress.
(e) The development of nanosensors: Nanotechnology can increase the sensitivity and specificity of biosensors (Acharya and Pal, 2020).

Nanomaterials in agriculture focus on reducing product losses and optimizing nutrient management (Suriya Prabha et al., 2022). The application of nanotechnology in agriculture is still under study, but it is expected to bring changes due to its large-scale applications (Acharya and Pal, 2020).

Nanotechnology has been applied to different aspects of food production, from agriculture to food processing and transportation. However, many studies are focusing on in vitro and in vivo toxicity of nanomaterials. These studies were conducted to apply nanomaterials to products consumed by people (He et al., 2019).

This chapter aims to describe the current concepts on essential aspects of agricultural nanotechnology, including a description of nanotechnology, NPs-microorganisms, and NPs-plant interactions, the use of NPs as fertilizers, and possible and possible alternatives to solve problems that still affect crop production.

2 Nanotechnology in agriculture

Nanotechnology is an upcoming technology with numerous applications in various sectors (Rathore and Mahesh, 2021). Nanotechnology is defined as the fusion, design, characterization, and use of tools and systems with size variations from 1 to 100 nm. It has a variety of applications in different fields, including science (Muthukrishnan, 2022; Neme et al., 2021; Nisha Raj et al., 2021). The prefix "nano" comes from Greek word, which is: "nanos," and the meaning is "very little." The small dimensions of the NPs give them a large specific surface area and reactivity (Salem et al., 2023). NPs are commonly classified into organic and inorganic. Organic NPs contain carbon, while inorganic ones contain metals, magnets, and semiconductors components (Zorraquín-Peña et al., 2020).

Researchers from all over the world are concentrating on nanotechnology, generating paradigms for the treatment of diseases (Thipe et al., 2022). Initially, these applications focused on physics and chemistry, but in recent years it has been extended to different areas, one of them being agriculture. Reports have shown that NPs can solve challenges related to agriculture and can be positive in resolving crop problems because NPs can be applied at different stages of crop development (Bartolucci et al., 2022; Hoang et al., 2022; Juárez-Maldonado et al., 2021; Muthukrishnan, 2022). In the last decade, metallic NPs have gained attention due to their applications in disease-directed drug design (Thipe et al., 2022).

Agriculture-linked nanotechnology can result in accurate tools for pest diagnosis and treatment (Zobir et al., 2021). The application of nanotechnology in agriculture has led to the production of agrochemicals, including nanopesticides, nanofertilizers, nanofungicides, and nanoinsecticides (Nisha Raj et al., 2021). Agrochemicals are essential because they defend plants against pests and diseases and provide fertilizer (Machado et al., 2022). However, conventional agrochemicals are dangerous for the environment and food chains. The application of nanotechnology can increase the quality and quantity of yields, reduce environmental hazards and promote a sustainable ecosystem (Nath et al., 2019). Furthermore, nanotechnology has given rise to new agricultural products with a higher potential to manage crop problems (Worrall et al., 2018).

The aforementioned products are functional because the use of NPs can increase the tolerance of plants and some beneficial microorganisms for plants to different stress models, increasing their ability to produce secondary metabolites (Juárez-Maldonado et al., 2021). The use of NPs has been carried out with the help of fewer chemical products, improves shelf life, increases the solubility of poorly water-soluble pesticides, and reduces the use of toxic solvents, leading to less addition of chemical substances in the soil (Bartolucci et al., 2022; Neme et al., 2021; Worrall et al., 2018).

NPs can act as nanocarriers to encapsulate, bind or trap the active particles to produce a potent formulation. Silica NPs are used for this purpose. The shape, structure, and size of these NPs can be controlled. Chitosan NPs are also used for this purpose. They are biodegradable, biocompatible, nontoxic compounds with antifungal, antimicrobial, and antiinflammatory properties. NPs can also be used as carriers for fungicides (Kutawa et al., 2021). Nanomaterials can affect different cellular compartments in living organisms, promoting dissimilar cellular interactions depending on their free energy. NPs can help microorganisms react under stress conditions (Juárez-Maldonado et al., 2021). The compact size of NPs is an advantage because it allows them to cross biological membranes, allowing interaction with proteins and, as an effect, causing metabolic changes (Hoang et al., 2022).

The use of nanotechnology can also make it possible the control of plant diseases (Castro-Restrepo, 2017). NPs provide the microorganism with a large area to which it can adhere. They can facilitate hydrolysis. NPs can rapidly penetrate the cell wall of microorganisms and affect or improve their metabolism (Baniamerian et al., 2021). An example is magnetite, which can enhance the production rate of some microorganisms because it can act as an electron conductor or decrease cell viability in others (Juárez-Maldonado et al., 2021).

Shielding NPs are materials with a range of 10–100 nm. NPs can be used in different parts of the plant to defend against pathogens (Kutawa et al., 2021). NPs can be used to protect plants via two mechanisms: they can provide crop protection, or they can be used as carriers for pesticides and can be applied by spraying or dipping to tissues, seeds, or roots. NPs used as carriers provide several benefits, such as increased shelf life, better feed solubility, reduced toxicity, and increased site-specific uptake by target pests. Nanopesticides can also resist environmental pressures, causing a reduction in the number of applications and reducing their cost (Worrall et al., 2018).

The use of NPs has many potentials, and they have shown more pronounced results than the possible risks they can cause (Juárez-Maldonado et al., 2021). The use of NPs in agriculture points to the progress of clean, safe, and ecological nanomaterials, using biocompatible and nontoxic solvents, natural biodegradable and biocompatible matrices, and sustainable and energy-efficient processes (Pascoli et al., 2019). The application of nanotechnology in agriculture is being studied in plant chemical delivery, nanosensors, seed monitoring, agrochemical release, and gene transfer. Although the application of NPs in agriculture has many advantages, they also have some limitations, such as the generation of waste and the use of toxic solvents (Kutawa et al., 2021).

3 Microorganisms in agriculture and their interaction with nanoparticles

Crops are affected by numerous biotic and abiotic factors. Biotic factors involve diseases caused by viruses, bacteria, fungi, nematodes, and insects, decreasing crop yields by 20%–30%. Fungal pathogens are responsible for millions of losses per year. One of the most devastating diseases is *Fusarium* wilt, produced by *Fusarium oxysporum* and altering different crops, affecting their nutritional values (Ashraf et al., 2022; Devi et al., 2022). This disease causes the loss of 10%–15% of crops per year (Khanna et al., 2022). Furthermore, this fungus is difficult to manage due to its survival in the soil, has a long incubation period, and can cause disease with a low inoculum level (Jamil et al., 2022).

Another example is *Ralstonia solanacearum*, the second most devastating plant pathogen worldwide. This bacterium is responsible for important agricultural disease, which affects more than 250 vegetable crops on Earth (Xu et al., 2022). This microorganism conquers the lateral roots of the plant and, through bacterial colonization, invades the host

(Zaki et al., 2022). *R. solanacearum* survives in soil, water, and infected plant tissues for a long time, which represents difficulties in controlling this pathogen (Xu et al., 2022).

To combat the aforementioned diseases, chemical treatments have been applied. However, these products accumulate in the food chain and damage soil microorganisms (Devi et al., 2022). In response to this problem, sustainable agriculture based on the depletion of the use of agrochemicals in crops has been proposed (Keswani et al., 2019). It has been proven that some species of fungi and bacteria can promote plant growth and reduce the causes of diseases that affect crops. Thus, microorganisms have multiple functions in agriculture. Therefore, some of them can be used in biological control to combat plant diseases and pesticides. The production of secondary metabolites determines this antagonistic activity. Microbial secondary metabolites are low molecular mass products and enzymes produced in the last growth phase by secondary metabolism (Keswani et al., 2019). These metabolites can also be used to biocontrol different plant pathogens (Falade et al., 2021).

An example is the bacterial genus *Bacillus*, which is currently used as a biocontrol agent. *Bacillus* can produce siderophores, extracellular metabolites, enzymes that give *Bacillus* its antagonistic activity, and endospores that make it resistant to the environment (Devi et al., 2022). Nanoencapsulation is considered one of the best alternatives to protect these metabolites from environmental factors that can cause their destruction, influencing their ability to inactivate pathogens and their excellent performance (Kamaruzaman et al., 2022). Nanomaterials also have antibacterial activity against pathogenic microorganisms, commonly used in the food sector. There are two types of antimicrobial agents present in NPs, organic and inorganic (Kumar et al., 2020).

Environmental agents can biostimulate or inhibit microorganisms. Factors outside the acceptable ranges can cause the death of the organism. Otherwise, a response to environmental factors that cause adaptive modifications of the metabolic process can regulate the growth of microorganisms, causing a more efficient use of environmental resources. Among the environmental factors that affect microorganisms are nanomaterials of metals, semimetals, and nonmetal nanomaterials. Their use shows physical and chemical effects related to their surface properties and compositions (Juárez-Maldonado et al., 2021). Metal-containing nanocomposites are used to enhance plant growth and inhibit the presence of plant pathogens. Copper, zinc, or silver NPs are known for their cytotoxicity and their ability to change the dynamics of microbes present in the soil. Cupric oxide (CuO) NPs have shown the ability to inhibit the growth of microorganisms due to their antimicrobial and antiviral properties. In the last decade, studies have focused on the use of NPs to inhibit the presence of bacterial pathogens in plants (Luong et al., 2022).

The nanomaterial-microorganism interaction initially takes place through surfaces or interfaces. According to the properties of nanomaterials, there is an impact they have on cell walls. When nanomaterials meet the environment, their surface is immediately modified by interaction with the compounds (Juárez-Maldonado et al., 2021). The interaction between nanomaterials and proteins leads to the development of a protein-modified surface, that is, a protein layer that adheres to the surface of nanomaterials (Chetwynd and Lynch, 2020). The nature of the modified surface makes the collision of nanomaterials with an organism challenging to predict, so microorganisms have different reactions with NPs. The functional groups of the modified surface are interrelated with the receptor surfaces of the cell. The sizes, structure, and nature of nanomaterials define the properties related to their impact on cells (Juárez-Maldonado et al., 2021). The main reason NPs are considered an alternative to antibiotics is that

NPs can effectively prevent microbial drug resistance. The extended use of antibiotics has led to many public health risks. Therefore, the search for new effective bactericidal materials is essential. NPs have emerged as a promising approach to combat drug resistance. However, in some cases, NPs can also promote the emergence of bacterial resistance (Wang et al., 2017). Due to its antimicrobial properties, it has been widely used in medicine and biotechnology. Silver NPs have antimicrobial properties and have been used in biotechnology fields previously. Different studies have evaluated the efficacy of silver NPs in inhibiting the growth of pathogenic bacteria such as *Staphylococcus aureus, Streptococcus mutans, Streptococcus pyogenes, Escherichia coli,* and *Proteus vulgaris* (Zorraquín-Peña et al., 2020).

The structure of biofilms makes bacteria highly resistant to foreign chemicals. Previous reports have shown that NPs compromise biofilm integrity. Bacteria like *Bacillus subtilis* grow to a point where the edge of the biofilm periodically stops growing, allowing nutrients to flow into the biofilm's center. In this way, the bacteria in the nucleus can resist foreign substances. However, in this case, NPs can affect the rate of bacterial adhesion and biofilm formation, but the exact mechanism is not fully understood. Researchers have demonstrated that NPs can affect the metabolism of bacterial communities. Bacterial metabolism is an important activity for biofilms. The long-range electrical signaling of bacteria in a biofilm occurs through channels of potassium ions. Furthermore, the diffusion of potassium ions coordinates the metabolic activities of bacteria inside and outside the biofilm. NPs can also adhere to and diffuse into biofilms, leading to potential membrane disruption, increased lipid peroxidation, and DNA binding. Disruption of the normal functioning of these processes reduces the ability of bacteria to form biofilms (Wang et al., 2017).

As an alternative to control the sensitivity to environmental factors, microorganisms can be confined to a specific region of solid support. This technology is better known as immobilization (Lou et al., 2019). Some immobilization mechanisms involve the adsorption of cells (Fig. 1) onto or within a matrix that prevents cells from being released but is porous enough to allow the diffusion of substrates and products (Lapponi et al., 2022).

Different studies have shown that immobilization technology can provide microorganisms with higher cell density, adaptability, reusability, and tolerance to negative factors (Lou et al., 2019; Wang et al., 2020). However, standard bacterial immobilization has several

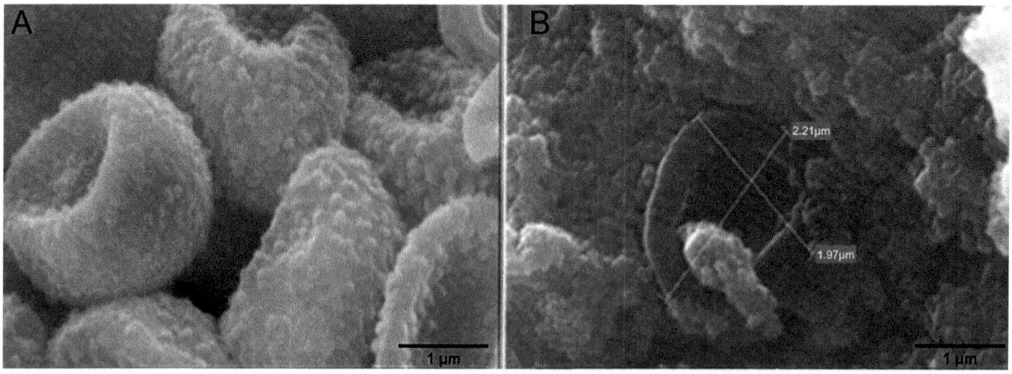

FIG. 1 Scanning electron microscope (SEM) microphotographs of (A) *Trichoderma* sp. free spores; (B) *Trichoderma* sp. spores immobilized on chitosan-coated magnetic magnetite nanoparticles. *Own elaboration.*

disadvantages. The most important is poor cell recovery. Therefore, magnetic NPs, especially iron oxide NPs, are the most widely used material for immobilization (Fig. 1) that allows magnetic separation of microorganisms (Wang et al., 2020).

The effect of NPs on stimulating the production of metabolites by microorganisms was demonstrated. Sun et al. (2021) reported that the use of $MnFe_2O_4$ NPs in anaerobic fermentation could provide a suitable environment for bacteria, causing better action of hydrogenase enzymes and electron transfer efficiency, improving bio-H_2 production using 400 mg/L of manganese ferrite NPs with the yield to 272.2 mL H_2/g glucose. This effect is related to the iron and manganese required for hydrogenase activity and microbial growth (Sun et al., 2021).

Zhong et al. (2020) also reported the efficiency of magnetite (Fe_3O_4) NPs in enhancing biohydrogen production. In batch experiments, the results revealed the maximum hydrogen yield of 12.97 mL H_2/g glucose with 50 mg/L hydrogens using 40–60 nm of magnetite. The abundance of butyrate-hydrogen-producing bacteria (*Clostridium*) decreased from 40.55% to 11.45%. In comparison, the quantity of ethanol-hydrogen-producing bacteria (*Acetanaerobacterium* and *Ethanoligenens*) increased from 19.62% to 33.35% with NPs, corroborating that the fermentation changed from butyrate type to ethanol type. Magnetite NPs contributed to hydrogen production, creating a more favorable environment to enhance the bacteria activity and proliferation (Zhong et al., 2020). These data show promising opportunities to use NPs to interact with microorganisms and influence metabolite production. Therefore, NPs can be considered control agents. However, there is not enough information about the mechanisms of this behavior and the environmental effect of NPs on plants. Some reports have indicated that the presence of NPs can cause changes in soil microbial communities and their interaction with plants (Ramírez-Valdespino and Orrantia-Borunda, 2021).

4 Nanobiofungicides

Agriculture plays an essential role in providing food and being a food source for many countries. About 86% of people living in rural areas depend on agricultural cultivation (Kutawa et al., 2021). Most fungi have a mutualistic relationship with plants and are saprotrophs, and they aren't plant parasites (Cardoso, 2020). Although few fungi are considered phytopathogens, and they are responsible for devastating epidemics in crop plants (El-Baky and Amara, 2021). To be considered a pathogen, a microorganism needs to be able to cause disease and completes its life cycle in the host plant (Cardoso, 2020). Fungal pathogens cause 70%–80% of yield losses. There are one to five million species classified under the "fungi" kingdom. Around 80%–90% of spray fungicides are lost after their application (Kutawa et al., 2021).

Pesticides are chemical substances that are used to eradicate insects, plants, rodents, and microorganisms that affect plants (Falck et al., 2020). Fungicides are a particular type of pesticide. Their main objective is to prevent or eradicate fungal diseases in plants or seeds. Commercial agriculture depends on the consistent application of chemical fungicides to protect plants against pathogenic fungi. Most fungicides have been reported to be of low toxicity. Even though some fungicides are responsible for human health problems (Gupta, 2018). The use of chemical fungicides is related to toxic effects on the living system, human and animal health, and the environment. Fungal phytopathogens are becoming resistant to chemical fungicides. Therefore, researchers are looking for a nontoxic environmentally friendly alternative to control fungal plant diseases (El-Baky and Amara, 2021). Some fungicides such as alkyldithiocarbamic, halogenated substituted monocyclic aromatics, carbamic acid derivatives, ferbam, mancozeb, and maneb metabolites, HCB, benzimidazoles, chloroalkylthiodicarboximides, and tridemorph are responsible for the development of toxicity and oncogenesis (Gupta, 2018).

All countries of the world use fungicides in plants to control pathogenic fungi (Cardoso, 2020). Fungicides are predominantly used on fruits and vegetables. They make up more than 35% of the pesticides on the market. The main demand for fungicides is in Europe, where the significant applications are for grains, cereals, fruits, and vegetables. In the United States, the main fungicide groups are dithiocarbamates, chloronitriles, demethylation inhibitors, and strobilurins (Zubrod et al., 2019). A commonly used group of pathogens is carboxamide fungicides. The active ingredient in these fungicides is fluxapyroxad. Strobilurins fungicides are another group of fungicides, which are natural chemical structures isolated from the genera *Strobilurus*. Another group of fungicides is the demethylation inhibitors which contain triazole fungicides (Cardoso, 2020). Fungicides are essential for the treatment of fungal diseases in global agricultural production. The use of fungicides is considered essential to ensure the world's food supply. However, fungicides can have devastating effects on the environment and be highly toxic for some organisms (Zubrod et al., 2019). Although synthetic fungicides have proven to be effective in controlling fungal diseases, they have multiple disadvantages (Rodrigo et al., 2022). Fungicides can be toxic to nontarget organisms because they act on fundamental biological processes that aren't specific to fungi. Fungicides aren't just applied in agriculture. They can also be used in urban areas and private spaces. Consequently, fungicides can cause potential risks to the environment to these applications. In addition, fungicides also cause indirect effects, which are defined as effects mediated by another organism group that has been directly affected; fungicides affect other microorganisms (Zubrod et al., 2019).

Various copper and organic fungicides are commonly used on crops to control plant diseases, resulting in soil, surface, and groundwater pollution (Ortega et al., 2022). Most seed-treated fungicides have a systemic action, so they can be absorbed into plant tissue where they protect against pests. Although soil-applied fungicides can attack soilborne pathogens, they have the potential to persist on the plant for several months (Zubrod et al., 2019). Government review presents a particular interest in the development of microbial fungicides. These biofungicides are safer than synthetic chemical fungicides (El-Baky and Amara, 2021). Fungicides are essential for the treatment of fungal diseases in global agricultural production. The use of fungicides is considered essential to ensure the world's food supply. However, fungicides can have devastating effects on the environment and be highly toxic for some organisms (Zubrod et al., 2019). Although synthetic fungicides have proven to be effective in controlling fungal diseases, they have multiple disadvantages (Rodrigo et al., 2022). Fungicides can be toxic to nontarget organisms because they act on fundamental biological processes that aren't specific to fungi (Zubrod et al., 2019).

The most widely used method for managing fungal diseases is chemical fungicides. These toxic compounds leave toxic residues in plants. Chemical fungicides pollute the environment. As an alternative, biocontrol agents to control fungal diseases decrease the ecological effects on the environment (El-Mehy et al., 2022). Unfortunately, agricultural practices often involve the misuse of pesticides, leading to the leaching of these chemicals into the soil and groundwater contamination (Zambito Marsala et al., 2020). An example of the incorrect daily use of fungicides is the black pods' cacao disease caused by *Phytophthora* species that causes devastating damage in production regions around the world (Nyadanu et al., 2019). In Nigeria, the most effective and widely used product to combat this disease is a copper-based fungicide. But the downside is that farmers in this country commonly misuse this fungicide, causing elemental damage to the environment (Sowunmi et al., 2019).

Many fungi species have shown that they can produce multiple benefits in host plants and protect them against fungal diseases. Table 1 shows the categories in which fungal metabolites can be classified. The category with the highest percentage of compounds is Polyketide, with 53%. This category is conformed up of 15 chemical substances, with isocoumarin being

TABLE 1 Some categories of fungal metabolites, indicating examples and the percentage of compounds in each category.

Categories of fungal metabolites	Examples	Percentage (%) of compounds
Polyketide	Isocumarain Furan derivative Acetophenone derivative Anthraquinone Aromatic polyketide Cytochalasin derivative Dihydroisocoumarin Diphenyl ether Furofurandione Halogenated Cyclopentenone Isochroman Naptho-α-pyrone Phenylethanoid Polyene Xanthene	53
Terpenoid	Cyclohexanoid Meroterpenoid Monoterpenoid Pyrrolidinone derivative Sesquiterpene	13
Nonribosomal peptide	Dilactone Epidithiodoketopoperazines Lactone Nonapeptide complex	10
Phenols	*P*-Quinone Methyl ketone Phenyl ether	7
Aliphatic compound	Fatty acid	5
Alkaloid	Alkaloid Chaetoglobosin Cholinated benzophenone	7
Others	Inorganic acid Inorganic compound	5

1. Types of nanohybrid fungicides

the most predominant followed by terpenoid, with 13%. The next category is nonribosomal peptides, with 10%. Continued by phenol and alkaloid, with 7% of each one. And the categories with a lower percentage of compounds are aliphatic compounds and "others," with 5% each. The category "others" is integrated by inorganic acids and inorganic compounds. Fungi can be used to develop new bioproducts that can reduce the use of chemical fungicides. In recent years, endophytic fungi have become an alternative to organic agriculture. These fungi are found within plant tissue, resulting in a closer relationship with hosts. Endophytic fungi produce secondary metabolites with different properties that make them have multiple applications in various industries. A wide range of produced metabolites has antimicrobial properties (Rodrigo et al., 2022).

Biofungicides are an attractive option for the control of fungal pathogens. Biofungicides are environmentally safe, affordable, and persistent in the field (dos Santos Gomes et al., 2021). An example of this case is corn wilt produced by *Magnaporthiopsis maydis*. Early blight is a potato disease caused by *Alternaria* species, especially *A. solani*. Fungicides are regularly used to control this disease. A disadvantage of the use of fungicides is that *A. solani* is developing resistance against this type of pesticide (Wolters et al., 2021). Corn wilt is a specific disease that affects maize production in Israel and Egypt. The genus *Trichoderma* is a biocontrol agent used to minimize infectious plant diseases. Degani et al. (2021) studied the effect of *Trichoderma asperellum* controlling maize wilt disease. The results showed that *T. asperellum* produced 6-pentyl-α-pyrone as a secondary metabolite. This metabolite was identified as a highly potent antifungal against plant pathogens (Degani et al., 2021).

NPs can be used to combat plant diseases due to their antimicrobial properties (Avila-Quezada et al., 2022). Nanotechnology used in agriculture is new. NPs can have many applications against infections. They can be used directly or as carriers for pesticides. Living organisms such as yeast, plants, or bacteria can be used to synthesize NPs. This synthesis of NPs has advantages over the traditional methods, like less energy consumption and avoiding the use of toxic solvents (Ali et al., 2020). Although, previous studies have shown that some metal NPs have antifungal activity. The application of nanomaterials as antifungal agents has shown promising results due to their high surface area/volume ratio. Phytochemical-derived nanostructured antifungal agents are a good option against plant pathogenic fungi (Osonga et al., 2022). NPs are crucial to suppress fungal development (Kutawa et al., 2021). Studies have shown that the adhesion of NPs with microbial cell membranes occurs due to the electrostatic attraction between the cell membrane of the microorganisms. The charge of the NP is also affected. The NPs affect the membrane structure, which causes depolarization, affects the membrane permeability, and damages the cell structure, causing its death. The effect of NP on the cell membrane also affects the internal cell content, affecting the production of proteins, enzymes, and metabolites of microorganisms (Ali et al., 2020).

There are studies of the antifungal activities of NPs. The most studied carriers are silica, chitosan, and polymer mixes, which show promising results (Kutawa et al., 2021). In recent years, the development of nanopesticides has been applied in agricultural practices. An alternative for the synthesis of NPs is the green synthesis mediated by microorganisms. This synthesis can occur extracellularly or intracellularly. For extracellular synthesis, the color change can be easily observed. The culture can be centrifugated and mixed with an aqueous metallic salt solution to obtain the product. For intracellular synthesis, NPs are collected by ultrasound, centrifugation, and washing (Ali et al., 2020). Table 2 provides some examples of NPs that represent green nanotechnology applications in fungal management.

TABLE 2 Plant extracts used to synthesize nanoparticles.

Nanoparticle	Green synthesis	Against phytopathogenic fungi
Ag NPs	*Alfalfa* sprouts Extract of *Punica granatum* peels *Ziziphora tenuior* leaves extract	*Rhizoctonia solani* *Sclerotinia sclerotiorum* *Alternaria alternata* *Botrytis cinerea* *Macrophomina phaseolina* *Curvularia lunata*
ZnO NPs	*Parthenium hysterophorus* leaves extract *Camellia sinensis* leaves extract	*Aspergillus fumigatus* *Aspergillus flavus* *Aspergillus niger*
Au NPs	Fresh leaves extract or Flowers of *Magnolia kobus, Diospyros kaki, Azadirachta indica, Mentha piperita, alfalfa, Helianthus annus, Moringa oleifera,* and *Artemisa dracunculus*	*Aspergillus flavus* *Aspergillus niger* *Puccinia graminis*
Cu NPs	Leaf extract of *Magnolia, Euphorbia nivulia* stem latex, *Carica papa,* and *Aloe vera*	*Fusarium oxysporum* *Fusarium culmorum* *Fusarium graminearum*

Based on El-Baky, N.A., Amara, A.A.A.F., 2021. Recent approaches towards control of fungal diseases in plants: an updated review. J. Fungi 7 (11), 900. https://doi.org/10.3390/JOF7110900. Article is licensed under an open access Creative Commons CC BY 4.0 license with permission from MDPI.

Plant extracts are considered nonphytotoxic and effectively control plant pathogens (dos Santos Gomes et al., 2021). Essential oils of *Origanum compactum* are used for the chemical composition of some fungicides due to their antifungal properties. Lucarini et al. (2022) apply a phytochemical study with these essential oils. Phytochemical screening was performed by gas chromatography-mass spectrometry. The main components of the essential oils were carvacrol, thymol, gamma-terpinene, and *o*-cymene, at 38.73%, 31.46%, 11.11%, and 9.07%, respectively. The essential oils were tested against *Candida albicans, Aspergillus flavus, Aspergillus niger*, and *F. oxysporum*. The results revealed that the vital oils inhibit 100% of the growth of these fungi (Lucarini et al., 2022).

In vitro, studies have been carried out using plant extracts as reducing agents for the synthesis of NPs. Plant materials used for NP synthesis are more beneficial than microbial or chemical methods because they do not have microbial or chemical effects. Plant extracts can be obtained from different parts (Ali et al., 2020). Although a wide range of fungi was tested for the efficacy of nanofungicides, only a few plants were assayed (Worrall et al., 2018). Most of the mentioned information was obtained through laboratory studies. It is crucial to apply all this information in in vivo studies to obtain knowledge about how plants and pathogens react under the use of biofungicides or nanobiofungicides (Ali et al., 2020).

Therefore, NPs are used to inhibit microbial growth and enhance metabolism. In addition, there is great interest in the ability of microorganisms to synthesize NPs. Three methods are currently known to synthesize nanomaterials: physical, chemical, and biological. The biological route includes the use of microorganisms (Ramírez-Valdespino and Orrantia-Borunda, 2021). Fungi are used to synthesize NPs and have shown advantages over other microorganisms. These advantages are due to their ease of propagation, economic viability, and the presence of mycelium that provides a large surface area (Salem et al., 2023) An example is

TABLE 3 Nanoparticles synthesized by the genus *Trichoderma* and microorganisms controlled by them.

NPs	Size (nm) and shape	Inhibited microorganisms
Ag	4–182 nm, spherical, ellipsoidal, irregular, oval, anisotropic, prism	*Candida* sp., *Aspergillus* sp., *Fusarium* sp., *E. coli, Pseudomonas aeruginosa, Staphylococcus aureus, Sclerotinia sclerotiorum, Klebsiella pneumonia, Alternaria alternata, Pyricularia oryzae, Streptococcus thermophiles, Bacillus subtilis, Aedes aegypti, Sclerotium rolfsii, Fusarium oxysporum, F. solani, F. moniliforme, F. semitectum, F. roseum, Rhizoctonia solani, Salmonella typhimurium*
ZnO	8–200 nm, spherical, fan/bouquet structure, hexagonal wurtzite, and peaks	*A. alternata, P. oryzae, Sclerotinia sclerotiorum, Xanthomonas oryzae*
Au	5–124 nm, spherical, hexagonal, pentagonal	*Bacillus subtilis, Pseudomonas aeruginosa, Serratia* sp., *E. coli, Aspergillus niger, A. flavus, A. fumigatus, Klebsiella granulomatis, P. syringae, P. aeruginosa, S. aureus, Shigella sonnei*
Se	20–312.5 nm, spherical, pseudo-spherical, hexagonal, irregular	*Fusarium* sp., *Alternaria* sp., *Sclerospora graminicola*
Si and Cu	55.5 nm (Si) and 56 nm (Cu), irregular spherical	*Poria hypolateritia, Phomopsis theae*
Au and Ag	10–75 nm, triangular and spherical	*Phomopsis theae*
α-Fe_2O_3	207 nm	*Sclerotinia sclerotiorum*
Chitosan	300 nm, spherical	*F. oxysporum, R. solani, S. rolfsii*
TiO_2	74.4–87.5 nm, spherical	*Helicoverpa armigera*

Based on Ramírez-Valdespino, C.A., Orrantia-Borunda, E., 2021. Trichoderma *and nanotechnology in sustainable agriculture: a review. Front. Fungal Biol. 2, 764675. Article is licensed under an open access Creative Commons CC BY 4.0 license with permission from Frontiers.*

Trichoderma, a microorganism commonly used for its antimicrobial activity, which is considered a component of the biosynthesis of NPs. *Trichoderma* produces secondary metabolites and enzymes that help convert ions into elemental metals. Studies have shown that *Trichoderma* has the potential for NP biosynthesis (Table 3), related to its resistance to metals, high pressure, and extracellular protein production (Ramírez-Valdespino and Orrantia-Borunda, 2021). The NPs synthesized by *Trichoderma* were active as inhibitors against different phytopathogenic microorganisms (Table 3).

Therefore, NPs can inhibit some microorganisms, stimulate the metabolism and proliferation of others, as well as synthesize NPs.

5 Nanostructured fertilizers and pesticides

Horticultural plants have their own physiological and morphological characteristics but require better nutrients from the soil to enhance their development. Fertilization is the most successful method and provides crops with a subsistence of macro and micronutrients (Sayed and Ouis, 2022). Fertilizer is a source of soil nutrients that promotes plant growth and expands productivity (Tarafder et al., 2020). The use of bio-based fertilizers and pesticides has intensified to replace environmentally harmful agrochemicals. The intensive use of agrochemical fertilizers has led to many adverse effects on the ecosystem, such as low adsorption capacity of nutrients by crops and high deprivation of chemical compounds in groundwater (Shalaby et al., 2022; Zhu et al., 2022). Chemical fertilizers are nitrogen, phosphorus, and potassium, which are necessary for plant growth (Elnahal et al., 2022). The enclosure of chemical and biological fertilizers in NPs is advantageous due to its potential for the controlled release of substances and protection of active substances against premature degradation (Machado et al., 2022).

In addition, factors related to the development of resistance of pathogens to fungicides have contributed to the need to reduce dependence on chemicals (Kalwani et al., 2022). As a result, this increases soil fertility, solubilizes insoluble phosphates in the soil, and generates chemicals impacting plant growth (Sayed and Ouis, 2022). *Bacillus* biofertilizer, a nontoxic multifunctional fertilizer, could be used due to functional bacteria to improve soil fertility and quality and reduce the toxicity of heavy metals and other pollutants (Zhu et al., 2022). The use of *Bacillus* in the control of plant pathogens involves the suppression of pathogens to favor plants, which is an alternative to chemical pesticides and fungicides on organic farms. Nanoencapsulated bacterial biocontrol agents and bacteria can stimulate plants' resistance by activating host defense mechanisms (El-Mehy et al., 2022). Nanoscale formulation improves delivery and better stability and controlled dispersion of nanopesticides and nanofertilizers. Furthermore, the nanometric size of the particles leads to efficacy compared to conventionally formulated products (Sun et al., 2019).

Nanofertilizers have a remarkable effect on the beneficial phytomicrobiome. Depending on the period of exposure to nanofertilizers in the soil, their function is to strengthen the microbial community. However, prolonged exposure can alter the structure of the soil microbial population (Kalwani et al., 2022). Furthermore, the effect of nanofertilizers depends not only on the number of minerals present in the ground but also on their availability for plants (Nath et al., 2019). Kumar et al. (2022) compared nitrogen and zinc nanofertilizers with chemical fertilizers. The results showed that nanofertilizer applications improve the average yield. The result was 5.35% higher in wheat, 24.24% higher in sesame, 4.2% higher in pearl millet, and 8.4% higher in mustard yield. The obtained results suggest that nanotechnology intervention with organic agriculture can minimize the amount of chemical fertilizers used to improve crop production (Kumar et al., 2022).

According to the United States Environmental Protection Agency, a pesticide is a general name for agricultural chemicals: herbicide, insecticide, fungicide, nematicide, and rodenticide (United States Environmental Protection Agency, 2007). However, a better explanation of these products is that they are chemical substances used to attack pests or prevent plant diseases. Some of the standard components of these pesticides are organophosphates, chlorinated hydrocarbons, carbamates, and carbamide derivatives. However, the conventional formulation of pesticides has limitations, including organic solvent content, low dispersibility, and the ability to remain in the soil. Therefore, researchers are focusing on developing nontoxic alternatives for pest and disease control (Abdollahdokht et al., 2022). Their function is to prevent or destroy any pest (Zobir et al., 2021). Furthermore, microorganisms produce biopesticides, which are more effective than synthetic chemical pesticides (Abdollahdokht et al., 2022).

Nanopesticides consist of small NPs with pesticidal properties that can rapidly degrade in the soil (Castro-Restrepo, 2017). Nanobiopesticides are a positive trend in green agriculture because they are safer for plants and generate less pollution (Elnahal et al., 2022). The term nano pesticides refer to two types of them. The first type relates to pesticides with nanoscaled ingredients. This type includes powder pesticides and microemulsion pesticides and consists of the use of NPs as pesticides (Sun et al., 2019). Nanoscale formulations are characterized by a controlled and targeted release of active ingredients. Therefore, they can produce the same effects at a lower dose than their bulk counterparts and are less toxic to nontargeted organisms, offering a powerful tool to improve agricultural production under changing climatic conditions (Kralova and Jampílek, 2021). Their efficiency is improved through polymers, active particles, metal oxides, and micelles. However, the shape, the size range, surface characteristics of NPs, the conditions in which they can be used, and release conditions must be defined before the application (Elnahal et al., 2022).

The second type refers to nanopesticides covered with nanomaterials on their surface. This type of pesticide usually improves the function of the original pesticides and protects them from external factors (Sun et al., 2019). Pesticide nanoencapsulation is a recent method of encapsulating the substances in nanomaterials for release and control. This nanoencapsulation protects bioactive substances against premature degradation (Abdollahdokht et al., 2022). However, NPs are qualified to alter the properties of encapsulated substances, so it is necessary to evaluate the effects of the encapsulated products for their safe use (Pascoli et al., 2019).

Constant changes in atmospheric composition lead to climate changes, accompanied by prolonged drought, massive storms, and other extreme weather events. The increasingly harmful effects of abiotic stresses and the increase in plant diseases and pests lead to significant crop losses. To combat these problems, new products that do not harm the environment must be used (Kralova and Jampílek, 2021). There are 9420 commercialized nanoproducts, but only 229 of them are being applied in agriculture. Nanotechnology is still being explored for agrochemical applications. Critical evaluations must be applied before their use is considered safe. Some researchers have proposed that NPs can be considered safe for the environment, but others suggested that the application of nanomaterials should be thoroughly tested for environmental effects (An et al., 2022).

6 Interaction nanoparticles and plants

The use of nanotechnology in agriculture has shown that plants can have different interactions depending on the type of NP they are exposed to (Sun et al., 2019). The NP can be applied to plants above or below ground to control plant diseases. Applied NPs can locally inhibit the translocation of phytopathogens through the plant system (Ali et al., 2020).

NPs can act as plant growth enhancers, or they can damage plants. For example, silver NPs are commonly used to make insecticides and pesticides. Labeeb et al. (2022) studied the impact of different concentrations of silver NPs in peas (*Pisum sativum* L.). The results showed that NPs caused cell wall and mitochondria malformations, plasmolysis and vacuolization in root cells, and damage at the level of mitochondria and chloroplasts. Furthermore, these effects were proportional to concentration. That is, they increased as the concentration of NPs increased. Again, in the presence of silver NPs, genomic variation in DNA was detected (Labeeb et al., 2022).

Xie et al. (2022) investigated the effect of ceric dioxide (CeO_2) NPs in different fields of action, such as on plant growth and soil microbial communities. The study was carried out in the rhizosphere of cucumber seedlings and the surrounding bulk soil, with cerium chloride ($CeCl_3$) as a control to identify the effect of NPs. The results showed that cerium accumulated significantly in cucumber tissue after exposure to CeO_2 NPs. In the roots, 5.3% of the accumulated cerium has been transformed into Ce^{3+}. However, this transformation could occur before root uptake, since 2.5% of CeO_2 NPs were found altered in the soil near the rhizosphere, indicating the critical role of rhizosphere chemistry in this transformation (Xie et al., 2022).

In addition to the above cases showing the adverse effects of NPs on plants, NPs can also benefit them. For example, NPs can be a sustainable alternative to salinity stress in plants and promote plant health. Salinity stress is a nonbiological pressure that affects plants and negatively affects crops around the world. NPs can stimulate plant growth and reduce salinity stress. The results have shown that NPs, for example, silver NPs, under appropriate use, regulate salinity tolerance in various plants by changing hormone concentrations, antioxidant enzyme activities, ion homeostasis, gene expression, and defense systems (Etesami et al., 2021).

Plants may also be a prospective alternative for NPs synthesis. For example, *Abrus precatorius* plant extracts can act as a reducing agent to form copper oxide (CuO) NPs. These NPs may have applications in the medical and environmental fields (Kavitha et al., 2022).

Plant-NP interactions are difficult to predict, and multiple factors must be considered. For example, NP size, chemical composition, and surface functionalization are essential, but environmental components must also be contemplated (Landry et al., 2019).

7 Conclusion

Nanotechnology is a relatively new area of science, which has been useful in solving various problems related to agriculture and food production. Nanometric systems have been useful to inhibit phytopathogenic microorganisms, and to enhance beneficial microorganisms for plants. The effects of beneficial microorganisms for plants immobilized in NPs, as well as the interaction between nanofertilizers and nanopesticides with plants, are still under investigation, but the results may be of great importance for the formulation of new treatments. The development of nanofungicides could bring some potentialities, such as higher efficacy and bioavailability of fungicides, reduced toxicity, and higher solubility of poorly water-soluble fungicides, and may also target delivery of actives and precise release and improve the useful life of the active compounds. Literature data analysis shows that nanotechnology can be applied in agriculture, but most companies are still afraid to apply this technology to their market products. Although the production of nanoagrochemicals is at an early stage of development, this application is expected to improve the efficiency of agrochemicals and reduce environmental pollution. More research is required to facilitate its large-scale application.

Acknowledgments

The authors thank the Mexican National Council of Humanities, Sciences and Technologies (CONAHCYT) for the financial support of the PDCPN project 2013-01-213844, Sandra Pacios Michelena Ph.D. scholarship, Liliana Sofía Farías Vázquez M.Sc. scholarship, and the financial support under the program "Investigadores e Investigadoras por México CONAHCYT" (Project No. 729) and FONCYT-Coahuila.

Conflicts of interest

The authors declare that they have not had any conflicts of interest.

References

Abdollahdokht, D., Gao, Y., Faramarz, S., Poustforoosh, A., Abbasi, M., Asadikaram, G., Nematollahi, M.H., 2022. Conventional agrochemicals towards nano-biopesticides: an overview on recent advances. Chem. Biol. Technol. Agric. 9, 13. https://doi.org/10.1186/s40538-021-00281-0.

Acharya, A., Pal, P.K., 2020. Agriculture nanotechnology: translating research outcome to field applications by influencing environmental sustainability. NanoImpact 19, 100232. https://doi.org/10.1016/J.IMPACT.2020.100232.

Ali, M.A., Ahmed, T., Wu, W., Hossain, A., Hafeez, R., Mahidul, M., Masum, I., Wang, Y., An, Q., Sun, G., Li, B., 2020. Nanomaterials review advancements in plant and microbe-based synthesis of metallic nanoparticles and their antimicrobial activity against plant pathogens state key laboratory for managing biotic and chemical threats to the quality and safety of agro-products. Nanomaterials 10 (6), 1146. https://doi.org/10.3390/nano10061146.

An, C., Sun, C., Li, N., Huang, B., Jiang, J., Shen, Y., Wang, C., Zhao, X., Cui, B., Wang, C., Li, X., Zhan, S., Gao, F., Zeng, Z., Cui, H., Wang, Y., 2022. Nanomaterials and nanotechnology for the delivery of agrochemicals: strategies towards sustainable agriculture. J. Nanobiotechnol. 20, 11. https://doi.org/10.1186/s12951-021-01214-7.

Ashraf, H., Anjum, T., Riaz, S., Batool, T., Naseem, S., Li, G., 2022. Sustainable synthesis of microwave-assisted IONPs using *Spinacia oleracea* L. for control of fungal wilt by modulating the defense system in tomato plants. J. Nanobiotechnol. 20 (1), 8. https://doi.org/10.1186/S12951-021-01204-9.

Avila-Quezada, G.D., Golinska, P., Rai, M., 2022. Engineered nanomaterials in plant diseases: can we combat phytopathogens? Appl. Microbiol. Biotechnol. 106 (1), 117–129. https://doi.org/10.1007/S00253-021-11725-W.

Baniamerian, H., Ghofrani-Isfahani, P., Tsapekos, P., Alvarado-Morales, M., Shahrokhi, M., Angelidaki, I., 2021. Multicomponent nanoparticles as means to improve anaerobic digestion performance. Chemosphere 283, 131277. https://doi.org/10.1016/j.chemosphere.2021.131277.

Bartolucci, C., Scognamiglio, V., Antonacci, A., Fraceto, L.F., 2022. What makes nanotechnologies applied to agriculture green? Nano Today 43, 101389. https://doi.org/10.1016/j.nantod.2022.101389.

Cardoso, F.C., 2020. Invited Review: applying fungicide on corn plants to improve the composition of whole-plant silage in diets for dairy cattle. Appl. Anim. Sci. 36 (1), 57–69. https://doi.org/10.15232/AAS.2019-01905.

Castro-Restrepo, D., 2017. Nanotecnología en la agricultura. Bionatura 2 (3), 384–389. https://doi.org/10.21931/RB/2017.03.03.9.

Chetwynd, A.J., Lynch, I., 2020. The rise of the nanomaterial metabolite corona, and emergence of the complete corona. Environ. Sci. Nano 7 (4), 1041–1060. https://doi.org/10.1039/C9EN00938H.

Degani, O., Khatib, S., Becher, P., Gordani, A., Harris, R., 2021. *Trichoderma asperellum* secreted 6-pentyl-α-pyrone to control *Magnaporthiopsis maydis*, the maize late wilt disease agent. Biology 10 (9), 897. https://doi.org/10.3390/BIOLOGY10090897.

Devi, N.O., Tombisana Devi, R.K., Debbarma, M., Hajong, M., Thokchom, S., 2022. Effect of endophytic *Bacillus* and arbuscular mycorrhiza fungi (AMF) against *Fusarium* wilt of tomato caused by *Fusarium oxysporum* f. sp. *lycopersici*. Egypt. J. Biol. Pest Control 32, 1. https://doi.org/10.1186/S41938-021-00499-Y.

dos Santos Gomes, A.C., da Silva, R.R., Moreira, S.I., Vicentini, S.N.C., Ceresini, P.C., 2021. Biofungicides: an eco-friendly approach for plant disease management. In: Encyclopedia of Mycology. vol. 2, pp. 641–649, https://doi.org/10.1016/B978-0-12-819990-9.00036-6.

El-Baky, N.A., Amara, A.A.A.F., 2021. Recent approaches towards control of fungal diseases in plants: an updated review. J. Fungi 7 (11), 900. https://doi.org/10.3390/JOF7110900.

El-Mehy, A.A., El-Gendy, H.M., Aioub, A.A.A., Mahmoud, S.F., Abdel-Gawad, S., Elesawy, A.E., Elnahal, A.S.M., 2022. Response of faba bean to intercropping, biological and chemical control against broomrape and root rot diseases. Saudi J. Biol. Sci. 29 (5), 3482–3493. https://doi.org/10.1016/J.SJBS.2022.02.032.

Elnahal, A.S.M., El-Saadony, M.T., Saad, A.M., Desoky, E.-S.M., El-Tahan, A.M., Rady, M.M., Abuqamar, S.F., El-Tarabily, K.A., 2022. Correction: The use of microbial inoculants for biological control, plant growth promotion, and sustainable agriculture: a review. Eur. J. Plant Pathol. 162 (4), 10077. https://doi.org/10.1007/S10658-022-02472-3.

Etesami, H., Fatemi, H., Rizwan, M., 2021. Interactions of nanoparticles and salinity stress at physiological, biochemical and molecular levels in plants: a review. Ecotoxicol. Environ. Saf. 225, 112769. https://doi.org/10.1016/J.ECOENV.2021.112769.

Falade, A.O., Adewole, K.E., Ekundayo, T.C., 2021. Aptitude of endophytic microbes for production of novel biocontrol agents and industrial enzymes towards agro-industrial sustainability. Beni-Suef Univ. J. Basic Appl. Sci. 10, 21. https://doi.org/10.1186/s43088-021-00146-3.

Falck, A.J., Mooney, S., Bearer, C.F., 2020. Adverse exposures to the fetus and neonate. In: Fanaroff and Martin's Neonatal-Perinatal Medicine, 14, pp. 239–259. https://www.clinicalkey.com/#!/content/book/3-s2.0-B9780323567114000146.

Gupta, P.K., 2018. Toxicity of fungicides. In: Veterinary Toxicology: Basic and Clinical Principles, third ed. Academic Press, pp. 569–580 (Chapter 45) https://doi.org/10.1016/B978-0-12-811410-0.00045-3.

He, X., Deng, H., Hwang, H.m., 2019. The current application of nanotechnology in food and agriculture. J. Food Drug Anal. 27 (1), 1–21. https://doi.org/10.1016/J.JFDA.2018.12.002.

Hoang, A.S., Cong, H.H., Shukanov, V.P., Karytsko, L.A., Poljanskaja, S.N., Melnikava, E.v., Mashkin, I.A., Nguyen, T.H., Pham, D.K., Phan, C.M., 2022. Evaluation of metal nano-particles as growth promoters and fungi inhibitors for cereal crops. Chem. Biol. Technol. Agric. 9 (1), 12. https://doi.org/10.1186/S40538-021-00277-W.

Hoque, M.N., Saha, S.M., Imran, S., Hannan, A., Seen, M.M.H., Thamid, S.S., Tuz-zohra, F., 2022. Farmers' agrochemicals usage and willingness to adopt organic inputs: watermelon farming in Bangladesh. Environ. Chall. 7, 100451. https://doi.org/10.1016/J.ENVC.2022.100451.

Jamil, F.N., Hashim, A.M., Yusof, M.T., Saidi, N.B., 2022. Analysis of soil bacterial communities and physicochemical properties associated with *Fusarium* wilt disease of banana in Malaysia. Sci. Rep. 12, 999. https://doi.org/10.1038/s41598-022-04886-9.

Juárez-Maldonado, A., Tortella, G., Rubilar, O., Fincheira, P., Benavides-Mendoza, A., 2021. Biostimulation and toxicity: the magnitude of the impact of nanomaterials in microorganisms and plants. J. Adv. Res. 31, 113–126. https://doi.org/10.1016/j.jare.2020.12.011.

Kalwani, M., Chakdar, H., Srivastava, A., Pabbi, S., Shukla, P., 2022. Effects of nanofertilizers on soil and plant-associated microbial communities: emerging trends and perspectives. Chemosphere 287 (2), 132107. https://doi.org/10.1016/J.CHEMOSPHERE.2021.132107.

Kamaruzaman, N.H., Mohd Noor, N.N., Radin Mohamed, R.M.S., Al-Gheethi, A., Ponnusamy, S.K., Sharma, A., Vo, D.V.N., 2022. Applicability of bio-synthesized nanoparticles in fungal secondary metabolites products and plant extracts for eliminating antibiotic-resistant bacteria risks in non-clinical environments. Environ. Res. 209, 112831. https://doi.org/10.1016/J.ENVRES.2022.112831.

Kavitha, K., Arockia John Paul, J., Kumar, P., Archana, J., Faritha Begam, H., Karmegam, N., Biruntha, M., 2022. Impact of biosynthesized CuO nanoparticles on seed germination and cyto-physiological responses of *Trigonella foenum-graecum* and *Vigna radiata*. Mater. Lett. 313, 131756. https://doi.org/10.1016/J.MATLET.2022.131756.

Keswani, C., Singh, H.B., Hermosa, R., García-Estrada, C., Caradus, J., He, Y.W., Mezaache-Aichour, S., Glare, T.R., Borriss, R., Vinale, F., Sansinenea, E., 2019. Antimicrobial secondary metabolites from agriculturally important fungi as next biocontrol agents. Appl. Microbiol. Biotechnol. 103 (23–24), 9287–9303. Springer https://doi.org/10.1007/s00253-019-10209-2.

Khanna, A., Raj, K., Kumar, P., Wati, L., 2022. Antagonistic and growth-promoting potential of multifarious bacterial endophytes against *Fusarium* wilt of chickpea. Egypt. J. Biol. Pest Control 32, 17. https://doi.org/10.1186/s41938-022-00516-8.

Kralova, K., Jampílek, J., 2021. Nanotechnology as effective tool for improved crop production under changing climatic conditions. In: Biobased Nanotechnology for Green Applications, pp. 463–512, https://doi.org/10.1007/978-3-030-61985-5_17.

Kumar, P., Mahajan, P., Kaur, R., Gautam, S., 2020. Nanotechnology and its challenges in the food sector: a review. Mater. Today Chem. 17, 100332. https://doi.org/10.1016/J.MTCHEM.2020.100332.

Kumar, A., Singh, K., Verma, P., Singh, O., Panwar, A., Singh, T., Kumar, Y., Raliya, R., 2022. Effect of nitrogen and zinc nanofertilizer with the organic farming practices on cereal and oil seed crops. Sci. Rep. 12, 6938. https://doi.org/10.1038/s41598-022-10843-3.

Kutawa, A.B., Ahmad, K., Ali, A., Hussein, M.Z., Abdul Wahab, M.A., Adamu, A., Ismaila, A.A., Gunasena, M.T., Rahman, M.Z., Hossain, M.I., 2021. Trends in nanotechnology and its potentialities to control plant pathogenic fungi: a review. Biology 10 (9), 881. https://doi.org/10.3390/biology10090881.

Labeeb, M., Badr, A., Haroun, S.A., Mattar, M.Z., El-Kholy, A.S., 2022. Ultrastructural and molecular implications of ecofriendly made silver nanoparticles treatments in pea (*Pisum sativum* L.). J. Genet. Eng. Biotechnol. 20, 5. https://doi.org/10.1186/s43141-021-00285-1.

Landry, M.P., Dispenza, C., Ambrosone, A., Sanzari, I., Leone, A., 2019. Nanotechnology in plant science: to make a long story short. Front. Bioeng. Biotechnol. 7, 120. https://doi.org/10.3389/fbioe.2019.00120.

Lapponi, M.J., Méndez, M.B., Trelles, J.A., Rivero, C.W., 2022. Cell immobilization strategies for biotransformations. Curr. Opin. Green Sustain. Chem. 33, 100565. https://doi.org/10.1016/J.COGSC.2021.100565.

Lou, L., Huang, Q., Lou, Y., Lu, J., Hu, B., Lin, Q., 2019. Adsorption and degradation in the removal of nonylphenol from water by cells immobilized on biochar. Chemosphere 228, 676–684. https://doi.org/10.1016/J.CHEMOSPHERE.2019.04.151.

Lucarini, M., Virginia Nevárez-Moorillón, G., Abouelatta, A., Aimad, A., Bourhia, M., Mohammad Salamatullah, A., Moussaoui, E.A., Salamatullah, M.A., Alyahya, K.H., Abdali Youness, E., Sanae, R., el Moussaoui, A., Alzahrani, A., Khalil Alyahya, H., Albadr, N.A., Nafidi, H.-A., Ouahmane, L., Mohamed, F., 2022. Chemical composition and antifungal, insecticidal and repellent activity of essential oils from *Origanum compactum* Benth. Used in the Mediterranean diet. Front. Plant Sci. 13, 798259. https://doi.org/10.3389/fpls.2022.798259.

Luong, H.T., Nguyen, C.X., Thuong, B., Lam, T., Nguyen, T.-H., Dang, Q.-L., Lee, J.-H., Hur, H.-G., Nguyen, H.T., Ho, C.T., 2022. Antibacterial effect of copper nanoparticles produced in a Shewanella-supported non-external circuit bioelectrical system on bacterial plant pathogens. RSC Adv. 12, 4428. https://doi.org/10.1039/d1ra08187j.

Machado, T.O., Grabow, J., Sayer, C., de Araújo, P.H.H., Ehrenhard, M.L., Wurm, F.R., 2022. Biopolymer-based nanocarriers for sustained release of agrochemicals: a review on materials and social science perspectives for a sustainable future of agri- and horticulture. Adv. Colloid Interf. Sci. 303, 102645. https://doi.org/10.1016/J.CIS.2022.102645.

Muthukrishnan, L., 2022. An overview on the nanotechnological expansion, toxicity assessment and remediating approaches in Agriculture and Food industry. Environ. Technol. Innov. 25, 102136. https://doi.org/10.1016/j.eti.2021.102136.

Nath, A., Molnár, M.A., Albert, K., Das, A., Bánvölgyi, S., Márki, E., Vatai, G., 2019. Agrochemicals from nanomaterials—synthesis, mechanisms of biochemical activities and applications. Compr. Anal. Chem. 84, 263–312. https://doi.org/10.1016/BS.COAC.2019.04.004.

Neme, K., Nafady, A., Uddin, S., Tola, Y.B., 2021. Application of nanotechnology in agriculture, postharvest loss reduction and food processing: food security implication and challenges. Heliyon 7 (12), e08539. https://doi.org/10.1016/j.heliyon.2021.e08539.

Nisha Raj, S., Anooj, E.S., Rajendran, K., Vallinayagam, S., 2021. A comprehensive review on regulatory invention of nano pesticides in Agricultural nano formulation and food system. J. Mol. Struct. 1239, 130517. https://doi.org/10.1016/J.MOLSTRUC.2021.130517.

Nyadanu, D., Lowor, S.T., Akrofi, A.Y., Adomako, B., Dzahini-Obiatey, H., Akromah, R., Awuah, R.T., Kwoseh, C., Adu-Amoah, R., Kwarteng, A.O., 2019. Mode of inheritance and combining ability studies on epicuticular wax production in resistance to black pod disease in cacao (*Theobroma cacao* L.). Sci. Hortic. 243, 34–40. https://doi.org/10.1016/J.SCIENTA.2018.07.002.

Ortega, P., Sánchez, E., Gil, E., Matamoros, V., 2022. Use of cover crops in vineyards to prevent groundwater pollution by copper and organic fungicides. Soil column studies. Chemosphere 303 (1), 134975. https://doi.org/10.1016/j.chemosphere.2022.134975.

Osonga, F.J., Eshun, G., Kalra, S., Yazgan, I., Sakhaee, L., Ontman, R., Jiang, S., Sadik, O.A., 2022. Influence of particle size and shapes on the antifungal activities of greener nanostructured copper against *Penicillium italicum*. ACS Agric. Sci. Technol. 2 (1), 42–56. https://doi.org/10.1021/acsagscitech.1c00102.

Pascoli, M., Jacques, M.T., Agarrayua, D.A., Avila, D.S., Lima, R., Fraceto, L.F., 2019. Neem oil based nanopesticide as an environmentally-friendly formulation for applications in sustainable agriculture: an ecotoxicological perspective. Sci. Total Environ. 677, 57–67. https://doi.org/10.1016/J.SCITOTENV.2019.04.345.

Ramírez-Valdespino, C.A., Orrantia-Borunda, E., 2021. *Trichoderma* and nanotechnology in sustainable agriculture: a review. Front. Fungal Biol. 2, 764675. https://doi.org/10.3389/ffunb.2021.764675.

Rathore, A., Mahesh, G., 2021. Public perception of nanotechnology: a contrast between developed and developing countries. Technol. Soc. 67, 101751. https://doi.org/10.1016/J.TECHSOC.2021.101751.

Rodrigo, S., García-Latorre, C., Santamaria, O., Suárez, A., 2022. Metabolites produced by fungi against fungal phytopathogens: review, implementation and perspectives. Plants 11 (1), 81. https://doi.org/10.3390/plants11010081.

Salem, S.S., Hammad, E.N., Mohamed, A.A., El-Dougdoug, W., 2023. A comprehensive review of nanomaterials: types, synthesis, characterization, and applications. Review, Biointerface Res. Appl. Chem. 13 (1), 41. https://doi.org/10.33263/BRIAC131.041.

Sayed, E.G., Ouis, M.A., 2022. Improvement of pea plants growth, yield, and seed quality using glass fertilizers and biofertilizers. Environ. Technol. Innov. 26, 102356. https://doi.org/10.1016/J.ETI.2022.102356.

Shalaby, T.A., Bayoumi, Y., Eid, Y., Elbasiouny, H., Elbehiry, F., Prokisch, J., El-Ramady, H., Ling, W., 2022. Can nanofertilizers mitigate multiple environmental stresses for higher crop productivity? Sustainability 14 (6), 3480. https://doi.org/10.3390/su14063480.

Sowunmi, F.A., Famuyiwa, G.T., Oluyole, K.A., Aroyeun, S.O., Obasoro, O.A., 2019. Environmental burden of fungicide application among cocoa farmers in Ondo state, Nigeria. Sci. African 6, 207. https://doi.org/10.1016/J.SCIAF.2019.E00207.

Sun, Y., Liang, J., Tang, L., Li, H., Zhu, Y., Jiang, D., Song, B., Chen, M., Zeng, G., 2019. Nano-pesticides: a great challenge for biodiversity? Nano Today 28, 100757. https://doi.org/10.1016/J.NANTOD.2019.06.003.

Sun, H., Shen, J., Hu, M., Zhang, J., Cai, Z., Zang, L., Zhang, F., Ji, D., 2021. Manganese ferrite nanoparticles enhanced biohydrogen production from mesophilic and thermophilic dark fermentation. Energy Rep. 7, 6234–6245. https://doi.org/10.1016/j.egyr.2021.09.070.

Suriya Prabha, A., Angelin Thangakani, J., Renuga Devi, N., Dorothy, R., Nguyen, T.A., Senthil Kumaran, S., Rajendran, S., 2022. Nanotechnology and sustainable agriculture. In: Denizli, A., Nguyen, T.A., Rajendran, S., Yasin, G., Nadda, A.K. (Eds.), Nanosensors for Smart Agriculture, pp. 25–39 (Chapter 2) https://doi.org/10.1016/B978-0-12-824554-5.00016-1.

Tarafder, C., Daizy, M., Morshed Alam, M., Ripon Ali, M., Jahidul Islam, M., Islam, R., Sohel Ahommed, M., Aly Saad Aly, M., Zaved Hossain Khan, M., 2020. Formulation of a hybrid nanofertilizer for slow and sustainable release of micronutrients. ACS Omega 5 (37), 23960–23966. https://doi.org/10.1021/acsomega.0c03233.

Thipe, V.C., Karikachery, A.R., Çakılkaya, P., Farooq, U., Genedy, H.H., Kaeokhamloed, N., Phan, D.H., Rezwan, R., Tezcan, G., Roger, E., Katti, K.v., 2022. Green nanotechnology—an innovative pathway towards biocompatible and medically relevant gold nanoparticles. J. Drug Deliv. Sci. Technol. 70, 1773–2247. https://doi.org/10.1016/J.JDDST.2022.103256.

United States Environmental Protection Agency, 2007, November. System of Registries. US EPA. Recovered April 11, 2022, from https://sor.epa.gov/sor_internet/registry/termreg/searchandretrieve/glossariesandkeywordlists/search.do;jsessionid=PqkgdpGVjEUePkuowu8KGlp3ZUfES12zIBZTLO-dgoE6mUhSDlkt!1954464232?details=&vocabName=Ag%20101%20Glossary&filterTerm=pesticide&checkedAcronym=false&checkedTerm=false&hasDefinitions=false&filterTerm=pesticide&filterMatchCriteria=Contains.

Usman, M., Farooq, M., Wakeel, A., Nawaz, A., Cheema, S.A., Rehman, H.u., Ashraf, I., Sanaullah, M., 2020. Nanotechnology in agriculture: current status, challenges and future opportunities. Sci. Total Environ. 721, 137778. https://doi.org/10.1016/J.SCITOTENV.2020.137778.

Wang, L., Hu, C., Shao, L., 2017. The antimicrobial activity of nanoparticles: present situation and prospects for the future. Int. J. Nanomedicine 12, 1227. https://doi.org/10.2147/IJN.S121956.

Wang, Q., Wu, X., Jiang, L., Fang, C., Wang, H., Chen, L., 2020. Effective degradation of Di-n-butyl phthalate by reusable, magnetic Fe3O4 nanoparticle-immobilized *Pseudomonas* sp. W1 and its application in simulation. Chemosphere 250, 126339. https://doi.org/10.1016/J.CHEMOSPHERE.2020.126339.

Wolters, P.J., Wouters, D., Kromhout, E.J., Huigen, D.J., Visser, R.G.F., Vleeshouwers, V.G.A.A., 2021. Qualitative and quantitative resistance against early blight introgressed in potato. Biology 10 (9), 892. https://doi.org/10.3390/BIOLOGY10090892.

Worrall, E.A., Hamid, A., Mody, K.T., Mitter, N., Pappu, H.R., 2018. Nanotechnology for plant disease management. Agronomy 8 (12), 285. https://doi.org/10.3390/agronomy8120285.

Xie, C., Guo, Z., Zhang, P., Yang, J., Zhang, J., Ma, Y., He, X., Lynch, I., Zhang, Z., 2022. Effect of CeO2 nanoparticles on plant growth and soil microcosm in a soil-plant interactive system. Environ. Pollut. 300, 118938. https://doi.org/10.1016/J.ENVPOL.2022.118938.

Xu, C., Zhong, L., Huang, Z., Li, C., Lian, J., Zheng, X., Liang, Y., 2022. Real-time monitoring of *Ralstonia solanacearum* infection progress in tomato and *Arabidopsis* using bioluminescence imaging technology. Plants Methods 18, 7. https://doi.org/10.1186/s13007-022-00841-x.

Zaki, A.G., Hasanien, Y.A., El-Sayyad, G.S., 2022. Novel fabrication of SiO_2/Ag nanocomposite by gamma irradiated *Fusarium oxysporum* to combat *Ralstonia solanacearum*. AMB Express 12, 25. https://doi.org/10.1186/s13568-022-01372-3.

Zambito Marsala, R., Capri, E., Russo, E., Bisagni, M., Colla, R., Lucini, L., Gallo, A., Suciu, N.A., 2020. First evaluation of pesticides occurrence in groundwater of Tidone Valley, an area with intensive viticulture. Sci. Total Environ. 736, 139730. https://doi.org/10.1016/J.SCITOTENV.2020.139730.

Zhong, D., Li, J., Ma, W., Xin, H., 2020. Magnetite nanoparticles enhanced glucose anaerobic fermentation for bio-hydrogen production using an expanded granular sludge bed (EGSB) reactor. Int. J. Hydrog. Energy 45 (18), 10664–10672. https://doi.org/10.1016/j.ijhydene.2020.01.095.

Zhu, Y., Lv, X., Song, J., Li, W., Wang, H., 2022. Application of cotton straw biochar and compound *Bacillus* biofertilizer decrease the bioavailability of soil cd through impacting soil bacteria. BMC Microbiol. 22, 35. https://doi.org/10.1186/S12866-022-02445-W.

Zobir, S.A.M., Ali, A., Adzmi, F., Sulaiman, R., Ahmad, K., 2021. A review on Nanopesticides for plant protection synthesized using the supramolecular chemistry of layered hydroxide hosts. Biology 10 (11), 1077. https://doi.org/10.3390/biology10111077.

Zorraquín-Peña, I., Cueva, C., Bartolomé, B., Moreno-Arribas, M.V., 2020. Silver nanoparticles against foodborne bacteria. Effects at intestinal level and health limitations. Microorganisms 8 (1). https://doi.org/10.3390/microorganisms8010132. MDPI AG.

Zubrod, J.P., Bundschuh, M., Arts, G., Brühl, C.A., Imfeld, G., Knäbel, A., Payraudeau, S., Rasmussen, J.J., Rohr, J., Scharmüller, A., Smalling, K., Stehle, S., Schulz, R., Schäfer, R.B., 2019. Fungicides: an overlooked pesticide class? Environ. Sci. Technol. 53 (7), 3347–3365. https://doi.org/10.1021/ACS.EST.8B04392/SUPPL_FILE/ES8B04392_SI_001.PDF.

CHAPTER

3

Chitosan-based agronanofungicides: A sustainable alternative in fungal plant diseases management

Ayat F. Hashim[a], *Khamis Youssef*[b,c], *Farah K. Ahmed*[d], *and Mousa A. Alghuthaymi*[e]

[a]Fats and Oils Department, Food Industries and Nutrition Research Institute, National Research Centre, Giza, Egypt [b]Plant Pathology Research Institute, Agricultural Research Center (ARC), Giza, Egypt [c]Agricultural and Food Research Council, Academy of Scientific Research and Technology, Cairo, Egypt [d]Biotechnology English Program, Faculty of Agriculture, Cairo University, Giza, Egypt [e]Biology Department, Science and Humanities College, Shaqra University, Alquwayiyah, Saudi Arabia

1 Introduction

Chitosan-based biodegradable nanomaterials (NMs) are nontoxic, biodegradable, and antifungal, making them ideal for agricultural applications. Nanoparticles, nanogels, and nanocomposites have become widespread in agriculture due to their antimicrobial and plant growth-boosting properties (Kumaraswamy et al., 2018). Chitosan-based nanoparticles inhibited foodborne bacteria significantly, notably in fruits and vegetables. There are few publications on the use of chitosan nanoparticles to treat plant diseases caused by fungal pathogens (Brunel et al., 2013; Abd El-Aziz et al., 2018). Nanotechnology-derived chitosan formulations, either alone or in combination with other pesticides, showed considerable potential against phytopathogenic fungi and phytophagous insect pests (Al-Dhabaan et al., 2018; Mishra et al., 2022). Furthermore, the chitosan nanodelivery system was loaded with agrochemicals as the active agent, resulting in the formation of chitosan-agrochemical nanoparticles with high efficacy and potency, as the active ingredient can reach the target cell or plant parts more effectively within a defined time frame (Duhan et al., 2017). Furthermore,

https://doi.org/10.1016/B978-0-443-23950-2.00002-3

the developed solutions may improve uptake while reducing agrochemical leaching and runoff, which can be harmful to human and the environment (Maluin and Hussein, 2020). The effectiveness of chitosan-based nanofungicides in addressing real-world agricultural problems should be assessed in comparison to conventional agrochemicals. As a result, this chapter discusses some recent findings about the use of nanochitosan-based products in fungal plant disease management, with a focus on its use as a plant resistance inducer against fungal diseases. The routes of action of chitosan nanoparticles against a range of plant pathogenic fungi are also being investigated.

2 Synthesis and characterization of chitosan-based nanocomposites

There are multiple phases in nanocomposite materials. These materials might include elements with changing phase domains, at least one of which would be a continuous phase and the other would have nanoscale dimensions (Ashfaq et al., 2020). These hybrid nanomaterials can be produced by cosynthesizing or impregnating multiple inorganic and organic components (Winter et al., 2020). Chemical techniques such as in situ polymerization, interfacial polymerization, and interfacial polycondensation can be used to synthesize chitosan-based nanocomposites. Examples of physicochemical processes include centrifugal extrusion-spheronization, fluid bed coating, complex coacervation, spray cooling, and spray-drying are examples of physical approaches (Fig. 1).

2.1 Gold-chitosan nanocomposites (Au-Cs-NCs)

Dananjaya et al. (2017) prepared a chitosan-gold nanocomposite by dissolving a chitosan solution (0.2% w/v) in acetic acid at 65°C. Subsequently, they added tetrachloroauric (III) acid trihydrate (10 mM) dropwise into the previously prepared solution. The mixture was then heated to 90°C and agitated with a magnetic stirrer for 15 min. The change of mixture color from colorless to purple indicates the gold nanoparticles (AuNPs) formation. The obtained suspension was filtered and washed three times using distilled water and then dried in a vacuum oven at 60°C for 6 h. The resulting product was characterized using a UV-vis spectrometer to obtain the optical absorption spectrum of the AuNPs. Particle size distribution and zeta potential were investigated by Zetasizer. Using a field emission transmission electron microscope, the particle size and morphology were displayed (FE-TEM). In addition, a field emission scanning electron microscope (FE-SEM) coupled with energy-dispersive X-ray spectroscopy was utilized to confirm the presence of AuNPs in the composite. Inductively coupled plasma atomic emission spectroscopy (ICP-AES) was used to detect the amount of Au in the prepared nanocomposite. Thermal gravimetric analysis (TGA) determines the thermal stability of a nanocomposite. Lipşa et al. (2020) produced chitosan-stabilized gold nanoparticles using various concentrations of tetrachloroauric acid solution (25, 50, and 75 g/mL) and chitosan solution (0.1 mg/mL). The different ratios of tetrachloroauric acid solution were mixed with chitosan solution and then treated with an ultrasonic field for 20 min at 55°C. The formation of AuNP in chitosan was confirmed by the change of the transparent color of the initial solution to shades of purple-pink (a few days later). The more AuNP

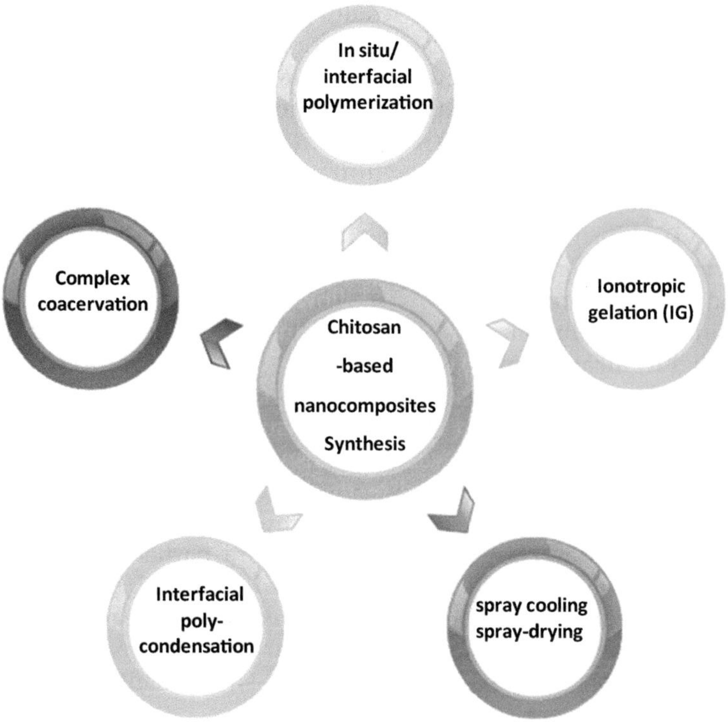

FIG. 1 Synthesis of chitosan-based nanocomposites using a variety of techniques.

concentration there is, the higher the intensity of the color is. Atomic force microscopy (AFM), fluorescence, and UV-vis spectra measurements were used to characterize the produced nanoparticles. Dynamic light scattering (DLS) was used for the measurement of average particle size and the zeta potential.

2.2 Silver-chitosan nanocomposites (Ag-Cs-NCs)

Artunduaga Bonilla et al. (2021) prepared nanocomposites based on silver and chitosan. Using a magnetic stirrer, solutions of cetyltrimethylammonium bromide (CTAB, 0.01 M), cysteine, and silver nitrate ($AgNO_3$, 0.01 M) were blended and stirred. The final volume was completed to 20 mL by adding deionized water. Then, the suspension was sonicated for 2 h after adding polyvinyl alcohol (0.02%, w/v). After the formation of silver nanoparticles (AgNPs), the suspension turned red-orange (Artunduaga Bonilla et al., 2015). The prepared AgNPs (298 μg/mL) were incorporated into chitosan (1%, w/v) at pH 12. The mixture was sonicated at 150 W using five cycles of 5 s pulses. The suspension was stirred at 300 rpm for 1 h at room temperature. Finally, the pH of the suspension was adjusted to 7.5 using sodium hydroxide (NaOH). The amber flask was used to keep the produced nanocomposites at room temperature.

The physicochemical properties of the resulting nanocomposites were detected using a spectrophotometer to observe the AgNPs spectrum in the UV-vis region. DLS and Zeta potential analyses were performed to determine the average hydrodynamic size and surface charge of the nanomaterial. A high-resolution continuum source graphite furnace atomic absorption spectrometer (HR-CS GF AAS) was used to measure the amount of silver. Transmission electron microscopy (TEM) and X-ray diffraction (XRD) were used to assess the size, shape, and crystallinity of the AgNPs. Baka et al. (2019) also synthesized silver chitosan nanocomposites via a biological method using the *Escherichia coli* D8 supernatant in the presence of a reducing agent (sunlight). While chitosan (0.25 g) was dissolved in 2% acetic acid, magnetic stirring was performed for 15 min, followed by sonication. The mixture was then filtered to obtain a clear solution. The chitosan amine sites were activated by changing the solution pH to slightly alkaline (pH 8) using ammonia. In the biosynthesis of AgNPs, $AgNO_3$ (1.5 mM) was added to 40 mL of CS solution. After that, 1 mL of bacterial supernatant was quickly added and stirred for 90 min at 70°C. The suspension was centrifuged for 15 min at a speed of 10,000 rpm. The residue was then rinsed with distilled water to eliminate any excess solution before being redistributed in the same amount of water. Dananjaya et al. (2017) developed and studied chitosan silver nanocomposites using transmission electron microscopy, X-ray diffraction, and UV-vis absorbance spectra, particle size distribution, zeta potential, and thermal stability analysis methods. Chowdappa et al. (2014) used chitosan as a matrix to form silver nanoparticles at 95°C. By detecting the peak at 415–420 nm with a UV-vis spectrometer, the production of silver nanoparticles was proven (characteristic peak for silver nanoparticles). The fabricated chitosan-silver nanocomposite exhibited a hydrodynamic range diameter of 495–616 nm.

2.3 Zinc-chitosan nanocomposites (Zn-Cs-NCs)

Wang et al. (2004) developed a chitosan nanocomposite with five distinct Zn concentrations (molar ratio of 4:1, 2:1, 1:1, 0.5:1, and 0.25:1 compared to chitosan singly). Analytical techniques including FT-IR and X-ray diffraction were used to illustrate the interaction between Zn and the molecules of chitosan. Du et al. (2009) produced and examined the chitosan solution for enhanced antibacterial activity after including Zn^{2+} granules. In brief, the $ZnSO_4$ solution was made by mixing chitosan solution (dissolved in 1% (v/v) acetic acid and 1% (w/v) TPP) with 0.3% (w/v) Zn granules. The study's findings showed that raising the quantity of Zn^{2+} considerably increased the nanocomposite's zeta potential value, which in turn increased the composite's antibacterial activity. Abd-Elsalam et al. (2018) impregnated chitosan with a metal organosol to create a Zn-Cs-nanocomposite. Four gram of chitosan in 120 mL of toluene and 0.5 g of Zn granules were commonly utilized in preparation. 1:50 was the molar ratio of metal to solvent. Various analytical instruments, including SEM, TEM, and X-ray fluorescence (XRF), were used to examine the produced nanocomposite for its physicochemical characteristics. Dananjaya et al. (2018) dissolved 30 mM zinc nitrate hexahydrate ($Zn(NO_3)_2 \cdot 6H_2O$) in 60 mL of deionized water with constant stirring at room temperature. Then, 20 mL of deionized water was used to dissolve 60 mM cyclohexylamine ($C_6H_{13}N$), and the resultant solution was added to the Zn solution previously mentioned. The reaction mixture was then added to 80 mL of deionized water, which was continuously stirred for 48 h.

The resulting white precipitate followed filtering, distilled water washing, and 24h of vacuum drying at room temperature. The dried precipitate was combined with 300mL of water and continuously agitated for 12h to eliminate contaminants before being dried once more under vacuum at room temperature. After that, the product was calcined in air for 3h at 500°C. To produce ZnO-C NCs, the synthesized ZnO NPs (1g) were first dissolved in 100mL of 1% (v/v) acetic acid. Next, 750mg of low molecular weight chitosan was added to the aforementioned solution. After 20min of sonication, 1M of NaOH was gradually added to the mixture until the pH of the solution was below 10. The solution was then heated for 3h at 65°C in a water bath. The precipitate was then filtered, repeatedly rinsed with distilled water, and dried in an oven for 6h at 50°C.

2.4 Copper-chitosan nanocomposites (Cu-Cs-NCs)

The copper-chitosan nanocomposites were prepared using a metal-vapor technique in three main steps. Briefly, in the first step, copper-toluene and copper acetone organosol were prepared. Then the chitosan powder was impregnated with organosol. In the last step, the solvent was removed, and the metal-carrying chitosan powder was dried in a vacuum at 60°C (Rubina et al., 2017). Jaiswal et al. (2012) added copper sulfate to a chitosan solution and then added NaOH to prepare **Cu-Cs-NCs**. The size range of the generated copper sulfate particles was 700–750nm.

2.5 Bimetallic (zinc-copper) chitosan nanocomposites (Zn-Cu-Cs-NCs)

Under magnetic stirring, a TPP solution of varying concentrations is gradually poured into a chitosan solution. The solution is kept at room temperature overnight for cross-linking. After the reaction is complete, the chitosan solution becomes a colloidal gel. The pH solution stayed in the range of 3.5–4, depending on the chitosan and TPP concentration. Separating nanoparticles can be done through alternate centrifugation and ultrasonication, followed by two to three washes with distilled water to a pH of around 4.5–5. Finally, the material can be frozen and stored for future use. By adding copper sulfate and zinc sulfate solutions to the synthesis of chitosan nanoparticles, copper and zinc-loaded chitosan nanoparticles can be produced. In this method, chitosan nanoparticles are formed mainly through the interaction of positively charged amino groups on chitosan and negative charges on TPP (Saharan et al., 2013).

The developed nanocomposites are characterized by average size, polydispersity index (PDI), zeta potential using DLS, functional group analysis using FT-IR, and internal and surface morphology by SEM and TEM (Saharan et al., 2013). Kaur et al. (2015) developed chitosan-metal nanocomposites by using copper sulfate hydrate (5mmolL^{-1}) and zinc acetate (5mmolL^{-1}) as precursors. In an acidic solution of chitosan (acetic acid in distilled water, 1% v/v), copper and zinc ions were reduced to nanoparticles after being agitated for 12h and subjected to sonication at 1.5kW for 30min. The suspension was then centrifuged at 4°C for 20min at 12,000rpm. The pellet was centrifuged once more, resuspended in distilled water, and freeze-dried (Lyophilized). The freeze-dried chitosan-metal nanocomposites were either used directly or suspended in distilled water for characterization and other research. For elemental analysis of nanoparticles, energy dispersive spectroscopy (SEM-EDS) is the most preferred method that provides information about the elemental composition of nanoparticles and their distribution. Atomic absorption spectroscopy (AAS) can be used to study the release profile or encapsulation efficacy of metals in copper and zinc chitosan nanoparticles (Jaiswal et al., 2012). Al-Dhabaan et al. (2017) synthesized chitosan-bimetallic nanocomposites based on chitosan and zinc oxide (ZnO) and copper oxide (CuO) solutions. Chitosan (0.5 g) was dissolved in acetic acid (2%) to prepare a CS solution. Na_2SO_4 solution (20%) was added dropwise to the chitosan solution while being stirred and sonicated simultaneously. Under sonication with stirring, synthesized ZnO and CuO nanoparticles (0.1 g) were added to the CS solution. The suspension was centrifuged and washed several times. The nanocomposite was then obtained and dried at 60°C.

3 Chitosan nanocomposites applications in fungal plant diseases management

The irreversible contamination of soil, water, and air caused by the excessive and illogical use of synthetic fungicides has worried us, leading to the development of resistance in bacteria and disrupting the biosphere. Therefore, the primary objective to replace or limit the use of synthetic fungicides in agriculture for crop protection is the search for compounds that are biodegradable and favorable to the environment. In this scenario, nanobiotechnology appears to be a boon for the synthesis of ecofriendly, biocompatible, and safe fungicides. These fungicides will not only improve the health of the soil and the defense system of plants, but they will also assist in the acquisition of healthy food for a population that is continually growing. Chitosan is being investigated as a potential new-generation smart material to be utilized in agriculture, particularly for plant protection. Chitosan is one of the biomaterials and biopolymers that are already accessible. This chapter provides an overview of the numerous chitosan nanocomposites (NCs) that have been tested in the lab and applied in the field to prevent fungal disease in various crops (Fig. 2).

3.1 Antifungal activity

The most harmful pathogens for agricultural commodities are fungi, which are regarded as one of the most widespread (Shuping and Eloff, 2017; Junior et al., 2019). Producing mycotoxins, seriously harms fruits, cereals, and other food products before and after harvest. A significant amount of the world's agricultural products go to waste because of fungi attacks during the current food emergency (da Cruz Cabral et al., 2013). There have been reports of 300 fungus metabolites being hazardous to both humans and animals (Nesic et al., 2014). One of the sclerotia-forming infections that affect a variety of crops worldwide is *Rhizoctonia solani*. Diseases are brought on by *Alternaria alternata*, a fungal pathogenic class that affects a variety of crops globally. Many conventional fungicides have been extensively used worldwide to address these issues. However, many fungicides have limitations on their application for a variety of environmental and health reasons (Gikas et al., 2022). For instance, certain fungicides are so toxic that they cannot be used to treat seeds or to protect fruits and other agricultural products from fungal growth after harvest. Numerous crops have been explored

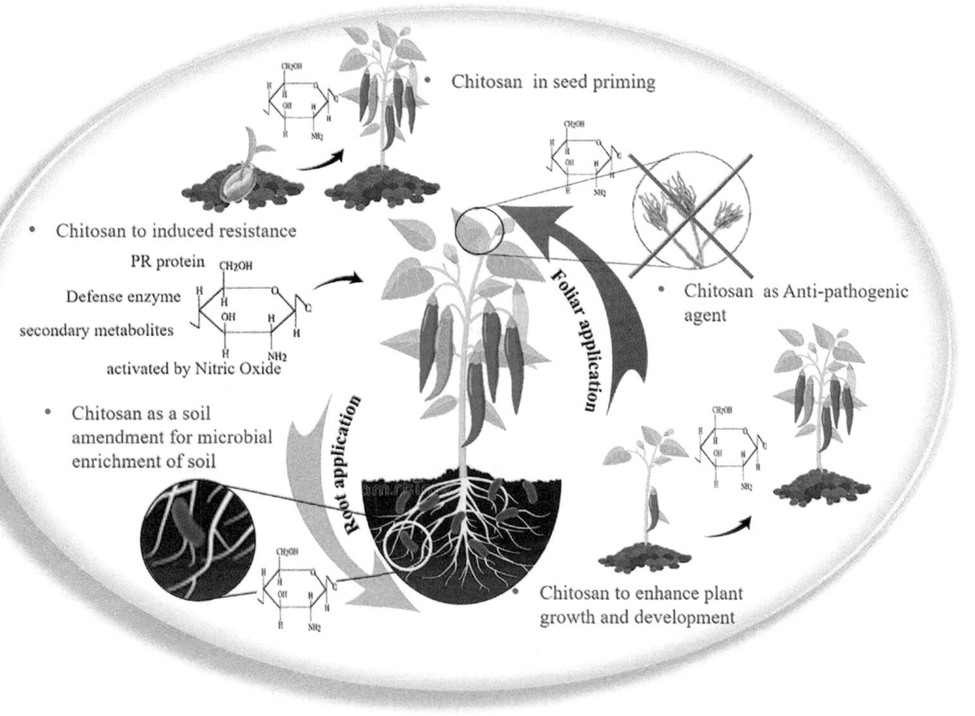

FIG. 2 Chitosan is a natural substance that has the potential to be used in the management of plant diseases, as well as antifungal activity, the induction of resistance, and antifungal activity mechanisms in plant pathology. *From Saberi-Riseh, R.S., Hassanisaadi, M., Vatankhah, M., Babaki, S.A., Barka, E.A., 2022. Chitosan as potential natural compound to manage plant diseases. Int. J. Biol. Macromol. with permission from Elsevier.*

using biopolymers such as chitin, chitosan, cellulose, and alginate as an alternative. Chitosan has stood out among them thanks to its exceptional antifungal and physicochemical capabilities (Salahuddin et al., 2018). The degree of deacetylation, molecular weight, and pH of the produced chitosan solution, as well as the source of chitosan, all affect the antifungal action of the substance. Inhibiting fungal growth at many growth phases, including mycelial growth, sporulation, spore viability, germination, and the synthesis of fungal virulence factors, has been effectively evaluated for chitosan. Overall, the effectiveness of chitosan does not only depend on the chitosan formulation but also on the fungus category and the type of treatment (Youssef and Hashim, 2020; Youssef et al., 2020).

For improved antifungal properties, chitosan has been used as a single ingredient or combined with additional components, particularly metal particles (Table 1). Due to the properties of inorganic and organic materials that function together to conduct the desired activity, nanocomposites have been widely used (Adnan et al., 2018). Nanoparticulate materials are typically added to long-chain or short-chain polymeric matrices to form nanocomposites. The resulting nanocomposites display enhanced features that cannot be found in any of the parts alone. The properties of the polymer are most likely to be greatly increased by

TABLE 1 Chitosan nanocomposites are used against various fungal pathogens.

No.	NCs	Application	References
1.	Chitosan NPs	In vivo antifungal activity against *Sclerospora graminicola*	Siddaiah et al. (2018)
2.	Cu-chitosan NPs	In vitro antifungal activity against *Alternaria alternata, Macrophomina phaseolina,* and *Rhizoctonia solani*	Saharan et al. (2013)
3.	Cu-chitosan NPs	In vitro antifungal activity against *Fusarium solani*	Vokhidova et al. (2014)
4.	Chitosan NPs	In vitro antifungal activity against various chili fungal disease	Chookhongkha et al. (2012)
5.	Ag-chitosan NPs	In vitro botryticidal activity against gray mold (*Botrytis cinerea*) in strawberry	Moussa et al. (2013)
6.	Ag-chitosan NPs	In vitro antifungal activity against the mycelial growth of *P. expansum*	Alghuthaymi et al. (2021)
7.	Ag-chitosan NPs	In vitro antifungal activity against *Aspergillus flaw* and *Aspergillus tennis*	Mathew and Kuriakose (2013)
8.	Silica-chitosan NPs	In vitro antifungal activity against *Phomopsis asparagi*	Cao et al. (2016)
	OCS, $nSiO_2$	In vitro antifungal activity of oligochitosan (OCS), hybrid with nanosilica ($nSiO_2$) against *Phytophthora infestans*	Nguyen et al. (2019).
9.	Chitosan-Pepper Tree (*Schisms mode*) essential oil (CS-EO) NPs	In vitro antifungal activity against *Aspergillus parasiticus* spores	Luque-Alcaraz et al. (2016)
10.	*Mentha piperita* essential oils in chitosan-cinnamic acid nanogel	In vivo antifungal activity against *Aspergillus flavus* in tomato during postharvest storage	Beyki et al. (2014)
	Zataria multiflora essential oils in chitosan nanoparticles	In vitro and in vivo botryticidal activity against gray mold (*Botrytis cinerea*) in strawberries at postharvest stage	Mohammadi et al. (2015)
	Chitosan boehmite-alumina nanocomposites films and thyme oil	Inhibited *Monilinia laxa,* which caused brown rot infection during postharvest storage of peaches	Cindi et al. (2015)
11.	Thiadiazole-functionalized chitosan derivatives	In vitro antifungal activity against *Colletotrichum lagenarium, Phomopsis asparagi,* and *Monilinia fructicola*	Li et al. (2013)
12.	Fungicide zineb (Zi) and chitosan-Ag nanoparticles	In vitro antifungal activity against *Neoscytalidium dimidiatum,* which caused brown rot disease in dragon fruit during postharvest storage	Ngoc and Nguyen (2018)
13.	Chitosan-Thyme-oregano, thyme-tea tree, and thyme-peppermint EO mixtures	In vitro antifungal activity against *Aspergillus niger, Aspergillus flavus, Aspergillus parasiticus,* and *Penicillium chrysogenum,* reducing their growth by 51%–77% in rice plants during postharvest storage	Hossain et al. (2019)

TABLE 1 Chitosan nanocomposites are used against various fungal pathogens—cont'd

No.	NCs	Application	References
14.	Chitosan-thymol nanoparticles	In vitro antifungal activity against the mycelial growth of *Botrytis cinerea* in blueberries and tomato cherries during postharvest storage	Medina et al. (2019)
15.	Chitosan-*Cymbopogon martinii* essential oil	In vitro antifungal activity against *Fusarium graminearum*, which causes *Fusarium* head blight disease in maize during postharvest storage	Kalagatur et al. (2018)

the addition of nanoparticles (Fahmy et al., 2020). Such nanocomposites are currently employed extensively in food processing, insect detection and management, food health screening, water treatment, disease detection, drug delivery systems, and enhancement of sustainable agriculture (Idumah et al., 2020; Wei et al., 2020). Similar to this, polymer composites serve as fertilizers by improving nutrient uptake and reducing soil toxicity (Guha et al., 2020). Furthermore, nanocomposites can be used as sensors and antimicrobial agents to extend the shelf life of food products (Usman et al., 2020).

Melo et al. (2020) demonstrated that the application of chitosan as a gel, nanoscale particles, or nanocomposite, can control the growth and viability of the strawberry phytopathogenic fungi. Elsherbiny et al. (2022) produced innovative chitosan-based nanocomposites as ecofriendly pesticide carriers for *Pythium debaryanum* and *Fusarium oxysporum*. Chitosan-based composites incorporated with nanomaterials and/or natural antimicrobials are very useful in improving shelf life and maintaining the quality of postharvest produce (Xing et al., 2019). El-Mohamedy et al. (2019) used chitosan nanoparticles as bio-nanopesticides against fungal diseases exploited for the delivery of agrochemicals. According to Lipşa et al. (2020), the usage of AuNPs-chitosan nanocomposites in agricultural settings offers a promising substitute for the conventional fungicides that are typically used to control plant infections like *F. oxysporum*. They proved that the antifungal properties of AuNPs-chitosan nanocomposites against two different strains of the widely distributed *F. oxysporum* impact tomato yield. CS-AuNPs exhibited antifungal activity against both strains of *F. oxysporum* by reducing the colony diameter (Lipşa et al., 2020). Chitosan significantly enhanced AuNPs antifungal activities (Dananjaya et al., 2017).

Kaur et al. (2012) added AgNPs to chitosan and then investigated the resulting nanocomposite formulation as an antifungal agent against isolates of *R. solani*, *Aspergillus flavus*, and *A. alternata*. The AgNPs, chitosan nanoparticles, and Ag-Cs-CNs were examined at various development stages. At all developmental stages, Ag-Cs-CNs showed good antifungal activity against all three fungi pathogens. Chowdappa et al. (2014) revealed good antifungal activity for the Ag-Cs nanocomposite against *Colletotrichum gloeosporioides*. The in vivo assay revealed that anthracnose was significantly inhibited by nanoformulation. Dananjaya et al. (2017) investigated the antifungal activities of Ag-Cs-CNs against *F. oxysporum* species complex. Al-Zubaidi et al. (2019) biosynthesized silver nanoparticles using an *Aspergillus niger* fungal isolate and evaluated them as a safe antifungal agent against plant pathogenic fungi. A disk-diffusion approach was used to explore the antifungal effects of oligochitosan (OCS), nanosilica ($nSiO_2$), and OCS/$nSiO_2$ hybrid materials. The results showed that the nanohybrid materials were more resistant to the fungus *Phytophthora infestans* than the individual components, and an $OCS_2/nSiO_2$ hybrid material concentration of 800 mg/L was the lowest concentration at which the material completely inhibited *P. infestans* growth, as measured using an agar dilution method (Nguyen et al., 2019). Mahmoud (2021) studied the antifungal activity and antiaflatoxin B1 of AgNPs and Ag-CS-NCs against aflatoxigenic isolates of *A. flavus*. The results showed notable antifungal activity and potency in thwarting aflatoxin B1 (AB1) production. Fig. 3 shows the antifungal activity of Ag-Cs NCs against *A. flavus* collected from feed samples (Alghuthaymi et al., 2021). Du et al. (2009) composited Zn^{2+} granules into the chitosan solution and tested it for enhanced antibacterial activity. Abd-Elsalam et al. (2018) demonstrated excellent antifungal activity against *R. solani* using a Zn-Cs nanocomposite. Kaur et al. (2018) evaluated chitosan-zinc oxide nanocomposites for their antifungal effectiveness against Fusarium wilt (caused by *F. oxysporum* f. sp. *ciceri* in chickpea). Cs-ZnO not only helped to reduce illness by about 40% after use but also promoted chickpea growth. Jaiswal et al. (2012) discovered a significant inhibitory effect of copper chitosan nanocomposite against fungal infections. Additionally, Cu-chitosan nanocomposite was demonstrated as a substantial crop growth enhancer. Through the formation of two Cu-CS nanocomposites, Rubina et al. (2017) established new

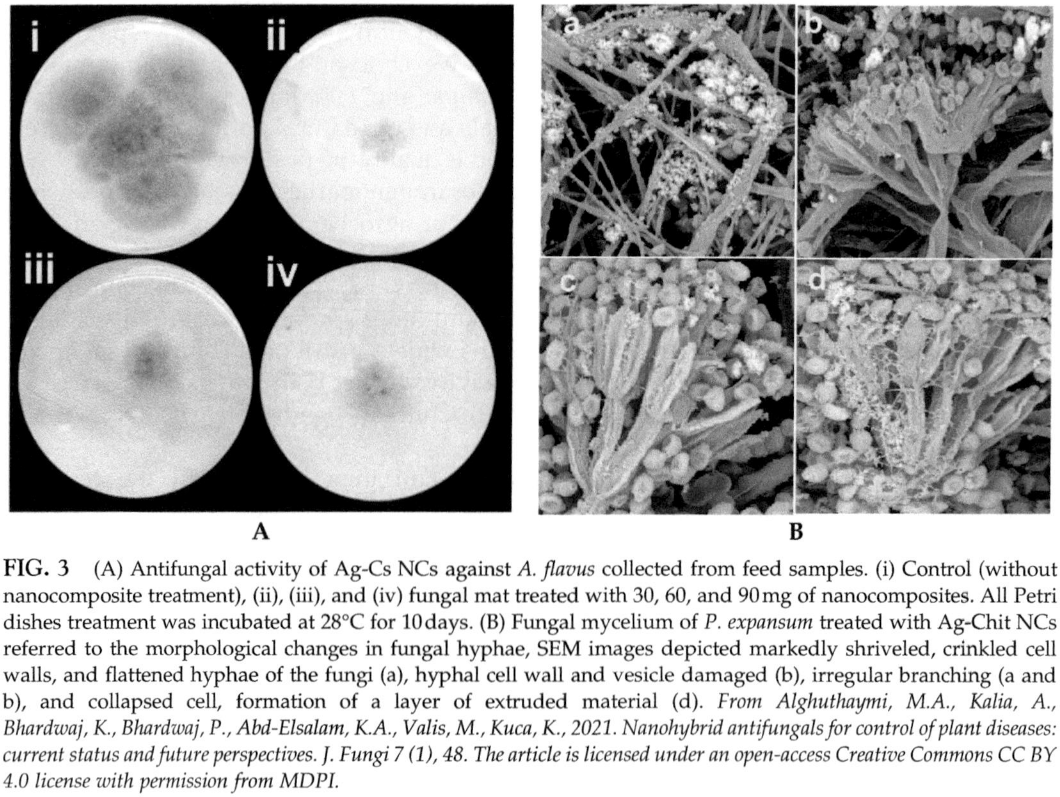

FIG. 3 (A) Antifungal activity of Ag-Cs NCs against *A. flavus* collected from feed samples. (i) Control (without nanocomposite treatment), (ii), (iii), and (iv) fungal mat treated with 30, 60, and 90 mg of nanocomposites. All Petri dishes treatment was incubated at 28°C for 10 days. (B) Fungal mycelium of *P. expansum* treated with Ag-Chit NCs referred to the morphological changes in fungal hyphae, SEM images depicted markedly shriveled, crinkled cell walls, and flattened hyphae of the fungi (a), hyphal cell wall and vesicle damaged (b), irregular branching (a and b), and collapsed cell, formation of a layer of extruded material (d). *From Alghuthaymi, M.A., Kalia, A., Bhardwaj, K., Bhardwaj, P., Abd-Elsalam, K.A., Valis, M., Kuca, K., 2021. Nanohybrid antifungals for control of plant diseases: current status and future perspectives. J. Fungi 7 (1), 48. The article is licensed under an open-access Creative Commons CC BY 4.0 license with permission from MDPI.*

methods for the control of sclerotium-forming plant pathogenic fungus. Abd-Elsalam et al. (2018) studied Cu-Cs, Zn-Cs, and Cs-bimetallic nanocomposites for their antifungal activity in both in vitro and in vivo at doses of 30, 60, and 100 μg/mL against two anastomosis groups of *R. solani* for control of cotton seedling damping-off. CuNPs, ZnONPs, and Zn-Cu-Cs-NCs were examined for their antifungal activity against the three plants' pathogenic fungi: *A. alternata*, *R. solani*, and *Botrytis cinerea* (Al-Dhabaan et al., 2017). Abd-Elsalam et al. (2020) demonstrated that copper polymer nanocomposites have outstanding effectiveness in inhibiting the growth of Aflatoxigenic *A. flavus*.

The use of polyethylene terephthalate punnets containing thyme oil and wrapped in chitosan/boehmite nanocomposite lidding films significantly reduced the incidence and severity of brown rot caused by *Monilinia laxa* in artificially inoculated peach fruit (cv. Kakawa) held at 25°C for 5 days and also significantly reduced brown rot occurrence at lower temperatures (Cindi et al., 2015). Cu-chitosan NCPs are synthesized and described, and their antifungal properties against *Fusarium verticillioides*, which cause premature flowering stalk rot (PFSR) in maize, are evaluated. Cu-chitosan NCs have exceptional potential as antifungal agents against PFSR of maize in both pot and field conditions (Choudhary, 2017). Chitosan NPs were produced from low-molecular-weight chitosan with a higher degree of acetylation and tested for their efficacy against the downy mildew disease of pearl millet caused by *Sclerospora graminicola* (Siddaiah et al., 2018).

3.1.1 Antifungal mechanisms

A promising nanofungicide should demonstrate activity that is similar to or greater than that of the bulk metal at relatively lower concentrations. It is also desirable to understand the problems with phytotoxicity and ecotoxicity caused by the release of metal ions. The antifungal activity of nanomaterials is carried out in a variety of ways (Fig. 4). Chitin, lipids, phospholipids, and polysaccharides are typically used in the construction of the fungal cell wall and cell membrane, with mannoproteins, 1,3-D-glucan, and 1,6-D-glucan proteins having a particular predominance (Patil and Chandrasekaran, 2020). Three different processes lead to the internalization of nanomaterials: (i) direct internalization of nanoparticles within the cell wall, (ii) specific receptor-mediated adsorption followed by internalization, and (iii) internalization of nanomaterials via ion transport proteins (Kalia et al., 2020a,b). The *N*-acetylglucosamine (*N*-acetyl-D-glucose-2-amine) production in the cell wall of fungi may be affected by the potential of nanomaterials to block the enzyme-glucan synthase after internalization. Abnormalities such as increased cell wall thickening, liquefaction of the cell membrane, dissolution or disorganization of the cytoplasmic organelles, hypervacuolization, and detachment of the cell wall from the cytoplasmic components may result from enzyme inhibition (Alghuthaymi et al., 2021).

Chitosan has been demonstrated to be a superior metal chelating or encapsulating agent for micro- or nanosize particles of Cu or Zn separately (Saharan et al., 2013; Rajasekaran and Santra, 2015). In the context of this, Cu- and Zn-based chitosan NPs can function in plants as both antimicrobial agents and growth enhancers (Du et al., 2009; Brunel et al., 2013; Saharan et al., 2013, 2015). Crop plants need transition metals like copper, zinc, and iron for a variety of purposes (Vasconcelos, 2014). Cu has a unique function in the biological system as a cofactor of various enzymes via the electron transport chain and redox reaction (Badawy and Rabea,

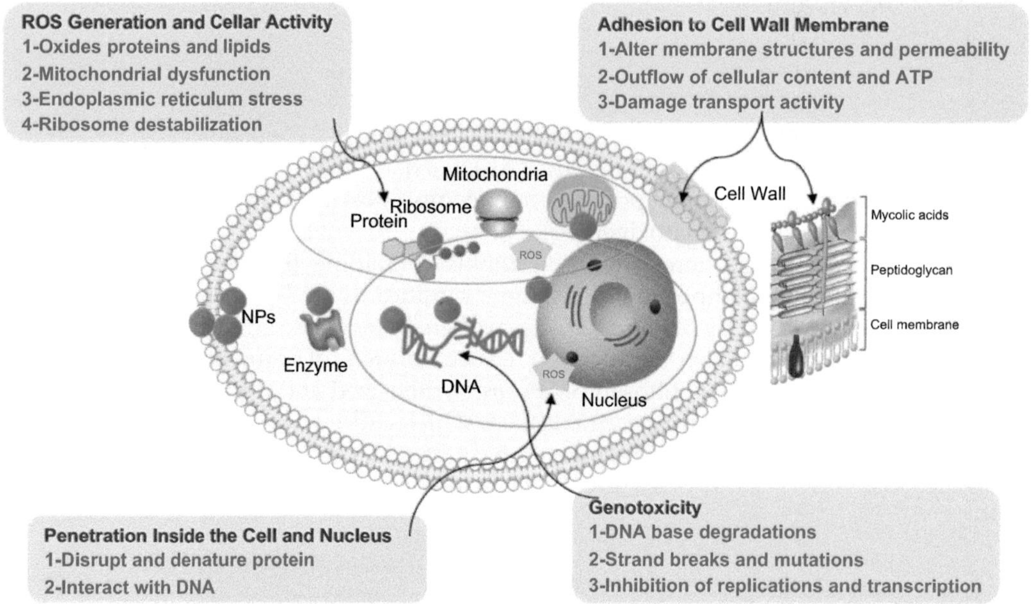

FIG. 4 Possible mechanistic illustration of hybrid nanomaterials' antifungal activity. *From Alghuthaymi, M.A., Kalia, A., Bhardwaj, K., Bhardwaj, P., Abd-Elsalam, K.A., Valis, M., Kuca, K., 2021. Nanohybrid antifungals for control of plant diseases: current status and future perspectives. J. Fungi 7 (1), 48. The article is licensed under an open-access Creative Commons CC BY 4.0 license with permission from MDPI.*

2011; Rajasekaran and Santra, 2015). Saharan et al. (2016) fabricated the Cu-chitosan nanoformulation and then treated it to maize seedlings. By increasing the overall protein content in the germinating seeds and generating α-amylase activity and protease enzymes, the nanoformulation showed promising benefits for seedling growth.

Similar to this, Zn is a crucial part of several enzymes, including oxidoreductases, transferases, and hydrolases. It plays a role in the production of chlorophyll and carbohydrates. It is also a part of zinc finger proteins, which are known to bind to DNA and RNA to regulate gene activity. Reactive oxygen species (ROS) are produced; membrane disintegration, chromosomal abnormalities, and other effects are all caused by excess zinc. Furthermore, many agrochemicals that are used in agriculture for crop protection or improvement historically contain significant amounts of Cu and Zn. Zinc oxide, silver, copper oxide, aluminum oxide, silicon dioxide, and other metal-based nanoparticles are more or less hazardous to plants, including rice and maize (Yang et al., 2015). The ROS production and antimicrobial activity of nanostructures are primarily influenced by the particle size, shape, surface charges, and surface groups. Smaller nanostructures can easily get through biological barriers like cell membranes. As particle size increases, the number of uptake decreases (Dizaj et al., 2014; Sathe et al., 2017). Additionally, superoxide radicals and other ROS can be produced when metal ions from nanomaterials, such as Ag^{+}, Cu^{+}, and Zn^{2+} ions, react with molecular oxygen. These ROS then have antimicrobial properties (Joe et al., 2018).

At the molecular level, the nanomaterials interact with different biomolecules and assemble into complexes with them, resulting in structural deformation of the biomolecules, inactivation of the catalytic proteins, and abnormalities of the nucleic acids, such as DNA breakage and chromosomal aberrations (Shoeb et al., 2013). The mechanism by which nanocomposite has antifungal action highly depends on reactive oxygen species. Metal ions cause ROS to be released, which damages biomolecules and causes cell death. Additionally, Lipovsky et al. (2011) discovered that the enhanced expression of lipid peroxidation is a clear sign of ROS generation to confirm the role of ROS in antifungal action. Ascorbate peroxidase, glutathione dismutase, and other stress-related enzymes were up or downregulated in fungi after exposure to nanomaterials (Yu et al., 2020).

Antimicrobial nanoparticles can electrostatically attach to microbial cell walls and modify the membrane potential, impairing respiration, balancing transport, interrupting energy transfer, and causing cell lysis, which can ultimately result in cell death (Pelgrift and Friedman, 2013). ROS are produced by nanomaterials and are the basis of another mechanism for nanostructured antimicrobials. Some nanostructures can trigger inflammation, which has the potential to chemically or catalytically transform less toxic oxidants like superoxide and hydrogen peroxide into more reactive free radicals like hydroxyl radicals, depending on the physicochemical properties of nanomaterials as well as the types of bacteria. A variety of semiconductor nanomaterials can also produce ROS when they interact with light, one of the environmental factors, as a result of the reaction between excited electron holes and water molecules, which is hazardous to bacteria. Significant oxidative stress produced by ROS can result in DNA/RNA damage, lipid peroxidation, protein modification, and enzyme inhibition, all of which can induce microbial cell death (Pan et al., 2010; Sathe et al., 2016). Direct inhibition of vital microbial enzymes, the formation of nitrogen reactive species (NRS), and the triggering of programmed cell death are additional impacts of antimicrobial nanomaterials (Dizaj et al., 2014). Dananjaya et al. (2017) showed that the chitosan-silver nanocomposite significantly damaged the membranes, disrupted the mycelium surface, increased membrane permeability, and even caused cell disintegration. The efficiency of the mechanism in action can be enhanced by the small size of the chitosan-based nanomaterials due to their high surface area that comes in contact with fungal pathogens. The small size can also enhance the uptake and increase of the penetrated and permeated chitosan on the thick coat of seeds, plant tissues, as well as the cell membranes of pathogens, hence resulting in better elicitations of plant immunity and defense response activities.

3.2 Biocontrol agents encapsulated with chitosan NPs

Micro- and nanocapsule research began around 1950 and grew quickly in the 1970s. Biocontrol agent encapsulation promotes agricultural plant growth and boosts harvestable yields by improving the plant's resistance to diseases. The delayed release of bioactive chemicals from the capsule ensures the bacteria's performance and survival while safeguarding them from adverse environmental elements and circumstances for extended periods (Saberi-Riseh et al., 2022). Chitin and its derivative chitosan are important encapsulation agents and can be employed as carriers for bioencapsulated microbial agents (Lakkis, 2016). The substance used to formulate the biocontrol agent has a substantial impact on its survival and effectiveness in eradicating pests and pathogens. Polymers such as alginate, starch, chitosan, and others have been studied in the development of biocontrol agents (Vemmer and Patel, 2013). Encapsulation enhances the viability, stability, and controlled release of bacteria while also reducing contamination during transportation and storage. Chitosan can be used as a carrier for microbial encapsulation and has proven to be an effective formulation component (Saberi-Riseh et al., 2021). For example, under both in vitro and in vivo conditions, chitosan nanoencapsulation of *Zataria multiflora* essential oil greatly reduced disease severity and incidence in strawberries infected with *B. cinerea*, the primary agent of gray mold disease (Mohammadi et al., 2015). Chitosan nanoparticles encapsulated with *Cymbopogon martinii* essential oil (CMEO) have antifungal and antimycotoxin efficacy against postharvest pathogens. Under in vitro circumstances, *Fusarium graminearum* was found in maize grains. According to the study, Ce-CMEO-NPs have stronger antifungal and antimycotoxin properties than CMEO because the regulated release of fungicidal components maintains antifungal activity. The Ce-CMEO-NPs could be a highly effective mycobiocide for protecting agricultural products from toxic fungi (Kalagatur et al., 2018). The antimicrobial activity of a chitosan/citral nanoemulsion against *A. niger* and *Rhizopus stolonifera* was dramatically increased (Marei et al., 2018). Foliar spraying and treating melon seeds with zinc nanoparticles encapsulated in chitosan effectively controlled Curvularia leaf spot disease, increased crop production, and enriched the grain with zinc micronutrients (Choudhary et al., 2019). Chitosan nanoparticles containing thiamine improved chickpea seedling growth and resistance to *F. oxysporum* f. sp. *ciceri* (Muthukrishnan et al., 2019). Saberi-Riseh and Moradi-Pour (2021) evaluated the possibility of encapsulating *Streptomyces fulvissimus* Uts22 in spray-dried chitosan/gellan microbeads. The effects of *S. fulvissimus* Uts22 chitosan/gellan microcapsules supplemented with ZnO nanoparticles on bacterial survival, efficiency in boosting wheat growth factors, and regulation of *Gaeumannomyces graminis* were also examined. As a result, encapsulating *S. fulvissimus* in a biodegradable coating (chitosan/gellan gum) can boost the bacterium's survival rate and efficiency by gradually releasing it. Furthermore, the ZnO nanoparticles used in encapsulation enhance plant growth. Because the bacteria are not out of reach of the plant, the slow release of germs from the capsule results in better control of plant pathogens. Garlic essential oil (GEO) was encapsulated in chitosan nanoparticles (NPCH) with sodium tripolyphosphate (TPP) to improve its stability and assess its efficacy as an antifungal coating agent on various cereal seeds. Coating seeds with GEO-NCPH is an intriguing alternate strategy for combating three fungal diseases. Furthermore, they demonstrated growth-promoting effects on wheat, oat, and barley by enhancing emergence, shoot, and root fresh weight (Mondéjar-López et al., 2022). The efficacy of using chitosan for biocontrol bacteria encapsulation needs further research, particularly in terms of the preparation and production of new formulations based on biodegradable polymers, as well as their ability to increase the quantity and quality of agricultural products while controlling pests and plant diseases (Saberi-Riseh et al., 2022). Table 2 shows the use of biological agents encapsulated in chitosan to protect against plant diseases.

3.3 Induce resistance

Chitosan-based fungicides have been shown to stimulate preharvest and postharvest defense systems in vegetables and fruits against viruses and pathogenic bacteria, insects,

TABLE 2 Some examples of biological agents encapsulated in chitosan to protect against plant pathogens.

Biological agent	Particle type	Pathogen	Encapsulation method	Results	Reference
Bacillus licheniformis	Alginate/chitosan	*Sclerotium rolfsii*	Nanoparticles beads	Suppressed the blight disease on chili plants	Panichikkal et al. (2021)
Streptomyces fulvissimus	Chitosan/gellan gum	*Gaeumannomyces graminis* var. *tritici*	Spray drying	Decreased the wheat take-all disease	Saberi-Riseh et al. (2021)
Bacillus sp. BK17	Chitosan/alginate	*Fusarium oxysporum*	Coated seed	Suppressed chili damping off	Suryanto et al. (2012)
Enterobacter sp. BK15	Chitosan/alginate	*Fusarium oxysporum*	Coated seed	Suppressed chili damping off	Suryanto et al. (2012)
Bacillus cereus B-71 (BMI)	Chitosan	*Radopholus similis*	BMI loaded nanoparticles	Promoting protection against *R. similis* in banana	Ureña-Saborío et al. (2017)
Bacillus sp. B-72 (BMI)	Chitosan	*Radopholus similis*	BMI loaded nanoparticles	Promoting protection against *R. similis* in banana	Ureña-Saborío et al. (2017)
Bacillus thuringiensis SER-217	Chitosan	*Radopholus similis*	BMI loaded nanoparticles	Promoting protection against *R. similis* in banana	Ureña-Saborío et al. (2017)
Pseudomonas syringae (Harpin pss)	Chitosan	*Rhizoctonia solani*	Ionic gelation	Inducing plant defense enzymes and reducing *R. solani* infection in tomato plants	Nadendla et al. (2018)
Cymbopogon martinii essential oil	Chitosan	*Fusarium graminearum*	Nanoparticles beads	Antifungal and antitoxins	Kalagatur et al. (2018)
Garlic oil	Chitosan		Nanoparticles beads	Promoting protection against three pathogenic fungi	Mondéjar-López et al. (2022)

Modified from Saberi-Riseh, R., Tamanadar, E., Hajabdollahi, N., Vatankhah, M., Thakur, V.K., Skorik, Y.A., 2022. Chitosan microencapsulation of rhizobacteria for biological control of plant pests and diseases: recent advances and applications. Rhizosphere 100565 with permission from Elsevier.

nematodes, fungi, and abiotic stressors. Furthermore, chitosan nanoparticles have coating capability, which increases the biophysical properties and lifespan of postharvest crops (Mehta et al., 2022). Chitosan is a possible elicitor chemical that induces plant innate immunity by activating defense-related responses, providing plants with stress tolerance (Stasińska-Jakubas and Hawrylak-Nowak, 2022). Secondary biomolecules, for example, produce various kinds of elicitors, which stimulate the plant's immunological response to stress. Similarly, chitosan is widely known as an elicitor that primes for signaling defense in response to various abiotic and biotic stressors (Katiyar et al., 2015). Cucumber plants treated with chitosan develop resistance to biotic stress by upregulating phytohormones and the biosynthesis of defense enzymes (Jogaiah et al., 2020). In another study, chitosan nanoparticle treatment induced the activity of defense-related enzymes and antioxidant enzymes, increasing phenolics and nitric oxide levels. This shows that chitosan nanoparticles play an important role in plant immunity via the nitric oxide signaling pathway. These chitosan-based nanoparticles develop complex defense mechanisms that protect the plant against disease and pest infection (Chandra et al., 2015).

Transcriptomic study has recently revealed that chitosan-based nanomaterials induce defensive gene expression in a variety of physiological processes, including alterations in hormone metabolism, heat shock proteins, protein reprogramming, and SAR activation. Chitosan nanoformulations can decrease crop diseases by directly acting on pathogens or pests and triggering some processes, including ROS and SAR production. They may also indirectly decrease diseases by enhancing crop nutrition and stimulating systemic and induced plant defense signaling pathways. Gene expression studies related to growth and secondary metabolite overproduction by chitosan nanoparticles should be the key goal of further studies in host-pathogen interactions.

As stand-alone products, chitosan NMs (Chit-NMs) can inhibit pathogenic fungi, bacteria, viruses, nematodes, and insects. They can also be combined with other engineered nanomaterials (ENMs) to form nanocomposites, which will increase both their function as nanocides and the gradual delivery of bioactive nutrients or chemicals to plants. Chit-NMs were synthesized in an anionic protein solution derived from the fungus *Penicillium oxalicum*. Later, the biosynthesized Chit-NMs were tested for antifungal activity against the pathogenic fungi *Pyricularia grisea, Alternaria solani*, and *F. oxysporum*. The Chit-NMs strongly suppressed the in vitro development of all three harmful fungi while also enhancing chickpea seed germination, seed vigor index, and biomass (Sathiyabama and Parthasarathy, 2016). Verticillium dahlia, a pathogenic fungus, was evaluated with oleoyl-chitosan nanocomposites (Xing et al., 2017). Another study found that bio-nanocomposite (Chit-NCs) chitosan with pepper tree essential oil has antifungal potency against pathogenic *Aspergillus parasiticus* (Adisa et al., 2019); the efficacy of the Chit-NCs was dependent on the individual strength of the components. Several in vitro studies on the antipathogenic effects of Chit-NCs propose a mechanism based on ROS formation and SAR activation, emphasizing that these Chit-NCs directly elucidate microbicidal activities against microbial pathogens and insect pests.

Furthermore, it has been reported that chitosan regulates several genes in plants, boosting particular pathways involved in plant defense. The enzyme phenylalanine ammonia-lyase (PAL) catalyzes the conversion of L-phenylalanine to trans-cinnamic acid and ammonium, which is a vital step in initiating metabolism in plants (Hyun et al., 2011) Chitosan also promotes plant immunity and defense by stimulating defense-related enzymes such as PAL, polyphenol-oxidase, catalase, and peroxidase (Lopez-Moya et al., 2019). The enzyme polyphenol oxidase aids in the catalysis of phenolic compounds during lignin formation. As a result, an increase in these enzymes might be read as an increase in lignin synthesis, which contributes to the creation of cell wall structure, thereby establishing a barrier for pathogen penetration (Li and Zhu, 2013). Similarly, substantial inhibition of *Physalospora piricola* and *Alternaria kikuchiana* spore germination and mycelial growth with elevated peroxidase activity in chitosan-treated pear has been found (Pacheco et al., 2008).

Peaches treated with chitosan had considerably higher peroxidase-gene expression than untreated peaches (Ma et al., 2013). Awadalla and Mahmoud used a novel chitosan derivative (carboxymethyl chitosan) to stimulate phytoalexins and induce Fusarium wilt resistance in cotton seeds (Awadalla and Mahamoud, 2005). Chitinase and β-1,3-glucanase can operate as catalytic converters in the hydrolysis of chitin and β-D-glucans present in fungal cell walls and insect exoskeletons, respectively. As a result, it destroys the fungal cell wall and inhibits fungal development in the host plant (Xing et al., 2015). A substantial increase in defense enzymes was seen in finger millet plants treated with Cu-chitosan NCs, both qualitatively and quantitatively. Blast disease suppression corresponds strongly with enhanced defense enzymes in Cu-chitosan NCs-treated finger millet plants (Sathiyabama and Manikandan, 2018). Chitosan nanoparticles (ChNP) have the potential to be an efficient biocontrol agent for the rice sheath blight pathogen (ShB) caused by *R. solani*. All of the defense enzymes had much higher enzyme-specific activity than the chemical control. It is a powerful plant immunity booster that can be used in place of commercially available chemical fungicides (Divya et al., 2020). When compared to the control treatment, chitosan nanoparticles prolonged latent and incubation times while decreasing infection type, pustule size, and several pustules. Salicylic acid treatment was less successful than chitosan nanoparticles in lowering the development of wheat leaf rust disease, which could be attributed to the fact that salicylic acid required more time and a higher dosage before inducing resistance (Elsharkawy et al., 2022).

Chitosan-based nanoformulations can deliver chemical and biopesticides efficiently. The primary benefit of chitosan-encapsulated nanopesticide carriers is that they play a key role due to their unique slow-release mechanisms with minimal evaporation and are environmentally friendly (Yu et al., 2021). The route of action of chitosan and plant innate immunity-induced activity is likely more complicated than indicated above, and more research is needed.

3.4 Chitosan nanocomposites against postharvest diseases

The growing public concern about the human health and environmental risks associated with high levels of pesticide use has sparked a lot of interest in developing nonpolluting control methods. Resistance has been frequently detected in several fungal pathogens exposed to fungicides applied to control postharvest diseases (Jacometti et al., 2010). Among alternative control means, chitosan and its nanocomposites have been widely used to manage the causal agents responsible for postharvest diseases in a wide range of horticultural crops. The shelf-life of fruits and vegetables has been extended using chitosan nanostructure (Meena et al., 2020). Chitosan and silicon oxide were investigated for their ability to stop postharvest weight loss and fungus disease in "Valencia Late" oranges (Beltrán et al., 2021).

In grapes, chitosan and silica-chitosan-copper nanoparticles reduced Botrytis growth in vitro by 38% and 43%, respectively, when used at a 4 mg/mL concentration. Under artificial fungal infection, chitosan, and silica chitosan-copper nanoparticles applied at 3 g/L were enough to reduce the percentage of rotted berries by 50% and 54%, respectively. Regarding natural infection, chitosan and silica-chitosan-copper NPs at 3 g/L were sufficient to reduce the percentage of rotted berries by 54% and 45%, respectively (Hashim et al., 2019). Similarly, chitosan/Ag/ZnO blend films were tested against several microorganisms such as *Bacillus subtilis*, *E. coli*, *Staphylococcus aureus*, *Penicillium* spp., *Aspergillus* spp., and *Rhizopus* spp. (Li et al., 2010). To prevent postharvest diseases and extend the shelf-life of table grapes, the chitosan-TiO_2 film (50–80 nm) was used (Zhang et al., 2017).

In strawberry, silver-chitosan nanocomposites were proposed as capable fungicidal members for the prevention of gray mold disease. This composite was effective for inhibiting pathogen growth, with a minimum inhibition concentration of 125 μg/mL (Moussa et al., 2013). The encapsulation of essential oil of *Z. multiflora* into chitosan nanoparticles (225–175 nm) permitted a release of the oil for 40 days, and the use of nanocapsules reduced disease severity and incidence of gray mold in strawberries (Mohammadi et al., 2015). Similarly, in peach fruits, chitosan-rice starch nanocomposite decreased *E. coli* and *S. aureus* and improved the quality of peaches, including the shelf-life period (Kaur et al., 2017). Chitosan/cellulose nanofibril bionanocomposites were effectively prepared and used to preserve the postharvest quality of strawberries (Resende et al., 2018).

In citrus fruits, clay-chitosan nanocomposite was prepared at different ratios of clay/chitosan and was tested against green mold caused by *Penicillium digitatum* on "Valencia Late" sweet orange. In in vitro tests, complete inhibition of the pathogen was achieved at 20 μg/mL for clay/chitosan (1:0.5), clay/chitosan (1:1), and clay/chitosan (1:2). Under artificial fungal infection, complete inhibition of the disease was obtained, and a high reduction of lesion development was reported for clay/chitosan at 1:2 ratio at 20 μg/mL. Scanning electron microscopy (SEM) showed that treatment with this nanocomposite caused severe collapse and irregular branching of hyphae in the apical part. In addition, clay-chitosan nanocomposite caused degradation of DNA of the tested pathogen at 20 μg/mL (Youssef and Hashim, 2020).

In tangerine fruits, the combination of montmorillonite was capable of improving the water vapor barrier parameters of chitosan coating and consequently reduced mass loss. Chitosan or chitosan/montmorillonite was used to coat tangerine fruits, and then stored at 10°C for 11 days; chitosan coating inhibited the decay rate of fruits only at the first 5 days. Nevertheless, when chitosan/montmorillonite containing 1% was used to coat tangerine fruits, a lower decay percentage, lower weight loss, higher contents of total soluble solids (TSS), and titratable acidity (TA) were noted (Xu et al., 2018).

In mango fruits, chitosan-silver nanocomposite exhibited a higher antifungal effect against the conidial germination of *C. gloeosporioides* the causal agent of anthracnose. In this regard, complete inhibition of spore germination was noted when chitosan-silver nanocomposite was applied at 100 μg/mL. Also, in mango cv. Alphonso, field experiments showed that *C. gloeosporioides* were significantly suppressed by chitosan-silver nanocomposite (Chowdappa et al., 2014).

4 Large-scale applications challenges

The bioactive compounds integrated into chitosan films demonstrated great potential for use in extending the shelf-life and maintaining the quality of food products, as well as reducing postharvest fungi and foodborne bacteria in the food system. However, the majority of the data provided consists of manufacturing employing casting procedures, which are mostly used on a laboratory scale. The casting process involves dissolving the polysaccharide in a suitable solvent, which for chitosan is typically an acetic acid solution, and incorporating the active compound, plasticizer, and nanofiller of interest at the same time, followed by pouring the resulting mixed dispersion onto an inert surface to evaporate the solvent and obtain the thin film. As a result, one issue in the use of chitosan is translating this laboratory-size approach to an industrial scale or finding other production methods soon (Souza et al., 2020). Pesticides, fertilizers, and carriers for plant nutrients still rarely use chitosan nanocomposites. The fact that chitosan-based products have received less focus and research is the reason for this problem. To create such new ecofriendly pesticides, chitosan polymeric nanoparticles should be further investigated. More research is needed to fully understand the interactions between chitosan and bioactive compounds to improve the efficacy of the bioactive integrated agents. The standardization of methods from chitosan production through to its industrial uses needs to take the spotlight. Additionally, it is necessary to look into the use of innovative and practical sources for the isolation of chitosan. Numerous studies have documented the manufacture of chitosan nanocomposite on a lab scale recently. Transferring these products from the laboratory to large-scale production facilities and finally to farmers, meanwhile, remains a challenging issue. Regarding potential threats to product development and farmer acceptance, industries continue to be wary of chitosan and its products. Research should concentrate on the creation of more practical, industrially acceptable chitosan synthesis techniques and chitosan-based products such as nanoformulations and fertilizers. This might make it possible for farmers and businesses to use chitosan.

5 Future trends

Due to a major lack of understanding of their bioactivity and modes of action against pathogenic microorganisms, as well as plant protection and growth, chitosan-based NMs have not yet been widely employed in agriculture. A series of investigations have shown that chitosan NMs, when compared to its bulk form, act prudently on harmful bacteria while also inducing plant antioxidant/defense responses. Chitosan-based NMs are emerging as a valuable biomaterial for usage as a biopesticide to replace synthetic pesticides, according to recent review literature results. We are certain that NMs produced by chitosan mimic natural elicitation and have a significant impact on plant protection and growth. Chitosan NMs have been used as seed treatments and foliar applications in a variety of crops. However, it is worthwhile to investigate the influence of chitosan-based NMs on soil microbiota through soil application. More research has demonstrated their role in eliciting plant defense responses such as defense genes/enzymes, antioxidant enzymes, total phenolics, and so on. More research on the toxicity of these formulations before the final phases of application is needed to take use of the antimicrobial interaction mechanisms, notably against the antimicrobial mechanism in plants. Furthermore, more data is required to determine the actual movement mechanism of chitosan-based agronanofungicides in plants, where the uptake, translocation, and transportation of the nanochitosan may be influenced by particle size, morphology, surface charge, solubility, bioavailability, plant types, and effective exposure concentrations.

6 Conclusion

Natural products, such as chitosan, can be viewed as viable alternatives for controlling plant diseases and reducing pesticides' negative effects on humans and the environment. Because of its outstanding thermal, mechanical, and electrical capabilities, chitosan and its nanostructures are the most viable contenders to serve as a significant filling agent for composite applications. Chitosan's chemical structure contains amino groups (—NH_2), which gives it unique and desirable features in several agricultural systems, including food conservation and food security through the manufacture of biodegradable edible coatings and films. Numerous potential applications indicate that future research and opportunities for chitosan-based nanocomposites are anticipated to greatly expand in a variety of fields and products available for purchase. Several alternative control methods are currently proposed to reduce and/or substitute chemical pesticides. This chapter not only discusses a novel environmentally friendly substance with unique synergistic antifungal actions that can replace current harmful agrochemicals, but it also discusses a new platform for future research in green agricultural applications. Chitosan biopolymer possesses adaptable physicochemical features that allow it to be converted into an intelligent nanochitosan product with the assistance of other bioactive substances. The use of chitosan and its nanostructures holds great potential for the management of plant diseases and/or the agents that cause them. As a result, we anticipate that fungicides of the next generation will exhibit the following characteristics: (1) action that is multitargeted and uses many modes to stimulate immune responses in plants; (2) demonstration of direct antifungal activity; and (3) gradual and regulated release of active component to ensure timely and long-lasting effects in the crop. As a resistance inducer, it activates plant defense mechanisms through a series of biochemical and molecular events, as well as increasing plant immunity against pathogenic fungi during the pre- and postharvest stages. To validate the precise pathways of chitosan nanostructures, additional research on biochemical and molecular features under field conditions is necessary. Chitosan NCs, therefore, have a significant potential for the development of new generations of fungicides that may be cost-effective, favorable to the environment, and give the biosphere a minimal amount of chemical burden.

References

Abd El-Aziz, A.R.M., Al-Othman, M.R., Mahmoud, M.A., Shehata, S.M., Abdelazim, N.S., 2018. Chitosan nanoparticles as a carrier for *Mentha longifolia* extract: synthesis, characterization and antifungal activity. Curr. Sci., 2116–2122.

Abd-Elsalam, K.A., Vasil'kov, A., Said-Galiev, E.E., Rubina, M.S., Khokhlov, A.R., Naumkin, A.V., Shtykova, E.V., Alghuthaymi, M.A., 2018. Bimetallic blends and chitosan nanocomposites: novel antifungal agents against cotton seedling damping-off. Eur. J. Plant Pathol. 151, 57–72.

Abd-Elsalam, K.A., Alghuthaymi, M.A., Shami, A., Rubina, M.S., Abramchuk, S.S., Shtykova, E.V., Yu Vasil'kov, A., 2020. Copper-chitosan nanocomposite hydrogels against aflatoxigenic *Aspergillus flavus* from dairy cattle feed. J. Fungi 6, 112.

Adisa, I.O., Pullagurala, V.L.R., Peralta-Videa, J.R., Dimkpa, C.O., Elmer, W.H., Gardea-Torresdey, J.L., White, J.C., 2019. Recent advances in nano-enabled fertilizers and pesticides: a critical review of mechanisms of action. Environ. Sci. Nano 6 (7), 2002–2030.

Adnan, M.M., Dalod, A.R.M., Balci, M.H., Glaum, J., Einarsrud, M.A., 2018. In situ synthesis of hybrid inorganic-polymer nanocomposites. Polymers 10, 1129.

Al-Dhabaan, F.A., Shoala, T., Ali, A.A.M., Alaa, M., Abd-Elsalam, K., 2017. Chemically-produced copper, zinc nanoparticles and chitosan–bimetallic nanocomposites and their antifungal activity against three phytopathogenic fungi. Int. J. Agric. Technol. 13, 753–769.

Al-Dhabaan, F.A., Mostafa, M., Almoammar, H., Abd-Elsalam, K.A., 2018. Chitosan-based nanostructures in plant protection applications. In: Nanobiotechnology Applications in Plant Protection. Springer, Cham, pp. 351–384.

Alghuthaymi, M.A., Rajkuberan, C., Rajiv, P., Kalia, A., Bhardwaj, K., Bhardwaj, P., Abd-Elsalam, K.A., Valis, M., Kuca, K., 2021. Nanohybrid antifungals for control of plant diseases: current status and future perspectives. J. Fungi 7, 48.

Al-Zubaidi, S., Al-Ayafi, A., Hayam Abdelkader, H., 2019. Biosynthesis, characterization and antifungal activity of silver nanoparticles by *Aspergillus niger* isolate. J. Nanotechnol. Res. 1, 023–036.

Artunduaga Bonilla, J.J., Paredes Guerrero, D.J., Sánchez Suárez, C.I., Ortiz López, C.C., Torres Sáez, R.G., 2015. In vitro antifungal activity of silver nanoparticles against fluconazole-resistant *Candida* species. World J. Microbiol. Biotechnol. 31, 1801–1809.

Artunduaga Bonilla, J.J., Honorato, L., Cordeiro de Oliveira, D.F., Gonçalves, R.A., Guimarães, A., Miranda, K., Nimrichter, L., 2021. Silver chitosan nanocomposites as a potential treatment for superficial candidiasis. Med. Mycol. 59, 993–1005.

Ashfaq, M., Talreja, N., Chuahan, D., Srituravanich, W., 2020. Polymeric nanocomposite-based agriculture delivery system: emerging technology for agriculture. Genet. Eng. Glimpse Tech. Appl., 1–16.

Awadalla, O.A., Mahamoud, G., 2005. New chitosan derivatives induced resistance to *Fusarium* wilt disease through phytoalexin (Gossypol) production. Sains Malays. 34, 141–146.

Badawy, M.E.I., Rabea, E.I., 2011. A biopolymer chitosan and its derivatives as promising antimicrobial agents against plant pathogens and their applications in crop protection. Int. J. Carbohydr. Chem. 1, 1–29.

Baka, Z.A., Abou-Dobara, M.I., El-Sayed, A.K.A., El-Zahed, M.M., 2019. Synthesis, characterization and antimicrobial activity of chitosan/Ag nanocomposite using *Escherichia coli* D8. Sci. J. Damietta Faculty Sci. 9, 1–6.

Beltrán, R., Otesinova, L., Cebrián, N., Zornoza, C., Breijo, F., Garmendia, A., Merle, H., 2021. Effect of chitosan and silicon oxide treatments on postharvest Valencia Late (*Citrus × sinensis*) fruits. J. Plant Sci. Phytopathol. 5, 065–071.

Beyki, M., Zhaveh, S., Khalili, S.T., Rahmani-Cherati, T., Abollahi, A., Bayat, M., Tabatabaei, M., Mohsenifar, A., 2014. Encapsulation of *Mentha piperita* essential oils in chitosan–cinnamic acid nanogel with enhanced antimicrobial activity against *Aspergillus flavus*. Ind. Crop. Prod. 54, 310–319.

Brunel, F., Gueddari, N.E.E., Moerschbacher, B.M., 2013. Complexation of copper (II) with chitosan nanogels: toward control of microbial growth. Carbohydr. Polym. 92, 1348–1356.

Cao, L., Zhang, H., Cao, C., Zhang, J., Li, F., Huang, Q., 2016. Quaternized chitosan-capped mesoporous silica nanoparticles as nanocarriers for controlled pesticide release. Nanomaterials 6 (7), 126.

Chandra, S., Chakraborty, N., Dasgupta, A., Sarkar, J., Panda, K., Acharya, K., 2015. Chitosan nanoparticles: a positive modulator of innate immune responses in plants. Sci. Rep. 5, 15195.

Chookhongkha, N., Sopondilok, T., Photchanachai, S., 2012, February. Effect of chitosan and chitosan nanoparticles on fungal growth and chilli seed quality. In: I International Conference on Postharvest Pest and Disease Management in Exporting Horticultural Crops-PPDM2012 973, pp. 231–237.

Choudhary, M.K., 2017. Development and Evaluation of Cu Chitosan Nanocomposite for Its Antifungal Activity against Post Flowering Stalk Rot (PFSR) Disease of Maize Caused by *Fusarium verticillioids* (Sheldon) (Ph.D. thesis). Maharana Pratap University of Agriculture and Technology, Udaipur. 79 pages.

Choudhary, R.C., Kumaraswamy, R.V., Kumari, S., Sharma, S.S., Pal, A., Raliya, R., Biswas, P., Saharan, V., 2019. Zinc encapsulated chitosan nanoparticle to promote maize crop yield. Int. J. Biol. Macromol. 127, 126–135.

Chowdappa, P., Gowda, S., Chethana, C.S., Madhura, S., 2014. Antifungal activity of chitosan-silver nanoparticle composite against *Colletotrichum gloeosporioides* associated with mango anthracnose. Afr. J. Microbiol. Res. 8, 1803–1812.

Cindi, M.D., Shittu, T., Sivakumar, D., Bautista-Baños, S., 2015. Chitosan boehmite alumina nanocomposite films and thyme oil vapour control brown rot in peaches (*Prunus persica* L.) during postharvest storage. Crop Prot. 72, 127–131.

da Cruz Cabral, L., Pinto, V.F., Patriarca, A., 2013. Application of plant derived compounds to control fungal spoilage and mycotoxin production in foods. Int. J. Food Microbiol. 166, 1–14.

Dananjaya, S.H.S., Udayangani, R.M.C., Chulhong, O., Nikapitiya, C., Lee, J., De Zoysa, M., 2017. Green synthesis, physio-chemical characterization and anti-candidal function of a biocompatible chitosan gold nanocomposite as a promising antifungal therapeutic agent. RSC Adv. 7, 9182–9193.

Dananjaya, S.H.S., Kumar, R.S., Yang, M., Nikapitiya, C., Lee, J., De Zoysa, M., 2018. Synthesis, characterization of ZnO-chitosan nanocomposites and evaluation of its antifungal activity against pathogenic *Candida albicans*. Int. J. Biol. Macromol. 108, 1281–1288.

Divya, K., Thampi, M., Vijayan, S., Varghese, S., Jisha, M.S., 2020. Induction of defence response in *Oryza sativa* L. against *Rhizoctonia solani* (Kuhn) by chitosan nanoparticles. Microb. Pathog. 149, 104525.

Dizaj, S.M., Lotfipour, F., Barzegar-Jalali, M., Zarrintan, M.H., Adibkia, K., 2014. Antimicrobial activity of the metals and metal oxide nanoparticles. Mater. Sci. Eng. C Mater. Biol. Appl. 44, 278–284.

Du, W.L., Niu, S.S., Xu, Y.L., Xu, Z.R., Fan, C.L., 2009. Antibacterial activity of chitosan tripolyphosphate nanoparticles loaded with various metal ions. Carbohydr. Polym. 75, 385–389.

Duhan, J.S., Kumar, R., Kumar, N., Kaur, P., Nehra, K., Duhan, S., 2017. Nanotechnology: the new perspective in precision agriculture. Biotechnol. Rep. 15, 11–23.

El-Mohamedy, R.S., El-Aziz, M.E.A., Kamel, S., 2019. Antifungal activity of chitosan nanoparticles against some plant pathogenic fungi in vitro. Agric. Eng. Int. 21, 201–209.

Elsharkawy, M.M., Omara, R.I., Mostafa, Y.S., Alamri, S.A., Hashem, M., Alrumman, S.A., Ahmad, A.A., 2022. Mechanism of wheat leaf rust control using chitosan nanoparticles and salicylic acid. J. Fungi 8 (3), 304.

Elsherbiny, A.S., Galal, A., Ghoneem, K.M., Salahuddin, N.A., 2022. Novel chitosan-based nanocomposites as ecofriendly pesticide carriers: synthesis, root rot inhibition and growth management of tomato plants. Carbohydr. Polym. 282.

Fahmy, H.M., Salah Eldin, R.E., Abu Serea, E.S., Gomaa, N.M., AboElmagd, G.M., Salem, S.A., Elsayed, Z.A., Edrees, A., Shams-Eldin, E., Shalan, A.E., 2020. Advances in nanotechnology and antibacterial properties of biodegradable food packaging materials. RSC Adv. 10, 20467–20484.

Gikas, G.D., Parlakidis, P., Mavropoulos, T., Vryzas, Z., 2022. Particularities of fungicides and factors affecting their fate and removal efficacy: a review. Sustainability 2022 (14), 4056.

Guha, T., Gopal, G., Kundu, R., Mukherjee, A., 2020. Nanocomposites for delivering agrochemicals: a comprehensive review. J. Agric. Food Chem. 68, 3691–3702.

Hashim, A.F., Youssef, K., Abd-Elsalam, K.A., 2019. Ecofriendly nanomaterials for controlling gray mold of table grapes and maintaining post-harvest quality. Eur. J. Plant Pathol. 154, 377–388.

Hossain, F., Follett, P., Salmieri, S., Vu, K.D., Fraschini, C., Lacroix, M., 2019. Antifungal activities of combined treatments of irradiation and essential oils (EOs) encapsulated chitosan nanocomposite films in in vitro and in situ conditions. Int. J. Food Microbiol. 295, 33–40.

Hyun, M.W., Yun, Y.H., Kim, J.Y., Kim, S.H., 2011. Fungal and plant phenylalanine ammonia-lyase. Mycobiology 39 (4), 257–265.

Idumah, C.I., Zurina, M., Ogbu, J., Ndem, J.U., Igba, E.C., 2020. A review on innovations in polymeric nanocomposite packaging materials and electrical sensors for food and agriculture. Compos. Interfaces 27, 1–72.

Jacometti, M.A., Wratten, S.D., Walter, M., 2010. Review: alternatives to synthetic fungicides for *Botrytis cinerea* management in vineyards. Aust. J. Grape Wine Res. 16, 154–172.

Jaiswal, M., Chauhan, D., Sankararamakrishnan, N., 2012. Copper chitosan nanocomposite: synthesis, characterization, and application in removal of organophosphorous pesticide from agricultural runoff. Environ. Sci. Pollut. Res. 19, 2055–2062.

Joe, A., Park, A.H., Kim, D.J., Lee, Y.J., Jhee, K.H., Sohn, Y., Jang, E.S., 2018. Antimicrobial activity of ZnO nanoplates and its Ag nanocomposites: insight into an ROS-mediated antibacterial mechanism under UV light. J. Solid State Chem. 267, 124–133.

Jogaiah, S., Satapute, P., De Britto, S., Konappa, N., Udayashankar, A.C., 2020. Exogenous priming of chitosan induces upregulation of phytohormones and resistance against cucumber powdery mildew disease is correlated with localized biosynthesis of defense enzymes. Int. J. Biol. Macromol. 162, 1825–1838.

Junior, O.J.C., Youssef, K., Koyama, R., Ahmed, S., Dominguez, A.R., Mühlbeier, D.T., Roberto, S.R., 2019. Control of gray mold on clamshell-packaged 'benitaka' table grapes using sulphur dioxide pads and perforated liners. Pathogens 8, 271.

Kalagatur, N.K., Nirmal Ghosh, O.S., Sundararaj, N., Mudili, V., 2018. Antifungal activity of chitosan nanoparticles encapsulated with *Cymbopogon martinii* essential oil on plant pathogenic fungi *Fusarium graminearum*. Front. Pharmacol. 9, 610.

Kalia, A., Abd-Elsalam, K.A., Kuca, K., 2020a. Zinc-based nanomaterials for diagnosis and management of plant diseases: ecological safety and future prospects. J. Fungi 6, 222.

Kalia, A., Sharma, S.P., Kaur, H., Kaur, H., 2020b. Novel nanocomposite-based controlled-release fertilizer and pesticide formulations: prospects and challenges. In: Abd-Elsalam, K.A. (Ed.), Multifunctional Hybrid Nanomaterials for Sustainable Agri-Food and Ecosystem. Elsevier, Amsterdam, pp. 99–134.

Katiyar, D., Hemantaranjan, A., Singh, B., 2015. Chitosan as a promising natural compound to enhance potential physiological responses in plant: a review. Indian J. Plant Physiol. 20 (1), 1–9.

Kaur, P., Thakur, R., Choudhary, A., 2012. An in vitro study of the antifungal activity of silver/chitosan nanoformulations against important seed borne pathogens. Int. J. Sci. Technol. Res. 1, 83–86.

Kaur, P., Thakur, R., Barnela, M., Chopra, M., Manujab, A., Chaudhury, A., 2015. Synthesis, characterization and in vitro evaluation of cytotoxicity and antimicrobial activity of chitosan–metal nanocomposites. J. Chem. Technol. Biotechnol. 90, 867–873.

Kaur, M., Kalia, A., Thakur, A., 2017. Effect of biodegradable chitosan–rice-starch nanocomposite films on post-harvest quality of stored peach fruit. Starch 69, 1600208.

Kaur, P., Duhan, J.S., Thakur, R., 2018. Comparative pot studies of chitosan and chitosan-metal nanocomposites as nano-agrochemicals against fusarium wilt of chickpea (*Cicer arietinum* L.). Biocatal. Agric. Biotechnol. 14, 466–471.

Kumaraswamy, R.V., Kumari, S., Choudhary, R.C., Pal, A., Raliya, R., Biswas, P., Saharan, V., 2018. Engineered chitosan-based nanomaterials: bioactivities, mechanisms and perspectives in plant protection and growth. Int. J. Biol. Macromol. 113, 494–506.

Lakkis, J.M. (Ed.), 2016. Encapsulation and Controlled Release Technologies in Food Systems. John Wiley & Sons.

Li, S.J., Zhu, T.H., 2013. Biochemical response and induced resistance against anthracnose (*Colletotrichum camelliae*) of camellia (*Camellia pitardii*) by chitosan oligosaccharide application. For. Pathol. 43 (1), 67–76.

Li, L.H., Deng, J.C., Deng, H.R., Liu, Z.L., Li, X.L., 2010. Preparation, characterization and antimicrobial activities of chitosan/Ag/ZnO blend films. Chem. Eng. J. 160, 378–382.

Li, Q., Ren, J., Dong, F., Feng, Y., Gu, G., Guo, Z., 2013. Synthesis and antifungal activity of thiadiazole functionalized chitosan derivatives. Carbohydr. Res. 373, 103–107.

Lipovsky, A., Nitzan, Y., Gedanken, A., Lubart, R., 2011. Antifungal activity of ZnO nanoparticles-the role of ROS mediated cell injury. Nanotechnology 22.

Lipşa, F.-D., Ursu, E.-L., Ursu, C., Ulea, E., Cazacu, A., 2020. Evaluation of the antifungal activity of gold–chitosan and carbon nanoparticles on *Fusarium oxysporum*. Agronomy 10, 1143.

Lopez-Moya, F., Suarez-Fernandez, M., Lopez-Llorca, L.V., 2019. Molecular mechanisms of chitosan interactions with fungi and plants. Int. J. Mol. Sci. 20 (2), 332.

Luque-Alcaraz, A.G., Cortez-Rocha, M.O., Velázquez-Contreras, C.A., Acosta-Silva, A.L., Santacruz-Ortega, H.D.C., Burgos-Hernández, A., Argüelles-Monal, W.M., Plascencia-Jatomea, M., 2016. Enhanced antifungal effect of chitosan/pepper tree (*Schinus molle*) essential oil bionanocomposites on the viability of *Aspergillus parasiticus* spores. J. Nanomater. 2016.

Ma, Z., Yang, L., Yan, H., Kennedy, J.F., Meng, X., 2013. Chitosan and oligochitosan enhance the resistance of peach fruit to brown rot. Carbohydr. Polym. 94 (1), 272–277.

Mahmoud, M.A., 2021. Characterization and antifungal efficacy of biogenic silver nanoparticles and silver–chitosan nanocomposites. Egypt. J. Phytopathol. 49, 68–79.

Maluin, F.N., Hussein, M.Z., 2020. Chitosan-based agronanochemicals as a sustainable alternative in crop protection. Molecules 25 (7), 1611.

Marei, G.I.K., Rabea, E.I., Badawy, M.E., 2018. Preparation and characterizations of chitosan/citral nanoemulsions and their antimicrobial activity. Appl. Food Biotechnol. 5 (2), 69–78.

Mathew, T.V., Kuriakose, S., 2013. Photochemical and antimicrobial properties of silver nanoparticle-encapsulated chitosan functionalized with photoactive groups. Mater. Sci. Eng. C 33 (7), 4409–4415.

Medina, E., Caro, N., Abugoch, L., Gamboa, A., Díaz-Dosque, M., Tapia, C., 2019. Chitosan thymol nanoparticles improve the antimicrobial effect and the water vapour barrier of chitosan-quinoa protein films. J. Food Eng. 240, 191–198.

Meena, M., Pilania, S., Pal, A., Mandhania, S., Bhushan, B., Kumar, S., Gohari, G., Saharan, V., 2020. Cu-chitosan nano-net improves keeping quality of tomato by modulating physio-biochemical responses. Sci. Rep. 10, 21914.

Mehta, M.R., Biradar, S.P., Mahajan, H.P., Bankhele, R.R., Hivrale, A.U., 2022. Chitosan and chitosan-based nanoparticles in horticulture: past, present and future prospects. In: Role of Chitosan and Chitosan-Based Nanomaterials in Plant Science. Academic Press, pp. 453–474.

Melo, N.F.C.B., de Lima, M.A.B., Stamford, T.L.M., Galembeck, A., Flores, M.A., de Campos Takaki, G.M., da Costa Medeiros, J.A., Stamford-Arnaud, T.M., Montenegro Stamford, T.C., 2020. In vivo and in vitro antifungal effect of fungal chitosan nanocomposite edible coating against strawberry phytopathogenic fungi. Int. J. Food Sci. Technol. 55, 3381–3391.

Mishra, K.K., Subbanna, A.R.N.S., Rajashekara, H., Paschapur, A.U., Singh, A.K., Jeevan, B., Maharana, C., 2022. Chitosan and chitosan-based nanoparticles for eco-friendly management of plant diseases and insect pests: a concentric overview. In: Role of Chitosan and Chitosan-Based Nanomaterials in Plant Sciences. Academic Press, pp. 435–451.

Mohammadi, A., Hashemi, M., Hosseini, S.M., 2015. Nanoencapsulation of *Zataria multiflora* essential oil preparation and characterization with enhanced antifungal activity for controlling *Botrytis cinerea*, the causal agent of gray mould disease. Innov. Food Sci. Emerg. Technol. 28, 73–80.

Mondéjar-López, M., Rubio-Moraga, A., López-Jimenez, A.J., Martínez, J.C.G., Ahrazem, O., Gómez-Gómez, L., Niza, E., 2022. Chitosan nanoparticles loaded with garlic essential oil: a new alternative to tebuconazole as seed dressing agent. Carbohydr. Polym. 277, 118815.

Moussa, S.H., Tayel, A.A., Alsohim, A.S., Abdallah, R.R., 2013. Botryticidal activity of nanosized silver-chitosan composite and its application for the control of gray mold in strawberry. J. Food Sci. 78 (10), M1589–M1594.

Muthukrishnan, S., Murugan, I., Selvaraj, M., 2019. Chitosan nanoparticles loaded with thiamine stimulate growth and enhances protection against wilt disease in Chickpea. Carbohydr. Polym. 212, 169–177.

Nadendla, S.R., Rani, T.S., Vaikuntapu, P.R., Maddu, R.R., Podile, A.R., 2018. HarpinPss encapsulation in chitosan nanoparticles for improved bioavailability and disease resistance in tomato. Carbohydr. Polym. 199, 11–19.

Nesic, K., Ivanovic, S., Nesic, V., 2014. Fusarial toxins: secondary metabolites of fusarium fungi. In: Whitacre, D. (Ed.), Reviews of Environmental Contamination and Toxicology. vol. 228. Springer, Cham.

Ngoc, U.T.P., Nguyen, D.H., 2018. Synergistic antifungal effect of fungicide and chitosan-silver nanoparticles on *Neoscytalidium dimidiatum*. Green Proc. Synth. 7, 132–138.

Nguyen, T.N., Huynh, T.N., Hoang, D., Nguyen, D.H., Nguyen, Q.H., Tran, T.H., 2019. Functional nanostructured oligochitosan–silica/carboxymethyl cellulose hybrid materials: synthesis and investigation of their antifungal abilities. Polymers 11 (4), 628.

Pacheco, N., Larralde-Coron, C.P., Sepulveda, J., Trombottoc, S., Domardc, A., Shirai, K., 2008. Evaluation of chitosans and *Pichia guillermondii* as growth inhibitors of *Penicillium digitatum*. Int. J. Biol. Macromol. 43, 20–26.

Pan, X., Welti, R., Wang, X., 2010. Quantitative analysis of major plant hormones in crude plant extracts by high-performance liquid chromatography–mass spectrometry. Nat. Protoc. 5, 986–992.

Panichikkal, J., Puthiyattil, N., Raveendran, A., Nair, R.A., Krishnankutty, R.E., 2021. Application of encapsulated *Bacillus licheniformis* supplemented with chitosan nanoparticles and rice starch for the control of *Sclerotium rolfsii* in *Capsicum annuum* (L.) seedlings. Curr. Microbiol. 78 (3), 911–919.

Patil, S., Chandrasekaran, R., 2020. Biogenic nanoparticles: a comprehensive perspective in synthesis, characterization, application and its challenges. J. Genet. Eng. Biotechnol. 18.

Pelgrift, R.Y., Friedman, A.J., 2013. Nanotechnology as a therapeutic tool to combat microbial resistance. Adv. Drug Deliv. Rev. 65 (13–14), 1803–1815.

Rajasekaran, P., Santra, S., 2015. Hydrothermally treated chitosan hydrogel loaded with copper and zinc particles as a potential micronutrient-based antimicrobial perspectives. Front. Plant Sci. 5, 616.

Resende, N.S., Gonçalves, G.A.S., Reis, K.C., Tonoli, G.H.D., Boas, E.V.B.V., 2018. Chitosan/cellulose nanofibril nanocomposite and its effect on quality of coated strawberries. J. Food Qual. 2018, 1727426.

Rubina, M.S., Vasil'kov, A.Y., Naumkin, A.V., Shtykova, E.V., Abramchuk, S.S., Alghuthaymi, M.A., Abd-Elsalam, K.A., 2017. Synthesis and characterization of chitosan–copper nanocomposites and their fungicidal activity against two sclerotia-forming plant pathogenic fungi. J. Nanostruct. Chem. 7, 249–258.

Saberi-Riseh, R., Moradi-Pour, M., 2021. A novel encapsulation of *Streptomyces fulvissimus* Uts22 by spray drying and its biocontrol efficiency against *Gaeumannomyces graminis*, the causal agent of take-all disease in wheat. Pest Manag. Sci. 77 (10), 4357–4364.

Saberi-Riseh, R., Skorik, Y.A., Thakur, V.K., Moradi Pour, M., Tamanadar, E., Noghabi, S.S., 2021. Encapsulation of plant biocontrol bacteria with alginate as a main polymer material. Int. J. Mol. Sci. 22 (20), 11165.

Saberi-Riseh, R., Tamanadar, E., Hajabdollahi, N., Vatankhah, M., Thakur, V.K., Skorik, Y.A., 2022. Chitosan microencapsulation of rhizobacteria for biological control of plant pests and diseases: recent advances and applications. Rhizosphere, 100565.

Saharan, V., Mehrotra, A., Khatik, R., Rawal, P., Sharma, S.S., Pal, A., 2013. Synthesis of chitosan based nanoparticles and their in vitro evaluation against phytopathogenic fungi. Int. J. Biol. Macromol. 62, 677–683.

Saharan, V., Sharma, G., Yadav, M., Choudhary, M.K., Sharma, S.S., Pal, A., Raliya, R., Biswas, P., 2015. Synthesis and in vitro antifungal efficacy of Cu-chitosan nanoparticles against pathogenic fungi of tomato. Int. J. Biol. Macromol. 75, 346–353.

Saharan, V., Kumaraswamy, R.V., Choudhary, R.C., Kumari, S., Pal, A., Raliya, P., Biswas, P., 2016. Cu-chitosan nanoparticle mediated sustainable approach to enhance seedling growth in maize by mobilizing reserved food. J. Agric. Food Chem. 64, 6148–6155.

Salahuddin, N., Elbarbary, A., Allam, N.G., Hashim, A.F., 2018. Chitosan modified with 1,3,4-oxa(thia)diazole derivatives with high efficacy to heal burn infection by *Staphylococcus aureus*. J. Bioact. Compat. Polym. 33, 254–268.

Sathe, P., Laxman, K., Myint, M.T.Z., Dobretsov, J., Richter, S., 2017. Bioinspired nanocoatings for biofouling prevention by photocatalytic redox reactions. Sci. Rep. 7 (1), 3624.

Sathe, P., Richter, J., Myint, M.T., Dobretsov, S., Dutta, J., 2016. Self-decontaminating photocatalytic zinc oxide nanorod coatings for prevention of marine microfouling: a mesocosm study. Biofouling 32, 383–395.

Sathiyabama, M., Manikandan, A., 2018. Application of copper-chitosan nanoparticles stimulate growth and induce resistance in finger millet (*Eleusine coracana* Gaertn.) plants against blast disease. J. Agric. Food Chem. 66 (8), 1784–1790.

Sathiyabama, M., Parthasarathy, R., 2016. Biological preparation of chitosan nanoparticles and its in vitro antifungal efficacy against some phytopathogenic fungi. Carbohydr. Polym. 151, 321–325.

Shoeb, M., Singh, B.R., Khan, J.A., Khan, W., Singh, B.N., Singh, H.B., Naqvi, A.H., 2013. ROS-dependent anticandidal activity of zinc oxide nanoparticles synthesized by using egg albumen as a biotemplate. Adv. Nat. Sci. Nanosci. Nanotechnol. 4.

Shuping, D.S.S., Eloff, J.N., 2017. The use of plants to protect plants and food against fungal pathogens: a review. Afr. J. Tradit. Complement. Altern. Med. 14, 120–127.

Siddaiah, C.N., Prasanth, K.V.H., Satyanarayana, N.R., Mudili, V., Gupta, V.K., Kalagatur, N.K., Satyavati, T., Dai, X.-F., Chen, J.Y., Mocan, A., Singh, B.P., Srivastava, R.K., 2018. Chitosan nanoparticles having higher degree of acetylation induce resistance against pearl millet downy mildew through nitric oxide generation. Sci. Rep. 8 (1), 2485.

Souza, V.G., Pires, J.R., Rodrigues, C., Coelhoso, I.M., Fernando, A.L., 2020. Chitosan composites in packaging industry—current trends and future challenges. Polymers 12 (2), 417.

Stasińska-Jakubas, M., Hawrylak-Nowak, B., 2022. Protective, biostimulating, and eliciting effects of chitosan and its derivatives on crop plants. Molecules 27 (9), 2801.

Suryanto, D., Indarwan, A., Munir, E., 2012. Examination of chitinolytic bacteria in alginate-chitosan encapsulation on chili seed against damping off caused by *Fusarium oxysporum*. Am. J. Agric. Biol. Sci. 7 (4), 461–467.

Ureña-Saborío, H., Madrigal-Carballo, S., Sandoval, J., Vega-Baudrit, J.R., Rodríguez-Morales, A., 2017. Encapsulation of bacterial metabolic infiltrates isolated from different *Bacillus* strains in chitosan nanoparticles as potential green chemistry-based biocontrol agents against *Radopholus similis*. J. Renew. Mater. 5 (3–4), 290–299.

Usman, M., Farooq, M., Wakeel, A., Nawaz, A., Cheema, S.A., Ur Rehman, H., Ashraf, I., Sanaullah, M., 2020. Nanotechnology in agriculture: current status, challenges and future opportunities. Sci. Total Environ. 721, 137778.

Vasconcelos, M.W., 2014. Chitosan and chitooligosaccharide utilization in phytoremediation and biofortification programs: current knowledge and future perspectives. Front. Plant Sci. 5, 616.

Vemmer, M., Patel, A.V., 2013. Review of encapsulation methods suitable for microbial biological control agents. Biol. Control 67 (3), 380–389.

Vokhidova, N.R., Sattarov, M.E., Kareva, N.D., Rashidova, S.S., 2014. Fungicide features of the nanosystems of silkworm (*Bombyx mori*) chitosan with copper ions. Microbiology 83 (6), 751–753.

Wang, X., Du, Y., Liu, H., 2004. Preparation, characterization and antimicrobial activity of chitosan–Zn complex. Carbohydr. Polym. 56, 21–26.

Wei, X., Wang, X., Gao, B., Zou, W., Dong, L., 2020. Facile ball-milling synthesis of CuO/biochar nanocomposites for efficient removal of reactive red 120. ACS Omega 5, 5748–5755.

Winter, J., Nicolas, J., Ruan, G., 2020. Hybrid nanoparticle composites. J. Mater. Chem. B 8, 4713–4714.

Xing, K., Zhu, X., Peng, X., Qin, S., 2015. Chitosan antimicrobial and eliciting properties for pest control in agriculture: a review. Agron. Sustain. Dev. 35 (2), 569–588.

Xing, K., Liu, Y., Shen, X., Zhu, X., Li, X., Miao, X., Feng, Z., Peng, X., Qin, S., 2017. Effect of O-chitosan nanoparticles on the development and membrane permeability of *Verticillium dahliae*. Carbohydr. Polym. 165, 334–343.

Xing, Y., Li, W., Wang, Q., Li, X., Xu, Q., Guo, X., Bi, X., Liu, X., Shui, Y., Lin, H., Yang, H., 2019. Antimicrobial nanoparticles incorporated in edible coatings and films for the preservation of fruits and vegetables. Molecules 24, 1695.

Xu, D., Qin, H., Ren, D., 2018. Prolonged preservation of tangerine fruits using chitosan/montmorillonite composite coating. Postharvest Biol. Technol. 143, 50–57.
Yang, Z., Chen, J., Dou, R., Gao, X., Mao, C., Wang, L., 2015. Assessment of the phytotoxicity of metal oxide nanoparticles on two crop plants, maize (*Zea mays* L.) and rice (*Oryza sativa* L.). Int. J. Environ. Res. Public Health 12, 15100–15109.
Youssef, K., Hashim, A.F., 2020. Inhibitory effect of clay/chitosan nanocomposite against *Penicillium digitatum* on citrus and its possible mode of action. Jordan J. Biol. Sci. 13, 349–355.
Youssef, K., Roberto, S.R., Tiepo, A.N., Constantino, L.V., de Resende, J.T.V., Abo-Elyousr, K.A.M., 2020. Salt solution treatments trigger antioxidant defense response against gray mold disease in table grapes. J. Fungi 6, 179.
Yu, Z., Li, Q., Wang, J., Yu, Y., Wang, Y., Zhou, Q., Li, P., 2020. Reactive oxygen species-related nanoparticle toxicity in the biomedical field. Nanoscale Res. Lett. 15.
Yu, J., Wang, D., Geetha, N., Khawar, K.M., Jogaiah, S., Mujtaba, M., 2021. Current trends and challenges in the synthesis and applications of chitosan-based nanocomposites for plants: a review. Carbohydr. Polym. 261, 117904.
Zhang, X., Xiao, G., Wang, Y., Zhao, Y., Su, H., Tan, T., 2017. Preparation of chitosan-TiO_2 composite film with efficient antimicrobial activities under visible light for food packaging applications. Carbohydr. Polym. 169, 101–107.

CHAPTER

4

Gum nanocomposites for postharvest fungal disease control in fruits

Jéssica de Matos Fonseca, Amanda Galvão Maciel, and Alcilene Rodrigues Monteiro

Department of Chemical and Food Engineering, Federal University of Santa Catarina, Florianópolis, Brazil

1 Introduction

Fungal diseases can be devastating, increasing postharvest losses and decreasing food quality. They can affect all tissues of plants, including fruit, leaves, stems, roots, and flowers, causing their rot and death. *Colletotrichum* spp, *Alternaria* spp., *Botrytis cinerea*, *Penicillium* spp, *Fusarium oxysporum*, *Aspergillus* spp., *Sclerotinia sclerotiorum*, *Rhizopus* spp., and *Mucor* spp. are common fungi deteriorating fruit (Chen et al., 2021a; Matrose et al., 2021; Thambugala et al., 2020).

The Food and Agriculture Organization of the United Nations (FAO) estimates that approximately 33% of all food produced worldwide, almost 1.3 billion tons of food, is lost. More than 45% of losses are fruit and vegetables, of which 54% occur during harvesting and postharvest and 46% during processing, distribution, and consumption. Annually, the total food loss generates a financial loss of US$ 750 billion (Santos et al., 2020). The postharvest losses occur due to inadequate handling, packaging, transportation, and contamination, which is facilitated by the previous factors. The highest contaminants of harvested food are fungi. Molds and yeasts are responsible for almost 20% of the postharvest losses per year (Davies et al., 2021; Government of Western Australia, 2018). Thus, identifying symptoms, fungi metabolism, etiology, and epidemiology is essential to provide control solutions for sustainable production, environmental impact reduction, and food safety.

Fungicides are the most common solutions for controlling fungi disease in harvested food. Although they exhibit an effective fungistatic or fungicide action, their accumulative toxic residues pose several risks to human health and environmental contamination. Due to this,

https://doi.org/10.1016/B978-0-443-23950-2.00016-3

emerging and sustainable technologies have been developed to reduce or replace toxic pesticide applications. Among these technologies are biological control by antagonists, natural active compounds, nonthermal technologies (modified atmosphere, ultrasound, UV light, cold plasma, active coatings, and films and ozonated and electrolyzed water), and combined technologies (Davies et al., 2021).

In this context, nanocomposites based on gums and active compounds have been applied as a postharvest technology to prevent fungal diseases, mainly in fruit. Gums such as Arabic (Ali et al., 2021), xanthan (Wani et al., 2021), shellac (Ma et al., 2021), carboxymethylcellulose (CMC) (Deng et al., 2021), and carrageenan (Wani et al., 2021) exhibit excellent properties to form films and coatings with good mechanical and barrier properties (Kadzińska et al., 2019; Md Nor and Ding, 2020). In the case of coatings, these characteristics are essential to avoid the penetration of fungi into food. Gum matrices can be added with antifungal properties by incorporating active compounds such as essential oils (Alizadeh-Sani et al., 2020; Cao and Song, 2019; Kawhena et al., 2020; Rastegar and Atrash, 2021; Wani et al., 2021; Xue et al., 2019), plant extracts (Al-Maqtari et al., 2021; Fan et al., 2021; Yang et al., 2019), metal oxides (Alizadeh-Sani et al., 2020; Joshy et al., 2020; Kanikireddy et al., 2020), resins (Pobiega et al., 2021), pigments (flavonoids) (Sucheta et al., 2019), and organic acids (Ali et al., 2020; Ma et al., 2021). In addition, these materials can be associated with biocontrol agents (antagonists) (Deng et al., 2021).

This chapter discusses the antifungal properties of gum-based nanocomposites incorporated with different antimicrobial agents and their application in postharvest.

2 Fungal diseases in postharvest

2.1 Influence on postharvest losses

Fungi diseases affect several plant organs, causing the deteriorating and rotting of fruit, stems, leaves, and roots. The fungal contamination of these products has caused significant economic losses. Both quantitative and qualitative losses occur during harvest and postharvest, affecting the production and supply of food worldwide (Fausto et al., 2019).

During the postharvest, tons of food are lost due to inadequate handling, packaging, and transportation. However, most losses are related to the fugal incidence, which is also accelerated by factors cited before. Mold and yeast are responsible for almost 20% of the postharvest losses per year (Davies et al., 2021; Government of Western Australia, 2018). Postharvest losses consist of a problem faced over decades by scientists. With the population increase, it reaches almost 8 billion people and increases as the food demand and the increase of the hungry worldwide. *Alternaria alternata*, *C. gloeosporioides*, *Botrysti cinerea*, *Penicillium* are among the ten top fungica using economic agriculture loss worldwide and the most targeted in the consulted papers (Youssef et al., 2019). Other many fungi contributing to harvest and postharvest losses are presented in Table 1. The microbial contamination of harvested food can occur over processing steps due to pest attacks and inadequate handling, transportation, and storage.

TABLE 1 Summary of common postharvest pathogens, associated diseases, and symptoms on fresh fruit.

Pathogen genera	Disease	Symptoms	Selected fruit host
Alternaria	Fruit rot, dark stop, sooty mold	Circular, dry, firm, and shallow lesions, covered with dark, olive green to black surface mycelial growth.	Apple, pear, peach, plum, cherry, papaya, grapes
Botrytis	Gray mold	Grayish-colored soft, mushy spots on leaves, stems, flowers, and on produce.	Strawberry, cherry, grapes, tomatoes, apples, pears, peach
Colletotrichum	Anthracnose, bitterrot	Small, slightly sunken, dark yellow spots	Banana, avocado, apple, guava, mango, strawberry, papaya, passion fruit
Diplodia	Stem-end rot, fingerrot, crown rot, stalkrot	White/gray mold on ears, generally starting at the base of ears and progressing toward the tips	Citrus fruit, avocado, mango, banana
Monilinia	Brown rot, blossom, and twig blight	Capable of forming thick, compact, and complex stromata	Peaches, apples, sweet cherries
Penicillium	Blue rot, green and blue mold, core rot	It starts as a soft, light-colored spot that spreads swiftly on the surface and deeply into the fruit tissue, with blue-green coremial fruiting bodies appearing on the surface	Pome and stone fruit, citrus fruit
Rhizopus	Watery white rot, soft rot	A water-soaked appearance in fruit tissues with brownish-gray to blackish-gray colonies.	Apple, pears, peach, plum, grapes, strawberry
Mucor	Juicy soft rot, mucorrot	Typically originates at wounds in the cuticle or the stem or calyx end of the fruit. At early infection stages, mycelium is long, coarse white or gray with pin-shaped dark spore heads that could beseen on fruit wound sites, splits, or lenticels	Tomato, strawberry, grapes
Sclerotinia	Soft watery rot	White fuzzy mycelial growth with dark lesions on the fruit surface containing black sclerotia.	Citrus fruit, carrot
Geotrichum	Sour rot	It appears as water-soaked lesions on the fruit surface and extends deep into the tissue. After that, the surface of diseased lesions could be covered with fluffy, white mycelium with a sour odor.	Citrus fruit, carrot, oriental melon, tomato, cucumber, potato, pumpkin
Phomopsis	Stem-end rot	The pathogen produces a dark, circular lesion at the fruit peduncle (stem-end) with defined edges, which spreads relatively slowly but penetrates deeply into the flesh. The fruit becomes soft and lighter in color.	Citrus fruit, mango, kiwifruit

Reprinted from Matrose, N.A., Obikeze, K., Belay, Z.A., Caleb, O.J., 2021. Plant extracts and other natural compounds as alternatives for post-harvest management of fruit fungal pathogens: a review. Food Biosci. 41, 100840 with permission from Elsevier.

TABLE 2 Some synthetic fungicides currently used in fruit and vegetable harvest and postharvest.

Fungicides group	Active compounds	Commodities application	Control actions	References
Benzimidazole	Benomyl Thiabendazole	Papaya	Anthracnose (*Colletotrichum gloeosporioides*)	Bautista-Baños et al. (2013)
Benzimidazole	Carbendazim	Citrus, beans, wheat	Motley (*Elsinore australis*), Anthracnose (*Colletotrichum Gloeosporioides*); Fusarium head blight (*Fusarium graminearum*); Blotch (*Stagonosporanodorum*) Septoria leaf spot or speckled leaf blotch (*Septoria tritici*)	Chung (2011); Gupta (2017)
Fuberidazole	Benzimidazole	Not specified	Multiple fungal pathogens in postharvest	Bradshaw et al. (2021)

2.2 Traditional control methods

Traditionally, the antifungal treatment in food harvest uses synthetic chemical pesticides that are toxic and unfriendly to the environment (Table 2). They can act as fungistatic and fungicides to (1) prevent sporulation, (2) cure horticulture injuries (avoiding fungi penetration and mycelial growth), and (3) eradicate vegetative cells and sporulation when there are already viable symptoms.

Foods treated with pesticides present residual toxic compounds that the consumers have ingested. They can trigger cancer, allergy, diabetes, and other human diseases. Although the consumption of organic fruits and vegetables has increased in the last decades, most of the population still does not have access to them, mainly in emerging countries, due to the high price. According to the fungicide committee (FRAC-1), fungicides are highly dangerous to human health, and fungi are more resistant to synthetic fungicides (Bradshaw et al., 2021). Fungicides from the benzimidazole family have shown side effects such as skeletal malformations, high mortalities, and anomalies in rats (Gupta, 2017).

Because of this issue, several emerging and sustainable methods have been researched to reduce toxic pesticide application. These methods include the application of antagonist agents (biological control), natural active compounds, nonthermal technologies (e.g., modified atmosphere, ultrasound, UV light, cold plasma, active coatings, and films, ozonated and electrolyzed water), and combined technologies (Table 1) (Davies et al., 2021). Among these possibilities, the gums have been highlighted as biopolymer matrices for incorporating antimicrobial compounds to prevent fungal diseases.

3 Gums

3.1 General properties and food applications

Gums are polysaccharides extracted and derived from natural sources, exhibiting interesting physicochemical and rheological properties for various applications in the food, pharmaceutical, and cosmetics industries. They are explored in the development of new technologies because they are abundant in nature, nontoxic, cheap, and ecologically correct compounds. Its main characteristics are being hydrophilic, resistant to organic solvents, hydrating in water, forming gels, and other functional properties such as emulsification, stabilization, and cryo-protection, enabling several applications (Amin et al., 2021; Nazarzadeh et al., 2019).

The functional properties of gums depend on intrinsic and extrinsic factors such as raw material composition, molecular mass, degree of branching and ionization of the molecule, pH, ionic strength, interaction in a specific system, temperature, and concentration. Therefore, each gum will have distinct physical and chemical characteristics (Badui, 2006). For example, gums with a linear chemical structure have a higher volume and viscosity than branched polymers of the same molecular weight and concentration. On the other hand, branched gums form gels more easily and are highly stable, as their branches prevent intermolecular interactions in a system (Rana et al., 2011).

The fields of application of these polysaccharides are diverse, including the development of films, encapsulating agents, nanocarriers, nanocomposites, and edible coatings. They are used as thickeners, binders, suspenders, stabilizers, and emulsifiers in bakery products, ice cream, confectionery, and sauces (Taheri and Jafari, 2019). Some works have reported antimicrobial and nutraceutical properties of some gums, indicating positive effects on human health, with health benefits mainly related to its properties as dietary fiber (Bashir et al., 2018; Hamdani et al., 2018; Phillips and Phillips, 2011). Commercial gums commonly used in foods as additives classified according to the international numbering system (INS) or the standard number in European Commission (E) are xanthan gum (E414), locust bean gum (E140), carrageenan (E407), and Arabic (E414). Other examples of commercial gums are ghatti (E419), karaya (E416), tragacanth (E413), guar (E412), gellan (E418), cellulose (E466), cassia (E427), carob bean (E410) (Codex Alimentarius Commission, 2021).

The gums increase the water vapor resistance of edible coatings and reduce fruit respiration, delaying ripening and senescence. Sensorial properties such as firmness and color are preserved, and the fruit shelf life is extended. Murmu and Mishra (2018) studied the effect of cosmetic coating based on gum arabic, sodium caseinate, cinnamon, and lemongrass essential oil on guava. Results showed higher total antioxidant activity and DPPH radical scavenging activity, good ascorbic acid retention, and higher phenolic content than the uncoated samples. The shelf life ranged from 7 days (control sample) to 40 days (coated samples). In the study carried out by Jafari et al. (2018) on coating based on gum tragacanth for improving the quality of fresh-cut apples, the contents of ascorbic acid and phenolic compounds were significantly maintained during storage of apple slices coated with gum compared to uncoated fruits. Insignificant differences were observed between the treatment and the control at the storage end.

Another alternative widely chosen for gums is their use as encapsulant agents due to water solubility, flexibility, availability of reactive sites for molecular interactions, and ease of use (Taheri and Jafari, 2019). Encapsulation aims to protect a compound and carry out its controlled release, applied in active and intelligent packaging to improve antibacterial and food preservation performance (Zhang et al., 2021a). Khoshakhlagh et al. (2018) studied the nanoencapsulation feasibility of D-limonenein *Akyssum homolocarpum* seed gum and obtained favorable results for encapsulation.

Gums have great potential in developing food packaging as they are a cheaper source, from renewable sources, biodegradable, and with interesting film-forming properties. Rahman et al. (2021) combined guar gum and cross-linked chitosan without plasticizers for film formation and obtained a film with high mechanical strength, lower water solubility, and hydrophobic nature. Barreiros et al. (2022) developed film-forming suspensions based on xanthan gum incorporated with clove, cinnamon, and oregano essential oils to act as an antimicrobial in milk. The suspensions showed adequate viscosity at low concentrations and pseudoplastic behavior and remained stable after 30 days of storage at room temperature.

Nanotechnology has been playing an important role in encompassing several segments, and the use of gums has gained space due to their unique properties. Janani et al. (2020) developed an active film with nanocomposites based on tragacanth gum, polyvinyl alcohol, ZnO nanoparticles, and ascorbic acid for antioxidant action in foods. The addition of the gum increased the film degradation rate in the soil. Kanikireddy et al. (2020) studied the influence of nanocomposites based on CMC, guar gum, and silver on the strawberry shelf-life. Nanocomposites displayed antimicrobial activity, including against fungi. It was observed that guar gum has high mechanical strength stability at high temperatures and pH ranges, being a good option for nanocomposites.

Despite all its advantages, there are limiting factors in the use of gums. Microbial contamination is one of the limiting factors. It happens due to the moisture content of the gums being, in most cases, equal to or greater than 10%, which favors the appearance of microorganisms. The uncontrolled hydration rate, change in viscosity during storage (usually reducing), and dependence on seasonal factors also make difficult the application of these polysaccharides (Ahmad et al., 2019).

3.2 Types of gums and characteristics

Gums can be classified based on their origin, shape, charge, and chemical structure. Depending on the source, gums can be categorized as microbial gums, plant exudate gums (trees, land plants, marine plants), or seed gums, with plants being the primary sources (Taheri and Jafari, 2019).

3.2.1 *Microbial gums*

Microbial gums are listed by the US Food and Drug Administration (FDA) as generally recognized as safe (GRAS) and have a wide range of physical and chemical properties suitable for various pharmaceutical, medicinal, and food applications. Microbial gums are high molecular weight polysaccharides composed of sugar residues linked by linear or branched, nontoxic glycosidic bonds (Alizadeh-Sani et al., 2020). They are produced by enzymes from microorganisms, such as bacteria, yeasts, algae, and fungi, through submerged aerobic fermentation in two types: exopolysaccharides and capsular polysaccharides.

These gums are diverse in structure, function, and producing strains but have carbohydrate components (e.g., glucose, mannose), noncarbohydrates (e.g., pyruvate, acetate), and uronic acid in their composition. The most significant advantages of microbial gums are their high potential for industrial-scale production, microbiological stability, adhesion, cohesion, wettability, transparency, and mechanical properties, enabling their application as films or coatings. However, there is a need for high-technology equipment and specific resources for their production, making it expensive (Abu Elella et al., 2021).

Xanthan gum is the most widely used fermentation product by *Xanthomonas campestris* bacteria. It is a branched nonionic polysaccharide composed of *D*-glucose, *D*-mannose, and *D*-glucuronic acid linked by a pentasaccharide unit. It is mainly applied as a stabilizer, thickener, or emulsifier due to its pseudoplastic behavior, high viscosity even in small concentrations, good stability in acidic media, solubility in both hot and cold water, and high yield value (Tahir et al., 2019; Wani et al., 2021).

Dextran is the first microbial polysaccharide commercially applied, consisting of linear and branched α-1,3 α-1,3 glycosidic bonds. It is produced by the bacteria *Leuconostocmesenteroides* and *Streptococcus mutans*. It has the characteristic of not forming a gel in its original state, being more applied in pharmaceutical products due to its antithrombotic and plasma volume expansion properties (Ng et al., 2020; Su et al., 2021).

Gellan gum is a linear repeating polysaccharide of two *D*-glucose, one *L*-rhamnose, and one D-glucuronic acid. It is anionic, a product of fermentation by the bacteria *Sphingomonas elodea*. Its main characteristic is its ability to gel, despite the need for a high temperature to gel, and it appears in two forms: high and low acetylation gel gum, but its use is limited due to the lack of specific sites for cell adhesion (Ng et al., 2020). It has excellent acid and enzymatic hydrolysis but has low stability and low mechanical strength, so it is usually used with another mixed polysaccharide (Zia et al., 2018).

Other more recent examples of microbial gums include gellan, pullulan, curdlan, cellulose and gellan, tara, and spruce gum (Abu Elella et al., 2021; Alizadeh-Sani et al., 2019; Ng et al., 2020).

3.2.2 Plant exudate gums

The gums exuded from plants are those secreted by trees, terrestrial plants (leaves, stems), and aquatic plants, produced through mechanisms against mechanical or microbial injury, resulting in a "tear," the exudate. The exudates usually have colors ranging from white to amber and gray to dark brown. It is obtained by the gummosis process, which involves the fragmentation of vegetable cellulose (Hamdani et al., 2019). This gum undergoes natural drying, becoming rigid and glassy. Examples of gum from trees are gum arabic, gum karaya, gum ghatti, gum tragacanth, cashew gum, and Persian gum. In chemically modified gums, we have pectin and cellulose. Gums originating from seaweeds are present in the cell wall of various algae or intracellular regions as food reserve materials such as carrageenan gum, alginates, and agar (Ahmad et al., 2019).

Gum Arabic is mainly extracted from wild trees located in Sudan and the Sahel (Verma and Quraishi, 2021). It is a heterogeneous biopolymer composed of glycoproteins and polysaccharides of galactose and arabinose, a polysaccharide chain with low or without protein. The other fraction comprises molecules of greater molecular mass constituting part of its structure. One of the characteristics of gum arabic is its high dissolution under agitation in water, which is a peculiar property among food gums (Al-Baadani et al., 2021; Musa et al., 2019; Yun et al., 2019).

Karaya gum, also known as sterculia gum, is an exudate with a color ranging from white to gray produced by Sterculia trees native to India, Senegal, Mali, Sudan, and Pakistan. It is obtained from cracks developed in the tree trunk, formed by the gummy process result of biotic and abiotic stresses (Prasad et al., 2022; Raj et al., 2021). The polysaccharide has a long and branched-chain structure and a high molecular mass, containing from 13% to 26% galactose and 15%–30% rhamnose. The presence of acetyl groups results in their insolubility in water and swelling ability. Uptake water ability is an exciting feature of karaya gum, where it absorbs water and increases about seventy times its size, forming a viscous solution (Le Cerf et al., 1990).

Gum tragacanth is extracted from trees of the genus Astragalus, grown in Iran, India, and Turkey. It is a dry exudate that varies from white to brown and has a mild flavor (Nejatian et al., 2020). Chemically, this gum exhibits residues of arabinose, glucose, xylose, rhamnose, galacturonic acid, and fucose, a complex mixture of heteropolysaccharides with branched acids (Nazarzadeh et al., 2019). It has higher viscosity at pH 5 and 6, decreasing with the reduction of pH due to decreased dissociation of carboxylic groups, good thermal stability, and excellent solubility. Compared to other gums, such as locust beans and guar, tragacanth gum has a higher water absorption capacity (Silva et al., 2017).

Cashew gum is obtained from cashew trees *Anacardium occidentale* L., found in Brazil and India (Loureiro et al., 2021; Oliveira et al., 2018). This gum is a heteropolysaccharide consisting of β-*D*-galactopyranose, α-*L*-rhamnopyranose, and β-*D*-glucurin and is used as an emulsifier, thickener, binder, and stabilizer agent (Koyyada and Orsu, 2021). Its viscosity is low compared to other common gums due to its highly branched structure, and it presents pseudoplastic behavior even at low concentrations (less than 1 wt%). However, an increase in viscosity is observed proportional to gum concentration (Botrel et al., 2017).

Exudates from marine plants have three main colloids from algae: agar, alginate, and carrageenan. The agar comes from the cell wall of red algae of the Gracilariaceae, Gelidiacea, Pterocladiaceae family; it has a mixture of agarose, the gelling fraction, and agaropectin, the nongelling fraction. Its chain is linear and composed of 3,6-anhydro-*L*-galactose and *D*-galactose units linked by alternating α-(1, 3) and β-(1, 4) glycosidic bonds (Chen et al., 2021b; Li et al., 2014). Its use is 90% aimed at the food industry and has properties as a gelling agent, stabilizer, and thickener (Mahmud, 2021).

Alginate, an anionic copolymer, is present in brown algae of the species *Ascophyllum nodosum*, *Laminaria hyperborean*, *Laminadigitata*, *Macrocystis pyrifera*, and some bacteria of the genera *Pseudomonas* and *Azotobacter*. It is composed of blocks of (1,4)-linked-β-*D*-mannuronate and α-L-guluronate residues (Williams et al., 2017). Higher proportions of G-Block result in stiff and dense gels, while M-Block is the opposite, with flexible and porous gels. Its viscosity increases with hydrogen bonds, and a high molecular weight improves the physical qualities of the gels (Mahmud, 2021).

Carrageenan is extracted from red algae (*Rhodophyta class*) and is the third most used polysaccharide in the food industry, after gelatin and starch. Carrageenan is not soluble in all organic solvents, such as alcohols and ketones (Mahmud, 2021). Its chemical structure consists of galactose and 3,6-anhydrogalactose linked by alternating β-1,4 and α-1,3 glycosidic bonds (Machado et al., 2019).

Pectin and cellulose are examples of gums that have undergone a chemical modification that results in hydrocolloids with gelling properties. Pectin is a biocompatible, straight-chain polysaccharide consisting of D-galacturonic acid molecules linked in chains by α-(1–4) glycosidic bond carboxyl/hydroxyl side chains and neutral sugars (Baran, 2018; Naqash et al., 2017). Three structures are considered primary in the conformation of pectin: homogalacturonan, rhamnogalacturonan I, and rhamnogalacturonan II, and their presence is associated with the specific physical characteristics of pectin, like linear structuring in pectin results in higher viscosity (Kumar et al., 2020).

Cellulose is the most abundant biopolymer on Earth, with high hydrophilicity and molecular weight. Its structure is composed of β-1,4-linked anhydro-*D*-glucose units and exhibits reducing and nonreducing end groups, facilitating functionalization and physical interaction (Bai et al., 2020). Methylcellulose (MC), carboxymethylcellulose (CMC), and hydroxypropylmethylcellulose (HPMC) are examples of modified cellulose from etherification reactions (displacement of hydroxyl groups) (He et al., 2021; Jia et al., 2016).

3.2.3 *Seeds gums*

The seed gums are also from a vegetable source but obtained from the seed grains containing surface cells containing proteins, fibers, gums, and mucilage. Examples of gums extracted from seeds are guar gum, locust bean gum, and tamarind gum.

Guar gum comes from the ground endosperm of *Cyamopsis tetragonolobus*, cultivated in Asia, Africa, and South America. This gum is a nonionic, hydrophilic, nonhygroscopic galactomannan polysaccharide (Jana et al., 2019), and it can form hydrogen bonds. Its structure is composed of (1–4) β-*D*-mannopyranosyl moieties linked with branch points of α-*D*-galactopyranosyl moieties linked by (1–6) bonds. Guar has good thickening, emulsifying properties, nontoxicity over a wide pH range, high thermal stability, and film-forming ability (Dehghani Soltani et al., 2021; Sharma et al., 2018).

Locust bean gum, composed of 1:4 ratio of galactose and mannose, is extracted from the seeds of pods, representing only 10% of the seed weight. The seed is from the carob tree (*Ceratonia siliqua* L.) native to the Mediterranean region (Brassesco et al., 2021). As a food additive, it improves the texture of meat products, beverages, dairy products, baked goods, jellies, and fish (Barak and Mudgil, 2014).

Tamarind gum, from tamarind seeds, is a polysaccharide containing glucose, galactose, and xylose monomers, corresponding to 65% of the grain. It is a neutral, nonionic, light yellow compound. Water solubility above 80°C can retain water, gelatinize and form filmogenic solutions. It has its main functions as a stabilizer and emulsifier. In addition, it can uptake water, gelatinize and compose film-forming dispersions (Premalatha et al., 2016).

Most of these gums presented exhibit characteristics that can be improved or new functions attributed by association with other materials such as biopolymers and natural active compounds to form blends and nanocomposites. These nanomaterials have a high potential for antifungal application.

4 Antifungal properties of gum-based nanocomposites

Natural materials such as essential oils, plant extracts, natural resins, organic acids, biopolymers, and biocontrol agents have been incorporated into gum-based materials to improve mechanical and gas barrier properties and provide antifungal activity (Deng et al., 2021; Fan et al., 2021; Kumar et al., 2021; Pobiega et al., 2021; Wani et al., 2021). Recent applications of gum-based nanocomposite as antifungal material for fruit and vegetables are presented in Table 3.

TABLE 3 Recent gum-based nanocomposite coatings developed to prevent fruit and vegetable fungal diseases.

Gum	Nanocomposite association	Fruit or vegetable application	Fungal inhibition	Main results	Reference
Carboximethyl cellulose (CMC)	*Cryptococcus laurentii* FRUC DJ1 (yeasts)	Grapefruit	*Penicillium digitatum*	Biocontrol of *Penicillium digitatum* by yeasts. Extension of grapefruit shelf life over 28 days of storage: the CMC-loaded yeasts induced the defense enzyme activity activities (e.g., chitinase, β-1,3-glucanase, peroxidase, polyphenol oxidase, and phenylalanine ammonia-lyase), and preserved weight, soluble solids, ascorbic acid, and titratable acidity.	Deng et al. (2021)
Cashew gum polysaccharide/ polyvinyl alcohol(PVA)	Cell wall degrading enzyme from *Trichoderma asperellum*	Strawberries	*Penicillium* sp.	Cashew gum/PVA edible coatings containing cell wall degrading enzymes from *Trichoderma asperellum* (e.g., chitinase, N-acetylglucosaminidase, and β-1,3-glucanase) were used to reduce *Penicillium* sp. growth on strawberries and preserve physicochemical properties stored at room temperature during 5 days. The visible microbial growth on the fruit was characterized as brown spots and softening in the injured zone. The fruit was inoculated with *Penicillium* sp. and coated with composites. Strawberries coated with cashew gum/PVA and cashew gum/PVA/enzymes showed a uniform appearance, lower weight loss, and color variation. They did not show signs of visible fungal decomposition during the storage period, while uncoated strawberries were entirely vulnerable to the deteriorative symptoms.	Moreira et al. (2020)
Gum Arabic, sodium alginate	Citral essential oil	Banana	*Fusariumpseudocircinatum*	Citral essential oil was encapsulated into Gum Arabic/sodium alginate by spray drying to improve its water solubility and promote antifungal activity on fruit. The spray-dried microcapsules presented sizes of 1–17 μm, 42.48% of encapsulation efficiency, and thermal stability between 37 and 148°C. They reduced the infection of bananas by *Fusarium pseudocircinatum* by 56% after 9 days of storage at 25°C.	Calderón-Santoyo et al. (2022)

Gum Arabic	Cactus pear stem, moringa and henna leaf extracts	Guava	Molds	Guavas were coated with Gum Arabic incorporated with cactus pear stem, moringa, and henna leaf extracts and stored at ambient and refrigeration for 7, 14, and 21 days for reducing rot ratio and extending the shelf life. The coating based on Gum Arabic (10% w/v) and moringa extract (10% w/w, gum) exhibited the highest values for maintaining firmness, total soluble solids, total sugars, and total antioxidant activity.	El-Gioushy et al. (2022)
Gum Arabic	Lemongrass Essential Oil and Pomegranate Peel Extract	Pomegranate Whole Fruit and Arils	Grey mold (*Botrytis cinerea*) and blue mold (*Penicillium* spp.)	Pomegranates were coated with blank Gum Arabic dispersion and Gum Arabic containing Lemongrass Essential Oil or Pomegranate Peel Extract. Coatings containing oil and extract reduced weight loss, respiration rate, and fungi incidence over the 6 weeks of storage at 5°C and 95% RH.	Kawhena et al. (2022)
Gum Arabic, xanthan and carrageenan	Lemongrass essential oil	Strawberry	Yeasts and molds	Strawberries were coated with nanocomposites based on different gums, packed into polyethylene pouches, and refrigerated at 4°C for 12 days. All coatings preserved the physicochemical and physiological properties of the fruit compared to the control. However, carrageenan-based nanocomposite was more effective in reducing the enzymatic activities of cellulase, pectin methylesterase, and β-galactosidase. It also achieved the maximum reduction in polygalacturonase activity at the end of storage. Carrageenan also preserved a higher fraction of anthocyanins and was more efficient in inhibiting psychrophilic bacteria, yeasts, and molds.	Wani et al. (2021)
Gum Arabic	None	Apricot	Not directly investigated	Apricots coated with gum Arabic were stored at 20 ± 1°C for 8 days. They exhibited lower weight loss, disease incidence, malondialdehyde concentration, and reduced hydrogen peroxide compared to the control.	Ali et al. (2021)

Continued

TABLE 3 Recent gum-based nanocomposite coatings developed to prevent fruit and vegetable fungal diseases—cont'd

Gum	Nanocomposite association	Fruit or vegetable application	Fungal inhibition	Main results	Reference
				Gum Arabic also preserved the total phenols, ascorbic acid, soluble solids content, titrable acidity, and antioxidant activity of fruit. The enzymatic activities of cellulases, pectin methylesterases, and PG [3] were suppressed by gum Arabic. Sensory analysis indicated the consumer preference for apricot coated with gum Arabic because of flavor and taste preservations and better visual appearance and overall quality.	
Gum Arabic	*Conyza bonariensis* (L.) Cronquist essential oil	Banana	*Colletotrichum musae*	Coatings based on Gum Arabic and *Conyza bonariensis* (L.) Cronquistessential oil effectively delayed the anthracnosis development and physicochemical alterations in bananas during the 5 days of storage at room temperature. Blank Gum Arabic was not capable of inhibiting fungal growth. The antifungal action of essential oil caused damage to the cytoplasmic membrane and disturbances of enzymatic activity of conidia, which could be related to cellular dysfunction, nonviability, and death.	Lundgren et al. (2022)
Konjac glucomannan	Pomegranate peel extract	Fresh-cut kiwifruit and green bell pepper	Yeasts and molds	Fresh-cut kiwifruit and green bell pepper were coated with nanocomposite, packed into polymeric film, and stored at 10°C for 8 days. Synergistic effects of gum and plant extract preserved the chlorophyll content, total phenolics, ascorbic acid, and antioxidant activity for both samples coated with nanocomposite compared to control. The nanocomposite also exhibited a higher antifungal effect than an antibacterial effect for mesophiles. The weight loss of kiwifruit was reduced by the gum effect, whereas for the pepper, this reduction was a consequence of a synergistic nanocomposite effect as the firmness loss reduction.	Fan et al. (2021)

Pullulan	Chitosan and pomegranate peel extract	Green bell pepper	Not directly investigated	Green bell peppers stored at room temperature (23 ± 3°C, 40%–45% RH) and cold (4 ± 1°C, 90%–95% RH) had their ripening delayed by pullulan-based nanocomposite coating. The nanocomposite exhibited antioxidant activity, preserved firmness and weight, and reduced enzymic browning and variations of color, soluble solids, titratable acidity, and pH.	Kumar et al. (2021)
Pullulan	Propolis extract	blueberry (*Vaccinium corymbosum*) fruit	*Botrytis cinerea*, *Colletotrichum gloeosporioides*, *Fusarium solani*, and *Penicilliumchrysogenum* molds	Pullulan-propolis nanocomposite also exhibited inhibition properties against bacteria (*E. coli*, *S. enteridis*, *L. monocytogenes*). The count of microorganisms was reduced by 3–4.5 logarithmic cycles after 21 days of storage at 16°C and 58%–63% RH. The fruit weight loss was reduced, and soluble solids content was preserved compared with control.	Pobiega et al. (2021)
Shellac	Tannic acid	Mango	*Colletotrichum gloeosporioides*, *Phomopsis mangiferae*	Extension of shelf life for approximately 10 days: preservation of firmness, weight, total polyphenolics, ascorbic acid, and reduction in ethylene evolution, respiration rate, color change, and activity of deteriorating enzymes (*e.g.* polyphenol oxidase, peroxidase, polygalacturonase, β-1,3-glucanase, and chitinase).	Ma et al. (2021)
Tragacanth gum	*Eremurus* extract	Sweet cherry fruit	Molds	The coating based on gum and gum/*Erasmus* extracts reduced the fungal decay index on sweet cherry fruit equally, whereas fruit control exhibited the highest rate of fungal decay.	Esmaeili et al. (2022)

These nanocomposites are prepared from colloidal dispersion systems such as nano- and microemulsions and suspensions and applied as coatings, films, and nanoparticles. Several researchers have studied the structural, antioxidant, and antifungal properties of gum-based nanocomposite coatings in vitro as film form, prepared by the casting method (Al-Maqtari et al., 2021; Cao and Song, 2019; Mahmood et al., 2020; McDaniel et al., 2019).

The coating forms a thin dried layer (film) on the fruit or vegetable that plays several functions: protecting fruit and vegetable against some physical injuries such as minor impacts and chilling; improving their visual appearance, avoiding contact of their surface with insects, larvae, and microorganisms; reducing gas exchanges (e.g. transpiration, respiration, ethylene evolution, aromatic losses) and providing active properties (nanocomposites) (Monteiro Fritz et al., 2019). These benefits delay biochemical reactions in vegetal tissue, increasing the shelf-life of fruit and vegetables (Ali et al., 2021).

4.1 Gums combined with other biopolymers

Mechanical and gas barrier properties of gums can be improved by association with other biopolymers. Sood and Saini (2022) investigated the effect of pectin incorporation on the mechanical, structural, morphological, and thermal properties of edible composite films based on red pomelo peel pectin, casein, and egg albumin. All properties were improved as pectin incorporation. Morphological analysis of the composite films revealed that pectin increased the composite matrix compactness due to increased polymer chain interactions. Pectin-albumin-casein exhibits better properties than individual formulations.

Kumar et al. (2021) developed antioxidant and antimicrobial nanocomposites based on gum pullulan, chitosan, and pomegranate peel extract to delay the ripeness of green bell peppers. The chitosan antimicrobial activity has been attributed to the cell permeability changes caused by interactions between cationic amino groups of chitosan chains and negative charges of the microbial cell membrane. These interactions cause the leakage of electrolytes and proteins (Severino et al., 2015). Pomegranate peel extract is composed of phenolics such as gallic acid, ellagic acid, punicalagin A, punicalagin B, and other hydrolyzable tannins, displaying natural antioxidant action that inhibits microbial growth (Kumar et al., 2021). Green bell peppers were coated using the dip-coating method. The nanocomposite preserved the firmness and weight and reduced enzymic browning and variations of color, soluble solids, titratable acidity, and pH of green bell peppers.

Zhang et al. (2021a,b) developed multilayer films from an antimicrobial nanocomposite layer based on chitosan and thyme essential oil sandwiched by two gellan layers. The multilayer films were prepared by casting method, and thymi oil was incorporated into chitosan as emulsion and nanoemulsion. There was a synergistic effect between the antimicrobial activities of chitosan and essential oil.

McDaniel et al. (2019) have reported the development of antifungal films based on a mix of biopolymers. The authors prepared pullulan, xanthan, and locust bean gum-based blends incorporated with essential oils (cinnamaldehyde, eugenol, and thymol) encapsulated in nanoemulsions. Films containing cinnamaldehyde oil exhibited the widest inhibition zones against *Alternaria alternata*, *Aspergillus niger*, and *Rhizopus stolonifera*.

Other authors have also reported the improvement of barrier and mechanical properties of other gum-based blends and composites to incorporate active compounds compared with a single gum-based matrix. Some of these biopolymer mixings are carboxymethylcellulose—guar gum (Kanikireddy et al., 2020), gum Arabic—methylcellulose (Kawhena et al., 2020), Kardi gum—Farsi gum (Shahbazi and Shavisi, 2020), alginate—guar gum (Rastegar and Atrash, 2021), and *Alyssum homalocarpum* seed gum—chitosan (Monjazeb Marvdashti et al., 2020).

4.2 Gums combined with active compounds

The active ingredients incorporated into gum matrices to provide antifungal actions to nanocomposites include essential oils (Alizadeh-Sani et al., 2020; Cao and Song, 2019; Kawhena et al., 2020; Rastegar and Atrash, 2021; Wani et al., 2021; Xue et al., 2019), plant extracts (Al-Maqtari et al., 2021; Fan et al., 2021; Yang et al., 2019), metal oxides (Alizadeh-Sani et al., 2020; Joshy et al., 2020; Kanikireddy et al., 2020), resins (Pobiega et al., 2021), pigments (flavonoids) (Sucheta et al., 2019), and organic acids (Ali et al., 2020; Ma et al., 2021). The antifungal action of these ingredients is mainly characterized by antioxidant properties and the generation of reactive oxygen species (ROS), causing damage to several cell structures (Fig. 1).

4.2.1 *Essential oils and plant extracts*

Essential oils such as cinnamaldehyde, eugenol, cumin, thymol, lemongrass (Mahmood et al., 2020; McDaniel et al., 2019; Wani et al., 2021), and plant extracts such as pomegranate, roselle, aloe, *Pulicaria jaubertti* (Al-Maqtari et al., 2021; Fan et al., 2021; Kawhena et al., 2020; Yang et al., 2019) can provide antimicrobial and antioxidant properties to harvested food. Both are widely prepared as nanoemulsions for food applications (Almasi et al., 2020; de Cenobio-Galindo et al., 2019; Noori et al., 2018; Severino et al., 2015; Shetta et al., 2019; Xiong et al., 2020; Yazdan-Bakhsh et al., 2020). The cytotoxic activity of essential oils against fungi depends on the oil composition and fungus species. The essential oil hydrophobicity favors its diffusion across the microbial cell membrane, changing the ion channels and depolarizing the mitochondria membrane. The cell pH is decreased, increasing its permeability, unbalancing the ion transport, and causing the loss of energy by the membrane and apoptosis or necrosis of the cell (Bakkali et al., 2008; Vellido-Perez et al., 2021).

The antioxidant is used to prevent mainly lipid peroxidation by the following mechanisms: removal of initiating oxidation radicals, avoiding chain inhibition; peroxide decomposition, avoiding the reconversion in initiating oxidation radicals; stopping the chain reaction; reduction in oxygen concentration and chain bonds catalysts starting chain bonds such as metal ions (Jamróz and Kopel, 2020).

In general, essential oils are poorly water soluble and easily oxidized when exposed to light, high temperatures, and oxygen. Thus, encapsulation in nanoemulsions can be an efficient method to prevent their active properties and avoid the generation of undesirable flavors compromising the sensorial food quality (El-Messery et al., 2020; Gumus and Gharibzahedi, 2021; Lamarra et al., 2020; Shetta et al., 2019).

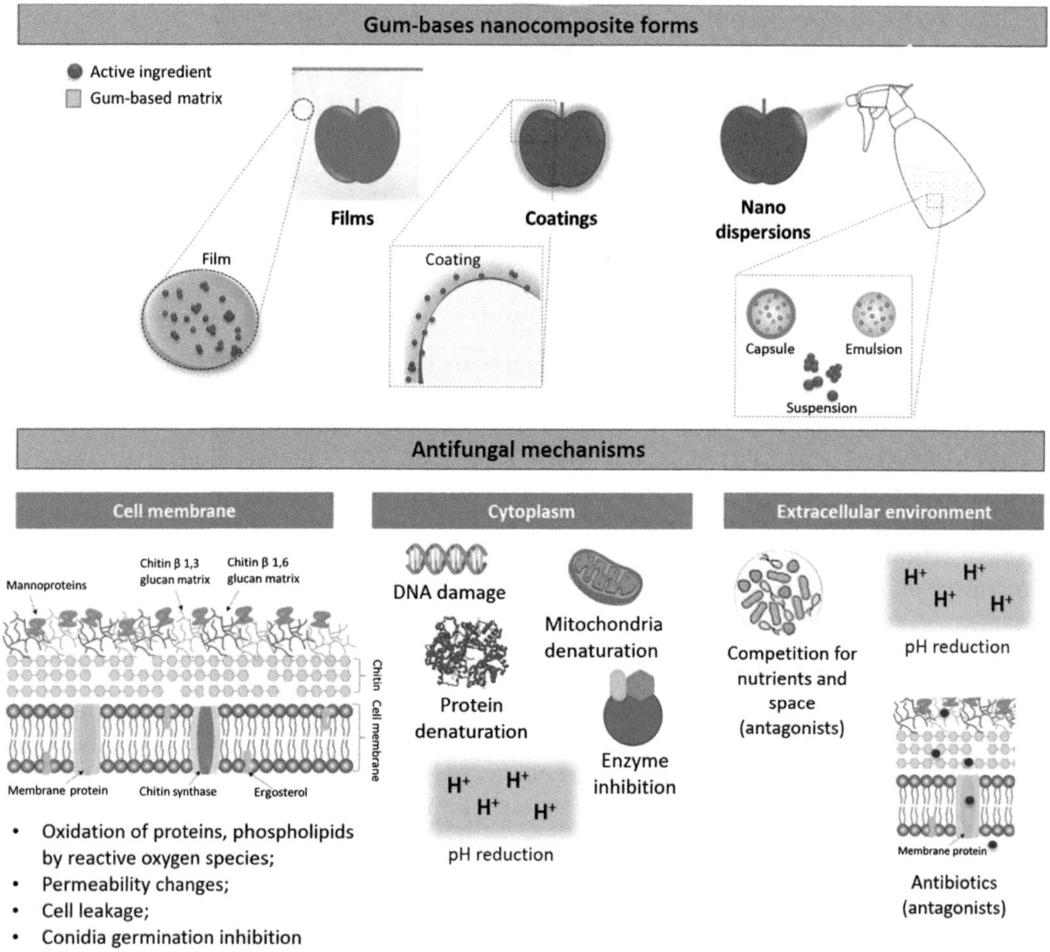

FIG. 1 Antifungal mechanisms of gum-based nanocomposites on harvested food.

Pullulan-xanthan-locust bean gum-based blends developed by McDaniel et al. (2019) showed higher antifungal activity when incorporated with essential oils, especially cinnamaldehyde, prepared by encapsulation in nanoemulsions. The nanosized emulsions droplets improved the essential oil homogenization in aqueous solutions, and the encapsulation protected them against oxidation and hydrolysis.

Xue et al. (2019) affirmed that the type and nature of essential oils affect the structural, antioxidant, and antimicrobial properties and the essential oil release profile of gum-essential oil nanocomposites due to different chemical and physical interactions between oil molecules and gum structure. Authors incorporated encapsulated essential oils (oregano, lemon, fruit of *Amomum tsaoko* Crevost et Lemaire, and grapefruit) in soy protein isolate—gum acacia edible films and evaluated their structure, active properties, and release profile in water/

ethanol (50:50). Films containing grapefruit essential oil showed better structural properties (water vapor permeability, higher tensile strength, and glass transition temperature). In contrast, films containing lemon and oregano essential oils showed the highest radical scavenging and antimicrobial activity. The lemon essential oil was the fastest released from the gum matrix, which justified its highest ability to scavenge radicals. Oregano essential oil showed an intermediate release, higher than grapefruit oil. Its highest antimicrobial activity was attributed to its chemical nature, volatility, and interaction with gum.

4.2.2 *Organic acids*

Organic acids exhibit carboxylic acid groups (R-COOH) in their structures. These compounds are natural preservatives and work as acidulants, antioxidants, and antimicrobials. Some examples of these acids are ascorbic, citric, acetic, lactic, benzoic, propanoic, and sorbic and their derived salts. Their preservative action is mainly characterized by microbial growth stagnation or microbial destruction, whose efficiency depends on the relation between the pH of the food and pK_a of the acid. Weak acids are more efficient antimicrobial agents than strong acids due to their lower dissociation degree (Jay et al., 2006).

When $pH < pK_a$, the acid is highly undissociated. Therefore, it is more soluble in lipids and diffuses easily across the cell membrane. Inside of cytoplasm ($pH \sim 7.0$), acid molecules (HA) dissociate in H^+ and A^-, acidifying the media. The H^+ protons cause the denaturation of proteins to peptides, also inactivating metabolic enzymes. The nutrient transport is also affected due to the permeability and activity changes of the cell membrane, alterations of calcium microchannels carrying ions (ionic disbalance), and reduction of ATP generation due to the reduction of ion transport and ion loss. Besides protein denaturation, toxic anions can be accumulated in the cytoplasm, and enzymes can be inactivated by complexation with metallic ions (chelates) (Lindsay, 2017).

On the other hand, when $pH > pK_a$, the acid molecules are highly dissociated and lower soluble in lipids. Thus, they diffuse slowly across the phospholipid membrane, minimizing deleterious damage in the cytoplasm (Lindsay, 2017).

Higher hydrophobic acids such as benzoic ($pK_a = 4.18$) and sorbic ($pK_a = 4.76$) and derived salts, characterized by long carbon chains, are more effective on mold than other acids, especially sorbic acid, due to their higher interaction with the fungal cell membrane. Sorbic acid inhibits dehydrogenase enzymes, and its metabolization in the human body produces CO_2 and H_2O (Lindsay, 2017).

The microorganism survival in media different from neutral pH depends on their ability to reverse the media pH. Fungi are more adapted to acid media than bacteria. The activation of the enzymes decarboxylases and deaminases synthesize amino and acid groups at $pH \sim 4{,}0$ and $pH \sim 8{,}0$, respectively, enabling media neutralization (Jay et al., 2006).

Ma et al. (2021) have reported the synergistic effect between shellac gum and tannic acid to delay the ripening of mangoes. The nanocomposite coating reduced respiration rate, physical changes in firmness, color, weight loss, lipid peroxidation, and fungal growth on fruit. Ali et al. (2020) obtained higher preservation of apples coated by composites based on Arabinoxylan and β-glucan stearic acid ester than shellac. The fungal growth in Arabinoxylan-β-glucan stearic acid ester composites was 12.2% higher than in shellac.

4.2.3 Metal ions

Several nanoparticles exhibit antifungal properties useful to control phytopathogenic fungi in agriculture. Among them highlight Ag, Cu, Se, Ni, Mg, Fe, Zn, Pd, and Ti. These particles are synthesized as oxides from different methods and show different shapes and sizes. Nanoparticles present a high volume-to-surface area ratio, which increases the contact area between nanoparticles and fungal cells, contributing to the outstanding antifungal activity of these materials. The antifungal activity of metallic nanoparticles is influenced by several factors such as size distribution, shape, composition, crystallinity, agglomeration, surface chemistry, and fungal species (Cruz-Luna et al., 2021).

Possible metallic nanoparticle mechanisms on pathogenic fungi were described by Cruz-Luna et al. (2021) and Lakshmeesha et al. (2020) (Fig. 2). In the first mechanism, metallic ions (*e.g.*, Ag^{2+}, Cu^{2+}, Zn^{2+}, Ti^{4+}) are released by oxide nanoparticles and bind membrane proteins, affecting cell permeability. In a second mechanism, metallic nanoparticles inhibit the conidia germination, suppressing their development. In a third mechanism, metallic nanoparticles and released ions disrupt electron transport, cause protein oxidation, and alter membrane potential. In the fourth and fifth mechanisms, metallic ions also interfere with protein oxidative electron transport and affect the mitochondrial membrane potential, increasing gene transcription levels in response to oxidative stress.

This stress is related to ROS, such as superoxide ($O_2^{\bullet -}$), hydroperoxyl ($HOO^{\bullet}$), and hydroxyl ($^{\bullet}OH$), generated from electron/hole pairs, which are initiated by light incidence, high temperature, chemical attacks, and ionic dissociation. In the sixth mechanism, ROS triggers oxidation reactions catalyzed by the different metallic nanoparticles, causing deleterious damage to proteins, phospholipid bilayer of cell membranes and deoxyribonucleic acid

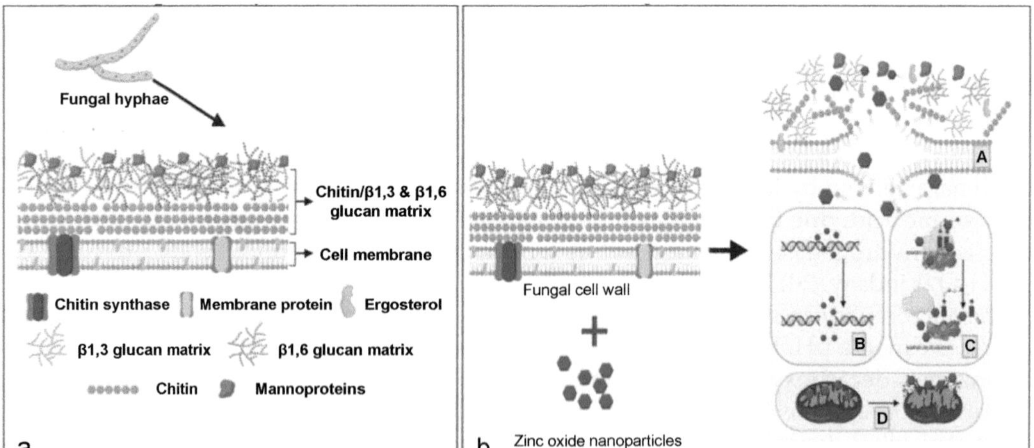

FIG. 2 Mechanism of antifungal activity of ZnO nanoparticles. (a) Fungal cell wall; (b) Mechanism of action; (A) Disruption of the fungal cell wall; (B) DNA damage; (C) Inhibition of protein synthesis; (D) Mitochondria damage. *Reprinted from Lakshmeesha, T.R., Murali, M., Ansari, M.A., Udayashankar, A.C., Alzohairy, M.A., Almatroudi, A., Alomary, M.N., Asiri, S.M.M., Ashwini, B.S., Kalagatur, N.K., Nayak, C.S., Niranjana, S.R., 2020. Biofabrication of zinc oxide nanoparticles from Melia azedarach and its potential in controlling soybean seed-borne phytopathogenic fungi. Saudi J. Biol. Sci. 27, 1923–1930 with permission from Elsevier.*

(DNA), and modifications in cell permeability, nutrient absorption, replication, and respiration. Finally, in the last mechanism, metallic ions have a genotoxic effect that destroys DNA, causing cell death (Cruz-Luna et al., 2021; de Fonseca et al., 2021).

Joshy et al. (2020) incorporated ZnO nanoparticles in xanthan-based coatings and tested them in apples to retard microbial growth. Uncoated apples, stored at room temperature, showed the highest weight loss over 5 days and tissue failure. Authors attributed this to the crack on the primary cell wall, facilitating attack by pathogens. Compared to the uncoated fruit, the nanocomposite coating preserved fruit freshness for more than a week.

Kanikireddy et al. (2020) prepared films based on CMC, Guar gum, Ag nanoparticles, and mentha leaves extract to package strawberries and prevent the growth of pathogenic microbes causing food-borne illnesses, such as *Escherichia coli, Bacillus subtilis, Salmonella typhi, Klebsiellapneumonia, Fusarium oxysporum, Staphylococus aureus, Candida albicans, Pseudomonas aeruginosa, Micrococcus liteus*, and *Salmonilla enterica*. The authors used inhibition zone tests and the viable cell count method to evaluate in vitro antimicrobial activity of the nanocomposite films. The films effectively inhibited bacteria and fungi (*Fusarium oxysporum, Candida albicans*). The packaging efficiency was evaluated by the shelf-life study of strawberries stored at room temperature for 6 days. Unpacked strawberries exhibited a weight loss of 90% after 6 days against 84% and 80% of strawberries packed in synthetic plastic and gum-based nanocomposite.

It has been known that the synthetic route of metal nanoparticles plays an essential role in their antifungal activity. It also generates preoccupation about precursors or surfactant residuals that are not easy to remove from the nanoparticles. Therefore, these residues can modify the surface chemistry of the nanoparticles and consequently influence their antifungal activity. Finally, another important factor is the accumulation and toxicity of nanoparticles in the human body (Cruz-Luna et al., 2021). Because of this, the migration and immobilization studies of metallic nanoparticles in films for packaging are crucial to ensure food safety.

4.3 Antagonistic agents

Utilizing living biological microbe agents to control postharvest diseases is a potential technology to replace traditional synthetic fungicides due to their lower toxicity to humans and environmental pollution (Matrose et al., 2021). This technology consists of a biocontrol action involving competition for space and nutrients, the synthesis of antifungal substances, and secondary metabolites (Davies et al., 2021; Hua et al., 2018).

Antagonistic agents can be yeasts, bacteria, and fungi. Some antifungal substances produced by antagonists are antibiotic substances (bacteriocins), cell wall deteriorating enzymes, which facilitate the infection of pathogens. In addition, they can induce host resistance. However, antagonists do not have a remarkable effect as fungicides. They are commonly associated with chemical fungicides or exogenous substances such as calcium chloride, salicylic acid, sodium bicarbonate, silicon, boron, and glycine betaine, to obtain satisfying fungi biocontrol (Hua et al., 2018).

Deng et al. (2021) and Moreira et al. (2020) induced the biocontrol of fungi *Penicillium digitatum* and *Penicillium* spp. in vitro and strawberries, using cell wall deteriorating enzymes produced by fungi *Cryptococcus laurentii* and *Trichoderma asperellum*, respectively (Table 3).

FIG. 3 Images of strawberries inoculated with *Penicillium* spp. (a) noncoated and coated with (b) blank cashew gum/polyvinyl alcohol blend and (c) blend containing cell wall deteriorating enzymes. *Reprinted from Moreira, B.R., Pereira-Júnior, M. a., Fernandes, K.F., Batista, K. a., 2020. An ecofriendly edible coating using cashew gum polysaccharide and polyvinyl alcohol. Food Biosci. 37. with permission from Elsevier.*

Cryptococcus laurentii was immobilized in CMC-based coatings to produce deteriorating enzymes chitinase, β-1,3-glucanase, peroxidase, polyphenol oxidase, and phenylalanine ammonia-lyase. The enzymes chitinase, N-acetylglucosaminidase, and β-1,3-glucanase were isolated from media containing *Trichoderma asperellum* and incorporated in cashew gum polysaccharide/polyvinyl alcohol-based coatings. Both antagonists reduced the fungi growth. Uncoated and coated strawberries inoculated with *Penicillium* spp. after 5 days of storage at room temperature are shown in Fig. 3.

5 Final considerations and conclusion

Individually, the edible gum coatings offer a protective barrier to fruit and vegetables against fungal contamination and avoid excessive transpiration, reducing weight loss, respiration rate, and sensorial alterations. Their barrier properties can be improved by forming blends and composites with other biopolymers and active agents. Incorporating nanosized active compounds in gum matrices also provides antifungal properties to the nanocomposites, making them high-potential materials to prevent fungal diseases in the harvested food. Among active compounds highlight plant extracts and essential oils because they are natural substances with antimicrobial and antioxidant properties. Organic acids could also be included in this group, but their antifungal efficiency is dependent on the pH value. On the other hand, metal ions are very efficient in inactivating microorganisms, but there are many discussions about their risks to human health and the environment.

Biocontrol agents (yeast, bacteria, and deteriorating enzymes) immobilized in gum matrices have been an excellent option to control pathogenic fungi growth. This area offers wide possibilities, such as antagonist microorganism inoculation for competing by nutrients and space, producing bacteriocins (antibiotics), and cell wall deteriorating enzymes. The last can be produced and isolated for food application.

Besides the coatings, gum-based nanocomposites have physical properties to use in other forms, such as capsules and films for packaging, which induces the intensive exploration of the potential of these materials.

References

Abu Elella, M.H., Goda, E.S., Gab-Allah, M.A., Hong, S.E., Pandit, B., Lee, S., Gamal, H., Rehman, A.U., Yoon, K.R., 2021. Xanthan gum-derived materials for applications in environment and eco-friendly materials: A review. J. Environ. Chem. Eng. 9, 104702.

Ahmad, S., Ahmad, M., Manzoor, K., Purwar, R., Ikram, S., 2019. A review on latest innovations in natural gums based hydrogels: preparations & applications. Int. J. Biol. Macromol. 136, 870–890.

Al-Baadani, H.H., Al-Mufarrej, S.I., Al-Garadi, M.A., Alhidary, I.A., Al-Sagan, A.A., Azzam, M.M., 2021. The use of gum Arabic as a natural prebiotic in animals: a review. Anim. Feed Sci. Technol. 274, 114894.

Ali, U., Basu, S., Mazumder, K., 2020. Improved postharvest quality of apple (Rich Red) by composite coating based on arabinoxylan and β-glucan stearic acid ester. Int. J. Biol. Macromol. 151, 618–627.

Ali, S., Akbar Anjum, M., Nawaz, A., Naz, S., Ejaz, S., Shahzad Saleem, M., Tul-Ain Haider, S., Ul Hasan, M., 2021. Effect of gum arabic coating on antioxidative enzyme activities and quality of apricot (*Prunus armeniaca* L.) fruit during ambient storage. J. Food Biochem. 45, 1–13.

Alizadeh-Sani, M., Ehsani, A., Moghaddas Kia, E., Khezerlou, A., 2019. Microbial gums: introducing a novel functional component of edible coatings and packaging. Appl. Microbiol. Biotechnol. 103, 6853–6866.

Alizadeh-Sani, M., Rhim, J.W., Azizi-Lalabadi, M., Hemmati-Dinarvand, M., Ehsani, A., 2020. Preparation and characterization of functional sodium caseinate/guar gum/TiO2/cumin essential oil composite film. Int. J. Biol. Macromol. 145, 835–844.

Al-Maqtari, Q.A., Mohammed, J.K., Mahdi, A.A., Al-Ansi, W., Zhang, M., Al-Adeeb, A., Wei, M., Phyo, H.M., Yao, W., 2021. Physicochemical properties, microstructure, and storage stability of Pulicaria jaubertii extract microencapsulated with different protein biopolymers and gum arabic as wall materials. Int. J. Biol. Macromol. 187, 939–954.

Almasi, H., Azizi, S., Amjadi, S., 2020. Development and characterization of pectin films activated by nanoemulsion and Pickering emulsion stabilized marjoram (*Origanum majorana* L.) essential oil. Food Hydrocoll. 99, 105338.

Amin, U., Khan, M.U., Majeed, Y., Rebezov, M., Khayrullin, M., Bobkova, E., Shariati, M.A., Chung, I.M., Thiruvengadam, M., 2021. Potentials of polysaccharides, lipids and proteins in biodegradable food packaging applications. Int. J. Biol. Macromol. 183, 2184–2198.

Badui, S., 2006. Salvador Badui Dergal, fourth ed. Química de los alimentos, México.

Bai, L., Huan, S., Zhao, B., Zhu, Y., Esquena, J., Chen, F., Gao, G., Zussman, E., Chu, G., Rojas, O.J., 2020. All-Aqueous Liquid Crystal Nanocellulose. vol. 14 ACS Publications, pp. 13380–13390. https://doi.org/10.1021/acsnano.0c05251.

Bakkali, F., Averbeck, S., Averbeck, D., Idaomar, M., 2008. Biological effects of essential oils—a review. Food Chem. Toxicol. 46, 446–475.

Barak, S., Mudgil, D., 2014. Locust bean gum: processing, properties and food applications—a review. Int. J. Biol. Macromol. 66, 74–80.

Baran, T., 2018. Pd (0) nanocatalyst stabilized on a novel agar/pectin composite and its catalytic activity in the synthesis of biphenyl compounds by Suzuki-Miyaura cross coupling reaction and reduction of o-nitroaniline. Carbohydr. Polym. 195, 45–52.

Barreiros, Y., de Meneses, A.C., Alves, J.L.F., Mumbach, G.D., Ferreira, F.A., Machado, R.A.F., Bolzan, A., de Araujo, P.H.H., 2022. Xanthan gum-based film-forming suspension containing essential oils: production and in vitro antimicrobial activity evaluation against mastitis-causing microorganisms. Lwt 153, 112470. https://doi.org/10.1016/j.lwt.2021.112470.

Bashir, M., Usmani, T., Haripriya, S., Ahmed, T., 2018. Biological and textural properties of underutilized exudate gums of Jammu and Kashmir, India. Int. J. Biol. Macromol. 109, 847–854.

Bautista-Baños, S., Sivakumar, D., Bello-Pérez, A., Villanueva-Arce, R., Hernández-López, M., 2013. A review of the management alternatives for controlling fungi on papaya fruit during the postharvest supply chain. Crop Prot. 49, 8–20.

Botrel, D.A., Borges, S.V., Yoshida, M.I., de Feitosa, J.P.A., de Fernandes, R.V.B., de Souza, H.J.B., de Paula, R.C.M., 2017. Properties of spray-dried fish oil with different carbohydrates as carriers. J. Food Sci. Technol. 54, 4181–4188.

Bradshaw, M.J., Bartholomew, H.P., Hendricks, D., Maust, A., Jurick, W.M., 2021. An analysis of postharvest fungal pathogens reveals temporal–spatial and host–pathogen associations with fungicide resistance-related mutations. 111, 1942–1951. https://doi.org/10.1094/PHYTO-03-21-0119-R.

Brassesco, M.E., Brandão, T.R.S., Silva, C.L.M., Pintado, M., 2021. Carob bean (*Ceratonia siliqua* L.): a new perspective for functional food. Trends Food Sci. Technol. 114, 310–322.

Calderón-Santoyo, M., Iñiguez-Moreno, M., Barros-Castillo, J.C., Miss-Zacarías, D.M., Díaz, J.A., Ragazzo-Sánchez, J.A., 2022. Microencapsulation of citral with Arabic gum and sodium alginate for the control of Fusarium pseudocircinatum in bananas. Iran. Polym. J. 31, 665–676. https://doi.org/10.1007/s13726-022-01033-z.

Cao, T.L., Song, K.B., 2019. Effects of gum karaya addition on the characteristics of loquat seed starch films containing oregano essential oil. Food Hydrocoll. 97, 105198.

Chen, T., Ji, D., Zhang, Z., Li, B., Qin, G., Tian, S., 2021a. Advances and strategies for controlling the quality and safety of postharvest fruit. Engineering 7, 1177–1184.

Chen, X., Fu, X., Huang, L., Xu, J., Gao, X., 2021b. Agar oligosaccharides: a review of preparation, structures, bioactivities and application. Carbohydr. Polym. 265, 118076.

Chung, K.R., 2011. Elsinoë fawcettii and Elsinoë australis: the fungal pathogens causing citrus scab. Mol. Plant Pathol. 12, 123–135.

Codex Alimentarius Commission, 2021. Class names and the international numbering system for food additives. Ind. High. Educ. 3, 1689–1699.

Cruz-Luna, A.R., Cruz-Martínez, H., Vásquez-López, A., Medina, D.I., 2021. Metal nanoparticles as novel antifungal agents for sustainable agriculture: current advances and future directions. J. Fungi 7, 1033.

Davies, C.R., Wohlgemuth, F., Young, T., Violet, J., Dickinson, M., Sanders, J.W., Vallieres, C., Avery, S.V., 2021. Evolving challenges and strategies for fungal control in the food supply chain. Fungal Biol. Rev. 36, 15–26.

de Cenobio-Galindo, A.J., Ocampo-López, J., Reyes-Munguía, A., Carrillo-Inungaray, M.L., Cawood, M., Medina-Pérez, G., Fernández-Luqueño, F., Campos-Montiel, R.G., 2019. Influence of bioactive compounds incorporated in a nanoemulsion as coating on avocado fruits (Persea americana) during postharvest storage: antioxidant activity, physicochemical changes and structural evaluation. Antioxidants 8, 500.

de Fonseca, J.M., dos Alves, M.J.S., Soares, L.S., de Moreira, R.F.P.M., Valencia, G.A., Monteiro, A.R., 2021. A review on TiO2-based photocatalytic systems applied in fruit postharvest: Set-ups and perspectives. Food Res. Int. 144, 110378.

Dehghani Soltani, M., Meftahizadeh, H., Barani, M., Rahdar, A., Hosseinikhah, S.M., Hatami, M., Ghorbanpour, M., 2021. Guar (*Cyamopsis tetragonoloba* L.) plant gum: from biological applications to advanced nanomedicine. Int. J. Biol. Macromol. 193, 1972–1985.

Deng, J., Li, W., Ma, D., Liu, Y., Yang, H., Lin, J., Song, G., Naik, N., Guo, Z., Wang, F., 2021. Synergistic effect of carboxymethylcellulose and Cryptococcus laurentii on suppressing green mould of postharvest grapefruit and its mechanism. Int. J. Biol. Macromol. 181, 253–262.

El-Gioushy, S.F., Abdelkader, M.F.M., Mahmoud, M.H., El Ghit, H.M.A., Fikry, M., Bahloul, A.M.E., Morsy, A.R., Lo'ay, A.A., Abdelaziz, A.M.R.A., Alhaithloul, H.A.S., Hikal, D.M., Abdein, M.A., Hassan, K.H.A., Gawish, M.S.,

2022. The effects of a gum arabic-based edible coating on guava fruit characteristics during storage. Coatings 12, 90. https://doi.org/10.3390/coatings12010090.

El-Messery, T.M., Altuntas, U., Altin, G., Özçelik, B., 2020. The effect of spray-drying and freeze-drying on encapsulation efficiency, in vitro bioaccessibility and oxidative stability of krill oil nanoemulsion system. Food Hydrocoll. 106.

Esmaeili, A., Jafari, A., Ghasemi, A., Gholamnejad, J., 2022. Improving postharvest quality of sweet cherry fruit by using tragacanth and eremurus. Int. J. Fruit Sci. 22, 370–382.

Fan, N., Wang, X., Sun, J., Lv, X., Gu, J., Zhao, C., Wang, D., 2021. Effects of konjac glucomannan/pomegranate peel extract composite coating on the quality and nutritional properties of fresh-cut kiwifruit and green bell pepper. J. Food Sci. Technol. 59, 228–238. https://doi.org/10.1007/s13197-021-05006-7.

Fausto, A., Rodrigues, M.L., Coelho, C., 2019. The still underestimated problem of fungal diseases worldwide. Front. Microbiol. 10, 214.

Government of Western Australia, 2018. Fruit and vegetable diseases | Agriculture and Food [WWW Document]. Fruit Veg. Dis. https://www.agric.wa.gov.au/diseases/fruit-and-vegetable-diseases?nopaging=1 (accessed 3.21.22).

Gumus, C.E., Gharibzahedi, S.M.T., 2021. Yogurts supplemented with lipid emulsions rich in omega-3 fatty acids: new insights into the fortification, microencapsulation, quality properties, and health-promoting effects. Trends Food Sci. Technol. 110, 267–279.

Gupta, P.K., 2017. Herbicides and fungicides. In: Reproductive and Developmental Toxicology. Academic Press, pp. 657–679.

Hamdani, A.M., Wani, I.A., Bhat, N.A., Masoodi, F.A., 2018. Chemical composition, total phenolic content, antioxidant and antinutritional characterisation of exudate gums. Food Biosci. 23, 67–74.

Hamdani, A.M., Wani, I.A., Bhat, N.A., 2019. Sources, structure, properties and health benefits of plant gums: a review. Int. J. Biol. Macromol. 135, 46–61.

He, X., Lu, W., Sun, C., Khalesi, H., Mata, A., Andaleeb, R., Fang, Y., 2021. Cellulose and cellulose derivatives: different colloidal states and food-related applications. Carbohydr. Polym. 255, 117334.

Hua, L., Yong, C., Zhanquan, Z., Boqiang, L., Guozheng, Q., Shiping, T., 2018. Pathogenic mechanisms and control strategies of Botrytis cinerea causing post-harvest decay in fruits and vegetables. Food Qual. Saf. 2, 111–119.

Jafari, S., Hojjati, M., Noshad, M., 2018. Influence of soluble soybean polysaccharide and tragacanth gum based edible coating to improve the quality of fresh-cut apple slices. J. Food Process. Preserv. 42, 1–8.

Jamróz, E., Kopel, P., 2020. Polysaccharide and protein films with antimicrobial/antioxidant activity in the food industry: A review. Polymers (Basel) 12 (6), 1289. https://doi.org/10.3390/polym12061289.

Jana, S., Maiti, S., Jana, S., Sen, K.K., Nayak, A.K., 2019. Chapter 7—Guar gum in drug delivery applications. In: Hasnain, M.S., Nayak, A.K. (Eds.), Natural Polysaccharides in Drug Delivery and Biomedical Applications. Academic Press, pp. 187–201. https://doi.org/10.1016/B978-0-12-817055-7.00007-8.

Janani, N., Zare, E.N., Salimi, F., Makvandi, P., 2020. Antibacterial tragacanth gum-based nanocomposite films carrying ascorbic acid antioxidant for bioactive food packaging. Carbohydr. Polym. 247, 116678.

Jay, J.M., Loessner, M.J., Golden, D.A., 2006. Chemical, biological, and physical methods. In: Heldman, D.R. (Ed.), Modern Food Microbiology. Springer, p. 241.

Jia, F., Liu, H., Zhang, G., 2016. Preparation of carboxymethyl cellulose from corncob. Procedia Environ. Sci. 31, 98–102.

Joshy, K.S., Jose, J., Li, T., Thomas, M., Shankregowda, A.M., Sreekumaran, S., Kalarikkal, N., Thomas, S., 2020. Application of novel zinc oxide reinforced xanthan gum hybrid system for edible coatings. Int. J. Biol. Macromol. 151, 806–813.

Kadzińska, J., Janowicz, M., Kalisz, S., Bryś, J., Lenart, A., 2019. An overview of fruit and vegetable edible packaging materials. Packag. Technol. Sci. 32, 483–495.

Kanikireddy, V., Varaprasad, K., Rani, M.S., Venkataswamy, P., Mohan Reddy, B.J., Vithal, M., 2020. Biosynthesis of CMC-Guar gum-Ag0 nanocomposites for inactivation of food pathogenic microbes and its effect on the shelf life of strawberries. Carbohydr. Polym. 236, 116053.

Kawhena, T.G., Tsige, A.A., Opara, U.L., Fawole, O.A., 2020. Application of gum arabic and methyl cellulose coatings enriched with thyme oil to maintain quality and extend shelf life of "acco" pomegranate arils. Plants 9, 1–20.

Kawhena, T.G., Opara, U.L., Fawole, O.A., 2022. Effects of gum arabic coatings enriched with lemongrass essential oil and pomegranate peel extract on quality maintenance of pomegranate whole fruit and arils. Foods 11, 593.

Khoshakhlagh, K., Mohebbi, M., Koocheki, A., Allafchian, A., 2018. Encapsulation of D-limonene in Alyssum homolocarpum seed gum nanocapsules by emulsion electrospraying: morphology characterization and stability assessment. Bioact. Carbohydr. Diet. Fibre 16, 43–52.

Koyyada, A., Orsu, P., 2021. Natural gum polysaccharides as efficient tissue engineering and drug delivery biopolymers. J. Drug Deliv. Sci. Technol. 63, 102431.

Kumar, M., Tomar, M., Saurabh, V., Mahajan, T., Punia, S., Contreras, M., Rudra, S.G., Kaur, C., Kennedy, J.F., 2020. Trends in Food Science & Technology Emerging trends in pectin extraction and its anti-microbial functionalization using natural bioactives for application in food packaging. Trends Food Sci. Technol. 105, 223–237.

Kumar, N., Pratibha, N., Ojha, A., Upadhyay, A., Singh, R., Kumar, S., 2021. Effect of active chitosan-pullulan composite edible coating enrich with pomegranate peel extract on the storage quality of green bell pepper. Lwt 138, 110435.

Lakshmeesha, T.R., Murali, M., Ansari, M.A., Udayashankar, A.C., Alzohairy, M.A., Almatroudi, A., Alomary, M.N., Asiri, S.M.M., Ashwini, B.S., Kalagatur, N.K., Nayak, C.S., Niranjana, S.R., 2020. Biofabrication of zinc oxide nanoparticles from Melia azedarach and its potential in controlling soybean seed-borne phytopathogenic fungi. Saudi J. Biol. Sci. 27, 1923–1930.

Lamarra, J., Calienni, M.N., Rivero, S., Pinotti, A., 2020. Electrospun nanofibers of poly(vinyl alcohol) and chitosan-based emulsions functionalized with cabreuva essential oil. Int. J. Biol. Macromol. 160, 307–318.

Le Cerf, D., Irinei, F., Muller, G., 1990. Solution properties of gum exudates from Sterculia urens (Karaya gum). Carbohydr. Polym. 13, 375–386.

Li, M., Li, G., Zhu, L., Yin, Y., Zhao, X., Xiang, C., Yu, G., Wang, X., 2014. Isolation and characterization of an Agaro-Oligosaccharide (AO)-hydrolyzing bacterium from the gut microflora of chinese individuals. PloS One 9.

Lindsay, R.C., 2017. Food addives. In: Damodaran, S., Parkin, K.L. (Eds.), Fennema's Food Chemistry. Taylor & Francis Group, Boca Raton, FL, p. 803.

Loureiro, K.C., Jäger, A., Pavlova, E., Lima-Verde, I.B., Štěpánek, P., Sangenito, L.S., Santos, A.L.S., Chaud, M.V., Barud, H.S., Soares, M.F.L.R., de Albuquerque-Júnior, R.L.C., Cardoso, J.C., Souto, E.B., da Mendonça, M.C., Severino, P., 2021. Cashew gum (*Anacardium occidentale*) as a potential source for the production of tocopherol-loaded nanoparticles: formulation, release profile and cytotoxicity. Appl. Sci. 11.

Lundgren, G.A., Braga, S.D.P., de Albuquerque, T.M.R., Rimá, Á., de Oliveira, K., Tavares, J.F., Vieira, W.A.D.S., Câmara, M.P.S., de Souza, E.L., 2022. Antifungal effects of *Conyza bonariensis* (L.) Cronquist essential oil against pathogenic Colletotrichum musae and its incorporation in gum Arabic coating to reduce anthracnose development in banana during storage. J. Appl. Microbiol. 132, 547–561.

Ma, J., Zhou, Z., Li, K., Li, K., Liu, L., Zhang, W., Xu, J., Tu, X., Du, L., Zhang, H., 2021. Novel edible coating based on shellac and tannic acid for prolonging postharvest shelf life and improving overall quality of mango. Food Chem. 354, 129510.

Machado, N.D., Fernández, M.A., Haring, M., Saldias, C., D'iaz, D.D., 2019. Niosomes encapsulated in biohydrogels for tunable delivery of phytoalexin resveratrol. RSC Adv. 9, 7601–7609.

Mahmood, A.-S., Rhim, J.-W., Azizi-Lalabadi, M., Hemmati-Dinarvand, M., Ehsani, A., 2020. Preparation and characterization of functional sodium caseinate/guar gum/TiO2/cumin essetial oil composite film. Int. J. Biol. Macromol. 145, 835–844.

Mahmud, N., 2021. Marine biopolymers: applications in food packaging. Processes 9 (12), 2245. https://doi.org/10.3390/pr9122245.

Matrose, N.A., Obikeze, K., Belay, Z.A., Caleb, O.J., 2021. Plant extracts and other natural compounds as alternatives for post-harvest management of fruit fungal pathogens: a review. Food Biosci. 41, 100840.

McDaniel, A., Tonyali, B., Yucel, U., Trinetta, V., 2019. Formulation and development of lipid nanoparticle antifungal packaging films to control postharvest disease. J. Agric. Food Res. 1, 100013.

Md Nor, S., Ding, P., 2020. Trends and advances in edible biopolymer coating for tropical fruit: a review. Food Res. Int. 134, 109208.

Monjazeb Marvdashti, L., Abdulmajid Ayatollahi, S., Salehi, B., Sharifi-Rad, J., Abdolshahi, A., Sharifi-Rad, R., Maggi, F., 2020. Optimization of edible Alyssum homalocarpum seed gum-chitosan coating formulation to improve the postharvest storage potential and quality of apricot (*Prunus armeniaca* L.). J. Food Saf. 40, 1–9.

Monteiro Fritz, A.R., de Matos Fonseca, J., Trevisol, T.C., Fagundes, C., Valencia, G.A., 2019. Active, eco-friendly and edible coatings in the post-harvest—a critical discussion. In: Polymers for Agri-Food Applications. Springer International Publishing, pp. 433–463.

Moreira, B.R., Pereira-Júnior, M.A., Fernandes, K.F., Batista, K.A., 2020. An ecofriendly edible coating using cashew gum polysaccharide and polyvinyl alcohol. Food Biosci. 37, 100722. https://doi.org/10.1016/j.fbio.2020.100722.
Murmu, S.B., Mishra, H.N., 2018. The effect of edible coating based on Arabic gum, sodium caseinate and essential oil of cinnamon and lemon grass on guava. Food Chem. 245, 820–828.
Musa, H.H., Ahmed, A.A., Musa, T.H., 2019. Chemistry, biological, and pharmacological properties of gum arabic. Ref. Ser. Phytochem., 797–814.
Naqash, F., Masoodi, F.A., Rather, S.A., Wani, S.M., Gani, A., 2017. Emerging concepts in the nutraceutical and functional properties of pectin—a review. Carbohydr. Polym. 168, 227–239.
Nazarzadeh, E., Makvandi, P., Tay, F.R., 2019. Recent progress in the industrial and biomedical applications of tragacanth gum: a review. Carbohydr. Polym. 212, 450–467.
Nejatian, M., Abbasi, S., Azarikia, F., 2020. Gum tragacanth: structure, characteristics and applications in foods. Int. J. Biol. Macromol. 160, 846–860.
Ng, J.Y., Obuobi, S., Chua, M.L., Zhang, C., Hong, S., Kumar, Y., Gokhale, R., Ee, P.L.R., 2020. Biomimicry of microbial polysaccharide hydrogels for tissue engineering and regenerative medicine—a review. Carbohydr. Polym. 241, 116345.
Noori, S., Zeynali, F., Almasi, H., 2018. Antimicrobial and antioxidant efficiency of nanoemulsion-based edible coating containing ginger (*Zingiber officinale*) essential oil and its effect on safety and quality attributes of chicken breast fillets. Food Control 84, 312–320.
Oliveira, M.A., Furtado, R.F., Bastos, M.S.R., Leitão, R.C., Benevides, S.D., Muniz, C.R., Cheng, H.N., Biswas, A., 2018. Performance evaluation of cashew gum and gelatin blend for food packaging. Food Packag. Shelf Life 17, 57–64.
Phillips, A.O., Phillips, G.O., 2011. Biofunctional behaviour and health benefits of a specific gum arabic. Food Hydrocoll. 25, 165–169.
Pobiega, K., Igielska, M., Włodarczyk, P., Gniewosz, M., 2021. The use of pullulan coatings with propolis extract to extend the shelf life of blueberry (*Vaccinium corymbosum*) fruit. Int. J. Food Sci. Technol. 56, 1013–1020.
Prasad, N., Thombare, N., Sharma, S.C., Kumar, S., 2022. Production, processing, properties and applications of karaya (Sterculia species) gum. Ind. Crop Prod. 177, 114467.
Premalatha, M., Mathavan, T., Selvasekarapandian, S., Monisha, S., Pandi, D.V., Selvalakshmi, S., 2016. Investigations on proton conducting biopolymer membranes based on tamarind seed polysaccharide incorporated with ammonium thiocyanate. J. Non Cryst. Solids 453, 131–140.
Rahman, S., Konwar, A., Majumdar, G., Chowdhury, D., 2021. Guar gum-chitosan composite film as excellent material for packaging application. Carbohydr. Polym. Technol. Appl. 2, 100158.
Raj, V., Lee, J.H., Shim, J.J., Lee, J., 2021. Recent findings and future directions of grafted gum karaya polysaccharides and their various applications: a review. Carbohydr. Polym. 258, 117687.
Rana, V., Rai, P., Tiwary, A.K., Singh, R.S., Kennedy, J.F., Knill, C.J., 2011. Modified gums : approaches and applications in drug delivery. Carbohydr. Polym. 83, 1031–1047.
Rastegar, S., Atrash, S., 2021. Effect of alginate coating incorporated with Spirulina, Aloe vera and guar gum on physicochemical, respiration rate and color changes of mango fruits during cold storage. J. Food Meas. Charact. 15, 265–275.
Santos, S.F.D., Cardoso, R.D.C.V., Borges, Í.M.P., Almeida, A.C.E., Andrade, E.S., Ferreira, I.O., Ramos, L.D.C., 2020. Post-harvest losses of fruits and vegetables in supply centers in Salvador, Brazil: analysis of determinants, volumes and reduction strategies. Waste Manag. 101, 161–170.
Severino, R., Ferrari, G., Vu, K.D., Donsì, F., Salmieri, S., Lacroix, M., 2015. Antimicrobial effects of modified chitosan based coating containing nanoemulsion of essential oils, modified atmosphere packaging and gamma irradiation against Escherichia coli O157:H7 and Salmonella Typhimurium on green beans. Food Control 50, 215–222.
Shahbazi, Y., Shavisi, N., 2020. Application of active Kurdi gum and Farsi gum-based coatings in banana fruits. J. Food Sci. Technol. 57, 4236–4246.
Sharma, G., Sharma, S., Kumar, A., Al-Muhtaseb, A.H., Naushad, M., Ghfar, A.A., Mola, G.T., Stadler, F.J., 2018. Guar gum and its composites as potential materials for diverse applications: a review. Carbohydr. Polym. 199, 534–545.
Shetta, A., Kegere, J., Mamdouh, W., 2019. Comparative study of encapsulated peppermint and green tea essential oils in chitosan nanoparticles: Encapsulation, thermal stability, in-vitro release, antioxidant and antibacterial activities. Int. J. Biol. Macromol. 126, 731–742.
Silva, C., Torres, M.D., Chenlo, F., Moreira, R., 2017. Rheology of aqueous mixtures of tragacanth and guar gums: effects of temperature and polymer ratio. Food Hydrocoll. 69, 293–300.

Sood, A., Saini, C.S., 2022. Red pomelo peel pectin based edible composite films: effect of pectin incorporation on mechanical, structural, morphological and thermal properties of composite films. Food Hydrocoll. 123, 107135.

Su, H., Zheng, R., Jiang, L., Zeng, N., Yu, K., Zhi, Y., Shan, S., 2021. Dextran hydrogels via disulfide-containing Schiff base formation: synthesis, stimuli-sensitive degradation and release behaviors. Carbohydr. Polym. 265, 118085.

Sucheta, Chaturvedi, K., Sharma, N., Yadav, S.K., 2019. Composite edible coatings from commercial pectin, corn flour and beetroot powder minimize post-harvest decay, reduces ripening and improves sensory liking of tomatoes. Int. J. Biol. Macromol. 133, 284–293.

Taheri, A., Jafari, S.M., 2019. Gum-based nanocarriers for the protection and delivery of food bioactive compounds. Adv. Colloid Interface Sci. 269, 277–295.

Tahir, H.E., Xiaobo, Z., Mahunu, G.K., Arslan, M., Abdalhai, M., Zhihua, L., 2019. Recent developments in gum edible coating applications for fruits and vegetables preservation: A review. Carbohydr. Polym. 224, 115141.

Thambugala, K.M., Daranagama, D.A., Phillips, A.J.L., Kannangara, S.D., Promputtha, I., 2020. Fungi vs. fungi in biocontrol: an overview of fungal antagonists applied against fungal plant pathogens. Front. Cell. Infect. Microbiol. 10, 718.

Vellido-Perez, J.A., Ochando-Pulido, J.M., Brito-de la Fuente, E., Martinez-Ferez, A., 2021. Novel emulsions-based technological approaches for the protection of omega-3 polyunsaturated fatty acids against oxidation processes—a comprehensive review. Food Struct. 27, 100175.

Verma, C., Quraishi, M.A., 2021. Gum Arabic as an environmentally sustainable polymeric anticorrosive material: recent progresses and future opportunities. Int. J. Biol. Macromol. 184, 118–134.

Wani, S.M., Gull, A., Ahad, T., Malik, A.R., Ganaie, T.A., Masoodi, F.A., Gani, A., 2021. Effect of gum Arabic, xanthan and carrageenan coatings containing antimicrobial agent on postharvest quality of strawberry: assessing the physicochemical, enzyme activity and bioactive properties. Int. J. Biol. Macromol. 183, 2100–2108.

Williams, P.R.A., Ampbell, K.E.T.C., Haraviram, H.E.G., Adrigal, J.U.L.M., 2017. Alginate-chitosan hydrogels provide a sustained gradient of sphingosine-1-phosphate for therapeutic. Angiogenesis 45, 1003–1014.

Xiong, Y., Li, S., Warner, R.D., Fang, Z., 2020. Effect of oregano essential oil and resveratrol nanoemulsion loaded pectin edible coating on the preservation of pork loin in modified atmosphere packaging. Food Control 114, 107226.

Xue, F., Gu, Y., Wang, Y., Li, C., Adhikari, B., 2019. Encapsulation of essential oil in emulsion based edible films prepared by soy protein isolate-gum acacia conjugates. Food Hydrocoll. 96, 178–189.

Yang, Z., Zou, X., Li, Z., Huang, X., Zhai, X., Zhang, W., Shi, J., Tahir, H.E., 2019. Improved postharvest quality of cold stored blueberry by edible coating based on composite gum arabic/roselle extract. Food Bioproc. Tech. 12, 1537–1547.

Yazdan-Bakhsh, M., Nasr-Esfahani, M., Esmaeilzadeh-Kenari, R., Fazel-Najafabadi, M., 2020. Evaluation of antioxidant properties of Heracleum Lasiopetalum extract in multilayer nanoemulsion with biopolymer coating to control oxidative stability of sunflower oil. J. Food Meas. Charact., 1014–1023. https://doi.org/10.1007/s11694-020-00691-y.

Youssef, K., de Oliveira, A.G., Tischer, C.A., Hussain, I., Roberto, S.R., 2019. Synergistic effect of a novel chitosan/silica nanocomposites-based formulation against gray mold of table grapes and its possible mode of action. Int. J. Biol. Macromol. 141, 247–258.

Yun, D., Cai, H., Liu, Y., Xiao, L., Song, J., Liu, J., 2019. Development of active and intelligent films based on cassava starch and Chinese bayberry (Myrica rubra Sieb. et Zucc.) anthocyanins. RSC Adv. 9, 30905–30916.

Zhang, L., Yu, D., Regenstein, J.M., Xia, W., Dong, J., 2021a. A comprehensive review on natural bioactive films with controlled release characteristics and their applications in foods and pharmaceuticals. Trends Food Sci. Technol. 112, 690–707.

Zhang, X., Liu, D., Jin, T.Z., Chen, W., He, Q., Zou, Z., Zhao, H., Ye, X., Guo, M., 2021b. Preparation and characterization of gellan gum-chitosan polyelectrolyte complex films with the incorporation of thyme essential oil nanoemulsion. Food Hydrocoll. 114, 106570.

Zia, K.M., Tabasum, S., Khan, M.F., Akram, N., Akhter, N., Noreen, A., Zuber, M., 2018. Recent trends on gellan gum blends with natural and synthetic polymers: a review. Int. J. Biol. Macromol. 109, 1068–1087.

CHAPTER

5

Nanoencapsulation of fungicides: New trend in plant disease control

Pallavi Nayak[a,b]

[a]Nuclear Medicine Unit, University Hospital Sant'Andrea, Rome, Italy [b]Department of Medical-Surgical Sciences and of Translational Medicine, Faculty of Medicine and Psychology, "Sapienza" University of Rome, Rome, Italy

1 Introduction

Fungi have evolved for 1 billion years and can be found in various habitats worldwide due to their adaptability and persistence. Some fungi are pathogenic, and humans have struggled to limit productivity losses since organized agriculture began. Many advances have been achieved in this struggle since the introduction of fungicides. Even though fungicides have significantly increased agricultural output, their overuse has health and environmental consequences (Tleuova et al., 2020). Animal pests account for 15%–18% of agricultural losses, whereas weeds and microbiological diseases account for 34.16% of crop losses. About 70%–80% of yield losses are attributed to fungi (Atiq et al., 2020). There are more than 1.5 million species of fungus, and these fungi pathogens, saprophytic and typically parasitic, cause various diseases in crops. Fungal diseases can cause significant reductions in crop yields worldwide every year (Worrall et al., 2018; Flood, 2010). The FAO of the United Nations estimates that around 969061 tons of fungicides were used globally in 2019 (FAOSTAT, 2019) (Fig. 1).

The production of agrochemicals with the required chemical composition and efficiency is made possible by nanotechnology, which has the potential to increase plant yield significantly (Wani et al., 2019). The global nanotechnology industry is expected to grow at a compound annual growth rate (CAGR) of 35.5% from 2021 to 2026, rising from $5.2 billion in 2021 to $23.6 billion by 2026. The North American nanotechnology industry is expected to grow at a CAGR of 34.5% between 2021 and 2026, rising from $1.6 billion in 2021 to $7.2 billion by 2026. The Asia-Pacific nanotechnology market is expected to increase at a CAGR of 37.6% between 2021 and 2026, rising from $1.2 billion in 2021 to $6.0 billion in 2026 (Global Nanotechnology Market Report, 2021–2026). Due to the delayed release of the chemical compounds,

Nanohybrid Fungicides
https://doi.org/10.1016/B978-0-443-23950-2.00019-9

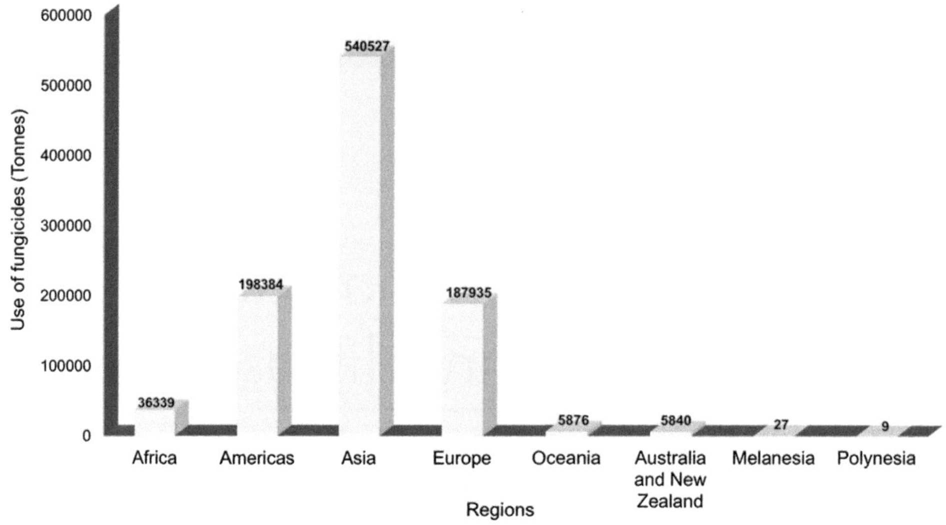

FIG. 1 FAOSTAT (2019) estimation on the use of fungicides worldwide.

nanopesticides specifically target pests while providing long-term crop protection (Golden et al., 2018; Sharma et al., 2017). Because of their small sizes, field-spraying effectiveness, improved droplet adhesion on plant surfaces, wettability, and quick absorption by the target, nanopesticides offer effective and environmentally benign benefits. Nanopesticides promote sustainable production by reducing the number of chemicals used, improving plant protection, and reducing hazardous residues (Deka et al., 2021). This chapter provides vital information on nanotechnology for researchers and the agrochemical industrial sector, mainly when using nanoencapsulation techniques to deliver fungicides to treat plant diseases.

2 Nanocarriers in the sustainable development of fungicides

In the current environmental scenario, pests and pathogens destroy approximately 20%–40% of crops yearly. Hazardous chemicals are frequently used to manage plant diseases, which are hazardous to environment and humans. In this context, nanotechnology makes pesticides innocuous by reducing toxicity, increasing the solubility of less water-soluble insecticides, and increasing shelf life. Additionally, it has a beneficial effect on the environment, particularly the soil. Nanoparticles (NPs) are small particles with sizes between 1 and 100 nm. In comparison to other organisms, the synthesis of NPs by fungi has proven advantageous. Fungi are beneficial due to their ease of isolation and ability to secrete extracellular enzymes (Sahoo et al., 2022; Suttee et al., 2019).

Furthermore, the process of metal ion reduction by fungi-secreted proteins was environmentl-friendly and less time-consuming. The effects of gold nanoparticles on *Candida albicans* and *S. cerevisiae* were studied by Das et al. in 2009. This study provided evidence on the mode of action that gold nanoparticles have on fungi. The results of the SEM analysis demonstrated that the interaction and activity of the NP caused the cell walls of the fungus to rupture (Das et al., 2009). In 2010, Shah and team found that the number of enzymes capable of digesting lignocellulose had reduced. Proteins and mitochondria were harmed due to the interaction with NPs (Shah et al., 2010). Fig. 2 depicts the lethal effects of NPs (Kumari and Khan, 2017; Prasad et al., 2016; Singh et al., 2014). Table 1 elaborates on the fabrication of nanocarrier-incorporated fungicides and their antifungal activities.

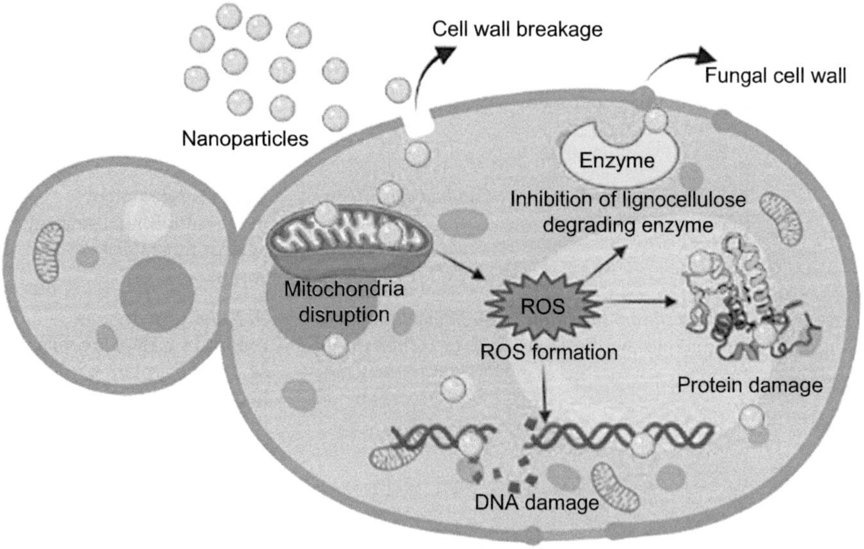

FIG. 2 Mechanism of antifungal action of NPs. *Created in Bio Render.com.*

TABLE 1 Nanocarrier incorporated fungicides and their antifungal activities.

Nanocarrier	Fabricated Fungicide	Fungus	Outcomes	Reference
Lignin–Chitosan Nanocarriers (NC)	Methacrylated lignin (ML)- Chitosan oligomers (COS)-NC loaded with extracts from *Rubia tinctorum, Silybum marianum, Equisetum arvense,* and *Urtica dioica*	*Neofusicoccum parvum*	*R. tinctorum* extract-loaded nanocarriers were very efficient against *N. parvum*	Sánchez-Hernández et al. (2022)
Liposomes	Chitosan-gum arabic-coated liposome-alizarin nanocarriers (CGL-Alz NCs)	*Candida albicans*	• Nanocarriers exhibited antibiofilm activity against *C. albicans* • An improved kinetic release mechanism of alizarin nanocarriers was found	Raj et al. (2022)

Continued

TABLE 1 Nanocarrier incorporated fungicides and their antifungal activities—cont'd

Nanocarrier	Fabricated Fungicide	Fungus	Outcomes	Reference
Nanocapsule	Boscalid@ZIF-67	*Botrytis cinerea*	On *Botrytis cinerea*, Boscalid@ZIF-67 had an antifungal effect that was 4–6 times stronger than the commercially available product	Zhang et al. (2022)
Micell	Surfactants were utilized as the host, and hexaconazole and dazomet were used as the guests in this experiment	*Ganoderma boninense*	An anionic surfactant-based system shows more outstanding inhibitory activities on the *Ganoderma pathogen* in comparison to a nonionic surfactant-based system	Mustafa et al. (2022)
AgNPs	Ag-NPmyc-loaded fluconazole (FLC)	*Candida spp.*	Demonstrate potent in vitro antifungal synergy against most *Candida* strains	Zainab et al. (2022)
AgNPs	*Purpureocillium lilacinum* fungus cells synthesized AgNPs	*A. flavus fungi* and *M. incognita*	*A. flavus* growth was reduced by 85%, juvenile *M. incognita* plant parasite hatching from eggs was reduced fourfold, and M. incognita nematode mortality was increased ninefold	Khan et al. (2022)
Zinc oxide nanoparticles (ZNONPs)	*Aspergillus niger* synthesized ZNONPs	*Alternaria solani*	ZNONPs have recently proven suitable fungicides, making them an excellent substitute for chemical fungicides	Singh et al. (2022)
AgNPs	Waste macrophytic biomass-incorporated AgNPs	*Pyriporia oryzea, Helminthisporium oryzea, Alternaria sp., Rhizoctonia solani*, and *Xanthomanas oryzae*	*Cymodocea serrulate–mediated AgNPs* (Cs-AgNP) exhibited effective antifungal activity against *Pyriporia oryzea, Helminthisporium oryzea*, and *Alternaria sp.*, and *Padina australis–mediated* AgNPs (Pa-AgNPs) showed potential activity against *Rhizoctonia solani* and *Xanthomanas oryzae*	Kailasam et al. (2022)

TABLE 1 Nanocarrier incorporated fungicides and their antifungal activities—cont'd

Nanocarrier	Fabricated Fungicide	Fungus	Outcomes	Reference
Copper nanoparticles (CuONPs)	*Pseudomonas fluorescens* and *Trichoderma viride*–mediated CuONPs	*P. parasitica*	Compared to CuONPs produced from *P. fluorescens*, *T. viride* CuONPs showed the highest percent growth inhibition	Sawake et al. (2022)
ZnONPs and CuONPs	Lemon peel extract-mediated ZnO NPs and CuO NPs	*Alternaria citri*	At $100\,mg\,ml^{-1}$, NPs prevented the growth of the fungus	Sardar et al. (2022)
Mesoporous Silica NPs	Prochloraz (Pro)-loaded bimodal mesoporous silica (BMMs) NPs	*R. solani*	Biological activity testing revealed that the NPs were efficient antifungal agents	Wu et al. (2022)
Graphene oxide	Graphene oxide–pyraclostrobin	*Sclerotinia sclerotiorum* and *Fusarium graminearum*	In a greenhouse, graphene oxide–pyraclostrobin nanocomposite was more effective than graphene oxide and pyraclostrobin alone against *Fusarium graminearum* and *Sclerotinia sclerotiorum*	Peng et al. (2022)
Sterosomes	Cymoxanil-loaded liposome	*Saccharomyces cerevisiae*	In a culture medium, sterosomes containing cyanoxanil reduced yeast cell development	Zhang et al. (2021)
Mesoporous silica nanoparticles (MSNs)	Ag+-polydopamine (PDA)@MSNs-eugenol (EU) NPs	*Botrytis cinerea*	*B. cinerea* mycelial growth is inhibited on detached and potted tomato leaves	Wang et al. (2021a,b)
Nanovaccine	Chitosan nanoparticles (CSNPs) loaded with d-limonene	*Botrytis cinerea*	Plant growth and development were downregulated, whereas stress response was upregulated	Vega-Vásquez et al. (2021)
Liposome	5ID liposomes coated with chitosan-gum arabic (CS-GA-5ID-LP-NPs)	*Botrytis cinerea*	Effective antifungal activity on *B. cinerea* at MIC values of 25 g/mL	Raj et al. (2021)
M.S.N.s	Embedding hydroxypropyl cellulose on hollow MSNPs that contain pyraclostrobin (PYR-HMS-HPC)	*Magnaporthe oryzae*	*Magnaporthe oryzae* was more resistant, and the fungicidal action was statistically significant from 7 to 21 days	Gao et al. (2021)
Self-assembled NPs	Self-assembled NPs loaded with fenhexamid (FHA) and polyhexamethylene biguanide (PHMB)	*Pseudomonas syringae pv. lachrymans*, *Botrytis cinerea*, and *Sclerotinia sclerotiorum*	Antimicrobial activity had a synergistic impact, and NPs showed greater efficacy against infections	Tang et al. (2021)

Continued

TABLE 1 Nanocarrier incorporated fungicides and their antifungal activities—cont'd

Nanocarrier	Fabricated Fungicide	Fungus	Outcomes	Reference
IONPs-coated chitosan (CS)	IONPs-CS-loaded miconazole (MCZ) or fluconazole (FLZ)	*C. albicans* and *C. glabrata*	Most effective treatments in reducing colony-forming units (CFUs)	Caldeirão et al. (2021)
Quaternary ammonium salt (Q)–modified mesoporous silica nanoparticles (MSN-Q NPs)	MSN-Q NPs loaded berberine hydrochloride (BH) and carboxylatopillar [5]arene (CP[5]A)	*Botrytis cinerea*	A novel fungicide nanoplatform could prevent mycelial growth and spore germination, controlling *B. cinerea*	Wang et al. (2021a,b)
Octahedral, iron-based metal-organic frameworks (MOFs) (NH2-Fe-MIL-101)	Diniconazole (Dini)-loaded NH2-Fe-MIL-101-MOFs	*Fusarium graminearum*	Showed promising bioactivities against the pathogenic fungus that causes wheat head scabs	Shan et al. (2020)
AgNPs	Rice leaf–extracted AgNPs	*Rhizoctonia solani*	Ag NP treatment completely inhibited the disease incidence at 20 μg/mL	Kora et al. (2020)
Iron NPs	*Calotropis procera* extract-mediated iron NPs	*Alternaria alternata*	Iron NPs showed 87.9% growth inhibition at 1.0 mg/mL concentration	Ali et al. (2020)
Magnetic graphene oxide	Magnetic graphene oxide that has been functionalized with zeolite imidazolate framework-8 (ZIF-8) (Fe3O4@APTES-GO/ZIF-8)	–	To perform well and produce satisfying results when analyzing trace triazole fungicides in complex matrices	Senosy et al. (2020)
Xylan-NPs	Fungicide-loaded xylan-based nanocarriers with toluene diisocyanate (TDI) as a crosslinking agent	*Phaeomoniella chlamydospora* (Pch) and *Phaeoacremonium minimum* (Pmi)	The growth of fungal mycelium was induced by empty xylan-based nanocarriers, indicating that xylan was degraded in the presence of fungi and emphasizing the degradation as a trigger for the release of a loaded agrochemical	Beckers et al. (2020)
Lignin-NPs	Boscalid, tebuconazole, pyraclostrobin, and azoxystrobin encased lignin NPs	*Phaeomoniella chlamydospora* (Pch) and *Phaeoacremonium minimum* (Pmi)	The biodegradable delivery vehicles made of lignin NPs are much sought-after for use in trunk injection treatments for the deadly fungus illness Esca	Machado et al. (2020)

TABLE 1 Nanocarrier incorporated fungicides and their antifungal activities—cont'd

Nanocarrier	Fabricated Fungicide	Fungus	Outcomes	Reference
CuNPs	Cu2+ is produced from copper(II) chloride dihydrate utilizing ascorbic acid as a green chemical reduction agent	*Fusarium oxysporum (F. oxysporum) and Phytophthora capsici (P. capsici)*	CuNPs are very effective at inhibiting fungi like *F. oxysporum* and *P. capsici.* At 30 ppm, *F. oxysporum* and *P. capsici* were stopped entirely after three days of incubation. After one day of incubation, they were stopped at 7.5 ppm	Pham et al. (2019)
Chitosan (Chit) NPs	Tripolyphosphate (TPP)-incorporated ChitNPs	*C. albicans*	TPP/ChitNPs decreased *C. Albicans* initial adhesion and biofilm formation	de Carvalho et al. (2019)
ChitNPs	Hexaconazole- and dazomet-loaded ChitNPs	*Ganoderma boninense*	On *G. boninense*, synthetic fungicides had the most potent antifungal effectiveness and the lowest half-maximal effective concentration (EC50)	Maluin et al. (2019a,b)
P-123 copolymeric micelles	Fluconazole alone and in conjunction with hypericin (Hyp) encapsulated P123 (FLU)	*Candida* species	P123-Hyp-PDI is a viable option for treating fungal infections and preventing biofilm formation and fungal dissemination in medical devices	Sakita et al. (2019)

2.1 Dendrimer

Dendrimers are three-dimensional (3D) globular polymeric nanostructures with a high monodispersity and multivalence (Mishra et al., 2021a,b,c; Saluja et al., 2021; Mishra et al., 2019; Mishra and Kesharwani, 2016). A dendrimer-based nanocarrier makes it possible for the pesticide to be carried into living cells in an effective manner, which dramatically boosts the cytotoxicity of the drug (Liu et al., 2015). The dendrimer can be broken down into three distinct sections: the core, the inner branches, and the lateral groups. Polyamidoamine, also known as PAMAM, is a branched-chain dendrimer that has amine branching and amide bridging. It is the dendrimer that finds the most widespread application. Its amine functional groups (for even-numbered PAMAMs) and carboxylates (for odd-numbered PAMAMs) aid in dendrimer dissolution in polar solvents (Thanh et al., 2018; Nguyen et al., 2017).

Thanh et al. encapsulated *Origanum majorana* L. essential oil in a polyamidoamine (PAMAM) G4.0 dendrimer to develop a potential nanocide against *Phytophthora infestans*. The antifungal activity of oil/G4 PAMAM was significantly higher than that of G4 PAMAM alone or marjoram essential oil. These findings suggested that the nanocide oil/G4 PAMAM helped the oil's antifungal properties to be strengthened and prolonged (Thanh et al., 2019).

Liu et al. used a water-soluble fluorescent perylenediimide-cored (PDI-cored) cationic dendrimer to encapsulate thiamethoxam, a hydrophobic pesticide. The findings revealed that dendrimer-based nanocarriers could effectively deliver drugs into living cells while significantly increasing the drug's cytotoxicity. Dendrimers have been used successfully as pesticide nanocarriers to increase the pesticide's cytotoxic effect (Liu et al., 2015).

2.2 Liposome

Liposomes are aqueous-cored bilayer vesicles. Because of this unique arrangement, liposomes can be used as a drug delivery system for hydrophobic and hydrophilic drugs. Liposomes also have biocompatibility, bioavailability, low toxicity, sustained release, and high drug load, among other advantages. Cholesterol is used in liposomes to improve membrane stiffness, stabilize vesicles, and regulate drug release rate (Mishra et al., 2021a,b,c; Zhang and Huang, 2020; Patil et al., 2019). Zhang et al. developed cymoxanil-loaded nonphospholipid liposomes by preparing nonphospholipid liposomes (sterosomes) with cholesterol and stearylamine as a pesticide nanocarrier. The results demonstrated that sterosomes containing cymoxanil suppressed the growth of yeast cells, which serve as model fungal targets. Sterosomes as nanocarriers increased the stability and efficacy of cymoxanil, presenting practical and cost-effective ways for developing innovative pesticide formulations (Zhang et al., 2021). Karny et al. constructed liposomes containing Fe and Mg to overcome the acute nutrient deficiency in tomato plant leaves. The findings demonstrate that nano-liposomes can penetrate foliage and migrate bidirectionally throughout the plant, utilizing the natural transportation mechanisms of a plant while producing no toxicity. Liposomes were found close to cell nuclei. After more than 48 h, they discharged their contents, showing that endogenous factors (e.g., cytoplasmic lipases) and osmotic instability promote liposomal membrane rupture. Extending nanotechnology to allow active substances to be delivered to plants can improve agricultural methodological approaches and outputs (Karny et al., 2018).

2.3 Niosomes

Niosomes are innovative vesicular drug delivery methods containing nonionic surfactant vesicles surrounding the solution. There are various advantages of using niosomes over standard medication delivery systems. Niosomes and liposomes are structurally similar in that they both have a bilayer. Niosomes are significantly more stable during the formulation and storage procedures than liposomes. Niosomes have the potential to improve bioavailability, drug degradation, drug insolubility, and volatility (Suttee et al., 2020). Aswin et al. used the transmembrane pH gradient method to create niosomes encapsulated with Indian herb oils, which were then tested for larvicidal properties of mosquitoes. The LC50 and LC90 values for encapsulated niosomes treated against *Aedes aegypti* larvae were 36, 59 ppm, 41, 79 ppm, and 43, 72 ppm, respectively, for neem oil, eucalyptus oil, and rosemary oil, respectively, and against *Culex quinquefasciatus* larvae were 33, 71 ppm and 35, 75 ppm. In light of these concerns, it was revealed that niosomes containing neem oil had a decreased rate of larval mortality (Aswin Jeno et al., 2021). To inhibit *Aspergillus flavus* growth and aflatoxin (AF) contamination of maize grains during storage, Garca-Daz et al. developed a niosome-encapsulated essential oil (EO) formulation. To investigate *Aspergillus flavus* growth control and AF generation by this fungus, the researchers used EOs derived from *Rosmarinus officinalis, Thymus vulgaris, Satureja montana, Origanum virens, O. majoricum,* and *O. vulgare*. All of the EOs evaluated modified the fungal growth rate and lengthened the lag phase at high concentrations. On the other hand, the EOs derived from *S. montana* and *O. virens* were the most effective at reducing fungal growth and AF generation at lower doses (García-Díaz et al., 2019).

2.4 Silver nanoparticles (AgNPs)

Silver nanoparticles (AgNPs) are among the most studied nanostructures in recent years, exhibiting more challenging and promising properties in various biomedical applications. Metallic nanoparticles (NPs) such as Ag and gold have been widely used in biomedical applications due to their physical, chemical, and biological properties (Burdușel et al., 2018; Faisal and Kumar, 2017). When AgNPs reach the nanoscale, they develop physicochemical properties that cause unusual biological responses. The unique properties of AgNPs allow them to be used as antibacterial, antiinflammatory, antifungal, and antiviral agents (Mishra et al., 2021a,b,c). Silver nanoparticles have "oligodynamic activity," which means they are effective against many microorganisms that cause infections such as bacteria, fungi, and viruses (Huang et al., 2018; Rudramurthy et al., 2016). Both human and plant diseases have been successfully treated with AgNPs by utilizing a variety of different biochemical mechanisms of antimicrobial action, such as DNA and membrane damage, interruption of electron transport and/or ATP synthesis, inhibition of protein synthesis, and ROS generation/induction of oxidative stress. Malandrakis et al. developed AgNP-loaded benzimidazoles (BEN-R), thiophene methyl (TM), and carbendazim (CARB) to inhibit *Monilia fructicola*. AgNPs inhibited mycelial growth in susceptible (BEN-S) and resistant (BEN-R) isolates. The combination of AgNPs and TM considerably boosted fungal toxic effects in vitro and on apple fruit, regardless of resistance phenotype, compared to the different treatments (BEN-R/S). The positive correlation between AgNPs and TM+AgNPs treatments indicates that the observed additive/synergistic action is likely due to increased AgNP activity/availability (Rai et al., 2017; Malandrakis et al., 2019). Malandrakis et al. developed AgNP-loaded benzimidazoles (BEN-R), thiophanare methyl (TM), and carbendazim (CARB) to inhibit *Monilia fructicola*. AgNPs successfully inhibited mycelial multiplication insensitive (BEN-S) and resistant isolates. The combination of AgNPs and TM considerably boosted fungal toxic effects in vitro and on apple fruit, regardless of resistance phenotype, compared to the different treatments (BEN-R/S). The positive correlation between AgNPs and TM+AgNPs treatments indicates that the observed additive/synergistic action is likely due to increased AgNP activity/availability (Malandrakis et al., 2020). Al-Otibi and his coworkers used *Malva parviflora L.* to biosynthesize the AgNPs. *M. parviflora* leaf extract and biosynthesized silver nanoparticles effectively inhibit the mycelial growth of *Helminthosporium rostratum, Fusarium solani, Fusarium oxysporum,* and *Alternaria alternata*. The most significant reduction in root length by biosynthesized nanoparticles was found against *H. rostratum* (88.6%). *M. parviflora* leaf extract, on the other hand, was the most effective against *F. solani* (65.3%) (Al-Otibi et al., 2021).

2.5 Carbon nanotube

Carbon nanotubes (CNTs) have been commonly recognized as the most interesting and effective nanocarriers for drug delivery and many potential applications due to their unique physicochemical features. In recent years, many researchers have been drawn to carbon nanotubes (CNTs) as drug delivery carriers. Carbon fullerenes rolled into cylindrical tubes are the third allotropic form of carbon fullerenes. To be integrated into biological systems, CNTs can be chemically modified or functionalized with therapeutically active chemicals, establishing stable covalent connections or supramolecular assemblies based on noncovalent interactions (Mishra et al., 2020; Mishra et al., 2018a,b). Because the exterior wall of pristine CNTs is inert, noncovalent and covalent changes to the CNT surface are required for improved biocompatibility and solubility. In the first approach, long-chain polymers are utilized, and in the second method, direct covalent alterations are used to modify CNT surfaces. Most current chemical methods processess try to create a strong link between CNTs and coupling chemicals. The functionalization of carbon nanotubes can lead to loss of mechanical strength and increased purifying time. CNTs are uniquely designed carriers for carrying modified genes and single-stranded RNAs (Mishra et al., 2021a,b,c). Fungicides are widely used in agriculture worldwide, and their pervasive presence poses a severe threat to the environment due to their hazardous nature. Li et al. developed a simple and effective pretreatment procedure to extract and purify the two target chiral fungicides (MWCNTs) using reversed-dispersive solid-phase extraction (r-DSPE) with multiwalled CNT. The fortification studies were carried out with three different types of MWCNTs (20–30 nm, 10–20 nm, and <8 nm, 10 mg). MWCNTs (10–20 nm, 10 mg) showed the highest purification efficiency for eliminating matrix co-extracts, resulting in the highest recovery values (Li et al., 2020). Garrido et al. designed a sensitive electrochemical sensor using a glassy carbon (GC) electrode modified by a mixture of MWCNT and -cyclodextrin (-CD) embedded in a polyaniline film to detect and determine the fungicide pyrimethanil in pome fruit (apples). The results reveal that the GC electrode modified with -CD/MWCNT conducts electrocatalytic pyrimethanil oxidation efficiently and with good reproducibility, repeatability, and stability (Garrido et al., 2016).

3 Nanoencapsulation techniques

Nanotechnology is being widely evaluated in various fields, including agriculture. In the realm of research, the nanoencapsulation of agrochemicals is becoming an increasingly attractive area. The ultimate advantage of nanoencapsulation is its ability to improve active compounds' stability, bioavailability, and activity prone to leaching, decomposition, and volatilization (Fig. 3) (Aphibanthammakit and Kasemwong, 2021). Nanoencapsulation is one method for producing nanoparticles (NPs) in agriculture and food production, specifically for protecting and controlling the delivery of valuable compounds such as pesticides, nutrients, or bioactive components. There are several methods for nanoencapsulation (Table 2). Which approach to employ is determined by the chemical and physical characteristics of the core and coating materials as well as the type of cell ingredients to be coated (Castro-Mayorga et al., 2020; Suganya and Anuradha, 2017; Yousefi et al., 2012).

3.1 Coacervation

It is a simple approach frequently employed to encapsulate tiny, reactive, or highly volatile compounds into nanoscale delivery devices. This approach comes in two varieties: aqueous and organic. Only lipophilic substances, including EOs and their bioactive components, may

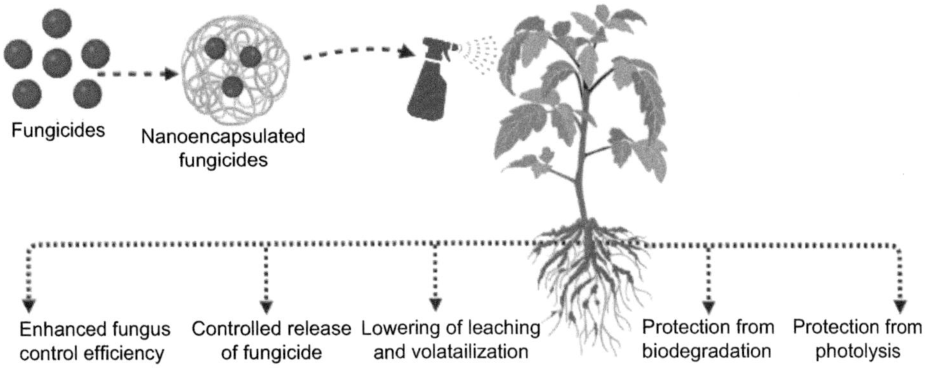

FIG. 3 A schematic diagram of nanoencapsulation in fungicide applications. *Created in Bio Render.com*

TABLE 2 Different nanoencapsulation techniques with their merits and demerits.

Encapsulation techniques	Principle	Merits	Demerits
Coacervation	The electrostatic deposition of a liquid-coating substance around the core material brings on entrapment	Heat-sensitive substances made at room temperature can be encapsulated using this method	• An expensive and complicated method • Toxic chemicals are utilized • Spheres' small size range • Complex coacervates are extremely unstable
Ionic gelation	Within an ionic solution, drops of coating material with dissolved core material are extruded The capsules are formed as a result of ionic contact	Avoid high temperature and pH levels, as well as organic solvents	• Primarily utilized in laboratories • Most capsules are highly porous, which encourages intense bursts of absorption
Emulsion phase Separation	A water-in-oil or oil-in-water emulsion inserts the core component into the polar or polar layer The emulsions are generated with a surfactant	• Emulsions can be used directly or indirectly while they are "wet" • It is possible to utilize polar, nonpolar (apolar), and amphiphilic substances	• It becomes unstable when subjected to external variables, such as heating and drying. • There are limited emulsifiers available for application
Simple extrusion	A mass of molten wall material is driven through a laboratory-sized die or a group of dies with the required cross-section in a desiccant liquid bath	• Any remaining core is washed outside • The wall material encircles the substance	• The capsule needs to be dried after being removed from the liquid bath

Continued

TABLE 2 Different nanoencapsulation techniques with their merits and demerits—cont'd

Encapsulation techniques	Principle	Merits	Demerits
	When the coating material comes into contact with liquids, it hardens and traps the active ingredients	• This method of low-temperature entrapment	• Capsules in the highly-dense carrier material are difficult to obtain
Spray drying	After dispersing the core material in an entrapment substance and atomizing and spraying the mixture into a chamber using a hot air desiccant, the core material is encapsulated in an entrapment substance	• Process cost is low • A diverse range of coating materials is available, as is a high level of encapsulation efficiency • The finished product has a high level of stability • Possibility of large-scale continuous production	• Highly temperature-sensitive substances can be degraded • It is not easy to control particle size • Small batch yields are moderate
Spray chilling	The only difference between air desiccant and spray drying is that air desiccant is cold	• Compounds that are sensitive to temperature can be encapsulated	• Controlling particle size is difficult • Small batch yields are moderate • Exceptional handling and storage conditions can be required
Lyophilization/ freeze drying	For encapsulation, a core and coating material are dissolved in an emulsion solution and then lyophilized	This method is helpful for encapsulating hazardous thermosensitive chemicals in aqueous solutions	• Processing time is long • Process costs are high • Expensive capsule storage and transportation
Fluidized bed coating	In this process, the coating material is sprayed through a nozzle into a fluidized bed of core material in a heated environment	• A low-cost technique that allows for customized capsule size distribution and product porosity	• Compounds degrade rapidly under high temperatures
Emulsion polymerization	The polymerization solution dissolves the core ingredient. Capsules can be made by polymerizing the monomers in an aqueous solution	• It is possible to create micro-nanocapsules with a narrow size distribution	• Capsule development is challenging to manage
Liposome entrapment	Phospholipids spontaneously disperse in an aqueous phase to form a liposome Core substance entrapment within a liposome	• Encapsulation can be done using either aqueous or lipid-soluble substances • Suitable for applications requiring a high level of water activity • Effective controlled delivery	• Typically applied on a laboratory scale

be encapsulated via aqueous-phase coacervation. Contrarily, the organic phase coacervation method involves using organic solvents while yet allowing the encapsulation of hydrophilic molecules. There are two types of aqueous-phase coacervation: simple and complicated. Complex coacervation is caused by the attraction forces of two oppositely charged colloids, often a protein (like whey protein or gelatin) plus polysaccharides (like alginate, gum arabic, and carrageenan to form a matrix around the active core compounds). High encapsulation efficiency, the requirement for the fewest encapsulating polymers, stability at extremely high temperatures, and the capacity to release components over an extended period are only a few advantages of the nano-encapsulated formulation created via complicated coacervation (de Souza Simoes et al., 2017; Bastos et al., 2020). Optimizing these conditions is essential to the process success because many factors affect it, including pH, ion concentration, the ratio of encapsulating wall materials to bioactive components, molecular weight, the solution's ionic strength, the chain flexibility of the polyelectrolytes, and charge density (Chaudhari et al., 2021). Oregano essential oil nanocapsules were produced by Shi et al. using a complex coacervation technique (OEO-NPs). The findings demonstrated that compared to the control, 2%, and 8% OEO-NPs treatments, % OEO-NPs exhibited a better fungistatic action and a lower decay rate. A slow-release fungistatic agent based on plant essential oils was developed in response to these discoveries (Shi et al., 2021). A coacervation method to encapsulate lambda-cyhalothrin. Gas chromatography and an FID detector showed 96% encapsulation effectiveness. (GC-FID). The release research results showed that, respectively, after 15, 30, and 180 min, 49%, 63%, and 89% of the encapsulated materials were released from their containers. The current research proposes a pesticide that can be used in the context of public health in a simple and cost-effective manner (Saghafi et al., 2021).

3.2 Ionotropic gelation

Ionotropic gelation (IG) uses electrostatic interactions between two ionic species to generate nano- and microparticles. A polymer must be present in at least one of the different species. When a drug or bioactive molecule is introduced to the procedure, it becomes trapped within the polymeric chains and is contained inside the nanoparticle or microparticle structure (Koukaras et al., 2012). This formulation enables regulated drug release and other advantages, including co-encapsulation of molecules, site-specific functionalization of the particles, and longer bioactivity times. Calvo et al. (1997) were the first to describe the IG procedure for nano-encapsulating proteins. The authors investigated the synthesis of chitosan (CS) nanoparticles in sodium tripolyphosphate (TPP). This reaction uses the cationic nature of CS in the presence of diluted acids and the polyanionic nature of TPP. Positively charged CS amino groups crosslink with negatively charged phosphate groups to create hydrogels. The nanoparticles develop spontaneously when the mixture is constantly stirred in a precise range of concentrations for each reagent (Pedroso-Santana and Fleitas-Salazar, 2020; Calvo et al., 1997). CS-NPs can be utilized as protectants (nano-pesticides) and carriers (fungicides, insecticides, herbicides, plant hormones, elicitors, and nucleic acids) to manage plant diseases. The use of CS-NPs as a delivery mechanism is relevant since it can load and safeguard the ingredients surrounding the environment before releasing them to the target site for absorption by the plants. Additionally, since NPs have the fundamental qualities of small size and large contact area, CS-NPs or CS-NP-loaded active substances can be easily penetrated and permeated into phytopathogen membranes or enhanced plant tissue uptake in increased control or defense response activity. As a result, these NPs can be utilized directly and indirectly to manage plant diseases (Hoang et al., 2022). CS-NPs using an emulsion-ionic gelation method and the essential oil extracted from *Pistacia atlantica subsp. Atlantica* (PAHEO) were developed under both in vitro and in vivo conditions, and the encapsulated PAHEO demonstrated potent antifungal activity against *Botrytis cinerea*. It also decreased the occurrence and severity of grey mold on strawberries during storage (Hesami et al., 2021). Maluin et al. fabricated CS-NPs fungicide by ion gelation method to target *Ganoderma boninense*. The in vitro fungicide release at pH 5.5 demonstrated that the fungicide from the NPs was released sustainably, with a release time of up to 86h. Smaller particle size results in a lower half-maximum effective concentration (EC50), indicating more vital, more potent antifungal activity against *G. boninense* in vitro (Maluin et al., 2019a,b).

3.3 Incorporation in layer-by-layer deposited nanolaminates

The layer-by-layer (LBL) method has seen widespread use in producing nanostructured thin films for various applications, including, chemical sensors, flame resistance, superhydrophobic drug delivery conductive films, antireflective coatings, and flame resistance, as well as antibacterial and biocompatible coatings. Because the films are made by continuously depositing solutions or dispersions of opposing charge polymers or nanoparticles onto a substrate, this method is highly user-friendly and flexible in its application. As a result of this, the films may be used for a wide variety of purposes. Nanocomposite films that are created utilizing the LBL approach have the potential to have extraordinarily alluring mechanical characteristics and shallow gas permeability values. This is possible when high-aspect-ratio nanofillers are used.

The traditional LBL method involves depositing polymers or nanoparticles onto the substrate; the time required to produce each pair of oppositely charged monolayers ranges from 5 to 20min depending on diffusion rates and adsorption of polyelectrolyte molecules and colloidal particles within the substrate. To expedite the LBL deposition procedure, Schlenoff et al. developed the spray-LBL approach in 2000. The monolayer forms following drainage and drying of polyelectrolyte solutions sprayed onto a vertical substrate; hence, the time for monolayer development is independent of molecular species diffusion. Spray-LBL has been shown to be capable of accelerating deposition while simultaneously producing homogeneous coatings with decreased surface roughness. Another LBL strategy with a short deposition time is the spin-LBL approach. The solutions or suspensions are deposited onto a spin-coater-attached substrate. The rotation speed generates a high centrifugal force and airflow at the surface, promoting rapid thinning and drying of the liquid and high layer uniformity. Because of the quick drying, the polymer chains cannot interpenetrate as they do in dipping-LBL, resulting in a highly stratified internal multilayer structure (Larocca et al., 2018; Schlenoff et al., 2000). Jin and his team fabricated iodopropynyl butylcarbamate (IPBC) fungicide using enlarged halloysite nanotubes (HNTs) with LBL assembly of polyelectrolyte multilayers. Prepared nanocomposites exhibited significant antifungal performance against three mold fungi (*Aspergillus niger*, *Trichoderma viride*, and *Penicillium citrinum*) and one stain fungus (*Botryodiplodia theobromae*) (Jin et al., 2019).

3.4 Emulsification–solvent evaporation

Evaporation of emulsions has long been used to make polymeric NPs from ready-to-use polymers. The procedure involves dissolving a polymer organic solution in water and evaporating the organic solvent. The polymer must first be dissolved in a suitable solvent (e.g., ethyl acetate, chloroform, or methylene chloride). The organic phase is poured into the continuous phase (aqueous phase), which includes a dissolved surfactant to give the emulsion stability. Emulsification is accomplished under strong shear force to minimize the size of the emulsion droplet. The ultimate particle size will be determined mainly by this method. Following emulsification, the system evaporates the organic solvent under a vacuum, resulting in polymer precipitation and NP production (Wang et al., 2016). Saleh et al. used a double emulsion/solvent evaporation method to develop fluconazole (FLZ)-loaded polymeric NPs. The results showed that loading FLZ in the alginate/chitosan-based polymeric matrix of NP4 significantly increased its antifungal activity against *C. albicans* (Saleh et al., 2022). Itraconazole (ITZ)-loaded Poly-(D,L-lactic-co-glycolic acid) (PLGA) nanoparticles (PLGA-NPs) to improve the antifungal activity stabilized by D-tocopherol polyethylene-glycol succinate-1000 (TPGS) were fabricated via nanoprecipitation and single emulsion solvent evaporation. The antifungal activity of ITZ against *Candida albicans* was preserved by drug-loaded PLGA-NPs (Alhowyan et al., 2019).

3.5 Polymer-based nanoencapsulation materials

Polymers and polymeric materials have a wide range of applications. An extensive study has been devoted to producing nanosized controlled-release medication formulations based on biodegradable polymers. The use of polymeric nanomaterials for pesticide delivery is a relatively new approach. Polymer nanocomposites (PNCs) are polymers with nanoparticles or nanofillers dispersed within the polymer matrix, and the active ingredients are typically encapsulated with polymer. Natural-source polymers are environmentally safe and biodegradable, produce no degradation byproducts, and are reasonably cheap. These characteristics effectively encapsulate active compounds' materials (Nuruzzaman et al., 2016).

3.5.1 Nanocapsules

A liquid/solid core is encapsulated in a polymeric shell in a polymeric nanocapsule. Polymeric nanocapsules have been more popular in the field of drug delivery applications in recent years due to the core–shell microstructure that they possess. When compared to polymeric nanospheres, the solid/oil core of nanocapsules has the potential to significantly increase drug-loading efficiency while simultaneously reducing the nanoparticle polymeric matrix content. In addition, the polymeric shell can insulate the encapsulated payload from the environment of the tissue, preventing the payload from being degraded or bursting out prematurely due to enzymes, temperature, or pH, among other things. Additionally, the polymeric shell can be functionalized by novel compounds interacting with specific proteins, enabling targeted drug delivery (Trindade et al., 2018). When wood products are exposed to dampness, they are highly prone to deterioration by degrading wood fungus. There is a significant interest in developing wood preservatives from renewable resources to prevent or minimize degradation. Andeme et al. constructed water-dispersible, double-shell lignin nanocapsules containing propiconazole fungicide. Regarding antifungal resistance, the novel preservative system performed better than its components. The increased antifungal activity of the double-shell lignin nanocapsule–propiconazole combination may have been brought on by the double-shell lignin nanocapsule's 74.94% radical quenching activity against DPPH (Andeme Ela et al., 2021).

Kumar et al. developed chitosan-carrageenan nanocomposites containing the fungicide mancozeb (CSCRG-M). In an in vitro study, nano CSCRG-1.5 suppressed *Alternaria solani* 83.1% and outperformed mancozeb (84.6%). At 1.0 and 1.5 ppm, the nanoformulation inhibited *Sclerotinia sclerotiorum* completely. In pot settings, mancozeb-loaded nanoparticles had a relatively greater in vivo disease control efficacy (79.4%) against *A. solani* in pathogenized plants than commercial fungicides (76.1%) (Kumar et al., 2021). Cui et al. used a water-oil-water double-emulsion approach to create a nanocapsule system to prevent and control rice sheath blight. The dual-functionalized pesticide nanocapsules demonstrated significantly improved foliar spread in comparison to more traditional formulations. This is significant for lowering the amount of pesticide that is lost on the leaves of the plant. The percent median effect concentration of the dual-functionalized pesticide nanocapsules against *Rhizoctonia solani* and the commercial formulation were 0.0082 and 0.0350 μg/mL, and the synergistic ratio was 2.088 and 0.917, respectively (Cui et al., 2020).

3.5.2 *Nanospheres*

Nanospheres are particles that range in size from 10 to 200 nm. Nanospheres protect drugs against enzymatic and chemical degradation, which can be amorphous or crystalline. The hydrophobic surfaces of these particles are susceptible to reticuloendothelial system opsonization and clearance. Biodegradable nanospheres include albumin nanospheres, modified starch nanospheres, gelatine nanospheres, polypropylene dextran, and polylactic acid nanospheres. Immune nanospheres and magnetic nanospheres are two different forms of nanospheres. Combining these two nanospheres can result in immuno-magnetic nanospheres, which could considerably improve targeting (Pachpute et al., 2019).

The potential of *Euphorbia wallichii* shoot extract gold nanospheres (AuNPs) was produced. AuNPs demonstrated significantly higher fungicidal efficacy against the investigated fungal strains than the shoot extract alone. At 30, 60, and 90 min of reaction time, the IC50 for 2,2-diphenyl-1-picrylhydrazyl scavenging was 31.52, 18.29, and 15.32 μg/mL, respectively (Ullah et al., 2021). An accurate and fast method for spotting fungicide residue in agricultural products must be developed to protect human health. An accurate and fast method for spotting fungicide residue in agricultural products must be designed to protect human health. AuNPs were synthesized by Nguyen et al. using a seed-mediated technique. The outcomes demonstrated that 53-nm-sized AuNPs were successfully produced, and their resonance peak was at 560 nm (Nguyen Thi Nhat et al., 2020). Ahuja et al. synthesized in situ sulfide nanospheres (MnS-NSs) using a sonochemical method. The effect of nanoscale manganese dioxide on the antifungal potential of rice mycopathogens *Drechslera oryzae, Curvularia lunata*, and *Fusarium verticillioides* was investigated. With a minimum ED50 value of 14 μg/ml, *C. lunata* showed the most significant percent inhibition (98.1.4), followed by *D. oryzae* and *F. verticillioides* with ED50 values of 17 and 18 μg/ml, respectively (Ahuja et al., 2019).

3.5.3 *Nanogels*

"Nanogel" (NGS) refers to nanosized particles formed physically or chemically by a crosslinked polymer. Crosslinked bifunctional networks containing a polyion and a nonionic polymer were initially used to transport polynucleotides (Ghaywat et al., 2021). Nanogels are standard formulations with diameters ranging from 20 to 200 nm (Ansari and Karimi, 2017; Sivaram et al., 2015). To ensure the three-dimensional (3D) structure, the volume percentage can be adjusted by adjusting the solvent quality and branching. Several features of nanogels contribute to their success as a delivery mechanism. They have excellent solubilization capability and low viscosity and can resist rigorous sterilization procedures (Saraogi et al., 2022; Anooj et al., 2021; Kaur et al., 2019; Vashist et al., 2018). Horvat et al. developed poly(glycidol)-based NGS using surfactant-free inverse nanoprecipitation. Breathing in lyophilized NGs in the drug solution, particles are loaded with the antifungal drug itraconazole. The results demonstrated that nanoprecipitation NGs provided efficient drug administration, improved fungicidal action against *Aspergillus Fumigatus*, and minimized adverse effects compared to the drug itself (Horvat et al., 2021). To prevent the fungus *Microsporum canis*, De Carvalho et al. developed a cinnamic acid–grafted chitosan nanogel that included the essential oils of *Syzygium aromaticum* and *Cinnamomum ssp*. Cinnamic acid was grafted onto chitosan nanogel, demonstrating antifungal action against *M. canis* in vitro, reducing up to 53.96% of its mycelial development (de Carvalho et al., 2021). Akbari and his team investigated *Gymnema sylvestre* essential oils (G. EOs) as they were encapsulated using chitosan and myristic acid, producing a nanogel with improved antifungal activity and oil stability against *C. albicans* strain. According to the findings, 4.68 μg/ml of G. OLNPs were needed to suppress biofilm development by 50%, while 18.07 μg/ml was required to do the same for 90% biofilm growth. Additionally, G. OLNP's cell absorption was significantly higher than free G. EOs. Akbari and his team investigated *Gymnema sylvestre* essential oils (G. EOs), as they were encapsulated using chitosan and myristic acid, producing a nanogel with improved antifungal activity and oil stability against *C. albicans* strain. According to the findings, 4.68 μg/ml of G. OLNPs was needed to suppress biofilm development by 50%, while 18.07 μg/ml was required to do the same for 90% biofilm growth. Additionally, G. OLNP's cell absorption was significantly higher than free G. EOs (Akbari et al., 2019).

3.6 Lipid-based nanoencapsulation techniques

3.6.1 *Nanoemulsion technique*

The thermodynamic instability between two immiscible liquids leads to the formation of a nanoemulsion, a colloidal dispersion of the two liquids. One of the liquids in a nanoemulsion produces the dispersed phase, whereas the other liquid produces the dispersing medium. A protective coating of emulsifier molecules covers each droplet in a nanoemulsion, and its diameter ranges from 10 to 200 nm. The nanoemulsions have droplet dimensions like the microemulsions ranging from <200 and, in some cases, <100 nm (Malode et al., 2022). De Melo et al. developed essential oils (EOs) loaded with nanoemulsions with *Lippia gracilis* and *Lippia sidoides* to test their antifungal effectiveness against *Lasiodiplodia theobromae*. Except for Nano-104, which showed fungicidal activity at a concentration of 10 mL^{-1}, the nanoemulsions could only slow the mycelial growth of the fungus. It was possible to observe that the Eos-loaded nanoemulsion showed more significant toxicity against *L. theobromae* (de Melo et al., 2022). Rafiee et al. produced denak EO–loaded nanoemulsion for *Penicillium digitatum*. Higher EO concentrations decreased mycelial development. The concentration improved EO's antifungal activity. EO nanoemulsion's MIC is $0.5\,mL^{-1}$. These results suggest that ultrasonication can cause denak EO. Nanoemulsion can cure *P. digitatum* and other postharvest illnesses. The concentration improved EO's antifungal activity. These findings show that ultrasonication can be utilized to generate a denak EO nanoemulsion that can cure *P. digitatum* and presumably other infections (Rafiee et al., 2022).

3.6.2 Solid lipid nanoparticle incorporation

SLNs were developed in 1991 to improve biocompatibility, storage stability, and drug degradation prevention. SLNs are colloidal carriers (50–1000 nm) made of solid lipids (high melting fat matrix) that overcome the shortcomings of polymeric nanoparticles and liposomes (e.g., lack of a suitable large-scale production method, phospholipid degradation, polymer degradation, and cytotoxicity; inferior stability, drug leakage, and fusion; sterilization issues; and high production cost). Several factors set SLNs apart from conventional colloidal carriers, including low toxicity, a sizable surface area, delayed drug release, better cellular absorption, and improved drug solubility and bioavailability. The formulation's matrix type and drug placement determine the drug release from SLNs. Hydrophilic and lipophilic bioactive substances can be added to SLNs constructed of biodegradable and biocompatible materials, making them suitable for precise and regulated drug administration. SLNs typically have a monolayer of phospholipids coating the hydrophobic solid core, where the medicine is generally disseminated or dissolved (Mishra et al., 2018a,b). SLNs typically have a monolayer of phospholipids coating the hydrophobic solid core, where the drug is usually given or dissolved (Mishra et al., 2018a,b). *Eucalyptus globulus* EOs (EGEO)-loaded SLNs were formulated to characterize the bactericidal and fungicide effect. The nanoparticles considerably decreased the minimum fungicidal concentration (MFC), minimum inhibitory concentration (MIC), and minimum bactericidal concentration (MBC) of EGEO. The cytotoxic action of EGEO was significantly boosted after being encapsulated using nano-lipo. As a consequence, utilizing nanocapsules may prove to be an effective method for improving both the biological efficacy and the stability of EOs (Azadmanesh et al., 2021). Sharma et al. used stearic acid and Pluronic F-68 to synthesize luliconazole-loaded SLNs. With a minimum inhibitory concentration of 6.25 μg/mL against *Candida albicans* (MTCC 227) and 12.5 μg/mL against *Aspergillus niger* (MTCC 8189), the SLNs demonstrated intense antifungal activity (Sharma et al., 2021).

4 Challenges

Despite the incorporation of nanotechnology into agricultural practices, there are a few issues that need to be resolved or managed. Antifungal management involves nanohybrid materials. This category includes elements such as silver, gold, copper, iron, graphene, and silica; polymers such as chitosan, PVC, and PLGA; and other organic molecules that are utilized in the production of composites or nano-hybrids. Because they need the utilization of pricey chemicals, reagents, and physical energy, the processes that result in the creation of nanohybrids are quite expensive. As a direct consequence of this, the nanohybrids that were produced may prove to be extremely powerful against phytopathogens. However, when these nanohybrids are used in the field, there is a possibility that they will migrate in a manner that was not intended for them. It is possible for them to penetrate the plant system or accumulate in the vegetative portions of the plant. Before developing any nanoformulation for antibacterial or fungicidal purposes, it is necessary to have an understanding of the effect that nanoparticles have on agricultural plants (Alghuthaymi et al., 2021).

Many researchers are of the opinion that nanoparticles might have a detrimental effect on the expansion and development of plants. Researchers Dimkpa et al. (2012) found that CuONPs might have an impact on the growth of wheat roots and shoots. In addition, the amount of chlorophyll present and the enzyme activity of peroxidase and catalase were reduced in plants that had been treated with CuONPs. In a similar manner, AgNPs at concentrations higher than 10 mg/L have the potential to interfere with the metabolic processes and cell defense systems of wheat (Vannini et al., 2014). TiO_2 treatment of *Oryza sativa* lowered biomass content, altered metabolite concentrations, and modified respiration pathways (Wu et al., 2017).

Similarly, CNT causes DNA damage and causes both the root and the shoot of a rice plant to shrink (Shen et al., 2010). The presence of nanoparticles in soil may not only have a detrimental effect on the plant life that grows there, but it may also have an effect on the bacteria that reside in the soil. All metallic/metal oxide nanoparticles suppress microbial abundance to varied degrees. Both CNTs and ZnONPs had an effect on the bacteria *Pseudomonas putida*, which is an essential component in the process of nitrogen recycling (Rai et al., 2018). Beneficial soil fungi were also speculated to have suffered damage at the hands of the same ZnONPs. Soil microbial biomass carbon (SMBC) was found to be extremely low in areas that included soil microorganisms that had been exposed to carbon nanotubes, AgNPs, NiNPs, and other nanoparticles. Ag-SiO_2 core–shell nanocomposites, which create reactive oxygen species, were found to limit the radial growth of a few hazardous plant fungi (ROS).

Another substantial barrier is the capacity of nanoparticles to avoid being ingested by organisms and instead accumulate at higher trophic levels. In tomato plants exposed to designed metal oxides, K concentrations increased while Mg, P, and S concentrations dropped, according to Vittori Antisari et al. (2015). It is still unclear what will happen to people when they eat such fruit. Because of this, there is still a big problem with using nanoparticles in agricultural applications, especially in wide fields.

5 Future perspectives

New nanomaterials and nanotechnology are rapidly gaining popularity in the agricultural and food industries. There are more commercially available nano-formulated pesticides and more publications on the worldwide market. Because of factors such as changed solubility, adhesion to surfaces, permeability across biological membranes, targeting, and controlled release, such products require a greater level of stability as well as efficiency and potency. These modified qualities allow for the use of lower pesticide dosages. Food items are now also included in packaging materials made of "smart" nanoparticles.

However, because there is widespread hostility to the use of pesticides, scientists are currently striving to develop biopesticides, which are compounds that are derived from natural sources and possess significant antipest qualities. These naturally occurring compounds have several limitations, such as a short action window and low stability, for example. As a result, nanotechnologies have been utilized in this sector, whether for the production of stable nanoformulations of biopesticides with effects that last for an extended period of time or for the "green" production of active inorganic nanomaterials with antibacterial, antifungal, and antiparasitic properties.

Only chemically based nanocomposite synthesis has been documented so far. To produce nanoscale silica, carbon, cellulose, chitosan polymers, and graphene, scientists are currently working to create nanocomposite using a green chemistry method that uses agri-waste products such as wheat whiskers, banana or orange peels, cotton or corn stalks, coconut or almond shells, corn silk, and rice husks. This method was developed in an effort to reduce the amount of waste produced by agriculture. Inorganic metal and metal oxide nanoparticles that have been manufactured in an environmentally friendly manner serve a variety of purposes in agriculture, including the regulation and mitigation of plant diseases caused by a wide variety of fungal phytopathogens (Saratale et al., 2018a,b,c). Using biological synthesis strategies can help reduce the amount of chemicals that are harmful to the environment that are required for the commercial manufacture of nanomaterials and their composites. At the same time, these strategies can improve the cost, amount of time, and energy requirements of the most common physical and chemical synthesis methods (Saratale et al., 2018a,b,c). The capability of biodegradable polymers to exert in plant antifungal activity and their ease of translocation through plant tissues are the two benefits that stand out the most when considering the use of biodegradable polymers in the synthesis of nanocomposites through the conjugation of metal and metal oxide nanoparticles. In light of the biocompatibility and environmental friendliness of biodegradable polymers, it is imperative that future research work encourage their use.

These novel organic nanomaterials, including nano-encapsulated biocidal essential oils, enzymes, or the reducing effects of plant extracts for synthesizing metal or metal oxide NPs, have a promising future. Farmers seek them out because they are very effective even in small doses, and the general population accepts them broadly since they are naturally active substances. Regulatory bodies take a proactive stance when approving applications. On the other hand, using these nanobiopesticides calls for particular handling and expertise. Furthermore, they cannot be employed in agricultural regions of the so-called third world due to their high cost.

6 Conclusion

In the field of agriculture, nanotechnology has been put to use to increase agricultural productivity while simultaneously improving crop quality by means of improvements in farming procedures, as indicated in the diagram. The use of engineered nanomaterials in sustainable agriculture has fundamentally altered the nature of global agriculture in terms of its capacity for innovation, rapid development, and large-scale production to meet projected levels of global food demand. As the accompanying image demonstrates, nanotechnology has been applied in agriculture with the goal of raising agricultural yields while simultaneously raising crop quality through the implementation of better farming techniques. In terms of innovation, quick expansion, and scale to satisfy the growing need for food in the future, the agricultural business has been altered by the use of engineered nanomaterials and their application in environmentally friendly agriculture. The prevention of environmental pollution is at the forefront of sustainable agriculture, and the use of nanomaterials ensures greater control as well as the conservation of plant production inputs. It is possible that nanotechnology will result in a shift in the current management strategy for plant pathogenic fungi. The creation of nanofungicides has the potential to increase the efficacy and bioavailability of fungicides, lower their toxicity, lengthen the shelf life of active activity, target delivery of actives, raise the solubility of fungicides with a low water solubility, and release precisely. Despite this, there are substantial holes in our knowledge regarding the ecotoxicity, safe limit, and absorption capacity of a wide variety of nanomaterials. As a result, further research is required to comprehend the behavior, destination, and interactions of altered agricultural inputs with biomacromolecules in biological systems and ecosystems.

References

Ahuja, R., Sidhu, A., Bala, A., 2019. Synthesis and evaluation of iron (ii) sulfide aqua nanoparticles (FeS-NPs) against Fusarium verticillioides causing sheath rot and seed discoloration of rice. Eur. J. Plant Pathol. 155 (1), 163–171.

Akbari, S., Bayat, M., Roudbarmohammadi, S., Hashemi, J., 2019. Chitosan nanogel design on Gymnema sylvestre essential oils to inhibit growth of Candida albicans biofilm and investigation of gene expression ALS1, ALS3. Period. Polytech. Chem. Eng. 63 (4), 569–581.

Alghuthaymi, M.A., Kalia, A., Bhardwaj, K., Bhardwaj, P., Abd-Elsalam, K.A., Valis, M., Kuca, K., 2021. Nanohybrid antifungals for control of plant diseases: current status and future perspectives. J. Fungi 7 (1), 48.

Alhowyan, A.A., Altamimi, M.A., Kalam, M.A., Khan, A.A., Badran, M., Binkhathlan, Z., Alkholief, M., Alshamsan, A., 2019. Antifungal efficacy of Itraconazole loaded PLGA-nanoparticles stabilized by vitamin-E TPGS: In vitro and ex vivo studies. J. Microbiol. Methods 161, 87–95.

Ali, M., Haroon, U., Khizar, M., Chaudhary, H.J., Munis, M.F.H., 2020. Facile single step preparations of phyto-nanoparticles of iron in Calotropis procera leaf extract to evaluate their antifungal potential against Alternaria alternata. Curr. Plant Biol. 23, 100157.

Al-Otibi, F., Perveen, K., Al-Saif, N.A., Alharbi, R.I., Bokhari, N.A., Albasher, G., Al-Otaibi, R.M., Al-Mosa, M.A., 2021. Biosynthesis of silver nanoparticles using Malva parviflora and their antifungal activity. Saudi J. Biol. Sci. 28 (4), 2229–2235.

Andeme Ela, R.C., Chipkar, S.H., Bal, T.L., Xie, X., Ong, R.G., 2021. Lignin–propiconazole nanocapsules are an effective bio-based wood preservative. ACS Sustain. Chem. Eng. 9 (7), 2684–2692.

Anooj, E.S., Charumathy, M., Sharma, V., Vibala, B.V., Gopukumar, S.T., Jainab, S.B., Vallinayagam, S., 2021. Nanogels: an overview of properties, biomedical applications, future research trends and developments. J. Mol. Struct. 1239, 130446.

Ansari, S., Karimi, M., 2017. Novel developments and trends of analytical methods for drug analysis in biological and environmental samples by molecularly imprinted polymers. TrAC - Trends Anal. Chem. 89, 146–162.

Aphibanthammakit, C., Kasemwong, K., 2021. Nanoencapsulation in agricultural applications. In: Handbook of Nanotechnology Applications. Elsevier, pp. 359–382.

Aswin Jeno, J.G., Maria Packiam, S., Nakkeeran, E., 2021. Contact in-vivo larvicidal toxicity and histological studies of Indian herb essential oils loaded niosomes against Aedes aegypti and Culex quinquefasciatus (Diptera: Culicidae). Int. J. Trop. Insect Sci., 1–15.

Atiq, M., Naeem, I., Sahi, S.T., Rajput, N.A., Haider, E., Usman, M., Shahbaz, H., Fatima, K., Arif, E., Qayyum, A., 2020. Nanoparticles: A safe way towards fungal diseases. Arch. Phytopathol. Pflanzenschutz. 53 (17–18), 781–792.

Azadmanesh, R., Tatari, M., Asgharzade, A., Taghizadeh, S.F., Shakeri, A., 2021. GC/MS profiling and biological traits of Eucalyptus globulus L. essential oil exposed to solid lipid nanoparticle (SLN). J. Essent. Oil-Bear. Plants 24 (4), 863–878.

Bastos, L.P.H., Vicente, J., dos Santos, C.H.C., de Carvalho, M.G., Garcia-Rojas, E.E., 2020. Encapsulation of black pepper (*Piper nigrum* L.) essential oil with gelatin and sodium alginate by complex coacervation. Food Hydrocoll. 102, 105605.

Beckers, S.J., Wetherbee, L., Fischer, J., Wurm, F.R., 2020. Fungicide-loaded and biodegradable xylan-based nanocarriers. Biopolymers 111 (12), e23413.

Burduşel, A.C., Gherasim, O., Grumezescu, A.M., Mogoantă, L., Ficai, A., Andronescu, E., 2018. Biomedical applications of silver nanoparticles: an up-to-date overview. Nanomater. 8 (9), 681.

Caldeirão, A.C.M., Araujo, H.C., Tomasella, C.M., Sampaio, C., dos Santos Oliveira, M.J., Ramage, G., Pessan, J.P., Monteiro, D.R., 2021. Effects of antifungal carriers based on chitosan-coated iron oxide nanoparticles on microcosm biofilms. Antibiotics 10 (5), 588.

Calvo, P., Remunan-Lopez, C., Vila-Jato, J.L., Alonso, M.J., 1997. Novel hydrophilic chitosan-polyethylene oxide nanoparticles as protein carriers. J. Appl. Polym. Sci. 63 (1), 125–132.

Castro-Mayorga, J.L., Cabrera-Villamizar, L., Balcucho-Escalante, J., Fabra, M.J., López-Rubio, A., 2020. Applications of nanotechnology in agry-food productions. In: Nanotoxicity. Elsevier, pp. 319–340.

Chaudhari, A.K., Singh, V.K., Das, S., Dubey, N.K., 2021. Nanoencapsulation of essential oils and their bioactive constituents: a novel strategy to control mycotoxin contamination in food system. Food Chem. Toxicol. 149, 112019.

Cui, J., Sun, C., Wang, A., Wang, Y., Zhu, H., Shen, Y., Li, N., Zhao, X., Cui, B., Wang, C., Gao, F., 2020. Dual-functionalized pesticide nanocapsule delivery system with improved spreading behavior and enhanced bioactivity. Nanomater. 10 (2), 220.

Das, S.K., Das, A.R., Guha, A.K., 2009. Gold nanoparticles: microbial synthesis and application in water hygiene management. Langmuir 25 (14), 8192–8199.

de Carvalho, F.G., Magalhaes, T.C., Teixeira, N.M., Gondim, B.L.C., Carlo, H.L., Dos Santos, R.L., de Oliveira, A.R., Denadai, Â.M.L., 2019. Synthesis and characterization of TPP/chitosan nanoparticles: Colloidal mechanism of reaction and antifungal effect on *C. albicans* biofilm formation. Mater. Sci. Eng. C 104, 109885.

de Carvalho, S.Y.B., Almeida, R.R., Pinto, N.A.R., de Mayrinck, C., Vieira, S.S., Haddad, J.F., Leitao, A.A., Guimaraes, L.G.D.L., 2021. Encapsulation of essential oils using cinnamic acid grafted chitosan nanogel: preparation, characterization and antifungal activity. Int. J. Biol. Macromol. 166, 902–912.

de Melo, J.O., Blank, A.F., de Souza Nunes, R., Alves, P.B., de Fátima Arrigoni-Blank, M., Gagliardi, P.R., do Nascimento-Júnior, A.F., Sampaio, T.S., Lima, A.D., de Castro Nizio, D.A., 2022. Essential oils of Lippia gracilis and Lippia sidoides chemotypes and their major compounds carvacrol and thymol: nanoemulsions and antifungal activity against Lasiodiplodia theobromae. Res., Soc. Dev. 11 (3). e36511326715.

de Souza Simoes, L., Madalena, D.A., Pinheiro, A.C., Teixeira, J.A., Vicente, A.A., Ramos, O.L., 2017. Micro-and nano bio-based delivery systems for food applications: in vitro behavior. Adv. Colloid. Interfac 243, 23–45.

Deka, B., Babu, A., Baruah, C., Barthakur, M., 2021. Nanopesticides: a systematic review of their prospects with special reference to tea pest management. Front. Nutr. 8.

Dimkpa, C.O., McLean, J.E., Latta, D.E., Manangón, E., Britt, D.W., Johnson, W.P., Boyanov, M.I., Anderson, A.J., 2012. CuO and ZnO nanoparticles: phytotoxicity, metal speciation, and induction of oxidative stress in sand-grown wheat. J. Nanopart. Res. 14 (9), 1–15.

Faisal, N., Kumar, K., 2017. Polymer and metal nanocomposites in biomedical applications. Biointerface Res. Appl. Chem. 7 (6), 2286–2294.

FAOSTAT, 2019. https://www.fao.org/faostat/en/#data/RP (Accessed on 01.04.2022).

Flood, J., 2010. The importance of plant health to food security. Food Security 2 (3), 215–231.

Gao, Y., Liu, Y., Qin, X., Guo, Z., Li, D., Li, C., Wan, H., Zhu, F., Li, J., Zhang, Z., He, S., 2021. Dual stimuli-responsive fungicide carrier based on hollow mesoporous silica/hydroxypropyl cellulose hybrid nanoparticles. J. Hazard. Mater. 414, 125513.

García-Díaz, M., Patiño, B., Vázquez, C., Gil-Serna, J., 2019. A novel niosome-encapsulated essential oil formulation to prevent Aspergillus flavus growth and aflatoxin contamination of maize grains during storage. Toxins 11 (11), 646.

Garrido, J.M.P.J., Rahemi, V., Borges, F., Brett, C.M.A., Garrido, E.M.P.J., 2016. Carbon nanotube β-cyclodextrin modified electrode as enhanced sensing platform for the determination of fungicide pyrimethanil. Food Control 60, 7–11.

Ghaywat, S.D., Mate, P.S., Parsutkar, Y.M., Chandimeshram, A.D., Umekar, M.J., 2021. Overview of nanogel and its applications. GSC biol. pharm. sci. 16 (1), 040–061.

Global Nanotechnology Market Report, 2021-2026 - Market Opportunities with Increasing Use of Nanotechnology in Building Materials, Research and Markets, Dublin, Nov. 30, 2021. https://www.prnewswire.com/news-releases/global-nanotechnology-market-report-2021-2026- - -market-opportunities-with-increasing-use-of-nanotechnology-in-building-materials-301433710.html (Accessed on 01 April 2022).

Golden, G., Quinn, E., Shaaya, E., Kostyukovsky, M., Poverenov, E., 2018. Coarse and nano emulsions for effective delivery of the natural pest control agent pulegone for stored grain protection. Pest Manag. Sci. 74 (4), 820–827.

Hesami, G., Darvishi, S., Zarei, M., Hadidi, M., 2021. Fabrication of chitosan nanoparticles incorporated with Pistacia atlantica subsp. kurdica hulls' essential oil as a potential antifungal preservative against strawberry grey mould. Int. J. Food Sci. 56 (9), 4215–4223.

Hoang, N.H., Le Thanh, T., Sangpueak, R., Treekoon, J., Saengchan, C., Thepbandit, W., Papathoti, N.K., Kamkaew, A., Buensanteai, N., 2022. Chitosan nanoparticles-based ionic gelation method: a promising candidate for plant disease management. Polymers 14 (4), 662.

Horvat, S., Yu, Y., Manz, H., Keller, T., Beilhack, A., Groll, J., Albrecht, K., 2021. Nanogels as antifungal-drug delivery system against aspergillus fumigatus. Adv. NanoBiomed. Res. 1 (5), 2000060.

Huang, W., Wang, C., Duan, H., Bi, Y., Wu, D., Du, J., Yu, H., 2018. Synergistic antifungal effect of biosynthesized silver nanoparticles combined with fungicides. Int. J. Agric. Biol. 20 (5), 1225–1229.

Jin, X., Zhang, R., Su, M., Li, H., Yue, X., Qin, D., Jiang, Z., 2019. Functionalization of halloysite nanotubes by enlargement and layer-by-layer assembly for controlled release of the fungicide iodopropynyl butylcarbamate. RSC Adv. 9 (72), 42062–42070.

Kailasam, S., Sundaramanickam, A., Tamilvanan, R., Kanth, S.V., 2022. Macrophytic waste optimization by synthesis of silver nanoparticles and exploring their agro-fungicidal activity. Inorg. Nano-Met. Chem., 1–10.

Karny, A., Zinger, A., Kajal, A., Shainsky-Roitman, J., Schroeder, A., 2018. Therapeutic nanoparticles penetrate leaves and deliver nutrients to agricultural crops. Sci. Rep. 8 (1), 1–10.

Kaur, M., Sudhakar, K., Mishra, V., 2019. Fabrication and biomedical potential of nanogels: An overview. Int. J. Polym. Mater. Polym. Biomater. 68 (6), 287–296.

Khan, M., Khan, A.U., Rafatullah, M., Alam, M., Bogdanchikova, N., Garibo, D., 2022. Search for effective approaches to fight microorganisms causing high losses in agriculture: application of *P. lilacinum* metabolites and mycosynthesised silver nanoparticles. Biomolecules 12 (2), 174.

Kora, A.J., Mounika, J., Jagadeeshwar, R., 2020. Rice leaf extract synthesized silver nanoparticles: An in vitro fungicidal evaluation against Rhizoctonia solani, the causative agent of sheath blight disease in rice. Fungal Biol. 124 (7), 671–681.

Koukaras, E.N., Papadimitriou, S.A., Bikiaris, D.N., Froudakis, G.E., 2012. Insight on the formation of chitosan nanoparticles through ionotropic gelation with tripolyphosphate. Mol. Pharm. 9 (10), 2856–2862.

Kumar, R., Najda, A., Duhan, J.S., Kumar, B., Chawla, P., Klepacka, J., Malawski, S., Kumar Sadh, P., Poonia, A.K., 2021. Assessment of antifungal efficacy and release behavior of fungicide-loaded chitosan-carrageenan nanoparticles against phytopathogenic fungi. Polymers 14 (1), 41.

Kumari, S., Khan, S., 2017. Synthesis and applications of nanofungicides: a next-generation fungicide. In: Fungal Nanotechnology. Springer, Cham, pp. 103–118.

Larocca, N.M., Bernardes Filho, R., Pessan, L.A., 2018. Influence of layer-by-layer deposition techniques and incorporation of layered double hydroxides (LDH) on the morphology and gas barrier properties of polyelectrolytes multilayer thin films. Surf. Coat. Technol. 349, 1–12.

Li, J., Dong, C., An, W., Zhang, Y., Zhao, Q., Li, Z., Jiao, B., 2020. Simultaneous enantioselective determination of two new isopropanol-triazole fungicides in plant-origin foods using multiwalled carbon nanotubes in reversed-dispersive solid-phase extraction and ultrahigh-performance liquid chromatography–tandem mass spectrometry. J. Agric. Food Chem. 68 (21), 5969–5979.

Liu, X., He, B., Xu, Z., Yin, M., Yang, W., Zhang, H., Cao, J., Shen, J., 2015. A functionalized fluorescent dendrimer as a pesticide nanocarrier: application in pest control. Nanoscale 7 (2), 445–449.

Machado, T.O., Beckers, S.J., Fischer, J., Müller, B., Sayer, C., de Araújo, P.H., Landfester, K. and Wurm, F.R., 2020. Bio-based lignin nanocarriers loaded with fungicides as a versatile platform for drug delivery in plants. Biomacromolecules 21 (7), 2755–2763.

Malandrakis, A.A., Kavroulakis, N., Chrysikopoulos, C.V., 2019. Use of copper, silver and zinc nanoparticles against foliar and soil-borne plant pathogens. Sci. Total Environ. 670, 292–299.

Malandrakis, A.A., Kavroulakis, N., Chrysikopoulos, C.V., 2020. Use of silver nanoparticles to counter fungicide-resistance in Monilinia fructicola. Sci. Total Environ. 747, 141287.

Malode, M.G.P., Chauhan, S.A., Bartare, S.A., Malode, L.M., Manwar, J.V., Bakal, R.L., 2022. A critical review on nanoemulsion: advantages, techniques and characterization. J. Appl. Pharm. Sci. 4 (3), 6–12.

Maluin, F.N., Hussein, M.Z., Yusof, N.A., Fakurazi, S., Idris, A.S., Zainol Hilmi, N.H., Jeffery Daim, L.D., 2019a. Preparation of chitosan–hexaconazole nanoparticles as fungicide nanodelivery system for combating Ganoderma disease in oil palm. Molecules 24 (13), 2498.

Maluin, F.N., Hussein, M.Z., Yusof, N.A., Fakurazi, S., Seman, I.A., Hilmi, N.H.Z., Daim, L.D.J., 2019b. Enhanced fungicidal efficacy on Ganoderma boninense by simultaneous co-delivery of hexaconazole and dazomet from their chitosan nanoparticles. RSC Adv. 9 (46), 27083–27095.

Mishra, V., Kesharwani, P., 2016. Dendrimer technologies for brain tumor. Drug Discov. 21 (5), 766–778.

Mishra, V., Bansal, K.K., Verma, A., Yadav, N., Thakur, S., Sudhakar, K., Rosenholm, J.M., 2018a. Solid lipid nanoparticles: Emerging colloidal nano drug delivery systems. Pharmaceutics 10 (4), 191.

Mishra, V., Kesharwani, P., Jain, N.K., 2018b. Biomedical applications and toxicological aspects of functionalized carbon nanotubes. Crit. Rev. Ther. Drug Carrier Syst. 35 (4), 293–330.

Mishra, V., Yadav, N., Saraogi, G.K., Tambuwala, M.M., Giri, N., 2019. Dendrimer based nanoarchitectures in diabetes management: an overview. Curr. Pharm. Des. 25 (23), 2569–2583.

Mishra, V., Singh, M., Nayak, P., Sriram, P., Suttee, A., 2020. Carbon nanotubes as emerging nanocarriers in drug delivery: an overview. Int. J. Pharm. Qual. Assur. 11 (3), 373–378.

Mishra, V., Nayak, P., Singh, M., Tambuwala, M.M., Aljabali, A.A., Chellappan, D.K., Dua, K., 2021a. Pharmaceutical aspects of green synthesized silver nanoparticles: A boon to cancer treatment. Anti-Cancer Agents Med. Chem. (Formerly Curr. Med. Chem.-Anti-Cancer Agents) 21 (12), 1490–1509.

Mishra, V., Singh, M., Nayak, P., 2021b. Smart functionalised-dendrimeric medicine in cancer therapy. In: Dendrimers in Nanomedicine. CRC Press, pp. 233–253.

Mishra, V., Singh, M., Mishra, Y., Charbe, N., Nayak, P., Sudhakar, K., Aljabali, A.A., Shahcheraghi, S.H., Bakshi, H., Serrano-Aroca, Á., Tambuwala, M.M., 2021c. Nanoarchitectures in management of fungal diseases: an overview. Appl. Sci. 11 (15), 7119.

Mustafa, I.F., Hussein, M.Z., Idris, A.S., Hilmi, N.H.Z., Ramli, N., Fakurazi, S., 2022. The effect of surfactant type on the physico-chemical properties of hexaconazole/Dazomet-Micelle nanodelivery system and its biofungicidal activity against ganoderma boninense. Colloids Surf. A Physicochem. Eng. Asp., 128402.

Nguyen Thi Nhat, H., Le, N.T.T., Phuong Phong, N.T., Nguyen, D.H., Nguyen-Le, M.T., 2020. Potential application of gold nanospheres as a surface plasmon resonance-based sensor for in-situ detection of residual fungicides. Sensors 20 (8), 2229.

Nguyen, T.L., Nguyen, T.H., Nguyen, C.K., Nguyen, D.H., 2017. Redox and pH responsive poly (amidoamine) dendrimer-heparin conjugates via disulfide linkages for letrozole delivery. Biomed. Res. Int. 2017, 8589212.

Nuruzzaman, M.D., Rahman, M.M., Liu, Y., Naidu, R., 2016. Nanoencapsulation, nano-guard for pesticides: a new window for safe application. J. Agric. Food Chem. 64 (7), 1447–1483.

Pachpute, T., Dwivedi, J., Shelke, T., Jeyabalan, G., 2019. Formulation and evaluation of mesalamine nanosphere tablet. J. Drug Deliv. Ther. 9 (4), 1045–1053.

Patil, A., Mishra, V., Thakur, S., Riyaz, B., Kaur, A., Khursheed, R., Patil, K., Sathe, B., 2019. Nanotechnology derived nanotools in biomedical perspectives: An update. Curr. Nanosci. 15 (2), 137–146.

Pedroso-Santana, S., Fleitas-Salazar, N., 2020. Ionotropic gelation method in the synthesis of nanoparticles/microparticles for biomedical purposes. Polym. Int. 69 (5), 443–447.

Peng, F., Wang, X., Zhang, W., Shi, X., Cheng, C., Hou, W., Lin, X., Xiao, X., Li, J., 2022. Nanopesticide formulation from pyraclostrobin and graphene oxide as a nanocarrier and application in controlling plant fungal pathogens. Nanomaterials (Basel) 12 (7), 1112.

Pham, N.D., Duong, M.M., Le, M.V., Hoang, H.A., 2019. Preparation and characterization of antifungal colloidal copper nanoparticles and their antifungal activity against Fusarium oxysporum and Phytophthora capsici. C. R. Chim. 22 (11–12), 786–793.

Prasad, R., Pandey, R., Barman, I., 2016. Engineering tailored nanoparticles with microbes: quo vadis? Wiley Interdiscip. Rev. Nanomed. Nanobiotechnol. 8 (2), 316–330.

Rafiee, S., Ramezanian, A., Mostowfizadeh-Ghalamfarsa, R., Niakousari, M., Saharkhiz, M.J., Yahia, E., 2022. Nanoemulsion of denak (Oliveria decumbens Vent.) essential oil: ultrasonic synthesis and antifungal activity against Penicillium digitatum. J. Food Meas. Charact. 16 (1), 324–331.

Rai, M., Ingle, A.P., Pandit, R., Paralikar, P., Gupta, I., Chaud, M.V., Dos Santos, C.A., 2017. Broadening the spectrum of small-molecule antibacterials by metallic nanoparticles to overcome microbial resistance. Int. J. Pharm. 532 (1), 139–148.

Rai, M., Gupta, I., Ingle, A.P., Biswas, J.K., Sinitsyna, O.V., 2018. Nanomaterials: What are they, why they cause ecotoxicity, and how this can be dealt with? In: Nanomaterials: Ecotoxicity, Safety, and Public Perception. Springer, Cham, pp. 3–18.

Raj, V., Raorane, C.J., Lee, J.H., Lee, J., 2021. Appraisal of chitosan-gum Arabic-coated bipolymeric nanocarriers for efficient dye removal and eradication of the plant pathogen Botrytis cinerea. ACS Appl. Mater. Interfaces 13 (40), 47354–47370.

Raj, V., Kim, Y., Kim, Y.G., Lee, J.H., Lee, J., 2022. Chitosan-gum arabic embedded alizarin nanocarriers inhibit biofilm formation of multispecies microorganisms. Carbohydr. Polym. 284, 118959.

Rudramurthy, G.R., Swamy, M.K., Sinniah, U.R., Ghasemzadeh, A., 2016. Nanoparticles: alternatives against drug-resistant pathogenic microbes. Molecules 21 (7), 836.

Saghafi, H., Zarkesh, F., Kooshyar, H., 2021. Coacervation process for preparation of encapsulate Lambda-Cyhalothrin. SRPH J. Fundam. Sci. 3 (2), 1–5.

Sahoo, B., Rath, S.K., Mahanta, S.K., Arakha, M., 2022. Nanotechnology mediated detection and control of phytopathogens. In: Bio-Nano Interface. Springer, Singapore, pp. 109–125.

Sakita, K.M., Conrado, P.C., Faria, D.R., Arita, G.S., Capoci, I.R., Rodrigues-Vendramini, F.A., Pieralisi, N., Cesar, G.B., Goncalves, R.S., Caetano, W., Hioka, N., 2019. Copolymeric micelles as efficient inert nanocarrier for hypericin in the photodynamic inactivation of Candida species. Future Microbiol. 14 (6), 519–531.

Saleh, N., Elshaer, S., Girgis, G., 2022. Biodegradable polymers-based nanoparticles to enhance the antifungal efficacy of fluconazole against *Candida albicans*. Curr. Pharm. Biotechnol. 23 (5), 749–757.

Saluja, V., Mishra, Y., Mishra, V., Giri, N., Nayak, P., 2021. Dendrimers based cancer nanotheranostics: an overview. Int. J. Pharm. 600, 120485.

Sánchez-Hernández, E., Langa-Lomba, N., González-García, V., Casanova-Gascón, J., Martín-Gil, J., Santiago-Aliste, A., Torres-Sánchez, S., Martín-Ramos, P., 2022. Lignin–Chitosan nanocarriers for the delivery of bioactive natural products against wood-decay phytopathogens. Agronomy 12 (2), 461.

Saraogi, G.K., Tholiya, S., Mishra, Y., Mishra, V., Albutti, A., Nayak, P., Tambuwala, M.M., 2022. Formulation development and evaluation of pravastatin-loaded nanogel for hyperlipidemia management. Gels 8 (2), 81.

Saratale, R.G., Benelli, G., Kumar, G., Kim, D.S., Saratale, G.D., 2018a. Bio-fabrication of silver nanoparticles using the leaf extract of an ancient herbal medicine, dandelion (Taraxacum officinale), evaluation of their antioxidant, anticancer potential, and antimicrobial activity against phytopathogens. Environ. Sci. Pollut. Res. 25 (11), 10392–10406.

Saratale, R.G., Karuppusamy, I., Saratale, G.D., Pugazhendhi, A., Kumar, G., Park, Y., Ghodake, G.S., Bharagava, R.N., Banu, J.R., Shin, H.S., 2018b. A comprehensive review on green nanomaterials using biological systems: Recent perception and their future applications. Colloids Surf. B Biointerfaces 170, 20–35.

Saratale, R.G., Saratale, G.D., Shin, H.S., Jacob, J.M., Pugazhendhi, A., Bhaisare, M., Kumar, G., 2018c. New insights on the green synthesis of metallic nanoparticles using plant and waste biomaterials: current knowledge, their agricultural and environmental applications. Environ. Sci. Pollut. Res. 25 (11), 10164–10183.

Sardar, M., Ahmed, W., Al Ayoubi, S., Nisa, S., Bibi, Y., Sabir, M., Khan, M.M., Ahmed, W., Qayyum, A., 2022. Fungicidal synergistic effect of biogenically synthesized zinc oxide and copper oxide nanoparticles against *Alternaria citri* causing citrus black rot disease. Saudi J. Biol. Sci. 29 (1), 88–95.

Sawake, M.M., Moharil, M.P., Ingle, Y.V., Jadhav, P.V., Ingle, A.P., Khelurkar, V.C., Paithankar, D.H., Bathe, G.A., Gade, A.K., 2022. Management of Phytophthora parasitica causing gummosis in citrus using biogenic copper oxide nanoparticles. J. Appl. Microbiol. 132 (4), 3142–3154.

Schlenoff, J.B., Dubas, S.T., Farhat, T., 2000. Sprayed polyelectrolyte multilayers. Langmuir 16 (26), 9968–9969.

Senosy, I.A., Guo, H.M., Ouyang, M.N., Lu, Z.H., Yang, Z.H., Li, J.H., 2020. Magnetic solid-phase extraction based on nano-zeolite imidazolate framework-8-functionalized magnetic graphene oxide for the quantification of residual fungicides in water, honey and fruit juices. Food Chem. 325, 126944.

Shan, Y., Xu, C., Zhang, H., Chen, H., Bilal, M., Niu, S., Cao, L., Huang, Q., 2020. Polydopamine-modified metal–organic frameworks, NH2-Fe-MIL-101, as pH-sensitive nanocarriers for controlled pesticide release. Nanomater. 10 (10), 2000.

Sharma, S., Singh, S., Ganguli, A.K., Shanmugam, V., 2017. Anti-drift nano-stickers made of graphene oxide for targeted pesticide delivery and crop pest control. Carbon 115, 781–790.

Sharma, M., Mundlia, J., Kumar, T., Ahuja, M., 2021. A novel microwave-assisted synthesis, characterization and evaluation of luliconazole-loaded solid lipid nanoparticles. Polym. Bull. 78 (5), 2553–2567.

Shen, C.X., Zhang, Q.F., Li, J., Bi, F.C., Yao, N., 2010. Induction of programmed cell death in Arabidopsis and rice by single-wall carbon nanotubes. Am. J. Bot. 97 (10), 1602–1609.

Shi, Z., Jiang, Y., Sun, Y., Min, D., Li, F., Li, X., Zhang, X., 2021. Nanocapsules of oregano essential oil preparation and characterization and its fungistasis on apricot fruit during shelf life. J. Food Process. Preserv. 45 (7), e15649.

Singh, D., Rathod, V., Ninganagouda, S., Hiremath, J., Singh, A.K., Mathew, J., 2014. Optimization and characterization of silver nanoparticle by endophytic fungi Penicillium sp. isolated from Curcuma longa (turmeric) and application studies against MDR E. coli and S. aureus. Bioinorg. Chem. Appl. 2014, 408021.

Singh, A., Gaurav, S.S., Shukla, G., Rani, P., 2022. Assessment of mycogenic zinc nano-fungicides against pathogenic early blight (Alternaria solani) of potato (*Solanum tuberosum* L.). Mater. Today: Proc. 49, 3528–3537.

Sivaram, A.J., Rajitha, P., Maya, S., Jayakumar, R., Sabitha, M., 2015. Nanogels for delivery, imaging and therapy. Wiley Interdiscip. Rev. Nanomed. Nanobiotechnol. 7 (4), 509–533.

Suganya, V., Anuradha, V., 2017. Microencapsulation and nanoencapsulation: a review. Int. J. Pharm. Clin. Res 9 (3), 233–239.

Suttee, A., Singh, G., Yadav, N., Barnwal, R.P., Singla, N., Prabhu, K.S., Mishra, V., 2019. A review on status of nanotechnology in pharmaceutical sciences. Int. J. Drug Deliv. Technol. 9 (1), 98–103.

Suttee, A., Mishra, V., Nayak, P., Singh, M., Sriram, P., 2020. Niosomes: potential nanocarriers for drug delivery. Int. J. Pharm. Qual. Assur. 11 (03), 389–394.

Tang, G., Tian, Y., Niu, J., Tang, J., Yang, J., Gao, Y., Chen, X., Li, X., Wang, H., Cao, Y., 2021. Development of carrier-free self-assembled nanoparticles based on fenhexamid and polyhexamethylene biguanide for sustainable plant disease management. Green Chem. 23 (6), 2531–2540.

Thanh, V.M., Nguyen, T.H., Tran, T.V., Ngoc, U.T.P., Ho, M.N., Nguyen, T.T., Chau, Y.N.T., Tran, N.Q., Nguyen, C.K., Nguyen, D.H., 2018. Low systemic toxicity nanocarriers fabricated from heparin-mPEG and PAMAM dendrimers for controlled drug release. Mater. Sci. Eng. C 82, 291–298.

Thanh, V.M., Bui, L.M., Bach, L.G., Nguyen, N.T., Thi, H.L., Hoang Thi, T.T., 2019. *Origanum majorana* L. essential oil-associated polymeric nano dendrimer for antifungal activity against Phytophthora infestans. Materials 12 (9), 1446.

Tleuova, A.B., Wielogorska, E., Talluri, V.P., Štěpánek, F., Elliott, C.T., Grigoriev, D.O., 2020. Recent advances and remaining barriers to producing novel formulations of fungicides for safe and sustainable agriculture. J. Control. Release 326, 468–481.

Trindade, I.C., Pound-Lana, G., Pereira, D.G.S., de Oliveira, L.A.M., Andrade, M.S., Vilela, J.M.C., Postacchini, B.B., Mosqueira, V.C.F., 2018. Mechanisms of interaction of biodegradable polyester nanocapsules with non-phagocytic cells. Eur. J. Pharm. Sci. 124, 89–104.

Ullah, R., Shah, S., Muhammad, Z., Shah, S.A., Faisal, S., Khattak, U., ul Haq, T., Akbar, M.T., 2021. In vitro and in vivo applications of Euphorbia wallichii shoot extract-mediated gold nanospheres. Green Process. Synth. 10 (1), 101–111.

Vannini, C., Domingo, G., Onelli, E., De Mattia, F., Bruni, I., Marsoni, M., Bracale, M., 2014. Phytotoxic and genotoxic effects of silver nanoparticles exposure on germinating wheat seedlings. J. Plant Physiol. 171 (13), 1142–1148.

Vashist, A., Kaushik, A., Vashist, A., Bala, J., Nikkhah-Moshaie, R., Sagar, V., Nair, M., 2018. Nanogels as potential drug nanocarriers for CNS drug delivery. Drug Discov. 23 (7), 1436–1443.

Vega-Vásquez, P., Mosier, N.S., Irudayaraj, J., 2021. Nanovaccine for plants from organic waste: D-limonene-loaded chitosan nanocarriers protect plants against *Botrytis cinerea*. ACS Sustain. Chem. Eng. 9 (29), 9903–9914.

Vittori Antisari, L., Carbone, S., Gatti, A., Vianello, G., Nannipieri, P., 2015. Uptake and translocation of metals and nutrients in tomato grown in soil polluted with metal oxide (CeO2, Fe3O4, SnO2, TiO2) or metallic (Ag, Co, Ni) engineered nanoparticles. Environ. Sci. Pollut. Res. 22 (3), 1841–1853.

Wang, Y., Li, P., Truong-Dinh Tran, T., Zhang, J., Kong, L., 2016. Manufacturing techniques and surface engineering of polymer-based nanoparticles for targeted drug delivery to cancer. Nanomater. 6 (2), 26.

Wang, C.Y., Jia, C., Zhang, M.Z., Yang, S., Qin, J.C., Yang, Y.W., 2021a. yA lesion microenvironment-responsive fungicide nanoplatform for crop disease prevention and control. Adv. Healthc. Mater., 2102617.

Wang, C.Y., Lou, X.Y., Cai, Z., Zhang, M.Z., Jia, C., Qin, J.C., Yang, Y.W., 2021b. Supramolecular nanoplatform based on mesoporous silica nanocarriers and Pillararene nanogates for fungus control. ACS Appl. Mater. Interfaces 13 (27), 32295–32306.

Wani, T.A., Masoodi, F.A., Baba, W.N., Ahmad, M., Rahmanian, N., Jafari, S.M., 2019. Nanoencapsulation of agrochemicals, fertilizers, and pesticides for improved plant production. In: Advances in Phytonanotechnology. Academic Press, pp. 279–298.

Worrall, E.A., Hamid, A., Mody, K.T., Mitter, N., Pappu, H.R., 2018. Nanotechnology for plant disease management. Agronomy 8 (12), 285.

Wu, B., Zhu, L., Le, X.C., 2017. Metabolomics analysis of TiO2 nanoparticles induced toxicological effects on rice (*Oryza sativa* L.). Environ. Pollut. 230, 302–310.

Wu, L., Pan, H., Huang, W., Hu, Z., Wang, M., Zhang, F., 2022. pH and redox dual-responsive mesoporous silica nanoparticle as nanovehicle for improving fungicidal efficiency. Materials 15 (6), 2207.

Yousefi, A.T., Ikeda, S., Mahmood, M.R., Rouhi, J., Yousefi, H.T., 2012. Encapsulation technology based on coacervation method to control release of manufactured nano particles. World Appl. Sci. J. 17 (4), 524–531.

Zainab, S., Hamid, S., Sahar, S., Ali, N., 2022. Fluconazole and biogenic silver nanoparticles-based nano-fungicidal system for highly efficient elimination of multi-drug resistant Candida biofilms. Mater. Chem. Phys. 276, 125451.

Zhang, Y., Huang, L., 2020. Liposomal delivery system. In: Nanoparticles for Biomedical Applications. Elsevier, pp. 145–152.

Zhang, Z., Yang, J., Yang, Q., Tian, G., Cui, Z.K., 2021. Fabrication of non-phospholipid liposomal nanocarrier for sustained-release of the fungicide cymoxanil. Front. Mol. Biosci. 8, 212.

Zhang, X., Tang, X., Zhao, C., Yuan, Z., Zhang, D., Zhao, H., Yang, N., Guo, K., He, Y., He, Y., Hu, J., 2022. A pH-responsive MOF for site-specific delivery of fungicide to control citrus disease of Botrytis cinerea. Chem. Eng. J. 431, 133351.

CHAPTER

6

Antifungal potential of nano- and microencapsulated phytochemical compounds and their impact on plant heath

Nasreen Musheer[a], *Anam Choudhary*[b], *Arshi Jamil*[b], *and Sabiha Saeed*[b]

[a]Glocal University, Saharanpur, Uttar Pradesh, India [b]Aligarh Muslim University, Aligarh, Uttar Pradesh, India

1 Introduction

Preharvest and postharvest infections caused by toxigenic fungi are emerging as serious health concerns to living organisms and the environment, which leads to the protected cultivation and storage of agricultural commodities (Villa et al., 2017). The toxigenic fungi belong to the genera *Aspergillus, Fusarium, Claviceps, Penicillium, Stachybotrys, Alternaria*, etc. They cause significant yield losses of approximately $45 billion worldwide every year (Makhuvele et al., 2020). For instance, *Rhizoctonia solani, Fusarium* spp., are *Phytophthora* spp. are reported worldwide (Abd-Elsalam and Alghuthaymi, 2015). The interaction of pathogenic fungi with plants disturbs the normal physiological and biochemical functions of plants (Vincent et al., 2020). Cell wall lytic enzymes, toxins, growth regulators, and their analogues are key factors that help fungal pathogens cause browning, yellowing, and necrosis in plants (Yin et al., 2016; Jajic et al., 2019; Yang et al., 2020). To control fungal diseases, conventional synthetic fungicides are being extensively used; these are nonbiodegradable and impose toxic residual effects by accumulating in plants, animals, soil, and water bodies (Villa et al., 2017). Many scientists have proposed the use of plant products in the management of fungal diseases. Several studies have reported that plant extracts provide better protection

https://doi.org/10.1016/B978-0-443-23950-2.00004-7

to the crop, soil, and environment as well as develop pathogen-resistant strains than chemical fungicides (Zaker, 2016). Plants are natural reservoirs of many secondary metabolites such as phenols, phenolic acids, quinones, flavones, and flavonoids, which are highly active to suppress phytopathogenic fungi.

A recent advance has been to introduce nanotechnology in the synthesis of plant-based nanoparticles to protect crops against phytopathogens (Vaculikova et al., 2016; Jampilek and Kralova, 2018; Ali et al., 2020b; Chengjun and Bing, 2020). These nanoparticles, which are termed "green nanoparticles," offer numerous mechanisms (Chengjun and Bing, 2020). Green nanoparticles have shown greater biocompatibility with plant systems to control fungal disease than synthetic fungicides, as they can easily react with plant metabolites such as enzymes, hormones, proteins, lipids, carbohydrates, and other organic acids (Nakkala et al., 2014; Chetan et al., 2022). For instance, phytonanofungicides could be used as carrier materials to resist the attacks of pests and pathogens (Suresh et al., 2018). It is widely used in the production of a new generation of agrochemicals such as fungicides, pesticides, herbicides, etc. (Jiang et al., 2021). Phytonanofungicides are capable of photocatalytic activity and degrade the residue toxicity of synthetic fungicides. Scientists have engineered cost-effective and nontoxic phytonanoparticles for various agricultural purposes such as plant breeding, crop production, crop protection, fertilizers, food product preservation, etc., by considering their desired size and shape with specific optical properties (Pestovsky and Martínez, 2017; Worrall et al., 2018; Camara et al., 2019; Jiang et al., 2021; Zobir et al., 2021). These phytonanoparticles could contribute to the sustainable intensification of agricultural production (Prasad et al., 2014; Katarzyna and Izabela, 2022). Nano-sized phytoextracts enable better uptake of minerals, fungicides, insecticides, and pesticides by plants than pure forms of phytoextracts (Chhipa, 2017; Kah et al., 2018; Din et al., 2018; Mohamed et al., 2021). They improve plant hormone delivery, seed germination, water management, transfer of target genes, nanobarcoding, nanosensors, and disease management strategies (Liu et al., 2015; Hayles et al., 2017; Goswami et al., 2020; Xiong et al., 2020). Phytonanofungicides are good alternatives to fungicides, minimizing the risk of fungicides such as residual ecotoxicity and pathogen resistance development (Angela et al., 2022; Avila et al., 2022).

Researchers are currently working to develop effective phytonanoparticles with high antifungal potential (Antunes and Cavaco, 2010; Alghuthaymi et al., 2021). Phytonanoparticles are also being used in combination with inorganic nanoparticles. Sulfur, silver, copper oxides, magnesium oxides, and zinc oxides, for example (Cruz-Luna et al., 2021; Jeevanandam et al., 2022), can improve fungicidal activity and reduce fungicide challenges (Abd-Elsalam and Alghuthaymi, 2015; Lowry et al., 2019; Khan et al., 2022). The effect of inorganic and phytoextract nanoparticles provides external and internal protection to plants against phytopathogens (Rai and Ingle, 2012; Abd-Elsalam and Alghuthaymi, 2015; Huang et al., 2018). Plant-based nanoparticles are potent for fungicides, pesticides, fertilizers, plant strengtheners, and phytostimulators (Sadeghi et al., 2017; Worrall et al., 2018). There are several advantages associated with the use of phytochemical-based nanobiocides; however, their commercialization for agricultural applications remains small due to inadequate knowledge of phytochemical extraction, short shelf life, soil properties, and weather factors.

In this chapter, we discuss the use of nanotechnology in the formulation of plant-based nanofungicides as a stringent dynamic approach, introducing plant materials to control various fungal diseases of many crops and reduce the ecotoxicological effects of synthetic chemicals with prolonged efficiency. Further, this chapter discusses the application of phytonanofungicides in the agriculture sector, which should be based on the advantages and disadvantages of using nanotechnology in plant systems. Finally, the chapter discusses the importance of phytonanofungicides over synthetic chemicals to develop a reliable, safe, and cost-effective solution for phytopathogen management for sustainable crop production.

2 Antifungal potential of phytochemicals

Plants are a rich repository of more than 10,000 secondary metabolites that can control pathogenic fungi and reduce the extensive use of fungicides with the goal of achieving sustainability in the agriculture system (Mohana and Raveesha, 2007; Haq et al., 2020). Now, scientists are investigating new plants with the aim of minimizing the negative effects of chemicals, developing alternative control of plant disease, and developing plant immunity inducers to eliminate plant pathogen infections (Peng et al., 2021; Arora et al., 2022). The plant immunity inducers activate the pathogen-associated molecular pattern (PAMP)-triggered immunity (PTI), which enables biochemical changes in the plant's cell membrane to prevent the virulent pathogen infection. The plant immunity inducers are classified as proteins, oligosaccharides, glycopeptides, lipids, lipopeptides, and other metabolites, which are associated with plant pattern recognition receptors to provide comprehensive disease resistance (Boutrot and Zipfel, 2017). Plant-derived protectants are an effective alternative to synthetic pesticides because they are rich in antifungal metabolites such as polyphenols, glycosides, quinones, terpenoids, saponins, allicin, flavonoids, and scopolamine, which all offer various defense mechanisms against fungal pathogens (Table 1). The extraction and characterization of desired bioactive compounds from plants for crop protection requires appropriate solvent extraction methods that include steam distillation, maceration, decoction, infusion, percolation, Soxhlet extraction, chromatography-assisted extraction, and spectroscopy-assisted extraction. The solvents commonly used to prepare phytoextracts are polar solvents (e.g., water, alcohols), intermediate polar solvents (e.g., acetone, dichloromethane), and nonpolar solvents (e.g., *n*-hexane, ether, chloroform) (Abubakar and Haque, 2020; Steglinska et al., 2022). The aqueous extract of *Matricaria chamomilla* L. is used as a capping and stabilizing agent for the greenly synthesized silver nanoparticles (AgNPs). Morphological and structural characterization of photo-induced biomolecule-capped AgNPs was done using microscopic and spectroscopic techniques such as TEM, SEM, EDX, XRD, and FTIR analysis. The phytomediated AgNPs were examined by TGA-DTG analysis for thermal stability and had photocatalytic activity against Rhodamine B (RB) that was examined for stability in the presence of ultraviolet (UV) light irradiation (Alshehri and Malik, 2020). Many previously published studies define the plant-based chemicals are found effective against various pathogens as mentioned in Table 1. Therefore, researchers used plant-derived nanoparticles to control phytopathogens and promote plant growth, as mentioned in Table 2.

Several technologies are being used to extract phytochemicals from plants parts such as the leaf, flower, fruit, and bark that have antifungal properties (Ehsan et al., 2020). Crude aqueous and ethanolic extracts of *Tithonia diversifolia* and *Byrum coronatum* showed considerable antifungal activities against *Penecillium atrovenetium*, *Aspergillus niger*, *Fusarium flocciferum*, and

TABLE 1 Mode of action of major phytochemicals to inhibit fungal growth.

Phytochemical class	Antifungal mechanism	References
Polyphenols	Disrupt cell wall, causing leakage of cytoplasmic content and inactivation of enzymes	Gillmeister et al. (2019)
Alkaloids	Malfunctioning of mitochondria has a direct impact on respiratory activity and enzymatic activity	El Hamdani et al. (2016)
Tannins	Affects the cell membrane integrity and permeability and inhibits enzyme synthesis	Congyi et al. (2019)
Flavonoids	Cell membrane disruption, mitochondria dysfunction, and interfere in cell wall formation, cell division, RNA, and protein synthesis	Saleh and Suresh (2020)
Coumarins	Degradation of eukaryotic DNA	Elgharbawy et al. (2020)
Lectins and polypeptides	Bind with chitin and inhibit the growth and germination of spores	Mohsen et al. (2018)
Allicin	Affect membrane permeability, inactive enzymes	Sarfraz et al. (2020)
Terpenoids	Hinders the formation and viability of hyphae and promotes morphological alterations	Elgharbawy et al. (2020)
Saponins	Damage the cell membrane and cause cellular materials to leach out, which ultimately leads to cell death	Longfei et al. (2018)
Peptides	Hinder cell wall synthesis	Elgharbawy et al. (2020)
γ-Terpinene, 1-phellandrene, γ-terpene, cuminaldehyde	Disrupt cell membrane, causing cytoplasmic leakage	Valarezo et al. (2021)
Linalool, β-caryophyllene	Affect cell membrane function and disrupt mitochondrial membrane potential	Beatovic et al. (2015)
Thymol, γ-terpinene, ρ-cymene	Hypopolarization of the cytoplasmic membrane and inhibit ATP biosynthesis	Alshaikh and Perveen (2021)
Carvacrol	Cytoplasmic membrane depolarized and cell membrane integrity disrupted	Kaskatepe et al. (2022)
Limonene and γ-terpinene	Cell membrane disruption induced cytoplasmic leakage	Valkova et al. (2021)
Menthol, menthone	Depolarization of the cytoplasmic membrane and cytolytic leakage	(Norouzi et al. (2021)
Eugenol, eugenyl acetate	Cytolytic leakage	Kaur et al. (2019)

TABLE 2 Plant-based nanomaterials used in plant growth and crop protection.

Phytonanoparticles	Mode of action	Pathogen/disease	References
Chitosan nanoparticles	Antifungal potential	Control *A. alternata*, *M. phaseolina* and *R. solani*	Saharan et al. (2013)
	Promote plant growth	Robusta coffee	Nguyen et al. (2013)
	Defense activity	Effective against Fusarium head blight	Kheiri et al. (2016)
	Plant promotion Antifungal activity	Improve seed germination and seedling vigor, activate defense response and show strong resistance against *Sclerospora graminicola*	Siddaiah et al. (2018)
Silver nanoparticle encapsulated *Zataria multiflora* essential oil (ZEO)	Antifungal potential	*Aspergillus ochraceus*, *Aspergillus flavus*, *Alternaria solani*, *Rhizoctonia solani*, and *Rhizopus stolonifera*	Nasseri et al. (2016)
CS NP encapsulated ZEO and incidence of *Botrytis*-inoculated strawberries	Decreased disease severity and incidence	Botrytis-inoculated strawberries	Mohammadi et al. (2015)
CS-cinnamic acid nanogel encapsulated *Mentha piperita* essential oils	Antifungal activity	*A. flavus*	Khalili et al. (2015)
Thyme oil sealed with CS/ boehmite nanocomposite	reduced the incidence and severity	Brown rot of peach fruits caused by *Monilinia laxa*	Beyki et al. (2014)
Adhatoda vasica leaf extract encapsulated in copper oxide	Inhibit fungal growth	*Aspergillus niger*	Bhavyasree and Xavier (2020)
Nanoencapsulated clove essential oil	Improve antifungal activity	*Aspergillus niger*	Hasheminejad et al. (2019)
Silver nanoparticles using extract of medicinal herb (*Ocimum kilimandscharicum*)	Antifungal activity	*Fusarium oxysporum* and *Colletotrichum gloeosporioides*	Singh et al. (2019)
Iron nanoparticle encapsulated with leaf extract of *Calotropis procera*	Antifungal potential	*Alternaria alternata*	Ali et al. (2020a)
Lemongrass and clove oil encapsulated into mesoporous silica nanoparticles (LGO-MSNPs)	Fungicidal efficacy	*Gaeumannomyces graminis*	Sattary et al. (2020)
Encapsulation of *Bunium persicum* (Boiss) essential oil into chitosan nanomatrix	Antifungal efficacy	Aflatoxin B1 (AFB1) contamination of stored masticatories	Singh et al. (2020)
Encapsulated oregano essential oil nanoparticles	Antifungal efficacy		Aguilar-Perez et al. (2021)

Continued

TABLE 2 Plant-based nanomaterials used in plant growth and crop protection.—Cont'd

Phytonanoparticles	Mode of action	Pathogen/disease	References
Cananga odorata essential oil encapsulated in chitosan nanoparticle	Antifungal efficacy	Aflatoxin B1 (AFB1) *Aspergillus flavus*	Upadhyay et al. (2021)
Biosynthesis of AgNPs from beech bark extract	Antifungal activity	*Candida albicans, guilliermondii*	Mare et al. (2021)
Green silver nanoparticle of *Portulaca oleracea*	Antifungal activity	*Curvuleria spicifera, Macrophomina phaseolina*, and *Bipolaris* sp.	Fatimah et al. (2022)
Alhagi graecorum AgNPs	Strong antifungal activity	*Candida* species	Hawar et al. (2022)

Geotrichum candidium (Liasu and Ayandele, 2008). Many studies reported that phytoextracts of *Azadirachta indica, Allium sativum, Callistemon rigidus, Capsicum annum, Datura inoxia, Lantana camara, Lawsonia enermis, Santalum album,* and *Terminalia theorlii* are significant to inhibit the fungal growth and spore germination of pathogenic fungi, namely *Altenaria alternata, A. niger, Fusarium moniliforme,* and *Trichoderma viride* (Pawar, 2011; Ehsan et al., 2020). Babu et al. (2008) determined the antifungal activity of leaf extracts of *Azadirachta indica, Eucalyptus globulus, Artemissia anunua, Ocimum sanctum,* and *Rheum emodi* against *Fusarium solani* of brinjal. Stem extracts of *Azadirachta indica, C. rigidus, C. annum, D. inoxia, L. camara, L. enermis, S. album,* and *T. theorlii* have considerably inhibited pathogenic fungi, *namely A. alternata, A. niger, F. moniliforme,* and *T. viride* (Pawar, 2011). Seint and Masaru (2011) studied the clove extract, neem leaf, and rosemary and pelargonium extracts that have inhibited the mycelial growth of rice phytopathogens *R. solani, Rhizoctonia oryzae, Rhizoctonia oryzae-sativae,* and *Sclerotium hydrophilum.* Lukman et al. (2016) reported that five plant leaf extracts of *E. globulus, Calotropis procera, Melia azedarach, Datura stramonium,* and *Acalypha indica* at 20% concentration caused maximum inhibition of 52.6%, 50.88%, 48.21%, and 47.42%, respectively, in the mycelial growth of *A. alternata,* whereas the lowest inhibition in the mycelial growth of *A. alternata* was recorded at 5% leaf extract concentration with all phytoextracts as compared to control. *Lantana camera, A. sativum,* and *Zingiber officinale* inhibited 80%, 54.44%, and 17.78% mycelial growth of *Alternaria brassicae,* respectively (Biswas and Tanmay, 2018). Garlic and eucalyptus extract at 5% and 10% concentrations gave a maximum 100% inhibition while Ashok Tulsi, Datura, and Neem at 10% concentration showed complete inhibition as compared to a 5% concentration (Yadav, 2019). The extracts of *Pistacia lentiscus, Stevia rebaudiana, A. sativum, C. procera, Azadirachta indica,* and *Z. officinale* were effective in inhibiting the Fusarium wilt disease of tomato plants (Ramirez et al., 2020; Poussio et al., 2021; Abo-Elyousr et al., 2022). Aqueous neem leaf extracts enhanced resistance to Fusarium wilt disease of the banana and promoted plant growth (Yi et al., 2021). Many studies reported that phytoextracts stimulate the plant defense response to resist pathogen infection, thus improving seed germination as well as increasing plant growth and production (Shafique et al., 2019; Ali et al., 2022). Naz et al. (2021) determined that extracts of *Jacaranda mimosifolia* induced defense-related enzymes to aid maize resistance against *Fusarium verticillioides.* Plant-derived oligosaccharides and polysaccharides can also serve as elicitors to activate the defense response against phytopathogens (Moenne and Gonzslez, 2021; Yang et al., 2022). Ghazanfar et al. (2019) reported that salicylic acid (SA) and botanical extracts were effective to manage the postharvest fungal disease of the tomato. Saberi et al. (2022) reported the phytoextracts trigger the expression of salicylic acid- and jasmonic acid-derived oligosaccharides and polysaccharides that help in inducing defense pathway against the fungal pathogens. Sernaite et al. (2020) reported the biofungicide potential of *Cinnamomum cassia* and *Syzygium aromaticum* extracts against *Botrytis cinerea* causing strawberry gray mold. Garlic extracts have potential as biopesticides that can protect the potato from the fungal attack of *Fusarium* sp., *R. solani*, *Colletotrichum coccodes*, *Alternaria* sp., and *Phoma exigua* during storage (Al-Baldawy et al., 2021; Steglinska et al., 2022). AgNPs ranging from 30 to 80nm using *Euporbia serpens* Kunth aqueous leaf extract suppressed the mycelial growth of *A. alternata*, *Fusarium gramium*, and *Candida albicans* (Ahmad et al., 2022).

Now, researchers are trying to reduce the extensive use of pesticides by proposing phytoextracts as good substitutes for synthetic fungicides in the management of crop diseases (Jepson et al., 2014). With increasing concern for human and animal health, scientists are engaged in extracting and testing plant-derived biochemicals to control fungal diseases.

3 Types of plant-based mediated nanoparticles and their effect on phytopathogens

With the importance of phytochemicals to suppress phytopathogens, plant-based nanoparticles are being synthesized physically, chemically, and biologically using various methods such as size reduction, homogenization under high pressure, sonication, precipitation, and solvent dispersion (Madeeha et al., 2020; Hano and Abbasi, 2022). Generally, nanoparticles are designed with sizes of 10–100nm for unique physiobiochemical properties (Hano and Abbasi, 2022). Nanoparticles act as protectants or are used in the production of effective agrochemicals such as fungicides, insecticides, growth stimulants, and germicides (Worrall et al., 2018; Shang et al., 2019). Natural polymers such as chitosan (CS) and silica are critically chosen to synthesize phytonanofungicides due to their high biodegradability, nonallergenicity, antifungal activity, and lack of danger to living organisms (Cota et al., 2013; Reddy et al., 2021). The chitosan has proven antifungal properties to control various fungal strains including *Rhizopus oryzae*, *A. niger*, and *A. alternata* (Samiyah et al., 2021). For example, *Mentha piperita* essential oil encapsulated with chitosan has inhibited the mycelia growth of *Aspergillus flavus* at an 800mg/mL concentration (Beyki et al., 2014). The chitosan can be absorbed into the fungal cell wall membranes at varying concentration levels to control many fungal diseases such as tomato fusarium wilt, botrytis grape bunch rot, and *Phyricularia* rice blast (Kashyap et al., 2015). However, it is less effective against bacteria. Beside absorbing into the fungal cell wall, it can adheres well to the plant parts such as leaves and stem, prolonging the contact time and facilitating the uptake of bioactive compounds (Malerba and Cerana, 2016). Plant extracts and essential oils were also used to maintain the thermal stability and increase the shelf life of chitosan-mediated nanoparticles (ChNPs). Hasheminejad et al. (2019) reported that the clove essential oil integrated chitosan nanoparticles improved the quality of pomegranate arils as well as extended the shelf life for 54days against fungal decay. The nanoparticles encapsulated with chitosan and essential oil of *Cymbopogon martinii* (Ce-CMEO-NPs) could control the postharvest fungal pathogen *Fusarium graminearum* (Kalagatur et al., 2018). Vanti et al. (2019) determined antifungal potential of chitosan encapsulated copper nanoparticles against phytopathogenic fungi such as *R. solani* and *Pythium aphanidermatum*, causing damping-off disease in seeds and young seedlings of vegetables crops and exhibiting plant growth promoting activity. The chitosan nanoemulsion formulation incorporated into *Cananga odorata* essential oil is used to inhibit aflatoxin B1 (AFB1) secreted by the fungal strain of *A. flavus* and increase the shelf life of stored grains (Upadhyay et al., 2021). The chitosan nanoparticle formulations were prepared using *Pelargonium graveolens* leaf extract, which has showed antifungal activity at 50 mg/mL against *B. cinerea* that causes strawberry leaf spot disease (El-Naggar et al., 2022). Elshaer et al. (2022) integrating chitosan nanoparticles with clotrimazole and *Vitis vinifera* juice extract inhibits the mycelial growth of *C. albicans* and *A. niger*.

Now, the implementation of nanoparticles in agriculture systems is gaining importance to deal with major problems such as crop production and protection, climatic factors, and nutrient management (Yaseen et al., 2018). To integrate different plant parts that possess antifungal potential, they are encapsulated with organic or inorganic nanocomposites called nanohybrids (Mishra et al., 2017). Nanohybrids are in demand due to their multifunctional abilities such as detection, detoxification, and management of pathogens (Kalia et al., 2020a). The use of metallic nanoparticles has been investigated enormously. Now, metals are being used with plants, bacteria, fungi, or yeast to synthesize green nanoparticles (Manoj, 2018; Hamza et al., 2019; Ali et al., 2020a; Kalia et al., 2020b; Jeevanandam et al., 2022). Phytopathogens were greatly suppressed by plant-mediated AgNPs encapsulated in flavonoids and phenolic molecules of plant extracts (Thajuddin and Mathew, 2020; Amruta et al., 2019; Rajaram et al., 2015). The integrated copper and chitosan nanoparticles have strong antifungal activity against the pathogens *A. alternata, Macrophomina phaseolina*, and *Rhizoctonia* in in vitro studies (Krishnaraj et al., 2012). AgNPs suppressed the mycelial growth of *A. alternata, Sclerotinia sclerotiorum, M. phseolina, R. solani, B. cinerea*, and *Curvularia lunata* (Krishnaraj et al., 2012). *Zataria multiflora* essential oil (ZEO) encapsulated in AgNPs) of 255 nm showed higher antifungal activity against *Aspergillus ochraceus* (at MIC 200 ppm), *A. flavus* (at MIC 200 ppm), *Alternaria solani* (at MIC 100 ppm), *R. solani* (at MIC 50 ppm), and *Rhizopus stolonifer* (at MIC 50 ppm) than pure essential oil (Nasseri et al., 2016). The particle size of chitosan encapsulation with ZEO nanoparticle ranged between 125–175 nm at 1500 ppm was significantly decreased both disease severity and incidence of *Botrytis*-inoculated strawberries (Mohammadi et al., 2015). The integration of chitosan, benzoic acid thyme essential oil (TEO) was produced antifungal properties carrying nanogel polymer (Khalili et al., 2015). The nanogel of chitosan, cinnamic acid and *M. piperita* essential oil mixture at 500 ppm MIC was found 4.2-fold more effective against *Aspergillus flavus* as compared to effect of pure essential oil under sealed condition (Beyki et al., 2014). Thyme oil encapsulated with a chitosan/boehmite nanoparticle has controlled the brown rot disease of peach fruits (cv. Kakawa) caused by *Monilinia laxa* (Cindi et al., 2015). *Adhatoda vasica* leaf extract encapsulated in copper oxide nanoparticles or carbon nanocomposite is effective in inhibiting the growth of *A. niger* (Bhavyasree and Xavier, 2020). Ali et al. (2020b) determined that the combination of *C. procera* (leaf extract) and iron nanoparticles could control the fungus *A. alternata*. However, other commonly used nanoparticles such as copper and titanium dioxide as well as gold have been used for crop production and plant protection in agriculture (Gogos et al., 2012; Sadeghi et al., 2017; Worrall et al., 2018). For example, phytonanofungicide formulation stability and biosafety are maintained using polymer poly (acrylic acid)-*b*-poly (butyl acrylate), polyvinylpyrrolidone, and polyvinyl alcohol. The leaf extract of *Vernonia amygdalina*-based nickel nanoparticles has been synthesized and determined its morphological characteristics using scanning electron microscopy and Fourier transform infrared spectroscopy (Habtemariam and Qomer, 2020). The encapsulated *Ligustrum lucidum* leaf extract into AgNPs at 170.20 μg/m has inhibited the conidial germination and colony growth of *Setosphaeria turcica* up to 50% (Weidong et al., 2020). Ecofriendly green magnetic iron nanoparticles were synthesized by using *Murraya koenigii* leaf extract (Sivakami et al., 2022).

4 Mode of action of phytonanoparticles in crop protection and production

Phytonanomaterials offer numerous mechanisms against fungal pathogens (Fig. 1). Nanomaterials enable rapid detection, allow target-specific delivery, and cure plant diseases (Avila et al., 2022). Scientists' efforts are necessary to understand the mechanisms of

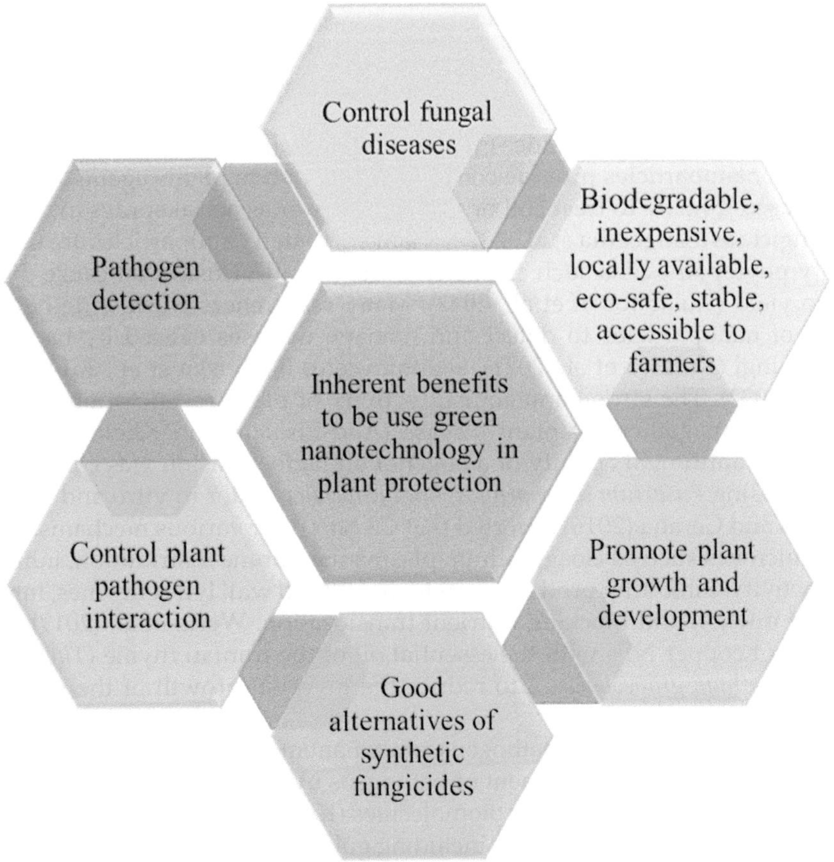

FIG. 1 Schematic representation of the inherent benefits of using green nanotechnology in plant protection. *Cited from Yifen, S., Kamrul, M.H., Golam, J.A., Mengqi, L., Hanqin, Y., Jie, Z., 2019. Applications of nanotechnology in plant growth and crop protection: a review. Molecules, 24.*

interaction of nanoparticles with the residing soil biota (Oliveira et al., 2015a, 2015b) and plants (Sadiko et al., 2014; Mishra and Singh, 2015; Patil and Chandrasekaran, 2020). Fungal cell walls and cell membrane compositions of chitin, lipids, phospholipids, and polysaccharides play important roles in the antifungal action of nanomaterials (Patil and Chandrasekaran, 2020). Silica-based nanoparticles are less expensive and more powerful, and they deliver target DNA in a plant system. Particle diameters ranging from 50 to 500 nm can easily enter xylem or phloem channels and be translocated throughout the plant system to control target pathogens (Machado et al., 2020). The interaction of pathogens with nanomaterials may disturb the structure of the pathogen biomolecule matrix, including catalytic enzymes deactivated, DNA breakage, and chromosomal aberrations (Patra et al., 2012; Shoeb et al., 2013).

Nanoparticles employ two different mechanisms to protect plants against phytopathogens: (a) act indirectly by inducing the defense response in plants against phytopathogens or (b) act directly as carriers for existing pesticides or other bioactive compounds to make them more effective with spray application, drenching, or soaking onto seeds, foliar tissue, or roots. Nanocarriers have several inherent benefits, including enhanced shelf life, improved solubility of water-soluble pesticides, increased stability of agrochemicals under environmental pressures (temperature and rain), reduced toxicity, and target-specific uptake (Hayles et al., 2017). Phytonanoparticles control fungal pathogen sporulation in four different ways: (1) inhibit spore germination, (2) interfere with sporulation and mycelial growth, (3) block water uptake, and (4) induce cytolytic leakage (Jampilek and Kralova, 2018; Chen et al., 2020). Bionanoparticles provide complete protection to plants against invading pathogens when used directly to treat soil or seed surfaces or as aerial sprays in comparison to chemical fungicides (Sangeetha et al., 2017). Plant-mediated nanoparticles are used to detect and identify phytopathogens such as bacteria, fungi, and viruses that cause diseases and reduce crop yield (Kulabhusan et al., 2022). Many researchers worldwide have reported the efficacy of nanoparticles to detect and manage diseases caused by bacteria (Rafael et al., 2022), fungi (Kutawa et al., 2021), actinomycetes (Golinska et al., 2014), and viruses (Rafael et al., 2022). The strong antagonistic activity of plant-based nanofungicides could be used in the management of plant disease (Abd-Elsalam and Khokhlov, 2015; Zahid et al., 2012). The antifungal activity of a eugenol oil nanoemulsion was confirmed against tomato wilt causing *Fusarium oxysporum* f. sp. *lycopersici* under in vitro and in vivo conditions. Malerba and Cerana (2016) reported that CS NPs offer various mechanisms to control pathogenic microbes such as isoagglutinin, plasma membrane distribution, inhibiting ATP synthetase activity, inhibiting production of toxin and cell wall lytic enzymes, inhibiting the expression of mRNA, and blocking nutrient translocation. Weria et al. (2019) studied the encapsulation of copper NPs with the essential oil of the Iranian thyme (*Thymus daenensis* L.) and dill (*Anethum graveolens* L.) to reduce the mycelial growth of the phytopathogen *Colletotrichum nymphaeae*.

In addition to controlling phytopathogens, phytonanomaterials enable the rapid, sensitive, and reliable diagnosis and management of pathogens by conjugation of phytonanoparticles with nucleic acids, proteins, and other biomolecules (Kashyap et al., 2015; Santos et al., 2019; Avila et al., 2022). Metallic nanoparticles including gold (Kulabhusan et al., 2022; Prasad et al., 2017) enable the rapid detection of pathogens. For example, nanogold immunosensor-based surface plasmon resonance (SPR) was used to detect *Tilletia indica*, a causal agent of the Karnal bunt disease of wheat (Singh et al., 2010). A copper NP-based electrochemical sensor was used as a detector (Wang et al., 2010) of the fungus *S. sclerotiorum* causing sclerotinia blight in oilseed crop. The chip-based hybridization of AgNPs (Schwenkbier et al., 2015) was used for the detection of the *Phytophthora* species.

Currently, plant-based hybrid nanoparticles have been shown to exhibit indistinguishable benefits in terms of improving fungicidal activity, promoting seed germination and plant growth, and offering environmental safety compared to the use of pure chemicals (Mishra et al., 2017; Shang et al., 2019). A hybrid nanomaterial is a composition of polymers/metals/organic molecules that act synergistically to control phytopathogens. For instance, lignin-based nanocarriers (NCs) have been recently investigated as a renewable, biodegradable, abundant, and inexpensive feedstock to develop sophisticated nanostructures (De Oliveira et al., 2014; Abd-Elsalam and Khokhlov, 2015; Shuaixuan et al., 2022). Min et al. (2009) demonstrated that AgNPs showed antifungal activity against *R. solani, S. sclerotium,* and *S. minor*. For mycotoxin detoxification, nanohybrid materials are used to detect, detoxify, and manage mycotoxin secretion in plant protection (Jampilek and Kralova, 2018). Hybrid nanoparticles can also detoxify the fungal toxin. To detoxify mycotoxin, Hamza et al. (2019) showed that the encapsulation of humic acid with β-glucan mannan lipid nanoparticles could detoxify the mycotoxin. Copper nanoparticles encapsulated with the essential oil of thyme and dill cause hyphal twisting and deformation in the mycelial growth of *C. nymphaeae* (Weria et al., 2019). The green synthesized copper NPs showed antifungal potential against *F. solani, F. oxysporum,* and *Nefofusicoccum* species (Pariona et al., 2019). Lemongrass essential oil-containing poly (lactic acid) nanocapsules control fungi that cause postharvest decay at an MIC dosage of 0.1% (*v*/v) against *Colletotrichum acutatum* and *Colletotrichum gloeosporioides* (Antonioli et al., 2020). Plant-mediated AgNPs suppress fungal pathogens such as *Aspergillus fumigates, A. niger, A. flavus, Trichophyton rubrum, C. albicans,* and *Penicillium* (Mansoor et al., 2021). Baldassarre et al. (2022) evaluated the fungistatic potential of pomegranate peel extract rich in phenolic content against *B. cinerea*.

4.1 Nanoparticle-plant interaction

Nanoparticle interaction with a plant induces several morphological and physiological changes in the plant, which leads to positive or negative effects on plants (Perez-de-Luque, 2017). The effect of NPs could be positive in terms of enhancing the seed germination rate, promoting plant growth, and improving plant metabolism or negative such as suppressing plant growth, affecting physiological process, etc. (Goswami et al., 2019). The nanoparticle reactivity with plants is driven by shape, size, chemical composition, and surface area of contact. Nanoparticles of 3.5–20nm are easily diffusible into the plant root through cytoplasmic membrane proteins, ionic channels, endocytosis, and inside the leaves through stomatal openings to show either the resistance or susceptibility of plants against phytopathogens, as shown in Fig. 2 (Chichiricco and Poma, 2015; Horejs, 2022). However, nanoparticles of size $\leq$43nm are suitable to penetrate the leaf with dorsal stomata (Eichert et al., 2008; Schreiber, 2005). Faraz et al. (2019) clarified the mechanisms of uptake, accumulation, and translocation of phytonanoparticles into plant cells, revealing the positive effects on the morphological and

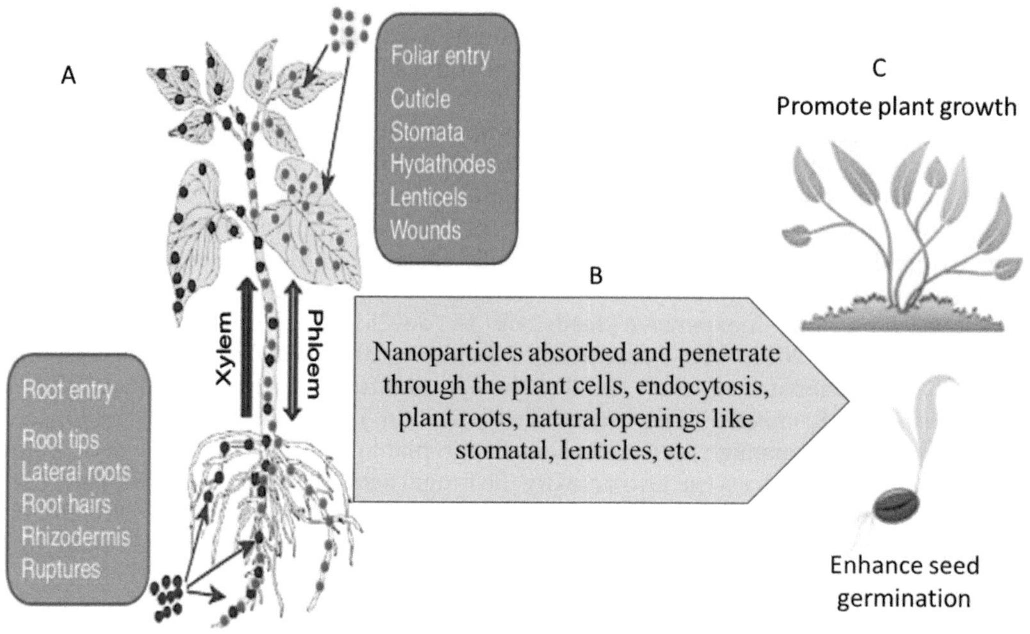

FIG. 2 Summary of nanoparticle interactions with plant: (A) Nanoparticle uptake by plants. (B) Nanoparticles absorbed and penetrate through the plant cells, endocytosis, plant roots, natural openings such as stomatal, lenticles, etc. (C) Enhance vegetation biomass and improve seed germination. *Cited from Perez-de-Luque, A., 2017. Interaction of nanomaterials with plants: what do we need for real applications in agriculture? Front. Environ. Sci., https://doi.org/10.3389/fenvs.2017.00012; Rai, M., Ingle, A., Pandit, R., Paralikar, P., Shende, S., Gupta, I., Biswas, J., Silva, S. 2018. Copper and copper nanoparticles: role in management of insect-pests and pathogenic microbes. Nanotechnol. Rev. 7. https://doi.org/10.1515/ntrev-2018-0031.*

physiological functions of the plants. The targeted delivery of the AgNP coated with protein into the leaf stomata and trichomes prevents pathogen entry into other parts of the plants (Spielman-Sun et al., 2020).

4.2 Nanoparticles-phytopathogens interaction

Nanostructured materials easily interact with microbes (Shabnam et al., 2021; Hawar et al., 2022) because of their excellent performance, selective adsorption of metal ions, operation over a broad range of ecological conditions (pH, ionic strength, temperature), and high biosorption capacity (Ferdous et al., 2012; Carapar et al., 2022). The degree of interaction of nanoparticles with phytopathogens depends on the nanoparticle porosity and concentration (Harish et al., 2022), and the microbial sensitivity of nanomaterials may cause cell death (Ghosh and Bera, 2021; Mansoor et al., 2021). Nanomaterials enter the fungal cell wall either directly or indirectly through specific receptor-mediated or protein-ion channels, as shown in

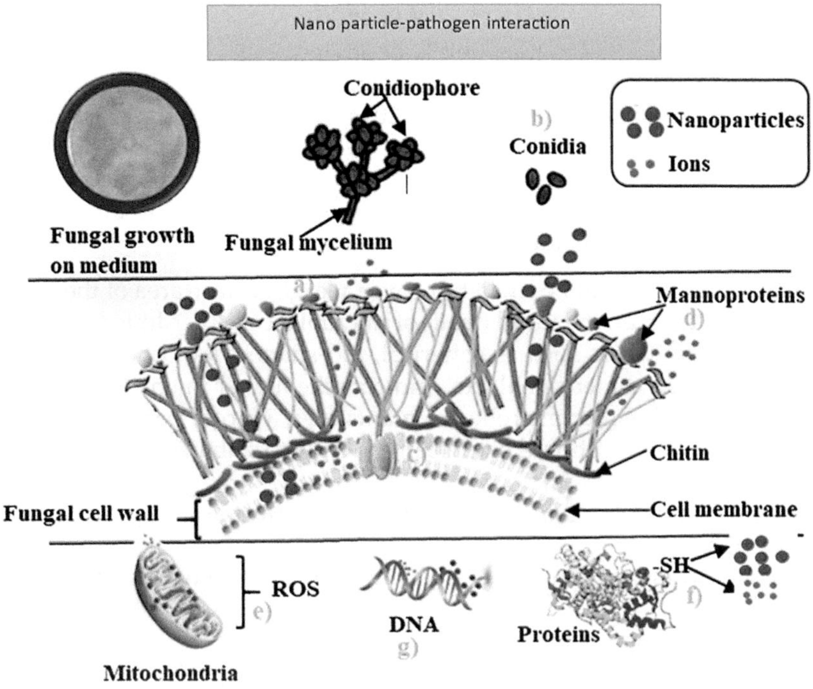

FIG. 3 The interaction and incorporation of different types of nanoparticles into the cell membrane and cellular components of a fungus occurs via possible mechanisms: (A) Metallic nanoparticle ions bind with membrane protein and interfere in cell permeability. (B) Nanoparticles bind with conidia and inhibit its germination. (C) Nanoparticles disrupt the fungal cell membrane, which affects the electron transport, protein oxidation, and membrane potential. (D) Interfering with protein electron transport system. (E) Affect the mitochondrial membrane function and inactivate replication and transcription process and increase the reactive oxygen species (ROS). (F) Increase the level of reactive oxygen species that oxidation catalysis of proteins, membranes, and deoxyribonucleic acid (DNA), thus, affects the uptake of mineral and water by fungi. (G) Metallic ions of nanoparticles having genotoxic potential, which causes fungal cell death. *Cited from Cruz-Luna, A.R., Cruz-Martinez, H., Vasquez-Lopez, A., Medina, D.I., 2021. Metal nanoparticles as novel antifungal agents for sustainable agriculture: current advances and future directions. J. Fungi 7, 1033. https://doi.org/10.3390/jof7121.*

Fig. 3 (Kalia et al., 2020a). The interaction of nanoparticles with pathogens causes better attachment and absorption into the microbial cell membrane as well as major structural and morphological changes, and hinders DNA replication and microbial cell death (Ghosh and Bera, 2021). Mansoor et al. (2021) demonstrated that smaller AgNPs could inhibit the growth of many fungi such as *A. fumigates*, *A. niger*, *A. flavus*, *T. rubrum*, *C. albicans*, and *Penicillium* by producing reactive oxygen species and free radicals, which affect the translation and transcription process, damage the proton pump and lipid peroxidation, and alter cell membrane permeability, ultimately leading to cell death. Tryfon et al. (2021) reported that CuZn and ZnO nanoflowers as nanofungicides that translocate from leaves to the root through the stem of *Lactuca sativa* provide protection against *B. cinerea* and *S. sclerotiorum* as well as improve the photosynthetic rate of the plant.

4.2.1 Factors affecting antifungal activity of phytonanofungicides

Different studies indicated that the stability of phytonanofungicides is achieved in terms of the biological and chemical properties of plant extracts to exhibit strong antifungal potential (Jeevanandam et al., 2018; Navya and Daima, 2016). The phytonanoparticle uptake, translocation, and accumulation in plants are affected by the physical factors of the environment such as reaction of temperature, pH value, pressure, time, particle size, and plant biometabolite concentration (Fig. 4). Thus, important factors need to be considered during preparation of phytonanofungicides (Khan et al., 2022):

- The nanoparticle size should be small but cover a large surface area of the phytopathogen, which is effective to cause the rapid death of microbial cells (Ferdous and Nemmar, 2020).
- Varied nanoparticle shapes such as spherical, rod-shaped, nanoshells, nanocages, nanowires, triangular, dimensional, etc., indicate the higher intensity of the interaction of the nanoparticles with either microbes or plants than a larger particle size.
- Viscosity and temperature should be chosen carefully to stabilize the antifungal properties of phytonanoparticles.
- The viscosity of nanobiofungicides is governed by surfactant, capping agents, water, or carrier components.
- Surfactant concentration should be high enough to facilitate nanoparticles diffusion into microbial cells.
- Force of attraction among nanoparticles.

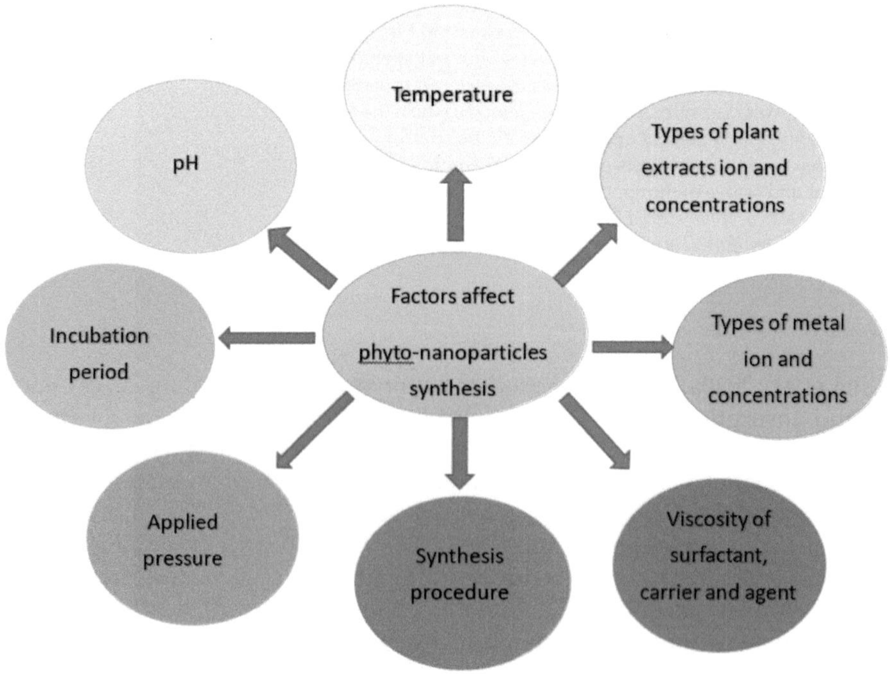

FIG. 4 Factors affecting the synthesis of phytonanoparticles.

- Reaction temperature is less than 100°C for the synthesis of phytonanoparticles.
- Pressure and pH values of the reaction medium should be maintained.
- The solution exposure times to light and storage conditions.

5 Advantages and disadvantages of using phytonanotechnology in crop protection

The application of nanotechnology in plant systems, called "phytonanotechnology," plays a crucial role in agriculture and general plant sciences. Phytonanotechnology benefits plant systems in various ways; however, special attention is devoted to transforming plant products with strong antifungal potential into nanoparticles for use in crop protection. Phytochemicals are the best natural source for producing ecosafe and biodegradable fungicides, which could be used either as protectants or as carriers for insecticides, fungicides, and herbicide formulations. The small size of phytonanoparticles covers a larger surface area and acts against the target pathogen of many agricultural crops. Utilization of plant-based microcarriers in modern agriculture systems might help to decrease the extensive use of fungicides and improve plant functions. Besides the antifungal potential of bionanoparticles, they induce a direct or indirect defense response in host plants against biotic and abiotic constraints. Thus, scientists have studied the introduction of plant-based nanofungicides in the agriculture system in a simple, non cost-effective manner to expose extra benefits such as high efficacy, durability, and ecofriendliness. Scientists ensure they should also have better leaching, evaporation, photolysis, chemical hydrolysis, and biodegradation properties than pure forms of phytochemicals (Abd-Elsalam and Alghuthaymi, 2015). Many studies reported that combined forms of phytochemicals and inorganic nanoparticles such as S, Ag, CuO, MgO, and ZnO possessed strong antifungal activity and allowed dsRNA for RNA interference (RNAi)-mediated protection in plant systems. Despite these promising perspectives, major obstacles include the nontarget site delivery of nanomaterials, which induces toxicity in plant cell structures (Chengjun and Bing, 2020). The phytonanoparticles used to combat plant pathogens are acceptable throughout the world, but the negative impact of nanoparticles on human and environmental health as well as the soil biota cannot be ignored, particularly in agriculture (Vande and Arai, 2012). There are bioengineering challenges associated with the application of phytonanoparticles inside plant cells and organelles to improve the plant's physiological functions or increase its resistance against pathogenic diseases (Ghorbanpour and Fahimirad, 2017; Ansari et al., 2020; Lowry et al., 2019). On the other hand, the production of plant-based nanoparticles is a cost-effective approach that requires proper lab conditions for their synthesis. Despite the benefits of green nanotechnology in crop protection systems, metallic-based phytonanomaterials may be toxic. This leads to the analysis of phytonanoparticles to test the rate of penetration and pathway of transport inside plant cells.

5.1 Challenges

This chapter addressed some challenges for the synthesis of phytonanoparticles and their introduction in the agriculture sector to protect crops against various fungal diseases. Nanotechnology is used to change the physical, chemical, and biological properties of antifungal plant extracts into something that is nanoscale. The major factors that may challenge the synthesis and antifungal properties of phytonanoparticles were discussed in this chapter, such as phytonanoparticle preparation and stabilization, which require specific protocols to maintain their size, shape, texture, and antifungal properties. The reaction temperature, incubation time, and pH of the solution also play significant roles in the synthesis and antifungal fungal activity of phytonanoparticles. There are issues associated with the use of nanotechniques in improving the quality of antifungal bioactive metabolites, which may be manipulated by using nanoprecipitation, physically applied pressure and temperature, solution, emulsion, or polymerization methods. These methods are expensive, time-consuming processes that require a well-trained person that can operate expensive machines to measure the quality of the synthesized plant-mediated metallic nanoparticles. The expanded cost of engineering nanodevices must be affordable and safer before releasing the technology into the environment. The types of solvent, surfactant, and capping agents also influence the morphology and rate of biosynthesis of the plant-mediated nanoparticles. Although nano-sized phytochemicals showed higher antifungal efficacy than pure phytochemicals, physical and chemical factors can affect the extraction and purification of phytonanoparticles. In addition, climatic change poses a risk of phytochemical thermal and photo degradation that hinders the shelf life and antifungal potential of phytonanochemicals under open field conditions. Thus, physical environmental factors pose difficulty in the application of phytonanoparticles in crop protection because increased atmospheric temperature and exposure to gases decrease the target-specific penetration and accumulation of phytonanoparticles. Thus, there is a need to understand all the issues associated with the implementation of phytonanotechnology in crop protection programs. We have addressed the major challenges to get plant-mediated nanotechnology used in the research field of crop protection.

6 Conclusion and future perspective

This chapter provides our insights into crop protection from the perspective of plant-based nanotechnology. Several researchers are focusing on introducing nanotechnology into crop protection. Current obstacles arise with the applications of phytonanotechnology; insights are used to tackle them in this dynamic research field. Concerning the ecotoxicological effects of synthetic fungicides, the use of phytonanotechnology in crop protection is an opportunistic and challenging strategy for improving plant health, controlling fungal pathogens, and reducing the overuse of synthetic fungicides. For better results from phytonanotechnology, researchers need to choose "smart plants" and amplify the antifungal potential of phytonanoparticles against devastating pathogens. Utilization of permissible antifungal bioactive compounds of plants in crop protection is an ecosafe and realistic approach for the management of phytopathogens in the future. Phytonanotechnology supports a higher level of sustainability for our planet in terms of being safe for humans, animals, and residing soil microbes than chemical-based synthesis of nanocomposites. Thus, researchers are focusing on evolving and fabricating nanocomposites through green chemistry that would be ecosafe in the management of various fungal diseases of plants. Applications of phytonanotechnology induce plant tolerance against biotic and abiotic factors as well as increasing crop production and protection. They also help in remediating environmental pollution to safeguard human society. Furthermore, future research in the 21st century is given to synthesized green nanohybrids in a field known as "green nanotechnology." These carry biological materials and other organic and inorganic constituents to control fungal pathogens that cause many plant diseases. Green nanohybrid particles containing botanical extracts and organic or inorganic metals work together to boost antifungal activity and plant growth. The most striking green nanohybrids are needed to meet the demands of growers, consumers, and environment activists. Researchers ensure that plant-based nanoparticles will not impose any adverse effects on plants, environments, or living organisms. With globalization, industrialization, and socialization, plant-based nanoformulations would be safe and ensure the proper delivery of phytochemicals. Because of public health concerns and legislative uncertainties, the antimicrobial potential of phytochemicals requires industrialization. Researchers will ensure that the encapsulated antimicrobial sources of phytonanoparticles have higher biodegradability, a longer shelf life, are feasible and easily accessible to farmers, and have a higher market value than pure phytoextracts.

References

Abd-Elsalam, K.A., Alghuthaymi, M.A., 2015. Nanobiofungicides: is it the next-generation of fungicides? J. Mater. Sci. Nanotechnol. 2 (2), 1–3.

Abd-Elsalam, K.A., Khokhlov, A.R., 2015. Eugenol oil nanoemulsion: antifungal activity against *Fusarium oxysporum* f. sp. *vasinfectum* and phytotoxicity on cottonseeds. Appl. Nanosci. 5, 255–265.

Abo-Elyousr, K.A.M., Ali, E.F., Sallam, N.M.A., 2022. Alternative control of tomato wilt using the aqueous extract of *Calotropis procera*. Horticulturae 8, 197. https://doi.org/10.3390/horticulturae8030197.

Abubakar, A.R., Haque, M., 2020. Preparation of medicinal plants: basic extraction and fractionation procedures for experimental purposes. J. Pharm. Bioallied Sci. 12 (1), 1–10. https://doi.org/10.4103/jpbs.JPBS_175_19.

Aguilar-Perez, K., Medina, D., Narayanan, J., Parra-Saldivar, R., Iqbal, H., 2021. Synthesis and nano-sized characterization of bioactive oregano essential oil molecule-loaded small unilamellar nanoliposomes with antifungal potentialities. Molecules 26, 2880.

Ahmad, N., Fozia, M.J., Zia Ul, H., Ijaz, A., Abdul, W., Zia Ul, I., Riaz, U., Ahmed, B., Mohamed, M.A., Fatma, M.E., Muhammad, Y.K., 2022. Green fabrication of silver nanoparticles using *Euphorbia serpens* Kunth aqueous extract, their characterization, and investigation of its *in-vitro* antioxidative, antimicrobial, insecticidal, and cytotoxic activities. BioMed. Res. Int., 11. https://doi.org/10.1155/2022/5562849. Article ID 5562849.

Al-Baldawy, M.S.M., Matloob, A.A.A.H., Almammory, M.K.N., 2021. Effect of plant extracts and biological control agents on *Rhizoctonia solani* Kuhn. IOP Conf. Ser. Earth Environ. Sci. 735, 012079.

Alghuthaymi, M.A., Rajkuberan, C., Rajiv, P., Kalia, A., Bhardwaj, K., Bhardwaj, P., Abd-Elsalam, K.A., Valis, M., Kuca, K., 2021. Nanohybrid antifungals for control of plant diseases: current status and future perspectives. J. Fungi 7, 48. https://doi.org/10.3390/jof7010048.

Ali, A.M., Ahmed, T., Wu, W., Hossain, A., Hafeez, R., Masum, M.I.M., Wang, Y., Qianli, A., Sun, G., Li, B., 2020a. Advancements in plant and microbe-based synthesis of metallic nanoparticles and their antimicrobial activity against plant pathogens. Nanomater 10, 1146. https://doi.org/10.3390/nano10061146.

Ali, M., Urooj, H., Maria, K., Hassan, J.C., Muhammad, F.H.M., 2020b. Facile single step preparations of phyto-nanoparticles of iron in *Calotropis procera* leaf extract to evaluate their antifungal potential against *Alternaria alternata*. Curr. Plant Biology 23. https://doi.org/10.1016/j.cpb.2020.100157.

Ali, E.F., Al-Yasi, H.M., Issa, A.A., Hessini, K., Hassan, F.A.S., 2022. Ginger extract and fulvic acid foliar applications as novel practical approaches to improve the growth and productivity of Damask Rose. Plan. Theory 11, 412.

Alshaikh, N.A., Perveen, K., 2021. Susceptibility of fluconazole resistant *Candida albicans* to thyme essential oil. Microorganisms 9, 2454. https://doi.org/10.3390/microorganisms9122454.

Alshehri, A.A., Malik, M.A., 2020. Phytomediated photo-induced green synthesis of silver nanoparticles using *Matricaria chamomilla* L. and its catalytic activity against rhodamine, B. Biomol. Ther. 10, 1604.

Amruta, S., Jaiprakash, S., Shampa, C., Ajay, V.S., Rajendra, P., Suresh, G., 2019. Helminthicidal and larvicidal potentials of biogenic silver nanoparticles synthesized from medicinal plant *Momordica charantia*. Med. Chem. 15, 781–789.

Angela, T., Marcello, L., Felisa, C., Christian, C., Luisa, D.M., Francesca, F., Donatella, P., D'Ambrosio, F., Ramundo, P., 2022. Nanotechnology-based green and efficient alternatives for the management of plant diseases. In: Balestra, G.M., Fortunati, E. (Eds.), Micro and Nano Technologies, Nanotechnology-Based Sustainable Alternatives for the Management of Plant Diseases. Elsevier, pp. 253–262, https://doi.org/10.1016/B978-0-12-823394-8.00014-7.

Ansari, M.H., Lavhale, S., Kalunke, R.M., Srivastava, P.L., Pandit, V., Gade, S., Yadav, S., Laux, P., Luch, A., Gemmati, D., 2020. Recent advances in plant nanobionics and nanobiosensors for toxicology applications. Curr. Nanosci. 16, 27–41.

Antonioli, G., Fontanella, G., Echeverrigaray, S., Delamare, A.P.L., Pauletti, G.F., Barcellos, T., 2020. Poly(lactic acid) nanocapsules containing lemongrass essential oil for postharvest decay control: *in-vitro* and *in-vivo* evaluation against phytopathogenic fungi. Food Chem. 326, 126997.

Antunes, M.D., Cavaco, A.M., 2010. The use of essential oils for post-harvest decay control. A review. Flavour Fragr. J. 25 (5), 351–366.

Arora, H., Sharma, A., Poczai, P., Sharma, S., Haron, F.F., Gafur, A., Sayyed, R.Z., 2022. Plant derived protectants in combating soil-borne fungal infections in tomato chilli. J. Fungi 8 (2), 213. https://doi.org/10.3390/jof8020213.

Avila, Q., Graciela, G., Patrycja, R.M., 2022. Engineered nanomaterials in plant diseases: can we combat phytopathogens? Appl. Microbiol. Biotechnol. 106, 1–13. https://doi.org/10.1007/s00253-021-11725-w.

Babu, J., Dar, M.A., Kumar, V., 2008. Bio-efficacy of plant extracts to control *Fusarium solani* f. sp. *melongenae* incitant of brinjal wilt. Global J. Biotech. Biochem. 3, 56–59.

Baldassarre, F., Vegaro, V., De Castro, F., Biondo, F., Suranna, P.G., Papadia, P., Fanizzi, P.F., Rongai, D., Ciccarella, G., 2022. Enhanced bioactivity of pomegranate peel extract following controlled release from $CaCO_3$ nanocrystals. Bioinorg. Chem. Appl., 16. https://doi.org/10.1155/2022/6341298. Article ID 6341298.

Beatovic, D., Krsti-Milosevic, D., Trifunovic, S., Siljegovic, J., Glamoclija, J., Ristic, M., Jelacic, S., 2015. Chemical composition, antioxidant and antimicrobial activities of the essential oils of twelve *Ocimum basilicum* L. cultivars grown in Serbia. Records Nat. Prod. 9, 62–75.

Beyki, M., Zhaveh, S., Khalili, S.T., Rahmani, C.T., Abollahi, A., Bayat, M., Tabatabaei, M., Mohsenifar, A., 2014. Encapsulation of *Mentha piperita* essential oils in chitosan-cinnamic acid nanogel with enhanced antimicrobial activity against *Aspergillus flavus*. Ind. Crop. Prod. 54, 310–319.

Bhavyasree, P.G., Xavier, T.S., 2020. Green synthesis of copper oxide/carbon nanocomposites using the leaf extract of *Adhatoda vasica* Nees, their characterization and antimicrobial activity. Heliyon 6, 03323.

Biswas, M.K., Tanmay, G., 2018. Evaluation of Phyto-extracts, biological agents and chemicals against the development of *Alternaria brassicae in-vitro* and *in-vivo*. Eur. J. Med. Plants 22 (3), 1. https://doi.org/10.9734/EJMP/2018/40412h.

Boutrot, F., Zipfel, C., 2017. Function, discovery, and exploitation of plant pattern recognition receptors for broad-spectrum disease resistance. Annu. Rev. Phytopathol. 55, 257–286. https://doi.org/10.1146/annurev-phyto080614-120106.

Camara, M.C., Campos, E.V.R., Monteiro, R.A., Do Espirito, S.P.A., De Freitas, P.P.L., Fraceto, L.F., 2019. Development of stimuli-responsive nano-based pesticides: emerging opportunities for agriculture. J. Nanobiotechnology 17, 1–19.

Carapar, I., Jurkovic, L., Pavicic-Hamer, D., Hamer, B., Lyons, D.M., 2022. Simultaneous influence of gradients in natural organic matter and abiotic parameters on the behavior of silver nanoparticles in the transition zone from freshwater to saltwater environments. Nanomater 12, 296. https://doi.org/10.3390/nano12020296.

Chen, J., Wu, L., Lu, M., Lu, S., Li, Z., Ding, W., 2020. Comparative study on the fungicidal activity of metallic MgO nanoparticles and macroscale MgO against soilborne fungal phytopathogens. Front. Microbiol. 11, 365. https://doi.org/10.3389/fmicb.2020.00365.

Chengjun, L.A., Bing, Y., 2020. Opportunities and challenges of phyto-nanotechnology. Environ. Sci.: Nano J. https://doi.org/10.1039/d0en00729c.

Chetan, P., Arpita, R., Suresh, G.A., Khusro, M.N., Islam, T., Bin Emran, S.E.L., Mayeen, U.K., David, A.B., 2022. Biological agents for synthesis of nanoparticles and their applications. J. King Saudi Univ. Sci. 34 (30), 101869. https://doi.org/10.1016/j.jksus.2022.101869.

Chhipa, H., 2017. Nanofertilizers and nanopesticides for agriculture. Environ. Chem. Lett. 15, 15–22.

Chichiricco, G., Poma, A., 2015. Penetration and toxicity of nanomaterials in higher plants. Nanomater 5 (2), 851–873.

Cindi, M.D., Shittu, T., Sivakumar, D., Bautista, B.S., 2015. Chitosan boehmite alumina nanocomposite films and thyme oil vapour control brown rot in peaches (*Prunus persica* L.) during postharvest storage. Crop Prot. 72, 127–131.

Congyi, Z.L., Mengying, A., Mebeaselassie, Z., Jiwu, L.J., 2019. Antifungal activity and mechanism of action of tannic acid against *Penicillium digitatum*. Physiol. Mol. Plant Pathol. 107. https://doi.org/10.1016/j.pmpp.2019.04.009.

Cota, A.O., Onofre, C.R.M., Burgos, H.A., Marina, E.B.J., Plascencia, J.M., 2013. Controlled release matrices and micro/nanoparticles of chitosan with antimicrobial potential: development of new strategies for microbial control in agriculture. J. Sci. Food Agric. 93, 1525–1536.

Cruz-Luna, A.R., Cruz-Martinez, H., Vasquez-Lopez, A., Medina, D.I., 2021. Metal nanoparticles as novel antifungal agents for sustainable agriculture: current advances and future directions. J. Fungi 7, 1033. https://doi.org/10.3390/jof7121.

De Oliveira, J.L., Campos, E.V.R., Bakshi, M., Abhilash, P.C., Fraceto, L.F., 2014. Application of nanotechnology for the encapsulation of botanical insecticides for sustainable agriculture: prospects and promises. Biotechnol. Adv. 32, 1550–1561.

Din, M.I., Nabi, A.G., Rani, A., Aihetasham, A., Mukhtar, M., 2018. Single step green synthesis of stable nickel and nickel oxide nanoparticles from *Calotropis gigantea*: catalytic and antimicrobial potentials. Environ. Nanotechnol. Monit. Manag. 9, 29–36.

Ehsan, H., Khan, A.M., Atiq, M., Ashraf, W., Yaseen, S., Bilal, A.M., Akbar, U., Fatima, I., Shahid, T., 2020. Efficacy of phytoextracts against fungal diseases of vegetables. J. Biodivers. Environ. Sci. 16 (4), 71–82.

Eichert, T., Kurtz, A., Steiner, U., Goldbach, H.E., 2008. Size exclusion limits and lateral heterogeneity of the stomatal foliar uptake pathway for aqueous solutes and water-suspended nanoparticles. Physiol. Plant 134 (1), 151–160. https://doi.org/10.1111/j.1399-3054.2008.01135.x. Epub 2008 May 20. PMID: 18494856.

El Hamdani, N.F.A., Najoie, F., Rabiaa, E., Ahmed, K.S., 2016. Antifungal activity of the alkaloids extracts from aerial parts of *Retama monosperma*. Res. J. Pharm. Biol. Chem. 7, 965–971.

Elgharbawy, A.A.M., Samsudin, N., Benbelgacem, F.F., Hashim, Z.H.Y.Y., Hamza, M.S., Santhanam, J., 2020. Phytochemicals with antifungal properties: cure from nature. Malaysian J. Microbiol. 16 (4), 324–345. https://doi.org/10.21161/mjm.190551.

El-Naggar, N.E.A., Saber, W.I.A., Zweil, A.M., et al., 2022. An innovative green synthesis approach of chitosan nanoparticles and their inhibitory activity against phytopathogenic *Botrytis cinerea* on strawberry leaves. Sci. Rep. 12, 3515. https://doi.org/10.1038/s41598-022-07073-y.

Elshaer, E.E., Elwakil, B.H., Eskandrani, A., Elshewemi, S.S., Olama, Z.A., 2022. Novel clotrimazole and *Vitis vinifera* loaded chitosan nanoparticles: antifungal and wound healing efficiencies. Saudi J. Biol. Sci. 29 (3), 1832–1841. https://doi.org/10.1016/j.sjbs.2021.10.041.

Faraz, A., Faizan, M., Sami, F., Siddiqui, H., Pichtel, J., Hayat, S., 2019. Nanoparticles: biosynthesis, translocation and role in plant metabolism. Nanatechnology 13 (3), 345–352. https://doi.org/10.1049/iet-nbt.2018.5251.

Fatimah, A., Shahad, A.A., Raedah, I.A., Abdulaziz, A.A., Rana, M.A., Hajar, F.A., Nadine, M.S.M., 2022. Comparative study of antifungal activity of two preparations of green silver nanoparticles from Portulaca oleracea extract. Saudi J. Biol. Sci. https://doi.org/10.1016/j.sjbs.2021.12.056.

Ferdous, Z., Nemmar, A., 2020. Health impact of silver nanoparticles: a review of the biodistribution and toxicity following various routes of exposure. Int. J. Mol. Sci. 21 (7), 2375. https://doi.org/10.3390/ijms21072375.

Ferdous, S., Ioannidis, M., Henneke, D.E., 2012. Effects of temperature, pH, and ionic strength on the adsorption of nanoparticles at liquid-liquid interfaces. J. Nanopart. Res. 14. https://doi.org/10.1007/s11051-012-0850-4.

Ghazanfar, M., Iqbal, Z., Ahmad, S., Qamar, M., 2019. Antifungal activity by resistance inducing salicylic acid and plant extracts against postharvest rots of tomato. Int. J. Biosci. 14, 264–274. https://doi.org/10.12692/ijb/14.4.264-274.

Ghorbanpour, M., Fahimirad, S., 2017. Medicinal Plants and Environmental Challenges. Springer, pp. 247–257.

Ghosh, S., Bera, T., 2021. Unraveling the mechanism of nanoparticles for controlling plant pathogens and pests. In: Advances in Nano-Fertilizers and Nano-Pesticides in Agriculture. Woodhead Publishing, https://doi.org/10.1016/B978-0-12-820092-6.00016-1.

Gillmeister, M., Silvia, B., Raschke, A., Geistlinger, J., Kabrodt, K., Baltruschat, H., Holger, B.D., Schellenberg, I., 2019. Polyphenols from rheum roots inhibit growth of fungal and oomycete phytopathogens and induce plant disease resistance. Plant Dis. 103, 1674–1684. https://doi.org/10.1094/PDIS-07-18-1168-RE.

Gogos, A., Knauer, K., Bucheli, T.D., 2012. Nanomaterials in plant protection and fertilization: current state, foreseen applications, and research priorities. J. Agric. Food Chem. 60, 9781–9792.

Golinska, P., Wypij, M., Ingle, A.P., Gupta, I., Dahm, H., Rai, M., 2014. Biogenic synthesis of metal nanoparticles from actinomycetes: biomedical applications and cytotoxicity. Appl. Microbiol. Biotechnol. 98, 8083–8097.

Goswami, P., Yadav, S., Mathur, J., 2019. Positive and negative effects of nanoparticles on plants and their applications in agriculture. Plant Sci. Today 6, 232. https://doi.org/10.14719/pst.2019.6.2.502.

Goswami, S.S.P., Rath, D., Thompson, S., Hedstrom, M., Annamalai, R., Pramanick, B.R., Ilic, S., Sarkar, S., Hooda, C.-A., Nijhuis, J., Martin, R.S., Williams, S.G., Venkatesan, T., 2020. Charge disproportionate molecular redox for discrete memristive and memcapacitive switching. Nat. Nanotechnol. 15, 380–389.

Habtemariam, B.A., Qomer, M., 2020. Plant extract mediated synthesis of nickel oxide nanoparticles. Mater. Int. 2 (2), 205–209. https://doi.org/10.33263/Materials22.205209.

Hamza, Z., El-Hashash, M., Aly, S., Hathout, A., Soto, E., Sabry, B., Ostroff, G., 2019. Preparation and characterization of yeast cell wall beta-glucan encapsulated humic acid nanoparticles as an enhanced aflatoxin B1 binder. Carbohydr. Polym. 203, 185–192.

Hano, C., Abbasi, B.H., 2022. Plant-based green synthesis of nanoparticles: production, characterization and applications. Biomolecules 12, 31. https://doi.org/10.3390/biom12010031.

Haq, I.U., Sarwar, M.K., Faraz, A., Latif, M.Z., 2020. Synthetic chemicals: major component of plant disease management. In: Haq, I.U., Ijaz, S. (Eds.), Plant Disease Management Strategies for Sustainable Agriculture through Traditional and Modern Approaches. Springer, Berlin/Heidelberg, Germany, pp. 53–81.

Harish, V., Tewari, D., Gaur, M., Yadav, A.B., Swaroop, S., Bechelany, M., Barhoum, A., 2022. Review on nanoparticles and nanostructured materials: bioimaging, biosensing, drug delivery, tissue engineering, antimicrobial, and agro-food applications. Nanomater 12, 457. https://doi.org/10.3390/nano12030457.

Hasheminejad, N., Khodaiyan, F., Safari, M., 2019. Improving the antifungal activity of clove essential oil encapsulated by chitosan nanoparticles. Food Chem. 275, 113–122.

Hawar, N.S., Al-Shmgani, H.S., Al-Kubaisi, A.Z., Sulaiman, M.G., Dewir, H.Y., Jesamine, R.J., 2022. Green synthesis of silver nanoparticles from *Alhagi graecorum* leaf extract and evaluation of their cytotoxicity and antifungal activity. J. Nanomater. https://doi.org/10.1155/2022/1058119.

Hayles, J., Johnson, L., Worthley, C., Losic, D., 2017. Nanopesticides: a review of current research and perspectives. In: New Pesticides and Soil Sensors. Academic Press, pp. 193–225. https://doi.org/10.1016/B978-0-12-804299-1.00006-0.

Horejs, C., 2022. Nanoparticle transport in plants. Nat. Rev. Mater. 7, 5. https://doi.org/10.1038/s41578-021-00403-y.

Huang, W., Wang, C., Duan, H., Bi, Y., Wu, D., Du, J., Yu, H., 2018. Synergistic antifungal effect of biosynthesized silver nanoparticles combined with fungicides. Int. J. Agric. Biol. 20, 1225–1229.

Jajic, I., Dudas, T., Krstovic, S., Krska, R., Sulyok, M., Bagi, F., et al., 2019. Emerging fusarium mycotoxins fusaproliferin, beauvericin, enniatins, and moniliformin in Serbian maize. Toxins 11, 357. https://doi.org/10.3390/toxins11060357.

Jampilek, J., Kralova, K., 2018. Benefits and potential risks of nanotechnology applications in crop protection. In: Abd-Elsalam, K.A., Prasad, R. (Eds.), Nanobiotechnology Applications in Plant Protection, Nanotechnology in the Life Sciences. Springer, Switzerland, pp. 189–246, https://doi.org/10.1007/978-3-319-91161-8_8. Chapter 8,.

Jeevanandam, J., Barhoum, A., Chan, Y.S., Dufresne, A., Danquah, M.K., 2018. Review on nanoparticles and nanostructured materials: history, sources, toxicity and regulations. Beilstein J. Nanotechnol. 9, 1050–1074. https://doi.org/10.3762/bjnano.9.98.

Jeevanandam, J., Kiew, F.S., Ansah, B.S., Lau, Y.S., Barhoum, A., Danquah, K.M., Rodrigues, J., 2022. Green approaches for the synthesis of metal and metal oxide nanoparticles using microbial and plant extracts. J. Nanosacle 7.

Jepson, P.C., Guzy, M., Blaustein, K., Sow, M., Sarr, M., Mineau, P., Kegley, S., 2014. Measuring pesticide ecological and health risks in West African agriculture to establish an enabling environment for sustainable intensification. Philos. Trans. R. Soc. B: Biol. Sci. 369, 1–18.

Jiang, M., Song, Y., Kanwar, M.K., et al., 2021. Phytonanotechnology applications in modern agriculture. J. Nanobiotechnol. 19, 430. https://doi.org/10.1186/s12951-021-01176-w.

Kah, R.S., Kookana, A.G., Bucheli, T.D., 2018. A critical evaluation of nanopesticides and nanofertilizers against their conventional analogues. Nat. Nanotechnol. 13, 677–684.

Kalagatur, N.K., Nirmal Ghosh, O.S., Sundararaj, N., Mudili, V., 2018. Antifungal activity of chitosan nanoparticles encapsulated with *Cymbopogon martinii* essential oil on plant pathogenic fungi *Fusarium graminearum*. Front. Pharmacol. 9, 610. https://doi.org/10.3389/fphar.2018.00610.

Kalia, A., Sharma, S.P., Kaur, H., Kaur, H., 2020a. Novel nanocomposite-based controlled-release fertilizer and pesticide formulations: prospects and challenges. In: Abd-Elsalam, K.A. (Ed.), Multifunctional Hybrid Nanomaterials for Sustainable Agri-Food and Ecosystem. Elsevier, Amsterdam, The Netherlands, pp. 99–134.

Kalia, A., Abd-Elsalam, K.A., Kuca, K., 2020b. Zinc-based nanomaterials for diagnosis and management of plant diseases: ecological safety and future prospects. J. Fungi 6, 222.

Kashyap, P.L., Xiang, X., Heiden, P., 2015. Chitosan nanoparticle-based delivery systems for sustainable agriculture. Int. J. Biol. Macromol. 77, 36–51.

Kaskatepe, B., Aslan Erdem, S., Ozturk, S., Safi Oz, Z., Subasi, E., Koyuncu, M., Vlainic, J., Kosalec, I., 2022. Antifungal and anti-virulent activity of *Origanum majorana* L. essential oil on *Candida albicans* and in-vivo toxicity in the *Galleria mellonella* larval model. Molecules 2022 (27), 663. https://doi.org/10.3390/molecules27030663.

Katarzyna, D., Izabela, M., 2022. The role of nanoparticles in sustainable agriculture. In: Chojnacka, K., Saeid, A. (Eds.), Smart Agrochemicals for Sustainable Agriculture. Academic Press, pp. 225–278, https://doi.org/10.1016/B978-0-12-817036-6.00007-8.

Kaur, K., Kaushal, S., Rani, R., 2019. Chemical composition, antioxidant and antifungal potential of clove (*Syzygium aromaticum*) essential oil, its major compound and its derivatives. J. Essent. Oil Bear. Plants 22, 1195–1217.

Khalili, S.T., Mohsenifar, A., Beyki, M., Zhaveh, S., Rahmani, C.T., Abdollahi, A., Bayat, M., Tabatabaei, M., 2015. Encapsulation of thyme essential oils in chitosan-benzoic acid nanogel with enhanced antimicrobial activity against *Aspergillus flavus*. LWT Food Sci. Technol. 60, 502–508. https://doi.org/10.1016/j.lwt.2014.07.054.

Khan, F., Shariq, M., Asif, M., Siddiqui, M.A., Malan, P., Ahmad, F., 2022. Green nanotechnology: plant-mediated nanoparticle synthesis and application. Nano 12, 673. https://doi.org/10.3390/nano12040673.

Kheiri, A., Jorf, S.A.M., Malihipour, A., Saremi, H., Nikkhah, M., 2016. Application of chitosan and chitosan nanoparticles for the control of *Fusarium* head blight of wheat (*Fusarium graminearum*) in vitro and greenhouse. Int. J. Biol. Macromol. 93, 1261–1272. https://doi.org/10.1016/j. ijbiomac.2016.09.072.

Krishnaraj, C., Ramachandran, R., Mohan, K., Kalaichelvan, P.T., 2012. Optimization for rapid synthesis of silver nanoparticles and its effect on phytopathogenic fungi. Spectrochim. Acta A 93, 9599.

Kulabhusan, P.K., Tripathi, A., Kant, K., 2022. Gold nanoparticles and plant pathogens: an overview and prospective for biosensing in forestry. Sensors 22, 1259. https://doi.org/10.3390/s22031259.

Kutawa, A.B., Ahmad, K., Ali, A., Hussein, M.Z., Abdul Wahab, M.A., Adamu, A., Ismaila, A.A., Gunasena, M.T., Rahman, M.Z., Hossain, M.I., 2021. Trends in nanotechnology and its potentialities to control plant pathogenic fungi: a review. Biology 10 (9), 881. https://doi.org/10.3390/biology10090881.

Liasu, M.O., Ayandele, A.A., 2008. Antimicrobial activity of aqueous and ethanolic extracts from *Tithonia diversifolia* and *Bryum coronatum* collected from Ogbomoso, Oyo state, Nigeria. Adv. Nat. Appl. Sci. 2, 31–34.

Liu, C.H., Chuang, Y.H., Chen, T.Y., Tian, Y., Li, H., Wang, M.K., Zhang, W., 2015. Mechanism of arsenic adsorption on magnetite nanoparticles from water: thermodynamic and spectroscopic studies. Environ. Sci. Technol. 49, 7726–7734.

Longfei, Y., Xin, L., Xinming, Z., Xuechao, F., Lili, Z., Tonghui, M., 2018. Antifungal effects of saponin extract from rhizomes of *Dioscorea panthaica* Prain et Burk against *Candida albicans*. Evid. Based Complement. Alternat. Med., 13. https://doi.org/10.1155/2018/6095307. Article ID 6095307.

Lowry, G.V., Avellan, A., Gilbertson, L.M., 2019. Opportunities and challenges for nanotechnology in the agri-tech revolution. Nat. Nanotechnol. 14, 517–522.

Lukman, A., Neha, P., Razia, K.Z., 2016. Antifungal potential of plant extracts against seed-borne fungi isolated from barley seeds (*Hordeum vulgare* L.). J. Plant Pathol. Microbiol. 7, 5. https://doi.org/10.4172/2157-7471.1000350.

Machado, T.O., Beckers, S.J., Fischer, J., Müller, B., Sayer, C., de Araujo, P., Landfester, K., Wurm, F.R., 2020. Bio-based lignin nanocarriers loaded with fungicides as a versatile platform for drug delivery in plants. Biomacromolecules 21 (7), 2755–2763. https://doi.org/10.1021/acs.biomac.0c00487.

Madeeha, N., Faiza, Z.G., Saad, H., Abdul, M., Sania, N., Joham, S.A., Muhammad, Z., 2020. Green and chemical syntheses of CdO NPs: a comparative study for yield attributes, biological characteristics, and toxicity concerns. ACS Omega 5 (11), 5739–5747. https://doi.org/10.1021/acsomega.9b03769.

Makhuvele, R., Kayleen, N., Sefater, G., Velaphi, C.T., Oluwafemi, A.A., Patrick, B.N., 2020. The use of plant extracts and their phytochemicals for control of toxigenic fungi and mycotoxins. Heliyon 6. https://doi.org/10.1016/j.heliyon.2020.e05291.

Malerba, M., Cerana, R., 2016. Chitosan effects on plant systems. Int. J. Mol. Sci. 17, 996.

Manoj, K., 2018. Role of microbes in plant protection using intersection of nanotechnology and biology. In: Abd-Elsalam, K.A., Prasad, R. (Eds.), Nanobiotechnology Applications in Plant Protection, Nanotechnology in the Life Sciences. vol. 5. Springer, Switzerland, Chapter, pp. 111–136, https://doi.org/10.1007/978-3-319-91161-8_5.

Mansoor, S., Zahoor, I., Baba, T.R., Padder, S.A., Bhat, Z.A., Koul, A.M., Jiang, L., 2021. Fabrication of silver nanoparticles against fungal pathogens. Front. Nanotechnol. 3, 679358. https://doi.org/10.3389/fnano.2021.679358.

Mare, A.D., Ciurea, C.N., Man, A., Mareş, M., Toma, F., Berţa, L., Tanase, C., 2021. In vitro antifungal activity of silver nanoparticles biosynthesized with beech bark extract. Plan. Theory 10 (10), 2153. https://doi.org/10.3390/plants10102153.

Min, J.S., Kim, K.S., Kim, S.W., Jung, J.H., Lamsal, K., Kim, S.B., Jung, M., Lee, Y.S., 2009. Effects of colloidal silver nanoparticles on sclerotium-forming phytopathogenic fungi. J. Plant Pathol. 25, 376–380.

Mishra, S., Singh, H.B., 2015. Biosynthesized silver nanoparticles as a nanoweapon against phytopathogens: exploring their scope and potential in agriculture. Appl. Microbiol. Biotechnol. 99, 1097–1107. https://doi.org/10.1007/s00253-014-6296-0.

Mishra, S., Keswani, C., Abhilash, P., Fraceto, L.F., Singh, H.B., 2017. Integrated approach of agri-nanotechnology: challenges and future trends. Front. Plant Sci. 8, 471.

Moenne, A., Gonzslez, A., 2021. Chitosan, alginate-carrageenan-derived oligosaccharides stimulate defense against biotic and abiotic stresses, and growth in plants: a historical perspective. Carbohydr. Res. 503, 108298.

Mohamed, T.E., Ameina, S.A., Manal, E.S., Najah, M., Albaqami, A.M.S., Amira, M.E.T., El-Sayed, M.D., Ahmed, S.M., Elnahal, A.A., Taia, A.A.E., Ayman, E.T., Ahmed, S.E., Ayman, M.H., 2021. Vital roles of sustainable nano-fertilizers in improving plant quality and quantity—an updated review. Saudi J. Biol. Sci. 28 (12), 7349–7359. https://doi.org/10.1016/j.sjbs.2021.08.032.

Mohammadi, A., Hashemi, M., Hosseini, S.M., 2015. Chitosan nanoparticles loaded with *Cinnamomum zeylanicum* essential oil enhance the shelf life of cucumber during cold storage. Postharvest Biol. Technol. 110, 203–213.

Mohana, D.C., Raveesha, K.A., 2007. Anti-fungal evaluation of some plant extracts against some plant pathogenic field and storage fungi. J. Agric. Technol. 4, 119–137.

Mohsen, F.E.S., Abbassy, M.A., Rabea, I.E., Abou-Tale, K.H., 2018. Isolation and antifungal activity of plant lectins against some plant pathogenic fungi. Alexandria Sci. Exch. J. 39 (1), 161–167.

Nakkala, J.R., Mata, R., Gupta, A.K., Sadras, S.R., 2014. Green synthesis and characterization of silver nanoparticles using *Boerhaavia diffusa* plant extract and their antibacterial activity. Ind. Crop. Prod. 52, 562–566.

Nasseri, M., Golmohammadzadeh, S., Arouiee, H., Jaafari, M.R., Neamati, H., 2016. Antifungal activity of Zataria multiflora essential oil-loaded solid lipid nanoparticles in vitro condition. Iran. J. Basic Med. Sci. 19, 1231–1237. https://doi.org/10.22038/ijbms.2016.7824.

Navya, P.N., Daima, H.K., 2016. Rational engineering of physicochemical properties of nanomaterials for biomedical applications with nanotoxicological perspectives. Nano Convergence 3, 1. https://doi.org/10.1186/s40580-016-0064-z.

Naz, R., Bano, A., Nosheen, A., et al., 2021. Induction of defense-related enzymes and enhanced disease resistance in maize against fusarium verticillioides by seed treatment with *Jacaranda mimosifolia* formulations. Sci. Rep. 11, 59. https://doi.org/10.1038/s41598-020-79306-x.

Nguyen, V.S., Dinh, M.H., Nguyen, A.D., 2013. Study on chitosan nanoparticles on biophysical characteristics and growth of Robusta coffee in green house. Biocatal. Agric. Biotechnol. 2, 289–294.

Norouzi, N., Alizadeh, F., Khodavandi, A., Jahangiri, M., 2021. Antifungal activity of menthol alone and in combination on growth inhibition and biofilm formation of *Candida albicans*. J. Herbal Med. 29, 100495. https://doi.org/10.1016/j.hermed.2021.100495.

Oliveira, H.C., Stolf, M.R., Martinez, C.B.R., Grillo, R., DeJesus, M.B., Fraceto, L.F., 2015a. Nanoencapsulation enhances the post-emergence herbicidal activity of atrazine against mustard plants. PLoS ONE 10, 0132971.

Oliveira, H.C., Stolf, M.R., Martinez, C.B.R., Sousa, G.F.M., Grillo, R., DeJesus, M.B., et al., 2015b. Evaluation of the side effects of poly (epsilon-caprolactone) nanocapsules containing atrazine toward maize plants. Front. Chem. 3, 61.

Pariona, N., Mtz-Enriquez, A., Rangel, D.S., Carrion, G., Delgado, P., Saito, G.R., 2019. Green-synthesized copper nanoparticles as a potential antifungal against plant pathogens. J. RSC Adv. 9, 18835–18843. https://doi.org/10.1039/c9ra03110crsc.li/rsc.

Patil, S., Chandrasekaran, R., 2020. Biogenic nanoparticles: a comprehensive perspective in synthesis, characterization, application and its challenges (review). J. Genet. Eng. Biotechnol. 18 (1), 67.

Patra, P., Mitra, S., Debnath, N., Goswami, A., 2012. Biochemical, biophysical, and microarray-based antifungal evaluation of the buffer-mediated synthesized nano zinc oxide: an *in vivo* and *in vitro* toxicity study. Langmuir 28, 16966–16978.

Pawar, B.T., 2011. Antifungal activity of some stem extracts against seed-borne pathogenic fungi. J. Phytol. 3, 49–51.

Peng, Y., Li, S.J., Yan, J., Tang, Y., Cheng, J.P., Gao, A.J., Yao, X., Ruan, J.J., Xu, B.L., 2021. Research Progress on phytopathogenic fungi and their role as biocontrol agents. Front. Microbiol. 12, 670135. https://doi.org/10.3389/fmicb.2021.670135.

Perez-de-Luque, A., 2017. Interaction of nanomaterials with plants: what do we need for real applications in agriculture? Front. Environ. Sci. https://doi.org/10.3389/fenvs.2017.00012.

Pestovsky, Y.S., Martínez, A.A., 2017. The use of nanoparticles and nanoformulations in agriculture. J. Nanosci. Nanotechnol. 17, 8699–8730.

Poussio, B.G., Abro, A.M., Syed, N.R., Khashkheli, I.M., Hajano, J.U., 2021. Eco-friendly management of tomato wilt disease caused by *Fusarium* sp. in Sindh Province, Pakistan. Int. J. Recycl. Organic Waste Agric. https://doi.org/10.30486/IJROWA.2021.1910530.1140.

Prasad, R., Kumar, V., Prasad, K.S., 2014. Nanotechnology in sustainable agriculture: present concerns and future aspects. Afr. J. Biotechnol. 13, 705–713.

Prasad, R., Gupta, N., Kumar, M., Kumar, V., Wang, S., Abd-Elsalam, K.A., 2017. Nanomaterials act as plant defense mechanism. In: Prasad, R., Kumar, V., Kumar, M. (Eds.), Nanotechnology. Springer, Singapore, pp. 253–269.

Rafael, M.C., Tamires, S.P., Murilo, H.M., Facure, D.M.S., Luiza, A.M., Luiz, H.C.M., Daniel, S.C., 2022. Current progress in plant pathogen detection enabled by nanomaterials-based (bio)sensors. Sensors Actuators Rep. 4. https://doi.org/10.1016/j.snr.2021.100068.

Rai, M., Ingle, A., 2012. Role of nanotechnology in agriculture with special reference to management of insect pests. Appl. Microbiol. Biotechnol. 94, 287–293.

Rajaram, K., Aiswarya, D.C., Sureshkumar, P., 2015. Green synthesis of silver nanoparticle using *Tephrosia tinctoria* and its antidiabetic activity. Mater. Lett. 138, 251–254.

Ramirez, P.G., Ramirez, D.G., Mejia, E.Z., Ocampo, S.A., Diaz, C.N., Rojas Martinez, R.I., 2020. Extracts of *Stevia rebaudiana* against *Fusarium oxysporum* associated with tomato cultivation. Sci. Hortic. (Amsterdam) 259, 108683.

Reddy, M.S.B., Ponnamma, D., Choudhary, R., Sadasivuni, K.K.A., 2021. Comparative review of natural and synthetic biopolymer composite scaffolds. Polymers 2021 (13), 1105. https://doi.org/10.3390/polym13071105.

Saberi, R.R., Gholizadeh, V.M., Ebrahimi-Zarandi, M., Skorik, Y.A., 2022. Alginate-induced disease resistance in plants. Polymers 14, 661. https://doi.org/10.3390/polym14040661.

Sadeghi, R., Rodriguez, R.J., Yao, Y., Kokini, J.L., 2017. Advances in nanotechnology as they pertain to food and agriculture: benefits and risks. Annu. Rev. Food Sci. Technol. 8, 467–492.

Sadiko, A., Du, N., Kariuki, V., Okello, V., Bushlyar, V., 2014. Current and emerging technologies for the characterization of nanomaterials. ACS Sustain. Chem. Eng. 2, 1707–1716.

Saharan, V., Mehrotra, A., Khatik, R., Rawal, P., Sharma, S.S., Pal, A., 2013. Synthesis of chitosan based-nanoparticles and their in vitro evaluation against phytopathogenic fungi. Int. J. Biol. Macromol. 62, 677–683.

Saleh, A.A.M., Suresh, M., 2020. Anti-fungal efficacy and mechanisms of flavonoids. Antibiotics 9, 45. https://doi.org/10.3390/antibiotics9020045.

Samiyah, S.A., Roop, S.B., Saleh, M.A., 2021. Antimicrobial activity of chitosan nanoparticles. Biotechnol. Biotechnol. Equip. 35 (1), 1874–1880. https://doi.org/10.1080/13102818.2022.2027816.

Sangeetha, J., Thangadurai, D., Hospet, R., Harish, E.R., Purushotham, P., Mujeeb, M.A., Shrinivas, J., David, M., Mundaragi, A.C., Thimmappa, A.C., Arakera, S.B., Prasad, R., 2017. Nanoagrotechnology for soil quality, crop performance and environmental management. In: Prasad, R., Kumar, M., Kumar, V. (Eds.), Nanotechnology. Springer Nature Singapore Pte Ltd, Singapore, pp. 73–97.

Santos, A.O., Vaz, A., Rodrigues, P., Veloso, A.C.A., Venancio, A., Peres, A.M., 2019. Thin films sensor devices for mycotoxins detection in foods: applications and challenges. Chem. Aust. 7, 3.

Sarfraz, M., Nasim, M.J., Jacob, C., Gruhlke, M.C.H., 2020. Efficacy of allicin against plant pathogenic fungi and unveiling the underlying mode of action employing yeast based chemogenetic profiling approach. Appl. Sci. 10, 2563. https://doi.org/10.3390/app10072563.

Sattary, M., Amini, J., Hallaj, R., 2020. Antifungal activity of the lemongrass and clove oil encapsulated in mesoporous silica nanoparticles against wheat's take-all disease. Pestic. Biochem. Physiol. 170, 104696.

Schreiber, L., 2005. Polar paths of diffusion across plant cuticles: new evidence for an old hypothesis. Ann. Bot. 95 (7), 1069–1073.

Schwenkbier, L., Pollok, S., Konig, S., Urban, M., Werres, S., Cialla-May, D., Weber, K., Popp, J., 2015. Towards on-site testing of phytophthora species. Anal. Meth. 7, 211–217.

Seint, S.A., Masaru, M., 2011. Effect of some plant extracts on rhizoctonia spp. and *Sclerotium hydrophilum*. J. Med. Plants Res. 5 (16), 3751–3757. http://www.academicjournals.org/JMPR.

Sernaite, L., Rasiukeviciue, N., Valiuskaite, A., 2020. The extracts of cinnamon and clove as potential biofungicides against strawberry grey mould. Plan. Theory 9, 613.

Shabnam, S., Mizanur, M.M., Chandan, S., Olubunmi, A., Mohammad, T.I., Oluyomi, S.A., 2021. Nanoparticles as antimicrobial and antiviral agents: a literature-based perspective study. Heliyon 7 (3). https://doi.org/10.1016/j.heliyon.2021.e06456.

Shafique, S., Shafique, S., Zameer, M., Asif, M., 2019. Plant defense system activated in chili plants by using extracts from *Eucalyptus citriodora*. Biocontrl Sci. 24, 137–144. https://doi.org/10.4265/bio.24.137.

Shang, Y., Kamrul, H.M., Ahammed, G.J., Li, M., Yin, H., Zhou, J., 2019. Applications of nanotechnology in plant growth and crop protection: a review. Molecules 24, 2558.

Shoeb, M., Singh, B.R., Khan, J.A., Khan, W., Singh, B.N., Singh, H.B., Naqvi, A.H., 2013. ROS-dependent anticandidal activity of zinc oxide nanoparticles synthesized by using egg albumen as a biotemplate. Adv. Nat. Sci. Nanosci. Nanotechnol. 4.

Shuaixuan, Y., Zhenru, G., Polycarp, C.O., Preston, C., Cyren, R., Feng, H., Jie, H., 2022. Green synthesis of nanoparticles: current developments and limitations. Environ. Technol. Innov. 26, 102336. https://doi.org/10.1016/j.eti.2022.102336.

Siddaiah, C.N., Prasanth, K.V.H., Satyanarayana, N.R., Mudili, V., Gupta, V.K., Kalagatur, N.K., Satyavati, T., Dai, X.-F., Chen, J.Y., Mocan, A., Singh, B.P., Srivastava, R.K., 2018. Chitosan nanoparticles having higher degree of acetylation induce resistance against pearl millet downy mildew through nitric oxide generation. Sci. Rep. 8 (1), 2485. https://doi.org/10.1038/s41598-017-19016-z.

Singh, S., Singh, M., Agrawal, V.V., Kumar, A., 2010. An attempt to develop surface plasmon resonance based immunosensor for Karnal bunt (*Tilletia indica*) diagnosis based on the experience of nano-gold based lateral flow immuno-dipstick test. Thin Solid Films 519, 1156–1159.

Singh, R., Gupta, A.K., Patade, V.Y., et al., 2019. Synthesis of silver nanoparticles using extract of *Ocimum kilimandscharicum* and its antimicrobial activity against plant pathogens. SN Appl. Sci. 1, 1652. https://doi.org/10.1007/s42452-019-1703-x Singh, A., Deepika, Chaudhari, A.K., Das, S.

Singh, V.K., Dwivedy, A.K., Shivalingam, R.K., Dubey, N.K., 2020. Assessment of preservative potential of *Bunium persicum* (Boiss) essential oil against fungal and aflatoxin contamination of stored masticatories and improvement in efficacy through encapsulation into chitosan nanomatrix. Environ. Sci. Pollut. Res. 27, 27635–27650.

Sivakami, M., Renuka Devi, K., Renuka, R., 2022. Phytomediated synthesis of magnetic nanoparticles by *Murraya koenigii* leaves extract and its biomedical applications. Appl. Phys. A Mater. Sci. Process. 128, 272. https://doi.org/10.1007/s00339-022-05437-9.

Spielman-Sun, E., Astrid, A., Garret, D.B., Emma, T.C., Ryan, V.T., Alvin, S.A., Gregory, V.L., 2020. Protein coating composition targets nanoparticles to leaf stomata and trichomes. Nanoscale. https://doi.org/10.1039/c9nr08100c.

Steglinska, A., Bekhter, A., Wawrzyniak, P., Kunicka-Styczynska, A., Jastrzabek, K., Fidler, M., Smigielski, K., Gutarowska, B., 2022. Antimicrobial activities of plant extracts against *Solanum tuberosum* L. phytopathogens. Molecules 27, 1579. https://doi.org/10.3390/molecules27051579.

Suresh, U., Murugan, K., Panneerselvam, C., Rajaganesh, R., Roni, M., Al-Aoh, H.A.N., Trivedi, S., Rehman, H., Kumar, S., Higuchi, A., 2018. Suaeda maritima-based herbal coils and green nanoparticles as potential biopesticides against the dengue vector *Aedes aegypti* and the tobacco cutworm *Spodoptera litura*. Physiol. Mol. Plant Pathol. 101, 225–235.

Thajuddin, N., Mathew, S., 2020. Phytonanotechnology: Challenges and Prospects. Elsevier Sci.

Tryfon, P., Kamou, N.N., Mourdikoudis, S., Karamanoli, K., Menkissoglu-Spiroudi, U., Dendrinou, S., C., 2021. CuZn and ZnO nanoflowers as nano-fungicides against *Botrytis cinerea* and *Sclerotinia sclerotiorum*: phytoprotection, translocation, and impact after foliar application. Materials 14, 7600. https://doi.org/10.3390/ma142476.

Upadhyay, N., Singh, V.K., Dwivedy, A.K., Chaudhari, A.K., Dubey, N.K., 2021. Assessment of nanoencapsulated *Cananga odorata* essential oil in chitosan nanopolymer as a green approach to boost the antifungal, antioxidant and in situ efficacy. Int. J. Biol. Macromol. 171, 480–490.

Vaculikova, E., Cernikova, A., Placha, D., Pisarcik, M., Dedkova, K., Peikertova, P., Devinsky, F., Jampilek, J., 2016. Cimetidine nanoparticles for permeability enhancement. J. Nanosci. Nanotechnol. 16, 7840–7843. https://doi.org/10.1166/jnn.2016.12562.

Valarezo, E., Gaona-Granda, G., Morocho, V., Cartuche, L., Calva, J., Meneses, M.A., 2021. Chemical constituents of the essential oil from Ecuadorian endemic species *Croton ferrugineus* and its antimicrobial, antioxidant and α-glucosidase inhibitory activity. Molecules 2021 (26), 4608. https://doi.org/10.3390/molecules26154608.

Valkova, V., Duranova, H., Galovicova, L., Stefanikova, J., Vukovic, N., Kacaniova, M., 2021. The *Citrus reticulata* essential oil: evaluation of antifungal activity against penicillium species related to bakery products spoilage. Potravinarstvo Slovak J. Food Sci. 15, 1112–1119. https://doi.org/10.5219/1695.

Vande, V.A.R., Arai, Y., 2012. Effect of silver nanoparticles on soil denitrification kinetics. Ind. Biotechnol. 8 (6), 358–364.

Vanti, G., Masaphy, S., Kurjogi, M., Chakrasali, S., Nargund, V., 2019. Synthesis and application of chitosan-copper nanoparticles on damping off causing plant pathogenic fungi. Int. J. Biol. Macromol. 156. https://doi.org/10.1016/j.ijbiomac.2019.11.179.

Villa, F., Cappitelli, F., Cortesi, P., Kunova, A., 2017. Fungal biofilms: targets for the development of novel strategies in plant disease management. Front. Microbiol. 8 (2017), 654.

Vincent, D., Rafiqi, M., Job, D., 2020. The multiple facets of plant-fungal interactions revealed through plant and fungal secretomics. Front. Plant Sci. 10, 1626. https://doi.org/10.3389/fpls.2019.01626.

Wang, Z., Wei, F., Liu, S.Y., Xu, Q., Huang, J.Y., Dong, X.Y., Yu, J.H., Yang, Q., Di Zhao, Y., Chen, H., 2010. Electrocatalytic oxidation of phytohormone salicylic acid at copper nanoparticles-modified gold electrode and its detection in oilseed rape infected with fungal pathogen *Sclerotinia sclerotiorum*. Talanta 80, 1277–1281.

Weidong, H., Minhui, Y., Haiming, D., Yaling, B., Xinxin, C., Haibing, Y., 2020. Synergistic antifungal activity of green synthesized silver nanoparticles and epoxiconazole against *Setosphaeria turcica*. J. Nanomater., 7. https://doi.org/10.1155/2020/9535432. Article ID 9535432.

Weria, W., Saadi, S., Jahanshir, A., Somaieh, H., Shima, Y., Filippo, M., 2019. Enhancement of the antifungal activity of thyme and dill essential oils against *Colletotrichum nymphaeae* by nano-encapsulation with copper NPs. Ind. Crop. Prod. 132, 213–225. https://doi.org/10.1016/j.indcrop.2019.02.031.

Worrall, E.A., Hamid, A., Mody, K.T., Mitter, N., Pappu, H.R., 2018. Nanotechnology for plant disease management. Agronomy 8, 285.

Xiong, Y., Dong, J., Huang, Z.Q., Xin, P., Chen, W., Wang, Y., Li, Z., Jin, Z., Xing, W., Zhuang, Z., Ye, J., Wei, X., Cao, R., Gu, L., Sun, S., Zhuang, L., Chen, X., Yang, H., Chen, C., Peng, Q., Chang, C.R., Wang, D., Li, Y., 2020. Single-atom Rh/N-doped carbon electrocatalyst for formic acid oxidation. Nat. Nanotechnol. 15, 390–397.

Yadav, J.K., 2019. Efficacy of plant extracts against *Alternaria brassicae* under *in-vitro* condition. J. Pharm. Phytochem. 8, 528–532.

Yang, J., Guo, W., Wang, J., Yang, X., Zhang, Z., Zhao, Z., 2020. T-2 toxin-induced oxidative stress leads to imbalance of mitochondrial fission and fusion to activate cellular apoptosis in the human liver 7702 cell line. Toxins 12, 43. https://doi.org/10.3390/toxins12010043.

Yang, B., Yang, S., Zheng, W., et al., 2022. Plant immunity inducers: from discovery to agricultural application. Stress Biol. 2, 5. https://doi.org/10.1007/s44154-021-00028-9.

Yaseen, T., Pu, H., Sun, D.W., 2018. Functionalization techniques for improving SERS substrates and their applications in food safety evaluation: a review of recent research trends. Trends Food Sci. Technol. 72, 162–174.

Yi, U., Zaharah, S.S., Ismail, I.S., Musa, H.M., 2021. Effect of aqueous neem leaf extracts in controlling *Fusarium* wilt, soil physicochemical properties and growth performance of Banana (*Musa* spp.). Sustainability 13 (22), 12335. https://doi.org/10.3390/su132212335.

Yin, Z., Ke, X., Kang, Z., Huang, L., 2016. Apple resistance responses against Valsa Mali revealed by transcriptomics analyses. Physiol. Mol. Plant Pathol. 93, 85–92. https://doi.org/10.1016/j.pmpp.2016.01.004.

Zahid, N., Ali, A., Manickam, S., Siddiqui, Y., Maqboo, M., 2012. Potential of chitosan-loaded nano-emulsions to control different *Colletotrichum* spp. and maintain quality of tropical fruits during cold storage. J. Appl. Microbiol. 113, 925–939.

Zaker, M., 2016. Natural plant products as eco-friendly fungicides for plant diseases control: a review. Agriculturists 14 (1), 134–141.

Zobir, S.A.M., Ali, A., Adzmi, F., Sulaiman, M.R., Ahmad, K., 2021. A review on nanopesticides for plant protection synthesized using the supramolecular chemistry of layered hydroxide hosts. Biology 10, 1077. https://doi.org/10.3390/biology101110.

CHAPTER

7

The antagonistic yeasts: Novel nano/biofungicides for controlling plant pathogens

Parissa Taheri[a], *Saeed Tarighi*[a], *and Farah K. Ahmed*[b]

[a]Department of Plant Protection, Faculty of Agriculture, Ferdowsi University of Mashhad, Mashhad, Iran [b]Biotechnology English Program, Faculty of Agriculture, Cairo University, Giza, Egypt

1 Introduction

Various plant diseases cause significant damage to both the quantity and quality of agricultural products, resulting in massive socioeconomic effects on human life on a global scale. Diseases caused by fungal pathogens, in particular, are increasingly recognized as a global threat to crop production and food safety. Currently, fungal diseases constitute 64%–67% of the total crop diseases reported universally (Fisher et al., 2012, 2018) and account for 20% of the losses at the production level and 10% at the postharvest level (Fisher et al., 2018). This could be derived from increasingly common agricultural practices, such as extensive monocultures and the use of a restricted number of plant cultivars, as well as increased global trade, which leads to disease spread over excessive distances (Fisher et al., 2012; Taheri et al., 2007).

For the management of plant diseases and crop protection against various pathogens, several control strategies can be used, which are currently mainly based on applying chemically hazardous fungicides. The continued use of chemical fungicides leads to the generation of fungicide-resistant fungal isolates in the fungal pathogen populations (Syed Ab Rahman et al., 2018) and, in the absence of other disease control methods, to the re-emergence of virulence (Fisher et al., 2018; Taheri and Flaherty, 2022). Application of fungicides seriously affects the microflora of different ecosystems, with destructive effects on beneficial microorganisms including epiphytic and endophytic fungi and bacteria, as well as several

https://doi.org/10.1016/B978-0-443-23950-2.00001-1

animal species that are important for the quality of various soil types (Syed Ab Rahman et al., 2018). The frequent use of these hazardous chemicals results in the persistence of their residues in the environment, which cause environmental pollution and toxicity for humans and animals (Dukare et al., 2018).

In recent decades, some different microorganisms, including yeasts, have been isolated and shown to protect plant tissues against various pathogens; some of them have been utilized to develop commercial products (Droby et al., 2009; Ferraz et al., 2019). Considerable success has been achieved by applying antagonistic microorganisms to control plant diseases, especially postharvest diseases and mycotoxin production, using various types of beneficial fungi (Dehghanpour-Farashah et al., 2019; Daroodi et al., 2021; Zhimo et al., 2016). The use of yeasts to control the postharvest fungal decay of several fruits has been extensively studied, and several examples of successful disease control exist (Kasfi et al., 2018; Raspor et al., 2010). Numerous studies have been focused on isolation, characterization, and identification of antagonistic yeasts and yeast-like fungi, including *Aureobasidium pullulans* (Bozoudi and Tsaltas, 2018; Gostinčar et al., 2014; Schena et al., 2003), *Candida membranifaciens* (Kasfi et al., 2018), *Meyerozyma (Pichia) guilliermondii* with anamorph of *Candida guilliermondii* (Kasfi et al., 2018; Lahlali et al., 2014; Raspor et al., 2010), *Metschnikowia fructicola*, *Metschnikowia pulcherrima* (Raspor et al., 2010), *Saccharomyces cerevisiae* (Nally et al., 2013; Parapouli et al., 2020), *Cryptococcus laurentii*, *Issatchenkia terricola*, *Candida incommunis*, and *Kluyveromyces thermotolerans*. In recent decades, some different microorganisms, including yeasts, have been isolated and shown to protect plant tissues against various pathogens; some of them have been utilized to develop commercial products (Droby et al., 2009; Ferraz et al., 2019). Considerable success has been achieved by applying antagonistic microorganisms to control plant diseases, especially postharvest diseases and mycotoxin production, using various types of beneficial fungi (Dehghanpour-Farashah et al., 2019; Daroodi et al., 2021; Zhimo et al., 2016). The use of yeasts to control the postharvest fungal decay of several fruits has been extensively studied, and several examples of successful disease control exist (Kasfi et al., 2018; Raspor et al., 2010).

2 Application of yeasts for biocontrol of postharvest diseases

Postharvest losses of various types of vegetables and fruits, including pome fruits, stone fruits, small fruits, and citrus species, are very important if the agricultural products are not in ideal conditions during handling, processing, and storage. Postharvest losses of up to 25% of the total crop production in industrialized countries and more than 50% in developing countries have been reported (Nunes, 2012). Significant levels of postharvest molds and rots caused by fungal pathogens belonging to several genera, such as *Aspergillus*, *Penicillium*, *Alternaria*, *Botrytis*, *Colletotrichum*, *Fusarium*, etc., can be directly attributable to the high amount of nutrients and water that the vegetables and fruits contain, together with low pH and reduced inherent defense responses after harvest. Nowadays, researchers are more focused on finding natural, eco-friendly ways to control plant diseases (Taheri and Flaherty, 2022). Especially, a huge effort is made to decrease the contamination of vegetables and fruits by some of the above-mentioned fungi, such as various species of *Aspergillus*, *Penicillium*,

Alternaria, and *Fusarium*, as producers of mycotoxins, which are harmful metabolites causing health problems in consumers (Maghsoodi and Taheri, 2021).

Control of postharvest diseases is traditionally performed via using large amounts of synthetic fungicides, with an urgent need to be replaced with biological control strategies that are safe for human health and environment. Application of yeasts as biocontrol agents against various types of pathogens associated with crops spoilage, such as postharvest molds and rots of fruits and vegetables, has been reported since the early 1960s, by demonstrating that the budding yeast *S. cerevisiae* is capable of producing toxins that kill other yeast species (Bevan and Makower, 1963).

Various species of yeasts and yeast-like fungi can be used for the control of phytopathogens, causing postharvest diseases. For instance, isolates of the saprophytic yeast-like ascomycetous fungus *A. pullulans* have antagonistic effects on phytopathogens and can be effective biocontrol agents not only in the field but also in postharvest phase (Bozoudi and Tsaltas, 2018). Postharvest storage diseases of grapes, caused by *Aspergillus* species and *Talaromyces rugulosus* causing decay of grapes in storage, can be successfully controlled via application of yeasts. Antagonistic yeasts have been isolated from the epiphytic and endophytic flora associated with grape berries and leaves from different regions, worldwide (Joubert and Doty, 2018; Kasfi et al., 2018). Yeast species have been screened for their microbial antagonism against *Aspergillus flavus*, *Aspergillus niger*, and *Aspergillus ochraceus*, the main species responsible for the accumulation of aflatoxin and ochratoxin in grape berries. Several yeast isolates, such as *C. membranifaciens* and *M. guilliermondii*, have been reported to be capable of showing significant inhibitory effects against *Aspergillus* species and their antagonistic activities were investigated using a dual culture assay, as a measurement of nutritional competition ability of the beneficial yeasts (Kasfi et al., 2018).

In order for sustainable agriculture to be effective, new biocontrol agents must be developed. Yeasts are microorganisms that live in many environments and can act as antagonists to numerous plant diseases. Yeasts colonize plant surfaces quickly, consume nutrients from a variety of sources, can survive in a wide temperature range, create no toxic metabolites, and have no negative effects on the final food product. Plant protection with microbiological agents can help reduce or eliminate the use of agrichemicals while also improving plant quality. As a result, more emphasis appears to be required for future studies on the use of yeasts to make green nanoparticles (NPs), which may trigger plant defense responses, develop resistance to infections, and boost plant growth (Fig. 1).

3 Yeasts as biocontrol agents of pathogens causing diseases on aerial parts of plants

Several yeast species belonging to various taxonomic groups of fungi have been characterized and reported to be capable of decreasing the progress of diseases caused by different types of pathogens and also promoting plant growth. Information related to the application of yeasts on various monocot and dicot plants to decrease the damages caused by various pathogens, including fungi, fungus-like organisms, bacteria, and viruses, was summarized in Table 1. As mentioned in this table, in vitro and in vivo assays revealed that several direct and indirect biocontrol mechanisms can be involved in the protective effects of beneficial

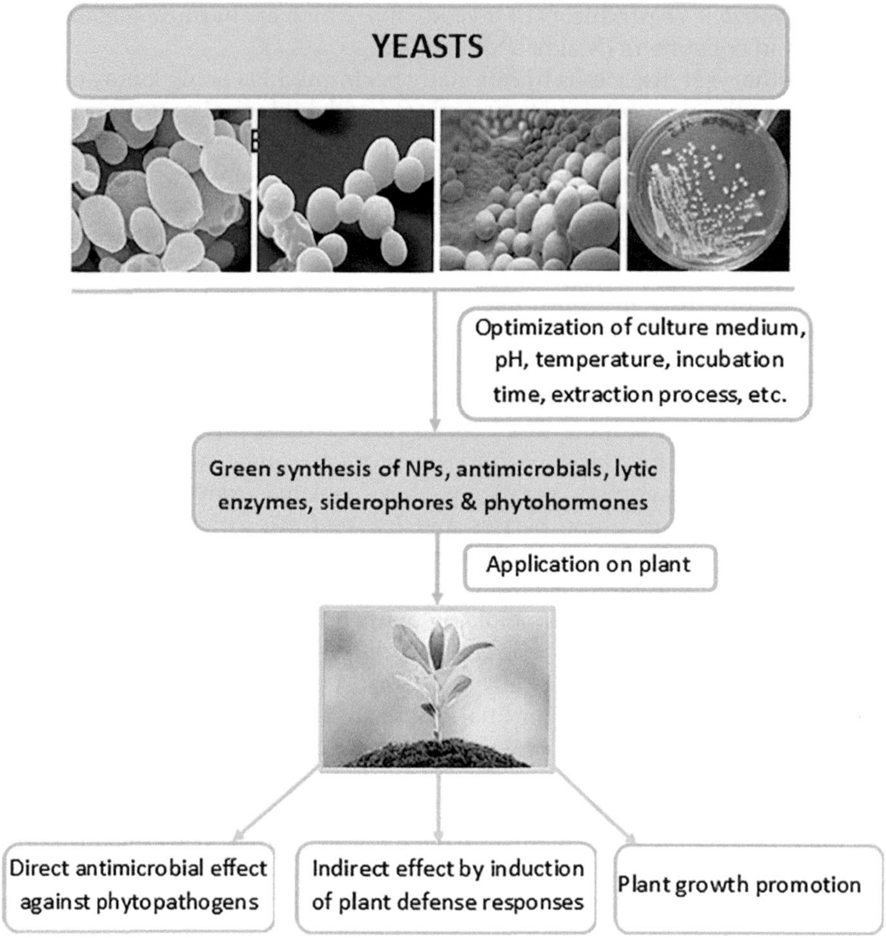

FIG. 1 Beneficial effects and applications using yeasts in agricultural systems.

TABLE 1 Application of yeasts for controlling diseases on aerial parts of plants.

Yeast	Pathogen	Plant	Conditions	Mechanism	References
Sporobolomyces roseus and *Cryptococcus laurentii* var. *flavescens*	*Colletotrichum graminicola*	*Zea mays*	In vivo	Reduction of infection via a reduced number of penetrations from the normally formed appressoria	Williamson and Fokkema (1985)
Anthracocystis flocculosa (*Pseudozyma flocculosa*)	Powdery mildew	Rose and cucumber	In vivo	Production of glycolipid material called Flocculosa	Thambugala et al. (2020)

TABLE 1 Application of yeasts for controlling diseases on aerial parts of plants—cont'd

Yeast	Pathogen	Plant	Conditions	Mechanism	References
Cryptococcus nodaensis and *Cryptococcus* sp.	*Gibberella zeae* (*Fusarium graminearum*)	Winter and spring wheat	In vivo	Effect on the ratio of nitrogen and carbon (C:N) and damage reduction	Khan et al. (2004)
Aureobasidium pullulans	*Fusarium culmorum* and *F. graminearum*	Wheat and maize	In vitro	Reduced sporulation	Luongo et al. (2005)
Saccharomyces cerevisiae	*Erysiphe betae*	Sugar beet	In vivo	Colonization on leaf surface of sugar beet and adhere itself around mycelium of *Erysiphe betae*	Ziedan and Farrag (2011)
Aureobasidium pullulans, Debaryomyces hansenii	*F. culmorum* and *F. avenaceum*	Winter wheat	In vitro	Inhibit the growth of *Fusarium* fungi through antibiosis and competition for nutrients, which reduced the production of *Fusarium* toxins	Wachowska et al. (2022)
Cryptococcus carnescens	*Fusarium poae*	Organic wheat	In vitro	Competition for nutrients and space, production of volatile metabolites, reduction of spore germination, mycelial growth inhibition, production of siderophores or extracellular lytic enzymes (chitinase and β-1,3-glucanase)	Podgórska-Kryszczuk et al. (2022)
Pseudozyma churashimaensis	*Xanthomonas axonopodis, Cucumber mosaic virus, Pepper mottle virus, Pepper mild mottle virus, Broad bean wilt virus*	Pepper	In vivo	Disease suppression by induced resistance	Lee et al. (2017)

Continued

TABLE 1 Application of yeasts for controlling diseases on aerial parts of plants—cont'd

Yeast	Pathogen	Plant	Conditions	Mechanism	References
Pichia kudriavzevii and *Issatchenkia terricola*	*Colletotrichum dematium, R. solani, A. alternata, A. niger, F. oxysporum*	Black Gram (*Vigna mungo* L.)	In vitro and in vivo	Production of indole acetic acid (IAA), siderophore, 1-amino cyclopropane-1-carboxylic acid deaminase (ACCD), and activating plant defense enzymes, promoting plant growth and health	Bright et al. (2022)
Candida oleophila	*Colletotrichum gloeosporioides*	Papaya	In vivo	Using it together with sodium bicarbonate reduced the pathogen growth and damage on fruit	Gamagae et al. (2003)
Pichia membranaefaciens	*Colletotrichum gloeosporioides*	Citrus	In vitro and in vivo	Secretion of hydrolytic enzymes (chitinase and β-1,3-glucanase)	Zhou et al. (2016)
Aureobasidium pullulans	*Phytophthora infestans*	Legume family (Fabaceae)	In vitro	Production of indole-3-acetic acid (IAA)	Ignatova et al. (2015)
Rhodotorula mucilaginosa	*Fusarium graminearum*	Legume family (Fabaceae)	In vitro	Production of indole-3-acetic acid (IAA)	Ignatova et al. (2015)
Pichia stipites	*Moniliophthora perniciosa*, causing Witches Broom disease	Cocoa	In vitro	Inhibition of the pathogen growth	Botha (2011)
Saccharomyces cerevisiae	*Fusarium oxysporum*	Sugar beet	In vivo	Reduced infection and penetration	Shalaby and El-Nady (2008)
Pseudozyma flocculosa	*Cladosporium cucumerinum*	–	In vitro	Production of two new antifungal fatty acids produced (9-heptadecenoic acid and 6-methyl-9-heptadecenoic acid)	Avis and Bélanger (2002)
Saccharomyces cerevisiae	*Colletotrichum acutatum*	Citrus	In vitro	Inhibition of mycelial growth	Díaz et al. (2020)
Candida (Pichia) guilliermondii	*Colletotrichum capsici*	Chili	In vitro	Inhibition of mycelial growth	Chanchaichaovivat et al. (2007)

TABLE 1 Application of yeasts for controlling diseases on aerial parts of plants—cont'd

Yeast	Pathogen	Plant	Conditions	Mechanism	References
Wickerhamomyces anomalus, Meyerozyma guilliermondii	*Colletotrichum gloeosporioides*	Papaya	In vitro	Synthesize two hydrolytic enzymes (chitinase and -1,3 glucanase)	Lima et al. (2013)
Cryptococcus infirmominiatus	*Monilinia fructicola*	Sweet cherry	In vitro	Inhibition of mycelial growth	Spotts et al. (2002)

yeasts against plant diseases. For instance, *A. pullulans* (belonging to Dothideomycetes) and *Debaryomyces hansenii* (belonging to Saccharomycetes) are capable of inhibiting the growth of *Fusarium* spp., pathogenic on small grain cereals such as wheat. Antibiosis and competition for nutrients have been reported as the biocontrol mechanisms involved in the inhibition of mycelial growth in *Fusarium* species, which leads to lower level of sporulation and reduction of toxins production by these toxigenic fungi (Wachowska et al., 2022). In addition, several other direct and indirect mechanisms of biocontrol, such as the production of hydrolytic enzymes (Lima et al., 2013; Podgórska-Kryszczuk et al., 2022), siderophores (Podgórska-Kryszczuk et al., 2022), glycolipids (Thambugala et al., 2020), and induction of plant defense responses via affecting phytohormone levels and resistance signaling (Bright et al., 2022; Lee et al., 2017) have been demonstrated to be involved in inhibiting growth and destructive effects of phytopathogens on various plant species (Table 1).

4 Application of yeasts for biocontrol of crown and root pathogens

Numerous species of soilborne pathogens are capable of causing crown and root rots on monocot and dicot plants, leading to huge yield losses in various geographic regions, every year. Several studies demonstrated the biocontrol effects of yeasts and yeast-like fungi against fungal pathogens and Oomycota, such as *Phytophthora* spp., causing crown and root rot diseases in plants (Table 2). The yeast-like fungus *A. pullulans* is a powerful biocontrol agent, not only against soilborne pathogens causing crown and root rot but also against different postharvest fungal pathogens. This beneficial microorganism is naturally present in the phyllosphere of various fruits and vegetables (Bozoudi and Tsaltas, 2018; Iqbal et al., 2021). *A. pullulans* can be used for biocontrol of plant diseases alone or in combination with other microorganisms or sustainable physical disease management strategies (Di Francesco et al., 2020; Zhang et al., 2010). Iqbal et al. (2021) demonstrated the efficacy of *A. pullulans* in biocontrol of *Phytophthora cactorum*, the causal agent of crown and root rot on strawberries in both in vitro and in vivo conditions. Studies by Matic et al. (2014) revealed beneficial effects of antagonistic yeasts in combination with thermotherapy as seed treatments to control soilborne fungi causing rice crown and root rot, such as *Fusarium fujikuroi*. Well-known fungal pathogens causing crown and root rot, such as *Rhizoctonia solani*, *Fusarium moniliforme*, *Fusarium oxysporum*, and *Fusarium culmorum*, can be successfully controlled by application of yeast

TABLE 2 Application of yeasts for controlling crown and root diseases on various plant species.

Yeast	Pathogen	Plant	Conditions	Mechanism	References
Candida saopaulonensis, *Cryptococcus laurentii*, and *Bullera sinensis*	*Rhizoctonia solani*	cowpea (*Vigna unguiculata*)	In vitro and in vivo	Competition, induction of enzymes such as peroxidase, catalase, and ascorbate peroxidase in cowpea	De Tenório et al. (2019)
Candida steatolytica and *Saccharomyces unispora*	*Fusarium oxysporum*	Bean (*Phaseolus vulgaris*)	In vivo	Antibiosis (production of inhibitory active compounds)	El-Mehalawy (2004)
Pichia guilliermondii, *Metschnikowia pulcherrima*, and *Sporidiobolus pararoseus*	*Fusarium fujikuroi*	Rice (*Oryza sativa*)	In vitro and in vivo	inhibited the mycelium growth and improved rice seed germination rate	Matic et al. (2014)
Candida valida, *Trichosporon asahii*, and *Rhodotorula glutinis*	*Rhizoctonia solani* AG 2-2	Sugar beet (*Beta vulgaris*)	In vitro and in vivo	Production of glucanase, inhibitory volatiles, hyphal plasmolysis, and degradation of fungal cell walls	El-Tarabily (2004)
Candida glabrata, *Candida maltosa*, *Candida slooffii*, *Rhodotorula rubra*, and *Trichosporon cutaneum*	*Cephalosporium maydis*	Maize (*Zea mays*)	In vivo	Production of antifungal metabolites, toxins, nonvolatile diffusible metabolites and growth factors	El-Mehalawy et al. (2004)
Candida orthopsilosis and *Rhodotorula mucilaginosa*	*F. oxysporum*	Wheat (*Triticum aestivum*)	In vitro and in vivo	Induction of plant defense responses, inhibition of mycelial growth and zearalenone production, production of fungal cell wall degrading enzymes (protease, chitinase, cellulase, pectinase, laccase, urease)	Marwa Abdel-Kareem et al. (2021)
Rhodotorula glutinis, *Candida sake*, *Debaryomyces hansenii*, and *Aureobasidium pullulans*	*Fusarium culmorum*	Durum wheat (*Triticum turgidum*) and bread wheat (*Triticum aestivum*)	In vitro and in vivo	Reduction of mycelial growth and DON mycotoxin production	Wachowska et al. (2020)
Wickerhamomyces anomalus	*Verticillium dahlia* and *Fusarium oxysporum*	Tomato (*Solanum lycopersicum*)	In vitro and in vivo	Inhibition of fungal spore germination, Inhibitory effect on mycelial growth of the	Fernandez-San Millan et al. (2021)

TABLE 2 Application of yeasts for controlling crown and root diseases on various plant species—cont'd

Yeast	Pathogen	Plant	Conditions	Mechanism	References
				pathogens, and in vivo, through their effect on reduction of fungal infections in tomato plants under	
Candida tropical, *Cryptococcus tephrensis*, and *Saccharomyces cerevisiae*	*Fusarium oxysporum* f. sp. *cucumerinum*	Cucumber (*Cucumis sativus*)	In vitro and in vivo	Production of antifungal metabolites, cell wall degrading enzymes, siderophores, HCN, antibiotics, auxin and gibberline, and phosphate solubilization	Kamel et al. (2016)
Torulaspora indica, *T. indica*, and *Wickerhamomyces anomalus*	*Rhizoctonia solani* and *Fusarium moniliforme*	Rice (*Oryza sativa*)	In vitro and in vivo	VOCs production (as the major biocontrol mechanism), competition for nutrients, leading to fungal mycelial growth inhibition and disease reduction	Into et al. (2020)

isolates belonging to *A. pullulans*, *D. hansenii*, *Wickerhamomyces anomalus*, *Candida sake* species via involvement of both direct and indirect mechanisms of biocontrol (Into et al., 2020; Wachowska et al., 2020).

5 Modes of yeast actions against phytopathogens

Understanding the mechanisms conferring biocontrol activity is the foundation for the successful development and application of yeasts as plant protection agents. Several direct and indirect mechanisms might be involved in the biocontrol effect of various yeast species against different types of phytopathogens, which are described below.

5.1 Secretion of hydrolytic enzymes by yeast species

Many enzymes have been reported which are secreted by yeasts in the lack of nutrients to meet the nutritional needs of the host cells and may lead to the destruction of the host cells. There are several studies on the secretion of enzymes such as chitinases, glucanases, proteases, and lipases by antagonistic yeasts and effect of these enzymes in bioinhibition activity of yeast species.

Secretion of chitin-degrading enzymes is very important for biological control agents due to the destruction of the fungal cell wall (Zhang et al., 2020). Chitinase production has been also reported in cultures during the interaction of *A. pullulans* with *Penicillium expansum* and also in apple wounds, in the interaction of phyllosphere yeasts *Tilletiopsis albescens* and *T. pallescens with* powdery mildew fungi such as *Podosphaera xanthii* (Urquhart and Punja, 2002), and in the interaction of *Candida saitoana* with *Botrytis cinerea*.

Glucans are the main components of fungal cell walls after chitin, and exoglucanases play a role in repairing the cell wall, cell adhesion, and resistance to lethal toxins. Studies aimed at reducing the effect of pathogens after harvesting products showed that the soil yeast *C. laurentii* reduced the growth of filamentous fungi in damaged fruits. This effect can be based on the interaction of chemical substances produced by the yeast, including hydrolytic enzymes such as glucanases, which affect the fungal cell walls and have fungicidal or fungistatic activity against the fungal pathogens (Lima et al., 1998).

Although proteases are important factors in the biocontrol effects of beneficial microorganisms, very scarce studies have been done on their role in the inhibitory potential of yeasts against phytopathogens. Protease activity has been reported in *Metschnikowia* and *Wickerhamomyces* genera, but further studies have not been done yet. Transcription of protease and glucanase genes of Saccharomycopsis yeast increases significantly during the hunting process, but there is still no applied research on it. In addition, lipolytic activity is often used in the screening of extracellular enzyme activities of yeasts and pseudo-yeasts. Several studies revealed the role of lipase in the bioinhibitory activity of yeast species, which can be the basis for novel and advanced studies in the field of biological control of plant diseases by the application of various beneficial yeasts (Freimoser et al., 2019).

5.2 Production of toxins and antimicrobial metabolites

Yeasts are not known as producers of secondary metabolites, which is one of the reasons why they often raise fewer biosafety concerns. Consequently, relatively few toxic molecules that may contribute to biocontrol activity have been described. Flocculosin is a low molecular weight cellobiose lipid produced by the biocontrol yeast *Pseudozyma flocculosa*. The ubiquitous yeast-like fungus, *A. pullulans* produces diverse polymers (e.g., pullulan, aubasidan-like exopolysaccharide, poly(β-L-malic acid)), lipids, volatiles, enzymes, and secondary metabolites. Some of these metabolites (e.g., aureobasidins, liamocins, 2 propylacrylic acid, 2-methylenesuccinic acid) confer antagonistic activity against bacteria or fungi. Toxin production provided a competitive advantage to *A. pullulans* under dry, oligotrophic conditions, whereas it had no effect (as compared to yeasts not producing toxins) on antagonistic activity in more humid environments. The most prominent toxins produced by many biocontrol yeast strains are proteinaceous killer toxins. These proteins were originally identified in *S. cerevisiae* and seem to mainly kill competing yeast species. Yeast killer toxins have thus been mainly studied with respect to the control of spoilage yeasts in the beverage and food industry or for medical applications. However, several of these toxins also inhibit or kill plant pathogenic fungi and were thus proposed for plant protection. Nevertheless, further investigations to evaluate the specificity of yeast toxins and assess their effects on other beneficial microorganisms (e.g., in the phyllosphere, in soil microbiota and, in the case of edible commodities, the human gut) are required, particularly in the light of a possible registration (Freimoser et al., 2019).

5.3 Mycoparasitism and predation

Mycoparasitism (or consumption of a fungus by yeast or another antagonist) is rarely described and poorly studied in yeasts. *P. guilliermondii* was shown to strongly adhere to hyphae of the plant pathogen *B. cinerea* which leads to hyphal collapse, presumably due to the secretion of hydrolytic enzymes, such as glucanases and chitinases. Similarly, the yeast-like Ustilaginomycetes *Pseudozyma aphidis* parasitizes the powdery mildew pathogen *P. xanthii* and also the necrotrophic fungal pathogen *B. cinerea*. The capability of predation has been reported for several yeasts belonging to Ascomycota and Basidiomycota. Predacious yeast species vary extensively in the optimal environmental conditions that favor predation. Some are inhibited by the presence of rich nitrogenous compounds or organic sulfur, whereas other species may be activated under the same conditions (Lachance et al., 2000). The genus *Saccharomycopsis*, comprising predacious yeasts directly feeding on their prey, has a biocontrol effect against different *Penicillium* species and can be applied to manage storage molds caused by these fungi on various fruits and vegetables (Freimoser et al., 2019).

5.4 Competition for space and nutrients

Competition for space is one of the most important mechanisms of biological control. The rapid growth of biocontrol agents helps a beneficial microorganism to occupy the space (the tissues of the host plant), before the pathogens colonize the host. Furthermore, some of the microbes occupy extra space with abundant amounts of secreted polysaccharides that have both direct (occupying space) and indirect (attachment inhibitors, growth inhibitors, etc.) functions on the growth of competitor pathogens (Bozoudi and Tsaltas, 2018).

Competition for nutrients, which can be easily investigated by dual culture assays on various culture media such as potato dextrose agar (PDA), is the main mechanism involved in the biocontrol effect of beneficial microorganisms like yeasts against phytopathogens (Gouka et al., 2022; Kasfi et al., 2018). The main nutrients, for instance, are sugars, amino acids, iron, and nitrogen compounds, which are involved in the growth, development, and reproduction of yeasts and pathogens. A complex network of signaling pathways, including G-proteins, G-protein coupled receptors, Ras, and various types of protein kinases, is involved in informing living cells on nutrient availability and influences several processes, such as transcriptional, translational, posttranslational, metabolic, and developmental phases (Taheri and Tarighi, 2011; Taheri, 2018). Therefore, cross-talk of these signaling pathways might be involved in the biocontrol of various plant pathogens by yeast species, which can be an interesting subject for more detailed studies in the future.

5.5 Induction of plant defense responses against pathogens

The induction of plant resistance against pathogens is an indirect mechanism of biocontrol that can be activated by treating plant tissues with natural chemical or biological plant protectants, such as yeast species (Fig. 1). Pattern recognition receptors, which are localized on the surface of plant cells, recognize microbe- or pathogen-associated molecular patterns (MAMPs or PAMPs), and intracellular nucleotide binding-leucine reach repeat (NB-LRR)-type R proteins subsequently detect effectors secreted by the pathogens. These two phases of resistance activation are known as MAMP- or PAMP-triggered immunity (MTI or PTI) and effector-triggered immunity (ETI) (Chisholm et al., 2006). The MAMPs are vital and conserved microbial structures in pathogenic, nonpathogenic, and beneficial microorganisms, including chitin and ergosterol as the main components of cell walls and membranes in higher fungi, respectively; bacterial lipopolysaccharides, which are glycolipid components of the outer membranes of gram-negative bacteria; and flagellin as the major structural component of flagella in bacteria (Zipfel and Felix, 2005).

Early plant defense responses related to the MTI mainly include ion fluxes across the plasma membrane, accumulation of reactive oxygen species (ROS), and reactive nitrogen species (RNS). Later MTI responses include strengthening plant cell walls via callose deposition, oxidative cross-linking of polymers and lignification, biosynthesis of antimicrobial compounds (such as phenolics, alkaloids, flavonoids, phytoalexins, terpenes), and activation of pathogenesis-related (PR) proteins, together with activating other defense components, including phytohormones signaling networks, mitogen-activated protein kinases (MAPKs), and regulating transcription factors which lead to transcriptional changes of several genes involved in plant physiology and defense responses (Taheri, 2018, 2022; Narusaka et al., 2015).

In *Saccharomyces pastorianus*, known as budding yeast, the yeast extract and a mannopeptide from the yeast invertase function as MAMP and induce plant defense responses. The yeast cell wall mainly consists of polysaccharides, including polymers of glucose (β-glucan) and also polymers of mannose (mannoproteins), which may act as MAMPs and might be involved in the induction of resistance in plant tissues (Klis et al., 2006). Using yeast cell wall extracts (YCWEs) via treating the yeast cell wall with cell wall degrading enzymes revealed that application of the YCWEs activates resistance signaling pathways, systemically. Plant treatment with a defense activator, known as Housaku Monogatari (HM) obtained from yeast cell wall extract revealed the efficacy of the HM in enhancing defense responses of *Arabidopsis thaliana* and *Brassica rapa* leaves against fungal and bacterial diseases via activating jasmonate (JA)/ethylene (ET) signaling and also triggering salicylic acid (SA) signal transduction pathway. These results suggest that the HM contains the MAMPs which activate plant defense responses against various types of phytopathogens (Narusaka et al., 2015).

6 Yeast-mediated synthesis of antifungal nanoparticles

Nanotechnology is a new field of science that can be used in all aspects of life, including the management of plant diseases (Elmer et al., 2018; Yadav et al., 2008). A nanoparticle (NP) is a very small particle with a size range between 1 and 100 nm. The NPs obtained from some materials, such as metal oxides, are capable of inducing eukaryotic cell death, and in prokaryotic cells, cell growth is inhibited by their cytotoxic effect (Silva, 2006). Nanoparticles of metal oxides, metalloids, nonmetals, and carbon have fungicidal and bactericidal activity, which can be used to control various phytopathogens. Many of these nanomaterials also have nutritional benefits for plants and can induce host plant resistance to several diseases. The NPs of CeO_2, CuO, MgO_2, SiO_2, and numerous other NPs, whose synthesis might be mediated by plant material, fungi, bacteria, and yeasts, can activate defense-related mechanisms in plants. Therefore, the application of NPs can be a novel and effective strategy for controlling plant diseases (Elmer et al., 2018).

Recently discovered is yeast's ability to produce metal (gold, silver, titanium, palladium, and selenium) NPs with varying dimensions, either extracellularly or intracellularly, via the function of enzymes involved in various mechanisms (Roychoudhury, 2020). These NPs have great importance as antifungal, antibacterial, antioxidant, and anticancer agents. Therefore, different yeast species can be considered major scaffolds for the green and environmentally safe synthesis of different NPs with numerous biological applications, especially in plant disease management. The synthesis of NPs by fungi, such as yeasts, is considered to be easier for stable production of NPs compared to NPs production by bacteria, due to the higher biomass production of yeasts in simple culture media, higher bioaccumulation of metabolites, higher tolerance to metals and uptake capability of metals, and high wall binding capability of metals (Kumar et al., 2011; Roychoudhury, 2020).

The yeast cells can be used not only for NP production but also as carriers of NPs. One of the advantages of using yeast cells as NP carriers is that a simple encapsulation mechanism is possible using only yeast cells, water, and reagents, with no requirement for stabilizers. Yeast cells are biomacromolecular microparticles with an envelope composed of chitin, glycoproteins, and -glucans on one hand, and microcapsules on the other, with the plasma membrane assisting in encapsulation so that yeast cells can encapsulate various NPs (Klis et al., 2006). When membrane-bound oxidoreductases are activated by increasing the pH, metal ions and NPs are reduced. Because of their strong nucleophilic and redox properties, quinones can also reduce metal ions, converting them to NPs via the application of yeasts (Salunke et al., 2015).

Despite the amount of knowledge on the industrial-scale production of consumer items by yeasts, few studies have looked into yeast cells' potential to biosynthesize metallic nanoparticles such as Ag NPs (Quester et al., 2013). Stabilized Ag NPs are typically produced from $AgNO_3$ through interaction with a reducing agent, followed by stabilizing agents for surface encapsulation/capping. As a result, βG was proposed for effective NM reduction and stabilization with numerous bioactive properties (Sen et al., 2013). The yeast *S. cerevisiae* model revealed the ability to synthesize fairly monodisperse silver nanoparticles extracellularly. High antifungal activity against fluconazole-susceptible and fluconazole-resistant *Candida albicans* strains (Niknejad et al., 2015). After being incubated with silver nitrate, 15 yeast isolates from termite digestive tracts were evaluated for their ability to create Ag/AgCl-NPs ($AgNO_3$). Two strains were particularly successful in producing nanoparticles, and the purified Ag/AgCl-NPs were studied using a number of physicochemical approaches (Eugenio et al., 2016). Green synthesis of AgNPs using *S. cerevisiae* biotransformations and examination of the sizes and forms of the NPs generated were evaluated. The biocatalyst was dried and freshly cultivated *S. cerevisiae*. Dried yeast produced few NPs, whereas freshly cultivated yeast produced a considerable number of them (Korbekandi et al., 2016). *M. guilliermondii* KX008616 was studied for the intracellular synthesis of

Ag/AgCl-NPs under aerobic and anaerobic conditions. The results revealed differences in size and form among Ag/AgCl-NPs generated aerobically or anaerobically. The Ag/AgCl-NPs were homogenous and spherical in aerobic conditions (Alamri et al., 2018). After removing yeast cells from *S. cerevisiae* cultures, the media demonstrated the ability to synthesize AgNPs from $AgNO_3$ without the use of any extra chemical agents. This method of producing biocidal AgNPs with restricted size ranges, crystalline structure, and spherical form employs a unique, scalable, and environmentally friendly chemical route (Kthiri et al., 2021).

Using yeast extract as reducing and capping agents, shape-controlled and well-dispersed Ag NPs were biosynthesized. When the Ag^+ solution was combined with the yeast extract, yeast micelles formed. Furthermore, the biomolecules improve the stability of the produced Ag NPs, allowing them to remain stable for more than a year without precipitation. The biomolecules in yeast extract operate as a capping agent and play an important role in determining the size distribution, shape, and morphology of Ag NPs throughout their production (Shu et al., 2020). Biomolecules in yeast extract play an important role in the production of Ag NPs by preventing aggregation. Biomolecule stabilizer aid in the prevention of redundant reactions between Ag NPs (Zada et al., 2018). Yeast β-glucan (βG) was extracted from *S. cerevisiae* and converted into nanoparticles (NPs) using an alkali/acid facile method. The novel Ag NPs biosynthesis with βG NPs and the combined βG-Ag NPs nanocomposites might be convincingly proposed as potent antibacterial possibilities with low potential toxicity (Elnagar et al., 2021).

The AgNPs are reported to have a strong antifungal effect against several destructive fungal phytopathogens, such as *Fusarium graminearum* (Jian et al., 2022), *R. solani* and *F. oxysporum* f. sp. *lycopersici* (Purohit et al., 2022). Mourato et al. (2011) produced gold (Au) NPs with diameter ranging from 30 to 100 nm in extremophilic yeast strains from acid mine drainage, grown in the presence of $HAuCl_4$ up to 0.09 mM. Also, AuNP production was reported by Gericke and Pinches (2006) using the yeast, *Pichia jadinii* through exposure to $HAuCl_4$. Antifungal activity of AuNP has been reported against several fungal phytopathogens causing pre- and postharvest diseases, such as *F. graminearum*, *F. oxysporum*, and various species of *Aspergillus* and *Penicillium* (Balakumaran et al., 2020). In addition to AgNPs and AuNPs, the yeasts can mediate the production of the NPs of several other elements. Numerous yeast species are not only capable of mediating NPs production but also could be used as NP carriers (Roychoudhury, 2020).

Baker's yeast was used to achieve yeast-mediated synthesis of TiO_2 NPs. The presence of reducing substances, such as glucose, lowers the redox potential, enhancing NP production. The production of Y-TiO_2 NPs was a cost-effective method for producing pure, very stable, anatase TiO_2 NPs with small particle sizes (Peiris et al., 2018). The *S. cerevisiae* strain was used to establish a new synthesis method for selenium sulfide NPs. The physicochemical properties of SeNPs as well as their antifungal action were examined (Fig. 2). The produced NPs significantly inhibited the growth of various dangerous fungi, including *Aspergillus*, *Candida*, and *Alternaria* (Asghari-Paskiabi et al., 2019). The biotransformation of selenite oxyanion into SeNPs as intracellular and extracellular deposits was accomplished using a novel isolated aquatic *Rhodotorula mucilaginosa* strain under resting cell strategy, resulting in a monodisperse preparation of spherical shape SeNPs with an average size of 83 nm (Ashengroph and Tozandehjani, 2022). SeNPs were created using a baker's yeast extract. The biosynthesized SeNPs derived from baker's yeast extract have antifungal properties against toxigenic fungi

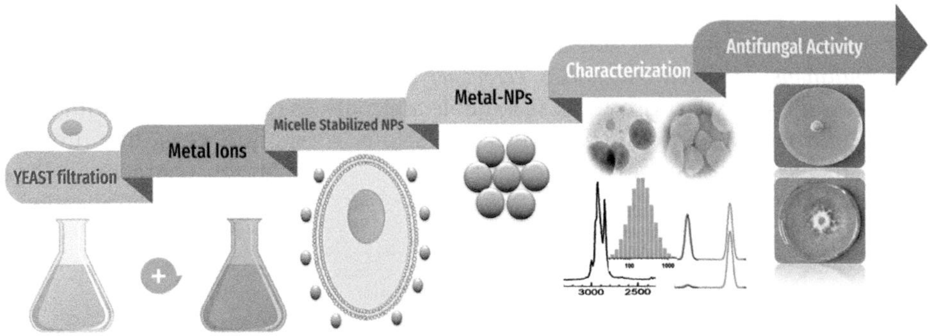

FIG. 2 A schematic representation of the efficiency of baker yeast end products in aerobic culture when confronted with diverse metal precursors. The physicochemical characteristics and antifungal activity of the metal NPs generated were studied.

such as *Aspergillus fumigatus* and *A. niger* and may be a useful efficacious antimicrobial agent in food preservation (Salem, 2022). A bio-mediated technique for synthesizing biocompatible barium carbonate nanoparticles ($nBaCO_3$) in living *S. cerevisiae* cells was presented based on biomimetic mineralization principles. This yeast cell method allows for the effective intracellular production of $BaCO_3$ NPs with strong monodispersibility. This biosynthesis approach, which uses a microorganism as a bioreactor, should be an appealing practice for the synthesis of additional insoluble carbonate nanoparticles since it is safe and reliable, allowing for the synthesis of nanomaterials with minimal costs and energy needs (Chang et al., 2021). Recent advances in the synthesis and applications of bionanomaterials produced through intracellular and/or extracellular biosynthesis processes employing different yeast and/or derived metabolites/secreted compounds. As a result, more research is needed to synthesize higher-monodispersity nanomaterials in a wider range of chemicals using the yeast synthesis technique. In this chapter, we provide a summary of the current research on the use of eukaryotes such as yeast and fungi in the biosynthesis of NPs and their applications (Boroumand Moghaddam et al., 2015).

7 Commercially produced yeasts as plant protectants and biocontrol agents

Nowadays, several yeast species are commercialized and yeast-based products are practically used to protect various plants, especially fruits and vegetables from destructive effects of postharvest fungal decays, and also for increasing the shelf-life of various agricultural crops and food products. Therefore, beneficial yeast species can be considered as good alternatives in the management of both preharvest and postharvest diseases (Droby et al., 2009; Ferraz et al., 2019). Isolates of *Candida* spp. have been used for manufacturing commercial yeast-based products, which can be used against various plant diseases, especially caused by fungal pathogens. For instance, "Aspire" and "Nexy" are commercial products of *Candida oleophila* isolates, which can be used to control postharvest diseases caused by several fungal species belonging to *Aspergillus*, *Penicillium*, *Botrytis*, and *Monilinia* genera. "Candifruit" is another commercial product obtained using the isolates of *C. sake*, which mainly can be used for treating pome fruits to protect them against storage molds caused by fungal pathogens such as *Rhizopus*, *Penicillium*, and *Botrytis*. Not only isolate *Candida* species, but also other yeasts, such as *A. pullulans*, *M. fructicola*, and *S. cerevisiae* (Pizzolitto et al., 2012) have been used for generating commercial yeast products as suitable alternatives to be replaced with hazardous chemical fungicides. More detailed studies on various yeasts and yeast-like fungi in both in vitro and in vivo conditions seem to be necessary to more accurately characterize these beneficial microorganisms and develop their commercialization and practical application as natural plant protectants against pathogens and pests.

8 Concluding remarks

Utilization of various yeasts as useful microorganisms is a sustainable strategy to reduce the destructive effects of environmental stresses, especially plant diseases caused by fungal pathogens. Epiphytic and endophytic yeast species, which have considerable advantages over filamentous fungi and bacteria, can be used in protecting plants against both biotic and abiotic stresses and also for plant growth promotion. Beneficial yeasts are easily cultured, stored, and applied to various tissues of various crops. It seems necessary to focus on more detailed research for determining mechanisms of crop protection via the application of yeasts in various plants, identifying beneficial metabolites produced by yeasts, their commercialization, and especially field trials to investigate the potential of yeasts and their products to be applied as biofertilizers and crop protectants in the environment. The green synthesis of metal NPs by beneficial yeasts, which has the advantages of nontoxicity, reproducibility in production, ease of scaling up, and well-defined morphology, will continue to become a common trend in NP production in the forthcoming years. The application of yeasts could be an ecologically friendly way to reduce the usage of hazardous synthetic fertilizers and other agrochemicals to control pathogens and water inputs in agricultural systems, leading to increased crop yield.

References

Alamri, S.A., Hashem, M., Nafady, N.A., Sayed, M.A., Alshehri, A.M., Alshaboury, G.A., 2018. Controllable biogenic synthesis of intracellular silver/silver chloride nanoparticles by *Meyerozyma guilliermondii* KX008616. J. Microbiol. Biotechnol. 28 (6), 917–930.

Asghari-Paskiabi, F., Imani, M., Rafii-Tabar, H., Razzaghi-Abyaneh, M., 2019. Physicochemical properties, antifungal activity and cytotoxicity of selenium sulfide nanoparticles green synthesized by *Saccharomyces cerevisiae*. Biochem. Biophys. Res. Commun. 516 (4), 1078–1084.

Ashengroph, M., Tozandehjani, S., 2022. Optimized resting cell method for green synthesis of selenium nanoparticles from a new *Rhodotorula mucilaginosa* strain. Process Biochem. 116, 197–205.

Avis, T.J., Bélanger, R.R., 2002. Mechanisms and means of detection of biocontrol activity of *Pseudozyma* yeasts against plant-pathogenic fungi. FEMS Yeast Res. 2 (1), 5–8.

Balakumaran, M.D., Ramachandran, R., Balashanmugam, P., Jagadeeswari, S., Kalaichelvan, P.T., 2020. Comparative analysis of antifungal, antioxidant and cytotoxic activities of mycosynthesized silver nanoparticles and gold nanoparticles. Adv. Perform. Mater. 37 (6). https://doi.org/10.1080/10667857.2020.1854518.

Bevan, E.A., Makower, M., 1963. The physiological basis of the killer character in yeast. In: Geerts, S.J. (Ed.), Proceedings of the Genetics Today, XIth International Congress of Genetics. vol. 1. Pergamon Press, Oxford, pp. 202–203.

Boroumand Moghaddam, A., Namvar, F., Moniri, M., Md. Tahir, P., Azizi, S., Mohamad, R., 2015. Nanoparticles biosynthesized by fungi and yeast: a review of their preparation, properties, and medical applications. Molecules 20 (9), 16540–16565.

Botha, A., 2011. The importance and ecology of yeasts in soil. Soil Biol. Biochem. 43 (1), 1–8.

Bozoudi, D., Tsaltas, D., 2018. The multiple and versatile roles of *Aureobasidium pullulans* in the vitivinicultural sector. Fermentation 4, 85.

Bright, J.P., Karunanadham, K., Maheshwari, H.S., Karuppiah, E.A., Thankappan, S., Nataraj, R., Pandian, D., Ameen, F., Poczai, P., Sayyed, R.Z., 2022. Seed-borne probiotic yeasts foster plant growth and elicit health protection in black gram (*Vigna mungo* L.). Sustainability 14 (8), 4618.

Chanchaichaovivat, A., Ruenwongsa, P., Panijpan, B., 2007. Screening and identification of yeast strains from fruits and vegetables: potential for biological control of postharvest chilli anthracnose (*Colletotrichum capsici*). Biol. Control 42 (3), 326–335.

Chang, Y., Chen, S., Liu, T., Liu, P., Guo, Y., Yang, L., Ma, X., 2021. Yeast cell route: a green and facile strategy for biosynthesis of carbonate nanoparticles. CrystEngComm 23 (26), 4674–4679.

Chisholm, S.T., Coaker, G., Day, B., Staskawicz, B.J., 2006. Host-microbe interactions: shaping the evolution of the plant immune response. Cell 124, 803–814.

Daroodi, Z., Taheri, P., Tarighi, S., 2021. Direct antagonistic activity and tomato resistance induction of the endophytic fungus *Acrophialophora jodhpurensis* against *Rhizoctonia solani*. Biol. Control 160, 104696.

De Tenório, D.A., Medeiros, E., Lima, C.S., da Silva, J.M., de Barros, J.A., Neves, R.P., Laranjeira, D., 2019. Biological control of *Rhizoctonia solani* in cowpea plants using yeast. Trop. Plant Pathol. 44, 113–119.

Dehghanpour-Farashah, S., Taheri, P., Falahati-Rastegar, M., 2019. Effect of polyamines and nitric oxide in *Piriformospora indica*-induced resistance and basal immunity of wheat against *Fusarium pseudograminearum*. Biol. Control 136, 104006. https://doi.org/10.1016/j.biocontrol.2019.104006.

Di Francesco, A., Di Foggia, M., Baraldi, E., 2020. *Aureobasidium pullulans* volatile organic compounds as alternative postharvest method to control brown rot of stone fruits. Food Microbiol. 87, 103395.

Díaz, M.A., Pereyra, M.M., Picón-Montenegro, E., Meinhardt, F., Dib, J.R., 2020. Killer yeasts for the biological control of postharvest fungal crop diseases. Microorganisms 8 (11), 1680.

Droby, S., Wisniewski, M., Macarisin, D., Wilson, C., 2009. Twenty years of postharvest biocontrol research: is it time for a new paradigm? Postharvest Biol. Technol. 52, 137–145.

Dukare, A.S., Paul, S., Nambi, V.E., Gupta, R.K., Singh, R., Sharma, K., et al., 2018. Exploitation of microbial antagonists for the control of postharvest diseases of fruits: a review. Crit. Rev. Food Sci. Nutr. 16, 1–16.

El-Mehalawy, A.A., 2004. The rhizosphere yeast fungi as biocontrol agents for wilt disease of kidney bean caused by *Fusarium oxysporum*. Int. J. Agric. Biol. 6, 310–316.

El-Mehalawy, A.A., Hassanein, N.M., Khater, H.M., Karam El-Din, E.A., Youssef, Y.A., 2004. Influence of maize root colonization by the rhizosphere actinomycetes and yeast fungi on plant growth and on the biological control of late wilt disease. Int. J. Agric. Biol. 6, 599–605.

Elmer, W., Ma, C., White, J., 2018. Nanoparticles for plant disease management. Curr. Opin. Environ. Sci. Health 6, 66–70.

Elnagar, S.E., Tayel, A.A., Elguindy, N.M., Al-saggaf, M.S., Moussa, S.H., 2021. Innovative biosynthesis of silver nanoparticles using yeast glucan nanopolymer and their potentiality as antibacterial composite. J. Basic Microbiol. 61 (8), 677–685.

El-Tarabily, K.A., 2004. Suppression of *Rhizoctonia solani* diseases of sugar beet by antagonistic and plant growth-promoting yeasts. J. Appl. Microbiol. 96, 69–75.

Eugenio, M., Müller, N., Frases, S., Almeida-Paes, R., Lima, L.M.T., Lemgruber, L., Farina, M., de Souza, W., Sant'Anna, C., 2016. Yeast-derived biosynthesis of silver/silver chloride nanoparticles and their antiproliferative activity against bacteria. RSC Adv. 6 (12), 9893–9904.

Fernandez-San Millan, A., Larraya, L., Farran, I., Ancin, M., Veramendi, M., 2021. Successful biocontrol of major postharvest and soil-borne plant pathogenic fungi by antagonistic yeasts. Biol. Control 160, 104683.

Ferraz, P., Cássio, F., Lucas, C., 2019. Potential of yeasts as biocontrol agents of the phytopathogen causing cacao witches' broom disease: is microbial warfare a solution? Front. Microbiol. 10, 1766.

Fisher, M.C., Henk, D.A., Briggs, C.J., Brownstein, J.S., Madoff, L.C., McCraw, S.L., et al., 2012. Emerging fungal threats to animal, plant and ecosystem health. Nature 484, 186–194.

Fisher, M.C., Hawkins, N.J., Sanglard, D., Gurr, S.J., 2018. Worldwide emergence of resistance to antifungal drugs challenges human health and food security. Science 360, 739–742.

Freimoser, F.M., Rueda-Mejia, M.P., Tilocca, B., Migheli, Q., 2019. Biocontrol yeasts: mechanisms and applications. World J. Microbiol. Biotechnol. 35 (10), 1–19.

Gamagae, S.U., Sivakumar, D., Wijeratnam, R.W., Wijesundera, R.L., 2003. Use of sodium bicarbonate and *Candida oleophila* to control anthracnose in papaya during storage. Crop Prot. 22 (5), 775–779.

Gericke, M., Pinches, A., 2006. Microbial production of gold nanoparticles. Gold Bull. 39, 22–28.

Gostinčar, C., Ohm, R.A., Kogej, T., Sonjak, S., Turk, M., Zajc, J., Zalar, P., Grube, M., Sun, H., Han, J., Sharma, A., Chiniquy, J., Ngan, C.Y., Lipzen, A., Barry, K., Grigoriev, I.V., Gunde-Cimerman, N., 2014. Genome sequencing of four *Aureobasidium pullulans* varieties: biotechnological potential, stress tolerance, and description of new species. BMC Genomics 15, 549.

Gouka, L., Vogels, C., Hansen, L.H., Raaijmakers, J.M., Cordovez1, V., 2022. Genetic, phenotypic and metabolic diversity of yeasts from wheat flag leaves. Front. Plant Sci. https://doi.org/10.3389/fpls.2022.908628.

Ignatova, L.V., Brazhnikova, Y.V., Berzhanova, R.Z., Mukasheva, T.D., 2015. Plant growth-promoting and antifungal activity of yeasts from dark chestnut soil. Microbiol. Res. 75, 78–83.

Into, P., Khunnamwong, P., Jindamoragot, S., Am-in, S., Intanoo, W., Limtong, S., 2020. Yeast associated with rice phylloplane and their contribution to control of rice sheath blight disease. Microorganisms 8, 362.

Iqbal, M., Jamshaid, M., Zahid, M.A., Andreasson, E., Vetukuri, R.R., Stenberg, J.A., 2021. Biological control of strawberry crown rot, root rot and grey mould by the beneficial fungus *Aureobasidium pullulans*. BioControl 66, 535–545.

Jian, Y., Chen, X., Ahmed, T., Shang, Q., Zhang, S., Ma, Z., Yin, Y., 2022. Toxicity and action mechanisms of silver nanoparticles against the mycotoxin-producing fungus *Fusarium graminearum*. J. Adv. Res. 38, 1–12.

Joubert, P.M., Doty, S.L., 2018. Endophytic yeasts: biology, ecology and applications. In: Pirttilä, A.M., Frank, A.C. (Eds.), Endophytes of Forest Trees. Forestry Sciences, vol. 86., https://doi.org/10.1007/978-3-319-89833-9.

Kamel, S.M., Morsy, E.M., Massoud, O.N., 2016. Potentiality of some yeast species as biocontrol agents against *Fusarium oxysporum* f. sp. *cucumerinum* the causal agent of cucumber wilt. Egypt. J. Biol. Pest Cont. 26 (2), 185–193.

Kasfi, K., Taheri, P., Jafarpour, B., Tarighi, S., 2018. Characterization of antagonistic microorganisms against *Aspergillus* spp. from grapevine leaf and berry surfaces. J. Plant Pathol. 100, 179–190.

Khan, N.I., Schisler, D.A., Boehm, M.J., Lipps, P.E., Slininger, P.J., 2004. Field testing of antagonists of Fusarium head blight incited by *Gibberella zeae*. Biol. Control 29 (2), 245–255.

Klis, F.M., Boorsma, A., De Groot, P.W.J., 2006. Cell wall construction in *Saccharomyces cerevisiae*. Yeast 23, 185–202.

Korbekandi, H., Mohseni, S., Mardani Jouneghani, R., Pourhossein, M., Iravani, S., 2016. Biosynthesis of silver nanoparticles using *Saccharomyces cerevisiae*. Artif. Cells Nanomed. Biotechnol. 44 (1), 235–239.

Kthiri, A., Hamimed, S., Othmani, A., Landoulsi, A., O'Sullivan, S., Sheehan, D., 2021. Novel static magnetic field effects on green chemistry biosynthesis of silver nanoparticles in *Saccharomyces cerevisiae*. Sci. Rep. 11 (1), 1–9.

Kumar, S.D., Karthik, L., Kumar, G., Rao, K.V., 2011. Biosynthesis of silver nanoparticles from marine yeast and their antimicrobial activity against multidrug resistant pathogens. Pharmacologyonline 3, 1100–1111.

Lachance, M.A., Pupovac-Velikonja, A., Natarajan, S., Schlag-Edler, B., 2000. Nutrition and phylogeny of predacious yeasts. Can. J. Microbiol. 46 (6), 459–505.

Lahlali, R., Hamadi, Y., Guilli, M.E., Jijakli, M.H., 2014. The ability of the antagonist yeast *Pichia guilliermondii* strain Z1 to suppress green mould infection in citrus fruit. Ital. J. Food Saf. 3, 4774.

Lee, G., Lee, S.H., Kim, K.M., Ryu, C.M., 2017. Foliar application of the leaf-colonizing yeast *Pseudozyma churashimaensis* elicits systemic defense of pepper against bacterial and viral pathogens. Sci. Rep. 7 (1), 1–3.

Lima, G., De Curtis, F., Castoria, R., De Cicco, V., 1998. Activity of the yeasts *Cryptococcus laurentii* and *Rhodotorula glutinis* against post-harvest rots on different fruits. Biocontrol Sci. Technol. 8 (2), 257–267.

Lima, J.R., Gondim, D.M., Oliveira, J.T., Oliveira, F.S., Gonçalves, L.R., Viana, F.M., 2013. Use of killer yeast in the management of postharvest papaya anthracnose. Postharvest Biol. Technol. 83, 58–64.

Luongo, L., Galli, M., Corazza, L., Meekes, E., Haas, L.D., Van Der Plas, C.L., Köhl, J., 2005. Potential of fungal antagonists for biocontrol of *Fusarium* spp. in wheat and maize through competition in crop debris. Biocontrol Sci. Technol. 15 (3), 229–242.

Maghsoodi, F., Taheri, P., 2021. Efficacy of *Althaea officinalis* leaf extract in controlling *Alternaria* spp. pathogenic on *Citrus*. Eur. J. Plant Pathol. 161, 799–813.

Marwa Abdel-Kareem, M., Zohri, A.N., Nasr, S.A.E., 2021. Novel marine yeast strains as plant growth-promoting agents improve defense in wheat (*Triticum aestivum*) against *Fusarium oxysporum*. J. Plant Dis. Prot. 128, 973–988.

Matic, S., Spadaro, D., Garibaldi, A., Gullino, M.L., 2014. Antagonistic yeasts and thermotherapy as seed treatments to control *Fusarium fujikuroi* on rice. Biol. Control 73, 59–67.

Mourato, A., Gadanho, M., Lino, A.R., Tenreiro, R., 2011. Biosynthesis of crystalline silver and gold nanoparticles by extremophilic yeasts. Bioinorg. Chem. Appl. 2011, 546074.

Nally, M.C., Pescea, V.M., Maturanoa, Y.P., Toroa, M.E., Combinab, M., Castellanos De Figueroa, L.I., Vazquez, F., 2013. Biocontrol of fungi isolated from sour rot infected table grapes by *Saccharomyces* and other yeast species. Postharvest Biol. Technol. 86, 456–462.

Narusaka, M., Minami, T., Iwabuchi, C., Hamasaki, T., Takasaki, S., Kawamura and Narusaka, Y., 2015. Yeast cell wall extract induces disease resistance against bacterial and fungal pathogens in *Arabidopsis thaliana* and *Brassica* crop. PLoS One. https://doi.org/10.1371/journal.pone.0115864.

Niknejad, F., Nabili, M., Ghazvini, R.D., Moazeni, M., 2015. Green synthesis of silver nanoparticles: advantages of the yeast *Saccharomyces cerevisiae* model. Curr. Med. Mycol. 1 (3), 17.

Nunes, C.A., 2012. Biological control of postharvest diseases of fruit. Eur. J. Plant Pathol. 133, 181–196.

Parapouli, M., Vasileiadis, A., Afendra, A.S., Hatzilouka, S., 2020. *Saccharomyces cerevisiae* and its industrial application. AIMS Microbiol. 6 (1), 1–31.

Peiris, M.M.K., Gunasekara, T.D.C.P., Jayaweera, P.M., Fernando, S.S.N., 2018. TiO2 nanoparticles from Baker's yeast: a potent antimicrobial. J. Microbiol. Biotechnol. 28 (10), 1664–1670.

Pizzolitto, R.P., Armando, M.R., Combina, M., Cavaglieri, L.R., Dalcero, A.M., Salvano, M.A., 2012. Evaluation of *Saccharomyces cerevisiae* strains as probiotic agent with aflatoxin B1 adsorption ability for use in poultry feedstuffs. J. Environ. Sci. Health B 47 (10), 933–941.

Podgórska-Kryszczuk, I., Solarska, E., Kordowska-Wiater, M., 2022. Biological control of *Fusarium culmorum*, *Fusarium graminearum* and *Fusarium poae* by antagonistic yeasts. Pathogens 11 (1), 86.

Purohit, A., Sharma, A., Ramakrishnan, R.S., Sharma, S., Kumar, A., Jain, D., Kushwaha, H.S., Maharjan, E., 2022. Biogenic synthesis of silver nanoparticles (AgNPs) using aqueous leaf extract of *Buchanania lanzan* Spreng and evaluation of their antifungal activity against phytopathogenic fungi. Bioinorg. Chem. Appl. 2022, 6825150.

Quester, K., Avalos-Borja, M., Castro-Longoria, E., 2013. Biosynthesis and microscopic study of metallic nanoparticles. Micron 54, 1–27.

Raspor, P., Miklic-Milek, D., Avbelj, M., Cadez, N., 2010. Biocontrol of grey mold disease on grape caused by *Botrytis cinerea* with autochthonous wine yeasts. Food Technol. Biotechnol. 48, 336–343.

Roychoudhury, A., 2020. Yeast-mediated green synthesis of nanoparticles for biological applications. Indian J. Pharmacol. Biol. Res. 8 (3), 26–31.

Salem, S.S., 2022. Bio-fabrication of selenium nanoparticles using Baker's yeast extract and its antimicrobial efficacy on food borne pathogens. Appl. Biochem. Biotechnol. 194 (5), 1898–1910.

Salunke, B.K., Sawant, S.S., Lee, S.I., Kim, B.S., 2015. Comparative study of MnO2 nanoparticle synthesis by marine bacterium *Saccharophagus degradans* and yeast *Saccharomyces cerevisiae*. Appl. Microbiol. Biotechnol. 99, 5419–5427.

Schena, L., Nigro, F., Pentimone, I., Ligorio, A., Ippolito, A., 2003. Control of postharvest rots of sweet cherries and table grapes with endophytic isolates of *Aureobasidium pullulans*. Postharvest Biol. Technol. 30, 209–220.

Sen, I.K., Mandal, A.K., Chakraborti, S., Dey, B., Chakraborty, R., Islam, S.S., 2013. Green synthesis of silver nanoparticles using glucan from mushroom and study of antibacterial activity. Int. J. Biol. Macromol. 62, 439–449.

Shalaby, M.E., El-Nady, M.F., 2008. Application of *Saccharomyces cerevisiae* as a biocontrol agent against Fusarium infection of sugar beet plants. Acta Biol. Szeged. 52 (2), 271–275.

Shu, M., He, F., Li, Z., Zhu, X., Ma, Y., Zhou, Z., Yang, Z., Gao, F., Zeng, M., 2020. Biosynthesis and antibacterial activity of silver nanoparticles using yeast extract as reducing and capping agents. Nanoscale Res. Lett. 15 (1), 1–9.

Silva, G.A., 2006. Neuroscience nanotechnology: progress, opportunities and challenges. Nat. Rev. Neurosci. 7, 65–74.

Spotts, R.A., Cervantes, L.A., Facteau, T.J., 2002. Integrated control of brown rot of sweet cherry fruit with a preharvest fungicide, a postharvest yeast, modified atmosphere packaging, and cold storage temperature. Postharvest Biol. Technol. 24 (3), 251–257.

Syed Ab Rahman, S.F., Singh, E., Pieterse, C.M.J., Schenk, P.M., 2018. Emerging microbial biocontrol strategies for plant pathogens. Plant Sci. 267, 102–111.

Taheri, P., 2018. Cereal diseases caused by *Fusarium graminearum*: from biology of the pathogen to oxidative burst-related host defense responses. Eur. J. Plant Pathol. 152, 1–20.

Taheri, P., 2022. Crosstalk of nitro-oxidative stress and iron in plant immunity. Free Radic. Biol. Med. 191, 137–149.

Taheri, P., Flaherty, J., 2022. Beneficial microorganisms in agriculture. Physiol. Mol. Plant Pathol. 117, 101777.

Taheri, P., Tarighi, S., 2011. Cytomolecular aspects of rice sheath blight caused by *Rhizoctonia solani*. Eur. J. Plant Pathol. 129, 511–528.

Taheri, P., Gnanamanickam, S., Höfte, M., 2007. Characterization, genetic structure, and pathogenicity of *Rhizoctonia* spp. associated with rice sheath diseases in India. Phytopathology 97, 373–383.

Thambugala, K.M., Daranagama, D.A., Phillips, A.J., Kannangara, S.D., Promputtha, I., 2020. Fungi vs. fungi in biocontrol: an overview of fungal antagonists applied against fungal plant pathogens. Front. Cell. Infect. Microbiol. 10, 604923.

Urquhart, E.J., Punja, Z.K., 2002. Hydrolytic enzymes and antifungal compounds produced by *Tilletiopsis* species, phyllosphere yeasts that are antagonists of powdery mildew fungi. Can. J. Microbiol. 48 (3), 219–229.

Wachowska, U., Stuper-Szablewska, K., Perkowski, J., 2020. Yeasts isolated from wheat grain can suppress *Fusarium* head blight and decrease trichothecene concentrations in bread wheat and durum wheat grain. Pol. J. Environ. Stud. 29 (6), 4345–4360.

Wachowska, U., Sulyok, M., Wiwart, M., Suchowilska, E., Kandler, W., Krska, R., 2022. The application of antagonistic yeasts and bacteria: an assessment of in vivo and under field conditions pattern of *Fusarium* mycotoxins in winter wheat grain. Food Control 138, 109039.

Williamson, M.A., Fokkema, N.J., 1985. Phyllosphere yeasts antagonize penetration from appressoria and subsequent infection of maize leaves by *Colletotrichum graminicola*. Neth. J. Plant Pathol. 91 (6), 265–276.

Yadav, S.K., Mohanpuria, P., Rana, N.K., 2008. Biosynthesis of nanoparticle: technological concepts and future applications. J. Nanopart. Res. 10, 507–517.

Zada, S., Ahmad, A., Khan, S., Yu, X., Chang, K., Iqbal, A., Ahmad, A., Ullah, S., Raza, M., Khan, A., Ahmad, S., 2018. Biogenic synthesis of silver nanoparticles using extracts of *Leptolyngbya* JSC-1 that induce apoptosis in HeLa cell line and exterminate pathogenic bacteria. Artif. Cells Nanomed. Biotechnol. 46 (sup3), S471–S480.

Zhang, D., Spadaro, D., Garibaldi, A., Gullino, M.L., 2010. Efficacy of the antagonist *Aureobasidium pullulans* PL5 against postharvest pathogens of peach, apple and plum and its modes of action. Biol. Control 54, 172–180.

Zhang, X., Li, B., Zhang, Z., Chen, Y., Tian, S., 2020. Antagonistic yeasts: a promising alternative to chemical fungicides for controlling postharvest decay of fruit. J. Fungi 6 (3), 158.

Zhimo, V.Y., Bhutia, D.D., Saha, J., 2016. Biological control of post harvest fruit diseases using antagonistic yeasts in India. J. Plant Pathol. 98 (2), 275–283.

Zhou, Y., Zhang, L., Zeng, K., 2016. Efficacy of *Pichia membranaefaciens* combined with chitosan against *Colletotrichum gloeosporioides* in citrus fruits and possible modes of action. Biol. Control 96, 39–47.

Ziedan, E.S., Farrag, E.S., 2011. Application of yeasts as biocontrol agents for controlling foliar diseases on sugar beet plants. J. Agric. Technol. 7 (6), 1789–1799.

Zipfel, C., Felix, G., 2005. Plants and animals: a different taste for microbes? Curr. Opin. Plant Biol. 8, 353–360.

CHAPTER

8

Antifungal activity of microbial secondary metabolites

Ragini Bodade and Krutika Lonkar

Department of Microbiology, Savitribai Phule Pune University, Pune, Maharashtra, India

1 Introduction

Fungi are the second most diverse group of organisms in ecosystems and are estimated to contain 2.2–3.8 million species based on next-generation sequencing. In the environment, it plays an essential role as decomposers, mutualists, and pathogens. Over the last few years, immunocompromised people undergoing human immunodeficiency virus (HIV) infections, cancer chemotherapy, and organ transplantation have mainly suffered from the opportunistic fungal infections caused by *Candida albicans*, *Pneumocystis jirovecii*, *Aspergillus fumigatus*, and *Cryptococcus neoformans*. Therefore, increasing numbers of virulent fungal infectious diseases threaten morbidity and mortality in human patients (Kaur et al., 2022). In recent years, it has been assessed that over 1 billion people suffer from serious fungal diseases or mycosis and more than 1.6 million people die annually from fungal diseases. Furthermore, pathogenic fungi pose the greatest threat to animals such as bats, amphibians, and reptiles (Kim, 2016; Aldholmi et al., 2019; Almeida et al., 2019; Wu et al., 2019). Moreover, around 19,000 fungi are known as crop plant pathogens and may survive as dormant but alive on both living and dead plant tissues until favorable conditions (Jain et al., 2019).

The term "antifungals" includes all chemical compounds (potassium iodide and zinc pyrithione), pharmacologic agents (polyenes, azoles, allylamines, thiocarbamates, octenidine, pirtenidine, and morpholines), and natural products (plant oils and honey) that are used to treat plant and human fungal infections (mycoses). Till the 1950s, human fungal infections were treated with chemotherapy using potassium iodide (effective in the treatment of sporotrichosis) and two useful polyenes only, i.e., nystatin and amphotericin B. Griseofulvin, the first line of drug introduced in 1959, and soon after Terbinafine, an allylamine antifungal, was approved by the FDA in 1996. Moreover, azoles, viz., miconazole, ketoconazole, fluconazole, itraconazole, voriconazole, and posaconazole, were introduced in between 1979 and 2006 and

Nanohybrid Fungicides
https://doi.org/10.1016/B978-0-443-23950-2.00005-9

more recently isavuconazonium in 2015. However, the success of treatment is gradually compromised due to antifungal drug resistance and its side effects like allergic reactions, nephrotoxicity (due to amphotericin B), and hepatotoxicity (due to echinocandins). Thus, the large number of unmet medical needs associated with invasive fungal diseases should stimulate the investigation of new antifungal compounds. Modifications of antifungals such as lipid formulations for polyenes aid to lower its toxicity, and triazoles for its wider spectrum of action have been introduced to tackle the problem to some extent. Moreover, the new target identification, validation, and drug action mechanism further aid in running sustainable antimicrobial drug discovery (Abu-Elteen and Hamad, 2017; McKeny et al., 2021).

In the same way, the agricultural industry is threatened worldwide by fungal pathogens producing plant diseases. One of the most common fungal plant pathogens, *Magnaporthe oryzae*, is responsible for rice blast disease (Eseola et al., 2021). Black Rust and brown rust diseases caused by *Puccinia graminis* f. sp. *tritici*, and *Puccinia triticina* are the most devastating infection to wheat crops, respectively (Figueroa et al., 2018). *Fusarium graminearum* is a highly destructive pathogen responsible for fusarium head blight to cereals and affects grain yield and quality. Moreover, it also induces mycotoxin-contaminated grain toxicity in humans (Gong et al., 2015). Wheat and barley, the world's topmost crops are affected by *Blumeria graminis* responsible for well-known powdery mildew disease (Troch et al., 2014). More than 200 different plants are infected by the grey mold *Botrytis cinerea* (Williamson et al., 2007). Moreover, several high-value crops, such as coffee, cacao, bananas, spices, mangos, and several nuts, are currently affected by fungal infections too. These phytopathogenic fungal diseases can cause enormous forfeiture in the yield and quality of crops, fruits, and other edible plant materials. Thus, fungal pathogens are not only a threat to human health but also to the global economy (Almeida et al., 2019) (Table 1).

Furthermore, all these fungal infections are controlled by chemical fungicides that may cause toxicity and health complications to humans and animals. In the 1990s, the importance of biocontrol agents has been increased to reduce the chemical fungicides usages and thereby their hazardous effects on human and animal health. Moreover, increased population growth and climate change can further elevate the incidence of fungal infections and their impact on arable land in the future. Thus, plant fungal infection management is important for increasing food production and poverty alleviation (Almeida et al., 2019; Rashad and Moussa, 2020). Compounds like secondary metabolites (SMs) are structurally diverse compounds secreted by microbes during their stationary growth phase. SMs from bacteria, fungi, actinomycetes, and endophytes are reported for therapeutic, cosmetics, nutraceutical, agricultural applications, and as food additives (Reddy et al., 2021).

In this chapter, we have highlighted microbial-sourced SMs exhibiting antifungal activity.

2 Secondary metabolites screening, production, purification, and characterization

The term SM was first coined by A. Kossel in 1891. Microbial SMs are synthesized by bacteria, fungi, and algae under nutrient depletion and environmental stress conditions. Further, they are found beneficial to microorganisms either by improving their growth or offering protection from competitive environments or predators. In a microbial cell, duplication or

TABLE 1 Some fungal diseases and their management using antifungal drugs.

Sr. no.	Host	Name of diseases	Pathogens	Antifungal drugs
1	Infection to humans	dermatophytosis, vaginal candidiasis, allergy, mycotoxicosis, invasive fungal infections (IFIs), Athlete's foot, Aspergillosis	*Pneumocystis jirovecii, Candida albicans, Aspergillus fumigatus* and *Cryptococcus neoformans, Fusarium* spp., *Scedosporium* spp., *Penicillium* spp., *Epidermophyton floccosum, Aspergillus fumigatus*	Amphotericin B, 5-Fluorocytosine, voriconazole, terbinafine, echinocandins
2	Infection to Plants	Leaf blight, Fusarium wilt, early blight, Wilt, Sheath blight, Rust	*Alternaria triticina, Fusarium oxysporum, Alternaria solani, Sclerotium rolfsii, Rhizoctonia solani, Puccinia graminis*	Morpholines (Fenpropimorph, tridemorph), strobilurins (azoxystrobin, fluoxystrobin), dithiocarbamates (mancozeb, maneb, zineb), Azole (prothioconazole, Metconazole)
3	Infection to Animals	Aspergillosis, Mucormycosis, Oral and gastrointestinal candidiasis, Peritonitis, Mastitis, Bronchopneumonia, Cryptococcosis, dermatophytosis, Histoplasmosis	*A. fumigatus, Aspergillus felis, Rhizopus, Mucor* sp., *Rhizomucor* sp., *Lichtheimia* sp., *C. albicans, C. catenulata, C. krusei, C. neoformans, C. glabrata*, and *C. gattii, Trichophyton simii, T. mentagrophytes, T. verrucosum, Microsporum canis, Histoplasma farciminosum*	Itraconazole, Ketoconazole, Fluconazole, Voriconazole, Posaconazole

mutations of primary metabolites genes can induce SM production pathways named mevalonic acid pathway (for terpenoids, steroids), shikimic acid pathway (for aromatic amino acids, alkaloids), acetate pathway (for polyketides, fatty acids), amino acid/peptide pathway, polyketide synthases (PKS) pathway, nonribosomal peptide synthetases (NRPS) pathway, hybrid (nonribosomal polyketide) synthetic pathway, and carbohydrate pathway (Spencer et al., 2023). Most of these SMs thus serve different functions such as antimicrobial agents against pathogens, toxins, metal-transporting agents (siderophores), sex hormones, and pigments (Fig. 1). The classical regulatory mechanisms for SMs production include autoinduction, carbon catabolite regulation, enzyme feedback regulation, and nitrogen/phosphate regulation. Many growth factors including nutrient exhaustion, pH, light intensity, temperature, redox status, and pheromones are responsible for SM induction (Vining, 2007; Thirumurugan et al., 2018; Barrios-González, 2018). All these SMs are produced by active synthesis machinery and encompass a single gene or a group of genes called biosynthetic gene cluster (BGC), arranged on chromosomal DNA. In addition to biosynthetic genes, other genes are involved in their expression control (regulatory genes), self-resistance, and export (permeability genes). In laboratory conditions, a single BGC or cluster of BCGs (superclusters)

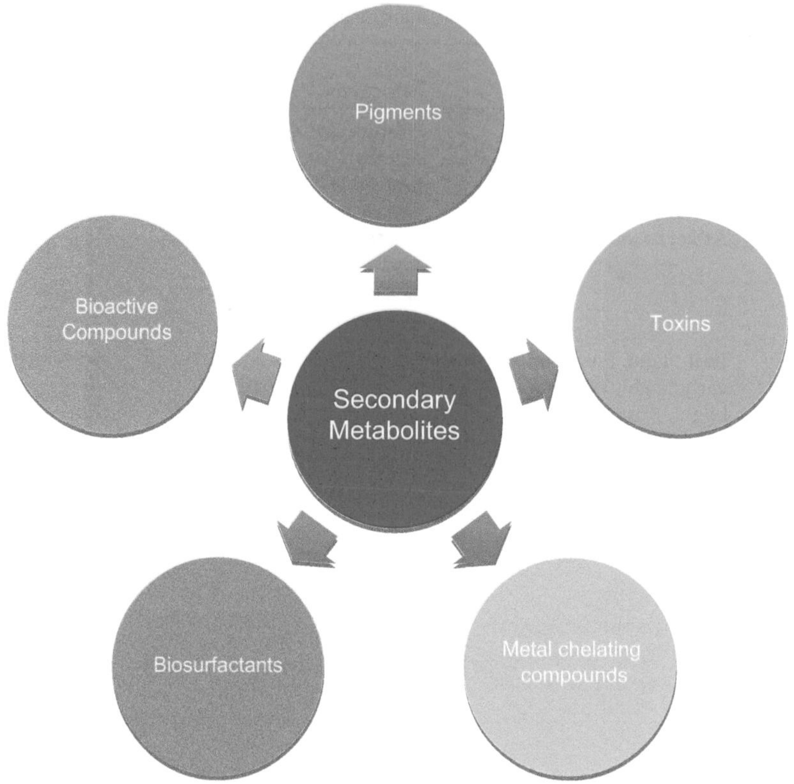

FIG. 1 Fungal secondary metabolites have shown promising antifungal properties and are employed in various applications, including agriculture, medicine, and natural product-based fungicides.

is possibly responsible for the production of a single or many (>100) structurally related molecular products with different biological activity. Thus, for the commercial production of SMs connection between the genomic diversity, the chemical diversity of their biosynthetic pathways as well as their regulation is required to understand (Martinet et al., 2019).

Microbes including bacteria, fungi, and yeast are known to produce diverse coloring pigments as SMs such as melanin, carotenoids, pyocyanin, prodigiosin, violacein, and lycopene. Most of them are applied in the pharmaceutical industry due to diverse biological activities like antimicrobial, antiulcerogenic, antioxidant, anti-leishmanial, anticancer, and enzyme modulation properties. Further, most of them are also used as replacements for artificial colorants in the food, textile, and cosmetic industries (Kalra et al., 2020).

A structurally diverse group of biosurfactants as SMs including glycolipids, lipopeptides, lipoproteins, phospholipids, neutral lipids, fatty acids, and polymeric macromolecules are also produced by microorganisms. Most of them have applications in the bioremediation of petroleum pollutants, heavy metals, and organic pollutants from soils and enhanced oil recovery. Biosurfactant has also achieved considerable importance in medicine due to their specific pharmacological activities such as antitumor, antimicrobial, and immunosuppressant properties (Soberón-Chávez and Maier, 2011). Siderophores are iron-chelating compounds involved in the growth promotion of organisms while toxins and antimicrobial compounds are found to suppress plant pathogens (Vining, 2007; Naz et al., 2022).

Confirmation of SMs producers is done by standard primary and secondary screening methods. However, product yield can be enhanced by strain improvement using classical (optimization of cultural conditions and mutagenesis), genetic recombination (conjugation, transformation, transduction, protoplast fusion), and genetic engineering methods. Also, co-culture techniques showed active SMs production with high yield and product novelty. Recently metagenomics approach allows bioactive compound identification through sequencing specific functional genes within metagenomics libraries (Barrios-Gonzalez et al., 2003). At the industrial scale level, SM is produced by submerged liquid fermentation (SmF) and Solid-state fermentation (SSF). SmF is routinely carried in the industry by batch and feed-batch methods under optimized cultural conditions with/without inducers. Various evidence confirmed the advantage of SSF for SM production by bacteria and fungi concerning the usage of agricultural residues, more product stability, cost-effective downstream processing, convenient fermenter size, and requirement of less energy. Whereas, SmF provides a controlled environment for pH, heat, and nutrient conditions. SMs, viz., toxins, antibiotics, bioactive compounds, and biosurfactants are produced by SSF using agriculture waste like wheat grains and bran, rice grains and husk, sweet potato, maize, cassava, corn, okara, peels, whole pomace, and bagasse. Various factors including pH, temperature, inoculum size, C-and N-source, metal ions, salinity, the growth stage of culture, fermentation time, water activity, and particle size revealed to affect SM productions by SSF (Barrios-Gonzalez et al., 2003; Kumar et al., 2021). Initial extraction of bioactive compounds is carried out by solvent extractions from cultured broth and dried fungal biomass. Water soluble compounds, viz., alkaloids, polyketides, shikimates, sugars, polyhydroxysteroids, amino acids, and saponins can be extracted using n-butanol, chloroform, ethyl acetate, acetone, methanol, ethanol, water, or a mixture of these solvents. Proteins and peptides can be extracted using dichloromethane (DCM), carbon tetrachloride, methanol, ethanol, acetone, butanol, and petroleum ether, while hydrocarbons, terpenes, and fatty acids can be extracted using carbon tetrachloride and hexane. Hydrophilic compounds are extracted effectively by cryo-crushing followed by distilled-water maceration. Moreover, other unconventional methods, viz., microwave-assisted extraction (MAE), supercritical fluid extraction (SFE), ultrasound-assisted extraction (UAE), subcritical water extraction (SWE), and pressurized liquid extraction (PLE) are also applied in the separation of bioactive compounds from microorganisms. Crud-dried extract powder was subjected to further purification by solvent partition and chromatography techniques and further screened for biological assay using each fraction (Duarte et al., 2012). Structural characterization of a purified bioactive compound by different techniques based on nuclear magnetic resonance (NMR) spectroscopy, X-ray crystallography (XRD), infrared (IR) spectroscopy, Fourier-transform infrared spectroscopy (FTIR), high-performance liquid chromatography (HPLC), gas chromatography-mass spectrometry (GC-MS), mass spectroscopy (MS), ultra-high-performance liquid chromatography (UPLC), and desorption electrospray ionization (DESI) tandem MS are applied routinely (Gomes et al., 2017).

Antifungal activity can be carried out by different screening methods for a therapeutic outcome, pathogenicity testing, and drug discovery. The antifungal activity of tested compounds can be detected by inhibiting fungal and yeast culture growth when treated with them. The most commonly reported methods for antifungal activity evaluation are disk diffusion, agar dilution, bioautography, time kill test, ATP bioluminescence, flow cytometric, ergosterol binding/quantitation assay, and poisoned food technique method (Scorzoni et al., 2017; Balouiri et al., 2016).

3 Mechanism of antifungal resistance and drug targets

In drug discovery, the antifungal drugs are designed against the fungal essential molecules or enzymes required for metabolic activity, thereby selectively preventing the fungal pathogens from producing diseases in humans and plants (Pimienta et al., 2023). All the drugs suppress fungal growth without inducing toxicity or inhibiting the host growth. Most of the invented antifungal drugs target the enzymes and essential molecules involved in cell wall/cell membrane synthesis, fungal DNA/RNA synthesis, protein synthesis, ATPase, and virulence factors (Kaur and Nobile, 2023). The endoplasmic reticulum acts as a site for several enzyme syntheses involved in the biosynthesis of cell wall/cell membrane components like ergosterol (Scorzoni et al., 2017; Ivanov et al., 2022) (Fig. 2).

The fungal cell wall contains important components such as glucan, Chitin, glycosylphosphatidylinositol (GPI), and mannoprotein, which makes it an ideal target for antifungal agents. Echinocandins (Echinocandins B from *Aspergillus nidulans)* inhibit β (1,3) D-glucan synthesis of the cell wall that leads to cell death, whereas synthesis of cell membrane components such as glycosphingolipids (GSLs), sterols, and ergosterol are effectively suppressed by some other antifungal agents. Polyene compounds such as amphotericin B (from *Streptomyces nodosus*), natamycin (from *Streptomyces lydicus)*, and nystatin (from *Streptomyces noursei*) bind effectively with ergosterol causing cell component leakages through membrane channels. Azoles target ergosterol biosynthesis and are of two types, viz., imidazole (bifonazole,

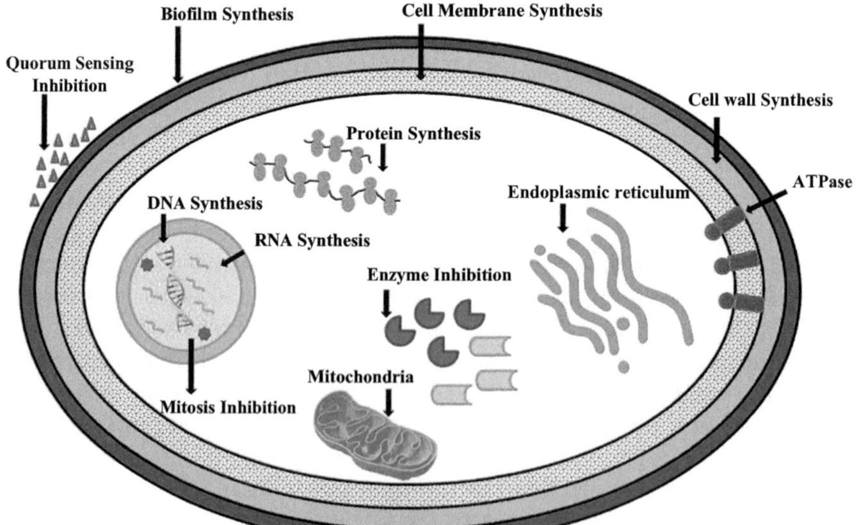

FIG. 2 Cellular targets of antifungal drugs.

clotrimazole, econazole, isoconazole, tioconazole, miconazole, butoconazole, fluconazole, and sulconazole) and derivatives of 1,2,4-triazole (terconazole) compounds. Fluconazole inhibits lanosterol 14α-demethylase essential for ergosterol biosynthesis. Allylamines inhibit the initial steps of fungal ergosterol biosynthesis by targeting the squalene epoxidase enzyme. They all are effective against dermatophyte infections. Proton ATPases of fungi cell membranes are also revealed as targets for antifungal agents. Omeprazole inhibits the *Saccharomyces cerevisiae* and *C. albicans* plasma membrane proton ATPase (Liu and Köhler, 2016). Nucleic acid (DNA and RNA) biosynthesis is targeted by 5-flurocytosin (5-FC). Fungal cell mitosis can be targeted by griseofulvin (from *Penicillium griseofulvum)* by hampering the microtubule functions. Fungal cellular protein synthesis can be inhibited by sordarines without affecting the host machinery. Some compounds such as arylamidine inhibit yeast mitochondrial functions. Some of the antifungal agents such as Piericidin A (from *Streptomyces mobaraensis*, *Streptomyces lividans*, and *Streptomyces anthocyanicus*) and glucopiericidin A (from *Streptomyces pactum*) are potential quorum sensing inhibitors (QSIs), while fungichromin (from *Streptomyces pandanus*) may inhibit the biofilm formation of *C. albicans* by downregulating the expression of the virulent genes. Fungichromin from *S. pandanus* MPS 702 inhibits cucumber downy mildew caused by *Pseudoperonospora cubensis*. Some antifungal compounds like streptochlorin (from *Streptomyces anulatus*) inhibit the catalytic activity of the monoamine oxidase enzyme. However, more efforts are required to search the new targets against evolving drug-resistant strains. Some drug resistance mechanisms adapted by cell includes target overexpression and alteration, drug sequestration, biofilm formation, chromosome aneuploidy, overexpression of drug efflux pumps, and blockage of drug entry. Consequently, current drug therapy is accelerated by drug combination/formulations, antimicrobial photodynamic therapy, enzyme therapy, drug repurposing/modification, drug delivery systems, and vaccines development (Cui et al., 2015; Scorzoni et al., 2017; Alam et al., 2022; Ivanov et al., 2022; Fan et al., 2019; Seyedmousavi et al., 2017). The structures of some antifungal drugs are as below (Fig. 3).

4 Antifungal SMs from bacteria

Bacterial SMs have a wide range of applications in agriculture and as therapeutics. Different types of bacterial-sourced antifungal metabolites are characterized as cyclic lipopeptides, linear lipopeptides, dihydroisocoumarins, alkaloids, pigments, and volatile organic compounds (VOCs), etc. *Chromobacterium* sp. NIIST (MTCC 5522) was isolated from clay mine acidic sediment and yielded a purple-blue violacein pigment. The pigment was revealed effective against several plants and human pathogenic fungi and yeast species such as *Cryptococcus gastricus*, *C. albicans*, *Penicillium expansum*, *Trichophyton rubrum*, *Fusarium oxysporum*, *Rhizoctonia solani*, and *Aspergillus flavus*. The promising activity was noted against *T. rubrum* and *P. expansum* at a minimal inhibitory concentration (MIC) of 2 μg/ml (Sasidharan et al., 2015). Pyocyanin pigment is a water-soluble blue-green color phenazine compound produced by *Pseudomonas aeruginosa* that has antifungal activity against *C. albicans*, *Candida glabrata*, *Candida krusei*, *Candida tropicalis*. and *Candida neoformans* at MIC 64 μg–40 μg/ml by arresting the electron transport both in vitro and in vivo conditions (Sudhakar and Karpagam, 2011). *Pseudomonas chlororaphis* GP 72 was isolated from green pepper rhizosphere and yielded two antifungal substances, viz.,

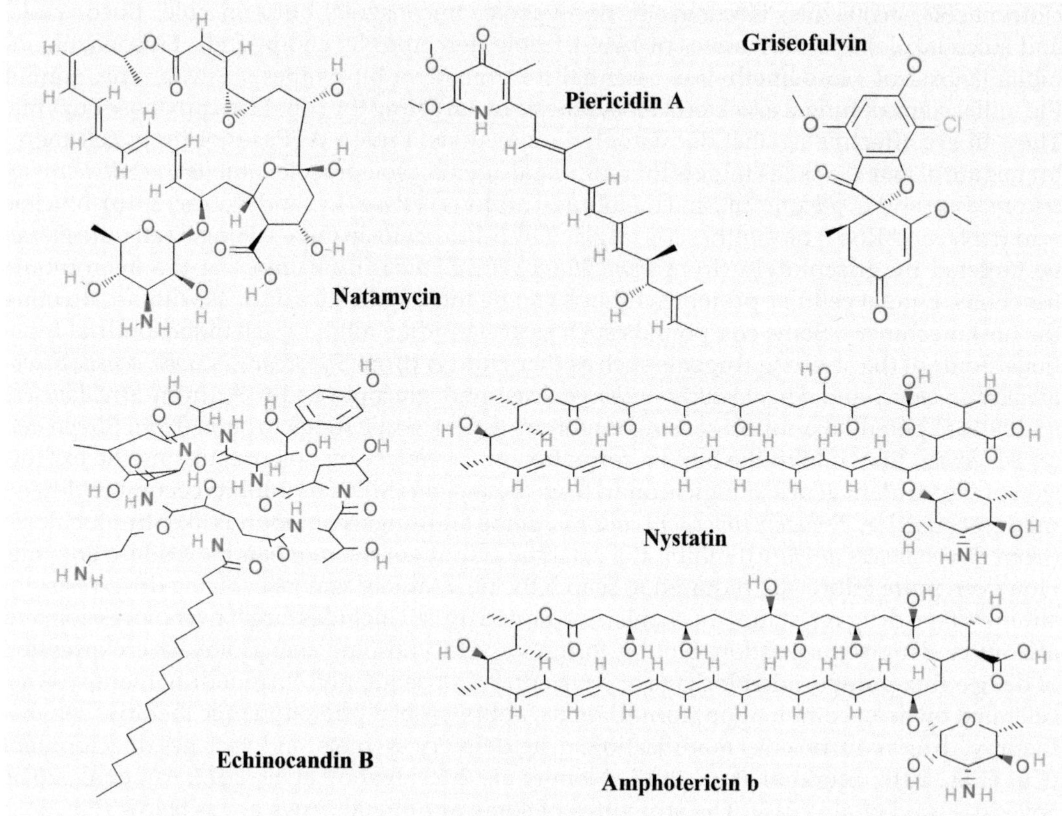

FIG. 3 Chemical structure of some common antifungal drugs.

phenazine-1-carboxylic acid and 2-hydroxyphenazine. The phytopathogens such as *Pythium ultimum, Colletotrichum lagenarium, Phytophthora capsici, Pythium aphanidermatum, Sclerotinia sclerotiorum, F. oxysporum.* f. sp. *cucumerinum, Carposina sasakii,* and *R. solani* were completely suppressed after 48 hrs by in vitro agar plug method (Liu et al., 2007a). Antifungal metabolites viz. Phenazine-1-carboxylic acid (PCA), Hydrogen cyanide (HCN), 2,4 diacetylphloroglucinol (DAPG), pyoluteorin (PLT), and pyrrolnitrin (PRN) were reported from the cultured broth of *Pseudomonas pseudoalcaligenes* P4 (Ayyadurai et al., 2007). In another study, *Pseudomonas fluorescens* JM-1 was isolated from maize rhizosphere to yield a natural phenolic compound 2,4 diacetylphloroglucinol (DAPG) and revealed to control the ear rot disease caused by *Fusarium moniliforme.* Further, its derivatives were equally found effective against *Penicillium digitatum* (29.7% inhibition) and *Penicillium italicum* (40.7% inhibition) at 25 μg/ml concentration (Gong et al., 2016; Mishra et al., 2022). In another study, a similar compound was extracted from *Lysobacter gummosus* (AB161361), and showed inhibitory activity against the fungal pathogen, *Batrachochytrium dendrobatidis* at IC_{50} value of 8.73 μM (Brucker et al., 2008). The bacteria *Photorhabdus akhurstii* sp. nov. 0813-124 was isolated from entomopathogenic nematodes and yielded two bioactive metabolites named glidobactin A and cepafungin I. Both the compounds showed antifungal activity at MIC values of 1.5 and 2.0 μg/mL against the plant pathogen *Colletotrichum gloeosporioides*, respectively (Tu et al., 2022). In another study antifungal compound, trans-cinnamic acid, from *Photorhabdus luminescens* revealed growth inhibition of *Fusicladium effusum* at a concentration of 148–200 μg/mL by agar plate method (Bock et al., 2014). Carotenoids are the most diverse yellow-orange-red colored pigments synthesized by bacteria and archea groups. Antifungal activity of yellow pigment from endolichenic *Bacillus* sp. DBS4 isolated from the lichen *Dirinaria aegialita* was analyzed against three fungi: *R. solani*, *F. oxysporum*, and *Sclerotium rolfsi*. The change in the hyphal morphology including length reduction, curling and lysis was noted (Dawoud et al., 2020). *Bacillus subtilis* CMB32 was isolated from soil collected from Gwangju, Korea and studied for the production of three antifungal lipopeptides biosurfactants viz. fengycin, iturin A, and surfactin A. Strain CMB32 has the strongest biosurfactant mediated antifungal activity against *C. gloeosporioides* followed by other fungal plant pathogens, viz., *Fusarium solani* KCTC 6328, *B. cinerea* KACC 40573, *F. oxysporum* KACC 40037, *R. solani* KACC 40151, and *P. capsici* KACC 40157 (Kim et al., 2010). In another study, similar bioactive lipopeptides such as iturins, fengycins, and surfactins were detected from the ethyl acetate extract of *Bacillus velezensis* DTU001.The crude extract showed antifungal activity against *Penicillium solitum*, *P. italicum*, *Penicillium sclerotigenum*, *Aspergillus uvarum*, *Aspergillus calidoustus*, *Cryptococcus neoformans*, *Penicillium ulaiense*, *Penicillium expansum*, *Talaromyces atroroseus*, and *Talaromyces amestolkiae* with up to 30 mm zone of inhibition (Devi et al., 2019). A dipeptide Bacilysin containing L-alanine and L-anticapsin amino acid from *B. subtilis*, *B. amyloliquefaciens*, and *B. pumilus* was found effective against *Erwinia amylovora*. Zwittermicin A 14, a potent amino polyol antifungal compound extracted from *Bacillus cereus* and *Bacillus thuringiensis* revealed against *Phytophthora medicaginis*. Other compounds kurstakin 18 and Bacthuricin (Bacteriocins), Bacilysocin (phospholipid), Mycosubtilin 6, and Fengycin 8 (Cyclic lipoheptapeptides) were also reported for their fungicidal activity (Sansinenea and Ortiz, 2011). Biosurfactant Bacilomycins L, a cyclic lipoheptapeptides was isolated from *B. amyloliquefaciens* K103 and showed antifungal activity against phytopathogen *R. solani* (Zhang et al., 2013a). In another study, *B. velezensis* PW192 has examined for lipopeptide type of biosurfactant named fengycin A and fengycin B. The antifungal activity of both compounds was received against *C. gloeosporioides* c1060 and *Colletotrichum musae* BCC 13080 at 1 mg concentration with a zone of inhibition 7.5 and 6.5, respectively (Jumpathong et al., 2022). *B. velezensis* CE 100 yields a cyclic tetrapeptide named cyclo-(prolyl-valyl-alanyl-isoleucyl) for protection against plant pathogen *C. gloeosporioides* at 1000 μg/mL concentration (Choub et al., 2021). *Bacillus licheniformis* A12 was isolated from a natural cave yield peptide A12- C. Most of the fungi named *Mucor plumbeus* CCM F 443, *Mucor mucedo* CECT 2653, and *Trichophyton mentagrophytes* CECT 2793 were highly sensitive to the purified antibiotic (Gálvez et al., 1993). A similar type of peptide fungicin M4 from *B. licheniformis* M4 was reported for inhibitory activity of *Microsporum canis*, *M. mucedo*, *M. plumbeus*, and *Sporothrix schenckii* (Lebbadi et al., 1994). Bacisubin, an antifungal protein from *B. subtilis* strain B-916 exhibited inhibitory activity against *R. solani* (IC_{50} 4.01 μM), *Alternaria oleracea* (IC_{50} 0.087 μM), *A. brassicae* (IC_{50} 0.055 μM), and *B. cinerea* (IC_{50} 2.74 μM) (Liu et al., 2007b). A Diazeniumdiolate compound named fragin, a metal chelating compound was reported from *Burkholderia cenocepacia* H111 for antifungal activity against *F. solani* with increased zone of inhibition from 20 to 80 μg/ml concentration (Jenul et al., 2018). In another study, antifungal activities of 2-hydroxymethyl-chroman-4-one produced by *Burkholderia* sp. MSSP exhibited good activities against phytopathogens such as *Pythium ultimum*, *Phytophthora capsici*, and *Sclerotinia sclerotiorum* at ED_{50} 54.99, 35.77, and 52.03 ppm, respectively (Kang et al., 2004). VOCs are also important SMs. Three strains of *Burkholderia ambifaria* showed phytopathogenic fungal inhibition through the emission of VOC like dimethyldisulfide (DMDS), dimethyltrisulfide (DMTS), S-methyl methanethiosulfonate, 4-octanone, 1-phenylpropan-1-one and 2-undecanone (Groenhagen et al., 2013). Another species *Burkholderia glumae* is a phytopathogen of panicle blight disease of rice. Toxoflavin, a bright yellow pigment from *B. glumae* revealed strong inhibitory action against *R. solani* growth, while other extracted purple and yellow-green pigments showed strong antifungal activity against *Collectotrichum orbiculare* (Karki et al., 2012). Phenyllactic acid (PLA) has been extracted from *Lactobacillus plantarum* and showed promising antifungal activity against the fungal strains belonging to *Aspergillus*, *Penicillium*, and *Fusarium*, at MIC_{90} ranging from 3.75 to 7.5 mg/ml (Lavermicocca et al., 2003). *Kribbella* sp. MI481-42F6 produced new antibiotics, namely kribellosides A-D inhibited *Saccharomyces cerevisiae* growth at MIC 3.12–100 μg/ml concentration (Igarashi et al., 2017). Epothilones, a macrolide from myxobacterium *Sorangium cellulosum* were reported initially for antifungal activity (Vaishnav and Demain, 2011).

5 Antifungal SMs from some yeast, filamentous fungi, and endophytes

Cryptocandin, a lipopeptide isolated from endophytic fungi *Cryptosporiopsis quercina* exhibited antifungal activity by inhibiting glucan components of the fungal cell wall of *C. albicans*, *Trichophyton mentagrophytes* and *Trichophyton rubrum* at MIC 0.03–0.07 μg/ml. Other similar lipopeptides named pneumocandins produced by *Pezicula* spp. and *Zalerion arboricola* and echinocandins by *Aspergillus species* were also reported (Strobel et al., 1999). Griseofulvin, a well-known spirocyclic benzofuran-3-one was isolated initially from *P. griseofulvum* in 1939 and considered as the first oral antifungal drug for the treatment of dermatomycoses. More than 400 griseofulvin analogues have been synthesized and used for drug screening later. Endophyte *Nigrospora oryzae* was isolated from leaves of *Emblica officinalis* and studied for griseofulvin production. The growth of pathogenic fungi *F. oxysporum*, *T. mentagrophytes*, and *Microsporum canis* was inhibited at MIC > 1.5 mg/ml concentration. It was also reported from *Xylaria* sp. F0010, for effective antifungal activity against *Magnaporthe grisea*, *Corticium sasaki*, *Puccinia recondite*, and *B. graminis* f. sp. Hordei, at a concentration of 50–150 μg/ml (Park et al., 2005b; Rathod et al., 2014; Bai et al., 2019). Varioxepine A, a novel 3*H*-oxepine-containing alkaloid was isolated from the marine algal-derived endophytic fungus *Paecilomyces variotii*. Varioxepine A exhibited potent inhibitory activity against the plant-pathogenic fungus *F. graminearum* (Zhang et al., 2014). Yeast *Rhodotorula babjevae* YS3 was isolated from an agricultural field of Assam, Northeast India produced sophorolipids and glycolipid biosurfactants for antifungal activity against *C. gloeosporioides* (MIC 62 μg/ml), *Fusarium verticilliodes* (MIC 125 μg/ml), *F. oxysporum* f. sp. *Pisi* (MIC 125 μg/ml), *Corynespora cassiicola*, (MIC > 2000 μg/ml), and *T. rubrum* at (MIC 1000 μg/ml) (Sen et al., 2017). Epipyrone A, a water-soluble polyene metabolite from *Epicoccum nigrum* fungal exudate inhibited the fungal spore germination of *Sclerotinia sclerotiorum* (MIC 0.2 mg/mL), *Aspergillus oryzae* (MIC 2 mg/mL), *S. cerevisiae* (MIC 0.03 mg/mL), and *C. albicans* (MIC 0.04 mg/mL). *E. nigrum* fungus produces other types of diverse SMs for antifungal activity like flavipin and orevactaene (Preindl et al., 2017; Lee et al., 2020). *Chaetopsina* sp. (CF-255912) was isolated from the leaf litter of the *Beilschmiedia tawa* tree. The fungal extract revealed antifungal activity against *C. krusei*, *C. parapsilosis*, *C. albicans*, *C. glabrata*, *C. tropicalis*, and *A. fumigatus* mediated through linoleyl sulfate, linolenyl sulfate, and oleyl sulfate. Two of them linoleyl sulfate and oleyl sulfate displayed significant antifungal activity at MIC 2–64 µg/ml (Crespo et al., 2017).

A cyclic depsilipopeptide Colisporifungin, and two linear peptides cavinafungins A and B, were extracted from the fungus *Colispora cavincola*. The linear lipopeptides cavinafungins A and cavinafungins B showed broad-spectrum antifungal activity against *Candida* species and filamentous fungi *A. fumigatus* at MIC values at 0.5–4 µg/mL and 8 µg/mL, respectively (Ortiz-Lopez et al., 2015). Citridone A, a pyridine alkaloid was originally isolated from culture both of *Penicillium* sp. FKI-1938. Citridone A and its synthetic derivatives revealed promising antifungal activity in addition to miconazole (60 nM) at 20µg/disk with a zone of inhibition for *C. albicans* at 15–21 mm (Fukuda et al., 2014). A novel azaphilone Sch 1385568 was extracted from *Aspergillus* sp. culture broth (SPRI-0814) and demonstrated for its antifungal activity against *S. cerevisiae* (PM503) at MIC of 32 mg/ml. Various azaphilones and hydrogenated azaphilones were also isolated from *Emericella* sp., *Penicillium* sp., *Phomopsis* sp., *Chaetomium* sp., *Pseudohalonectria* sp., and *Anuulohypoxylon* sp. (Yang et al., 2009).

Chaetoviridins, a class of orange-colored azaphilones pigments was reported from *Chaetomium globosum*. The culture showed strong antifungal activities due to chaetoviridins A and B against rice blast (*Magnaporthe grisea*) and wheat leaf rust (*Puccinia recondita*). Chaetoviridin A at 62.5 µg/ml suppressed the development of rice blast and wheat leaf rust by over 80% (Park et al., 2005a). In addition to *Chaetomium* sp., azaphilones are produced by *Monascus* sp. These Monascus pigments exist as cell-bound forms and are six known types with three colors: red (monascorubramine and rubropunctamine), orange (monascorubrin and rubropunctatin), and yellow (ankaflavin and monascin). Derivatives with L-Asp, D-Asp, L-Tyr, and D-Tyr exhibited effective MIC values >32–64 µg/ml against the filamentous fungi *A.niger*, *Penicillium citrinum*, and *C. albicans* (Kim et al., 2006; Agboyibor et al., 2018). Oosporein, an SM was extracted from the culture broth of *Verticillium psalliotae* and evaluated for antifungal activity against *Phytophthora infestans* and *Alternaria solani*, and *F. oxysporum*. Oosporein exhibited strong inhibitory activity against *P. infestans* at MIC 16 µm only (Nagaoka et al., 2004). Amphiol, an antifungal pigment extracted from the culture broth of the fungus *Pseudogymnoascus* sp. PF1464 showed promising growth inhibition of *Schizosaccharomyces pombe* (MIC 0.8 µg/mL), *C. albicans* (MIC 8 µg/mL), and *A. fumigatus* (MIC 16 µg/mL). Structurally the metabolite was revealed as a polyene macrolide that targets the ergosterol molecule of the cell wall (Fujita et al., 2021). A new yellow pigment, anishidiol as the antifungal substance was identified from *Aspergillus nishimurae* strain IFM58441. The pigment exhibited antifungal activity at 100 mg/disk against *A. fumigatus*, *A. niger*, *C. albicans*, and *C. neoformans* with the zone of inhibition 12, 16, 21, and 13 mm, respectively (Hosoe et al., 2011). A strain of *Curvularia lunata* has been characterized for the production of three anthraquinone derivatives: helminthosporin, chrysophanol, and cynodontin. Cynodontin (1,4,5,8-tetrahydroxy-3-methyl anthraquinone) showed potential antifungal activity (Lin and Xu, 2020). In another study, two compounds were identified as (−) 2,3,4-trihydroxybutanamide, and (−) erythritol from *Penicillium ochrochloron* and revealed strong antifungal activity against *C. albicans* at MIC 2.5µg/ml and 5µg/ml, respectively (Rančić et al.,

2006). Moreover, three new 3,4,6-trisubstituted α-pyrone derivatives, namely 6-(2′R-hydroxy-3′E,5′E-diene-1′-heptyl)-4-hydroxy-3-methyl-2H-pyran-2-one, 6-(2′S-hydroxy-5′E-ene-1′-heptyl)-4-hydroxy-3-methyl-2H-pyran-2-one, and 6-(2′S-hydroxy-1′-heptyl)-4-hydroxy-3-methyl-2H-pyran-2-one, together with one known compound trichodermic acid, were isolated from *Taxus media* roots associated endophytic culture of *P. ochrochloron*. All compounds displayed significant antifungal activity against *A. solani*, *Cercospora arachidicola* Hori, *Bipolaris carbonum* Wilson, and *F. graminearum* fungal strains at MIC 12.5 μg/ml (Zhao et al., 2019). The epipolythiodioxopiperazine (ETP) family of fungal SMs was produced well by *A. fumigatus*, *A. terreus*, *A. flavus*, and *A. niger*. One compound from the ETP group named "gliotoxin" has potent antifungal activity against *C. albicans* (MIC 2 μg/ml) and *C. neoformans* (MIC 4 μg/ml). Moreover, the other two compounds acetylgliotoxin and a derivative of hyalodendrin from the supernatant of *A. fumigatus* inhibited both pathogens at MIC 3.2 mg/ml (Coleman et al., 2011). In another study, gliotoxin from *Trichoderma virens* (ITC-4777) was found to be active against *Rhizoctonia bataticola* (ED_{50} 0.03 μg/ml), *Macrophomina phaseolina* (ED_{50} 1.76 μg/ml), *Pythium deharyanum* (ED_{50} 29.38 μg/ml), *Pythium aphanidermatum* (ED_{50} 12.02 μg/ml), *Sclerotium rolfsii* (ED_{50} 2.11 μg/ml), *R. solani* (ED_{50} 3.18 μg/ml) (Singh et al., 2005). Fungus *Bombardioidea anartia* isolated from deer dung revealed antifungal activity of two extracted compounds i.e. (E/Z)-Bombardolides A and D against *C. albicans* with zones of inhibition of 35 and 20 mm, respectively, by disk diffusion method (Hein et al., 2001). Three new antifungal 2(5H)- furanones named Appenolides A, B, and C, have been isolated from liquid cultures of the coprophilous fungus *Podospora appendiculata*. The compounds showed promising antifungal activity against *C. albicans* (zones of inhibition ranging from 12 to 14 mm at 150 μg/disk), *Sordaria fimicola* (54% growth inhibition with 150μg/disk), and *Ascobolus furfuraceus* (26% growth inhibition with 50μg/disk) (Wang et al., 1993). In another study, two new tetracyclic sesquiterpene lactones, decipienolides A-B, and Decipinin A were isolated from the coprophilous fungus *Podospora decipiens*. Only Decipinin A showed inhibitory activity against *F. verticillioides* at 250 μg/disk with a 22 mm zone of inhibition (Che et al., 2002). Fungus *Podospora anserine* also produces two new benzoquinones anserinones A and B. Anserinone A showed fungal growth inhibition against *S. fimicola* (60%) and *A. furfuraceus* (100%) at 200 μg/disk concentration, while anserinone B caused 50% and 37% growth inhibition against *S. fimicola* and *A. furfuraceus* fungi, respectively (Wang et al., 1997).

Roridin E and three novel isocoumarin derivatives Cercophorins A-C, decarboxycitrinone, and 4-acetyl-8-hydroxy-6-methoxy-5-methylisocoumarin, have been isolated from the coprophilous fungus *Cercophora areolate* for antifungal activity. Roridin E and decarboxycitrinone caused a 100% growth inhibition of both *S. fimicola* and *A. furfuraceus* fungi, while 4-acetyl-8-hydroxy-6-methoxy-5-methylisocoumarin and cercophorins A, B, and C caused 46%, 34%, 10%, and 51% growth reduction of *S. fimicola*, respectively (Whyte et al., 1996). Aromatic butenolides named gymnoascolide A have been extracted from the ascomycete *Gymnoascus reessii* and tested for antifungal activity against the pathogenic plant fungus *Septoria nodorum* at MIC 13 μg/ml (Clark et al., 2005). The two new fusicoccane diterpenes named Periconicin A and B were extracted from the endophytic fungus *Periconia* sp. OBW-15. Periconicin A showed potent inhibitory activity against *C. albicans*, *Trichophyton mentagrophytes*, and *T. rubrum*, with MIC in the range of 3.12–6.25 μg/ml (Shin et al., 2005). Endophytic fungi *Hormonema* sp. produce a triterpene glycoside "Enfumafungin" for antifungal activity against

C. albicans (MIC 0.25 μg/ml), *C. tropicalis* (MIC 0.5 μg/ml), and *A. fumigatus* (MIC < 0.03 μg/ml (Peláez et al., 2000). Metabolite 3-(5-oxo-2,5-dihydrofuran-3-yl) propanoic acid, a furan derivative was isolated from *Decaisnea insignis* plant-associated endophytic fungal culture *Aspergillus tubingensis* and exhibited effective antifungal activity against *F. graminearum* at MIC value of 16 μg/ml (Yang et al., 2019). Sambacide, a tetracyclic triterpene sulfate from *Fusarium sambucinum* B10.2 exhibited anticandidal activity at MIC 64 μg/ml (Dong et al., 2016). SMs from *Aspergillus* sp., viz., Yanuthones (meroterpenoids), Aspernigrins (NRP), Asperfuran (dihydrobenzofuran), Flufuran (furans), Avenaciolide (lactone), antafumicins A and B (alkenyl sulphates), Tensidols (furopyrrols,), and Canadensolide (γ-lactone) from *Penicillium* were reported for antifungal activity (Frisvad et al., 2018). Compactin (statins), a new antifungal metabolite and sophorolipids have been reported for antifungal activity from *Penicillium brevicompactum* and yeast (Vaishnav and Demain, 2011; Demain and Martens, 2017).

6 Antifungal SMs from actinomycetes

Actinomycetes are the treasure house of SMs with more than 140 identified genera. Among these, 75% of metabolites are alone produced by Streptomyces and the rest are from *Salinispora, Nocardiopsis, Actinoplanes, Saccharopolyspora, Amycolatopsis,* and *Micromonospora*. Most of them are isolated from terrestrial soil, forest soil, river water bodies, marine sponges, marine sediments, and sponges (Solecka et al., 2012; Shrivastava and Kumar, 2018). Diverse SMs have been identified from actinomycetes such as ansamycins, aminoglycosides, lactones, oligosaccharides, macrolides, anthracyclines, polycyclic ethers, peptides, polyenes, and terpenoids (Al-Fadhli et al., 2022).

Streptomyces genus alone produces a wide range of SM for antifungal activity, viz., piericidin-A1, nigericin, fungichromin, cyphomycin, mycangimycin, streptochlorin, natalamycin A, daryamide B, actinomycin, thailandin B, antifungalmycin, amphotericin B, aureofacin, candicidin, nystatin, oligomycin, frontalamide A, frontalamide B, bafilomycin B1, and C1. Further, some of the compounds like aflatoxins, Kasugamycin, streptomycin, and polyoxin D from *Streptomyces* sp. are used as fungicides for plant disease management (Alam et al., 2022). Various antifungal compounds are secreted by the actinomycetes, viz., Amphotericin B a polyene antibiotic was isolated from *Streptomyces nodosus* in 1959. It is the choice of drug for infectious *C. albicans* and *A. fumigatus* (Lemke et al., 2005). *S. nodosus* NRRLB-2371 yielded a corifungin, a sodium salt of AmB also showed promising antifungal activity against various *Candida* sp. (MICs 0.125–1 μg/mL) and *Aspergillus* sp. (MIC 0.5–2 μg/mL). Further, fungi *Coccidioides, Histoplasma capsulatum, Blastomyces,* and *Paracoccidioides*, were inhibited in the range of MIC 0.5–2 μg/ml. Other fungal pathogens Alternaria, Scedosporium, and Paecilomyces are also get inhibited during in vitro assay (Victoria Castelli et al., 2017). Another polyene macrolide metabolite named nystatin produced by *S. noursei* ATCC 11455 showed inhibitory activity against a broad range of filamentous fungi and yeast (Brautaset et al., 2000). A tetraene macrolide named natamycin from *Streptomyces natalensis, Streptomyces chattanogenesis, Streptomyces gilvosporeus,* and *Streptomyces lydicus* got Food and Drug Administration (FDA) approval for the treatment of fungal infections and as food additives (Elsayed et al., 2019). The cyclic lipopeptide,

Neopeptins A and B, were reported from the *Streptomyces* sp. K710 inhibits the fungal mycelial growth of *Alternaria mali, Botrytis cinerea, Glomerella cingulate, Colletotrichum lagenarium,* and *Cochliobolus miyabeanus* at MIC 4–250µ concentration. The target was identified as β-1,3-glucan and proteoheteroglycan synthesis of the cell wall (Satomi et al., 1982; Ubukata et al., 1986). Moreover, neopeptin A and B from strain *Streptomyces* sp. KNF2047 also showed antifungal activity at MIC in the range of 128–512 µg/ml against *A. mali, Didimella bryoniae, Cladosporium cucumerinum, C. lagenarium, B. cinerea* and *Magnaporthe grisea* (Kim et al., 2007). The antifungal peptide "globopeptin," was isolated from *Streptomyces* sp. strain MA-23 that inhibits the chitin synthetase of the fungal cell wall and thus preventing the growth of *Mucor racemosus* KF-223, *Pyricularia oryzae* KF-180, and *Botrytis cinerea* KF-241 at MIC 0.2–0.78 µg/ml (Tanaka et al., 1987; Shrivastava and Kumar, 2018). *Streptomyces lavendulae* H698 SY2 secreted a cyclic tetrapeptide antibiotic, glomecidin, and was revealed to exhibit growth inhibitory activity of plant pathogenic fungi such as *G. cingulate* (0.78 µg/ml), *C. gloeosporioides* (0.78 µg/ml), and *C. lagenarium* (1.56 µg/ml). Moreover, *Streptomyces lavendulae* H698 SY2, produced ileumycin antibiotic that has stronger antifungal activity than glomecidin (Kunihiro and Kaneda, 2003). Actinobacteria *Streptomyces morookaense* AM25 was isolated from guarana (*Paullinia cupana*) bulk soil and assayed for antifungal activity against plant pathogen *C. gloeosporioides* with 42.7% inhibition at 0.1 mg by disc diffusion method. The compound was revealed as a novel cyclic peptide "gloeosporiocide" (Vicente dos Reis et al., 2019). Reveromycins A and B, a spiroacetal polyketide compounds extracted from *Streptomyces* sp. 3–10 effectively inhibited the mycelial growth of *B. cinerea, Mucor hiemails, Rhizopus stolonifer,* and *Sclerotinia sclerotiorum* at more than 97% at 5 µg/ml concentrations (Lyu et al., 2017). Novonestmycins A and B were identified as 32-membered macrolide compounds from *Streptomyces phytohabitans* HBERC-20821. Both the compounds showed inhibitory activity against phytophathogenic fungi *Corynespora cassiicola, R. solani,* and *Septoria nodorum,* with MIC at 0.78, 0.39, and 0.78 µg/ml (Wan et al., 2015). Most of the studied macrolides comprised around 26 groups, viz., ammocidins, neomaclafungins, bafilomycins, spinosyns, rosaramicins, and tiacumicins with diverse biological activities (Al-Fadhli et al., 2022). Bafilomycins are a family of polyene macrolides containing a 16-member lactone ring. Sixteen bafilomycin antibiotics were identified as bafilomycins A_1, A_2, B_1, B_2, C_1, and C_2 from *Streptomyces griseus* sp. *Sulphurus*, bafilomycins D and E from *Streptomyces griseus* Tu 2599, bafilomycins F-J from *Streptomyces* spp., bafilomycin K from *Streptomyces flavotricini* Y12-2 and hygrobafilomycin from both *Streptomyces varsoviensis* and *Streptomyces cavourensis* YY01-17. Hygrobafilomycin revealed antifungal activity against *Pyricularia oryzae, B. cinerea,* and *R. solani* at MIC 16, 31, and 125 µg/mL, respectively while, bafilomycins K showed medium activities against *P. oryzae* and *C. lagenarium* with MICs of 31 and 62 mg ml1, respectively (Xu et al., 2013; Zhang et al., 2011). Faeriefungin, a new polyene antibiotic from *Streptomyces griseus* var. autotrophicus revealed to control root rot and stem wilt of *Asparagus officinalis* L. caused *by F. oxysporum* f. sp asparagi (FOA) (Smith et al., 1990). *Streptomyces* sp. IA1, isolated from Saharan soil produces actinomycin D metabolite that effectively reduces fusarium wilt disease by almost 60% (Toumatia et al., 2015). Saquayamycins are identified as inhibitors for the farnesyl-protein transferase, an enzyme involved in cell division. The Saquayamycins A and C were extracted from *Streptomyces* spp. PAL114 for antifungal activity against *S. cerevisiae* (MIC 30 µg/ml), *C. albicans* (MIC 50 µg/ml), *Aspergillus carbonarius* (MIC 75 µg/ml), *A. flavus* (MIC 100 µg/ml),

Fusarium culmorum (MIC 75 μg/ml) and *Penicillium glabrum* (MIC 75 μg/ml) (Aouiche et al., 2014). *Streptomyces* sp. MK-30 isolated from the soil of a tea field yields a new bafilomycin analogue named makinolide against *S. cerevisiae* (NBRC2376) and *M. hiemalis* (NBRC 9405) with a zone of inhibition 12 and 15 mm at 15 μg/ml concentration (Kodani et al., 2012). Malayamycin is a perhydrofuropyran C-nucleoside extracted from *Streptomyces malaysiensis*. The compound was revealed to control the growth and sporulation of the pathogen *Stagonospora nodorum* at 50 μg/ml and 10 μg/ml concentration, respectively (Li et al., 2008). *Streptomyces* sp. No. AC-69 culture produced lipopeptin A and was revealed to inhibit cell wall glycan synthesis of *Piricularia oryzae* at 150 μg/ml concentration (Nishii et al., 1981). Enduspeptides A-F (1–6), the six new cyclic octa depsipeptides were isolated from coal mine-derived *Streptomyces* sp. strain. Compounds 1, 2, and 3 exhibited promising antifungal activities against *C. glabrata* (ATCC 90030) with IC_{50} values of 5.33 ± 1.56, 1.72 ± 0.71, and 8.13 ± 1.98 μg/mL, respectively (Chen et al., 2017). *Streptomyces zhaozhouensis* CA-185989 was reported for polycyclic tetramic acid macrolactams derivative. Metabolite isoikarugamycin inhibited the growth of *C. albicans* and *A. fumigatus* at MIC 2–4 and 4–8 μg/ml, whereas 28-N-methylikaguramycin and ikarugamycin showed inhibition at MIC 4 μg/ml against *C. albicans* and at MIC value range of 4–8 μg/ml against *A. fumigatus* respectively (Lacret et al., 2014). Polyene macrocyclic lactam named "sceliphrolactam" was extracted from *Sceliphron caementarium* associated *Streptomyces* sp. and displayed potent antifungal activity against amphotericin B-resistant *C. albicans* at MIC 4 μg/ml (Oh et al., 2011). Metabolite 4′ phenyl-1-napthyl-phenyl acetamide from the cultured broth of *Streptomyces* sp. DPTB16 exhibited antifungal activity against *C. albicans* (Zone of Inhibition 25.05 ± 0.81 mm) followed by *A. niger* (Zone of Inhibition 13.6 ± 1.23 mm), *A. flavus* (Zone of Inhibition 13.66 ± 0.54 mm), *Mucor* sp. (Zone of Inhibition 13.27 ± 0.44 mm), *A. fumigatus* (Zone of Inhibition 10.8 ± 0.49 mm) (Dhanasekaran et al., 2008). Soil-derived isolate *Streptomyces lavenduligriseus*, was investigated for the production of polyene macrolides compound 1–4, and revealed to inhibit *C. albicans* growth at MIC 6.25, 6.25, 200, and 200 μg/ ml, respectively. Compound I was identified as filipin III (Yang et al., 2016). In another study, actinomycete *Streptomyces* sp. SY1965 culture was isolated form the Mariana Trench sediment and extracted for streptothiazolidine A, streptodiketopiperazines A and B. All the compounds showed antifungal activity against *C. albicans* at MIC values of 47, 42, and 42 μg/ml, respectively. Other extracted compounds viz. [2-hydroxy-1-(hydroxymethyl) ethyl]-2-methoxybenzamide, salicylamide, and 4-hydroxymethyl benzoate, exhibited weak antifungal activity against *C. albicans* at MIC 43, 38, and 38 μg/mL, respectively. Spoxazomicin C showed stronger anti-candida activity at MIC 10 μg/ml (Yi et al., 2020). Actinomycetes *Streptomyces* sp. CS was isolated from plant tissue of *Manglietia hookeri* yielding Naphthomycin L and Naphthomycin A using oatmeal agar. Both the compounds showed inhibitory activity of *F. moniliforme* at MIC 300 and 100 μg/mL (Yang et al., 2012). Kasugamycin from *Streptomyces kasugaensis* targets the protein synthesis of rice blast-causing fungal pathogens *Pyricularia oryzae* (Umezawa et al., 1965). *Streptomyces cacaoi* var. asoensis yield polymixin B and D, which inhibit the rice sheath blast disease caused by *R. solani* by affecting cell wall synthesis enzyme i.e. chitin synthase. It also affects the pathogenic fungi causing diseases in vegetables, fruits, and ornamental crops (Isono et al., 1967). Mildiomycin was another highly active antifungal metabolite obtained from *Streptoverticillium rimofaciens* for the control of several powdery mildews affected crops. It effectively controls the powdery mildew of barley plants at a concentration of 31.2–62.5 ppm (Iwasa et al., 1978). *Streptomyces hygroscopicus* var. geldanus and *S. griseus*, produce geldanamycin, a new antifungal agent against *R. Saloni*-induced root rot of pea (Rothrock and Gottlieb, 1984). Another macrodiolide compound named gopalamicin, was reported from *Streptomyces hygroscopicus* inhibits the fungal pathogen, viz., *A. fumigatus, F. oxysporum, A. flavus, F. moniliforme, A. niger, Pythium ultimum, C. albicans, Phialophora graminicola, A. solani* and *Leptosphaeria korrae* at 12–16 μg/ ml (Nair et al., 1994). A new antifungal nucleoside compound named tubercidin was produced by *Streptomyces viola-ceoniger* and revealed effective against phytopathogen *Phytophthora capsici* and *Magnaporthe grisea*. The *P. capsici* blight in *Capsicum annuum* was completely controlled at 500 μg/ml (Hwang and Kim, 1995). Guanidylfungin A, geldanamycin, and nigericin, from *Streptomyces violaceusniger* YCED-9, revealed effective against *Sclerotinia homeocarpa* and *R. saloni* (Trejo-Estrada et al., 1998). *Streptomyces diastaticus* produces macrolide antibiotics Oligomycins A and C, exhibited strong inhibitory activity against *A. niger, A. alternata, P. capsici*, and *Botrytis cinerea* at MIC 2–10 μg/ml concentration (Yang et al., 2010). *Streptomyces padanus* strain PMS-702 revealed antifungal activity against pathogenic *R. solani* AG-4, due to its polyene macrolide, fungichromin, at MIC 72 μg/ml (Shih et al., 2003; Fan et al., 2019). Antimycin A_{17} from a *Streptomyces* sp. GAAS7310 was reported for significant antimicrobial activity against fungal pathogens, viz., *Curvularia lunata* (Wakker) Boed, *Rhizopus nigrtcans* Ehrb, *Alternaria solani* (E. et M.), *Cladosporium* sp., *Sclerotinia sclerotiorum* (Lib.) de Bary, and *Colletotrichum nigrum* EL. *et* Halst. (Chen et al., 2005). Isolate *Streptomyces cavourensis* subsp. cavourensis inhibited *C. gloeosporioides*, a fungal pathogen of pepper plants. The inhibitory activity was due to a combined effect of lytic enzymes (chitinase, lipase, β-1,3-glucanase, and protease) and 2-furancarboxaldehyde (Lee et al., 2012). In another study, antifungal components bafilomycins B1 and C1 from *S. cavourensis* NA4 were reported as potential biocontrol agents for soilborne fungal pathogens viz. *Fusarium* spp., *R. solani*, and *B. cinerea* (Pan et al., 2015). Resistomycin was isolated from a termite-associated *Streptomyces canus* BYB02 culture and investigated for antifungal activity against pathogen *Valsa mali* and *Magnaporthe grisea* at IC_{50} 1.1 μg/ml and IC_{50} 3.8 μg/ml, respectively (Zhang et al., 2013b). Ansacarbamitocins a new class of maytansinoids from *Amycolatopsis* CP2808 showed modest activity against plant pathogens like *Septoria tritici, Erisiphe graminis*, and *Puccinia recondite* by in vivo testing (Snipes et al., 2007). Formamicin from *Saccharothrix* sp. strain MK27-91F2 was reported for its strong antifungal activity against *P. oryzae, P. digitatum, R. solani*, and *Diaporthe citri*, at MIC values ranging from 0.39 to 1.56 μg/ml (Igarashi et al., 1997). *Trichoderma* spp. is a rich source of SMs responsible for the production of more than a hundred compounds such as pyrones, terpenoids, steroids, polyketides siderophores, and epipolythiodioxopiperazines (ETPs). Gliotoxin and gliovirin from ETP class of peptides are applicable for the control of Rhizoctonia and oomycete pathogens. The volatile compound 6-Pentyl pyrone (6-PP) is reported to have antifungal and plant growth-promoting activities (Mukherjee et al., 2012).

Rustmicin from *Micromonospora narashinoensis* strain 980-MC1 displayed potent fungicidal activity against the wheat stem rust fungus *C. neoformans* (MIC < 1 ng/ml), *C. tropicalis* MY1012 (MIC 0.03 μg/ml) and *C. albicans* MY1055 (MIC 6.25 μg/ml). The mechanism of action includes inhibition of fungal biosynthesis of inositol phosphoceramide, and sphingolipids synthesis. Four rustmicin analogues, neorustmicins A-D from *Micromonospora chalcea* exhibited significant antifungal activity against *Puccinia grannnis* f. sp. Tritici at 0.2, 1, 4, and 5 μg/ml, respectively.

Primycin from *Micromonospora paleriensis* displayed antifungal activities against *C. albicans*, at MIC of 0.25 μg/ml. Chemically the compound was revealed as a 36-membered marginolactone macrolide with guanidine and arabinose moieties as the side chain (Al-Fadhli et al., 2022). *Micromonospora* sp. M39 showed antifungal activity against the rice blast pathogen *P. oryzae* MPO 293. The characterized crude extract identified the antifungal metabolites viz. 2,3-dihydroxybenzoic acid, cervinomycin A1, phenylacetic acid, and cervinomycin A2 (Ismet et al., 2004). An Oligomycin type of macrolide named Neomaclafungins A-I was isolated from the broth of *Actinoalloteichus* sp. and displayed significant antifungal activity against *Trichophyton mentagrophytes* ATCC 9533 between MIC 1 and 3 μg/ml. The structural activity relationship (AR) study revealed that the substitution at C-24 and the absence of ketones at the lactone ring were critical for better antifungal activity (Sato et al., 2012). Mathemycin A, a new antifungal macrolactone from *Actinomycete* sp. HIL Y-8620959 showed antifungal activity against *Phytophthora infestans* (MIC 7.8mg/L), followed by *A. mali* and *B. cinerea* (MIC 15.6 mg/L) (Nadkarni et al., 1998). *Actinomadura roseola*, yielded an anthracycline compound daunomycin, which showed promising inhibitory activity against *R. solani* and *S. cerevisiae* at MIC 10 μg/ml (Kim et al., 2000).

7 Antifungal SMs from a marine microorganism

The World's surface is merely covered by 70% of the ocean with a wide diversity of organisms. Near about 1.5 million species are estimated, out of which only 465 species are identified. A plethora of research has been carried out on the extraction of SMs from these marine organisms and their evaluation of their biological activity. According to the modified Mayer's chemical classification of the marine environment, the diverse natural products are belonging to six groups viz. sugars, peptides, terpenes, alkaloids, polyketides, and shikimates. Some highlighted compounds extracted from Marine fungi are penicillin, caspofungin, mevinolin, and fingolimod from Penicillium, Aspergillus, Phomopsis, Emericella, Epicoccum, Exophiala, and Paraphaeospaeria. Most of them are proven as anticancer, antimicrobial, antiplasmodial, and antiinflammatory agents. Moreover, they act as biocontrol agents in agriculture and as PGPR in host–microbiome interactions (Gomes et al., 2017).

The fungus *Trichoderma koningii*, isolated from marine mud of the South China Sea, yields new polyketide viz. 7-O-methylkoninginin D, four derivatives of trichodermaketones (A–D) and koninginins (A, D, E, F). Compound trichodermaketones A showed synergistic antifungal activity against *C. albicans* with 0.05 μg/ml ketoconazole (Song et al., 2010). Similarly, in another study monodictyquinone A, an anthraquinone was isolated from the *Anthocidaris crassispina*-associated marine-derived *Monodictys* sp. and revealed inhibition of *C. albicans* (zone of inhibition 7 mm) at 2.5 μg/ml (El-Beih et al., 2007). Awajanoran, a dihydrobenzofuran derivative from the marine fungus *Acremonium* sp. AWA16-1 showed antifungal activity against *C. albicans* with a zone of inhibition of 15 mm at 50 μg/disk (Jang et al., 2006).

Six new macrocyclic trichothecenes, named roridin Q, 12′-hydroxyroridin E, and 2′,3′-deoxyroritoxin D, were isolated from the marine-derived fungus *Myrothecium roridum* TUF 98F42, and roridin R, isororidin E, roridins A and H from *Myrothecium* sp. TUF 02F6. Out of all, Compound 2′,3′-deoxyroritoxin D, showed inhibitory activity against *S. cerevisiae* at 1 μg/disc with inhibition zone 12.2 mm (Xu et al., 2006). A new isoprenyl phenyl ether compound 3-hydroxy-4-(3-methylbut-2-enyloxy) benzoic acid methyl ester from a mangrove unknown fungus exhibited strong inhibitory activity against *F. oxysporum* at MIC value, of 12.5 μg/mL (Shao et al., 2007).

A halotolerant fungal strain (THW-18) identified as *Alternaria raphani* was isolated from marine sediment and produced three new cerebrosides named alternarosides A–C, and a new diketopiperazine alkaloid, alternarosin A. All the compounds showed weak anticandidal activity at MIC 1–4 caused by *Puccinia graminis f.* sp. *tritici*, g/ml (Wang et al., 2009). A marine fungal isolate *Chromocleista* sp. produced p-hydroxyphenopyrrozin with mild antifungal activity against *C. albicans* at MIC 25 μg/ml (Park et al., 2006). Mangrove *Ceriops Tagal*-associated endophytic fungi *Cytospora* sp., produced a new biscyclic sesquiterpene viz. seiricardine D, and eight known metabolites, xylariterpenoid A, xylariterpenoid B, and regiolone, ß-sitosterol, 4-hydroxy phenethyl alcohol, (22E, 24R)5, 8-epidioxy-5a, 8a-ergosta-6,22E-dien-3ß-ol, (22E, 24R) 5, 8-epidioxy-5a, 8a-ergosta-6,9(11), 22-trien-3 ß-ol, and stigmast-4-en-3-one. All compounds except 22-trien-3 ß-ol and ß-sitosterol showed weak antifungal activity against *Magnaporthe oryzae* at MIC 181.1–839.6 μM (Deng et al., 2020). Two lactone derivatives viz. penicipyrone and penicilactone were extracted from the sea fan *Annella* sp., associated endophytic fungus *Penicillium* sp. PSU-F44 and revealed antifungal activity against clinical isolate *Microsporum gypseum* SH-MU-4 at MIC 320 and 160 μg/ml, respectively (Trisuwan et al., 2009). Fungus *Fusarium* sp. FH-146 was isolated from driftwood and yielded three polyketides viz. neofusapyrone, fusapyrone, and deoxyfusapyrone. All three compounds were demonstrated for antifungal activity against *A. clavatus* at MIC 6.25 μg/ml, 25 μg/ml, and 3.12 μg/ml, respectively. The SAR studies suggested the requirement of the -OH group in deoxyfusapyrone over fusapyrone for enhanced activity (Hiramatsu et al., 2006). The marine-derived fungus *Stagonosporopsis cucurbitacearum* produced four new 4-hydroxy-2-pyridone alkaloids named Didymellamide A-D in the fermentation broth. out of which compound didymellamide A revealed good antifungal activity against azole-resistant and sensitive *C. albicans*, *C. neoformans*, and *C. glabrata* at MIC 1.6–3.1 μg/ml (Haga et al., 2013). Marine sponge *Halichondria japonica*-associated fungi *Phoma* sp. Q60596 demonstrated for antifungal activity of lactone metabolite against *C. albicans* ATCC10231 (IC_{80}s 6.25 μg/ml), *C. neoformans* TIMM0362 (IC_{80}s 1.56 μg/ml), *A. fumigatus* TIMM1776 (IC_{80}s 12.5 μg/ml) and *S. cerevisiae* YFC805 (IC_{80}s 1.56 μg/ml) by a microdilution method (Nagai et al., 2002). Polyene macrolides named marinisporolides A and B were isolated from *Marinispora* sp. CNQ-140 for antifungal activity against *C. albicans* at MIC 22 μM (Kwon et al., 2009). Another new macrolactam derivative from the marine actinomycete HF-11225 named nivelactam B exhibited inhibitory activity against *Sclerotinia sclerotiorum* at 100 μg/ml with a zone of inhibition 9 mm (Chen et al., 2018). A macrolide antibiotic viz. N-acetylborrelidin B and borrelidin were extracted from the marine *Streptomyces mutabilis* sp. MII and displayed antifungal activity against *S. cerevisiae*, *C. albicans* (Zone of inhibition between 14 and 22 mm), and *A. niger* (Zone of inhibition 10 mm) by agar diffusion method (Hamed et al., 2018). *Streptomyces* strain CNQ-085 was isolated from marine sediment and extracted for daryamides A and B. Both the compounds daryamides A and B exhibited antifungal activity against *C. albicans* at MIC values of 62.5 and 125 μg/ml, respectively (Asolkar et al., 2006). Metabolites pentadecanoic acid, N-hexadecanoic acid, and tetradecanoic acid from the crud extract of *Streptomyces* isolate showed inhibitory activity against *Talaromyces marneffei* yeast at MICs $\leq$ 0.03–0.25 μg/ ml (Sangkanu et al., 2021). Metabolite C 1,6-

dihydroxyphenazine-5,10-dioxide (iodinin) was extracted from *Streptosporangium* sp. DSMZ 45942 showed anticandidal activity against *C. glabrata* and *C. albicans* at MIC 0.36 µg/ml (Sletta et al., 2014). *Streptomyces xinghaiensis* SCSIO S15077 isolated from the South China Sea sediment produced compounds viz. Tunicamycin A, tunicamycin B, tunicamycin C, tunicamycin E, streptovirudin D_2, tunicamycin X, and tunicamycin C_3. All seven compounds exhibited antifungal activity against *C.albicans* ATCC 96901 and *C. albicans* CMCC (F) 98001 at MIC values between 2 and 32 µg/ml (Zhang et al., 2020a). *Bruguiera gymnorrhiza*-associated endophytic *Streptomyces albidoflavus* strain was assessed for antifungal activity. Metabolite Antimycin A18 with an acetoxy group at C-8 was purified and tested against *Colletotrichum lindemuthianum, B. cinerea, A. solani,* and *Magnaporth grisea* at MIC 0.01, 0.06, 0.03, and 0.20 µg/ml, respectively (Yan et al., 2010). In two studies, polyene macrolactam Heronamides and its derivatives were extracted from marine-derived *Streptomyces* sp., and forest soil isolates *Actinoalloteichus cyanogriseus* IFM 11549. The metabolites exhibited potent antifungal activity against fission yeast cells at MIC 0.50 µM through binding to phospholipids of the cell membrane (Sugiyama et al., 2014; Fujita et al., 2016).

New dilactone-tethered pseudo-dimeric peptides named Mohangamide A, and B, were reported from marine *Streptomyces* sp. against pathogenic *C. albicans*. Both peptides mohangamide A and B displayed strong inhibitory activity against *C. albicans* at IC_{50} 4.4 µM and IC_{50} 20.5 µM, respectively (Bae et al., 2015). *Aspergillus* sp. Isolated from Waikiki Beach of Honolulu, produced a novel prenylated indole alkaloid, i.e., waikialoid that showed anticandidal activity at MIC IC_{50} 1.4 µM (Wang et al., 2012). Two novel cyclic hexapeptides viz. sclerotides A and B containing anthranilic acid and dehydroamino acid units were isolated from the marine-derived halotolerant *Aspergillus sclerotiorum* PT06-1 strain and confirmed for antifungal activity against *C. albicans* at MIC 7.0 and 3.5 µM, respectively (Zheng et al., 2009). Antifungal compound mevinolinic acid methylester from *Aspergillus terreus* was recorded against *C. albicans* with the zone of inhibition 11 mm by plate assay method (Adpressa and Loesgen, 2016). Turbinmicin-producing strain *Micromonospora* sp. WMMC-415 was isolated from the ascidian *Ecteinascidia turbinate* for antifungal activity against *Candida auris* and *A. fumigatus* at MIC 0.25 µg/ml and 0.03 µg/ml, respectively (Zhang et al., 2020b). Marine fungus *Asteromyces cruciatus* revealed bis-N-norgliovictin mediated antifungal activity against *Mycotypha microspora* (13.5 mm total inhibition), *Eurotium rubrum* (4 mm total inhibition), and *Microbotryum violaceum* (13 mm total inhibition) at 50 µg/ml (Gulder et al., 2012). A marine bacterium *B. licheniformis* isolated from a sediment sample collected at Ieodo, Republic of Korea's southern reef was reported for antifungal activity. Two purified glycolipids ieodoglucomide C and ieodoglycolipid exhibited antifungal activity against *A. niger, R. solani, B. cinerea, C. acutatum,* and *C. albicans* at MIC 0.03–0.05 µM (Tareq et al., 2015). *B. amyloliquefaciens* SH-B10 from deep-sea sediment was reported for two anti-fungal lipopeptides, viz., C16 fengycin A and fengycin B containing aminobutyric acid at position 6 of the peptide backbone. Both the compounds revealed antifungal activity against phytopathogens viz. *F. oxysporum* f. sp. *cucumerinum, F. graminearum, F. oxysporum* f. sp. *vasinfectum, F. oxysporum* f. sp. *cucumis melo* L., and *F. graminearum* f. sp. zea mays L. with a zone of inhibition between 14 and 22 mm at 2 mg/ml concentration (Chen et al., 2010). A marine-sediment-derived bacterium *Bacillus velezensis* SH-B74 was reported for cyclic lipopeptides (CLPs) named *anteiso*-C_{15} $Ile_{2,7}$ surfactin production. The peptide was revealed to control the appressoria formation of rice blast pathogen *Magnaporthe oryzae* up to 23.4% (±8.6), and

11.3% (±3.4) at 10 and 50 μM concentration, respectively (Ma et al., 2020). Haliangicin is a beta-methoxy acrylate antibiotic with a conjugated tetraene moiety that was isolated from the marine myxobacterium *Haliangium luteum*, and exhibited antifungal activity against a wide spectrum of pathogenic fungi like *A. niger* AJ117374, *A. Fumigatus* AJ-117190, *B. cinerea* AJ 117140, *M. hiemalis* AJ 117396, *P. ultimum* IFO32210, and *Saprolegnia parasitica* IFO8978 at MIC 0.1–12.5 μg/ml (Fudou et al., 2001). *Aneurinibacillus* sp. YR247 was isolated from the deep-sea sediment of Sagami Bay, Japan, and studied for antifungal activity against the *Aspergillus brasiliensis* NBRC9455 at 5mg/ml crude extract with a zone of inhibition of 20.3 mm. The purified compound was identified as a gramicidin-like novel antifungal peptide (Kurata et al., 2017). A cold-adapted *Pseudomonas* strain GWSMS-1 was isolated from marine sediment and showed significant inhibition of *Verticillium dahlia* CICC 2534 and *F. oxysporum* f. sp. *cucumerinum* CICC 2532 mediated by crude chitinase enzyme (Liu et al., 2019). An antifungal compound 9,10-dihydrophenanthrene-2-carboxylic acid was isolated from a marine-derived bacterium *Pseudomonas putida* and was found active against *C. albicans* at MIC value of 20μg/ml (Uzair et al., 2018). Overall, mostly the *Streptomyces* sp., *Aspergillus* sp., and *Bacillus* sp. can produce diverse SMs and thus be labeled as treasures (Table 2).

TABLE 2 Some secondary metabolite extracted for antifungal activity.

Source	Secondary metabolite	Antifungal activity against pathogens	References
Bacteria			
Chromobacterium sp. NIIST (MTCC 5522)	Violacein	*T. rubrum* and *P. expansum*,	Sasidharan et al. (2015)
P. aeruginosa	Pyocyanin	*C. albicans*, *C. krusei*, *C. glabrata*, *C. tropicalis* and *C. neoformans*	Sudhakar and Karpagam (2011)
P. chlororaphis GP 72	Phenazine-1-Carboxylic Acid, 2-Hydroxyphenazine	*C. lagenarium*, *P. capsici*, *P. aphanidermatum*, *P. ultimum*, *S. sclerotiorum*, *F. oxysporum*. f. sp. *cucumerinum*, *C. sasakii* and *R. solani*	Liu et al. (2007a)
P. pseudoalcaligenes P4	Phenazine-1-Carboxylic Acid (PCA), 2,4 Diacetylphloroglucinol (DAPG), Hydrogen Cyanide (HCN), Pyoluteorin (PLT,) Pyrrolnitrin (PRN)	*R. solani*, *M. grisea*, *S. oryzae*	Ayyadurai et al. (2007)
P. fluorescens JM-1	2,4 Diacetyl Phloroglucinol (DAPG)	*F. moniliforme*	Mishra et al. (2022)
L. gummosus (AB161361),	2,4 Diacetyl Phloroglucinol (DAPG)	*B. dendrobatidis*	Brucker et al. (2008)
P. akhurstii sp. nov. 0813-124	Glidobactin A, Cepafungin I	*C. gloeosporioides*	Tu et al. (2022)
P. luminescens	Trans-Cinnamic Acid	*F. effusum*	Bock et al. (2014)

TABLE 2 Some secondary metabolite extracted for antifungal activity—cont'd

Source	Secondary metabolite	Antifungal activity against pathogens	References
Bacillus sp. DBS4	Carotenoids	*R. solani*, *F. oxysporum*, and *Sclerotium rolfsi*	Dawoud et al. (2020)
B. subtilis CMB32	Iturin A, Fengycin, Surfactin A	*C. gloeosporioides*, *F. solani* KCTC 6328, *B. cinerea* KACC 40573, *F. oxysporum* KACC 40037, *R. solani* KACC 40151, and *P. capsici* KACC 40157	Kim et al. (2010)
B. velezensis DTU001	Iturins, Fengycins, Surfactins	*P. sclerotigenum*, *P. solitum*, *P. italicum*, *A. uvarum*, *A. calidoustus*, *C. neoformans*, *P. ulaiense*, *P. expansum*, *T. atroroseus* and *T. amestolkiae*	Devi et al. (2019)
B. subtilis, *B. amyloliquefaciens*, *B. pumilus*	Bacilysin	*E. amylovora*	Sansinenea and Ortiz (2011)
B. amyloliquefaciens K103	Bacilomycins L	*R. solani*	Zhang et al. (2013a)
B. velezensis PW192	Fengycin A, Fengycin B	*C. gloeosporioides* c1060 and *C. musae* BCC 13080	Jumpathong et al. (2022)
B. velezensis CE 100	Cyclic Tetrapeptide Cyclo-(Prolyl-Valyl-Alanyl-Isoleucyl)	*C. gloeosporioides*	Choub et al. (2021)
B. licheniformis A12	Peptide A12- C	*M. plumbeus* CCM F 443, *M. mucedo* CECT 2653, and *T. mentagrophytes* CECT 2793	Gálvez et al. (1993)
B. licheniformis M4	Fungicin M4	*Microsporum canis*, *Mucor mucedo*, *Mucor plumbeus* and *Sporothrix schenckii*	Lebbadi et al. (1994)
B. subtilis strain 916	Bacisubin	*R. solani*, *A. oleracea*, *A. brassicae* and *B. cinerea*	Liu et al. (2007b)
B. cenocepacia H111	Fragin	*F. solani*	Jenul et al. (2018)
Burkholderia sp. MSSP	2-Hydroxymethyl-Chroman-4-One	*P. ultimum*, *P. capsici* and *S. sclerotiorum*	Kang et al. (2004)
B. ambifaria	Dimethyldisulfide (DMDS), Dimethyltrisulfide (DMTS), S-Methyl Methanethiosulfonate, 4-Octanone, 1-Phenylpropan-1-One And 2-Undecanone	phytopathogenic fungi	Groenhagen et al. (2013)
B. glumae	Toxoflavin	*R. solani*	Karki et al. (2012)
L. plantarum	Phenyllactic Acid (PLA)	*A.*, *Penicillium*, and *Fusarium*	Lavermicocca et al. (2003)

Continued

TABLE 2 Some secondary metabolite extracted for antifungal activity—cont'd

Source	Secondary metabolite	Antifungal activity against pathogens	References
Kribbella sp. MI481-42F6	Kribellosides A-D	*Saccharomyces cerevisiae*	Igarashi et al. (2017)
S. cellulosum	Epothilones,	Antifungal activity	Vaishnav and Demain (2011)
Fungi			
C. quercina	Cryptocandin	*C. albicans*, *T. mentagrophytes* and *T. rubrum*	Strobel et al. (1999)
Pezicula spp & *Zalerion arboricola*	Pneumocandins	*Antifungal*	
Aspergillus species	Echinocandins	*Antifungal*	
N. oryzae, *Xylaria* sp. F0010, *Chromocleista* sp.	Griseofulvin	*F. oxysporum*, *T. mentagrophytes*, *M. canis*, *M. grisea*, *Corticium sasaki*, *P. recondite*, *B. graminis*	Park et al. (2005b); Rathod et al. (2014); Bai et al. (2019)
P. variotii	Varioxepine A	*F. graminearum*	Zhang et al. (2014)
R. babjevae YS3	Sophorolipids, Glycolipid Biosurfactants	*C. gloeosporioides* *F. verticilliodes*, *F. oxysporum*, *C. cassiicola*, *T. rubrum*	Sen et al. (2017)
E. nigrum	Flavipin, Orevactaene, Epipyrone A	*S. sclerotiorum*, *A. oryzae*, *S. cerevisiae*, *C. albicans*	Preindl et al. (2017); Lee et al. (2020)
Chaetopsina sp.	Alkenyl Sulphates, Linoleyl Sulfate, Linolenyl Sulphate, Oleyl Sulphate.	*C. glabrata*, *C. krusei*, *C. parapsilosis*, *C. tropicalis*, *C. albicans*, and *A. fumigatus*	Crespo et al. (2017)
C. cavincola	Colisporifungin, Cavinafungins A and B	*Candida* sp., *A. fumigatus*	Ortiz-Lopez et al. (2015)
Penicillium sp. FKI-1938	Citridone A	*C. albicans*	Fukuda et al. (2014)
Aspergillus sp.	Azaphilone	*S. cerevisiae*	Yang et al. (2009)
C. globosum F0142	Chaetoviridins, Chaetoviridins A and B	*M. grisea*, *P. recondita*	Park et al. (2005a)
Monascus sp.	Azaphilone	*A. niger*, *P. citrinum*, and *C. albicans*	Kim et al. (2006)
V. psalliotae	Oosporein	*P. infestans* and *A. solani*, *F. oxysporum*	Nagaoka et al. (2004)
Pseudogymnoascus sp. PF1464	Amphiol	*S. pombe*, *C. albicans*, *A. fumigatus*	Fujita et al. (2021)
A. nishimurae strain IFM58441	Anishidiol	*A. fumigatus*, *A. niger*, *C. albicans*, and *C. neoformans*	Hosoe et al. (2011)

TABLE 2 Some secondary metabolite extracted for antifungal activity—cont'd

Source	Secondary metabolite	Antifungal activity against pathogens	References
C. lunata	Chrysophanol, Helminthosporin, Cynodontin,	Antifungal activity	Lin and Xu (2020)
P. ochrochloron	(−) 2, 3, 4-Trihydroxybutanamide, (−) Erythritol	*C. albicans*	Rančić et al. (2006)
P. ochrochloron	6-(2′R-Hydroxy-3′E,5′E-Diene-1′-Heptyl)-4-Hydroxy-3-Methyl-2H-Pyran-2-One, 6-(2′S-Hydroxy-5′E-Ene-1′-Heptyl)-4-Hydroxy-3-Methyl-2H-Pyran-2-One, And 6-(2′S-Hydroxy-1′-Heptyl)-4 -Hydroxy-3-Methyl-2H-Pyran-2-One, Trichodermic Acid	*C. arachidicola* Hori, *A. solani*, *B. carbonum* Wilson and *F. graminearum*	Zhao et al. (2019)
A. fumigatus, *A. terreus*, *A. flavus*, and *A. niger*	Gliotoxin, Acetylgliotoxin, Derivative of Hyalodendrin	*C. albicans* and *C. neoformans*	Coleman et al. (2011)
T. virens (ITC-4777)	Gliotoxin	*R. bataticola*, *P. deharyanum*, *P. aphanidermatum*, *M. phaseolina*, *S. rolfsii*, *R. solani*	Singh et al. (2005)
B. anartia	(E/Z)-Bombardolides A And D	*C. albicans*	Hein et al. (2001)
P. appendiculata	Appenolides A, B, and C	*C. albicans*, *S. fimicola*, *A. furfuraceus*	Wang et al. (1993)
P. decipiens	Sesquiterpene Lactones, Decipienolides A-B, and Decipinin A	*F. verticillioides*	Che et al. (2002)
P. anserine	Anserinones A and B	*S. fimicola* and *A. furfuraceus* fungi	Wang et al. (1997)
C. areolate	Cercophorins A-C, Decarboxycitrinone, 4-Acetyl-8-Hydroxy-6-Methoxy-5-Methylisocoumarin, Roridin E	*S. fimicola* and *A. furfuraceus*	Whyte et al. (1996)
G. reessii	Gymnoascolide A	*S. nodorum*	Clark et al. (2005)
Periconia sp. OBW-15	Periconicin A and B,	*C. albicans*, *T. mentagrophytes*, and *T. rubrum*	Shin et al. (2005)
Hormonema sp.	Enfumafungin	*C. albicans*, *C. tropicalis*, *A. fumigatus*	Peláez et al. (2000)
A. tubingensis	3-(5-Oxo-2,5-Dihydrofuran-3-Yl) Propanoic Acid	*F. graminearum*	Yang et al. (2019)
F. sambucinum	Sambacide	*C. albicans*	Dong et al. (2016)

Continued

TABLE 2 Some secondary metabolite extracted for antifungal activity—cont'd

Source	Secondary metabolite	Antifungal activity against pathogens	References
Penicillium sp.	Yanuthones (Meroterpenoids), Aspernigrins (NRP), Asperfuran (Dihydrobenzofuran), Flufuran (Furans), Avenaciolide (Lactone), Antafumicins A, Antafumicins A And B, Tensidols (Furopyrrols), Canadensolide	Antifungal activity	Frisvad et al. (2018)
P. brevicompactum	Compactin	Antifungal activity	Vaishnav and Demain (2011)
Actinomycetes			
S. nodosus	Amphotericin B	*C. albicans* and *A. fumigatus*	Lemke et al. (2005)
S. nodosus NRRLB-2371	Corifungin	*Candida* sp., *Aspergillus* sp.	Victoria Castelli et al. (2017)
S. noursei ATCC 11455	Nystatin	Antifungal activity	Brautaset et al. (2000)
S. natalensis, *S. chattanogenesis*, *S. gilvosporeus*, S. *lydicus*	Natamycin	Antifungal activity	Elsayed et al. (2019)
Streptomyces sp. K710 *Streptomyces* sp. KNF2047	Neopeptins A and B	*A. mali*, *B. cinerea*, *G. cingulate*, *C. lagenarium* and *C. miyabeanus*	Satomi et al. (1982); Ubukata et al. (1986)
Streptomyces sp. strain MA-23	Globopeptin	*M. racemosus* KF-223, *P. oryzae* KF-180, and *B. cinerea* KF-241	Tanaka et al. (1987)
S. lavendulae H698 SY2	Glomecidin, Ileumycin	*G. cingulate*, *C. gloeosporioides*, and *C. lagenarium*	Kunihiro and Kaneda (2003)
S. morookaense AM25	Gloeosporiocide"	*C. gloeosporioides*	Vicente dos Reis et al. (2019)
Streptomyces sp. 3–10	Reveromycins A and B	*B. cinerea*, *M. hiemails*, *R. stolonifer*, and *S. sclerotiorum*	Lyu et al. (2017)
S. phytohabitans HBERC-20821	Novonestmycins A and B	*C. cassiicola*, *R. solani* and *S. nodorum*,	Wan et al. (2015)
S. griseus Tu 2599 *S. flavotricini* Y12-2, *S. varsoviensis* and *S. cavourensis* YY01-17	Bafilomycins A_1, A_2, B_1, B_2, C_1, and C_2, D, E, F, J, K, Hygrobafilomycin	*P. oryzae*, *B. cinerea*, and *R. solani*, *C. lagenarium*	Xu et al. (2013), Zhang et al. (2011)
S. griseus var. autotrophicus	Faeriefungin	*F. oxysporum* f. sp asparagi (FOA)	Smith et al. (1990)

TABLE 2 Some secondary metabolite extracted for antifungal activity—cont'd

Source	Secondary metabolite	Antifungal activity against pathogens	References
Streptomyces sp. IA1	Actinomycin D	Antifungal activity	Toumatia et al. (2015)
Streptomyces spp. PAL114	Saquayamycins	*C. albicans, S. cerevisiae, A. carbonarius, A. flavus, F. culmorum, P. glabrum*	Aouiche et al. (2014)
Streptomyces sp. MK-30	Makinolide	*S. cerevisiae* and *M. hiemalis*	Kodani et al. (2012)
S. malaysiensis	Malayamycin	*S. nodorum*	Li et al. (2008)
Streptomyces sp. No. AC-69	Lipopeptin A	*P. oryzae*	Nishii et al. (1981)
Streptomyces sp.	Enduspeptides A-F	*C. glabrata*	Chen et al. (2017)
S. zhaozhouensis CA-185989	Ikarugamycin, 28-N-Methylikaguramycin	*C. albicans* and *A. fumigatus*	Lacret et al. (2014)
Streptomyces sp.	Sceliphrolactam	*C. albicans*	Oh et al. (2011)
Streptomyces sp. DPTB16	4′ Phenyl-1-Napthyl-Phenyl Acetamide	*C. albicans, A. niger, A. flavus, Mucor* sp., *A. fumigatus*	Dhanasekaran et al. (2008)
S. lavenduligriseus	Filipin III	*C. albicans*	Yang et al. (2016)
Streptomyces sp. SY1965	Streptothiazolidine A, Streptodiketopiperazines, A B, Spoxazomicin C	*C. albicans*	Yi et al. (2020)
Streptomyces sp. *CS*	Naphthomycin L and Naphthomycin A	*F. moniliforme*	Yang et al. (2012)
S. kasugaensis	Kasugamycin	*P. oryzae*	Umezawa et al. (1965)
S. cacaoi var. asoensis	Polymixin B and D	*R. solani*	Isono et al. (1967)
S. rimofaciens	Mildiomycin	Antifungal activity	Iwasa et al. (1978)
S. hygroscopicus var. geldanus and *S. griseus*	Geldanamycin	*R. solani*	Rothrock and Gottlieb (1984)
S. hygroscopicus	Gopalamicin,	*A. fumigatus, F. oxysporum, A. flavus, F. moniliforme, A. niger, P. ultimum, C. albiccans, P. graminicola, A. solani* and *L. korrae*	Nair et al. (1994)
S. viola-ceoniger	Tubercidin	*P. capsici* and *M.grisea.*	Hwang and Kim (1995)
S. violaceusniger YCED-9	Guanidyl Fungin A, Geldanamycin, Nigericin	*S. homeocarpa* and *R. saloni*	Trejo-Estrada et al. (1998)

Continued

TABLE 2 Some secondary metabolite extracted for antifungal activity—cont'd

Source	Secondary metabolite	Antifungal activity against pathogens	References
S. diastaticus	Oligomycins A and C	*A. niger*, *A. alternata*, *P. capsici*, and *B. cinerea*	Yang et al. (2010)
S. padanus PMS-702	Fungichromin	*R. solani* AG-4	Shih et al. (2003) Fan et al. (2019)
Streptomyces sp. GAAS7310	Antimycin A_{17}	*C. lunata* (Wakker) Boed, *R. nigrtcans* Ehrb, *R. nigrtcans* Ehrb, *A. solani* (E. et M.), *Cladosporium* sp., *S. sclerotiorum* (Lib.) de Bary, and *C. nigrum* EL. *et* Halst	Chen et al. (2005)
S. cavourensis subsp	2-Furancarboxaldehyde	*C. gloeosporioides*	Lee et al. (2012)
S. cavourensis NA4	Bafilomycins B1 and C1	*Fusarium* spp., *R. solani*, and *B. cinerea*	Pan et al. (2015)
S. canus BYB02	Resistomycin	*V. mali* and *M. grisea*	Zhang et al. (2013b)
Amycolatopsis CP2808	Ansacarbamitocins	*S. tritici*, *E. graminis*, *P. recondite*	Snipes et al. (2007)
Saccharothrix sp. strain MK27-91F2	Formamicin	*P. oryzae*, *R. solani*, *P. digitatum* and *D. citri*	Igarashi et al. (1997)
Trichoderma spp.	6-Pentyl Pyrone (6-PP), Gliotoxin, Gliovirin	Antifungal activity	Mukherjee et al. (2012)
M. narashinoensis	Rustmicin	*C. neoformans*, *C. tropicalis*, *C. albicans*	Al-Fadhli et al. (2022)
M. chalcea	Neorustmicins A-D	*P. grannnis*	
M. paleriensis	Primycin	*C. albicans*	
Micromonospora sp. M39	2,3-Dihydroxybenzoic Acid, Phenylacetic Acid, Cervinomycin A1 and A2	*P. oryzae* MPO 293	Ismet et al. (2004)
Actinoalloteichus sp. NPS702	Neomaclafungins A-I	*T. mentagrophytes*	Sato et al. (2012)
Actinomycete sp. HIL Y-8620959	Mathemycin A	*P. infestans*, *A. mali*, *B. cinerea*	Nadkarni et al. (1998)
A. roseola	Daunomycin	*P. capsici*, *R. solani* and *S. cerevisiae*	Kim et al. (2000)
Marine microorganisms			
T. koningii	Trichodermaketones A	*C. albicans*	Song et al. (2010)
Monodictys sp.	Monodictyquinone A	*C. albicans*	El-Beih et al. (2007)
Acremonium sp. AWA16-1	Awajanoran	*C. albicans*	Jang et al. (2006)

TABLE 2 Some secondary metabolite extracted for antifungal activity—cont'd

Source	Secondary metabolite	Antifungal activity against pathogens	References
M. roridum TUF 98F42	2′,3′-Deoxyroritoxin D	*S. cerevisiae*	Xu et al. (2006)
A. raphani	Alternarosides A–C, Alternarosin A	*Candida* sp.	Wang et al. (2009)
Chromocleista sp.	*P*-Hydroxyphenopyrrozin	*C. albicans*	Park et al. (2006)
Cytospora sp.,	22-Trien-3 ß-Ol and ß-Sitosterol	*M. oryzae*	Deng et al. (2020)
Penicillium sp. PSU-F44	Penicipyrone and Penicilactone	*M. gypseum*	Trisuwan et al. (2009)
Fusarium sp. FH-146	Neofusapyrone, Fusapyrone, Deoxyfusapyrone	*A. clavatus*	Hiramatsu et al. (2006)
S. cucurbitacearum	Didymellamide A-D	*C. neoformans*, *C. albicans*, *C. glabrata*	Haga et al. (2013)
Phoma sp. Q60596	Lactone Metabolite	*C. albicans*, *C. neoformans*, *A. fumigatus*, *S. cerevisiae*	Nagai et al. (2002)
Marinispora sp.	Marinisporolides A and B	*C. albicans*	Kwon et al. (2009)
Actinomycete HF-11225	Nivelactam B	*S. sclerotiorum*	Chen et al. (2018)
S. mutabilis sp. MII	N-Acetylborrelidin B, Borrelidin	*S. cerevisiae*, *C. albicans*, *A. niger*	Hamed et al. (2018)
Streptomyces strain CNQ-085	Daryamides A and B	*C. albicans*	Asolkar et al. (2006)
Streptomyces sp.	N-Hexadecanoic Acid, Tetradecanoic Acid, Pentadecanoic Acid	*T. marneffei*	Sangkanu et al. (2021)
Streptosporangium sp. DSMZ 45942	C 1,6-Dihydroxyphenazine-5,10-Dioxide (Iodinin)	*C. glabrata* and *C. albicans*	Sletta et al. (2014)
S. xinghaiensis SCSIO S15077	Tunicamycin E (1), Tunicamycin B (2), Tunicamycin X (3), Tunicamycin A (4), Streptovirudin D_2 (5), Tunicamycin C (6), Tunicamycin C_3	*C. albicans*	Zhang et al. (2020a)
S. albidoflavus	Antimycin A18	*C. lindemuthianum*, *B. cinerea*, *A. solani* and *M. grisea*	Yan et al. (2010)
Streptomyces sp., and *Actinoalloteichus cyanogriseus* IFM 11549	Heronamides	Antifungal activity	Sugiyama et al. (2014), Fujita et al. (2016)
Streptomyces sp.	Mohangamide A, and B	*C. albicans*	Bae et al. (2015)

Continued

TABLE 2 Some secondary metabolite extracted for antifungal activity—cont'd

Source	Secondary metabolite	Antifungal activity against pathogens	References
Aspergillus sp.	Waikialoid	*Candida* sp.	Wang et al. (2012)
A. sclerotiorum PT06-1	Sclerotides A and B	*C. albicans*	Zheng et al. (2009)
A. terreus	Mevinolinic Acid Methylester	*C. albicans*	Adpressa and Loesgen (2016)
Micromonospora sp. WMMC-415	Turbinmicin	*C. auris* and *A. fumigatus*	Zhang et al. (2020b)
Asteromyces cruciatus	Bis-N-Norgliovictin	*M. microspore, E. rubrum, M. violaceum*	Gulder et al. (2012)
B. licheniformis	Ieodoglucomide C and Ieodoglycolipid	*A. niger, R. solani, B. cinerea, C. acutatum,* and *C. albicans*	Tareq et al. (2015)
B. amyloliquefaciens	Fengycin A, Fengycin B	*F. oxysporum* and F. *graminearum*	Chen et al. (2010)
B. velezensis SH-B74	*Anteiso*-C_{15} $Ile_{2,7}$ Surfactin	*M.oryzae*	Ma et al. (2020)
H. luteum	Haliangicin	*A. niger, A. Fumigatus, B. cinerea, M. hiemalis, P. ultimum, S. parasitica*	Fudou et al. (2001)
Aneurinibacillus sp. YR247	Gramicidin-Like Novel Antifungal Peptide	*A. brasiliensis*	Kurata et al. (2017)
Pseudomonas strain GWSMS-1	Chitinase	*V. dahlia* and *F. oxysporum*	Liu et al. (2019)
P. putida	9,10-Dihydrophenanthrene - 2 - Carboxylic Acid	*C. albicans*	Uzair et al. (2018)

8 Future perspectives and conclusion

Human health is increasingly threatened by fungal infections, ranging from superficial to invasive. The prevalence of systemic fungal infections has increased significantly in the last decade. Moreover, a limited number of antifungal drugs are currently available for the treatment of life-threatening fungal infections. Antimicrobial resistance (AMR) is still increasing at an alarming rate and can be the next global pandemic. Since fungi are metabolically similar to mammalian cells, few pathogen-specific targets are available, making antifungal research even more challenging. Other challenges include the development of diagnostics for pathogen detection and antimicrobial susceptibility testing. Therefore, an urgent medical need to develop new antifungal agents with new mechanisms of action is required. Using combined expertise from microbiology, genomics, metabolomics, natural product chemistry, and pharmacology, we can enrich the pipeline with valuable pharmacological leads that may ultimately contribute to the worldwide search for clinically useful antifungal drugs. New analogues or templates in which SMs serve as lead compounds will lead to the discovery and design of new drugs. Moreover, the co-culture method using more than one microbe will initiate the novel bioactive compounds for fungal pathogens. Overall, new strategies for antifungal therapy, target identification, and rational drug design technologies can significantly accelerate the process of new antifungal development, reducing cure time or providing a better quality of life to patients. Moreover, improving existing molecules, developing new formulations, and alternative therapy for prevention and treatment are important for treating fungal infections and increasing quality of life. Also, the rise in fungicide-resistant pathogenic fungi has led to an increase in the use of antifungals in agriculture in recent years. Thus, the need to search and screen new antifungal agents is a priority. Considerable progress can be possible in this field, by using bioinformatics and nanotechnology approaches to identify and develop novel antifungal compounds.

Acknowledgments

The authors are thankful to the Department of Science and Technology Science and Engineering Research Board (DST-SERB project EEQ/2019/000469), New Delhi, India, for providing financial assistance to conduct the project on bioactive compounds from endophytes. The authors are thankful to the Head of the Department of Microbiology, Savitribai Phule Pune University, Pune, MS, India, for providing the necessary facility to conduct the DST-SERB project.

Conflict of interest

No conflict of interest.

References

Abu-Elteen, K.H., Hamad, M., 2017. Antifungal agents for use in human therapy. In: Fungi: Biology and Applications. John Wiley & Sons, Inc, pp. 299–332.

Adpressa, D.A., Loesgen, S., 2016. Bioprospecting chemical diversity and bioactivity in a marine-derived *Aspergillus terreus*. Chem. Biodivers. 13 (2), 253–259.

Agboyibor, C., Kong, W.B., Chen, D., Zhang, A.M., Niu, S.Q., 2018. Monascus pigments production, composition, bioactivity, and its application: a review. Biocatalysis Agri. Biotechnol. 16, 433–447.

Al-Fadhli, A.A., Threadgill, M.D., Mohammed, F., Sibley, P., Al-Ariqi, W., Parveen, I., 2022. Macrolides from rare actinomycetes: structures and bioactivities. Int. J. Antimicrob. Agents, 106523.

Alam, K., Mazumder, A., Sikdar, S., Zhao, Y.M., Hao, J., Song, C., Wang, Y., Sarkar, R., Islam, S., Zhang, Y., Li, A., 2022. Streptomyces: the biofactory of secondary metabolites. Front. Microbiol. 13, 968053.

Aldholmi, M., Marchand, P., Ourliac-Garnier, I., Le Pape, P., Ganesan, A., 2019. A decade of antifungal leads from natural products: 2010–2019. Pharmaceuticals 12 (4), 182.

Almeida, F., Rodrigues, M.L., Coelho, C., 2019. The still underestimated problem of fungal diseases worldwide. Front. Microbiol. 10, 214.

Aouiche, A., Bijani, C., Zitouni, A., Mathieu, F., Sabaou, N., 2014. Antimicrobial activity of saquayamycins produced by *Streptomyces* spp. PAL114 isolated from a Saharan soil. J. Mycol. Méd. 24 (2), e17–e23.

Asolkar, R.N., Jensen, P.R., Kauffman, C.A., Fenical, W., 2006. Daryamides A–C, weakly cytotoxic polyketides from a marine-derived actinomycete of the genus *Streptomyces* strain CNQ-085. J. Nat. Prod. 69 (12), 1756–1759.

Ayyadurai, N., Naik, P.R., Sakthivel, N., 2007. Functional characterization of antagonistic fluorescent *pseudomonads* associated with rhizospheric soil of rice (*Oryza sativa* L.). J. Microbiol. Biotechnol. 17 (6), 919–927.

Bae, M., Kim, H., Moon, K., Nam, S.J., Shin, J., Oh, K.B., Oh, D.C., 2015. Mohangamides A and B, new dilactone-tethered pseudo-dimeric peptides inhibiting *Candida albicans* isocitrate lyase. Org. Lett. 17 (3), 712–715.

Bai, Y.B., Gao, Y.Q., Nie, X.D., Tuong, T.M.L., Li, D., Gao, J.M., 2019. Antifungal activity of griseofulvin derivatives against phytopathogenic fungi in vitro and in vivo and three-dimensional quantitative structure-activity relationship analysis. J. Agric. Food Chem. 67 (22), 6125–6132.

Balouiri, M., Sadiki, M., Ibnsouda, S.K., 2016. Methods for in vitro evaluating antimicrobial activity: a review. J. Pharm. Anal. 6 (2), 71–79.

Barrios-González, J., 2018. Secondary metabolites production: physiological advantages in solid-state fermentation. In: Current Developments in Biotechnology and Bioengineering, pp. 257–283.

Barrios-Gonzalez, J., Fernandez, F.J., Tomasini, A., 2003. Microbial secondary metabolites production and strain improvement. Indian J. Biotechnol. 2 (3), 322–333.

Bock, C.H., Shapiro-Ilan, D.I., Wedge, D.E., Cantrell, C.L., 2014. Identification of the antifungal compound, trans-cinnamic acid, produced by *Photorhabdus luminescens*, a potential biopesticide against pecan scab. J. Pestic. Sci. 87 (1), 155–162.

Brautaset, T., Sekurova, O.N., Sletta, H., Ellingsen, T.E., Strøm, A.R., Valla, S., Zotchev, S.B., 2000. Biosynthesis of the polyene antifungal antibiotic nystatin in *Streptomyces noursei* ATCC 11455: analysis of the gene cluster and deduction of the biosynthetic pathway. Chem. Biol. 7 (6), 395–403.

Brucker, R.M., Baylor, C.M., Walters, R.L., Lauer, A., Harris, R.N., Minbiole, K.P., 2008. The identification of 2, 4-diacetylphloroglucinol as an antifungal metabolite produced by cutaneous bacteria of the salamander *Plethodon cinereus*. J. Chem. Ecol. 34 (1), 39–43.

Che, Y., Gloer, J.B., Koster, B., Malloch, D., 2002. Decipinin A and decipienolides A and B: new bioactive metabolites from the coprophilous fungus *Podospora decipiens*. J. Nat. Prod. 65 (6), 916–919.

Chen, G., Lin, B., Lin, Y., Xie, F., Lu, W., Fong, W.F., 2005. A new fungicide produced by a *Streptomyces* sp. GAAS7310. J. Antibiot. 58 (8), 519–522.

Chen, L., Wang, N., Wang, X., Hu, J., Wang, S., 2010. Characterization of two anti-fungal lipopeptides produced by *Bacillus amyloliquefaciens* SH-B10. Bioresour. Technol. 101 (22), 8822–8827.

Chen, Y., Liu, R.H., Li, T.X., Huang, S.S., Kong, L.Y., Yang, M.H., 2017. Enduspeptides AF, six new cyclic depsipeptides from a coal mine derived *Streptomyces* sp. Tetrahedron 73 (5), 527–531.

Chen, H., Cai, K., Yao, R., 2018. A new macrolactam derivative from the marine actinomycete HF-11225. J. Antibiot. 71 (4), 477–479.

Choub, V., Maung, C.E.H., Won, S.J., Moon, J.H., Kim, K.Y., Han, Y.S., Cho, J.Y., Ahn, Y.S., 2021. Antifungal activity of cyclic tetrapeptide from *Bacillus velezensis* CE 100 against plant pathogen *Colletotrichum gloeosporioides*. Pathogens 10 (2), 209.

Clark, B., Capon, R.J., Lacey, E., Tennant, S., Gill, J.H., Bulheller, B., Bringmann, G., 2005. Gymnoascolides A–C: aromatic butenolides from an Australian isolate of the soil ascomycete *Gymnoascus reessii*. J. Nat. Prod. 68 (8), 1226–1230.

Coleman, J.J., Ghosh, S., Okoli, I., Mylonakis, E., 2011. Antifungal activity of microbial secondary metabolites. PloS One 6 (9), e25321.

Crespo, G., Gonzalez-Menendez, V., de la Cruz, M., Martin, J., Cautain, B., Sanchez, P., Pérez-Victoria, I., Vicente, F., Genilloud, O., Reyes, F., 2017. Antifungal long-chain alkenyl sulphates isolated from culture broths of the fungus *Chaetopsina* sp. Planta Med. 234 (06), 545–550.

Cui, J., Ren, B., Tong, Y., Dai, H., Zhang, L., 2015. Synergistic combinations of antifungals and anti-virulence agents to fight against *Candida albicans*. Virulence 6 (4), 362–371.

Dawoud, T.M., Alharbi, N.S., Theruvinthalakal, A.M., Thekkangil, A., Kadaikunnan, S., Khaled, J.M., Almanaa, T.N., Sankar, K., Innasimuthu, G.M., Alanzi, K.F., Rajaram, S.K., 2020. Characterization and antifungal activity of the yellow pigment produced by a *Bacillus* sp. DBS4 isolated from the lichen *Dirinaria agealita*. Saudi J. Biol. Sci. 27 (5), 1403–1411.

Demain, A.L., Martens, E., 2017. Production of valuable compounds by molds and yeasts. J. Antibiot. 70 (4), 347–360.

Deng, Q., Li, G., Sun, M., Yang, X., Xu, J., 2020. A new antimicrobial sesquiterpene isolated from endophytic fungus *Cytospora* sp. from the Chinese mangrove plant Ceriops tagal. Nat. Prod. Res. 34 (10), 1404–1408.

Devi, S., Kiesewalter, H.T., Kovács, R., Frisvad, J.C., Weber, T., Larsen, T.O., Kovács, Á.T., Ding, L., 2019. Depiction of secondary metabolites and antifungal activity of *Bacillus velezensis* DTU001. Synth. Syst. Biotechnol. 4 (3), 142–149.

Dhanasekaran, D., Thajuddin, N., Panneerselvam, A., 2008. An antifungal compound: 4′ phenyl-1-napthyl–phenyl acetamide from *Streptomyces* sp. DPTB16. Facta Universitatis Ser. Med. Biol. 15 (1), 7–12.

Dong, J.W., Cai, L., Li, X.J., Duan, R.T., Shu, Y., Chen, F.Y., Wang, J.P., Zhou, H., Ding, Z.T., 2016. Production of a new tetracyclic triterpene sulfate metabolite sambacide by solid-state cultivated *Fusarium sambucinum* B10. 2 using potato as substrate. Bioresour. Technol. 218, 1266–1270.

Duarte, K., Rocha-Santos, T.A., Freitas, A.C., Duarte, A.C., 2012. Analytical techniques for discovery of bioactive compounds from marine fungi. TrAC Trends Anal. Chem. 34, 97–110.

El-Beih, A.A., Kawabata, T., Koimaru, K., Ohta, T., Tsukamoto, S., 2007. Monodictyquinone A: a new antimicrobial anthraquinone from a sea urchin-derived fungus *Monodictys* sp. Chem. Pharm. Bull. 55 (7), 1097–1098.

Elsayed, E.A., Farid, M.A., El-Enshasy, H.A., 2019. Enhanced Natamycin production by Streptomyces natalensis in shake-flasks and stirred tank bioreactor under batch and fed-batch conditions. BMC Biotechnol. 19 (1), 1–13.

Eseola, A.B., Ryder, L.S., Osés-Ruiz, M., Findlay, K., Yan, X., Cruz-Mireles, N., Molinari, C., Garduño-Rosales, M., Talbot, N.J., 2021. Investigating the cell and developmental biology of plant infection by the rice blast fungus *Magnaporthe oryzae*. Fungal Genet. Biol. 154, 103562.

Fan, Y.T., Chung, K.R., Huang, J.W., 2019. Fungichromin production by *Streptomyces padanus* PMS-702 for controlling cucumber downy mildew. Plant Pathol. J. 35 (4), 341.

Figueroa, M., Hammond-Kosack, K.E., Solomon, P.S., 2018. A review of wheat diseases—a field perspective. Mol. Plant Pathol. 19 (6), 1523–1536.

Frisvad, J.C., Møller, L.L., Larsen, T.O., Kumar, R., Arnau, J., 2018. Safety of the fungal workhorses of industrial biotechnology: update on the mycotoxin and secondary metabolite potential of *Aspergillus niger, Aspergillus oryzae,* and *Trichoderma reesei*. Appl. Microbiol. Biotechnol. 102 (22), 9481–9515.

Fudou, R., Iizuka, T., Yamanaka, S., 2001. Haliangicin, a novel antifungal metabolite produced by a marine myxobacterium 1. Fermentation and biological characteristics. J. Antibiot. 54 (2), 149–152.

Fujita, K., Sugiyama, R., Nishimura, S., Ishikawa, N., Arai, M.A., Ishibashi, M., Kakeya, H., 2016. Stereochemical assignment and biological evaluation of BE-14106 unveils the importance of one acetate unit for the antifungal activity of polyene macrolactams. J. Nat. Prod. 79 (7), 1877–1880.

Fujita, K., Ikuta, M., Nishimura, S., Sugiyama, R., Yoshimura, A., Kakeya, H., 2021. Amphiol, an antifungal fungal pigment from *Pseudogymnoascus* sp. PF1464. J. Nat. Prod. 84 (4), 986–992.

Fukuda, T., Shimoyama, K., Nagamitsu, T., Tomoda, H., 2014. Synthesis and biological activity of Citridone A and its derivatives. J. Antibiot. 67 (6), 445–450.

Gálvez, A., Maqueda, M., Martínez-Bueno, M., Lebbadi, M., Valdivia, E., 1993. Isolation and Physico-chemical characterization of an antifungal and antibacterial peptide produced by *Bacillus licheniformis* A12. Appl. Microbiol. Biotechnol. 39 (4), 438–442.

Gomes, A.R., Duarte, A.C., Rocha-Santos, T.A.P., 2017. Analytical techniques for discovery of bioactive compounds from marine fungi. In: Mérillon, J.M., Ramawat, K. (Eds.), Fungal Metabolites. Reference Series in Phytochemistry. Springer, Cham, pp. 415–434.

Gong, L., Jiang, Y., Chen, F., 2015. Molecular strategies for detection and quantification of mycotoxin-producing *Fusarium* species: a review. J. Sci. Food Agric. 95 (9), 1767–1776.

Gong, L., Tan, H., Chen, F., Li, T., Zhu, J., Jian, Q., Yuan, D., Xu, L., Hu, W., Jiang, Y., Duan, X., 2016. Novel synthesized 2,4-DAPG analogs: antifungal activity, mechanism, and toxicology. Sci. Rep. 6 (1), 1–9.

Groenhagen, U., Baumgartner, R., Bailly, A., Gardiner, A., Eberl, L., Schulz, S., Weisskopf, L., 2013. Production of bioactive volatiles by different *Burkholderia ambifaria* strains. J. Chem. Ecol. 39 (7), 892–906.

Gulder, T.A., Hong, H., Correa, J., Egereva, E., Wiese, J., Imhoff, J.F., Gross, H., 2012. Isolation, structure elucidation and total synthesis of lajollamide A from the marine fungus *Asteromyces cruciatus*. Mar. Drugs 10 (12), 2912–2935.

Haga, A., Tamoto, H., Ishino, M., Kimura, E., Sugita, T., Kinoshita, K., Koyama, K., 2013. Pyridone alkaloids from a marine-derived fungus, *Stagonosporopsis cucurbitacearum*, and their activities against azole-resistant *Candida albicans*. J. Nat. Prod. 76 (4), 750–754.

Hamed, A., Abdel-Razek, A.S., Frese, M., Wibberg, D., El-Haddad, A.F., Ibrahim, T.M., Kalinowski, J., Sewald, N., Shaaban, M., 2018. N-Acetylborrelidin B: A new bioactive metabolite from *Streptomyces mutabilis* sp. MII. Z. Naturforsch. 73 (1–2), 49–57.

Hein, S.M., Gloer, J.B., Koster, B., Malloch, D., 2001. Bombardolides: new antifungal and antibacterial γ-lactones from the coprophilous fungus *Bombardioidea anartia*. J. Nat. Prod. 64 (6), 809–812.

Hiramatsu, F., Miyajima, T., Murayama, T., Takahashi, K., Koseki, T., Shiono, Y., 2006. Isolation and structure elucidation of neofusapyrone from a marine-derived *Fusarium* species, and structural revision of fusapyrone and deoxyfusapyrone. J. Antibiot. 59 (11), 704–709.

Hosoe, T., Mori, N., Kamano, K., Itabashi, T., Yaguchi, T., Kawai, K.I., 2011. A new antifungal yellow pigment from *Aspergillus nishimurae*. J. Antibiot. 64 (2), 211–212.

Hwang, B.K., Kim, B.S., 1995. In-vivo efficacy and in-vitro activity of tubercidin, an antibiotic nucleoside, for control of *Phytophthora capsici* blight in *Capsicum annuum*. Pestic. Sci. 44 (3), 255–260.

Igarashi, M., Kinoshita, N., Ikeda, T., Nakagawa, E., Hamada, M., Takeuchi, T., 1997. Formamicin, a novel antifungal antibiotic produced by a strain of *Saccharothvix* sp. I. Taxonomy, production, isolation and biological properties. J. Antibiot. 50 (11), 926–931.

Igarashi, M., Sawa, R., Yamasaki, M., Hayashi, C., Umekita, M., Hatano, M., Fujiwara, T., Mizumoto, K., Nomoto, A., 2017. Kribellosides, novel RNA 5′-triphosphatase inhibitors from the rare actinomycete *Kribbella* sp. MI481-42F6. J. Antibiot. 70 (5), 582–589.

Ismet, A., Vikineswary, S., Paramaswari, S., Wong, W.H., Ward, A., Seki, T., Fiedler, H.P., Goodfellow, M., 2004. Production and chemical characterization of antifungal metabolites from *Micromonospora* sp. M39 isolated from mangrove rhizosphere soil. World J. Microbiol. Biotechnol. 20 (5), 523–528.

Isono, K., Nagatsu, J., Kobinata, K., Sasaki, K., Suzuki, S., 1967. Studies on polyomixins, antifungal antibiotics: Part V. Isolation and characterization of polyoxins C, D, E, F, G, H and I. Agric. Biol. Chem. 31 (2), 190–199.

Ivanov, M., Ćirić, A., Stojković, D., 2022. Emerging antifungal targets and strategies. Int. J. Mol. Sci. 23 (5), 2756.

Iwasa, T., Suetomi, K., Kusaka, T., 1978. Taxonomic study and fermentation of producing organism and antimicrobial activity of mildiomycin. J. Antibiot. 31 (6), 511–518.

Jain, A., Sarsaiya, S., Wu, Q., Lu, Y., Shi, J., 2019. A review of plant leaf fungal diseases and its environment speciation. Bioengineered 10 (1), 409–424. https://doi.org/10.1080/21655979.2019.1649520.

Jang, J.H., Kanoh, K., Adachi, K., Shizuri, Y., 2006. New dihydrobenzofuran derivative, awajanoran, from marine-derived *Acremonium* sp. AWA16-1. J. Antibiot. 59 (7), 428–431.

Jenul, C., Sieber, S., Daeppen, C., Mathew, A., Lardi, M., Pessi, G., Hoepfner, D., Neuburger, M., Linden, A., Gademann, K., Eberl, L., 2018. Biosynthesis of fragin is controlled by a novel quorum sensing signal. Nat. Commun. 9 (1), 1–13.

Jumpathong, W., Intra, B., Euanorasetr, J., Wanapaisan, P., 2022. Biosurfactant-producing *Bacillus velezensis* PW192 as an anti-fungal biocontrol agent against *Colletotrichum gloeosporioides* and *Colletotrichum musae*. Microorganisms 10 (5), 1017.

Kalra, R., Conlan, X.A., Goel, M., 2020. Fungi as a potential source of pigments: harnessing filamentous fungi. Front. Chem. 8, 369.

Kang, J.G., Shin, S.Y., Kim, M.J., Bajpai, V., Maheshwari, D.K., Kang, S.C., 2004. Isolation and anti-fungal activities of 2-hydroxymethyl-chroman-4-one produced by *Burkholderia* sp. MSSP. J. Antibiot. 57 (11), 726–731.

Karki, H.S., Shrestha, B.K., Han, J.W., Groth, D.E., Barphagha, I.K., Rush, M.C., Melanson, R.A., Kim, B.S., Ham, J.H., 2012. Diversities in virulence, antifungal activity, pigmentation and DNA fingerprint among strains of *Burkholderia glumae*. PloS One 7 (9), e45376.

Kaur, P., Kaur, S., Sood, A.S., 2022. Opportunistic fungal infections in immunocompromised patients of a tertiary care center, Amritsar, India. Nat. J. Lab. Med. 11 (1), MO30–MO33.

Kaur, J., Nobile, C.J., 2023. Antifungal drug-resistance mechanisms in *Candida biofilms*. Curr. Opinion Microbiol. 71, 102237.

Kim, J.Y., 2016. Human fungal pathogens: why should we learn? J. Microbiol. 54 (3), 145–148.

Kim, B.S., Moon, S.S., Hwang, B.K., 2000. Structure elucidation and antifungal activity of an anthracycline antibiotic, daunomycin, isolated from *Actinomadura roseola*. J. Agric. Food Chem. 48 (5), 1875–1881.

Kim, C., Jung, H., Kim, Y.O., Shin, C.S., 2006. Antimicrobial activities of amino acid derivatives of Monascus pigments. FEMS Microbiol. Lett. 264 (1), 117–124.

Kim, Y.S., Kim, H.M., Chang, C., Hwang, I.C., Oh, H., Ahn, J.S., Kim, K.D., Hwang, B.K., Kim, B.S., 2007. Biological evaluation of neopeptins isolated from a *Streptomyces* strain. Pest Manag. Sci. 63 (12), 1208–1214.

Kim, P.I., Ryu, J.W., Kim, Y.H., Chi, Y.T., 2010. Production of biosurfactant lipopeptides iturin A, fengycin, and surfactin A from *Bacillus subtilis* CMB32 for control of *Colletotrichum gloeosporioides*. J. Microbiol. Biotechnol. 20 (1), 138–145.

Kodani, S., Murao, A., Hidaki, M., Sato, K., Ogawa, N., 2012. Isolation and structural determination of a new macrolide, makinolide, from the newly isolated *Streptomyces* sp. MK-30. J. Antibiot. 65 (6), 331–334.

Kumar, V., Ahluwalia, V., Saran, S., Kumar, J., Patel, A.K., Singhania, R.R., 2021. Recent developments on solid-state fermentation for production of microbial secondary metabolites: Challenges and solutions. Bioresour. Technol. 323, 124566.

Kunihiro, S., Kaneda, M., 2003. Glomecidin, a novel antifungal cyclic tetrapeptide produced by *Streptomyces lavendulae* H698 SY2. J. Antibiot. 56 (1), 30–33.

Kurata, A., Yamaura, Y., Tanaka, T., Kato, C., Nakasone, K., Kishimoto, N., 2017. Antifungal peptidic compound from the deep-sea bacterium *Aneurinibacillus* sp. YR247. World J. Microbiol. Biotechnol. 33 (4), 1–8.

Kwon, H.C., Kauffman, C.A., Jensen, P.R., Fenical, W., 2009. Marinisporolides, polyene-polyol macrolides from a marine actinomycete of the new genus Marinispora. J. Org. Chem. 74 (2), 675–684.

Lacret, R., Oves-Costales, D., Gómez, C., Díaz, C., De la Cruz, M., Pérez-Victoria, I., Vicente, F., Genilloud, O., Reyes, F., 2014. New ikarugamycin derivatives with antifungal and antibacterial properties from *Streptomyces zhaozhouensis*. Mar. Drugs 13 (1), 128–140.

Lavermicocca, P., Valerio, F., Visconti, A., 2003. Antifungal activity of phenyllactic acid against molds isolated from bakery products. Appl. Environ. Microbiol. 69 (1), 634–640.

Lebbadi, M., Galvez, A., Maqueda, M., Martínez-Bueno, M., Valdivia, E., 1994. Fungicin M4: a narrow spectrum peptide antibiotic from *Bacillus licheniformis* M-4. J. Appl. Bacteriol. 77 (1), 49–53.

Lee, S.Y., Tindwa, H., Lee, Y.S., Naing, K.W., Hong, S.H., Nam, Y., Kim, K.Y., 2012. Biocontrol of Anthracnose in Pepper Using Chitinase, β-1,3 Glucanase, and 2-Furancarboxaldehyde Produced by *Streptomyces cavourensis* SY224. J. Microbiol. Biotechnol. 22 (10), 1359–1366.

Lee, A.J., Cadelis, M.M., Kim, S.H., Swift, S., Copp, B.R., Villas-Boas, S.G., 2020. Epipyrone A, a broad-spectrum antifungal compound produced by *Epicoccum nigrum* ICMP 19927. Molecules 25 (24), 5997.

Lemke, A., Kiderlen, A.F., Kayser, O., 2005. Amphotericin B. Appl. Microbiol. Biotechnol. 68 (2), 151–162.

Li, W., Csukai, M., Corran, A., Crowley, P., Solomon, P.S., Oliver, R.P., 2008. Malayamycin, a new Streptomyces antifungal compound, specifically inhibits sporulation of *Stagonospora nodorum* (Berk) Castell and Germano, the cause of wheat glume blotch disease. Pest Manag. Sci. 64 (12), 1294–1302.

Lin, L., Xu, J., 2020. Fungal pigments and their roles associated with human health. J. Fungus. 6 (4), 280.

Liu, N.N., Köhler, J.R., 2016. Antagonism of fluconazole and a proton pump inhibitor against Candida albicans. Antimicrob. Agents Chemother. 60 (2), 1145–1147.

Liu, H., He, Y., Jiang, H., Peng, H., Huang, X., Zhang, X., Thomashow, L.S., Xu, Y., 2007a. Characterization of a phenazine-producing strain *Pseudomonas chlororaphis* GP72 with broad-spectrum antifungal activity from green pepper rhizosphere. Curr. Microbiol. 54 (4), 302–306.

Liu, Y., Chen, Z., Ng, T.B., Zhang, J., Zhou, M., Song, F., Lu, F., Liu, Y., 2007b. Bacisubin, an antifungal protein with ribonuclease and hemagglutinating activities from *Bacillus subtilis* strain B-916. Peptides 28 (3), 553–559.

Liu, K., Ding, H., Yu, Y., Chen, B., 2019. A cold-adapted chitinase-producing bacterium from Antarctica and its potential in biocontrol of plant pathogenic fungi. Mar. Drugs 17 (12), 695.

Lyu, A., Liu, H., Che, H., Yang, L., Zhang, J., Wu, M., Chen, W., Li, G., 2017. Reveromycins A and B from *Streptomyces* sp. 3–10: antifungal activity against plant pathogenic fungi in vitro and in a strawberry food model system. Front. Microbiol. 8, 550.

Ma, Z., Zhang, S., Zhang, S., Wu, G., Shao, Y., Mi, Q., Liang, J., Sun, K., Hu, J., 2020. Isolation and characterization of a new cyclic lipopeptide surfactin from a marine-derived *Bacillus velezensis* SH-B74. J. Antibiot. 73 (12), 863–867.

Martinet, L., Naômé, A., Deflandre, B., Maciejewska, M., Tellatin, D., Tenconi, E., Smargiasso, N., De Pauw, E., van Wezel, G.P., Rigali, S., 2019. A single biosynthetic gene cluster is responsible for the production of bagremycin antibiotics and ferroverdin iron chelators. MBio 10 (4), e01230-19.

McKeny, P.T., Nessel, T.A., Zito, P.M., 2021. Antifungal antibiotics. In: Stat Pearls [Internet]. StatPearls Publishing, Treasure Island (FL).

Mishra, J., Mishra, I., Arora, N.K., 2022. 2,4-Diacetylphloroglucinol producing *Pseudomonas fluorescens* JM-1 for management of ear rot disease caused by *Fusarium moniliforme* in *Zea mays* L. 3 Biotech. 12 (6), 1–15.

Mukherjee, P.K., Horwitz, B.A., Kenerley, C.M., 2012. Secondary metabolism in Trichoderma—a genomic perspective. Microbiology 158 (1), 35–45.

Nadkarni, S.R., Mukhopadhyay, T., Bhat, R.G., Gupte, S.V., Sachse, B., 1998. Mathemycin A, a new antifungal macrolactone from *Actinomycete* sp. HIL Y-862095 I. Fermentation, isolation, physico-chemical properties and biological activities. J. Antibiot. 51 (6), 579–581.

Nagai, K., Kamigiri, K., Matsumoto, H., Kawano, Y., Yamaoka, M., Shimoi, H., Watanabe, M., Suzuki, K., 2002. YM-202204, a new antifungal antibiotic produced by marine fungus *Phoma* sp. J. Antibiot. 55 (12), 1036–1041.

Nagaoka, T., Nakata, K., Kouno, K., 2004. Antifungal activity of oosporein from an antagonistic fungus against *Phytophthora infestans*. Z. Naturforsch. 59 (3–4), 302–304.

Nair, M.G., Chandra, A., Thorogood, D.L., Ammermann, E., Walker, N., Kiehs, K., 1994. Gopalamicin, an antifungal macrodiolide produced by soil actinomycetes. J. Agric. Food Chem. 42 (10), 2308–2310.

Naz, R., Khushhal, S., Asif, T., Mubeen, S., Saranraj, P., Sayyed, R.Z., 2022. Inhibition of bacterial and fungal phytopathogens through volatile organic compounds produced by Pseudomonas sp. In: Secondary Metabolites and Volatiles of PGPR in Plant-Growth Promotion. Springer, Cham, pp. 95–118.

Nishii, M., Isono, K., Izaki, K., 1981. Inhibition of microbial cell wall synthesis by lipopeptin A. Agric. Biol. Chem. 45 (4), 895–902.

Oh, D.C., Poulsen, M., Currie, C.R., Clardy, J., 2011. Sceliphrolactam, a polyene macrocyclic lactam from a wasp-associated *Streptomyces* sp. Org. Lett. 13 (4), 752–755.

Ortiz-Lopez, F.J., Monteiro, M.C., Gonzalez-Menendez, V., Tormo, J.R., Genilloud, O., Bills, G.F., Vicente, F., Zhang, C., Roemer, T., Singh, S.B., Reyes, F., 2015. Cyclic colisporifungin and linear cavinafungins, antifungal lipopeptides isolated from *Colispora cavincola*. J. Nat. Prod. 78 (3), 468–475.

Pan, H.Q., Yu, S.Y., Song, C.F., Wang, N., Hua, H.M., Hu, J.C., Wang, S.J., 2015. Identification and characterization of the antifungal substances of a novel *Streptomyces cavourensis* NA4. J. Microbiol. Biotechnol. 25 (3), 353–357.

Park, J.H., Choi, G.J., Jang, K.S., Lim, H.K., Kim, H.T., Cho, K.Y., Kim, J.C., 2005a. Antifungal activity against plant pathogenic fungi of chaetoviridins isolated from *Chaetomium globosum*. FEMS Microbiol. Lett. 252 (2), 309–313.

Park, J.H., Choi, G.J., Lee, S.W., Lee, H.B., Kim, K.M., Jung, H.S., Jang, K.S., Cho, K.Y., Kim, J.C., 2005b. Griseofulvin from *Xylaria* sp. strain F0010, an endophytic fungus of *Abies holophylla* and its antifungal activity against plant pathogenic fungi. J. Microbiol. Biotechnol. 15 (1), 112–117.

Park, Y.C., Gunasekera, S.P., Lopez, J.V., McCarthy, P.J., Wright, A.E., 2006. Metabolites from the marine-derived fungus *Chromocleista* sp. isolated from a deep-water sediment sample collected in the Gulf of Mexico. J. Nat. Prod. 69 (4), 580–584.

Peláez, F., Cabello, A., Platas, G., Díez, M.T., del Val, A.G., Basilio, A., Martán, I., Vicente, F., Bills, G.F., Giacobbe, R.A., Schwartz, R.E., 2000. The discovery of enfumafungin, a novel antifungal compound produced by an endophytic *Hormonema* species biological activity and taxonomy of the producing organisms. Syst. Appl. Microbiol. 23 (3), 333–343.

Pimienta, D.A., Cruz Mosquera, F.E., Palacios Velasco, I., Giraldo Rodas, M., Oñate-Garzón, J., Liscano, Y., 2023. Specific focus on antifungal peptides against azole resistant *Aspergillus fumigatus*: current status, challenges, and future perspectives. J. Fungi 9 (1), 42.

Preindl, J., Schulthoff, S., Wirtz, C., Lingnau, J., Fürstner, A., 2017. Polyunsaturated C-glycosidic 4-hydroxy-2-pyrone derivatives: Total synthesis shows that putative orevactaene is likely identical with epipyrone A. Angew. Chem. Int. Ed. 56 (26), 7525–7530.

Rančić, A., Soković, M., Karioti, A., Vukojević, J., Skaltsa, H., 2006. Isolation and structural elucidation of two secondary metabolites from the filamentous fungus *Penicillium ochrochloron* with antimicrobial activity. Environ.Toxicol. Pharmacol. 22 (1), 80–84.

Rashad, Y.M., Moussa, T.A., 2020. Biocontrol agents for fungal plant diseases management. In: Cottage Industry of Biocontrol Agents and Their Applications. Springer, Cham, pp. 337–363.

Rathod, D.P., Dar, M.A., Gade, A.K., Rai, M.K., Baba, G., 2014. Griseofulvin producing endophytic *Nigrospora oryzae* from Indian *Emblica officinalis* Gaertn: a new report. Austin J. Biotechnol. Bioeng. 1 (3), 1–5.

Reddy, S., Sinha, A., Osborne, W.J., 2021. Microbial secondary metabolites: recent developments and technological challenges. In: Volatiles and Metabolites of Microbes. Academic Press, pp. 1–22. https://doi.org/10.1016/B978-0-12-824523-1.00007-9.

Rothrock, C.S., Gottlieb, D., 1984. Role of antibiosis in antagonism of *Streptomyces hygroscopicus* var. geldanus to *Rhizoctonia solani* in soil. Can. J. Microbiol. 30 (12), 1440–1447.

Sangkanu, S., Rukachaisirikul, V., Suriyachadkun, C., Phongpaichit, S., 2021. Antifungal activity of marine-derived actinomycetes against *Talaromyces marneffei*. J. Appl. Microbiol. 130 (5), 1508–1522.

Sansinenea, E., Ortiz, A., 2011. Secondary metabolites of soil *Bacillus* spp. Biotechnol. Lett. 33 (8), 1523–1538.

Sasidharan, A., Sasidharan, N.K., Amma, D.B.N.S., Vasu, R.K., Nataraja, A.V., Bhaskaran, K., 2015. Antifungal activity of violacein purified from a novel strain of *Chromobacterium* sp. NIIST (MTCC 5522). J. Microbiol. 53 (10), 694–701.

Sato, S., Iwata, F., Yamada, S., Katayama, M., 2012. Neomaclafungins A–I: oligomycin-class macrolides from a marine-derived actinomycete. J. Nat. Prod. 75 (11), 1974–1982.

Satomi, T., Kusakabe, H., Nakamura, G., Nishio, T., Uramoto, M., Isono, K., 1982. Neopeptins A and B, new antifungal antibiotics from *Actinomyces malachiticus*. Agric. Biol. Chem. 46 (10), 2621–2623.

Scorzoni, L., de Paula Silva, A.C., Marcos, C.M., Assato, P.A., de Melo, W.C., de Oliveira, H.C., Costa-Orlandi, C.B., Mendes-Giannini, M.J., Fusco-Almeida, A.M., 2017. Antifungal therapy: new advances in the understanding and treatment of mycosis. Front. Microbiol. 8, 36.

Sen, S., Borah, S.N., Bora, A., Deka, S., 2017. Production, characterization, and antifungal activity of a biosurfactant produced by *Rhodotorula babjevae* YS3. Microb. Cell Fact. 16 (1), 1–14.

Seyedmousavi, S., Rafati, H., Ilkit, M., Tolooe, A., Hedayati, M.T., Verweij, P., 2017. Systemic antifungal agents: current status and projected future developments. In: Human Fungal Pathogen Identification. Humana Press, New York, NY, pp. 107–139.

Shao, C., Guo, Z., Peng, H., Peng, G., Huang, Z., She, Z., Lin, Y., Zhou, S., 2007. A new isoprenyl phenyl ether compound from mangrove fungus. Chem. Nat. Compd. 43 (4), 377–380.

Shih, H.D., Liu, Y.C., Hsu, F.L., Mulabagal, V., Dodda, R., Huang, J.W., 2003. Fungichromin: a substance from *Streptomyces padanus* with inhibitory effects on *Rhizoctonia solani*. J. Agric. Food Chem. 51 (1), 95–99.

Shin, D.S., Oh, M.N., Yang, H.C., Oh, K.B., 2005. Biological characterization of periconicins, bioactive secondary metabolites, produced by *Periconia* sp. OBW-15. J. Microbiol. Biotechnol. 15 (1), 216–220.

Shrivastava, P., Kumar, R., 2018. Actinobacteria: Eco-friendly candidates for control of plant diseases in a sustainable manner. In: New and Future Developments in Microbial Biotechnology and Bioengineering. Elsevier, pp. 79–91.

Singh, S., Dureja, P., Tanwar, R.S., Singh, A., 2005. Production and antifungal activity of secondary metabolites of *Trichoderma virens*. Pestic. Res. J. 17 (2), 26–29.

Sletta, H., Degnes, K.F., Herfindal, L., Klinkenberg, G., Fjærvik, E., Zahlsen, K., Brunsvik, A., Nygaard, G., Aachmann, F.L., Ellingsen, T.E., Døskeland, S.O., 2014. Anti-microbial and cytotoxic 1,6-dihydroxy-phenazine-5,10-dioxide (iodinin) produced by *Streptosporangium* sp. DSM 45942 isolated from the fjord sediment. Appl. Microbiol. Biotechnol. 98 (2), 603–610.

Smith, J., Putnam, A., Nair, M., 1990. In vitro control of Fusarium diseases of *Asparagus officinalis* L. with a Streptomyces or its polyene antibiotic. Faeriefungin. J. Agric. Food Chem. 38 (8), 1729–1733.

Snipes, C.E., Duebelbeis, D.O., Olson, M., Hahn, D.R., Dent Iii, W.H., Gilbert, J.R., Werk, T.L., Davis, G.E., Lee-Lu, R., Graupner, P.R., 2007. The ansacarbamitocins: polar ansamitocin derivatives. J. Nat. Prod. 70 (10), 1578–1581.

Soberón-Chávez, G., Maier, R.M., 2011. Biosurfactants: a general overview. In: Soberón-Chávez, G. (Ed.), Biosurfactants. Microbiology Monographs. vol. 20. Springer, Berlin, Heidelberg, pp. 1–11.

Solecka, J., Zajko, J., Postek, M., Rajnisz, A., 2012. Biologically active secondary metabolites from Actinomycetes. Open Life Sci. 7 (3), 373–390.

Song, F., Dai, H., Tong, Y., Ren, B., Chen, C., Sun, N., Liu, X., Bian, J., Liu, M., Gao, H., Liu, H., 2010. Trichodermaketones A–D and 7-O-Methylkoninginin D from the marine fungus *Trichoderma koningii*. J. Nat. Prod. 73 (5), 806–810.

Spencer, A.C., Brubaker, K.R., Garneau-Tsodikova, S., 2023. Systemic fungal infections: a pharmacist/researcher perspective. Fungal Biol. Rev. 44, 100293.

Strobel, G.A., Miller, R.V., Martinez-Miller, C., Condron, M.M., Teplow, D.B., Hess, W.M., 1999. Cryptocandin, a potent antimycotic from the endophytic fungus *Cryptosporiopsis* cf. quercina. Microbiology 145 (8), 1919–1926.

Sudhakar, T., Karpagam, S., 2011. Antifungal efficacy of pyocyanin produced from bioindicators of nosocomial hazards. In: International Conference on Green technology and environmental Conservation (GTEC-2011), December. IEEE, pp. 224–229.

Sugiyama, R., Nishimura, S., Matsumori, N., Tsunematsu, Y., Hattori, A., Kakeya, H., 2014. Structure and biological activity of 8-deoxyheronamide C from a marine-derived *Streptomyces* sp.: Heronamides target saturated hydrocarbon chains in lipid membranes. J. Am. Chem. Soc. 136 (14), 5209–5212.

Tanaka, Y., Hirata, K., Takahashi, Y., Iwai, Y., Omura, S., 1987. Globopeptin, a new antifungal peptide antibiotic. J. Antibiot. 40 (2), 242–244.

Tareq, F.S., Lee, H.S., Lee, Y.J., Lee, J.S., Shin, H.J., 2015. Ieodoglucomide C and ieodoglycolipid, new glycolipids from a marine-derived bacterium *Bacillus licheniformis* 09IDYM23. Lipids 50 (5), 513–519.

Thirumurugan, D., Cholarajan, A., Raja, S.S., Vijayakumar, R., 2018. An introductory chapter: secondary metabolites. Second metab—sources Appl., 1–21.

Toumatia, O., Yekkour, A., Goudjal, Y., Riba, A., Coppel, Y., Mathieu, F., Sabaou, N., Zitouni, A., 2015. Antifungal properties of an actinomycin D-producing strain, *Streptomyces* sp. IA1, isolated from a Saharan soil. J. Basic Microbiol. 55 (2), 221–228.

Trejo-Estrada, S.R., Sepulveda, I.R., Crawford, D.L., 1998. In vitro and in vivo antagonism of *Streptomyces violaceusniger* YCED9 against fungal pathogens of turfgrass. World J. Microbiol. Biotechnol. 14 (6), 865–872.

Trisuwan, K., Rukachaisirikul, V., Sukpondma, Y., Phongpaichit, S., Preedanon, S., Sakayaroj, J., 2009. Lactone derivatives from the marine-derived fungus *Penicillium* sp. PSU-F44. Chem. Pharm. Bull. 57 (10), 1100–1102.

Troch, V., Audenaert, K., Wyand, R.A., Haesaert, G., Höfte, M., Brown, J.K., 2014. Formae speciales of cereal powdery mildew: close or distant relatives? Mol. Plant Pathol. 15 (3), 304–314.

Tu, P.W., Chiu, J.S., Lin, C., Chien, C.C., Hsieh, F.C., Shih, M.C., Yang, Y.L., 2022. Evaluation of the antifungal activities of *Photorhabdus akhurstii* and its secondary metabolites against phytopathogenic *Colletotrichum gloeosporioides*. J. Fungus. 8 (4), 403.

Ubukata, M., Uramoto, M., Uzawa, J., Isono, K., 1986. Structure and biological activity of neopeptins A, B, and C, inhibitors of fungal cell wall glycan synthesis. Agric. Biol. Chem. 50 (2), 357–365.

Umezawa, H., Okami, Y., Hashimoto, T., Suhara, Y., Hamada, M., Takeuchi, T., 1965. A new antibiotic, kasugamycin. J. Antibiot. Series A 18 (2), 101–103.

Uzair, B., Bano, A., Niazi, M.B.K., Khan, F., Mujtaba, G., 2018. In vitro antifungal activity of 9, 10-dihydrophenanthrene-2-carboxylic acid isolated from a marine bacterium: *Pseudomonas putida*. Pak. J. Pharm. Sci. 31 (6), 2733–2736.

Vaishnav, P., Demain, A.L., 2011. Unexpected applications of secondary metabolites. Biotechnol. Adv. 29 (2), 223–229.

Vicente dos Reis, G., Abraham, W.R., Grigoletto, D.F., de Campos, J.B., Marcon, J., da Silva, J.A., Quecine, M.C., de Azevedo, J.L., Ferreira, A.G., de Lira, S.P., 2019. Gloeosporiocide, a new antifungal cyclic peptide from *Streptomyces morookaense* AM25 isolated from the Amazon bulk soil. FEMS Microbiol. Lett. 366 (14), fnz175.

Victoria Castelli, M., Gabriel Derita, M., Noeli Lopez, S., 2017. Novel antifungal agents: a patent review (2013-present). Expert Opin. Ther. Patent. 27 (4), 415–426.

Vining, L.C., 2007. Roles of secondary metabolites from microbes. In: Ciba Foundation Symposium 171-Secondary Metabolites: their Function and Evolution: Secondary Metabolites: Their Function and Evolution: Ciba Foundation Symposium 171. John Wiley & Sons, Ltd., Chichester, UK, pp. 184–198.

Wan, Z., Fang, W., Shi, L., Wang, K., Zhang, Y., Zhang, Z., Wu, Z., Yang, Z., Gu, Y., 2015. Novonestmycins A and B, two new 32-membered bioactive macrolides from *Streptomyces phytohabitans* HBERC-20821. J. Antibiot. 68 (3), 185–190.

Wang, Y., Gloer, J.B., Scott, J.A., Malloch, D., 1993. Appenolides AC: three new antifungal furanones from the coprophilous fungus *Podospora appendiculata*. J. Nat. Prod. 56 (3), 341–344.

Wang, H.J., Gloer, K.B., Gloer, J.B., Scott, J.A., Malloch, D., 1997. Anserinones A and B: new antifungal and antibacterial benzoquinones from the coprophilous fungus *Podospora anserina*. J. Nat. Prod. 60 (6), 629–631.

Wang, W., Wang, Y., Tao, H., Peng, X., Liu, P., Zhu, W., 2009. Cerebrosides of the halotolerant fungus *Alternaria raphani* isolated from a sea salt field. J. Nat. Prod. 72 (9), 1695–1698.

Wang, X., You, J., King, J.B., Powell, D.R., Cichewicz, R.H., 2012. Waikialoid A suppresses hyphal morphogenesis and inhibits biofilm development in pathogenic *Candida albicans*. J. Nat. Prod. 75 (4), 707–715.

Whyte, A.C., Gloer, J.B., Scott, J.A., Malloch, D., 1996. Cercophorins A–C: Novel antifungal and cytotoxic metabolites from the coprophilous fungus *Cercophora areolata*. J. Nat. Prod. 59 (8), 765–769.

Williamson, B., Tudzynski, B., Tudzynski, P., Van Kan, J.A., 2007. *Botrytis cinerea*: the cause of grey mould disease. Mol. Plant Pathol. 8 (5), 561–580.

Wu, B., Hussain, M., Zhang, W., Stadler, M., Liu, X., Xiang, M., 2019. Current insights into fungal species diversity and perspective on naming the environmental DNA sequences of fungi. Mycology 10 (3), 127–140.

Xu, J., Takasaki, A., Kobayashi, H., Oda, T., Yamada, J., Mangindaan, R.E., Ukai, K., Nagai, H., Namikoshi, M., 2006. Four new macrocyclic trichothecenes from two strains of marine-derived fungi of the genus *Myrothecium*. J. Antibiot. 59 (8), 451–455.

Xu, W., Zhang, D., Si, C., Tao, L., 2013. Antifungal macrolides from *Streptomyces cavourensis* YY01-17. Chem. Nat. Compd. 49, 988–989.

Yan, L.L., Han, N.N., Zhang, Y.Q., Yu, L.Y., Chen, J., Wei, Y.Z., Li, Q.P., Tao, L., Zheng, G.H., Yang, S.E., Jiang, C.X., 2010. Antimycin A18 produced by an endophytic *Streptomyces albidoflavus* isolated from a mangrove plant. J. Antibiot. 63 (5), 259–261.

Yang, S.W., Chan, T.M., Terracciano, J., Loebenberg, D., Patel, M., Gullo, V., Chu, M., 2009. Sch 1385568, a new azaphilone from *Aspergillus* sp. J. Antibiot. 62 (7), 401–403.

Yang, P.W., Li, M.G., Zhao, J.Y., Zhu, M.Z., Shang, H., Li, J.R., Cui, X.L., Huang, R., Wen, M.L., 2010. Oligomycins A and C, major secondary metabolites isolated from the newly isolated strain *Streptomyces diastaticus*. Folia Microbiol. 55 (1), 10–16.

Yang, Y.H., Fu, X.L., Li, L.Q., Zeng, Y., Li, C.Y., He, Y.N., Zhao, P.J., 2012. Naphthomycins L–N, ansamycin antibiotics from *Streptomyces* sp. CS. J. Nat. Prod. 75 (7), 1409–1413.

Yang, J., Yang, Z., Yin, Y., Rao, M., Liang, Y., Ge, M., 2016. Three novel polyene macrolides isolated from cultures of *Streptomyces lavenduligriseus*. J. Antibiot. 69 (1), 62–65.

Yang, X.F., Wang, N.N., Kang, Y.F., Ma, Y.M., 2019. A new furan derivative from an endophytic *Aspergillus tubingensis* of *Decaisnea insignis* (Griff.) Hook. f. & Thomson. Nat. Prod. Res. 33 (19), 2777–2783.

Yi, W., Qin, L., Lian, X.Y., Zhang, Z., 2020. New antifungal metabolites from the mariana trench sediment-associated actinomycete *Streptomyces* sp. SY1965. Mar. Drugs 18 (8), 385.

Zhang, D.J., Wei, G., Wang, Y., Si, C.C., Tian, L., Tao, L.M., Li, Y.G., 2011. Bafilomycin K, a new antifungal macrolide from *Streptomyces flavotricini* Y12-26. Antibiotics 64 (5), 391–393.

Zhang, B., Dong, C., Shang, Q., Cong, Y., Kong, W., Li, P., 2013a. Purification and partial characterization of bacillomycin L produced by *Bacillus amyloliquefaciens* K103 from lemon. Appl. Biochem. Biotechnol. 171 (8), 2262–2272.

Zhang, Y.L., Li, S., Jiang, D.H., Kong, L.C., Zhang, P.H., Xu, J.D., 2013b. Antifungal activities of metabolites produced by a termite-associated *Streptomyces canus* BYB02. J. Agric. Food Chem. 61 (7), 1521–1524.

Zhang, P., Mandi, A., Li, X.M., Du, F.Y., Wang, J.N., Li, X., Kurtan, T., Wang, B.G., 2014. Varioxepine A, a 3 H-oxepine-containing alkaloid with a new oxa-cage from the marine algal-derived endophytic fungus *Paecilomyces variotii*. Org. Lett. 16 (18), 4834.

Zhang, S., Gui, C., Shao, M., Kumar, P.S., Huang, H., Uu, J., 2020a. Antimicrobial tunicamycin derivatives from the deep sea-derived *Streptomyces xinghaiensis* SCSIO S15077. Nat. Prod. Res. 11, 1499–1504.

Zhang, F., Zhao, M., Braun, D.R., Ericksen, S.S., Piotrowski, J.S., Nelson, J., Peng, J., Ananiev, G.E., Chanana, S., Barns, K., Fossen, J., 2020b. A marine microbiome antifungal targets urgent-threat drug-resistant fungi. Science 370 (6519), 974–978.

Zhao, T., Xu, L.L., Zhang, Y., Lin, Z.H., Xia, T., Yang, D.F., Chen, Y.M., Yang, X.L., 2019. Three new α-pyrone derivatives from the plant endophytic fungus *Penicillium ochrochloronthe* and their antibacterial, antifungal, and cytotoxic activities. J. Asian Nat. Prod. Res. 21 (9), 851–858.

Zheng, J., Zhu, H., Hong, K., Wang, Y., Liu, P., Wang, X., Peng, X., Zhu, W., 2009. Novel cyclic hexapeptides from marine-derived fungus, *Aspergillus sclerotiorum* PT06-1. Org. Lett. 11 (22), 5262–5265.

CHAPTER

9

Fungal metabolites as novel plant pathogen antagonists

Jagriti Singh, Shweta Mishra, and Vineeta Singh

Department of Biotechnology, Institute of Engineering and Technology, Dr. A.P.J. Abdul Kalam Technical University, Lucknow, Uttar Pradesh, India

1 Introduction

Plant pathogens including pests are one of the major concerns for agricultural fields as they directly affect the productivity of infected areas, which in turn affects the economy of a country. In addition, the remains of pathogen infections such as toxins in crop areas and postharvest bins are also causing a threat to humans and livestock (Rizzo et al., 2021; Stoev, 2015). Various approaches ranging from physical methods (heat treatment, ultraviolet irradiation, refrigeration) to chemical (chemically synthesized biocontrol agents) to genetically engineered (development of tolerant cultivars through plant breeding, genetically modified plants) are being used to treat plant diseases (Thambugala et al., 2020). These strategies have offered significant improvement in overall crop productivity and quality. However, scientists are still searching for promising and natural approaches to control plant diseases. Microbial biocontrol agents (BCAs) are mostly fungal or bacterial strains that are isolated from the phyllosphere (surface or the apoplast of the leaf), endosphere (the internal area of plant tissue where endophytic microbes live), or rhizosphere (microbes residing in the soil of the root area of plants), and they can play an important role in controlling plant pathogenic organisms (Helepciuc and Todor, 2023). BCAs prevent pathogens from infecting host plants. The mechanisms involved in the regulation of these pathogens are most probably those that act primarily on pathogens (Bonaterra et al., 2022). Based on their nature, BCAs are categorized into four groups: (1) macroorganisms (e.g., predators, insects, nematodes), (2) microorganisms (e.g., bacteria, fungi, viruses), (3) chemical mediators such as pheromones, and (4) natural metabolites originating from plant or animal sources (Masi et al., 2018a,b). Biopesticides that are used for pest regulation are either the whole microorganism or a product of biological origin, most commonly the microbial metabolites produced during their growth and development

https://doi.org/10.1016/B978-0-443-23950-2.00012-6

phase (Kumar et al., 2021). Based on natural origin, biopesticides are intended to be more ecofriendly. Moreover, Masi et al. (2018b) mentioned that the half-life of biopesticides is usually shorter than that of chemical pesticides, but their activity spectrum makes them promising agents. Natural products with a new mechanism of action also serve as models for the development of synthetic or semisynthetic pesticides that have a significant impact on pest/pathogen control. However, stepwise research, development, and regulation are still required to increase the effective solutions in the global market of BCAs (Kumar et al., 2021).

Currently used strategies of pest management are based on synthetic solutions, and these are now showing a range of side effects after long-term use. Nature, being the diversity reservoir itself, is able to provide some solution to any problem. Hence, studies are ongoing to find some natural alternates with fewer side effects. Fungi are heterotrophic eukaryotes characterized by a chitinous cell wall, no phagotrophic capabilities, and cell organizations ranging from unicellular to complex macroscopic structures (Masi et al., 2018a,b). Fungi can produce many bioactive secondary metabolites belonging to different classes of natural products (Keller, 2019). Under different stress conditions, fungi are also able to modify these secondary metabolites to sustain themselves. These fungal metabolites play a very important role to combat plant and human pathogens. A number of these natural compounds with antagonistic activity against plant and human pathogens are further subjected to the development of semisynthetic derivatives with enhanced selectivity and stability toward natural products and reduced toxicity for potential practical applications (Newman and Cragg, 2020).

Fungi, which have relatively higher reproduction rates (sexual and asexual), shorter generation times, and specific target structures, showed potential for the application of fungal BCAs against plant pathogens. Furthermore, fungi can survive without a host in the environment and as per the requirement, it can switch from parasitic mode to saprotrophic mode to maintain sustainability. It's possible that endophytic fungi, which spend all or part of their lives inhabiting the healthy tissues of the host plant without inflicting any disease, have defense systems that protect plants from phytopathogenic fungi (Alam et al., 2021). This specific symbiotic relationship between plants and endophytes has specific biochemical metabolic pathways, as a result of which they produce some special metabolites with a specific biological activity in the medicinal and agrochemical fields (Sharma et al., 2021; Strobel, 2003). This might be due to the genetic recombination occurring between the endophytic fungi and the plant (Tan and Zou, 2001). The purpose of this chapter is to enlighten readers about the metabolites produced by fungal populations, which may serve as novel plant pathogen antagonists.

2 Plant pathogens and complexity of treatment

Plant pathology is a branch of science that deals with the detailed study of pathogens causing plant diseases, their mechanism, and also how plants can survive under unfavorable conditions or how plants will survive in the presence of disease-causing parasite bacteria. From the outside, this branch appears to be quite fascinating, but within it is still very challenging to regulate or advance without upsetting the balance of the ecosystem (Jeger et al., 2021). Plant diseases are the result of intricate interactions between pathogens, plants, and the environment (Sosa-Gómez and Moscardi, 1998). So, proper advancement of this branch can lead to options for appropriate farming and the development of nutrient-rich plants in cultivated regions. One of the biggest dangers to society's ability to develop sustainably is plant disease, which affects the staple crops of rice, wheat, maize, and potato and causes production losses of 13%–22% per year, costing the US economy billions of dollars (He et al., 2021). Of the average total loss, 14.1% is estimated to be due to disease, 10.2% to insects, and 12.2% to weeds (Secretariat et al., 2021). Because developed nations take greater precautions to defend against these factors, these losses are often lower in developed nations than in developing nations. Hence, it is necessary to revisit and incorporate current plant pathology principles into a broader quantitative framework that can be used across disciplines so that it can contribute significantly to our understanding of how and when diseases occur.

Pathogens may be viruses, bacteria, fungi, nematodes, or parasites that affect the various activities of the plant, which can be easily seen after a certain period of infections. These plant pathogens can attack different plant parts such as the leaf, stem, root, vascular system, or fruit and give rise to disease conditions. Factors responsible for the development of mechanisms of virulence vary from case to case; however, some of the important factors are: (1) an arrangement of plant genetic material, (2) and transmission and (3) modes (s) and broadcast and the life history (4) conservational influences in the different plant (Nishiguchi and Kobayashi, 2011). To achieve a certain level of infection, there must be several factors such as: (A) surface receptors of the plant, (B) the ability of a plant-based pathogen to differentially achieve full contamination manifesting as an overload of plant defenses in the plant, (C) plant pathogens interfere with the metabolic pathways of plant secondary metabolite production, and (D) they interfere in the biological, genetic, and various functions of the plant under the influence of different factors of contamination.

Microbes reside in a plant atmosphere in a community where more than one species interacts with each other, affecting the plant environment and vice versa. Results of these synergistic interactions sometimes lead to disease conditions commonly called a "complex disease" (Lamichhane and Venturi, 2015). Tomato pith necrosis is an example of this coinfection that arises either alone or due to synergistic interactions with eight bacterial pathogens, *Pseudomonas cichorii*, *Pseudomonas corrugata*, *Pseudomonas viridiflava*, *Pseudomonas mediterranea*, *Pseudomonas fluorescens*, *Pectobacterium atrosepticum*, *Pectobacterium carotovorum*, and *Dickeya chrysanthemi* (Lamichhane and Venturi, 2015). Similarly, there are also reports for fungal-fungal, viral-viral, and mixed microbial interactions that are responsible for various plant diseases. Some of them are summarized in Table 1.

These synergistic interactions of microbes create hurdles in the control of pathogens. As the interaction mechanism that microbes adopt in the community is not known, it is therefore difficult to design their control mechanism.

3 Fungal activities helpful in controlling plant pathogens

Humanity faces increasing difficulty as bacteria gain resistance to antimicrobial treatments, and natural products continue to offer vital scaffolds for the creation of substitute biologically active compounds. There are some factors involved in plant disease prevention such as understanding the mechanism of infection, selecting the right plants for the desired activity, using disease-resistant varieties, crop rotation, using different sanitizing tools,

TABLE 1 Some common plant diseases and their pathogens.

Disease	Plant affected	Pathogens	Reference
Bacterial community interaction			
Pith necrosis	Tomato	*Pseudomonas corrugate, P. mediterranea, P corrugata, P. marginalis*	Moura et al. (2005), Kůdela et al. (2010)
Wilt	Mulberry	*Enterobacter asburiae, Enterobacter* sp.	
Leaf spot	Sugar beet	*Xanthomonas* sp.	Mbega et al. (2012)
Head rot	Broccoli	*P. marginalis, Erwinia carotovora, P. fluorescens, P. viridiflava*	Canaday et al. (1991)
Fungal community interaction			
Sooty blotch and flyspeck	Apple	*Zygophiala* sp, *Microcyclospora, Microcyclosporella*	Batzer et al. (2008)
Sigatoka	Banana	*Mycosphaerella fijiensis, M. musicola, M. eumusae*	Arzanlou et al. (2007)
Root rot	Cassava	*Fusarium* sp., *Botryodiplodia theobromae, Armillaria* sp.	Bandyopadhyay et al. (2006)
Bare patch disease	Cereals	*Rhizoctonia* sp.	Roberts and Sivasithamparam (1986)
Virus community interactions			
Corn lethal necrosis	Corn	*Maize chlorotic mottle virus (MCMV) and wheat streak mosaic virus (WSMV)*	Niblett and Claflin (1978)
Cowpea stunt	Cowpea	*Cucumber mosaic virus (CMV) and black-eye cowpea mosaic virus (BCMV)*	Kline et al. (1997)
Blackberry yellow vein disease (BYVD).	Blackberry	*Tobacco ringspot virus, raspberry bush dwarf virus, and crinivirus*	Martin et al. (2013)
Motley dwarf	Carrot	*Carrot red leaf luteovirus and carrot mottle umbravirus*	Watson et al. (1998)
Mixed colony infections			
Gummy stem blight and black rot	Pumpkin	*Didymella bryoniae, Pectobacterium carotovorum, Pseudomonas viridiflava, P. syringae*, and *X. cucurbitae*	Grube et al. (2011)
Brown apical necrosis	Walnut	*Fusarium, Alternaria, Cladosporium, Colletotrichum, Phomopsis*, and *Xanthomonas*	Belisario et al. (2002)
Root rot	*Panax notoginseng*	*Alternaria* sp., *Cylindrocarpon* sp., *Fusarium* sp., *Phytophthora cactorum, Phoma herbarum, Rhizoctonia solani, Pseudomonas* sp., and *Ralstonia* sp.	Miao et al. (2006)
Root rot	Sugar beet	*Leuconostoc mesenteroides* subsp. *dextranicum, Lactobacillus, Gluconobacter*	Strausbaugh and Gillen (2008), Strausbaugh and Eujayl (2012)

maintaining the soil moisture, and providing proper mulching, pruning, and fertilizing. In addition, some specific compounds are also used to reduce infections in plants (Combarros-Fuertes et al., 2020; Pusztahelyi et al., 2015). These bioactive compounds can be classified on the basis of causing agents such as insecticides (act on insects), herbicides (act on herbs), fungicides (act on fungi), algaecides (act on algae), bactericides (act on bacteria), rodenticides (act on rodents), larvicides (act on larvae), repellents, virucides (act on virus), nematicides (act on nematodes), etc. Some pant pathogens as well as their hosts, antagonists, specific compounds, and the procedure of activity are shown in Table 2.

TABLE 2 Plant pathogens, their hosts, and antagonists with procedures.

Pathogens	Host	Antagonists	Procedure	Metabolites	Reference
Rhizoctonia solani	Bean plants	*Gliocladium* sp.	In-vitro cultivation and greenhouse	Polyketides, nonribosomal peptides	Bazgir et al. (1992)
Athelia rolfsii (*Sclerotium rolfsii*)	Groundnut	*Trichoderma harzianum*	Cultivation in greenhouse	Polyketides, alkaloids, terpenoids, peptaibols	Asghari and Myee (1992)
Fusarium solani	Apple plants	*T. koningii*, *T. viride*, *T. harzianum*, and *T. virens* (*Gliocladium virens*)	Cultivation in greenhouse	Naphthoquinone and aza-anthraquinone	Karampour and Okhovat (1992)
Colletotrichum coccodes	Potato plants	*Trichoderma* spp.	In vitro cultivation	Polyketides, alkaloids, terpenoids, and peptaibols	Okhovat et al. (1994)
R. solani	Rice plants	*T. koningii*, *T. viride*, *T. harzianum*, and *T. virens*	Field cultivation	Polyketides, nonribosomal peptides	Izadyar and Padasht (1994)
Phytophthora erythroseptica	Potato plants	*T. harzianum*, *T. viride*	In vitro cultivation	GEMs	Zafari et al. (1994)
Sclerotinia sclerotiorum	Eggplants	*T. reesei*, *T. hamatum*, *T. longibrachiatum*, *T. koningii*, *T. viride*, *T. virens*, and *Gliocladium* sp.	In vitro cultivation	Epipolythiodioxopiperazines	Amir-Sadeghi et al. (1994)
H. schachtii	Beet plants	*Paecilomyces fumosoroseus*	Cultivation in greenhouse	Hydroxyproline	Alizadeh et al. (2020)
*F. o. f.*sp. *cucumerinum*	Cucumber plants	*T. harzianum*	Cultivation in greenhouse	Acetylcholine, serotonin	Peyghami and Nishabouri (1998)
H. schachtii	Sugar beet plants	*P. farinosus*	In vitro cultivation	Glutathione	Ahmadi et al. (1996)

1. Types of nanohybrid fungicides

3.1 Bactericides

These are chemical agents that help protect plants from bacterial infection. Some of the drug molecules are also used in agricultural crops. Fungal-derived secondary metabolites plan an important role in this direction. Biological management of plant diseases is mostly achieved by the following principal antagonistic activities: (1) direct antagonism: Lytic/some nonlytic mycoviruses, including *Trichoderma virens*, *Ampelomyces quisqualis*, *Lysobacter enzymogenes*, and *Pasteuria penetrans*, and (2) indirect antagonism: identification of molecular patterns associated with infections by interaction with the fungal cell walls (Köhl et al., 2019; Masi et al., 2018a,b). De Sá et al. (2022) reported around 213 secondary metabolites isolated from the cultures of the *Neosartorya* genus an endophytic, marine, and terrestrial fungus with diverse biological activity. The most harmful plant diseases that afflict both annual crops and woody perennials are vascular wilt diseases brought on by bacterial and fungal infections. Boutaj et al. (2022) reported *P. fluorescens* and *Trichoderma* sp. as the most promising biocontrol agents against vascular wilt diseases.

3.2 Fungicides

Fungal metabolites act as pesticides for most prevalent plant pathogens, and inhibit the growth of pathogenic fungi and their spores. Numerous studies have been conducted to create different management solutions that reduce crop losses caused by a pathogenic fungus assault on a field. There are some biochemical compounds such as ethylene bis dithiocarbamates (mancozeb, maneb, nabam, zineb) that are also used for this purpose (Masi et al., 2018a, b). Various fungal secondary metabolites that produce naturally in the environment have potential fungicidal activity against phytopathogenic fungi. Resistance developed in pathogens during the course of evolution makes the pathogen less susceptible to the fungicide's effect, making it less effective or even ineffective. Tóth et al. (2020) reported on the biofungicidal potential of *Neosartorya fischeri* antifungal protein (NFAP) and novel synthetic γ-core peptides, suggesting that these metabolites can be evaluated as potential crop preservatives. Fungicides can easily penetrate and move inside leaves, where they can exert their therapeutic effects and protect the plant tissue beyond just where the fungicide was initially applied. This is because they are developed to target specific particular enzymes or proteins produced by fungi while at the same time not affecting the plant tissue. Because the mode of action of these fungicides is so specific, hence minor genetic alterations in fungi might render them ineffective, and pathogen populations may develop resistance to subsequent administrations. For instance, the Trichoderma species compete for nitrogen or carbon, produce antibiotics and enzymes that have the ability to inhibit soil-borne phytopathogenic fungi by destroying the fungal cell wall, parasitize, and also produce auxin-like substances that promote plant development (El-Baky and Amara, 2021). In particular, the fungi of forest trees are a good source of bioactive metabolites that have been extensively studied (Litwin et al., 2020).

3.3 Weedicides

Weeds are undesirable plants growing in fields that create an issue in agricultural production because they decrease crop yields, affect the quality, and cause allergies through infected pollens (Harding and Raizada, 2015). Common weed management techniques, such as mechanical and chemical methods, have some downsides, such as being pricy, energy-intensive, difficult, having negative impacts on the environment, developing resistance, and potentially harming nontarget organisms. A weedicide is a chemical (natural and synthetic) sprayed to kill weeds on the fields (Radi and Banaei-Moghaddam, 2020). Commonly used weedicides include benefin, dithiopyr, dicamba, 2,4-D ethyl ester, glyphosate, fluazifop, neem extract, trifluralin, and others. Weedicides are diluted with water for the desired concentration and then sprayed in the fields with a sprayer. A herbicide containing chlorophenoxyacetic acid is used to defoliate broad-leaved flowers. It contains 2,4,5-trichlorophenoxyacetic acid, a synthetic auxin. In the current condition, new herbicides with safer toxicological profiles and innovative mechanisms of action are urgently needed. Actinomycetes are a group of microbes known to produce diverse bioactive compounds, some of which have strong biopesticidal activity (Abdallah et al., 2013; Amelia-Yap et al., 2022). The notable fungal genera *Colletotrichum*, *Phoma*, and *Sclerotinia* have drawn interest as potential bioherbicides (Harding and Raizada, 2015).

3.4 Herbicides

Herbicides are compounds used to manage undesirable plants, usually known as "weed killers." Generally 5%–40% of the applied herbicides are used for weed destruction and the rest remain in the field. This high concentration of herbicide can impact the cell division of emerging seedlings (Kozhuro et al., 2005). Some of the common herbicides that are used are glyphosate (Roundup), acetochlor, atrazine, amitrole, and dinosep. Buffelgrass is a species of native African grass native that has led to an extremely dangerous condition in the Sonoran Desert of southern Arizona, where it increases the frequency of fires and has the potential to wipe out the saguaro forest ecosystem. Considering the fact that buffelgrass is difficult to eradicate with common chemical herbicides that cause significant harm to native vegetation, as an alternate solution to this situation, phytotoxic secondary metabolites of *Cochliobolus australiensis* and *Pyricularia grisea* were evaluated for their ability to be used as potential natural herbicides. Generally, the phytotoxic secondary metabolites are used to control the specific molecular targets associated with amino acid synthesis, shikimate pathway function, glutamine synthase, carotenoid synthesis, lycopene cyclase, phytoene desaturase, deoxy-D-xylulose phosphate synthase, chlorophyll synthesis, fatty acid synthesis, photosynthesis, etc. (Gaines et al., 2020).

3.5 Pesticides

Fungal insecticides are an alternative to conventional insecticides and are used in genetically modified crops for controlling insect pests that are highly resistant to both Pest and genetically modified crops. *Beauveria bassiana* and *Metarhizium anisopliae* are entomopathogenic fungi that produce a number of toxic secondary metabolites that are able to infect sucking pests such as *Nezara viridula* (green vegetable bug) and *Creontiades* sp. (green and brown mirids). These pests generally affect the leaves of plants and cause severe damage within a short time. These pests cannot be eliminated by traditional controlling agents sprayed on plants. *B. bassiana* and other insect-pathogenic fungi infect their hosts through the cuticle rather than the digestive tract, where virulence factors are missing (Pedrini, 2022). Entomopathogenic fungi (EPF) have the potential to be used as biocontrol agents against herbivores, offering a more ecologically friendly alternative to conventional insect pest management methods. Secondary metabolites that are produced within the insect have different mode of actions. For example, such as beauverolides produced from *Beauveria tenella* and *Paecilomyces fumosoroseus* are involved in killing the host. Destruxins produced from *M. anisopliae* fungus and Oosporein, an abundant metabolite of *Beauveria caledonica*, mainly function as antimicrobials (Pedrini, 2022). However, they also promote infection, probably by reducing the number of insect hemocytes, with the consequent alteration of the humoral immune system.

4 Fungal metabolites as an approach to treat plant pathogen antagonists

To check the attacks of plant pathogens on crop fields, various chemical agents have long been used. However, this long use has its own drawbacks, such as the development of resistance among pathogens that makes their control more difficult, the accumulation of chemicals that can lead to toxicity for humans and animals, water pollution, etc. Therefore, as an alternate to these hazardous control strategies, the biological method of control is a biodegradable, more sustainable, and environmentally acceptable management method (Köhl et al., 2019). In biological control, plant disease or pathogen growth, this is controlled by other living organisms such as bacteria or fungi. Among the living sources, bacteria and fungi are well known to produce secondary metabolites under stress conditions (Lahlali et al., 2022). These are omnipresent and able to survive in extreme conditions. Secondary metabolites are not required for the normal growth of microbes, but still produce a sufficient amount to have other important roles. There is evidence showing that genes encoding various secondary metabolites are regulated according to the abiotic and/or biotic stressors to maintain fungal development. As a result of this regulation, either the loss or overproduction of specific secondary metabolites was observed that altered fungal development.

Fungal secondary metabolites (SMs) play important roles in the defense system of the host and its niche (Peng et al., 2021). Diverse products from these SMs are the immunosuppressant cyclosporin biosynthetic gene cluster (BGC)-2, the antibiotic penicillin BGC4, the cholesterol-reducing lovastatin BGC3, the mycotoxin trichothecene, the trichothecene T-2 toxin BGC6, and aflatoxin BGC7 (Rokas et al., 2020). Secondary metabolites are generally synthesized from diverse metabolic pathways where primary metabolite pools react with acyl-CoAs, resulting in the formation of initial building blocks for the synthesis of polyketide (aflatoxin), terpene (carotene), amino acids (nonribosomal peptides such as penicillin), etc. (Keller, 2019). Despite the fact that these tiny molecules are not necessary for fungal existence and development, their bioactive qualities make them extremely important to humans (medications, poisons, pigments), plants and animals (Adibah and Azzreena, 2019; Rokas et al., 2020). These secondary metabolites include several nonvolatile and volatile compounds such as ethylene, acetaldehyde, acetone, etc. (Yassin et al., 2022). Fungal and bacteria both are common inhabitants of soil; therefore, they are easy to handle for large-scale production. They are used to reduce the host-specific infections of plant pathogens. Bacteria such as *Bacillus*, *Pseudomonas*, and *Trichoderma* sp. are known as biological control agents.

Plants release some volatile compounds when damaged by herbivores. For instance, it is well known that herbivore-induced volatiles directly ward off small sucking and stinging insects. Aphids are discouraged by terpene mixtures, thrips are repelled by alkaloids such as nicotine, and whiteflies are kept at bay by volatile compounds found in green leaves such as aldehydes, alcohol derivatives, and esters. Terpene mixtures or volatiles from green leaves also deter whiteflies and egg-laying moths (Divekar et al., 2022; Matsuura and Fett-Neto, 2015). Similarly, mixtures of terpenes or other volatiles present in green leaves also repel egg-laying moths and butterflies. In addition, volatile substances have been proven to have antimicrobial properties through which they are able to prevent the cell growth of plant pathogens that generally enter though the wound site to colonize and affect plant tissues. Volatile terpene molecules, which are released in response to herbivory, have also been reported to be associated with the attraction of insect-eating predators. For instance, the accumulation of a group of metabolites with antifeedant action was discovered to be induced by the monoterpene myrcene in cotton. Allo-ocimene and linalool are examples of monoterpenes that have been found to increase the resistance to fungal and bacterial pathogens after application to *Arabidopsis thaliana* and rice. When *A. thaliana* was treated with myrcene or a mixture of ocimene isomers, it affected the expression of transcripts, transcription factors, and stress/defense-related genes. Similarly, homoterpenes (4,8-dimethyl-1,3,7-nonatriene (DMNT)), (Z)-3-hexenal (Z)-3-hexenyl acetate, and 3-hexenol have been reported to elevate the expression of genes related to defense systems in plants (Tyśkiewicz et al., 2022).

Nonvolatile metabolites such as polyphenolic compounds like flavonoids have long been known as plant defense weapons that play important roles as internal plant messengers. For example, the larvae of the moth species *Spodoptera frugiperda* are found to show a higher mortality rate when they grow on high concentrations of quercetin, a type of flavonoid. It is speculated that these quercetin molecules might inhibit the activity of mitochondrial ATPase and also interfere in some specific detoxification systems occurring in the insect gut site. Panigrahy et al. (2020) reported that flavonoid molecules are able to inhibit the P-glycoproteins present in the plasma membrane, and are responsible for auxin efflux into the apoplast. Beside this, flavonoids affect plant architecture by showing direct involvement in the polar auxin transport indirectly through modifying the auxin metabolism via regulatory pathways. Products of glycosylate hydrolysis are well known for their toxic and deterrent properties.

Natural predators or parasites such as leafhoppers, ladybugs, wasps, and moth larvae feed on various insects such as aphids, scale insects, mealybugs, whiteflies, and others. Microbes such as *Trichoderma* spp. can be introduced to provide protection to plants to keep them healthy (Matern et al., 2019). To reduce the invasion of birds and animals on crops, generally various obstacles such as scarecrows, traps, or nets are used. Besides this, several cover crops such as red clover, sweet clover, daikon radish, or vetch are successfully used to control weeds and provide assistance in fixing soil nitrogen. The introduction of cross-breeding and gene-modification techniques in various plants has improved their capacity for specific food combinations, high protein content, herbicide resistance, drought and salinity tolerance, and disease/insect resistance. The lack of a target pest while introducing natural predators, for instance, can harm crops because using natural predators does not guarantee the target insect's complete eradication. Further, the introduction of these natural predators can take some time, especially if the farm is big. Using cover crops to stop soil water loss has been successful. However, during winter it reduces the soil water. Over the years, genetically modified (GM) species have offered solutions to a number of crop protection problems. Due to their herbicide resistance, it is difficult to remove these GM plants from farms when they grow in unwanted regions (Kuhar et al., 2020).

Fungal strains have some basic advantages over bacteria, including a comparatively higher reproduction rate, short generation time, target specificity, and ability to change their living mode, such as parasitism to saprotrophic, in the absence of the host, suggesting their efficiency to maintain their sustainability (Thambugala et al., 2020). Further, fungal strains also show better resistance toward acidic conditions than bacteria; hence, fungal antagonists are expected to exert better antagonistic activity under these conditions (Tagawa et al., 2010). Therefore, it is important to explore more fungal antagonists to discover some novel antagonistic functions of fungal metabolites. An explanation for the nature of active compounds and their mode of action against pathogens varies from microbe to microbe. Mechanisms of inhibition are more complex, and may involve nutrient competition, antibiosis, mycoparasitism, and the production of cell-wall degrading enzymes (Yassin et al., 2022) (Fig. 1).

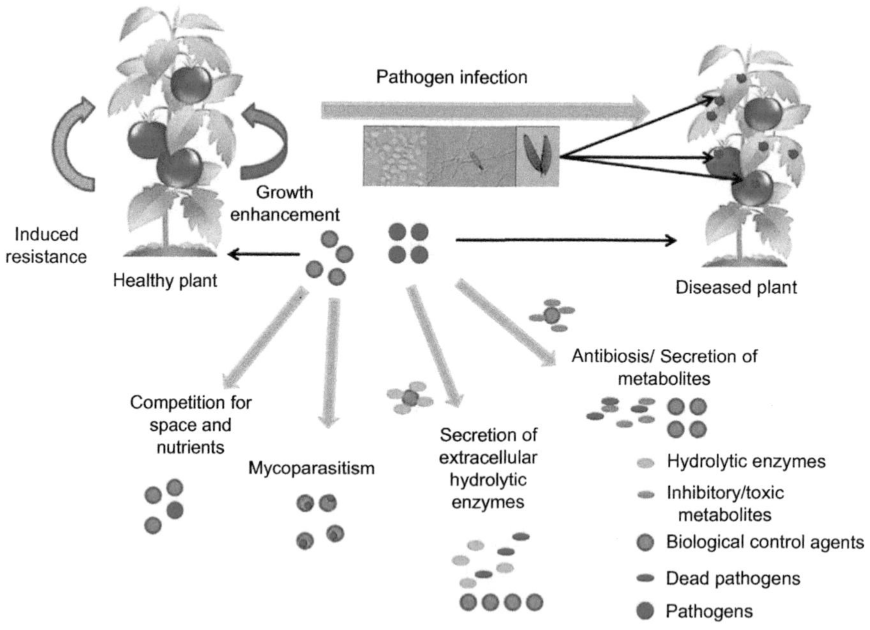

FIG. 1 Key mechanisms of action involved in the biological control of plant fungal pathogens. *Reproduced with permission from Thambugala, K.M., Daranagama, D.A., Phillips, A.J., Kannangara, S.D., Promputtha, I., 2020. Fungi vs. fungi in biocontrol: an overview of fungal antagonists applied against fungal plant pathogens. Front. Cell. Infect. Microbiol. 10, 604923. The present article is licensed under an open access Creative Commons CC BY license, Frontiers.*

5 Fungal secondary metabolites

Globally, there is a huge diversity and abundance of fungi. They are the world's second-largest group, trailing insects. According to recent estimates, there are around 1.5 million fungi on Earth. The total number of individuals described is currently around 100,000 species of fungi. Approximately 10% of the estimated 150,000 higher fungal species on Earth are currently known to science (Bérdy, 2005; Chen and Liu, 2017; Mueller and Schmit, 2007; Wasser, 2010). Fungi secondary metabolites can be released as volatiles from cells, excreted into the environment, incorporated into cell structural elements, or exist as cell-bound. Secondary metabolites are compounds biosynthesized by pathways that use primary metabolites as building blocks to assemble more complex molecules, such as terpenoids, polyketides, nuclear-encoded ribosomal peptides, nonribosomal peptides, and molecules of mixed biogenic root arising from hybrid pathways. There are distinct secondary metabolites found in both fungi and plants. Because these two eukaryotic domains have different basic metabolisms and different evolutionary origins for their unique biosynthetic pathways, it is theorized that these two eukaryotic domains have different metabolic activities (Bills and Gloer, 2016; Clevenger et al., 2017). The members of fungi produce different fungal metabolites under the influence of the environment that these fungal species received to survive (Schnarr et al., 2022). These metabolites can be divided into the following major chemical classes: polyketide, terpenoid, phenol, alkaloid, flavonoid, and peptide (Keller et al., 2005; Wang et al., 2022). Certain secondary metabolites, such as diterpenoid and gibberellin plant hormones, are synthesized by both plants and fungi, although via distinct enzyme pathways (Stoppacher et al., 2010). Gibberellic acids (GAs) are the best-known unequivocal example of convergent secondary metabolites between plant-inhabiting fungi and plants (Bömke and Tudzynski, 2009). Secondary metabolite biosynthesis may be strictly regulated. Their expression is frequently induced by environmental or chemical stimuli and is coordinated with the organism's development and morphogenesis (Lu et al., 2014; Brakhage, 2012). Some examples of the members of these classes are summarized in Fig. 2 and Table 3.

5.1 Polyketides

The most common and structurally diverse class of fungal secondary metabolites is polyketides, which are naturally occurring bioactive substances (Arasu et al., 2013). Their carbon scaffolds are produced from the polymerization of short-chain carboxylic acid units and possess a similar biosynthetic origin. The biosynthesis of these polyketide molecules requires a number of multifunctional enzymes known as polyketide synthases (PKSs), and based on the enzymatic routes and catalytic domains involved in their production, they can be categorized into four classes (Cox and Simpson, 2009; Shen, 2003). Type I PKSs are multidomain megaenzymes that have been described for the production of important fungal polyketides. This category of enzymes has a single ketosynthase (KS) domain and is connected to type III PKSs. Type I PKSs are the most well-known and abundant in fungal genomes, whereas just a few fungal strains have been documented for type III PKSs (Navarro-Muñoz and Collemare, 2020).

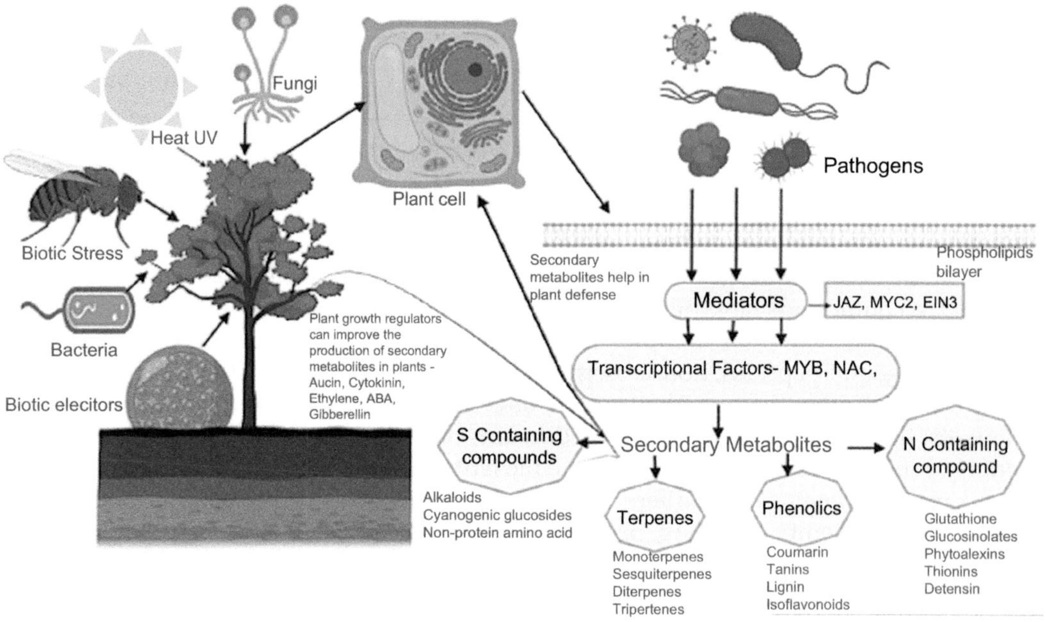

FIG. 2 Major classes of secondary metabolites playing important roles in the control of plant fungal pathogens.

5.2 Terpenoid

Terpenoids are a vast and diverse class of organic molecules found in nature. They are generally derived from isoprene, a five-carbon compound, and the polymers of isoprene known as terpenes. Higher members of fungi, such as ascomycota and basidiomycota, are well-known producers of a variety of natural products belonging to terpenoids. These are further classified according to their use including antibiotics, mycotoxins, antitumor agents, and phytohormones (Koczyk et al., 2021). Besides their applications, terpenoids can also be classified on the basis of the scaffolds from which they are derived, that is, solely from isoprenyl units or mixed in biosynthetic origin. The isoprenyl group is further divided on the basis of the number of isoprene units, that is, mono-, sesqui- (illudin S, deoxynivalenol), di- (gibberellin GA3, brassicicene A), or triterpenoids (helvolic acid), which contain two to six C-5 isoprene units. Carotenoids are an example of this group. The terpenoid of mixed biosynthetic origin includes meroterpenoids (secondary metabolites partially derived from terpenoid pathways such as pyripyropene and fumagillin), indole diterpenoids (secondary metabolites that have a common cyclic diterpene as a structural backbone such as paxillin), and prenylated aromatic natural products (Morrow, 2016; Schmidt-Dannert, 2015) (Fig. 3).

All fungal terpenoids are derived from isopentenyl diphosphate (IPP: 5C isoprenyl diphosphate intermediates) and dimethylallyl diphosphate (DMAPP). Both these intermediates are synthesized via the mevalonate pathway and acetyl-CoA is the parent substrate for biosynthesis. A number of enzymes such as thiolases (acetoacetyl-CoA thiolase), synthases (3-hydroxy-3-methylglutaryl CoA synthase), reductases (3-hydroxy-3-methylglutaryl CoA reductase), kinases (mevalonate kinase, phosphomevalonate kinase), decarboxylases

TABLE 3 Secondary metabolites produced by fungal endophytes.

Fungus	Source	Compound	Activity	Reference
Penicillium expansum	*Excoecaria agallocha*	Polyphenols, expansols A and B	Cytotoxic	Lu et al. (2010)
Curvularia lunata	*Niphates olemda*	Cytoskyrins	Antibacterial, anticancer agent	Brady et al. (2000)
Phoma medicaginis	*Medicago sativa, Medicago lupulina*	Brefeldin A	Antibacterial	Weber et al. (2004)
Phomopsis spp. PSU-D15	*Garcinia dulcis*	Phomoenamide	Antimycobacterial activity	Rukachaisirikul et al. (2008)
Muscodor albus	Cinnamomum zeylanicum	Volatile organic compounds	Antifungal, antibacterial	Ezra et al. (2004)
Pestaloptiopsis microspora	Taxus wallichiana	Paclitaxel	Anticancer	Strobel et al. (1996)
Chaetomium acuminata	*Edenia gomezpompa*	Naphthoquinone spiroketal	Allelochemical activity	Macías-Rubalcava et al. (2008)
Alternaria alternata	*Azadirachta indica*	Flavonoids	Bactericidal and antioxidant	Chatterjee et al. (2019)
Epicoccum sorghinum	*Annona senegalensis*	Phenolic	Antioxidant	Sibanda et al. (2018)
Aspergillus nidulans and *Aspergillus oryzae*	*Ginkgo biloba* L.	Phenolic and flavonoid	NR	Qiu et al. (2010)
Curvularia verruculosa	*Catharanthus roseus*	Vinblastine	Cytotoxicity against HeLa cell line	Parthasarathy et al. (2020)
Endophytic fungus	*Camptotheca acuminata*	Pentacyclic quinoline alkaloid camptothecin	Anticancer	Kusari et al. (2009)

(mevalonate diphosphate decarboxylase), isomerases (isopentenyl diphosphate isomerase), synthases (geranyl-, farnesyl-, or geranylgeranyl diphosphate synthase, terpene synthase, oxidosqualene synthase), transferases (prenyl transferase), and cyclases (terpene cyclase) are involved in the pathway and perform the required actions for the synthesis of a variety of structures (Schmidt-Dannert, 2015).

5.3 Phenol

Among plant secondary metabolites, phenols undoubtedly represent the largest group. Phenols range from the simplest molecules with a single aromatic ring to highly complicated polymeric compounds, sharing the presence of one or more phenolic groups as a common characteristic feature (Lattanzio, 2013). Because phenolic compounds and their derivatives are known to have a variety of biological properties, such as antibacterial, antioxidant,

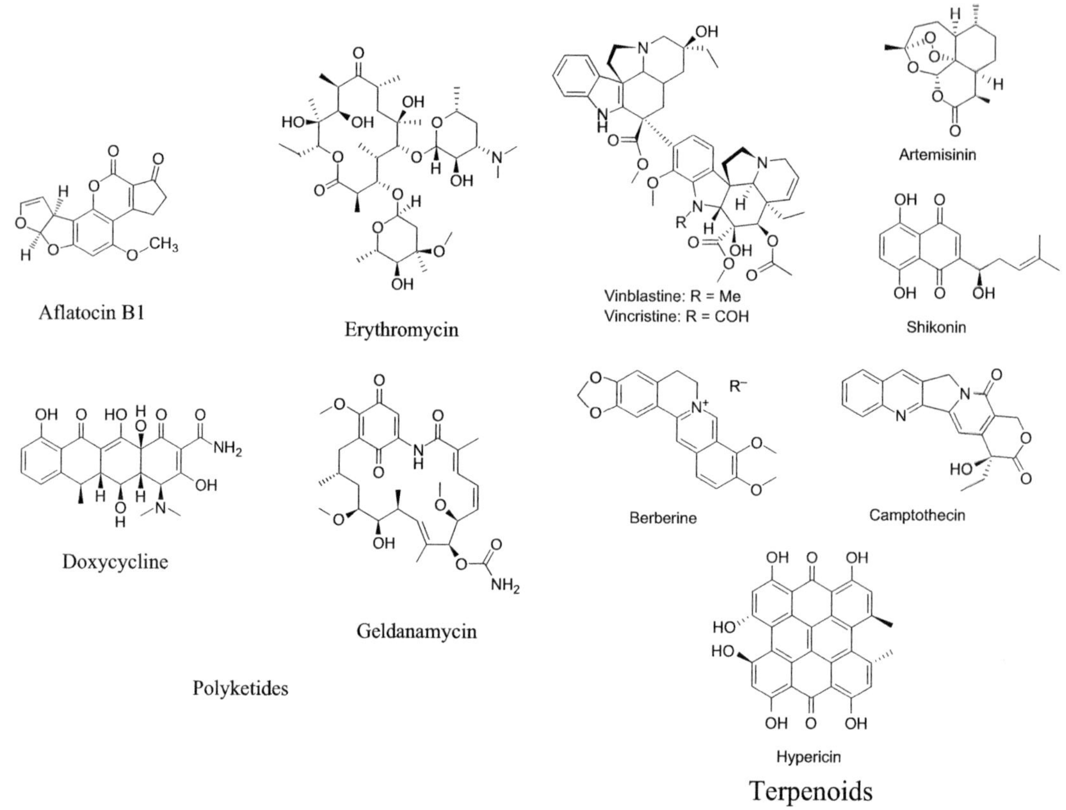

FIG. 3 Some polyketide and terpenoid fungal metabolites.

cytotoxic, and others, they are extremely important (Mutha et al., 2021). As a result of these activities, the food, chemical, pharmaceutical, cosmetic, and pharmaceutical industries all have high demand for phenolic secondary metabolites. Due to this importance, it has become very relevant to explore their structures and biological functions. The well-known shikimate pathway has been shown to biosynthesize phenolic compounds. Even though the shikimic acid pathway is more prominent in plants, it has also been described in microbes. The phenolic metabolites, which are initially produced by the plants through biosynthesis, can occasionally be found in the endophytic microbes of plants as a result of plant-microbe interactions. Therefore, a number of phenolic metabolites have been detected in the cultures of various endophytic microorganisms. In this direction, Lunardelli Negreiros de Carvalho et al. (2016) in their work covered the 124 secondary metabolites that are phenolic in nature and are produced by various species of endophytic fungi. Phenolic compounds can be categorized based on their chemical structure or whether they are biosynthetic. Phenolic compounds can be categorized into tannins, simple phenols, coumarins, chromones, flavonoids, xanthones, lignans, or stilbenes (Hussein and El-Anssary, 2019; Koo et al., 2022).

5.4 Simple phenols

Gallic acid is extensively present in plants and is the parent substance of gallotannin, whereas free phenols are uncommon in plants. *Inonotus obliquus*, a type of mushroom that lives on the tree trunks of *Betula* (birch), produces a diverse range of secondary metabolites, including phenolic compounds, melanins, and lanostane-type triterpenoids (Hwang et al., 2016). Gallic acid is primarily known for its astringent characteristics, but it has also been demonstrated to have antiviral, antibacterial, anticancer, antifungal, antiinflammatory, antianaphylactic, choleretic, antimutagenic, and bronchodilating properties in laboratory practices. Kanpiengjai et al. (2020) reported the coproduction of gallic acid and a novel cell-associated tannase by a pigment-producing yeast, *Sporidiobolus ruineniae*. The functional group such as aldehyde, hydroxyl, or carboxyl is responsible for variations in these phenolic substances, which include vanillin (phenolic aldehyde), and eugenol (phenolic phenylpropane) as well as salicylic, caffeic (phenolic acids), and ferulic acids. Hydroquinone is also one of the most widely distributed simple phenols, found in a wide range of plants and microbial strains. Riquelme et al. (2020) reported the de novo production of halogenated hydroquinone (drosophilin A) by the white-rot fungus *Phylloporia boldo*. Glycoside synthesis is ubiquitous, and the widespread glycoside coniferin and other phenolic cinnamic alcohol derivatives are lignin precursors. The antibacterial and diuretic effects of *Arctostaphylos uva-ursi* have been ascribed to its phenolic content, as have the pharmacological activities of numerous plants (Kumar et al., 2022a,b). A marine fungal strain, *Acremonium* sp., has the potential to produce some derivatives of hydroquinine (Abdel-Lateff et al., 2002). Similarly, the algicolous marine fungal strain *Wardomyces anomalus* is known to produce xanthone derivatives (Abdel-Lateff et al., 2003). Some of the important phenolic metabolites are mentioned in Fig. 4.

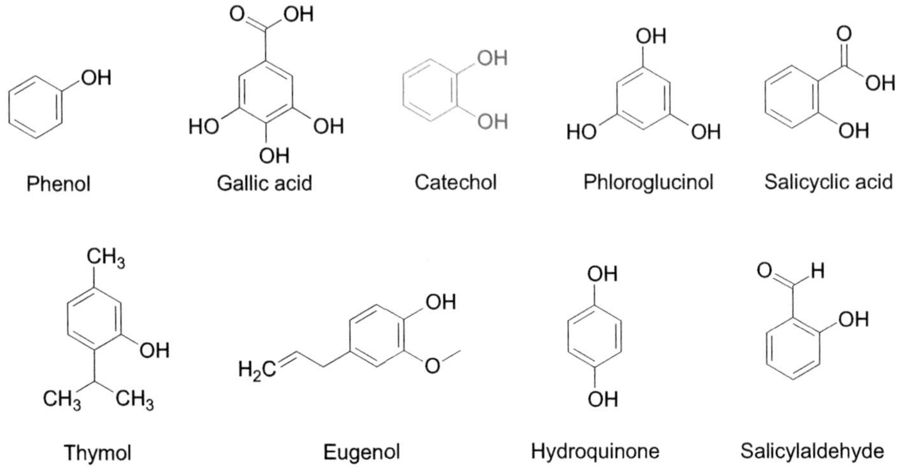

FIG. 4 Some important phenolic metabolites.

5.5 Carbohydrates

The constituents of carbohydrates are carbon, oxygen, and hydrogen, from which the last two are present in the same ratio as in water. Monosaccharides, disaccharides, oligosaccharides, and polysaccharides are the four chemical groups that constitute carbohydrates. Monosaccharides range from three to nine carbon atoms, but those with five and six carbon atoms (pentoses, $C_5H_{10}O_5$, and hexoses, $C_6H_{12}O_6$) are the ones that accumulate in plants in the greatest amounts. Depending on the number of carbohydrate units, monosaccharide condensation gives credence to several forms of carbohydrates. In addition to fulfilling essential biological and structural roles, carbohydrates in plants could indeed have therapeutic qualities, such as mucilage. Almost all plants and some microbes create slime, a viscous sticky substance that thickens plant membranes and serves as a protective layer. It is also used for seed germination as well as to preserve food and water. A polar glycoprotein and an exopolysaccharide saccharide make up its chemical composition. Mucilage is a sedative that is employed in medicine. The two major sources are cactus (and other succulent plants) and *Linum usitatissimum* (linseed). *Althaea officinalis* root mucilage extract, historically used to manufacture marshmallows, was used as a cough suppressant due to its calming effects. Due to its mucilaginous composition, the inner bark of *Ulmus rubra* (slippery elm) is also used as a sedative (Ameri et al., 2015). Currently, the suggested functions of mucilage in fungi include spores or mycelium adhesion to various surfaces and protection. In addition, the pathogenicity and environmental adaptability of *Hirsutella satumaensis* Aoki are influenced by the conidial mucilage, a natural film coating. Mucilage may also serve as storage for nutrients for fungus that are used during deprivation (Qu et al., 2017). Other polysaccharides such as β-D-glucans, galactose, mannose, arabinose, rhamnose, and fructose from basidiomycete fungi *Ganoderma lucidum* have been shown to play important therapeutic roles by exhibiting immunomodulation, antitumor, antiangiogenic, and cytotoxic effects (Shankar and Sharma, 2022). Although carbohydrates are primary metabolites, the pharmaceutical industries utilize those fungus byproducts quite effectively (Martinez et al., 2018). In addition, the *Pholiota microspora* polysaccharide has the major ability to reduce low-density lipoprotein and raise high-density cholesterol (Du et al., 2021).

5.6 Peptides

Proteins and peptides play a crucial role in SMs, where substances released from many different organisms share similarities in their metabolite products (Ng, 2004; Ramachander Turaga, 2020). Most edible mushrooms, including the *Pleurotus* species, *Agaricus bisporus*, *Volvariella volvacea*, and *Lentinus edodes*, contain 20%–30% protein by dry weight, which helps vegetarians to retain adequate levels of protein, vitamins, and fiber in their diets (Du et al., 2021). Among all organisms, the fungi have one of the richest and most complex reservoirs of nonribosomal peptides (NRPs). The filamentous ascomycota represent the pinnacle of this variety. A variety of naturally occurring peptide compounds with biological activity produced by the corresponding NRP megaenzymes are synthesized from either proteinogenic amino acids or nonproteinogenic amino acids by nonribosomal peptide synthetases (NRPSs). Iron transport, homeostasis, and storage as well as pathogenicity toward plants and animals are all regulated by siderophores generated from NRPS. The well-known drugs

TABLE 4 Some important fungal metabolites that are commercially used.

S. no.	Metabolite	Source	Commercialized products
1.	Penicillins G and V	*Penicillium rubens, Penicillium chrysogenum*	Amoxicillin, piperacillin, and ampicillin
2.	Cephalosporin C	*Acremonium chrysogenum*	Cephalexin, cephalothin, cefadroxil, and cephradine
3.	Pneumocandin B0	*Glarea lozoyensis*	Precursor of caspofungin (Cancidas)
4.	FR901379	*Coleophoma cylindrospora*	Precursor of micafungin (Mycamine, Fungiguard)
5.	Echinocandin B	*Aspergillus pachycristatus* and *Aspergillus* spp.	Precursor of anidulafungin (Eraxis)
6.	Cyclosporin A	*Tolypocladium inflatum*	Cyclosporine A
7.	Ergotamine	*Claviceps purpurea, Claviceps paspali,* and *Claviceps fusiformis*	Migril, Ergomar, Migranal, DHE 45, Cafergot, and Cafetrate
8.	Ergometrine (ergonovine)		Ergotrate, Ergovin, and Syntometrine LSD-25
9.	Ergocryptine		Parlodel
10.	PF1022A	*Rosellinia* sp.	Emodepside

penicillin, cyclosporine A, cephalosporin, and echinocandin are all biosynthesized by NRPSs, as indicated in Table 4 (Bills and Gloer, 2016; Keller et al., 2005). NRPSs have a modular structure similar to polyketides by sharing an assembly line thiotemplate approach to synthesizing peptides (Fischbach and Walsh, 2006). Both share an analogous set of three core functional domains: adenylation (A) domains, peptidyl carrier protein (thiolation (T)) domains, and condensation (C) domains. The adenylation domain is responsible for the selection and activation of amino acids, ketocarboxylic acid, or hydroxycarboxylic acid monomers; the thiolation domain for covalently anchoring the monomer; and the condensation domain catalyzes the addition of the monomer to the growing NRP chain (Zhang et al., 2020). The size of the peptide produced, whether the peptide is cyclized, and changes in the roles of the domains all contribute to the diversity of NRPs (Keller et al., 2005). Cyclosporins, ergopeptides, penicillins, and cephalosporins are a few examples derived from fungal peptides or depsipeptides that have attracted attention in the pharmaceutical field (Anke and Antelo, 2009) (Fig. 5). Linear polypeptide antibiotics such as peptaibols contain 5–20 amino acid residues with molecular weights ranging from 500 to 2200 Da that are formed in a microheterogeneous mixture (Szekeres et al., 2005). The peptaibol synthetase of *T. virens* is another example of an NRPS that generates a linear peptide (Wiest et al., 2002), which then synthesizes the cyclic undecapeptide cyclosporin by NRPS in *Tolypocladium niveum*. The drug is used as an immunosuppressive agent in treating patients after organ transplant surgery. A semisynthetic depsipeptide called emodepsin is used in veterinary medicine to treat helminths, and is obtained from PF1022A, a metabolite of a fungal endophyte from *Camellia japonica* (Von Samson-Himmelstjerna et al., 2005). The diketopiperazines

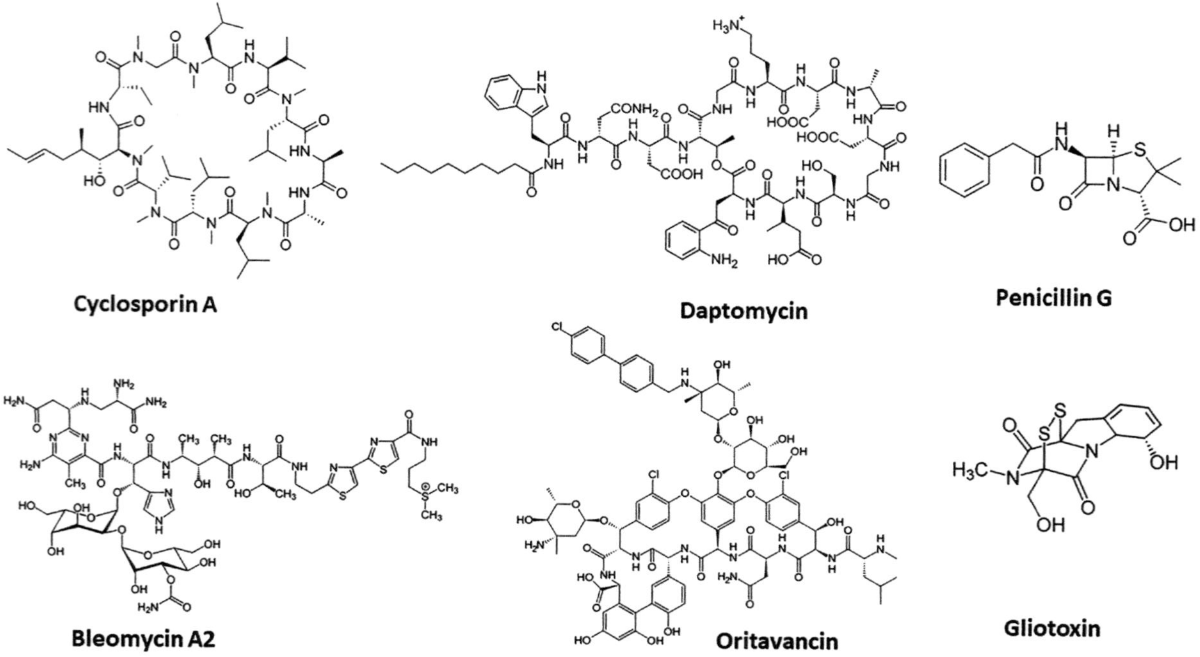

FIG. 5 Some important peptide secondary metabolites produced by fungal strains.

are a different class of cyclopeptides that is made up of two amino acid residues covalently linked with peptide bonds that also show antibacterial, amoebicidal, antiviral, antifungal, and immunosuppressive properties (Anke and Antelo, 2009). For instance, a diketopiperazine obtained from *Cordyceps sinensis* mycelium named cordysinin A exhibits antiinflammatory properties showing inhibitory effects on superoxide anion generation. Other diketopiperazines includes lepistamides A–C, diatretol from *Lepista sordida*, echinuline from *Lentinus strigellus*, emestrin, emestrins F and G from *Armillaria tabescens*, and neoechinulin A from *Xylaria euglossa* (Chen and Liu, 2017).

6 Potential of CRISPR/Cas9 system for enhancing plant disease resistance via editing of plant and fungal genome

With an increase in population, the demand for food also increases accordingly. With the Earth's population expected to reach 9.8 billion in 2050, there is a need for technological advancement in the agriculture area to meet food demand (Tripathi et al., 2022). Agriculture is generally challenged by climate change and emerging resistant strains of plant pathogens. These plant pathogens have host infection systems by which these pathogens can wreck plants in a brief time frame. In the past, severe famines such as the Irish potato famine occurred due to the infection of *Phytophthora infestans*, a pathogen belonging to the oomycetes. Similarly, the Bengal famine was caused by the rice brown spot fungal pathogen, *Cochliobolus miyabeanus*. Both led to drastic reductions in crop production. To reduce the occurrence of such events, generally various pesticides and cultivation practices are developed (Paul et al., 2021). However, as discussed earlier, these insecticides or pesticides are chemical in nature, so long-term use can result in environmental problems such as a reduction in soil fertility, resistance to plant pathogens, accumulation of chemicals in plant parts causing diseases after consumption of that part, etc. Hence, research is promoted by governments of various countries to reduce these undesirable side effects. The development of new and effective biopesticides and disease-resistant plant varieties are some outcomes of such research efforts (Pandey et al., 2021).

Techniques used for plant breeding play important roles in the development of disease-resistant plants. Currently, both conventional and molecular techniques are utilized for plant breeding to develop a plant variety with the desired characteristics (Liu et al., 2013). According to the requirements, different techniques are used for the development of new varieties of plants with better tolerance. Some conventional or traditional strategies used for plant breeding are pure line selection, pedigree, interspecific hybrids, and back-cross methods, whereas variation within multiple genes, marker-assisted breeding, tissue culture, and gene silencing are some of the molecular approaches that are commonly used (Paul et al., 2021). Conventional breeding techniques have played important roles in crop improvement over the decades, but there are some limitations of this technique, including: (i) labor cost; growing and examining crops over multiple generations is a time-consuming and labor-intensive process, (ii) inherent genetic variation inside plant populations, and (iii) undesired gene transfer; there is also the possibility of transferring nondesired genes at the side of preferred resistance genes or along the preferred gene. Genetic engineering has successfully reduced these limitations to an extent. These methods can introduce, remove, and modify specific genes with hardly any unintended modifications to the crop genome. Interspecies genetic material exchange is made possible via genetic engineering. Similarly, this technique allows the introduction of new genes into vegetatively propagated crops, as in the case of *Musa* sp. (banana), *Manihot esculenta* (cassava), and *Solanum tuberosum* (potato). In conventional transgenic methods using plant transformation methods, genes of interest are randomly introduced into the plant genome. However, current genome editing tools allow changes to the endogenous plant DNA at designated targets (Talakayala et al., 2022a, 2022b).

Based on nuclease enzyme activity, some recent advancements in breeding technology have been developed to overcome these limitations. Among these, meganucleases (MNs, also known as molecular DNA scissors), transcription-activator-like effector nucleases (TALENs), zinc-finger nucleases (ZFNs), clustered frequently interspaced palindromic repeats (CRISPR), and CRISPR-associated protein nine (Cas9) endonucleases are some of these advancements (Paul et al., 2021). These advancements allow genetic adjustments of gene targets (single or multiple) in desired plants more precisely than earlier techniques. The MNs, ZFNs, TALENs, and CRISPR/Cas9 are site-directed nucleases that specifically target DNA. Although MNs, ZFNs, and TALENs came before the introduction of CRISPR/Cas9, ZFNs and TALENs are site-specific endonucleases with a distinctive DNA-binding domain along with a cleavage domain as characteristic features. These are also successfully used in rice, maize, wheat, and tomato. However, due to complex construction methods, the large-scale application of these techniques in plants is limited (Tian et al., 2022; Hasan et al., 2021). On the other hand, due to the high-efficiency and ease of use, CRISPR/Cas9 is now becoming more popular. Numerous studies have been conducted on the application of CRISPR/Cas9 techniques to modify the host plant genome and fungal/oomycete pathogens for enhancing disease resistance (Jamil et al., 2020; Tian et al., 2022).

7 Delivery of CRISPR/Cas factor into higher plants

The transfer of CRISPR components into higher plant life due of its effectiveness and longevity in altering the genome, makes CRISPR/Cas as a recent molecular biology revolution. The transfer of CRISPR components into the plant genome plays a key role in the genome modification (Liang et al., 2020; Tian et al., 2022). The three significant CRISPR systems that have been successfully shown to edit the plant genome are Cas 9, Cas 12a, and Cas 12b (Zhang et al., 2021). The general workflow of gene editing technologies used to develop resistance in crops against plant diseases is summarized in Fig. 6. Schenke and Cai (2020) summarized several strategies that are used to get plant disease resistance through the CRISPR/Cas system. Some of them are:

- **(i)** Knock-out genes for susceptibility factor.
- **(ii)** Deletion, insertion, or modification of cis-elements in the promoter region.
- **(iii)** Specific coding region mutations.
- **(iv)** Plant surface receptor proteins' related amino acid alternations to evade secreted pathogen effectors.
- **(v)** Knock-out of negative regulators of plant defense responses.
- **(vi)** Modification of the primary defensive response regulators.

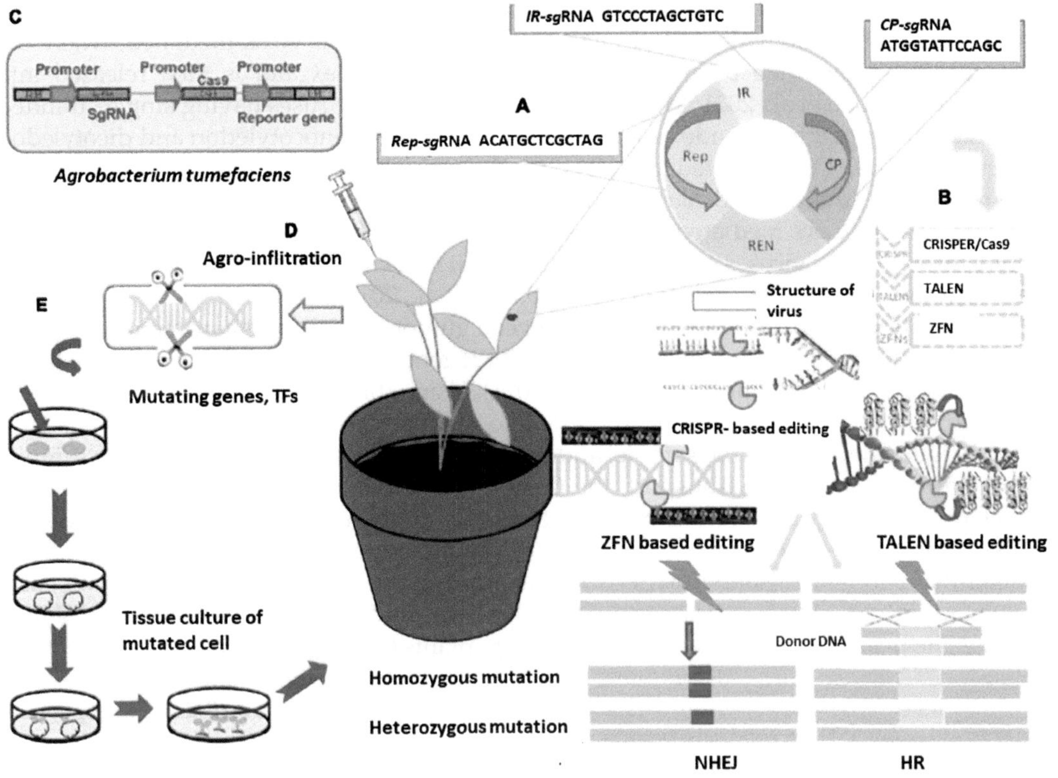

FIG. 6 Various gene editing technologies used to engineer disease resistance in plants. *Reproduced with permission from Mushtaq, M., Sakina, A., Wani, S.H., Shikari, A.B., Tripathi, P., Zaid, A., et al., 2019. Harnessing genome editing techniques to engineer disease resistance in plants. Front. Plant Sci. 10, 550. The present article is licensed under an open access Creative Commons CC BY license, Frontiers.*

Initially, *Agrobacterium tumefaciens* was used to deliver CRISPR/Cas components into dicotyledons, in addition to some monocotyledon plant cells to obtain a transgenic plant with the desired characteristics. Rice (*Oryza sativa* L.), sorghum (*Sorghum bicolor* L.), and tobacco (*Nicotiana tabacum* L.) are some examples in which to insert T-DNA for steady transgenic expression, *A. tumefaciens* was used to deliver CRISPR/Cas components into target cells (Kuluev et al., 2019). The Cas9 and gRNA are cloned in appropriate vectors to deliver into the specific plant genome by extraordinary techniques such as agrobacterium-mediated T-DNA shipping, PEG-mediated protoplast transformation, biolistic transformation, and particle bombardment (Tian et al., 2022). The most common tool for achieving genetic alterations is agrobacterium-mediated transformation. So far, this technique has successfully modified more than 20 plant species. Recently, CRISPR/Cas9 was used to demonstrate mutations in *Silene latifolia* protoplasts in *Agrobacterium rhizogenes*-mediated temporary experiments (Dalla Costa et al., 2020).

Another method of delivery of the CRISPR/Cas factor into higher plant cells is through plant viruses. The most widely used method of gene delivery is virus-mediated, where the viral genome is modified by the CRISPR/Cas9 gene complex and is then released into infected cells (Duan et al., 2021; Mushtaq et al., 2019). Geminiviruses having single-stranded DNA are known to infect a wide range of plants, including monocotyledon and dicotyledon (Baltes et al., 2014). Once inserted into a plant cell, the single-stranded DNA of geminiviruses becomes double-stranded and starts replicating actively. The virus-based gRNA delivery system for CRISPR/Cas9 mediated plant genome editing (VIGE) is an example of genome editing through plant viruses such as the cabbage leaf curl virus, the bean yellow dwarf virus, the wheat dwarf virus, etc. (Yin et al., 2015; Baltes et al., 2014).

The biolistics method, often known as the gene gun or particle bombardment method, is another common technique to introduce the CRISPR gene into plant cells for transient expression and stable transformation (Miller et al., 2021). Using this gene gun, it is possible to successfully introduce molecules of interest into plants. However, the consistency and reproducibility depends on the sample, and they can vary drastically from sample to sample. In bombardment-mediated transport, the Cas9/gRNA-RNP additives coated with gold or tungsten particles are transferred into plant cells by applying excessive strain, but due to excessive pressure, the enhancing performance can be reduced. In addition to this, another method that is utilized is one in which the Cas protein and gRNA(s), which are both parts of the CRISPR system, are preassembled in order to produce ribonucleoproteins (RNPs), and then the preassembled system is introduced to plants (Zhang et al., 2021). Direct delivery of CRISPR/Cas9 with the RNP complex allows for a reduction in the off-beam effect and prevents the incorporation of foreign DNA into the host genome (Gu et al., 2021). Nanoparticle and pollen magnetofection techniques are gaining momentum for transgenic free mutant flora. With the use of one-of-a-kind techniques, researchers are growing genome-edited smart crops (Duan et al., 2021).

Protoplasts, plant cells without cell walls, are frequently employed in crop breeding and plant science research because they may be transfected with great effectiveness using PEG-Ca^{2+} and electrophoresis without the need for a selection marker. DNA, RNA, and RNP are used as the genome editing reagents that can be introduced into protoplasts via transfection. Further, these engineered/edited protoplasts can be regenerated into plants. These are well explained experimentally in the case of the dicotyledonous species (Zhang et al., 2021).

8 Conclusion and future aspects

Fungi are a class of microbes that are known to produce a wide range of metabolites with various functions. In this chapter, we have discussed the applicability of fungal metabolites against various plant pathogens; in conclusion, we can say that fungal metabolites have emerged as promising and innovative plant pathogen antagonists with tremendous potential for disease management and sustainable agriculture. The fungal metabolites provide both direct and indirect methods of action, giving a diverse strategy for combating plant pathogens. Further, it was also found that these metabolites are playing an important tool for combating the social or economic problems arised by crop plant infections because of their power to directly limit pathogen growth and development as well as increase plant defenses and improve overall plant health. The diversity of fungal species and the metabolites produced by them, along with the development of advanced techniques in molecular biology and 3D printing, presents a vast array of possibilities for future research and development to uncover even more effective and environmentally friendly solutions for managing plant diseases.

References

Abdallah, M.E., Haroun, S.A., Gomah, A.A., El-Naggar, N.E., Badr, H.H., 2013. Application of actinomycetes as biocontrol agents in the management of onion bacterial rot diseases. Arch. Phytopathol. Plant Protect. 46 (15), 1797–1808.

Abdel-Lateff, A., König, G.M., Fisch, K.M., Höller, U., Jones, P.G., Wright, A.D., 2002. New antioxidant hydroquinone derivatives from the algicolous marine fungus *Acremonium* sp. J. Nat. Prod. 65 (11), 1605–1611.

Abdel-Lateff, A., Klemke, C., König, G.M., Wright, A.D., 2003. Two new xanthone derivatives from the algicolous marine fungus *Wardomyces anomalus*. J. Nat. Prod. 66 (5), 706–708.

Adibah, K.Z.M., Azzreena, M.A., 2019. Plant toxins: alkaloids and their toxicities. GSC Biol. Pharm. Sci. 62.

Ahmadi, A.R., Hedjaroude, C.A., Sharifi-Tehrani, A., Kheiri, A., Akiyani, A., 1996. First report on isolation and identification of *Paecilomyces farinosus* from *Heterodera schachtii* and its antagonistic effects on the eggs in Iran. In: 12th Iranian Plant Protection Congress, p. 354.

Alam, B., Lǐ, J., Gě, Q., Khan, M.A., Gōng, J., Mehmood, S., Yuán, Y., Gǒng, W., 2021. Endophytic fungi: from symbiosis to secondary metabolite communications or vice versa? Front. Plant Sci. 12, 3060.

Alizadeh, M., Vasebi, Y., Safaie, N., 2020. Microbial antagonists against plant pathogens in Iran: a review. Open Agric. 5 (1), 404–440.

Amelia-Yap, Z.H., Azman, A.S., AbuBakar, S., Low, V.L., 2022. Streptomyces derivatives as an insecticide: current perspectives, challenges and future research needs for mosquito control. Acta Trop., 106381.

Ameri, A., Heydarirad, G., Mahdavi Jafari, J., Ghobadi, A., Rezaeizadeh, H., Choopani, R., 2015. Medicinal plants contain mucilage used in traditional Persian medicine (TPM). Pharm. Biol. 53 (4), 615–623.

Amir-Sadeghi, S., Sharifi-Tehrani, A., Hejaroud, D.H.A., Okhovat, M., Rouhani, H., 1994. Investigation on antagonistic properties of *Trichoderma* spp. Against *Sclerotinia sclerotiorum* (Lib) debary the causal agent of Sclerotinia disease of eggplant. In: 11th Protection Congress Plant Iranian, p. 151.

Anke, H., Antelo, L., 2009. Cyclic peptides and depsipeptides from fungi. In: Esser, K. (Ed.), Physiology and Genetics: Selected Basic and Applied Aspects, 2009th ed. Springer, pp. 273–296.

Arasu, M.V., Al-Dhabi, N.A., Saritha, V., Duraipandiyan, V., Muthukumar, C., Kim, S.J., 2013. Antifeedant, larvicidal and growth inhibitory bioactivities of novel polyketide metabolite isolated from *Streptomyces* sp. AP-123 against *Helicoverpa armigera* and *Spodoptera litura*. BMC Microbiol. 13, 105.

Arzanlou, M., Abeln, E.C.A., Kema, G.H.J., Waalwijk, C., Carlier, J., de Vries, I., et al., 2007. Molecular diagnostics for the Sigatoka disease complex of banana. Phytopathology 97, 1112–1118.

Asghari, M.R., Myee, C.D., 1992. Comparative efficiency of management practices on stem and pod rot of groundnut. In: 10th Iranian Plant Protection Congress, p. 100.

Baltes, N.J., Gil-Humanes, J., Cermak, T., Atkins, P.A., Voytas, D.F., 2014. DNA replicons for plant genome engineering. Plant Cell 26, 151–163.

Bandyopadhyay, R., Mwangi, M., Aigbe, S.O., Leslie, J.F., 2006. Fusarium species from the cassava root rot complex in west Africa. Phytopathology 96, 673–676.

Batzer, J.C., Arias, M.M.D., Harrington, T.C., Gleason, M.L., Groenewald, J.Z., Crous, P.W., 2008. Four species of *Zygophiala* (Schizo-thyriaceae, Capnodiales) are associated with the sooty blotch and flyspeck complex on apple. Mycologia 100, 246–258.

Bazgir, A., Rouhani, H., Okhovat, M., 1992. The investigation of *Gliocladium* sp. effect on *Rhizoctonia solani* the causal agent of seedling death and bean seed rot. In: 10th Iranian Plant Protection Congress, p. 108.

Belisario, A., Maccaroni, M., Corazza, L., Balmas, V., Valier, A., 2002. Occurrence and etiology of brown apical necrosis on Persian (English) walnut fruit. Plant Dis. 86 (6), 599–602.

Bérdy, J., 2005. Bioactive microbial metabolites. J. Antibiot. (Tokyo) 58, 1–26.

Bills, G.F., Gloer, J.B., 2016. Biologically active secondary metabolites from the fungi. Microbiol. Spectr. 4 (6).

Bömke, C., Tudzynski, B., 2009. Diversity, regulation, and evolution of the gibberellin biosynthetic pathway in fungi compared to plants and bacteria. Phytochemistry 70 (15–16), 1876–1893.

Bonaterra, A., Badosa, E., Daranas, N., Francés, J., Roselló, G., Montesinos, E., 2022. Bacteria as biological control agents of plant diseases. Microorganisms 10 (9), 1759.

Boutaj, H., Meddich, A., Roche, J., Mouzeyar, S., El Modafar, C., 2022. The effects of mycorrhizal fungi on vascular wilt diseases. Crop Prot. 155, 105938.

Brady, S.F., Wagenaar, M.M., Singh, M.P., Janso, J.E., Clardy, J., 2000. The cytosporones, new octaketide antibiotics isolated from an endophytic fungus. Org. Lett. 14, 4043–4046.

Brakhage, A.A., 2012. Regulation of fungal secondary metabolism. Nat. Rev. Microbiol. 11 (1), 21–32.

Canaday, C.H., Wyatt, J.E., Mullins, J.A., 1991. Resistance to broccoli to bacterial soft rot caused by *Pseudomonas marginalis* and fluorescent *Pseudomonas* species. Plant Dis. 75, 715–720.

Chatterjee, S., Ghosh, R., Mandal, N.C., 2019. Production of bioactive compounds with bactericidal and antioxidant potential by endophytic fungus *Alternaria alternata* AE1 isolated from *Azadirachta indica* A. Juss. PLoS One. 14, e0214744.

Chen, H.P., Liu, J.K., 2017. Secondary metabolites from higher fungi. Prog. Chem. Org. Nat. Prod. 106, 1–201.

Clevenger, K.D., Bok, J.W., Ye, R., Miley, G.P., Verdan, M.H., Velk, T., et al., 2017. A scalable platform to identify fungal secondary metabolites and their gene clusters. Nat. Chem. Biol. 13 (8), 895–901.

Combarros-Fuertes, P., Fresno, J.M., Estevinho, M.M., Sousa-Pimenta, M., Tornadijo, M.E., Estevinho, L.M., 2020. Honey: another alternative in the fight against antibiotic-resistant bacteria? Antibiotics 9 (11), 774.

Cox, R.J., Simpson, T.J., 2009. Fungal type I polyketide synthases. Methods Enzymol. 459, 49–78.

Dalla Costa, L., Piazza, S., Pompili, V., Salvagnin, U., Cestaro, A., Moffa, L., et al., 2020. Strategies to produce T-DNA free CRISPRed fruit trees via *Agrobacterium tumefaciens* stable gene transfer. Sci. Rep. 10 (1), 1–14.

De Sá, J.D., Kumla, D., Dethoup, T., Kijjoa, A., 2022. Bioactive compounds from terrestrial and marine-derived fungi of the genus *Neosartorya*. Molecules 27 (7), 2351.

Divekar, P.A., Narayana, S., Divekar, B.A., Kumar, R., Gadratagi, B.G., Ray, A., et al., 2022. Plant secondary metabolites as defense tools against herbivores for sustainable crop protection. Int. J. Mol. Sci. 23 (5), 2690.

Du, X., Muniz, A., Davila, M., Juma, S., 2021. Egg white partially substituted with mushroom: taste impartment with mushroom amino acids, 5′-nucleotides, soluble sugars, and organic acids, and impact factors. ACS Food Sci. Technol. 1 (7), 1333–1348.

Duan, L., Ouyang, K., Xu, X., Xu, L., Wen, C., Zhou, X., et al., 2021. Nanoparticle delivery of CRISPR/Cas9 for genome editing. Front. Genet. 12, 673286.

El-Baky, N.A., Amara, A.A.A.F., 2021. Recent approaches towards control of fungal diseases in plants: an updated review. J. Fungi 7 (11), 900.

Ezra, D., Hess, W.M., Strobel, G.A., 2004. New endophytic isolates of *Muscodor albus*, a volatile-antibiotic-producing fungus. Microbiology 150, 4023–4031.

Fischbach, M.A., Walsh, C.T., 2006. Assembly-line enzymology for polyketide and nonribosomal peptide antibiotics: logic, machinery, and mechanisms. Chem. Rev. 106 (8), 3468–3496.

Gaines, T.A., Duke, S.O., Morran, S., Rigon, C.A., Tranel, P.J., Küpper, A., Dayan, F.E., 2020. Mechanisms of evolved herbicide resistance. J. Biol. Chem. 295 (30), 10307–10330.

Grube, M., Fürnkranz, M., Zitzenbacher, S., Huss, H., Berg, G., 2011. Emerging multi-pathogen disease caused by *Didymella bryoniae* and pathogenic bacteria on Styrian oil pumpkin. Eur. J. Plant Pathol. 131, 539–548.

Gu, X., Liu, L., Zhang, H., 2021. Transgene-free genome editing in plants. Front. Genome Editing 3.

Harding, D.P., Raizada, M.N., 2015. Controlling weeds with fungi, bacteria and viruses: a review. Front. Plant Sci. 6, 659.

Hasan, N., Choudhary, S., Naaz, N., Sharma, N., Laskar, R.A., 2021. Recent advancements in molecular marker-assisted selection and applications in plant breeding programmes. J. Genet. Eng. Biotechnol. 19 (1), 1–26.

He, D.C., He, M.H., Amalin, D.M., Liu, W., Alvindia, D.G., Zhan, J., 2021. Biological control of plant diseases: an evolutionary and eco-economic consideration. Pathogens 10 (10), 1311.

Helepciuc, F.E., Todor, A., 2023. Making the best of research investment in pathogens control through biocontrol. How is research correlated with agricultural microbial biological control product availability? PLoS Pathog. 19 (1), e1011071.

Hussein, R.A., El-Anssary, A.A., 2019. Plants secondary metabolites: the key drivers of the pharmacological actions of medicinal plants. In: Herbal Medicine. vol. 1.

Hwang, B.S., Lee, I.K., Yun, B.S., 2016. Phenolic compounds from the fungus *Inonotus obliquus* and their antioxidant properties. J. Antibiot. 69 (2), 108–110.

Izadyar, M., Padasht, F., 1994. Antagonistic activity of some micro-organisms against rice sheath blight. In: 11th Iranian Plant Protection Congress, p. 65.

Jamil, S., Shahzad, R., Ahmad, S., Fatima, R., Zahid, R., Anwar, M., et al., 2020. Role of genetics, genomics, and breeding approaches to combat stripe rust of wheat. Front. Nutr. 7, 173.

Jeger, M., Beresford, R., Bock, C., Brown, N., Fox, A., Newton, A., et al., 2021. Global challenges facing plant pathology: multidisciplinary approaches to meet the food security and environmental challenges in the mid-twenty-first century. CABI Agric. Biosci. 2 (1), 1–18.

Kanpiengjai, A., Khanongnuch, C., Lumyong, S., Haltrich, D., Nguyen, T.H., Kittibunchakul, S., 2020. Co-production of gallic acid and a novel cell-associated tannase by a pigment-producing yeast, *Sporidiobolus ruineniae* A45.2. Microb. Cell Factories 19, 1–12.

Karampour, F., Okhovat, M., 1992. Investigation on the effect of some isolates of *Trichoderma* and *Gliocladium* on the growth of *Fusarium solani* at lab conditions. In: 10th Iranian Plant Protection Congress, p. 143.

Keller, N.P., 2019. Fungal secondary metabolism: regulation, function and drug discovery. Nat. Rev. Microbiol. 17 (3), 167–180.

Keller, N.P., Turner, G., Bennett, J.W., 2005. Fungal secondary metabolism—from biochemistry to genomics. Nat. Rev. Microbiol. 3 (12), 937–947.

Kline, A.S., Anderson, E.J., Smith, E.B., 1997. Occurrence of cowpea stunt disease causing viruses on wild bean in Arkansas. Plant Dis. 81 (2), 231.

Koczyk, G., Pawłowska, J., Muszewska, A., 2021. Terpenoid biosynthesis dominates among secondary metabolite clusters in mucoromycotina genomes. J. Fungi 7 (4), 285.

Köhl, J., Kolnaar, R., Ravensberg, W.J., 2019. Mode of action of microbial biological control agents against plant diseases: relevance beyond efficacy. Front. Plant Sci. 845.

Koo, H.B., Hwang, H.S., Han, J.Y., Cheong, E.J., Kwon, Y.S., Choi, Y.E., 2022. Enhanced production of pinosylvin stilbene with aging of *Pinus strobus* callus and nematicidal activity of callus extracts against pinewood nematodes. Sci. Rep. 12 (1), 1–13.

Kozhuro, Y., Afonin, V., Maximova, N., 2005. Cytogenetic effect of herbicides on plant cells. BMC Plant Biol. 5, 1–2.

Kůdela, V., Krejzar, V., Pánková, I., 2010. *Pseudomonas corrugata* and *Pseudomonas marginalis* associated with the collapse of tomato plants in rockwool slab hydroponic culture. Plant Prot. Sci. 46, 1–11.

Kuhar, T.P., Hastings, P.D., Hamilton, G.C., VanGessel, M.J., Johnson, G.C., Wyenandt, C.A., Vuuren, V.M., 2020. 2020-2021 Mid-Atlantic Commercial Vegetable Recommendations.

Kuluev, B.R., et al., 2019. Delivery of CRISPR/Cas components into higher plant cells for genome editing. Russ. J. Plant Physiol. 66, 694–706.

Kumar, J., Ramlal, A., Mallick, D., Mishra, V., 2021. An overview of some biopesticides and their importance in plant protection for commercial acceptance. Plants 10 (6), 1185.

Kumar, S., Saini, R., Suthar, P., Kumar, V., Sharma, R., 2022a. Plant secondary metabolites: their food and therapeutic importance. In: Plant Secondary Metabolites. Springer, Singapore, pp. 371–413.

Kumar, G., Kumar, R., Gautam, G.K., Rana, H., 2022b. The phytochemical and pharmacological properties of *Catharanthus roseus* (Vinca). Sci. Prog. Res. 2 (1), 472–477. https://doi.org/10.52152/spr/2022.158.

Kusari, S., Zühlke, S., Spiteller, M., 2009. An endophytic fungus from *Camptotheca acuminata* that produces camptothecin and analogues. J. Nat. Prod. 72, 2–7.

Lahlali, R., Ezrari, S., Radouane, N., Kenfaoui, J., Esmaeel, Q., El Hamss, H., et al., 2022. Biological control of plant pathogens: a global perspective. Microorganisms 10 (3), 596.

Lamichhane, J.R., Venturi, V., 2015. Synergisms between microbial pathogens in plant disease complexes: a growing trend. Front. Plant Sci. 6, 385.

Lattanzio, V., 2013. Phenolic compounds: introduction. In: Ramawat, K.G., Mérillon, J.M. (Eds.), Natural Products: Phytochemistry, Botany and Metabolism of Akaloids, Phenolics and Terpenes. Springer, Berlin/Heidelberg, Germany, pp. 1543–1580.

Liang, Y.J., Ariyawansa, H.A., Becker, J.O., Yang, J.I., 2020. The evaluation of egg-parasitic fungi *Paraboeremia taiwanensis* and *Samsoniella* sp. for the biological control of *Meloidogyne enterolobii* on Chinese cabbage. Microorganisms 8 (6), 828.

Litwin, A., Nowak, M., Różalska, S., 2020. Entomopathogenic fungi: unconventional applications. Rev. Environ. Sci. Biotechnol. 19 (1), 23–42.

Liu, Y.J., Liu, J., Ying, S.H., Liu, S.S., Feng, M.G., 2013. A fungal insecticide engineered for fast per os killing of caterpillars has high field efficacy and safety in full-season control of cabbage insect pests. Appl. Environ. Microbiol. 79 (20), 6452–6458.

Lu, Z., Zhu, H., Fu, P., Wang, Y., Zhang, Z., Lin, H., Liu, P., Zhuang, Y., Hong, K., Zhu, W., 2010. Cytotoxic polyphenols from the marine-derived fungus *Penicillium expansum*. J. Nat. Prod. 73, 911–914.

Lu, M.Y.J., Fan, W.L., Wang, W.F., Chen, T., Tang, Y.C., Chu, F.H., Chang, T.T., Wang, S.Y., Li, M.Y., Chen, Y.H., Lin, Z.S., Yang, K.J., Chen, S.M., Teng, Y.C., Lin, Y.L., Shaw, J.F., Wang, T.F., Li, W.H., 2014. Genomic and transcriptomic analyses of the medicinal fungus *Antrodia cinnamomea* for its metabolite biosynthesis and sexual development. Proc. Natl. Acad. Sci. 111 (44).

Lunardelli Negreiros de Carvalho, P., de Oliveira Silva, E., Aparecida Chagas-Paula, D., Honorata Hortolan Luiz, J., Ikegaki, M., 2016. Importance and implications of the production of phenolic secondary metabolites by endophytic fungi: a mini-review. Mini Rev. Med. Chem. 16 (4), 259–271.

Macías-Rubalcava, M.L., Hernández-Bautista, B.E., Jiménez-Estrada, M., González, M.C., Glenn, A.E., Hanlin, R.T., Hernández-Ortega, S., Saucedo-Garcia, A., Muria-González, J.M., Anaya, A.L., 2008. Naphthoquinone spiroketal with allelochemical activity from the newly discovered endophytic fungus *Edenia gomezpompae*. Phytochemistry 69, 1185–1196.

Martin, R.R., MacFarlane, S., Sabandzovic, S., Quito, D., Poucdel, B., Tzanetakis, I.E., 2013. Viruses and virus diseases of *Rubus*. Plant Dis. 97, 168–182.

Martinez, D.A., Loening, U.E., Graham, M.C., 2018. Impacts of glyphosate-based herbicides on disease resistance and health of crops: a review. Environ. Sci. Eur. 30, 1–14.

Masi, D., Kumar, V., Garza-Reyes, J.A., Godsell, J., 2018a. Towards a more circular economy: exploring the awareness, practices, and barriers from a focal firm perspective. Prod. Plan. Control 29 (6), 539–550.

Masi, M., Nocera, P., Reveglia, P., Cimmino, A., Evidente, A., 2018b. Fungal metabolites antagonists towards plant pests and human pathogens: structure-activity relationship studies. Molecules 23 (4), 834.

Matern, A., Böttcher, C., Eschen-Lippold, L., Westermann, B., Smolka, U., Döll, S., et al., 2019. A substrate of the ABC transporter PEN3 stimulates bacterial flagellin (flg22)-induced callose deposition in *Arabidopsis thaliana*. J. Biol. Chem. 294 (17), 6857–6870.

Matsuura, H.N., Fett-Neto, A.G., 2015. Plant alkaloids: main features, toxicity, and mechanisms of action. Plant Toxins 2 (7), 1–5.

Mbega, E.R., Mabagala, R.B., Adriko, J., Lund, O.S., Wulff, E.G., Mortensen, C.N., 2012. Five species of Xanthomonads associated with bacterial leaf spot symptoms in tomato from Tanzania. Plant Dis. 96, 7602.

Miao, Z.Q., Li, S.D., Liu, X.Z., Chen, Y., Li, Y., Wang, Y., et al., 2006. The causal microorganisms of *Panax notoginseng* root rot disease. Sci. Agric. Sin. 39, 1371–1378.

Miller, K., Eggenberger, A.L., Lee, K., Liu, F., Kang, M., Drent, M., et al., 2021. An improved biolistic delivery and analysis method for evaluation of DNA and CRISPR-Cas delivery efficacy in plant tissue. Sci. Rep. 11 (1), 7695.

Morrow, G.W., 2016. Biosynthesis of alkaloids and related compounds. In: Bioorganic Synthesis. Oxford University Press.

Moura, M.L., Jacques, L.A., Brito, L.M., Mourao, I.M., Duclos, J., 2005. Tomato pith necrosis caused by *P-corrugata* and *P-mediterranea*: severity of damages and crop loss assessment. Acta Hort. 695, 365–372.

Mueller, G.M., Schmit, J.P., 2007. Fungal biodiversity: what do we know? What can we predict? Biodivers. Conserv. 16 (1), 1–5.

Mushtaq, M., Sakina, A., Wani, S.H., Shikari, A.B., Tripathi, P., Zaid, A., et al., 2019. Harnessing genome editing techniques to engineer disease resistance in plants. Front. Plant Sci. 10, 550.

Mutha, R.E., Tatiya, A.U., Surana, S.J., 2021. Flavonoids as natural phenolic compounds and their role in therapeutics: an overview. Future J. Pharm. Sci. 7 (1), 1–13.

Navarro-Muñoz, J.C., Collemare, J., 2020. Evolutionary histories of type III polyketide synthases in fungi. Front. Microbiol. 10, 3018.

Newman, D.J., Cragg, G.M., 2020. Natural products as sources of new drugs over the nearly four decades from 01/1981 to 09/2019. J. Nat. Prod. 83 (3), 770–803.

Ng, T.B., 2004. Peptides and proteins from fungi. Peptides 25 (6), 1055–1073.

Niblett, C.L., Claflin, L.E., 1978. Corn lethal necrosis-A new virus disease of corn in Kansas. Plant Dis. Rep. 62, 15–19.

Nishiguchi, M., Kobayashi, K., 2011. Attenuated plant viruses: preventing virus diseases and understanding the molecular mechanism. J. Gen. Plant Pathol. 77, 221–229.

Okhovat, M., Zafari, D.M., Karimi-Roosbahani, A.R., Rohani, 1994. Evaluation of antagonistic effects of Trichoderma on Colletotrichum coccodes isolated from potato. In: 11th Iranian Plant Protection Congress, p. 149.

Pandey, A.K., Sinniah, G.D., Babu, A., Tanti, A., 2021. How the global tea industry copes with fungal diseases–challenges and opportunities. Plant Dis. 105 (7), 1868–1879.

Panigrahy, N., Barik, M., Sahoo, N.K., 2020. Kinetics of phenol biodegradation by an indigenous *Pseudomonas citronellolis* NS1 isolated from coke oven wastewater. J. Hazard. Toxic Radioact. Waste 24 (3), 04020019.

Parthasarathy, R., Shanmuganathan, R., Pugazhendhi, A., 2020. Vinblastine production by the endophytic fungus *Curvularia verruculosa* from the leaves of *Catharanthus roseus* and its in vitro cytotoxicity against HeLa cell line. Anal. Biochem. 593, 113530.

Paul, N.C., Park, S.W., Liu, H., Choi, S., Ma, J., MacCready, J.S., et al., 2021. Plant and fungal genome editing to enhance plant disease resistance using the CRISPR/Cas9 system. Front. Plant Sci., 1534.

Pedrini, N., 2022. The entomopathogenic fungus *Beauveria bassiana* shows its toxic side within insects: expression of genes encoding secondary metabolites during pathogenesis. J. Fungi 8 (5), 488.

Peng, Y., Li, S.J., Yan, J., Tang, Y., Cheng, J.P., Gao, A.J., et al., 2021. Research progress on phytopathogenic fungi and their role as biocontrol agents. Front. Microbiol., 1209.

Peyghami, E., Nishabouri, M.R., 1998. Studying biological control of cucumber Fusarium wilt by *Trichoderma harzianum* Rifai. In: 13th Iranian Plant Protection Congress, p. 178.

Pusztahelyi, T., Holb, I.J., Pócsi, I., 2015. Secondary metabolites in fungus-plant interactions. Front. Plant Sci. 6, 573.

Qiu, M., Xie, R.S., Shi, Y., Zhang, H., Chen, H.M., 2010. Isolation and identification of two flavonoid-producing endophytic fungi from *Ginkgo biloba* L. Ann. Microbiol. 60 (1), 143–150.

Qu, J., Zou, X., Yu, J., Zhou, Y., 2017. The conidial mucilage, natural film coatings, is involved in environmental adaptability and pathogenicity of *Hirsutella satumaensis* Aoki. Sci. Rep. 7 (1).

Radi, H.C., Banaei-Moghaddam, A.M., 2020. Biological control of weeds by fungi: challenges and opportunities. Acta Sci. Microbiol. 3, 62–70.

Ramachander Turaga, V.N., 2020. Peptaibols: antimicrobial peptides from fungi. In: Bioactive Natural Products in Drug Discovery. Springer, Singapore, pp. 713–730.

Riquelme, C., Candia, B., Ruiz, D., Herrera, M., Becerra, J., Pérez, C., et al., 2020. The de Novo Production of Halogenated Hydroquinone Metabolites by the Andean-Patagonian White-Rot Fungus *Phylloporia boldo*.

Rizzo, D.M., Lichtveld, M., Mazet, J.A., Togami, E., Miller, S.A., 2021. Plant health and its effects on food safety and security in a One Health framework: four case studies. One Health Outlook 3 (1), 1–9.

Roberts, F.A., Sivasithamparam, K., 1986. Identity and pathogenicity of *Rhizoctonia* spp. associated with bare patch disease of cereals at a field site in Western Australia. Neth. J. Plant Pathol. 92, 185–195.

Rokas, A., Mead, M.E., Steenwyk, J.L., Raja, H.A., Oberlies, N.H., 2020. Biosynthetic gene clusters and the evolution of fungal chemodiversity. Nat. Prod. Rep. 37 (7), 868–878.

Rukachaisirikul, V., Sommart, U., Phongpaichit, S., Sakayaroj, J., Kirtikara, K., 2008. Metabolites from the endophytic fungus *Phomopsis* sp. PSU-D15. Phytochemistry 69 (3), 783–787.

Schenke, D., Cai, D., 2020. Applications of CRISPR/Cas to improve crop disease resistance: beyond inactivation of susceptibility factors. Iscience 23 (9).

Schmidt-Dannert, C., 2015. Biosynthesis of terpenoid natural products in fungi. Adv. Biochem. Eng. Biotechnol. 148, 19–61. https://doi.org/10.1007/10_2014_283.

Schnarr, L., Segatto, M.L., Olsson, O., Zuin, V.G., Kümmerer, K., 2022. Flavonoids as biopesticides—systematic assessment of sources, structures, activities and environmental fate. Sci. Total Environ., 153781.

Secretariat, I.P.P.C., Gullino, M.L., Albajes, R., Al-Jboory, I., Angelotti, F., Chakraborty, S., et al., 2021. Scientific Review of the Impact of Climate Change on Plant Pests. FAO on behalf of the IPPC Secretariat.

Shankar, A., Sharma, K.K., 2022. Fungal secondary metabolites in food and pharmaceuticals in the era of multi-omics. Appl. Microbiol. Biotechnol. 106 (9–10), 3465–3488.

Sharma, H., Rai, A.K., Dahiya, D., Chettri, R., Nigam, P.S., 2021. Exploring endophytes for in vitro synthesis of bioactive compounds similar to metabolites produced in vivo by host plants. AIMS Microbiol. 7 (2), 175.

Shen, B., 2003. Polyketide biosynthesis beyond the type I, II and III polyketide synthase paradigms. Curr. Opin. Chem. Biol. 7 (2), 285–295.

Sibanda, E., Mabandla, M., Chisango, T., Nhidza, A.F., Mduluza, T., 2018. Endophytic fungi associated with *Annona senegalensis*: identification, antimicrobial and antioxidant potential. Curr. Biotechnol. 7, 317–322.
Sosa-Gómez, D.R., Moscardi, F., 1998. Laboratory and field studies on the infection of stink bugs, *Nezara viridula, Piezodorus guildinii,* and *Euschistus heros* (Hemiptera: Pentatomidae) with *Metarhizium anisopliae* and *Beauveria bassiana* in Brazil. J. Invertebr. Pathol. 71 (2), 115–120.
Stoev, S.D., 2015. Foodborne mycotoxicoses, risk assessment and underestimated hazard of masked mycotoxins and joint mycotoxin effects or interaction. Environ. Toxicol. Pharmacol. 39, 794–809.
Stoppacher, N., Kluger, B., Zeilinger, S., et al., 2010. Identification and profiling of volatilmetabolites of the biocontrol fungus *Trichoderma atroviride* by HS-SPME-GC-MS. J. Microbiol. Methods 81, 187–193.
Strausbaugh, C.A., Eujayl, I.A., 2012. Influence of sugarbeet tillage systems on the Rhizoctonia-bacterial root rot complex. J. Sugar Beet Res. 49, 57–78.
Strausbaugh, C.A., Gillen, A.M., 2008. Bacteria and yeast associated with sugar beet root rot at harvest in the intermountain west. Plant Dis. 92, 357–363.
Strobel, G.A., 2003. Endophytes as sources of bioactive products. Microbes Infect. 5 (6), 535–544.
Strobel, G., Yang, X.S., Sears, J., Kramer, R., Sidhu, R.S., Hess, W.M., 1996. Taxol from *Pestalotiopsis microspora*, an endophytic fungus of *Taxus wallachiana*. Microbiology 142, 435–440.
Szekeres, A., Leitgeb, B., Kredics, L., Antal, Z., Hatvani, L., Manczinger, L., Vágvölgyi, C., 2005. Peptaibols and related peptaibiotics of *Trichoderma*. Acta Microbiol. Immunol. Hung. 52 (2), 137–168.
Tagawa, M., Tamaki, H., Manome, A., Koyama, O., Kamagata, Y., 2010. Isolation and characterization of antagonistic fungi against potato scab pathogens from potato field soils. FEMS Microbiol. Lett. 305 (2), 136–142.
Talakayala, A., Mekala, G.K., Malireddy, R.K., Ankanagari, S., Garladinne, M., 2022a. Manipulating resistance to mungbean yellow mosaic virus in greengram (*Vigna radiata* L): through CRISPR/Cas9 mediated editing of the viral genome. Front. Sustain. Food Syst. 276.
Talakayala, A., Ankanagari, S., Garladinne, M., 2022b. CRISPR-Cas genome editing system: a versatile tool for developing disease resistant crops. Plant Stress, 100056.
Tan, R., Zou, W., 2001. Endophytes: a rich source of functional metabolites. Nat. Prod. Rep. 18, 448–459.
Thambugala, K.M., Daranagama, D.A., Phillips, A.J., Kannangara, S.D., Promputtha, I., 2020. Fungi vs. fungi in biocontrol: an overview of fungal antagonists applied against fungal plant pathogens. Front. Cell. Infect. Microbiol. 10, 604923.
Tian, Q., Li, B., Feng, Y., Zhao, W., Huang, J., Chao, H., 2022. Application of CRISPR/Cas9 in rapeseed for gene function research and genetic improvement. Agronomy 12 (4), 824.
Tóth, L., Váradi, G., Boros, É., Borics, A., Ficze, H., Nagy, I., et al., 2020. Biofungicidal potential of *Neosartorya (Aspergillus) fischeri* antifungal protein NFAP and novel synthetic γ-core peptides. Front. Microbiol. 11, 820.
Tripathi, L., Dhugga, K.S., Ntui, V.O., Runo, S., Syombua, E.D., Muiruri, S., Wen, Z., Tripathi, J.N., 2022. Genome editing for sustainable agriculture in Africa. Front. Genome 4, 876697.
Tyśkiewicz, R., Nowak, A., Ozimek, E., Jaroszuk-Ściseł, J., 2022. Trichoderma: the current status of its application in agriculture for the biocontrol of fungal phytopathogens and stimulation of plant growth. Int. J. Mol. Sci. 23 (4), 2329.
Von Samson-Himmelstjerna, G., Harder, A., Sangster, N.C., Coles, G.C., 2005. Efficacy of two cyclooctadepsipeptides, PF1022A and emodepside, against anthelmintic-resistant nematodes in sheep and cattle. Parasitology 130 (3), 343–347.
Wang, S., Liu, Z., Wang, X., Liu, R., Zou, L., 2022. Mushrooms do produce flavonoids: metabolite profiling and transcriptome analysis of flavonoid synthesis in the medicinal mushroom *Sanghuangporus baumii*. J. Fungi 8 (6), 582.
Wasser, S.P., 2010. Current findings, future trends, and unsolved problems in studies of medicinal mushrooms. Appl. Microbiol. Biotechnol. 89 (5), 1323–1332.
Watson, M.T., Tian, T., Estabrook, E., Falk, B.W., 1998. A small RNA resembling the beet western yellows luteovirus ST9-associated RNA is a component of the California carrot motley dwarf complex. Phytopathology 88, 164–170.
Weber, R.W., Stenger, E., Meffert, A., Hahn, M., 2004. Brefeldin A production by Phoma medicaginis in dead precolonized plant tissue: a strategy for habitat conquest. Mycol. Res. 108, 662–671.
Wiest, A., Grzegorski, D., Xu, B.W., Goulard, C., Rebuffat, S., Ebbole, D.J., Bodo, B., Kenerley, C., 2002. Identification of peptaibols from *Trichoderma virens* and cloning of a peptaibol synthetase. J. Biol. Chem. 277 (23), 20862–20868.

Yassin, M.T., Mostafa, A.A.F., Al-Askar, A.A., 2022. In vitro antagonistic activity of *Trichoderma* spp. against fungal pathogens causing black point disease of wheat. J. Taibah Univ. Sci. 16 (1), 57–65.

Yin, K., Han, T., Liu, G., Chen, T., Wang, Y., Yu, A.Y., Liu, Y., 2015. A geminivirus-based guide RNA delivery system for CRISPR/Cas9 mediated plant genome editing. Sci. Rep. 5, 14926.

Zafari, D., Rouhani, H., Okhovat, M., Hejaroud, G.A., 1994. Biological control of *Phytophthora erythroseptica* by *Trichoderma* spp. In: 11th Iranian Plant Protection Congress, p. 156.

Zhang, L., Fasoyin, O.E., Molnár, I., Xu, Y., 2020. Secondary metabolites from hypocrealean entomopathogenic fungi: novel bioactive compounds. Nat. Prod. Rep. 37 (9), 1181–1206.

Zhang, Y., Iaffaldano, B., Qi, Y., 2021. CRISPR ribonucleoprotein-mediated genetic engineering in plants. Plant Commun. 2 (2), 100168.

CHAPTER

10

Trichoderma secondary metabolites for effective plant pathogen control

A.D. Lopes[a], W.R. Rivadavea[b], and G.J. Silva[c]

[a]Biotechnology Department, Post-Graduate Program in Biotechnology Applied to Agriculture, Paranaense University, Umuarama, Paraná, Brazil [b]Agronomy Department, Graduate program in Agronomy Engineering, Paranaense University, Umuarama, Paraná, Brazil [c]Biotechnology Department, Post-Graduate Program in Biotechnology Applied to Agriculture, Paranaense University, Toledo, Paraná, Brazil

1 Introduction

The *Trichoderma* fungal genus is composed of a large number of species that preferentially colonize fruiting bodies of other fungi and dead wood (Jaklitsch and Voglmayr, 2015). The most robust studies, to date, account for more than 320 catalogued species (Bissett et al., 2015; Qiao et al., 2018). All these *Trichoderma* sp. are exclusive parasites of fungi (Kubicek et al., 2019). However, most exhibit characteristics of antibiosis; competition for nutrients and space; release of Secondary Metabolites (SM); end up showing antagonistic behavior against a multitude of different pathogens, such as fungi, insects, nematodes and bacteria (Zhang et al., 2017a,b; Rodrigues et al., 2023, Gunjal, 2023). By improving the nutritional support of plants, they also help in tolerance to abiotic stresses (Sood et al., 2020; Fazeli-Nasab et al., 2022; Poveda and Eugui, 2022).

Fungi of *Trichoderma* genus have a multitude of SM, and with new molecular biology techniques, more and more species have been discovered, and consequently, more genes that produce these metabolites. A thorough review carried out by (Zhang et al., 2021) more than 200 newly described SM, which, if added to the metabolites discovered before that date, add up to more than 400 (Reino et al., 2008).

The great majority of these metabolites are concentrated in studies aimed at their identification. However, some of these already have their metabolic functions elucidated and

Nanohybrid Fungicides
https://doi.org/10.1016/B978-0-443-23950-2.00008-4

are in the testing phase concentrated. This chapter will address the most recent discoveries involving the production of SM of *Trichoderma* in the biocontrol of several species of pathogens important to agriculture, with a main focus on fungi and nematodes.

2 Trichoderma against nematodes

Nematodes represent a limiting factor for success in crops. More than 4100 nematode species have been described so far, with a group of genera considered as the main phytopathogens, and others, specific to a more limited scope of crops, both of which have major impacts on sparing important crops. It is estimated, according to Singh et al. (2015), that the damage caused by nematodes to plants, caused yield losses of 12.3% (US$ 157 billion) worldwide, more significant damage than that caused by bugs, about 70 billion dollars (Bradshaw et al., 2016).

These data, however, may be underestimated, most likely because farmers are often unsuspecting that these phytoparasites are present, as symptoms caused to plant are usually nonspecific, hampering to attribute crop losses to damage caused by nematodes (Jones et al., 2013; Siddique and Grundler, 2018). Additional losses may also be related to food quality and visual defects associated with infection symptoms (Palomares-Rius et al., 2017).

Among the phytonematodes of economic importance, we highlight the nematodes root-knot (RKNs; *Meloidogyne* spp.), which are the world's largest agricultural pests, causing an estimated US$ 118 billion in annual economic losses (McCarter, 2008). The *Megalaima incognita* species, extremely important, infects the roots of almost all cultivated plants, promotes the formation of massive cells in its roots, which block the absorption of nutrients and water and facilitate infection with opportunistic microorganisms (Jang et al., 2016).

The main difficulty encountered at root-knot nematodes control is the high reproduction rate and the short generation time of new organisms (Trudqill and Blok, 2001; Manzanilla-Lopez et al., 2004). The pathogen control carried out by chemical nematicides is generally used, however, its continuous use, in addition to toxicity, presents a loss of efficiency (Fuller et al., 2008; Ntalli and Caboni, 2012). The development of effective and safe alternatives is mainly due to the increased demand for environmental protection measures. In that regard, environmentally friendly farming practices have been prioritized (OECD, 2001). Therefore, the use of *Trichoderma* fungi and their SM, becomes a potential opportunity for the environment and control of plant diseases (Hyde and Soytong, 2008).

2.1 Interaction between *Trichoderma* fungi for nematodes control and their mechanisms of action

Secondary metabolites are compounds produced by the organism, not directly essential for growth (Keller et al., 2005; Hoffmeister and Keller, 2007; Osbourn, 2010), but that can perform different functions, including antimicrobial activity against important phytopathogens for agriculture (Daoubi et al., 2009). Fungi produce a wide diversity of these SM. In addition, they are known for their ability to produce a huge amount of antibiotics, enzymes, vitamins, organic acids and polysaccharides (Meyer, 2008). The most potent agents currently in use are produced by fungi of the genus *Trichoderma* (Harman et al., 2004; Ming et al., 2012). As will be reported in the topics below, SM such as gliotoxin, gliovirin, peptaibols, polyketides, terpenes, and pyrones are extremely effective against phytopathogenic organisms such as bacteria, yeast, fungi, and nematodes (Monte, 2001; Vizcaino et al., 2005; Vinale et al., 2008).

The investigation of fungal strains with potential antimicrobial activity in vivo, through greenhouse evaluations, is a laborious and time-consuming activity. In this way, in vitro tests for the selection of biocontrol antagonists are performed in order to demonstrate their mechanism of action (Carvalho Filho et al., 2008), i.e., the first screening in the selection of effective candidates (Jeyaseelan et al., 2014; Barari, 2016). An effective strategy for screening fungal species that produce compounds (SM) is in vitro evaluation. This assessment facilitates and speeds up the selection of specific microorganisms. This technique has been successfully used with other microorganisms, such as studies by (Becker et al., 1988) which selected strains of rhizobacteria producing nematicidal compounds, as control agents of *M. incognita*.

Recent research, such as that by Ali Khan et al. (2020), investigated, in vitro, how was the activity of SM from several *Trichoderma* strains, as *T. acitrinoviride* (T130), *T. afroharzianum* (32,233 and T52), *T. asperelloides* (T136), *T. koningiopsis* (T84), *T. longibrachiatum* (T161), *T. pseudoharzianum* (T113, T129 and T160) and *T. viridescens* (T196), cultured in different culture media, against eggs and juveniles (J2) of *M. incognita*. Abamacetin at a concentration of 10 μg/mL was used as a positive control.

The nematicidal activity of *Trichoderma* SM, which have a truly inhibiting power against the *M.incognita* eggs, was described only for the species *T. hamatum* (T21) and *T. viridae*. The greatest inhibition of egg hatching (71.6%) was observed in *T. viridae*, in STP medium, followed by 67.3% in solid medium. The hatching of eggs that had the second highest inhibition (59.2% and 54.7%) came from the metabolites produced by *T. hamatum* (T21) in STP and solid medium. The other strains were not effective, presenting no or few effects against nematodes. For J2 juveniles, the result was similar and the highest percentage of death in juveniles was accomplished by *T. viridae* and T. *hamatum* (T21) respectively. In this study, the *Trichoderma* spp. SM were not effective.

The same authors reported, as an effect of SM produced by *Trichoderma*, *M. incognita* J2 presented morphological variations, with emphasis on the results involving the species *T. viridae*. It is very likely that the internal organs were infected and destroyed, which resulted in a straightened and stiffened body, as observed by microscopic analysis. The same was observed for the positive control (Abamectin). SM of the same species (*T. viridae*) obtained in MOF medium showed less efficacy promoting in the J2 nematode, less rigidity and straightness during the treatment with MMK2, SuM medium and methanol, showing no effect of mortality on J2, and therefore no morphological changes at nematodes were induced.

Recently, Baazeem et al. (2021) evaluated SM of *T. hamatum* (FB10), in relation to nematicidal activity. For this, they cultivated fungal strains in different cultures (carrot broth, cornmeal broth, potato dextrose broth, modified potato dextrose broth and water). *M. incognita* (J2) and eggs were submitted for evaluation, from the counting of 100 eggs and 100 J2, after incubation for 72 h with SM at 100 mg/mL extracted in ethyl acetate. The results indicated that *T. hamatum* (FB10) showed great capability to hinder the hatchability of eggs (78 ± 2.6% in modified potato dextrose broth); great capability to hinder the J2 *M. incognita* (89 ± 2.5% in

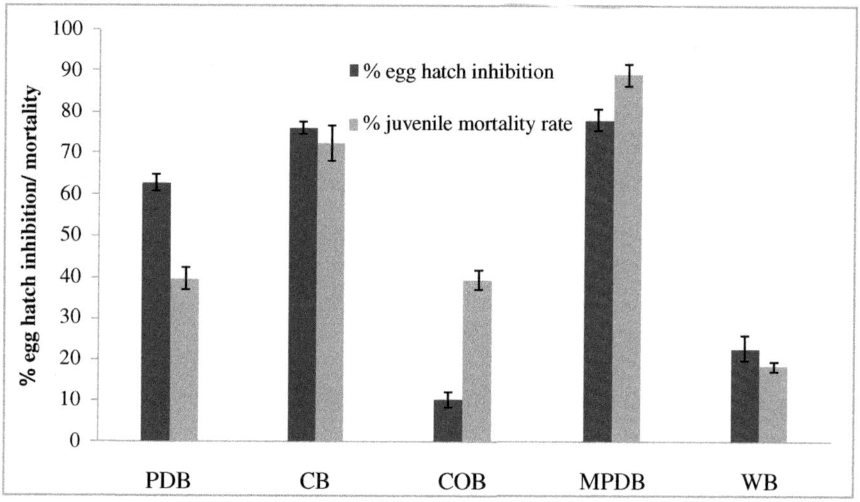

FIG. 1 Effect of SM from *T. hamatum* (FB10) cultured in various culture media and nematicidal activity (PDB: Potato Dextrose broth; CB: Carrot broth; COB: Cornmeal broth; MPD: Modified Potato Dextrose broth; WB: Water broth). *Reprinted from Baazeem, A., Almanea, A., Manikandan, P., Alorabi, M., Vijayaraghavan, P., Abdel-Hadi, A Yan et al. 2021. In Vitro antibacterial, antifungal, nematocidal and growth-promoting activities of Trichoderma hamatum FB10 and its SM. J. Fungi, 7, 331, 1–13. The present article is licensed under an open-access Creative Commons CC BY 4.0 license, MDPI.*

the same culture medium mentioned above). The carrot broth culture media showed 72.5 ± 4.4% mortality, while cornmeal broth showed lower efficiency when compared to other culture media (Fig. 1).

Khan et al. (2018) evaluated the ability of several species of *Trichoderma* to control *Megalaima incognita* in tomatoes, using the technique of exudates and cell suspension cultivation. In this study, the hatching of eggs was reduced and the mortality of J2 was accentuated, showing the efficiency of the SM of the studied strains in controlling *M. incognita*. In vivo tests verified, from the application of a culture containing fungus spores, that the J2 population was greatly affected. On the contrary, plant growth was higher, especially for plants colonized with *T. harzianum*. These results are promising as they confirm the potential as biological control agents against this nematode.

The nematode *M. incognita* is an important nematode that limits both indoor and outdoor cucumber production. Applications of *T. harzianum* and *T. viriens* in the soil and in seed treatment have significantly contributed to the decrease in cucumber losses due to problems related to nematodes and some opportunistic fungi (Nakkeeran et al., 2023).

Sahebani and Hadavi (2008) studied the effect of inoculating tomato seeds with *T. harzianum*, under greenhouse conditions. The authors evaluated several parameters such as some galls, the number of eggs per plant, establishment, development and reproduction of the plant. A reduction in all levels of disease caused by *M. javanica* was observed. Furthermore, in another study, when *T. harzianum* colonized plant roots, parasitism was prevented at several stages (invasion, galls and reproduction) (Martínez-Medina et al., 2017a, b).

The *Meloidogyne* genus females eggs are secreted wrapped in a sticky array, which form a bunch, considered as a defensive envelope with the function of protecting the eggs against pathogens, thus enabling their survival when in the soil. (2007). This sticky array makes a fundamental part in the task of joining the conidia of *Trichoderma* to *M. javanica*, explaining the parasitism caused by some species of *Trichoderma*. However, this mechanism of action cannot be extended to all species of the genus, since, for example, this matrix inhibited the growth of *T. harzianum*, when investigated by Poveda et al. (2020).

As previously reported in this review study, the genus *Trichoderma* is a powerful ally in the biological control against root-knot nematodes and nematodes that directly parasitize eggs and larvae. Studies suggest that the fungus begins to act against nematode eggs - consequently decreasing the number of J2 - when it releases extracellular enzymes capable of reducing cell wall components. These enzymes, mainly chitinases and proteases, have as their main action to degrade chitin and protein components, which are structures present in the cell wall of eggs.

As an example, we can highlight the promising results of *T. longibrachiatum* on *Heterodera avenae* eggs. The spores of *T. longibrachiatum* cover the eggs completely, releasing enzymes on them, and causing degradation, with consequent infiltration (Zhang et al., 2014). On the same fungus mentioned above, we found studies showing its action on females, J2 and again on *H. avenae* eggs (Zhang et al., 2017b). From these, we can verify that *T. longibrachiatum* has great potential to be used in the management of *H. avenae*.

The action of *Trichoderma* is not limited to the release of enzymes that will degrade the cell wall of pathogens. There are other mechanisms by which they act. In a study that aimed to verify the fungus-nematode interaction, Contina et al. (2017) used *T. harzianum* as a biomarker, for the species *Globoder pallida* (nematode known to infect potatoes). There was a reduction in nematode infection and reproduction, as cysts and J2 decreased. However, it did not affect eggs. As a method of action, *T. harzianum* established a connection (colonization) from the release of hyphae in the rhizoplane and in the rhizosphere where the potatoes were cultivated, allowing long-standing protection against this pathogen. This study demonstrated that, through direct interaction, *T. harzianum* fungi can control this type of nematode.

2.2 *Trichoderma* genus and resistance induction in nematodes

Filamentous fungi, in addition to acting as biological control agents, are capable of stimulating protection tools in plants against pathogens (Poveda et al., 2020), as is the case of the genus *Trichoderma* spp. which can stimulate the production of substances in plants that will be important for the defense against different types of pathogens. However, the mechanisms used by *Trichoderma* to help plant resistance against pathogens, especially when dealing with root-knot nematodes (RKNs), are a black box (Yan et al., 2021). The first authors to report the *Trichoderma*-mediated induction of systemic resistance were Calderón et al. (1993), in vine culture. Subsequently, Bigirimana et al. (1997) described the induction of resistance from soil treatment with *T. harzianum*, which increased bean resistance to important diseases caused by *Colletotrichum lindemuthianum* and *Botrytis cinerea*. Some studies have shown that *Trichoderma*, which colonize plant roots, can stimulate this protection against nematodes and other phytopathogens (Hermosa et al., 2012). The induction of resistance promoted by *Trichoderma* in several different species of plants can lead to changes in omic levels (Mukherjee et al., 2012).

The defense promoted by filamentous fungi systematically acts on the plant's immune response, that is, it prepares the plant's organisms, after priming, leaving the plant on alert against the attack of pathogens. This reduces the possibility of disease spreading (Mendoza-Mendoza et al., 2018). n ways of regulating resistance induction is signaling mediated by jasmonic acid/ethylene (JA/ET). In previous studies, Leonetti et al. (2014) observed that systemic signaling is downregulated in tomato roots when they are infected by *M. javanica*. However, when the roots are treated with *Trichoderma*, the JA/ET mediated response is induced. These findings corroborate the authors previously reported.

There are reports that the salicylic acid (SA) metabolic pathway integrates, together with JA, the induction of systemic defense caused by *T. virens* against *F. oxysporum* f.sp. *lycopersici* sp., in tomato plants (Martínez-Medina et al., 2017a,b; Jogaiah et al., 2018). The aforementioned authors, in their studies, showed that together, JA and SA orchestrate the plant's ability to resist or not resist nematode infection. It was evidenced that the plant/root-knot-nematode interaction is versatile. Even more versatile is the plant response, which is very different for each stage of infection. This shows plasticity in the induction of defense by *Trichoderma* (Fig. 2).

In the process of resistance induction, when the fungus *T. harzianum* comes into direct contact with plant roots, it stimulates defense responses, in the plant, mediated by SA. This initial

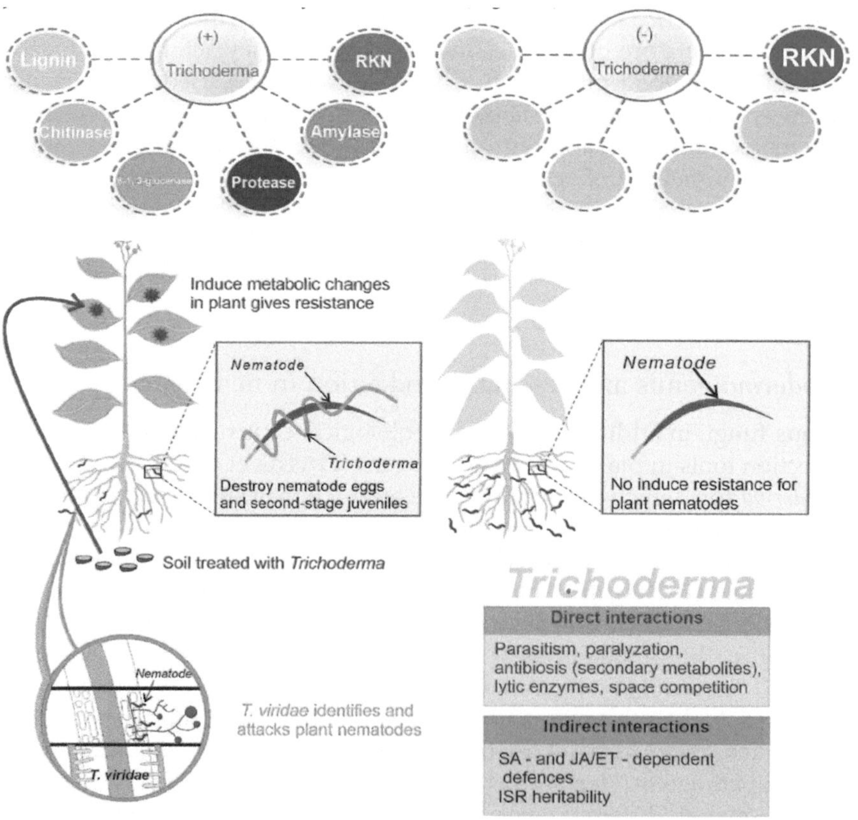

FIG. 2 Schematic of the types of resistance (direct and induced) against nematodes, from filamentous fungi in a hypothetical plant root.

defense is important. However, *M. incognita* afterwards manages to suppress it. At this moment, *Trichoderma* stimulates the expression of genes related to the metabolic pathway of JA, acting as an antagonist, and promoting the suppression of *M. incognita*-mediated defenses. This leads to reduced nematode populations. After the real establishment of the parasitism, *Trichoderma* again induces the expression of SA-mediated defenses, contributing to the prevention of new invasions (Martínez-Medina et al., 2017b).

The heritability of the fungus *T. atroviride* is represented by the association of plant growth after interaction with the fungus. This was confirmed by Medeiros et al. (2017). After the plant/fungus interaction, the metabolic defense pathways, dependent on SA and JA, are reduced. These interactions, both fungus/plant and nematode/plant, induce epigenetic changes that may be associated with small RNAs (Cabrera et al., 2016; Medeiros et al., 2017; Medina et al., 2017; Ruiz-Ferrer et al., 2018) in addition to already documented changes in plant methylomes (Hewezi et al., 2018). These processes are important both for the development of resistance and for symbiotic relationships.

As expected, some commercial formulations based on some *Trichoderma* strains, in recent studies, were able to induce resistance against *M. incognita* in tomatoes. In addition, they were able to promote an additive effect on the M1 1.2 resistance gene (Pocurull et al., 2020) These results open perspectives for integrated nematode management programs, with combined control programs. Interesting data were verified by Mukherjee et al. (2012), who found that high doses of *Trichoderma* inoculation can trigger an SA-mediated systemic resistance response, very similar to the response triggered by necrotrophic fungi. These data show that both the ISR and/or SAR pathways, and the genes that make their connections, can be induced by *Trichoderma* (Mendoza-Mendoza et al., 2018). As the data show, the defense response pathways may vary according to strains, *Trichoderma* species, pathogen and plant species (Brotman et al., 2012; Nawrocka and Małolepsza, 2013).

Yan et al. (2021) highlighted in their studies that *T. harzianum* can effectively extinguish, by 61.88%, the infestation of *M. incognita* (RKN) in *Solanum lycopersicum* L. plants, in addition to significantly reducing the levels of malondialdehyde, reactive oxygen species (ROS) and electrolyte leakage. After 75 days of inoculation with *M. incognita* (Table 1), the decrease in these markers was related to the accumulation of numerous SM, such as phenols, cellulose, flavonoids and lignin. In the same study, it was possible to verify that, in the same way, that SM

TABLE 1 Results of inoculation evaluation of *T. harzianum* and root-knot nematodes regarding to content of phenols, flavonoids, DPPH activity and cellulose in tomato roots.

Treatment	Phenols content ($mg\,g^{-1}$ FW)	Flavonoids content ($mg\,g^{-1}$ FW)	Lignin content ($mg\,g^{-1}$ FW)	Cellulose content ($mg\,g^{-1}$ FW)	DPPH activity
Control	2.31 ± 0.03	19.85 ± 1.22 d	1.23 ± 0.27c	30.12 ± 2.85 d	0.22 ± 0.03 d
RKN	2.69 ± 0.03c	25.63 ± 2.04c	1.41 ± 0.16c	38.56 ± 3.05c	0.34 ± 0.03c
TH	3.85 ± 0.04b	36.54 ± 3.17b	3.83 ± 0.41b	47.55 ± 5.13b	0.48 ± 0.04b
TH+RKN	4.75 ± 0.03 a	44.38 ± 5.12 a	4.96 ± 0.52 a	61.34 ± 4.95 a	0.62 ± 0.05 a

Means indicated by the same lower-case letters in the same column do not significantly differ according to Tukey's test at $P \leq 0.05$. *RKN: root-knot nematode: TH T.harzianum.*

and defense-related enzymatic activity was increased, the expression of associated genes (C4H, PAL, 4CL, LPO, CAD, CCOMT, G6PDH and Tpx1) also increased significantly mainly in the roots inoculated with *T. harzianum* + RKN. In addition, there was a significant increase in enzymes related to cell wall degradation of pathogens, such as β-1,3-glucanase (86.31%), chitinase (36.25%), protease (39, 78%) and amylase (17.06%)., in addition to SA and JA, a result not found in roots infected only with RKN.

All these results indicate that *T. harzianum* can be excellent sustainable alternatives in inducing systemic resistance against RKNs, as they activate important pathways, such as secondary metabolism, gene transcription of defense-related enzymes, etc.

3 *Trichoderma* × fungi

Trichoderma are excellent biofungicides. Recent data show that more than 60% of all biofungicides in the world are derived from *Trichoderma* (Abbey et al., 2019). The mechanisms of action by which phytopathogenic fungi are affected by *Trichoderma* SM are well described (Fig. 3). These affect spore germination, i.e., sporulation; degrade or prevent the formation and elongation of hyphae and mycelia; increase the release of enzymes and SM; produce toxins; cancel the development of reproductive structures; chelate nutrients, making it impossible for other microorganisms to use them, in addition to hindering the synthesis of mycotoxins from other fungi (Al-Ani, 2019).

There are numerous important SM in the biocontrol of pathogens that have been well characterized in several species of *Trichoderma* (Table 2). One of them, the first to be characterized (Brian, 1944), is gliotoxin (Dolan et al., 2015), a class of diketopiperazines, which internally has a disulfide bridge that acts by producing ROS, inactivating proteins (Gardiner et al., 2005). Some researchers show that the expression of gliotoxins in *Trichoderma* species may be linked to the antagonistic capacity of this species, as we will show below.

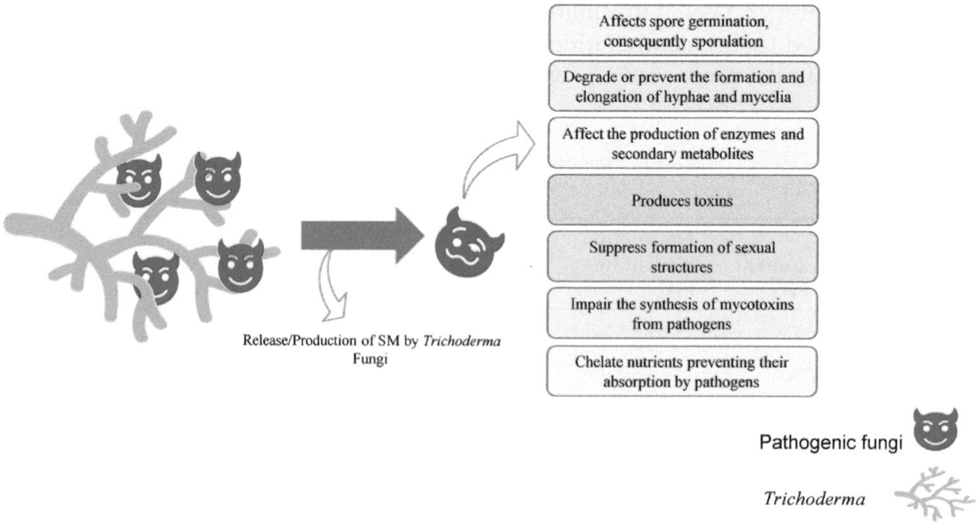

FIG. 3 Mechanisms of action of SM produced by *Trichoderma* on pathogenic fungi.

TABLE 2 Classification of SM that act in the biocontrol of pathogens, by species.

Secondary metabolite	Species	Phytopathogen	Reference
Gliotoxin	*Trichoderma virens* (HZA14); *T. virens* (T23)	*Phytophthora capsici; Sclerotium rolfsii*	Dolan et al. (2015), Tomah et al. (2020), Hua et al. (2021)
Gliovirins	*T. virens*	*P. palmivora, P. megakaria*	Pakora et al. (2018)
Peptaibols	*T. asperellum, T. longibrachiatum, T. viride, T. koningiopsis* and *T. gamsii*	*Alternaria alternata* (SZMC 16,085), *A. solani, Aspergillus niger, F. culmorum, F. graminearum, F. moniliforme, F. oxysporum, Fusarium solani* (SZMC 11467), *Micrococcus luteus* (SZMC 0264), *Staphylococcus aureus* (SZMC 0579), *Phoma cucurbitacearum* (SZMC 16088), *Pseudomonas aeruginosa* (SZMC 0568), *Rhizoctonia solani; Rhizoctonia solani* (SZMC6252J), *T. agressivum f. europaeum* (SZMC 1811), *T. gamsii* (SZMC 1656), *T. koningiopsis* (SZMC 12,500) and *T. pleuroti* (SZMC 12454).	Tamandegani et al. (2020), Gamal et al. (2022), Marik et al. (2018)
Pyrones	*T. asperellum*, (VM115) *T. koningiopsis*	*P. oryzae, Aspergillus fumigatus, Botrytis cinera*	Degani et al. (2021), Leylaie and Zafari (2018)
Harzianopiridone	*T. harzianum, T. harzianum* (CCTCC-SBW0162)	*R. solani, Sclerotinia scleritiorum, A. flavus; Botrytis cinerea; S. esculentum*	Dolatabad et al. (2019), Saravanakumar et al. (2018)
Azaphylones	*T. guizhouense*	*Fusarium oxysporum f.* sp. *cubense* race 4 (Foc4)	(Pang, 2020)

A study carried out with 15 *Trichoderma* isolates showed that 19.5% of them managed to suppress the development of *Phytophthora capsici*. In this study, *T. virens* (HZA14) was able to devastate and degrade the *P. capsici* colony and reduce the incidence of the disease by up to 64%. It was possible to identify the gliotoxin with high activity when *T. virens* (HZA14) acted against *P. capsici* (Tomah et al., 2020). Another study involving 10 *Trichoderma* strains against *Sclerotium rolfsii* showed that *T.virens* (T23) inhibited up to 70.2% of the pathogen's growth from an antagonistic effect. In this study, the *glyl-T* gene (glycotoxin pathway) was silenced, resulting in deficient production of this secondary metabolite, in addition to attenuating the antagonistic effect against *S. rolfsii* (Hua et al., 2021).

Gliovirins are members of the same class of gliotoxins and are described in several studies, acting against pathogens as *Phytophthora palmivora* and *P. megakaria* (Pakora et al., 2018). The biosynthetic pathway composed of 22 genes that lead to the production of gliovirins in *T.virens* was recently discovered (Sherkhane et al., 2017), opening new perspectives for the development of bioproducts based on gliovirins.

Another class of SM very well described in *Trichoderma* are the Peptaibols. These are a class of peptides with antibiotic properties. In nature there are more than 300 peptaibols catalogued and available in public databases (Whitmore and Wallace, 2004). A recent study showed that during the interaction between phytopathogenic fungi (*Alternaria solani*, *Fusarium culmorum*, *F. graminearum*, *F. moniliforme*, *F. oxysporum* species complex, and *Rhizoctonia solani*), with the fungi *T. asperellum* and *T. longibrachiatum*. *T. longibrachiatum* produced a significantly large and different amount of peptaibols, demonstrating that the interaction is important for the production of these metabolites in vivo (Tamandegani et al., 2020).

In a study conducted by Zhang et al. (2022), it was possible to identify 13 new peptaibols extracted from *T. longibrachiatum* Rifai DMG-3-1-1. These compounds were evaluated against three cell lines and showed high cytotoxicity against BV2 and MC F-7 cells. Furthermore, it was possible to identify moderate antibacterial activity against *Staphylococcus aureus* MRSA T144.

A study conducted with *T.viride* extracts, where there was confirmation by (UPLC/MS/MS) of the high concentration of peptaibols, proved to be able to promote the apoptotic effect of *Aspergillus niger*, through the increase of lipid peroxidation enzymes such as catalase, proteins and caspase-3 (suggesting mitochondrial action) (Gamal et al., 2022). Another study that aimed to elucidate the molecular structure of peptaibols from *T. gamsii* and *T. koningiopsis*, through HPLC-ESI-MS, released about 30 sequences that demonstrated broad antifungal and antibacterial activity against a range of microorganisms, including *Micrococcus luteus* (SZMC 0264), *S. aureus* (SZMC 0579), *Pseudomonas aeruginosa* (SZMC 0568), *Alternaria alternata* (SZMC 16085), *Fusarium solani* (SZMC 11467) species complex, *Rhizoctonia solani* (SZMC 6252J), *Phoma cucurbitacearum* (SZMC 16088), *T. agressivum f. europaeum* (SZMC 1811), *T. pleuroti* (SZMC 12454), *T. gamsii* (SZMC 1656) and *T. koningiopsis* (SZMC 12500) (Marik et al., 2018).

Another class of SM found in several species of *Tricoderms* are pyrones and their analogues. Pyrones are aromatic compounds and are well known for being responsible for the characteristic smell of coconut. They are reported in the literature for having great antimicrobial potential (Khan et al., 2020). In a recent study, researchers were able to isolate and identify, by GC–MS, the compound 6-Pentyl-α-pyrone, extracted from *T. asperellum*, which was responsible for inhibiting 100% of the growth of *Magnaportiopsis maydis*, an important pathogen in maize (Degani Yan et al., 2021). It is important to highlight that this was a compound present in another fungus of another species of *Trichoderma*, and this will be exposed in the next paragraph and next table (Table 3). In another study identified in the fungus *T. koningiopsis* (VM115), the production of 6-n-pentyl-6H-pyran-2-one (6PP), which was one of the most abundant metabolites was found. In this study, it was possible to prove a cytotoxic and fungicidal effect against *Aspergillus fumigatus*, *Botrytis cinerea* and *P. oryzae* (Leylaie and Zafari, 2018).

In a study conducted with the fungus *T. hamatum* (FB10), Baazeem et al. (2021) isolated and identified SM extract from ethyl acetate from GC/MS, and identified 6-pentyl-alpha-pyrone as the main compound of this strain, highlighted by the authors.

The previously reported study also performed biological activity against phytopathogenic fungi. The evaluated species were: *Alternaria radicina*, *A. citri*, *A. daucie*, *Rhizoctonia solani* and *S. sclerotiorum*. For these species, T. hamatum (FB10) had antagonistic activity (Table 4). The joint work (identification of compounds and biological activity) made it possible to effectively verify that the presence of the compound 6-pentyl-alpha-pyrone was possibly responsible for the effective inhibition of the fungi growth reported above.

TABLE 3 Analysis of compounds found from the fungal extract of *T. hamatum* from GC/MS.

Peak no	Chemical name	Chemical formula	Retention time (min)	Abundance (%)
1	Butyrolactone	$C_4H_6O_2$	7.371	63.51
2	Sulfurous acid, octyl 2-pentyl ester	$C_{13}H_{28}O_3S$	7.862	39.03
3	Ethanolic acid	$C_2H_4O_2$	7.896	27.07
4	2-Butoxyethyl acetate	$C_8H_{16}O_3$	14.298	87.02
5	Butanoic acid, Butyl ester	$C_8H_{16}O_2$	14.82	48.01
6	1-Hydroxy-2-propanone	C_6H_6O	15.96	79.66
7	3,5-Bis(1,1-dimethylethyl)phenol	$C_{14}H_{22}O$	17.25	83.65
8	6-Penthyl-alpha-pyrone	$C_{10}H_{14}O_2$	22.03	67.05
9	Hexadecanoic acid	$C_{20}H_{40}O_2$	22.87	82.69
10	2H-pyran-2-one	$C_6H_{10}O_3$	27.65	40.26
11	2,6-Dimethyl-naphthalene	$C_{12}H_{12}$	29.97	70.81
12	Hexadecane	$C_{16}H_{34}$	38.06	82.39
13	2-Octene	C_8H_{16}	49.2	72.7

Reprinted from Baazeem, A., Almanea, A., Manikandan, P., Alorabi, M., Vijayaraghavan, P., Abdel-Hadi, A., 2021. In Vitro antibacterial, antifungal, nematocidal and growth-promoting activities of Trichoderma hamatum FB10 and its SM. J. Fungi, 7, 331, 1–13. The present article is licensed under an open-access Creative Commons CC BY 4.0 license, MDPI.

TABLE 4 Antifungal activities of SM against fungal phytopathogens.

Fungal phytopathogens	MIC (μg/mL)	MBC (μg/mL)
S. sclerotiorum	63.5±7.25	120±10.5
R. solani	71±3.25	153±2.5
A. radicina	58.5±3.0	115.5±1.25
A. critri	60.5±5.5	110.2±3.75
A. dauci	65±3.75	122.5±3.0

Reprinted from Baazeem, A., Almanea, A., Manikandan, P., Alorabi, M., Vijayaraghavan, P., Abdel-Hadi, A., 2021. In Vitro antibacterial, antifungal, nematocidal and growth-promoting activities of Trichoderma hamatum FB10 and its SM. J. Fungi, 7, 331, 1–13. The present article is licensed under an open-access Creative Commons CC BY 4.0 license, MDPI.

Another recent study showed that crude extracts of *T. pinnatum* LS029-3 were effective against the fungus *Lasiodiplodia theobromae*, which causes postharvest rot in mangoes. In this crude extract, quantified by GC/MS, it was possible to identify the major compounds (Z)-13-docosenamide (ACROS), Hexanedioic acid, bis(2-ethylhexyl) ester (DOA) and (Z)-9-Octadecenamide (MSDS) (Zhan et al., 2023).

The secondary metabolite called harzianopiridone, a pyridone long reported in *Trichoderma* species, was recently found in a study that performed protoplast fusion between *T. harzianum* strains. In this study, in addition to the production of harzianopyridones in high concentrations, it was possible to control the phytopathogenic fungi *R.solani*, *Sclerotinia scleritiorum* and *A. flavus* (Dolatabad et al., 2019). Another study also with T. *harzianum* showed that the strain CCTCC-SBW0162 inhibited 90.6% of the production of the phytopathogenic fungus *Botrytis cinerea* and 80.7% of the disease in *S. esculentum*. This study, based on molecular docking, predicted that harzianopyridone, harzianolide and anthraquinone C could be responsible for this result (Saravanakumar et al., 2018).

Azaphylones are structurally diverse, bicyclic, oxygenated metabolites with a chiral quaternary center derived from polyketides (Osmanova et al., 2010). This compound is usually found in fungi and is associated with antibiosis. Researchers found that the fungus *T. guizhouense*, when fighting the fungus *Fusarium oxysporum* f. sp. *cubense* race 4 (Foc4), from the formation of aerial hyphae and release of hydrogen peroxide, protected itself from the same hydrogen peroxide that it released, producing an azaphylone called trigazafilone (Pang, 2020). These and several other SM of fungi of the genus *Trichoderma* are promising for the formulation of agricultural products with biocontrol action.

4 Challenges and future trends

The current status of all the studies mentioned in this chapter, and referring to the use of different species and strains of Trichoderma, confirm the potential of this genus in terms of antifungal and nematicide activity, revealing a positive correlation between biological control and the strains studied. However, the discussion about the potentiality of Trichoderma in the control of phytoparasites is not yet fully elucidated. Many authors have reported the in vitro *Trichoderma* activity against bacteria, fungi and nematodes. According to these studies, in addition to antimicrobial and nematicide activities, *Trichoderma* SM can act as plant growth promoters and influence the activity of soil enzymes, such as ureases, phosphatases, catalase and sucrase. These enzymes play important roles in soil fertility and the conversion of soil organic matter. Therefore, expanding research on this genus is essential in the process of prospecting new strains that are efficient in the biological control of phytopathogens, and essential to expand its use, thus representing a valid alternative to the use of chemicalsBiocontrol strategies for phytoparasitic nematodes constitute a valid alternative to toxic chemical nematicides, and therefore, much has been reported, especially with the use of filamentous fungi as biological control agents. These act mainly through two types of action mechanisms: those that include the production of SM (antibiosis), lytic enzymes and spatial competition by *Trichoderma*; and those that act based on the induction of plant defenses, such as SAR and ISR activation by *Trichoderma*.

The association of action mechanisms after treatment with fungi, with empirical data in the field, and the analysis of plant responses at a molecular level, can assist to identify and predict problems such as the presence of antagonists of the biological control agent in the soil, which could interfere in the rhizosphere biota and consequently in the control efficiency of these organisms. This perspective may allow more effective control of nematodes, in a given crop, for example. Furthermore, a more in-depth study of the molecular mechanisms of resistance induced by biological control agents would allow the direct manipulation of plants, taking into account, however, the synergistic effects of the SM produced by the biocontrol agents.

The results described in this chapter provide new insights into important mechanisms of resistance induced by *Trichoderma* species, such as *T. harzianum* against root-knot nematode, which may have potential implications for biological control. Despite the diversity of mechanisms by which fungi of the genus *Trichoderma* can act, there are still questions to be answered in future research. Recent studies, for example, highlight the occurrence of hereditary biocontrol after infection by biocontrol agents, possibly driven by epigenetics and mechanisms not yet well known, thus opening a new field of research with special applied interest.

Furthermore, the results described here highlight important Trichoderma SM that act especially against phytopathogenic fungi. Fungal diseases are the most devastating to major crops. The SM highlighted in this chapter show great possibilities for the development of biological products that can be effective against these fungal diseases, helping to reduce the use of chemicals, through the use of sustainable biological products.

5 Conclusion

The diversity of SM found in the most numerous species of *Trichoderma* opens up a range of possibilities for market exploration aimed at biological control against agricultural pathogens. Every day, due to the molecular biology techniques available, new metabolites, new microorganisms and new mechanisms of action are unveiled and these discoveries are extremely important in view of the great need to reduce the use of chemical molecules and fungicides in agriculture. In this review, we approach recent studies on the performance of the most studied species of *Trichoderma* against nematodes and phytopathogenic fungi, in addition to elucidating some mechanisms of action. We could verify that most of the studies involving SM of *Trichoderma* against nematodes were carried out with only a few species of *Trichoderma*. In addition, most studies report the effect of the fungus against nematodes, without considering which metabolites are associated with the action. In this way, given the enormous diversity of species already discovered, we can consider that many studies still need to be carried out and thus, the use of fungi of the genus *Trichoderma* in the biocontrol of nematodes is extremely promising. Regarding the action of fungi of the genus *Trichoderma* against phytopathogenic fungi, there are more studies with more species of *Trichoderma* and more species of phytopathogens. In addition, their studies also focus on elucidating which are the SM present in the phytopathogenic action. This is an important and necessary step for the production of metabolites at an industrial level for large-scale application in agriculture. In addition, these studies open up possibilities for the development of new prospects for SM, such as exposure of these microorganisms to other biotic or abiotic stresses. Finally, we understand and conclude that, in general, studies involving fungi of the genus *Trichoderma* and its SM can be used as a potential source for the isolation of bioactive SMs against nematodes and pathogenic fungi. The utilization of these SM could act as an effective, without chemicals, and eco-friendly disease management tool against plant pathogens, which must be increasingly explored in the face of climate change, which alters the entire balance of crops, and in the face of greater awareness of increasingly sustainable agriculture.

References

Abbey, J.A., Percival, D., Abbey, L., Asiedu, S.K., Prithiviraj, B., Schilder, A., 2019. Biofungicides as alternative to synthetic fungicide control of grey mould (Botrytis cinerea)–prospects and challenges. Biocontrol Sci. Tech. 29 (3), 207–228.

Al-Ani, L.K.T., 2019. Bioactive SM of *Trichoderma* spp. para o gerenciamento eficiente de fitopatógenos. In: Singh, H., Keswani, C., Reddy, M., Sansinenea, E., García-Estrada, C. (Eds.), SM of Plant Growth Promoting Rhizomicroorganisms. Springer.

Ali Khan, R.A., Najeeb, S., Mao, Z., Ling, J., Yang, Y., Li, Y., Xie, B., 2020. Bioactive SM from *Trichoderma* spp. against phytopathogenic bacteria and root-knot nematode. Microorganisms 8, 401.

Baazeem, A., Almanea, A., Manikandan, P., Alorabi, M., Vijayaraghavan, P., Abdel-Hadi, A., 2021. In vitro antibacterial, antifungal, Nematocidal and growth promoting activities of *Trichoderma hamatum* FB10 and its SM. J. Fungi 7, 331.

Barari, H., 2016. Biocontrol of tomato Fusarium wilt by *Trichoderma* species under in vitro and in vivo conditions. Cercet. Agron. Mold. 49, 91–98.

Becker, J.O., Zavaleta-Mejia, E., Colbert, S.F., Schroth, M.N., Weinhold, A.R., Hancock, J.G., Van Gundy, S.D., 1988. Effects of rhizobacteria on root-knot nematodes and gall formation. Phytopathology 78, 1466–1469.

Bigirimana, J., De Meyer, G., Poppe, J., Elad, Y., Höfte, M., 1997. Induction of systemic resistance on bean (*Phaseolus vulgaris*) by *Trichoderma harziamum*. Univ. Gent Fac. Landbouwwet. Meded. 62, 1001–1007.

Bissett, J., Gams, W., Jaklitsch, W., Samuels, G.J., 2015. Accepted *Trichoderma* names in the year 2015. IMA Fungus 6 (2), 263–295.

Bradshaw, C.J., Leroy, B., Bellard, C., Roiz, D., Albert, C., Fournier, A., et al., 2016. Massive yet grossly underestimated global costs of invasive insects. Nat. Commun. 7, 1–8.

Brian, P.W., 1944. Production of gliotoxin by *Trichoderma viride*. Nature 154, 667–668.

Brotman, Y., Lisec, J., Méret, M., Chet, I., Willmitzer, L., Viterbo, A., 2012. Transcript and metabolite analysis of the *Trichoderma*-induced systemic resistance response to *Pseudomonas syringae* in *Arabidopsis thaliana*. Microbiology 158, 139–146.

Cabrera, J., Barcala, M., García, A., Rio-Machín, A., Medina, C., Jaubert-Possamai, S., et al., 2016. Differentially expressed small RNAs in Arabidopsis galls formed by *Meloidogyne javanica*: a functional role for miR390 and its TAS 3-derived tasi RNAs. New Phytol. 209, 1625–1640.

Calderón, A.A., Zapata, J.M., Muñoz, R., Pedreño, M.A., Barceló, A.R., 1993. Resveratrol production as a part of the hypersensitivelike response of grapevine cells to an elicitor from *Trichoderma viride*. New Phytol. 124, 455–463.

Carvalho Filho, M.R., Menezes, J.E., Mello, S.C.M., Santos, R.P., 2008. Avaliação de Isolados de *Trichoderma* No Controle da Mancha Foliar do Eucalipto in Vitro e Quanto a Esporulação Em Dois Substratos Sólidos. Bol. Pesqui. Desenvolv. 225, 21.

Contina, J.B., Dandurand, L.M., Knudsen, G.R., 2017. Use of GFP-tagged *Trichoderma harzianum* as a tool to study the biological control of the potato cyst nematode *Globodera pallida*. Appl. Soil Ecol. 115, 31–37.

Daoubi, M., Pinedo-rivilla, C., Rubio, M.B., Hermosa, R., Monte, E., Aleu, J., Collado, I.G., 2009. Hemisynthesis and absolute configuration of novel 6-pentyl-2H-pyran-2-one derivatives from *Trichoderma* spp. Tetrahedron 65, 4834–4840.

Degani, O., Khatib, S., Becher, P., Gordani, A., Harris, R., 2021. *Trichoderma asperellum* secreted 6-Pentyl-α-Pyrone to control *Magnaporthiopsis maydis*, the maize late wilt disease agent. Biology 10 (9), 897.

Dolan, S.K., O'Keeffe, G., Jones, G.W., Doyle, S., 2015. Resistance is not futile: gliotoxin biosynthesis, functionality and utility. Trends Microbiol. 23, 419–428.

Dolatabad, H.K., Javan-Nikkhah, M., Safari, M., Golafaie, T.P., 2019. Effects of protoplast fusion on the antifungal activity of *Trichoderma* strains and their molecular characterisation. Arch. Phytopathol. Pflanzenschutz. 52 (17–18), 1255–1275.

Fazeli-Nasab, B., Shahraki-Mojahed, L., Piri, R., Sobhanizadeh, A., 2022. Trichoderma: melhorando o crescimento e a tolerância a estresses bióticos e abióticos nas plantas. In: Trends of Applied Microbiology for Sustainable Economy. Imprensa Acadêmica, pp. 525–564.

Fuller, V.L., Lilley, C.J., Urwin, P.E., 2008. Nematode resistance. New Phytol. 180, 27–44.

Gamal, M., Abou Zaid, M., Abou Mourad, I.K., Abd El Kareem, H., Gomaa, O.M., 2022. Peptaibol bioativo de *Trichoderma* viride induz apoptose em Aspergillus niger infectando tilápia em pisciculturas. Aquaculture 547, 737474.

Gardiner, D.M., Waring, P., Howlett, B.J., 2005. A classe epipolitiodioxopiperazina (ETP) de toxinas fúngicas: Distribuição, modo de ação, funções e biossíntese. Microbiology 151, 1021–1032.

Gunjal, A.B., 2023. Trichoderma: An eco-friendly biopesticide for sustainable agriculture. In: Organic Farming for Sustainable Development. Apple Academic Press, pp. 3–21.

Harman, G.E., Howell, C.R., Viterbo, A., Chet, I., Lorito, M., 2004. *Trichoderma* species–opportunistic avirulent plant symbionts. Nat. Rev. Microbiol. 2, 43–56.

Hermosa, R., Viterbo, A., Chet, I., Monte, E., 2012. Plant-beneficial effects of *Trichoderma* and of its genes. Microbiology 158, 17–25.

Hewezi, T., Pantalone, V., Bennett, M., Stewart, C.N., Burch-Smith, T.M., 2018. Phytopathogen-induced changes to plant methylomes. Plant Cell Rep. 37, 17–23.

Hoffmeister, D., Keller, N.P., 2007. Natural products of filamentous fungi: enzymes, genes, and their regulation. Nat. Prod. Rep. 24, 393–416.

Hua, L., Zeng, H., He, L., Jiang, Q., Ye, P., Liu, Y., Zhang, M., 2021. Gliotoxin is an important secondary metabolite involved in suppression of *Sclerotium rolfsii* by *Trichoderma virens* T23. Phytopathology 111, 1720–1725. https://doi.org/10.1094/PHYTO-09-20-0399-R.

Hyde, K.D., Soytong, K., 2008. The fungal endophyte dilemma. Fungal Divers 33, 173.

Jaklitsch, W.M., Voglmayr, H., 2015. Biodiversity of *Trichoderma* (Hypocreaceae) in southern Europe and Macaronesia. Stud. Mycol. 80, 1–87.

Jang, J.Y., Choi, Y.H., Shin, T.S., Kim, T.H., Shin, K.S., Park, H.W., 2016. Biological control of Meloidogyne incognita by Aspergillus niger F22 producing oxalic acid. PloS One 11, 156–230.

Jeyaseelan, E.C., Tharmila, S., Niranjan, K., 2014. Antagonistic activity of *Trichoderma* spp. and Bacillus spp. against *Pythium aphanidermatum* isolated from tomato damping. Arch. Appl. Sci. Res. 4, 1623–1627.

Jogaiah, S., Abdelrahman, M., Tran, L.S.P., Ito, S.I., 2018. Different mechanisms of *Trichoderma* virens-mediated resistance in tomato against Fusarium wilt involve the jasmonic and salicylic acid pathways. Mol. Plant Pathol. 19, 870–882.

Jones, J.T., Haegeman, A., Danchin, E.G., Gaur, H.S., Helder, J., Jones, M.G., et al., 2013. Top 10 plant-parasitic nematodes in molecular plant pathology. Mol. Plant Pathol. 14, 946–961.

Keller, J.M., Mcclellan-green, P.D., Kucklick, J.R., Keil, D.E., Peden-adams, M.M., 2005. Efects of organochlorine contaminants on loggerhead sea turtle immunity: comparison of a correlative field study and in vitro exposure experiments. Environ. Health Perspect. 114, 70–76.

Khan, M.R., Ahmad, I., Ahamad, F., 2018. Effect of pure culture and culture filtrates of *Trichoderma* species on root-knot nematode, Meloidogyne incognita infesting tomato. Indian Phytopathol. 71, 265–274.

Khan, R.A.A., Najeeb, S., Hussain, S., Xie, B., Li, Y., 2020. Bioactive SM from *Trichoderma* spp. against phytopathogenic fungi. Microorganisms 8 (6), 817.

Kubicek, C.P., Steindorff, A.S., Chenthamara, K., Manganiello, G., Henrissat, B., Zhang, J., Druzhinina, I.S., 2019. Evolution and comparative genomics of the most common *Trichoderma* species. BMC Genet. 20 (1), 1–24.

Leonetti, P., Costanza, A., Zonno, M., Molinari, S., Altomare, C., 2014. How fungi interact with nematode to activate the plant defence response to tomato plants. Commun. Agric. Appl. Biol. Sci. 79, 357–362.

Leylaie, S., Zafari, D., 2018. Antiproliferative and antimicrobial activities of SM and phylogenetic study of endophytic *Trichoderma* species from Vinca plants. Front. Microbiol. 9, 1484.

Manzanilla-Lopez, R.H., Kenneth, E., Bridge, J., 2004. Plant diseases caused by nematodes. In: Chen, Z.X., Chen, S.Y., Dickson, D.W. (Eds.), Nematology: Advanced and Perspectives. Nematode Management and Utilization. vol. 2. CAB International, Wallingford, UK, pp. 637–716.

Marik, T., Tyagi, C., Racić, G., Rakk, D., Szekeres, A., Vágvölgyi, C., Kredics, L., 2018. New 19-residue peptaibols from *Trichoderma* clade Viride. Microorganisms 6 (3), 85.

Martínez-Medina, A., Appels, F.V., Van Wees, S.C., 2017a. Impact of salicylic acid-and jasmonic acid-regulated defences on root colonization by *Trichoderma* harzianum T-78. Plant Signal. Behav. 12, 1345404.

Martínez-Medina, A., Fernandez, I., Lok, G.B., Pozo, M.J., Pieterse, C.M., Van Wees, S.C., 2017b. Shifting from priming of salicylic acid-to jasmonic acid-regulated defences by *Trichoderma* protects tomato against the root knot nematode Meloidogyne incognita. New Phytol. 213, 1363–1377.

McCarter, J.P., 2008. Nematology: Terra incognita no more. Nat. Biotechnol. 26, 882–884.

Medeiros, H.A., de Araújo Filho, J.V., De Freitas, L.G., Castillo, P., Rubio, M.B., Hermosa, R., et al., 2017. Tomato progeny inherit resistance to the nematode *Meloidogyne javanica* linked to plant growth induced by the biocontrol fungus *Trichoderma atroviride*. Sci. Rep. 7, 40216.

Medina, C., da Rocha, M., Magliano, M., Ratpopoulo, A., Revel, B., Marteu, N., et al., 2017. Characterization of microRNAs from Arabidopsis galls highlights a role for miR159 in the plant response to the root-knot nematode Meloidogyne incognita. New Phytol. 216, 882–896.
Mendoza-Mendoza, A., Zaid, R., Lawry, R., Hermosa, R., Monte, E., Horwitz, B.A., et al., 2018. Molecular dialogues between *Trichoderma* and roots: role of the fungal secretome. Fungal Biol. Rev. 32, 62–85.
Meyer, V., 2008. Genetic engineering of filamentous fungi-Progress, obstacles and future trends. Biotechnol. Adv. 26, 177–185.
Ming, Q., Han, T., Li, W., Zhang, Q., Zhang, H., Zheng, C., 2012. Tanshinone II a and tanshinone I production by *Trichoderma atroviride* D16, an endophytic fungus in *Salvia miltiorrhiza*. Phytomedicine 19, 330–333.
Monte, E., 2001. Understanding *Trichoderma*: between agricultural biotechnology and microbial ecology. Int. Microbiol. 4, 1–4.
Mukherjee, M., Mukherjee, P.K., Horwitz, B.A., Zachow, C., Berg, G., Zeilinger, S., 2012. *Trichoderma*–plant–pathogen interactions: advances in genetics of biological control. Indian J. Microbiol. 52, 522–529.
Nakkeeran, S., Sreenayana, B., Saravanan, R., Renukadevi, P., Vinodkumar, S., 2023. Harnessing the antifungal and Antinemic potential of Trichoderma in integrated crop management system for mitigating fungal nematode complex in cucumber. In: Integrated Pest Management in Diverse Cropping Systems. Apple Academic Press, pp. 485–516.
Nawrocka, J., Małolepsza, U., 2013. Diversity in plant systemic resistance induced by *Trichoderma*. Biol. Control 67, 149–156.
Ntalli, N.G., Caboni, P., 2012. Botanical nematicides: a review. J. Agric. Food Chem. 60, 9929–9940.
OECD, 2001. Agricultural Policies in OECD Countries: Monitoring and Evaluation. OECD, Paris, France.
Osbourn, A., 2010. Gene clusters for secondary metabolic pathways: an emerging theme in plant biology. Plant Physiol. 1, 154.
Osmanova, N., Schultze, W., Ayoub, N., 2010. Azaphilones: a class of fungal metabolites with diverse biological activities. Phytochem. Rev. 9 (2), 315–342.
Pakora, G.A., Mpika, J., Kone, D., Ducamp, M., Kebe, I., Nay, B., Buisson, D., 2018. Inhibition of Phytophthora species, agents of cocoa black pod disease, by SM of *Trichoderma* species. Environ. Sci. Pollut. Res. 25 (30), 29901–29909.
Palomares-Rius, J.E., Escobar, C., Cabrera, J., Vovlas, A., Castillo, P., 2017. Anatomical alterations in plant tissues induced by plant-parasitic nematodes. Front. Plant Sci. 8, 1987.
Pang, G., et al., 2020. Azaphilones biosynthesis complements the defence mechanism of *Trichoderma guizhouense* against oxidative stress. Environ. Microbiol. https://doi.org/10.1111/1462-2920.15246.
Pocurull, M., Fullana-Pons, A.M., Ferro, M., Valero, P., Escudero, N., Saus, E., et al., 2020. Commercial formulates of *Trichoderma* induce systemic plant resistance to Meloidogyne incognita in tomato and the effect is additive to that of the Mi-1.2 resistance gene. Front. Plant Sci. 10, 1–10. https://doi.org/10.3389/fmicb.2019.03042.
Poveda, J., Eugui, D., 2022. Combined use of Trichoderma and beneficial bacteria (mainly Bacillus and Pseudomonas): development of microbial synergistic bio-inoculants in sustainable agriculture. Biol. Control 105100.
Poveda, J., Abril-Urias, P., Escobar, C., 2020. Biological control of plant-parasitic nematodes by filamentous Fungi inducers of resistance: *Trichoderma*, mycorrhizal and endophytic fungi. Front. Microbiol. 11, 992.
Qiao, M., Du, X., Zhang, Z., Xu, J., Yu, Z., 2018. Three new species of soil-inhabiting *Trichoderma* from Southwest China. MycoKeys 44, 63.
Reino, J.L., Guerrero, R.F., Hernández-Galán, R., Collado, I.G., 2008. SM from species of the biocontrol agent *Trichoderma*. Phytochemistry 7, 89–123.
Rodrigues, A.O., May De Mio, L.L., Soccol, C.R., 2023. Trichoderma as a powerful fungal disease control agent for a more sustainable and healthy agriculture: recent studies and molecular insights. Planta 257 (2), 1–15.
Ruiz-Ferrer, V., Cabrera, J., Martinez-Argudo, I., Artaza, H., Fenoll, C., Escobar, C., 2018. Silenced retrotransposons are major rasiRNAs targets in Arabidopsis galls induced by *Meloidogyne javanica*. Mol. Plant Pathol. 19, 2431–2445.
Sahebani, N., Hadavi, N., 2008. Biological control of the root-knot nematode Meloidogyne javanica by *Trichoderma* harzianum. Soil Biol. Biochem. 40, 2016–2020.
Saravanakumar, K., Lu, Z., Xia, H., Wang, M., Sun, J., Wang, S., Chen, J., 2018. Triggering the biocontrol of Botrytis cinerea by *Trichoderma harzianum* through inhibition of pathogenicity and virulence related proteins. Front. Agric. Sci. Eng. 5 (2), 271–279.
Sherkhane, P.D., Bansal, R., Banerjee, K., Chatterjee, S., Oulkar, D., Jain, P., Mukherjee, P.K., 2017. Descoberta orientada pela genômica do cluster de genes da biossíntese de gliovirina no fungo benéfico da planta *Trichoderma* virens. ChemistrySelect 2 (11), 3347–3352.

Siddique, S., Grundler, F.M., 2018. Parasitic nematodes manipulate plant development to establish feeding sites. Curr. Opin. Microbiol. 46, 102–108.

Singh, S., Singh, B., Singh, A.P., 2015. Nematodes: a threat to sustainability of agriculture. Procedia Environ. Sci. 29, 215–216.

Sood, M., Kapoor, D., Kumar, V., Sheteiwy, M.S., Ramakrishnan, M., Landi, M., Sharma, A., 2020. *Trichoderma*: the "secrets" of a multitalented biocontrol agent. Plan. Theory 9 (6), 762.

Tamandegani, P., Marik, T., Zafari, D., Balázs, D., Vágvölgyi, C., Szekeres, A., Kredics, L., 2020. Changes in peptaibol production of *Trichoderma* species during in vitro antagonistic interactions with fungal plant pathogens. Biomol. Ther. 10 (5), 730.

Tomah, A.A., Abd Alamer, I.S., Li, B., Zhang, J.Z., 2020. A new species of *Trichoderma* and gliotoxin role: a new observation in enhancing biocontrol potential of *T. virens* against *Phytophthora capsici* on chili pepper. Biol. Control 145, 104–261.

Trudqill, D.L., Blok, V.C., 2001. Apomictic, polyphagous root-knot nematodes: exceptionally successful and damaging biotrophic root pathogens. Annu. Rev. Phytopathol. 39, 53–77.

Vinale, F., Sivasithamparam, K., Ghisalberti, E.L., Woo, S.L., Nigro, M., Marra, R., 2008. *Trichoderma* SM active on plants and fungal pathogens. Open Mycol. J. 8, 127–139.

Vizcaino, J.A., Sanz, L., Cardoza, R.E., Monte, E., Gutierrez, S., 2005. Detection of putative peptide synthetase genes in *Trichoderma* species. Application of this method to the cloning of a gene from *T. harzianum* CECT 2413. FEMS Microbiol. Let. 244, 139–148.

Whitmore, L., Wallace, B., 2004. The Peptaibol database: a database for sequences and structures of naturally occurring peptaibols. Nucleic Acids Res. 32, 593–594.

Yan, Y., et al., 2021. *Trichoderma harzianum* induces resistance to root-knot nematodes by increasing secondary metabolite synthesis and defense-related enzyme activity in *Solanum lycopersicum* L. Biol. Control 158, 1–11.

Zhan, X., Khan, R.A.A., Zhang, J., Chen, J., Yin, Y., Tang, Z., Liu, T., 2023. Control of postharvest stem-end rot on mango by antifungal metabolites of *Trichoderma pinnatum* LS029-3. Sci. Hortic. 310, 111696.

Zhang, S., Gan, Y., Xu, B., Xue, Y., 2014. The parasitic and lethal effects of *Trichoderma longibrachiatum* against *Heterodera avenae*. Biol. Control 72, 1–8.

Zhang, J., Chen, G.Y., Li, X.Z., Hu, M., Wang, B.Y., Ruan, B.H., Zhou, H., Zhao, L.X., Zhou, J., Ding, Z.T., et al., 2017a. Phytotoxic, antibacterial, and antioxidant activities of mycotoxins and other metabolites from *Trichoderma* sp. Nat. Prod. Res. 31, 2745–2752.

Zhang, S., Gan, Y., Ji, W., Xu, B., Hou, B., Liu, J., 2017b. Mechanisms and characterization of *Trichoderma longibrachiatum* T6 in suppressing nematodes (*Heterodera avenae*) in wheat. Front. Plant Sci. 8, 1491.

Zhang, J.L., Tang, W.L., Huang, Q.R., Li, Y.Z., Wei, M.L., Jiang, L.L., Zhang, X.X., 2021. *Trichoderma*: a treasure house of structurally diverse SM with medicinal importance. Front. Microbiol., 2037.

Zhang, S.H., Zhao, X., Xu, R., Yang, Y., Tang, J., Yue, X.L., 2022. Eleven-residue peptaibols isolated from *Trichoderma longibrachiatum* Rifai DMG-3-1-1 and their structure-activity relationship. Chem. Biodivers. 19 (9), e202200627.

CHAPTER

11

Exploring biological control strategies for managing *Fusarium* mycotoxins

Mirza Abid Mehmood[a], *Areeba Rauf*[a], *Muhammad Ashfaq*[a], *and Furqan Ahmad*[b]

[a]Plant Pathology, Institute of Plant Protection, Muhammad Nawaz Shareef University of Agriculture, Multan, Pakistan [b]Institute of Plant Breeding and Biotechnology, Muhammad Nawaz Shareef University of Agriculture, Multan, Pakistan

1 Introduction

Due to an increase in global population over the next 30 years, the world will require a 50%–70% increase in food production to preclude global food insecurity (Mottaleb et al., 2022). Food insecurity is a severe threat in developing nations, especially in Africa, where a huge population is facing a shortage of food and the condition is predicted to worsen in the future (Sasson, 2012; Wudil et al., 2022; Odeya et al., 2022). The aim of feeding these people securely must be achieved in a domain with decreasing cultivable land, less and costly fossil reserves, progressively inadequate water supplies, social instability, economic uncertainty, and a highly dynamic climate situation. Furthermore, the consequences of plant diseases cannot be overstated. A major challenge to global food accessibility and security is the influence of fungal infections and unexpected deviations of reputable pathogens on agriculturally indispensable crops. An estimated 8.5% of the world's population suffered significantly from diseases affecting our major agricultural crops (Fisher et al., 2012; Anand and Rajeshkumar, 2022). Without effective management of these diseases, the objective of serving the world's rising population will be impossible to achieve toxin-producing pathogens, their secondary metabolites having distinct chemical structures as well as biological properties that possess a broad range of harmful effects on humans as well as cattle, posed to be a hazard to food security (Chukwudi et al., 2021; Zain, 2011; De Ruyck et al., 2015). The chapter aims to examine the bioactive agents against various infections and their toxins.

https://doi.org/10.1016/B978-0-443-23950-2.00014-X

2 Management of plant diseases

Plant diseases can be efficiently managed by employing various methods, which includes growing resilient cultivars, practice of efficient crop alternation patterns, and using chemical as well (Corkley et al., 2022). Due to the harmful effects of chemicals, used to protect the plant, human, and animal health, the European Union has encouraged exploration of alternate and environmentally friendly options such as the use of various pest management strategies as well as the practice of biological control agents (BCAs) (Bronzwaer et al., 2022). In plant pathology, biocontrol, also refers as biological control, aims to use microbes to preclude the invasion or development of harmful plant pathogens (Pal and Gardener, 2006).

Beneficial microorganisms that can combat plant diseases and safeguard the plant are termed biocontrol agents (Weller, 1988; Abdallah et al., 2018; El-Tarabily and Sivasithamparam, 2006b; Punja and Utkhede, 2003).

3 Mycotoxins

Practically, all crops contain mycotoxins, which are produced under specific environmental conditions during or after the microbial invasion of the plant (Chukwudi et al., 2021; Marin et al., 2013). Short- and long-term exposure to mycotoxins can have a variety of harmful consequences on a variety of organisms (Zain, 2011; De Ruyck et al., 2015; Ferrigo et al., 2016; Marin et al., 2013; Voss et al., 2007; Gurikar et al., 2022).

4 Various pathogenic genera producing mycotoxins

The most investigated plant pathogenic genera that can produce toxins are *Alternaria*, *Fusarium*, *Claviceps*, and *Aspergillus* spp. (Chukwudi et al., 2021; Ferrigo et al., 2016; Yazar and Omurtag, 2008; Pitt and Miller, 2017). A variety of food commodities such as cereals, beans, nuts, sugarcane, and sugar beet are affected by these plant pathogenic genera. Under field conditions, the genera that affect these foodstuffs are *Fusarium*, *Alternaria*, and *Claviceps* sp. On the other hand, storage (e.g., *Aspergillus* spp.) affects the food commodity. In wheat and maize, *Fusarium graminearum* is the most common pathogen, whereas maize is majorly affected by *Fusarium verticillioides* as well as groundnuts. Maize is also affected by *Aspergillus flavus*. On the other hand, other members of *Aspergillus* and *Fusarium* genera including *A. carbonarius*, *A. parasiticus*, *F. avenaceum*, *A. niger*, *F. proliferatum*, and *F. acuminatum*, as well as *Alternaria alternata* and *Claviceps purpurea* have little attention in research so far.

Not only these fungal toxins are detrimental to animals (mycotoxins), but they are also harmful to plants as well. They produce harmful toxins in plants well-known as phytotoxins. When the concentration of harmful mycotoxin surpasses the maximum allowable levels, these natural pollutants obstruct international trade and have a substantial impact on the global economy.

Despite the economic importance of these harmful mycotoxins produced by these toxin-producing plant pathogens, many research institutions overlook them while exploring biological management measures. At that stage, these investigations are insufficient for the BCAs' fungicidal effects, although the response of BCAs to the synthesis of mycotoxin is frequently neglected.

Keeping in view the significance of these toxins for human as well as animal health, this chapter mainly focuses on how BCAs affect the synthesis of mycotoxins by toxin-producing fungi, as well as the relevance of how mycotoxins impact both animal and human health. In the first section, we'll go through the various modes of action that BCAs can adopt. In the second section, the influence of BCAs on the production of key mycotoxins is discussed in detail. In conclusion, we have suggested some recommendations for future study as well as potential obstacles.

5 *Fusarium* genus

Fusarium is a phytopathogenic fungus, which is considered the most significant genera. Innumerable species of *Fusarium* have the capacity to contaminate several crops particularly cereals such as barley, wheat, oat along with maize, as explained in Table 1; that is considered as the predominant species varies depending on the species of crop, geographic region, and climatic conditions (Logrieco et al., 2002; van der Lee et al., 2015; Emran et al., 2022). As a result of *Fusarium* contamination by toxigenic fungi, toxins that are produced are primarily secondary metabolites capable of contaminating cereals naturally. These hazardous metabolites are the primary source of concern in cereals as well as foodstuffs based on cereals, culminating in food and feed contamination (Placinta et al., 1999). One of the prominent genera of pathogenic fungi, *Fusarium*, infects or contaminates food crops both before and after harvest, which has a detrimental impact on grain quality and quantity. Food and feed contaminated with toxic and pathogenic fungi harm health and cause economic losses (Navale and Vamkudoth, 2022). *Fusarium* spp. is also proficient in causing secondary losses such as seedling blight disease or reduction in seed germination, as well as losses incurred as a result of direct damage such as seedling foot and stalk rots. Moreover, in addition to ear rot disease of maize, various other diseases, such as head blight disease affecting the cereal crops, are the most significant diseases of cereals that is the reason for severe yield and quality reductions (Nganje et al., 2004; Munkvold, 2003).

During field trials, cohabitation of multiple species of *Fusarium* is common, although there may be a substantial percentage of detectable species (Logrieco et al., 2002), merely a few of the species are virulent, particularly under appropriate environmental circumstances. The configuration of species intricated in the disease complex of *Fusarium* is dynamic (Köhl et al., 2007). *F. verticillioides* species is the most predominant one that is also involved in FER with an incidence of 100% under favorable conditions. Nevertheless, *F. proliferatum* as well as *Fusarium subglutinans* are the most significant causative species of *Fusarium*. *F. verticillioides* is the most important species, which is obtained from infected maize globally (Munkvold and Desjardins, 1997). FER occurs in warmer and drier environments as compared to GER, particularly after the course of pollination (Parsons and Munkvold, 2012). Over the period, *F. verticillioides* has been found to predominate among *Fusarium* isolates in different countries, e.g., Africa (Fandohan et al., 2003), Asia (Mohammadi et al., 2016), Europe

TABLE 1 Most important *Fusarium* mycotoxins in food/feed worldwide.

Source	*Fusarium* sp.	Mycotoxin	Europe	North Asia	South East Asia	Central south America	North America	Africa[b]/Middle East	Ref.
Maize, wheat, barley	*F. culmorum* *F. graminearum*	Deoxynivalenol	Persistent	Persistent	Persistent	Infrequent	Persistent	Persistent	Habschied et al. (2021), Perrone et al. (2020)
Maize	*F. proliferatum* *F. verticillioides*	Fumonisins (B1,B2,B3)	Persistent	Very frequent	Very frequent	Persistent	Very frequent	Very frequent	
Wheat, maize	*F. culmorum* *F. graminearum* *F. crookwellense*	Zearalenone	Persistent	Infrequent	Persistent	Persistent	Persistent	Infrequent	
Maize, barley, Wheat, rice, oats	*F. poae* *F. langsethiae* *F. sporotrichioides*	T-2/HT-2 toxins	Persistent	Persistent	Infrequent	Infrequent	Infrequent	Infrequent	

Reproduced with permission from Perrone, G., Ferrara, M., Medina, A., Pascale, M., Magan, N., 2020. Toxigenic fungi and mycotoxins in a climate change scenario: ecology, genomics, distribution, prediction and prevention of the risk. Microorganisms 8 (10), 1496.

(Balconi et al., 2014; Cao et al., 2014), and America (Jurjevic et al., 2005; Stumpf et al., 2013), with up to 90% level of colonization. Sometimes *F. verticillioides* gets confused with *F. subglutinans* that shares the identical ecological habitat as *F. verticillioides* and consequently competes for resources and space.

6 Types of *Fusarium* mycotoxins and their biocontrol activity

Mycotoxin, a class of noxious secondary metabolites produced by a variety of fungal species, has a wide range of biological and chemical properties (Navale and Vamkudoth, 2022) as explained in Fig. 1. Trichothecenes, Zearalenones, and Fumonisins are the most common toxin-producing groups by *Fusarium* (Figlan and Mwadzingeni, 2022; Navale and Vamkudoth, 2022). Other mycotoxins, notably, ENs, MON, BEA, and FUSP can, nevertheless, be found in conjunction with the other toxins produced by *Fusarium* (Jestoi, 2008). Though cereal fusariosis has a substantial impact on crop yield. Numerous *Fusarium* species release a variety of toxins that can reach levels that turn out to be unsafe to humans as well as animals (Kumar et al., 2022). Mycotoxin combinations vary depending on the species (Thrane et al., 2004; Kokkonen et al., 2010) and strain as well (Sewram et al., 2005; Varga et al., 2015). The toxigenic quality of an affected agricultural crop by *Fusarium* is not only determined by the prevailing infectious species, but also by the less significant species.

6.1 Trichothecenes

Trichothecenes comprise of massive array of metabolites. This type of mycotoxin encompasses an epoxide that is responsible for the mutagenic effects. *Fusarium* spp. is capable of producing Trichothecenes that are found worldwide in every cereal-growing region.

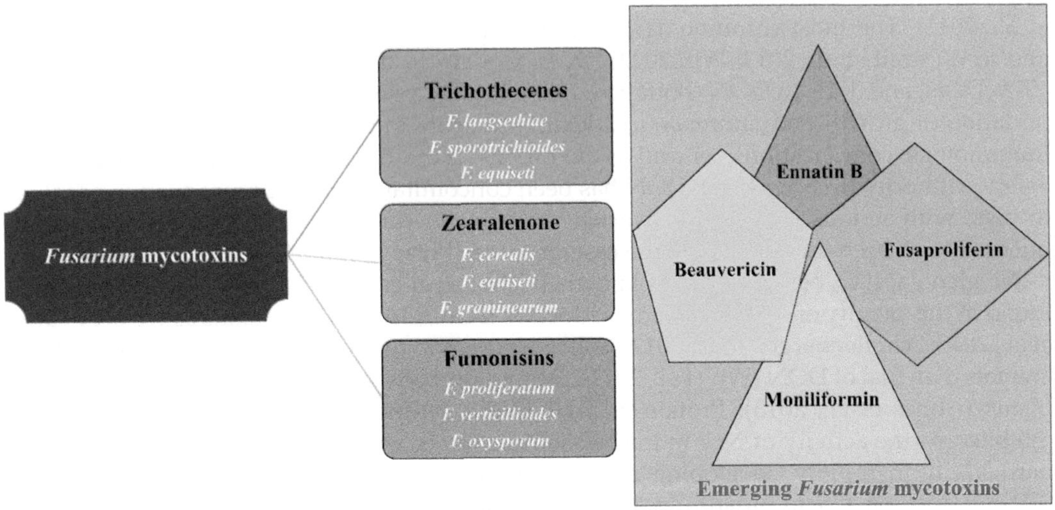

FIG. 1 Different types of *Fusarium* mycotoxins.

This toxin is classified into two types: (A) and (B), which arc recognized based on the existence of distinctive functional groups, located in the backbone structure of Trichothecene position (C-8) (Shank et al., 2011; Abbas et al., 2013; Chen et al., 2022). Toxins that are included in the group A are: (T-2), (HT-2) toxins, diacetoxy (DAS) and Monoacetoxy scirpenol (MAS), as well as Neoolaniol (NEO). The most prevalent member of the group B is Deoxynivalenol (DON), Nivalenol (NIV), 3-AcetylDON, 15-AcetylDON, and Fusarenone as well (Koch, 2004).

Various species belonging to *Fusarium* such as *F. langsethiae*, *F. sporotrichioides*, *F. equiseti*, and *F. poae* are adaptive to form Trichothecenes belonging to type A. Whereas, Trichothecenes can be found in type B group, produced by *Fusarium culmorum* as well as *F. graminearum*. Nonetheless, recently found strains of *F. graminearum* can produce novel Trichothecenes (called NXs) in wheat with a similar structure to that of type A (Kelly et al., 2015). For that reason, a stringent division among different sorts of Trichothecenes belonging to particular species of *Fusarium* is ineffective for systematic separation. Both the DON and NIV are included in the prevalent Trichothecenes (type B) found in the kernels affected by *Fusarium*. It is perceived that *F. graminearum* complex species also capable of producing DON are found all over the world (Ward et al., 2008).

Deoxynivalenol (DON), Nivanelol (NIV), and T-2 Toxin are by far the most prominent members. The biochemical significance of DON for the growth of fungus and its development is unknown (Patel et al., 2022). Nevertheless, it might be able to cause a fungal infection and colonization as well as function as a factor of virulence (Audenaert et al., 2013). Repression of cellular protein formation in animals by DON causes feed refusal and vomiting; on the other hand, NIV has the potential to cause genotoxicity and leucopenia in persistent exposure (Chukwudi et al., 2021; Zain, 2011; Marin et al., 2013). One of the lethal toxins known as T-2 toxin causes immune cells to apoptose (Audenaert et al., 2013). As a result of the complicated life cycle of *Fusarium* spp., majorly two application tactics have been used by scientists to biologically manage the disease. One is treating crop remains with an antagonist and the other one is treating wheat ears during anthesis (Goswami and Kistler, 2004). However, populations with a substantial incidence of NIV-producers have been documented (Gale et al., 2011). The most common Trichothecene in cereals and maize is consumed as food and feed (Streit et al., 2013; Wu, 2022). DON subsists in different forms such as 3-AcDON, 15-AcDON, and 3, 15-AcDON (Bottalico, 1998). It can cause long-term consequences such retardation of growth and anorexia, in addition to acute intoxication with purging (emesis), immunotoxic complications, as well as alterations in brain neurochemicals (European Food Safety Authority (EFSA), 2004). DON has been concomitant to cause reproductive problems in pigs, with serious impacts on ovarian function (Cortinovis et al., 2014). Similar consequences have been anticipated in the case of cattle (Pizzo et al., 2015).

Nivalenol (NIV), belongs to well-known Trichothecenes (type B). This toxin is frequently found in the vicinity of some other toxins such as DON, DAS, and T2 toxins as well (Covarelli et al., 2015; Schollenberger et al., 2012). When it comes to toxin lethality, NIV is more lethal to animals than that of DON (Wu et al., 2012). On the contrary, the toxicity of DON is observed in plants (Abbas et al., 2013). Protein as well as the synthesis of nucleic acid is effectively inhibited by the activity of NIV in the cells of mammals, and it can also induce cell necrosis, particularly in rapidly developing and proliferating tissues like intestinal epithelial cells (EFSA Panel on Contaminants in the Food Chain (CONTAM), 2013; Alassane-Kpembi et al., 2015). Certain strains of *Fusarium cerealis*, *F. culmorum*, *F. graminearum* along with *F. poae* are capable of producing NIV as certain (di-acetylated derivatives), such as 4,

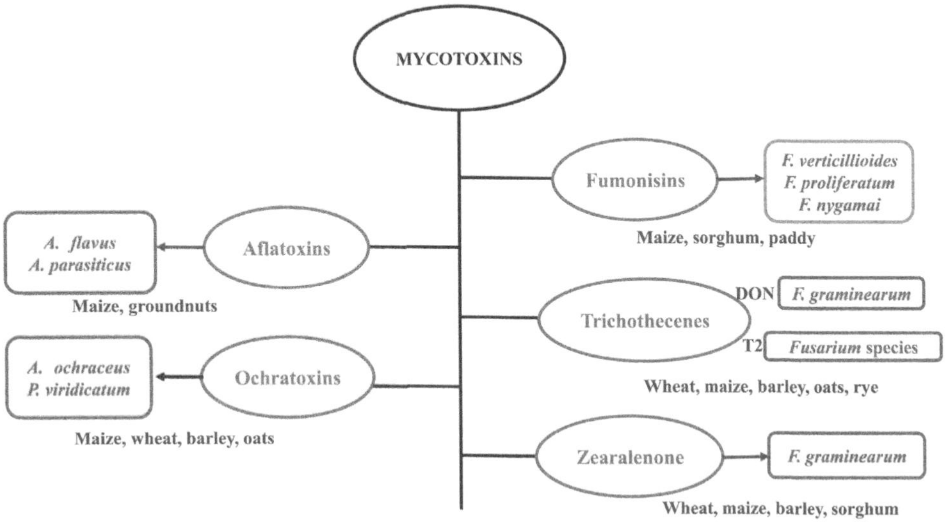

FIG. 2 Mycotoxins produced by various fungal species on different host plants. *Reproduced with permission from Deepa, N., Sreenivasa, M.Y., 2019. Sustainable approaches for biological control of mycotoxigenic fungi and mycotoxins in cereals. New and Future Developments in Microbial Biotechnology and Bioengineering. p. 150, Copyright (2022), with permission from Elsevier.*

15-AcNIV (Haratian et al., 2008). During relatively drier as well as warmer seasons, NIV has been reported in cereals (in comparison with DON toxin) owing to the ecology of the predominant producing species (Beyer et al., 2014). Overall adulteration in cereals due to NIV is less significant than DON (Lindblad et al., 2013; Van Der Fels-Klerx et al., 2012).

Numerous *Fusarium* spp. are capable of producing toxin T-2 as well as HT-2 as explained in Fig. 2. These are most frequently found in oats. Predominantly in Northern Europe, it was assumed that T-2 as well as HT-2 toxins are formed mostly due to the infection caused by various species of *Fusarium*. The immediate transformation of Toxin T-2 to HT-2 occurs in the body and owns comparable detrimental effects as that of T-2 with slight differences in strength (Schuhmacher-Wolz et al., 2010). Both of these toxins are dominant inducers of oxidative stress as well as inhibitors of nucleic acid, protein synthesis, and activity of mitochondria, making them significant pollutants for human as well as animal health (EFSA Panel on Contaminants in the Food Chain (CONTAM), 2011; Yang et al., 2016). Moreover, Diacetoxyscirpenol (DAS) can affect T-2 as well as HT-2 toxins (Lattanzio et al., 2013; Thrane et al., 2004). The most deadly diseases that impact wheat and maize worldwide are *Fusarium* Ear Rot (FER) as well as *Fusarium* Head Blight (FHB) (Comby et al., 2017; Baffoni et al., 2015; Hu et al., 2014). FHB has been mostly caused by three *Fusarium* species in recent years: *F. graminearum*, *F. culmorum*, and *F. avenaceum* (Jensen et al., 2016; Johansson et al., 2003). However, *F. verticillioides*, *F. proliferatum*, *F. subglutinans*, and *F. graminearum* are the primary cause of disease FER (Bacon et al., 2001; Samsudin et al., 2017). In contrast, FHB is most commonly seen as a multicomplex of species (Ferrigo et al., 2016; Audenaert et al., 2013). Each disease possesses a variety of negative effects on the crop, affecting yield as well as grain quality. *Fusarium* spp. produce around 180 forms of trichothecenes, which contaminate largely agricultural commodities like wheat, maize, and barley (Ferrigo et al., 2016; Yazar and Omurtag, 2008).

The majority of these trials utilized bacteria as BCAs that count on antibiosis to manage disease and level of DON mycotoxin. The outcome of the bio-control agents on NIV (Dawson et al., 2004) as well as T-2 toxin (Musyimi et al., 2012) has received little attention. *Trichoderma gamsii* 6085, a *Trichoderma* isolate, was chosen as a possible antagonist opposed to *F. culmorum* and *F. graminearum*. This strain was revealed to have a 92% detrimental effect on DON production by both pathogens (Matarese et al., 2012).

In a two-season field study on winter wheat, the effectiveness of several BCAs against ear blight and accompanying DON concentrations was assessed. The best results were displayed by the two *F. equiseti* strains. These strains were effective enough to reduce the mycotoxins induced by the activity of *F. culmorum* along with *F. graminearum* by 70% and 94% correspondingly (Tretiakova et al., 2022). When cereals were tested with *F. equiseti*, an insignificant amount of NIV was reported (Dawson et al., 2004). *Piriformospora indica* has recently demonstrated its potential in minimizing the incidence of infection induced by the *Fusarium* sp., *F. graminearum* in addition to the contamination of DON mycotoxin in wheat crop up to 70%–80%. Also, it was observed that *P. indica* potentially increased the accumulative grain weight of samples treated with *F. graminearum* by 54% (Rabiey and Shaw, 2016). *Paenibacillus polymyxa* as well as *Citrobacter* are considered novel endophytic bacteria that were capable of detoxifying DON under in vitro conditions. The enactment of a few of these has yet to be documented in the field as well as lab experiments (Mousa et al., 2015). In multiple field experiments, three *Cryptococcus* spp. strains were found to reduce the incidence of infection by 50%–60% on susceptible wheat. On the other hand, the level of DON was similar to that of the control (Khan et al., 2004). In the following study, the strain *Cryptococcus flavescens* was applied at the early stage of flowering resulting in no impact on the level of DON (Schisler et al., 2014).

F. graminearum is the fungus that produces DON in wheat crop (Pei et al., 2022). In wheat, this has been suppressed by utilizing fungal and yeast BCAs (Zhao et al., 2014; Hu et al., 2014; Palazzini et al., 2007, 2016) in addition to Maize (Mousa et al., 2015). For consecutive two seasons, *Bacillus subtilis* (RC 218) as well as *Brevibacillus* spp. (RC 263) were practiced at the stage of flowering (Palazzini et al., 2016). As a result, there was an extreme decline in DON level (Palazzini et al., 2007), though there was no perpetual decrease in disease occurrence. In a field experiment, Khan and Doohan examined three *Pseudomonas* strains, two *fluorescens* strains, and one *frederiksbergensis* strain, against *F. culmorum* as well as DON production in wheat and barley. DON contamination was found to be minimized by 12% and 21% in wheat and barley, respectively (Khan and Doohan, 2009). Due to their limited prevalence in crops, other forms of trichothecenes have not been studied as thoroughly as the toxins mentioned earlier. Under greenhouse conditions, variable findings for the toxin (T-2 toxin) were recorded after spraying the ears of vulnerable as well as resilient wheat cultivars with *Trichoderma* spp., *Penicillium* spp. *Epicoccum* spp., and *Alternaria* spp. Nevertheless, only *Alternaria* spp. is known to produce *Alternaria* toxins (Musyimi et al., 2012).

6.1.1 Zearalenone

Zearalenone, usually abbreviated as ZEA is a mycotoxin that appears to exist in four hydroxyl derivatives (Placinta et al., 1999). Despite its moderate to acute toxicity, ZEA is toxic to the liver, immune system, and cancer causing in a variety of mammals (EFSA Panel on

Contaminants in the Food Chain (CONTAM), 2011). Furthermore, ZEA and its metabolites have been demonstrated to compete for estrogen receptors in a variety of species, causing hyperestrogenism as well as sterility in livestock (Cortinovis et al., 2014). The following mycotoxin is predominant not only in the variable zone of America (Tralamazza et al., 2016), Europe (Edwards, 2011), and Asia (Wang et al., 2013), but also found in Africa (Zaied et al., 2012). When compared to the most significant Trichothecenes (DON), this particular toxin has a global distribution having variations in proportion and amount of contamination that are considerably lesser (Yazar and Omurtag, 2008). Since Zearalenone (ZEN) is a ubiquitous mycotoxin in food commodities particularly in cereals, it has received less attention in terms of control. ZEN is a strong mycoestrogen that competitively adheres to estrogen receptors, eliciting reproductive problems in farm animals as well as humans (Zain, 2011). Zearalenone as well as (α and β) zearalanol are other forms of ZEN that are frequently observed at slighter concentrations than ZEN.

Trichoderma isolates have recently been discovered to detoxify ZEN by converting it to reduced and sulfated forms (Tian et al., 2018). Under in vitro conditions, two isolates of *Trichoderma* in addition to six isolates of *Clonostachys* were evaluated against two isolates of *F. culmorum* in addition to *F. graminearum* and relatively parallel results were attained (Gromadzka et al., 2009). The effectiveness of these isolates performed in the greenhouse as well as fields has not been proven despite a high degree of ZEN reduction (above 96%) (Gromadzka et al., 2009). ZEN is reduced by *Clonostachys rosea* into less noxious compounds by the phenomenon of lactonohydrolase enzymatic alkaline hydrolysis under controlled conditions (Daguerre et al., 2014; Kosawang et al., 2014). This procedure was completely explained after the coding sequence of the relevant gene (zhd 101) was cloned. The gene was further expressed in *Schizosaccharomyces pombe* (Takahashi-Ando et al., 2002) and *Escherichia coli*. It was not expressed in *Saccharomyces cerevisiae*, due to the reason of weak detoxifying action against the toxin ZEN (Takahashi-Ando et al., 2004). However, it was proven in this approach that resistance of BCAs to mycotoxin is a key trait to assure efficiency as well as durability. It requires direct contact between BCAs and pathogen toxin. *C. rosea* has also been shown to be ZEN tolerant due to the existence of a large number of ATP that binds to cassette transporters (Karlsson et al., 2015).

6.1.2 Fumonisins

Fumonisins belong to the class of mycotoxins derived from the polyketide with a broad geographical prevalence, making them the most frequent mycotoxins found in maize across the globe (; Marasas, 1996; Yang et al., 2022). However, up to 13 *Fusarium* species are capable of producing the toxin Fumonisins (Marin et al., 2004). The most frequent species implicated with Fumonisin contamination are *F. verticillioides* and *F. proliferatum*. In animals, Fumonisins can induce serious diseases (Caloni and Cortinovis, 2010), including apoptosis as a result of membrane peroxidation of lipids (Garbetta et al., 2015). The intake of maize infected with this toxin has been linked to cancer of the esophagus, as well as neural tube abnormalities in humans (Alexander et al., 2012). Apparently Fumonisin types A, B, C, and P series (Bartók et al., 2006) are the four major classes of Fumonisins. B group comprises the highly active Fumonisins (FB1) and its different types (Bartók et al., 2010). The most prevalent fumonisin synthesized in maize is FB1, which poses significant toxicological concerns (El-Sayed et al., 2022). FB1 contributes 70%–80% to Fumonisins, compared to the other groups of Fumonisins. In cereals, the contamination with Fumonisin is the main concern around the world, especially *F. verticillioides* that is the principal Fumonisin producer. A higher prevalence of Fumonisin contamination is seen in temperate-warm climates (Gil-Serna et al., 2013). It is produced by *F. verticillioides* as well as *F. proliferatum*. Both can cause infection in maize (Ferrigo et al., 2016). This mycotoxin inhibits ceramide synthase in horses that results in neurological damage in them, pulmonary edema in pigs, also the possibility of hepatotoxicity and cancer of the esophagus in humans as well (Voss et al., 2007).

To manage this mycotoxin efficiently in maize, multiple experiments have been performed. Bacterial BCAs were used in every experiment conducted in the field (Chandra Nayaka et al., 2009; Pereira et al., 2007, 2010, 2011). While on the one hand, some of the other BCAs and fungi were only tested in vitro (Yates et al., 1999; Bacon et al., 2001; Samsudin and Magan, 2016; Samsudin et al., 2017). *Pseudomonas* and *Bacillus* rhizobacterial isolates from maize considerably reduced mycotoxin generation by 70%–100%. While, *Pseudomonas solanacearum* and *B. subtilis* in combination were incapable of altering FB1 concentration in another investigation (Cavaglieri et al., 2005). In maize field testing, treatment of seed with *Bacillus amyloliquefaciens* (Ba-S13) was adequate to lower the level of Fumonisins B1 (Pereira et al., 2007). This method was further confirmed by conducting a field trial for 2 years in a row with a similar bacterium, *B. amyloliquefaciens*, using two different treatments: inoculation of seeds before seeding and maize ears during flowering (Pereira et al., 2010). During a three-year research by Chandra Nayaka et al. (2009), *Pseudomonas fluorescens* secluded from the rhizosphere of maize exhibited a clear reduction in FB1 concentration as well as disease incidence after exposing it to *F. verticillioides*. The incidence of Fumonisins was reduced by 88% after treating the seed with a pure culture of *P. fluorescens* followed by spray treatment (Chandra Nayaka et al., 2009).

It was suggested that with the help of endophytic bacterium *B. subtilis*, FB1 synthesis can be prevented. This simple strategy was recommended to prevent the fungi from spreading vertically (Bacon et al., 2001). FB1 was reduced by 50% under greenhouse conditions (Bacon et al., 2001). When *Trichoderma viride* and *F. verticillioides* were co-inoculated in maize kernels, FB1 was reduced by 72%–85% depending on at which stage inoculation has been done (Yates et al., 1999). During storage, this fungus has been recommended as a postharvest agent to avert toxins from developing (Yates et al., 1999; Bacon et al., 2001). *C. rosea* has been proven to suppress the Fumonisins synthesis by *F. verticillioides* without degrading (Chatterjee et al., 2016). Based on temperature, a combination of the pathogen along with the *C. rosea* 016 in a ratio of 50:50 was sprayed at distinctive ripening stages of maize cobs, and FB1 was reduced by 60%–70%. These studies were carried out because *F. verticillioides* has been reported to attack maize during ripening in the appropriate conditions (Samsudin et al., 2017). A previously similar outcome was reported in milled maize agar at the same ratio (50:50/ *C. rosea* 016: pathogen) (Samsudin and Magan, 2016). We can say that utilizing antibiosis-based bacterial BCAs was more efficient in controlling FB1 in vitro as well as in field testing.

7 Emerging *Fusarium* toxins

In addition to the most frequent *Fusarium* toxins, several newly emerging *Fusarium* toxins have been identified in high quantities, and their contamination seems to be related to both temperature and crop type. Enniatins, Beauvericin, Fusaproliferin, and Moniliformin are among the newly discovered mycotoxins that should be investigated (Singh and Kumari, 2022). Several *Fusarium* species produce Enniatins (ENs) in addition to Beauvericin (BEA), which are found to infect cereals and their by-products as well (Fotso et al., 2002; Santini et al., 2012). The following toxins have identical chemical structures and hazardous dynamic effects, such as antibacterial, antimicrobial, insecticidal, and cytotoxic properties. They are also readily taken up into cellular membranes, altering the ionic equilibrium and affecting cell homeostasis (EFSA Panel on Contaminants in the Food Chain (CONTAM), 2014).

In Northern and Eastern Europe, the prevalence of enniatin in barley and wheat is frequently high (Lindblad et al., 2013; Bolechová et al., 2015), having contamination levels up to 100%, while the Mediterranean environment can also encourage the establishment of EN-producing toxic molds. In Spain, an investigation was conducted on cereals, indicating that the manifestation rate of cereals such as wheat, maize, and barley contaminated with EN were 89%, 62%, and 50%, respectively (Meca et al., 2010b). While, in countries like Morocco, at low levels, parallel high frequencies were found (Zinedine et al., 2011) in Tunisia (Oueslati et al., 2011).

Moniliformin (MON) is a highly polar molecule. Naturally, this toxin can be obtained in the form of sodium as well as potassium salts (Steyn et al., 1978). *F. proliferatum*, *F. avenaceum*, *F. verticillioides*, *F. subglutinans*, *F. chlamydosporum*, *Fusarium oxysporum*, and *F. tricinctum* are the *Fusarium* species most often related with the production of this toxin worldwide (Chelkowski et al., 1990; Logrieco et al., 2002). Predominantly, in the Nordic nations, other than ENs, *F. avenaceum* is the extremely important producer of MON (Jestoi, 2008). Due to the structural resemblance of (MON) to pyruvate, MON presumably impacts metabolic pathways involving pyruvate and inhibits the oxidation of tricarboxylic acid (TCA) cycle intermediates as well. This phenomenon leads to respiratory stress (Thiel, 1978). In addition to cereals, the toxin Moniliformin has been found in cereal products around the world (Peltonen et al., 2010), with erratic levels of frequency. Recently, samples of maize collected from northern Italian fields were found to contain levels as high as 2500 g/kg (Scarpino et al., 2013), with a 93% positive sample rate. MON levels were subsequently found to be higher in Nordic wheat and maize, but lower in the case of barley and oats (Uhlig et al., 2004; Sørensen et al., 2007). The contamination level of MON in Canadian durum wheat, soft wheat, rye, in addition to samples of oat were 75%, 56%, 33%, and 16%, respectively (Tittlemier et al., 2013).

7.1 Enniatin B

Enniatin (ENN) can be found in a variety of grains as well as their derivatives. The following toxin can also be found in a variety of other products such as fish, dried fruits, nuts, spices, chocolate, and coffee (Magazù et al., 2008; Covarelli et al., 2015; García-Moraleja et al., 2015; Vickers, 2017; Tolosa et al., 2017; Zinedine et al., 2017). However, various studies are demonstrating their existence in a variety of foods and feeds, as well as their toxic effects (Juan et al., 2013). Several ENNs analogs have been discovered. But in Europe, type A, A1, B, and B1 are the most common ENNs reported as natural pollutants in cereals (Ivanova et al., 2012). ENN B is the most investigated of the four ENNs described above, as it has been detected in some of the most unrefined as well as refined cereals from European countries. ENN B amounts in grains range from just a few grams/kilograms to more than milligram/kilogram (Jestoi, 2008).

7.1.1 *Fusarium incarnatum-equiseti* species complex (FIESC)

To manage insects, innumerable species of *Fusarium* might be practiced merely if these entomopathogenic species are accurately distinguished from that phytopathogenic strains (O'Donnell et al., 2012). Even though, various *Fusarium* spp. are renowned plant, insect, and human pathogens (Majumdar et al., 2008). Certain *Fusarium* spp., are exclusively known as insect pathogens and do not produce plant diseases. Though *F. proliferatum* can be adopted to manage *Thaumastocoris peregrinus*, it does not affect the other commodities such as tomato, sugarcane, and corn (de HC Maciel et al., 2021). *Fusarium* species are capable of producing secondary substances. These species are also capable of demonstrating insecticidal activities, such as Beauvericin, which aids the pathogen in controlling insects (Liuzzi et al., 2017; Sharma and Marques, 2018).

Fusarium species, including representatives of the FIESC (O'Donnell et al., 2018; Villani et al., 2016), are capable of producing BEA (Liuzzi et al., 2017), with an outcome of enhancing their potential for insect control well. One of the most commonly identified mycotoxins among FIESC entomopathogenic isolates followed by FUS was Beauvericin. Multiple mycotoxins are produced by FIESC species, notably Zearalenone (Avila et al., 2019), Moniliformin (O'Donnell et al., 2018), with Beauvericin (Desjardins, 2006). Trichothecenes and some of the other secondary substances such as Beavericin, Butenolide, Equisetin, Fusarochromanone, in addition to Zearalenone are also produced by FIESC members (Desjardins, 2006). *Fusarium* wilt is considered as the most communal disease in crops, and numerous species of *Fusarium* are phytopathogenic (Carnegie et al., 2022). On the other hand, the practice of nonvirulent isolates of *F. oxysporum* against *Fusarium* wilt disease has been intensively researched (Shishido et al., 2005; Edel-Hermann et al., 2011; Aimé et al., 2013).

7.1.2 Nonpathogenic *Fusarium oxysporum*

Fusarium caatingaense (URM 6782) and (URM 6778) were the prevalent pathogenic strains of *Fusarium* having pathogenic potential against *Dactylopius opuntiae* in Brazil (Carneiro-Leão et al., 2017; Diniz et al., 2022). The avirulent *F. oxysporum* (FO12) is considered one of the proficient bioactive agents of wilts produced by pathogen *Verticillium dahliae* (Varo et al., 2016). The efficiency of several biocontrol agents against pathogenic *Fusarium* species was demonstrated by other researchers (Kavitha and Nelson, 2013; Petrisor et al., 2017). They also demonstrated that nonpathogenic *Fusarium* isolates can be exploited to manage wilting and other diseases in different crops.

7.1.3 Entomopathogenic approach of FIESC

The capacity of FIESC species to serve as biologically active agents to minimize agricultural pests due to their entomopathogenic potential (Addario and Turchetti, 2011; Diniz et al., 2020) permits for cost and environmental savings, resulting in a much more consistent management strategy (Sahayaraj et al., 2011). *Fusarium* species, including members of the FIESC (Desjardins, 2006; Villani et al., 2016), are capable of producing BEA (Liuzzi et al., 2017), augmenting their efficacy for insect control. In Brazil, the substantial species having entomopathogenic potential against *D. opuntiae* in Brazil were *F. caatingaense* (URM 6782) and (URM 6778) (Carneiro-Leão et al., 2017). Insect-affected crops were found to be free of infections in other research on *Fusarium* species with entomopathogenic potential (de HC Maciel et al., 2021; Mikunthan and Manjunatha, 2006; Wenda-Piesik et al., 2009). The entomopathogenic potential of FIESC strains (Carneiro-Leão et al., 2017; Diniz et al., 2020; Gonçalves Diniz et al., 2020) as well as their capacity to produce toxins that aid in infection might also reduce their harmful effects in the environment (Santos et al., 2020). BEA is one such substance that has a resilient insecticidal activity against wide-ranging insect pests (Wang and Xu, 2012) along with no detrimental effects on human lives as well as the health of the other animals (Li et al., 2013).

7.2 Discovery of Fusaproliferin

Fusarium solani is considered one of the important species of *Fusarium*. The pathogen is responsible for causing multiple diseases. Other *Fusarium* spp. strain infections are quite infrequent (Hoque et al., 2018). Fusaproliferin (FUS), a mycotoxin initially obtained from Italian *F. proliferatum* strains, was termed "proliferin" (Randazzo et al., 1993). Later, after the chemical study of the ethyl acetate extract obtained from the *F. solani*, it was termed as "fusaproliferin" (Ritieni et al., 1995). In 1996, the utter chemistry of the compounds was validated by Santini et al. (2012). FUS is produced by *F. subglutinans* as well as some other 15 ex-type strains of *Fusarium* species (Fotso et al., 2002; Ritieni et al., 1999).

7.3 Beauvericin

Beauvericin (BEA) is a potential *Fusarium* mycotoxin that has been found in various food commodities around the world. It is a cyclic hexadepsipeptide produced by several toxin-producing fungi (Wang and Xu, 2012; Tao et al., 2015). Varying *Fusarium* spp. in different areas can induce BEA. For instance, in the United States and South Africa, the most common BEA-producing fungus is *Fusarium circinatum*. On the other hand *Fusarium sambucinum* as well as *F. subglutinans* are most common in Europe (Mallebrera et al., 2017). BEA is a natural toxin found in several bowls of cereal as well as cereal-based commodities as a mycotoxin (Shin et al., 2009; Juan et al., 2013). Various fungi are responsible for the production of this eminent toxin, mainly *Beauveria bassiana* (Hamill et al., 1969) as well as *Fusarium* spp. (Logrieco et al., 1998). BEA toxin belongs to the antibiotic family of Enniatins (ENNs), and it is comparable to the ENNs, structurally (Shin et al., 2009; Yoo et al., 2017). *B. bassiana* (Hamill et al., 1969; Peczyńska-Czoch et al., 1991) in addition to *Fusarium* spp. are among the fungi that are capable of producing BEA (Logrieco et al., 1998). BEA is an antibiotic that belongs to the Enniatins (ENNs) family and is structurally related to the ENNs produced by different species of *Fusarium* (Shin et al., 2009; Yoo et al., 2017). BEA is found naturally in cereals as well as cereal-based products. BEA is produced by numerous species of *Fusarium*. For the first time, BEA was obtained from the culture of *B. bassiana*, which is considered one of the important fungi with entomopathogenic properties (Hamill et al., 1969).

7.3.1 Biological property of Beauvericin (BEA)

This mycotoxin, interestingly, has an extensive range of biological attributes. It shows promising antibacterial as well as antifungal properties. It also enhances the effectiveness of other antifungal drugs (Zhang et al., 2007). BEA possesses a broad range of biological properties include antifungal, antibacterial, and insecticidal, along with nematicidal activities. All of these attributes are essential in the progression of medicine and chemicals as well. BEA is alleged to have the potential to be established as a pharmaceutical or a chemical as well due to its effectiveness as an anticancer, antibacterial, and insecticidal agent (Wang and Xu, 2012).

7.3.2 *Antifungal and antibacterial properties of Beauvericin*

BEA shows a very prominent activity against fungal pathogens. The application of the following toxin along with ketoconazole demonstrates an antifungal effect more than 100-fold higher than that by an application of a single toxin (Zhang et al., 2007). Numerous investigations have been explored to study the mechanism of BEA's antifungal activity (Mei et al., 2009; Tong et al., 2016; Pieterse et al., 2009). BEA's antifungal effectiveness is moderate on its own; however, it can be considerably enhanced when combined with ketoconazole or miconazole (Zhang et al., 2007; Fukuda et al., 2004). BEA has also been shown to have antifungal properties when tested on *Candida albicans*, it showed inhibitory activity (Hsu et al., 2013). The antibacterial activity of BEA has been confirmed against humans and animals. Such pathogenic bacteria includes both Gram-positive and Gram-negative (Castlebury et al., 1999; Nilanonta et al., 2000; Meca et al., 2010a; Santos et al., 2020; Hsu et al., 2013).

7.3.3 *Cytotoxic activity of Beauvericin*

BEA can damage DNA strands and alter mitochondrial membrane potential (Mallebrera et al., 2017). According to recent investigations, BEA cytotoxicity involves mitochondrial changes, apoptosis, and cell cycle disruptions (Manyes et al., 2018). A substantially greater proportion of apoptotic rate was observed. The findings of the investigations indicated that the following toxin might be neurotoxic, although nothing is confirmed regarding the mode of action (Žužek et al., 2016).

7.3.4 *Antimicrobial activity of Beauvericin*

BEA exhibits potent antibacterial properties against both types of bacteria (Gram-positive as well as Gram-negative) (Meca et al., 2010a; Nilanonta et al., 2000, 2002). BEA also inhibits a wide range of bacterial strains, including *Bacillus* spp., *Bifidobacterium adolescentis*, *Clostridium perfringens*, *Paenibacillus* spp., and *Peptostreptococcus* spp., without making a distinction between both types of bacteria (Santos et al., 2020; Castlebury et al., 1999). Based on its antibacterial action against plant pathogens, BEA could be used to manage nonfood crop infections and resolve drug resistance concerns (Santos et al., 2014; Tong et al., 2016).

7.3.5 *Insecticidal and nematicidal activity of Beauvericin*

Beauvericin serves as an eminent insecticide. Various model organisms such as *Calliphora erythrocephala*, *Aedes aegypti* (Grove and Pople, 1980), *Spodoptera frugiperda* (Fornelli et al., 2004), *Leptinotarsa decemlineata* (Gupta et al., 1991), and *Schizaphis graminum* were used to evaluate insecticidal activity (Ganassi et al., 2002). The insecticidal potential of BEA is quite promising. BEA possesses the insecticidal effect against a model organism, *Artemia salina*, initially demonstrated by Hamill et al. (1969). The activity of BEA has been also confirmed against various insects including *A. salina* (Hamill et al., 1969), *A. aegypti* (Wang and Xu, 2012),

S. graminum (Ganassi et al., 2002). *C. erythrocephala* (Grove and Pople, 1980), *Lygus* spp. (Leland et al., 2005), along with *S. frugiperda* (Fornelli et al., 2004). The nematicidal action of BEA is noteworthy. The nematicidal activity of *B. bassiana* culture filtrates against *Meloidogyne hapla* was studied (Liu et al., 2008). The filtrates significantly caused reductions in the nematode population density, along with the reduced development of gall and egg mass production by the tested nematode (Liu et al., 2008). Different cultures of *B. bassiana* with varying levels of activity against a similar nematode were observed (Zhao et al., 2013). In addition, a similar filtrate showed selective toxic levels against various nematodes (Kepenekci et al., 2017). Nematicidal activity of this toxin was also confirmed against *Bursaphelenchus xylophilus*, a pine wood nematode, along with *Caenorhabditis elegans* (Shimada et al., 2010). Octa-BEA, with antiparasitic efficacy against *Leishmania donovani* and *Trypanosoma cruzi* was recently discovered (Steiniger et al., 2017).

7.3.6 Management of insects by *Fusarium* spp.

Fusarium as well as some other well-established pathogenic genera are capable of interchanging between distinct life style stages (Van Kan et al., 2014; De Silva et al., 2017) and also act on nonplant hosts (Perez-Nadales et al., 2014; van Diepeningen and de Hoog, 2016) comprising several animals (O'Donnell et al., 2012, 2016; Torbati et al., 2018) and even some other fungi (Torbati et al., 2018). Among animals, insects are the predominant ones that are associated with *Fusarium* spp. Various species of *Fusarium* are pathogenic to different insect pest species. For instance, *F. oxysporum*, *F. equiseti*, *F. solani*, and *F. semitectum* are pathogenic to 8, 4, 14, and 5 insect species, respectively. Among various *Fusarium* spp., *F. proliferatum* was described as the prevalent one. Some entomopathogenic fungi produce insecticidal toxins that are capable of aiding the disease and ultimately cause faster host death (Goettel and Glare, 2010). Beauvericin is one of these metabolites, as it shows an effective toxic activity against a wide range of insect pests (Wang and Xu, 2012).

8 Mechanism of biocontrol agents

Antibiosis, competition, mycoparasitism, and activation/increase in plant defense are the main mechanisms of BCAs (Pal and Gardener, 2006; Palmieri et al., 2022). To combat the pathogen, BCAs frequently employ many modes of action as shown in Fig. 3. The existence of one dominating mode of action does not rule out the possibility of others.

(i) Secondary metabolites that might inhibit growth, weaken or kill pathogenic fungi are called antibiotics (Wang et al., 2007; Nagórska et al., 2007; Couillerot et al., 2009), e.g., degrading enzymes (El-Tarabily and Sivasithamparam, 2006a) and other proteins (Daguerre et al., 2014).

(ii) Two/more fungi can compete for similar nutrients that are necessary for their growth along with their development (Benítez et al., 2004; Alabouvette et al., 2009). This terms as competition.

It is also possible to compete in a way that excludes others by inhabiting the same habitat (Ehrlich, 2014; Abbas et al., 2006).

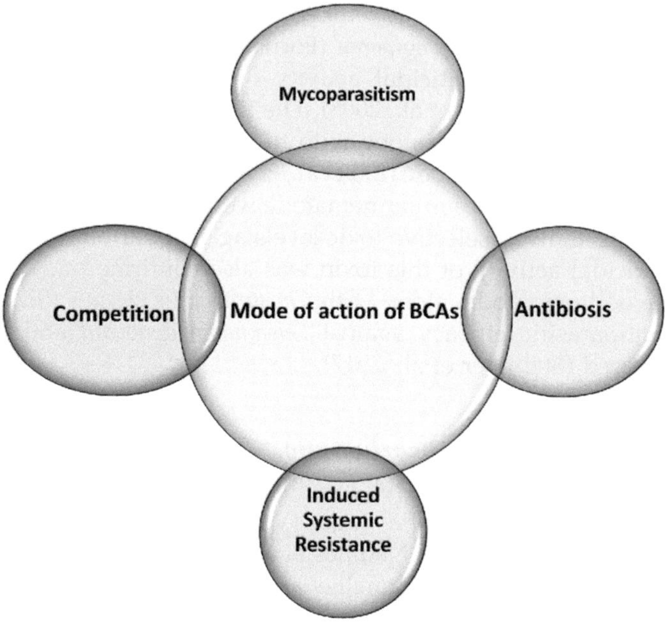

FIG. 3 Mechanism of biological control agents.

(iii) Mycoparasitism occurs when one fungus attacks the other one, resulting in the fatality of the host (Kim and Vujanovic, 2016; Howell, 2003; Blakeman and Fokkema, 1982).

(iv) A plant's resistance to infections can be improved by the introduction of helpful microorganisms into the plant, which can elicit local or systemic defense responses (Shoresh et al., 2010; Lugtenberg et al., 2016).

8.1 Antibiosis

Production of multiple enzymes, antifungal substances as well as antibiotics that are responsible for causing adverse effects on plant pathogens is a distinctive feature of diverse fungal BCAs. Example of such compound includes *Trichoderma* spp., *Clonostachys* spp. (Weller, 1988; Punja and Utkhede, 2003; Benítez et al., 2004; Hue et al., 2009). *Bacillus* spp., *Lactobacillus* spp., *Streptomyces* spp., and *Pseudomonas* spp. are the significant fungal and bacterial bioactive agents (Wang et al., 2007; Nagórska et al., 2007; Gupta et al., 2001; Zhao et al., 2014) along with yeast such as *Cryptococcus* spp., *Kluyveromyces* spp. and *Saccharomyces* spp. (El-Tarabily and Sivasithamparam, 2006a; Etcheverry et al., 2009). All of these BCAs have a resource of metabolites that target distinctive structures of the pathogen, thereby limiting its development or in some cases eradicating the pathogen.

8.1.1 Enzymatic hydrolyzation

Cell wall of fungus is a multifaceted structure comprised primarily of polymers of chitin as well as glucan. Molecules that intrude the cell wall have been identified for various BCAs. Peptaibols, which are linear oligopeptides stimulated by *Trichoderma* spp., can block beta-glucan synthase and thus avert the pathogen from rebuilding its cell wall (Vinale et al.,

2008). Different studies confirmed that the colony color of *A. flavus* was modified by culture filtrates from the isolate of *Trichoderma harzianum*, which had a clear effect on growth. According to the microscope examination, *A. flavus* morphology was altered by the improper formation of vesicles and multiple aberrant conidial heads suggesting deformation of the cell wall (Calistru et al., 1997). Extracellular enzyme production was also confirmed, although inhibition was linked to the carbon source available in the media (Calistru et al., 1997; Jeong et al., 2022).

8.1.2 Production of cell membrane-disrupting metabolites

Several BCAs have been found to produce antifungal metabolites that interfere with membrane structures. Lipopeptides, which are capable of interfering with the membrane, are the most significant class (Romero et al., 2007). Lipopeptides are efficient against numerous toxigenic genera of fungus, including *A. niger* as well as *Fusarium* spp. Inclusion of lipopeptides, iturin along with surfactin demonstrated that *Bacillus* spp. (P1 and P11) had significant antifungal action (Nagórska et al., 2007) against *A. flavus* (Veras et al., 2016). Similarly, *A. flavus* growth was entirely inhibited by *B. subtilis* BS119m, which was associated with its capacity to form an adequate amount of surfactin (Mohammadipour et al., 2009). Under field as well as controlled conditions, iturins that are produced by *B. amyloliquefaciens* were also studied in wheat (Crane et al., 2013). An inverse relation between iturin levels and the prevalence of *Fusarium* disease was perceived (Crane et al., 2013). Lipopeptide isolated from *B. subtilis* IB culture, fengycin, was found to inhibit *F. graminearum* (Wang et al., 2007).

8.1.3 Production of antifungal substances with antibiotic properties

Antibiotics have been well-known as an important associate in the fight against bacterial contamination. With respect to that, numerous fungicidal compounds have also been described. Due to wide-ranging efficacy against several fungal infections, Polyketide molecule 2,4-diacetylphloroglucinol (DAPG) generated by *P. fluorescens* has gained exceptional attention (Maurhofer et al., 2004; Mavrodi et al., 2001; Brazelton et al., 2008; Shanahan et al., 1992). The molecule was initially obtained from the rhizosphere of sugar beets by *Pseudomonas* strain F113 (Shanahan et al., 1992) and then from the rhizosphere of several crops by *Pseudomonas* strain F113 (De La Fuente et al., 2006). Antifungal properties of DAPG have been demonstrated against *Fusarium* as well as *Alternaria* spp. (Müller et al., 2016).

8.2 Nutrient competition

Competition for habitat or competitive exclusion occurs when multiple microorganisms compete for similar nutrients, mandatory for growth along with the production of secondary substances on the host plant. Whereas, competition for nutrients occurs in different microorganisms for the similar resource of macro- and micronutrients necessary for growth as well as the production of secondary metabolite (Pal and Gardener, 2006). *A. flavus* control (Ehrlich, 2014) is an eminent as well as promising example of competition for ecological habitat and nutrition as well. Some of the other toxin-producing organisms, such as *Fusarium pseudograminearum* (Lakhesar et al., 2010), *F. culmorum*, and *F. graminearum* (Dawson et al., 2004), have been shown to compete for nutrients. Atoxigenic *A. flavus* strains are effective BCAs in controlling toxin-producing *A. flavus* strains in cottonseed (Cotty, 1989, 1990; Cotty and

Bhatnagar, 1994), maize (Abbas et al., 2006; Atehnkeng et al., 2014; Dorner, 2009, 2010), and varying sorts of nuts (Zanon et al., 2013, 2016; Horn and Dorner, 2009; Mallikarjunaiah et al., 2017).

Currently, depending on the endemic region, several strains of nontoxic *A. flavus* are employed. Sometimes a combination of strains is exercised in the field by inoculating kernels of corn with GFP-labeled AF70 and wild-type AF36. This competitive exclusion idea was recently proven in situ. After viewing under UV, scientists discovered a reduction in population of up to 82% between the co-infected kernels with both fungi and the control ones inoculated exclusively with GFP-labeled AF70. Moreover, compared to the control, aflatoxins (AFs) study revealed a 73% reduction (Hruska et al., 2014).

A. flavus produces a variety of hazardous chemicals along with AFs. Certain strains of *A. flavus* are capable of producing the mycotoxin CPA, especially the atoxigenic strains, that mostly affect livestock's liver and muscles (Chang and Ehrlich, 2011; Uka et al., 2017). The commercially registered BCAs AF36, for example, has been proven for its CPA generation in cottonseeds while being efficacious against toxigenic *A. flavus*. As a result, researchers screened and tested novel strains that did not produce both toxins for the same crops as before (Abbas et al., 2011; Zhou et al., 2015; Mauro et al., 2015). Practicing nontoxic strains of *A. flavus* against other AFs-producing fungi like *A. parasiticus* was significantly less for the reason that *A. parasiticus* is not that active and does not appear as frequently in the soil as compared to *A. flavus* as well (Zanon et al., 2013).

When *Trichoderma* and *Clonostachys* spp. are administered before pathogen appearance, they compete for nutrients and niche (Punja and Utkhede, 2003; Verma et al., 2007). Once the available iron content is low, *Trichoderma* spp., particularly *T. harzianum*, creates siderophores (Daguerre et al., 2014). The iron source is accessible when siderophores chelate oxidized ferric ions (Fe^{3+}) (Benítez et al., 2004; Vinale et al., 2008; Altomare et al., 1999), permits *Trichoderma* spp. to contend for iron, necessary for the development of innumerable microorganisms (Benítez et al., 2004; Verma et al., 2007).

8.3 Mycoparasitism

Parasitic association directly between fungi with other fungal hosts is known as mycoparasitism (Benítez et al., 2004; Adeleke et al., 2022). The mycoparasitic relationship is facilitated by genes that are responsible for the formation of substances or enzymes primarily chitinases, in addition to proteases, which allow the fungal parasites to disintegrate and penetrate the host cells (Benítez et al., 2004; Howell, 2003; Harman et al., 2004). This technique is used by a variety of BCAs to compete against a variety of mycotoxigenic diseases. For instance, *Fusarium* spp., as well as *Trichoderma* spp., are ubiquitous mycoparasitic biological control agents available in the soil as well as on plants (Punja and Utkhede, 2003; Harman et al., 2004; Verena et al., 2011). The fungi are mostly biotrophic and interact with other fungi in a mycoparasitic manner; however, they also compete for niche and resources, improve systemic and localized plant resistance, and release secondary antifungal compounds (Howell, 2003; Verma et al., 2007). Mycoparasitic interaction between the *Trichoderma* spp. with *Fusarium* spp. resulted in the upregulation of certain chitinase-encoding genes (Verena et al., 2011;

Matarese et al., 2012). In an in vitro study, *T. viride* exhibited potent antagonisms against *F. verticillioides*, and this was verified by the fungal radial extension that is being suppressed by 46% after 6 days as well as 90% after 14 days of exposure (Yates et al., 1999).

T. harzianum did extremely well against *F. verticillioides* on rice by the phenomenon of mycoparasitism and demonstrated mutual antagonism through contact (Sempere and Santamarina, 2009). It was hypothesized that metabolites such as enzymes that degrade cell-wall, chitinases, and β-1,3 glucanases might be involved in the mechanism. Thus, in this investigation cryo scanning electron microscopic analyses provided evidence of mycoparasitism. Previously, the same experimental setup was used on rice with a similar BCA, but against *A. alternata*. In the result, similar findings were reported by Sempere and Santamarina (2007). When chitinases are produced by *Trichoderma* spp. that results in the degradation of the fungal cell wall, another chemical named exochitinase is secreted, and the pathogen is killed (Benítez et al., 2004). In wheat, *Trichoderma* spp. have been frequently studied as a biologically active microorganism against *F. graminearum* (Calistru et al., 1997; Dawson et al., 2004; Dal Bello et al., 2002; Inch and Gilbert, 2007; Schöneberg et al., 2015). In field testing, the production of perithecia in *F. graminearum* has been reduced by 70% by T-22 strain (Inch and Gilbert, 2007).

Another fungus that is well-known for its parasitic activity is *Clonostachys* with an encouraging BCA against various plant pathogens such as *F. graminearum*, *F. verticillioides*, *F. poae*, as well as *F. culmorum*. *Clonostachys* spp., on the other hand, are understudied in comparison with *Trichoderma*. *C. rosea* is the best studied within the *Clonostachys* spp. and has been interlinked to various mechanisms of action, including antibiosis (Hue et al., 2009), stimulation of resistance to plants (Roberti et al., 2008) as well as habitat and nutrient competition (Samsudin and Magan, 2016). Various antibiotics along with chitinases and glucanases are capable of degrading the cell wall. These antibiotics as well as enzymes are thought to be produced by *C. rosea*. *C. rosa* ACM941 has been found to form enzymes capable of hydrolyzing chitin and cell wall degradation of *F. culmorum* (Mamarabadi et al., 2009). Lately, by exploiting mycoparasitism approaches, *Sphaerodes* spp. have been identified as a possible source to control *Fusarium* spp., with encouraging results.

From the field of wheat along with asparagus, *Sphaerodes mycoparasitica* was obtained in combination with *Fusarium* spp. (Vujanovic and Goh, 2009), and it has been demonstrated to restrict *Fusarium* infection in both chemotypes (3-ADON, 15-ADON), as well as inhibit DON formation in vivo (Vujanovic and Goh, 2009, 2012). Both the strains of *P. fluorescens* (strains JP2034 and JP2175) were found to be capable of producing antifungal metabolites as well as chitinases, which can prevent the growth of *A. flavus* as well as *F. verticillioides*, respectively (Palumbo et al., 2007).

8.4 Induced systemic resistance

Rhizobacteria, as a plant growth-promoter, have been studied to improve systemic plant resistance, resulting in efficient protection against various diseases (Heil and Bostock, 2002; Choudhary et al., 2007; Pieterse et al., 2014). *P. fluorescens* is well distinguished to form substances responsible for plant growth and also interferes with plant signaling, such as IAA,

Gibberellins, and Cytokinins (Mantelin and Touraine, 2004). Antibiotics, volatile chemicals, and enzymes are also produced (Couillerot et al., 2009; Alimisup, 2012). The subsistence of pathogens such as *F. verticillioides* M1 triggers the synthesis of indole-3-acetic acid by *P. fluorescens* MPp4, which adds to its antagonistic activity (Hernández-Rodríguez et al., 2008). In barley, carbon diversion, as well as reduction in plant biomass as a result of infection by *F. graminearum* were inhibited by *P. fluorescens* CHA0 (Henkes et al., 2011).

P. fluorescens MKB158 exhibits antagonistic effect against *F. culmorum*. The following antagonistic activity was demonstrated earlier (Khan et al., 2006). Later, it was stated that the antagonistic activity is mediated by an indirect action involving the plant's systemic resistance (Khan et al., 2006). In addition to the production of lytic enzymes in wheat, *Lysobacter enzymogenes* strain C3 also serves as a biological control agent by inducing resistance to *F. graminearum* (Jochum et al., 2006). The presence of some fungal elicitors was confirmed by the effectual decline in the pathogen after treating the C3 broth cultures with heat to deactivate the cells of bacteria as well as lytic enzymes. Besides rhizobacteria, *T. harzianum* is a fungus that has also been demonstrated to stimulate plant development, upsurge nutrient accessibility, as well as improved resistance to fungal diseases through invasion of plant roots (Benítez et al., 2004; Vinale et al., 2008; Harman et al., 2004).

Trichoderma spp. have been studied extensively against various *Fusarium* spp., such as *F. verticillioides* (Danielsen and Jensen, 1999), *F. graminearum* (Schöneberg et al., 2015), and *A. flavus* (Aiyaz et al., 2015). In maize, *T. harzianum* has been illustrated to suppress *F. verticillioides* by activating systemic resistance through the ethylene and jasmonate signaling pathways (Ferrigo et al., 2014). Recently, different *Trichoderma* spp. been recognized as true endophytes such as *T. stromaticum*, *T. amazonicum*, *T. evansii*, *T. martiale*, *T. taxi*, and *T. theobromicola* after being found to colonize plant tissue farther from root, induce transcriptome alterations, and defend plants from numerous diseases along with abiotic pressures (Mukherjee et al., 2012). Colonization is another method for increasing plant resilience. Extensive research has been conducted to lessen plant diseases as well as mycotoxins in crops, and to discover endophytic bacteria that colonize plants (tissue) without affecting the plant (Porras-Alfaro and Bayman, 2011) to minimize the diseases and mycotoxins in crops (Mousa et al., 2015; Rabiey and Shaw, 2016; Rabiey et al., 2015; Waller et al., 2005; Deshmukh and Kogel, 2007). By activating plant defense responses, endophytes can improve the growth as well as the fitness of plant and also imparts protection against biotic and abiotic stressors.

It should be emphasized, however, that some of them are harmful to plants at different stages of their lifecycles or in different environments (Porras-Alfaro and Bayman, 2011). Some endophytes, such as *P. indica* (Varma et al., 1999; Sherameti et al., 2005) improves the host immunity against numerous fungal pathogens by improving the uptake of nutrient. In barley, to minimize the root rot disease, the efficacy of *P. indica* has been confirmed (Deshmukh and Kogel, 2007). This was corroborated by an in vitro study that showed no repression of the fungal growth of *F. graminearum* when co-inoculated with *P. indica*. *Epicoccum nigrum*, another endophyte, has been reported to have a biocontrol efficacy against a variety of plant pathogens (Ogórek and Plaskowska, 2011). However, its potential to combat diseases caused by mycotoxin-producing fungi has been understudied (Musyimi et al., 2012; Jensen et al., 2016).

9 In vitro and in vivo testing

Several BCAs have been evaluated under lab conditions against various strains of mycotoxigenic fungus. Under field conditions, however, not all of them were effective against mycotoxigenic fungus. For example, in the field study, to avoid infection caused by *F. culmorum*, 164 bacterial isolates out of 600 were examined on wheat crops. These isolates coupled with three strains of fluorescent *Pseudomonas* and *Pantoea* sp., offered a high level of control and sustainable results (Johansson et al., 2003).

Generally, variations in the performance of BCAs in controlled field conditions could be due to the impact of various field conditions such as weather-related parameters, soil properties, nutrient accessibility, and population of microbes. All of these factors could hamper the efficiency of the screening BCAs. Other important constraints to ensure BCA interaction against the pathogen that is not existent under in vivo studies comprise the method of BCA deliverance to the host by means of inoculation (spray or direct inoculation), transfer method by means of suspension of conidia or spore suspension, time of application, and application route. AF36 as well as Afla-Guard are commercial examples of BCAs, which are successfully applicable before harvesting of crops to minimize infections by aflatoxins in the United States (Hruska et al., 2014) as well as Polyversum, which is a recently approved commercial product in France.

It is critical to examine all the factors related to the application in the field since these constraints might produce dramatically different outcomes, which are not always tracked in many field experiments against various toxigenic diseases. For instance, in the case of wheat crops, *Streptomyces* sp. BN1 point inoculation failed to manage FHB, although the application of bacterial spores during blooming was quite effective (Jung et al., 2013). While in most field treatments, a suspension comprised of conidia as well as spore of the BCAs was applied, which might give fluctuating and unpredictable outcomes. In various crops, the effectual preparation of *C. rosea* ACM941 assured its capability in managing FHB disease under filed conditions (Xue et al., 2014). Treatment of the ears during a practical investigation with *Enterobacter hormaechei* and *B. amyloliquefaciens* produces extremely variable outcomes, whereas, in the management of *F. verticillioides* infection in maize along with toxin concentration, seed treatment produced more consistent results (Pereira et al., 2010).

Numerous bacterial strains such as *B. subtilis* were revealed to be more active in minimizing the root rot disease compared to the treated soybean seeds in the soil treatment (Zhang et al., 2009; Ouhaibi Ben Abdeljalil et al., 2022). The selection of BCAs may be misvalued if one or more of the above characteristics are left out. In rare circumstances, combining two or more BCAs during field testing to attain effective disease control may be necessary if they have a synergistic impact. For instance, in the case of durum wheat, the combination of *Lactobacillus plantarum* SLG17 and *B. amyloliquefaciens* FLN13 was more efficient in suppressing FHB (Baffoni et al., 2015).

9.1 Factors affecting the performance of BCAs

Although field experiments require a lot of energy and time, it is crucial to account for the method, application time, and effective dose. It is the best way to accurately assess the enactment of the designated BCAs and efficient control of mycotoxigenic fungal infection and mycotoxins. Legislation is a significant impediment to the commercialization of BCAs. BCAs are classified as Plant Protection Products/Pesticides under current European legislation, and as such, they must adhere to pesticide standards. This means that the method of action for each BCA must be recorded and their application must be balanced (Villaverde et al., 2014).

10 Findings and future outlooks

In spite of extensive investigation that has been executed to screen as well as select efficient BCAs for controlling mycotoxigenic infections and associated mycotoxins, there are still a few drawbacks to exercising BCAs. For example, the wide-ranging phenomenon of antagonism of some biologically active agents, such as *Trichoderma* spp., affecting multiple infectious fungi might also be detrimental to various beneficial organisms in the soil (Brimner and Boland, 2003), necessitating more research to discover target-specific BCAs. Using a biological control tactic to substitute synthetic chemicals is widely recommended, there are numerous concerns about the biological as well as environmental stability of BCAs. For instance, *A. flavus* population, is extremely diverse, including atoxigenic strains. This suggests that under specific environmental circumstances, nontoxic strains may intercross with toxic *A. flavus* and generate mycotoxins as an outcome (Ehrlich, 2014). Furthermore, it is unknown whether the atoxigenic strains will persist for an extensive period or what influence they will have on the soil microenvironment in the short and long term.

Reducing mycotoxins is a key element of managing mycotoxigenic pathogens, which makes finding an effectual biocontrol agent to limit fungal growth and toxin synthesis at the same time even more difficult. It is generally known that a single fungal pathogen can create numerous unrelated mycotoxins at the same time; for example, DON, as well as ZEN, is produced by *F. graminearum*, which comprises two distinct metabolic pathways. The majority of scientific research has been focused on controlling a single form of mycotoxin. As a result, selecting a single biocontrol agent capable of concurrently limiting the development of both toxins will be more valuable. It is an important factor that the designated BCAs should be resistant to mycotoxins (Karlsson et al., 2015), as this will ensure long-standing efficacy in the field.

It is important to make sure that, aside from controlling plant diseases, BCAs do not produce any hazardous compounds. *C. rosea*, for example, secretes gliotoxin, a human-toxic metabolite. Trichothecenes (Tri) genes have also been found in several *Trichoderma* strains, which produce proteins that are comparable to *Fusarium* Tri proteins (Tijerino et al., 2011; Keswani et al., 2014). This conveys that *Trichoderma* spp. and *Fusarium* spp. both produce trichothecenes toxins including T-2 toxin. Different *Trichoderma* spp. including *T. virens*, *T. harzianum*, along with *T. viride* produced gliotoxin in addition to viridian, which had a phytotoxic effect on wheat seed germination and human toxicity (Kim and Vujanovic, 2016). As a result, releasing such microbes into the environment could exacerbate food security and public fitness concerns. Furthermore, from an economic standpoint, it is vital to evaluate the entire budget of the application as well as the requirement for periodic reapplication of BCAs to ensure that expenses do not surpass present standards.

Varying BCAs with diverse modes of action, formulations, treatments, as well as their application intervals, were evaluated, indicating that finding a single BCA capable of reducing all regulated mycotoxins may be difficult (Abbas et al., 2017; Mancini and Romanazzi, 2014). Perhaps a mixture of various BCAs or fungicides could be used to combat this issue. To obtain the desired control, the application dose should be thoroughly examined. It has been established in prior studies that a suboptimal or sublethal fungicide application (Audenaert et al., 2010) can cause pathogens to produce mycotoxins as a stress response. Some mycotoxins can be modified by plants by changing their structure. These mycotoxins are termed "masked" or "modified" mycotoxins (Berthiller et al., 2013). One of the examples is conversion of DON into deoxynivalenol-3-glucoside (DON3G), which was found to be important in the defensive mechanism against various microorganisms. Furthermore, additional types of mycotoxins that pose health hazards, such as ENNs, emerging mycotoxins produced as a result of *Fusarium* spp. infection, have not been examined properly, and this demands further research (Ferrigo et al., 2016; Jestoi, 2008).

Instead of focusing solely on growth control, BCAs should be used to control both the growth as well as mycotoxin. Partially minimizing growth could result in a higher buildup of mycotoxins, lowering the worth of the harvested product, particularly grains as well as nuts. Pests inflict a lot of damage to ripening crops and make it easier for mycotoxigenic fungus to get in. Thus, the combination of BCAs that are based on entomogenous fungi with those that are effective in controlling toxigenic fungi could have a synergistic effect on the extent of toxin control. Discovering alternative BCAs with unusual mechanisms can help to manage toxic plant pathogens more effectively. *Enterobacter* spp., also known as a bacterial endophyte, was recently discovered to possess a diverse principle of action than initially assumed, blocking the invasion of *F. graminearum* by forming a physicochemical barrier. However, whether this mechanism of action can be used with other crops like maize as well as wheat is debatable (Mousa et al., 2016). Lastly, effective implementation of preharvest methods can reduce crop damage, and it does not completely certify food security as the attack of fungus can occur during storage or processing, mandating postharvest control.

References

Abbas, H.K., Zablotowicz, R.M., Bruns, H.A., Abel, C.A., 2006. Biocontrol of aflatoxin in corn by inoculation with non-aflatoxigenic *Aspergillus flavus* isolates. Biocontrol Sci. Technol. 16 (5), 437–449.

Abbas, H., Zablotowicz, R., Horn, B., Phillips, N., Johnson, B., Jin, X., Abel, C., 2011. Comparison of major biocontrol strains of non-aflatoxigenic *Aspergillus flavus* for the reduction of aflatoxins and cyclopiazonic acid in maize. Food Addit. Contam. 28, 198–208.

Abbas, H.K., Yoshizawa, T., Shier, W.T., 2013. Cytotoxicity and phytotoxicity of trichothecene mycotoxins produced by *Fusarium* spp. Toxicon 74, 68–75.

Abbas, H.K., Accinelli, C., Shier, W.T., 2017. Biological control of aflatoxin contamination in US crops and the use of bioplastic formulations of *Aspergillus flavus* biocontrol strains to optimize application strategies. J. Agric. Food Chem. 65, 7081–7087.

Abdallah, M.F., Ameye, M., De Saeger, S., Audenaert, K., Haesaert, G., 2018. Biological control of mycotoxigenic fungi and their toxins: an update for the pre-harvest approach. In: Mycotoxins—Impact and Management Strategies. IntechOpen.

Addario, E., Turchetti, T., 2011. Parasitic fungi on *Dryocosmus kuriphilus* in *Castanea sativa* necrotic galls. Bull. Insectol. 64 (2), 269–273.

Adeleke, B.S., Ayilara, M.S., Akinola, S.A., Babalola, O.O., 2022. Biocontrol mechanisms of endophytic fungi. Egypt. J. Biol. Pest Control 32 (1), 1–17.

Aimé, S., Alabouvette, C., Steinberg, C., Olivain, C., 2013. The endophytic strain *Fusarium oxysporum* Fo47: a good candidate for priming the defense responses in tomato roots. Mol. Plant-Microbe Interact. 26 (8), 918–926.

Aiyaz, M., Divakara, S.T., Nayaka, S.C., Hariprasad, P., Niranjana, S.R., 2015. Application of beneficial rhizospheric microbes for the mitigation of seed-borne mycotoxigenic fungal infection and mycotoxins in maize. Biocontrol Sci. Technol. 25 (10), 1105–1119.

Alabouvette, C., Olivain, C., Migheli, Q., Steinberg, C., 2009. Microbiological control of soil-borne phytopathogenic fungi with special emphasis on wilt-inducing *Fusarium oxysporum*. New Phytol. 184 (3), 529–544.

Alassane-Kpembi, I., Puel, O., Oswald, I.P., 2015. Toxicological interactions between the mycotoxins deoxynivalenol, nivalenol and their acetylated derivatives in intestinal epithelial cells. Arch. Toxicol. 89 (8), 1337–1346.

Alexander, J., Benford, D., Boobis, A., Eskola, M., Fink-Gremmels, J., Fürst, P., Heppner, C., Schlatter, J., van Leeuwen, R., 2012. Risk assessment of contaminants in food and feed. EFSA J. 10 (10), s1004.

Alimisup, M., 2012. Characterization and application of microbial antagonists for control of *Fusarium* head blight of wheat caused by *Fusarium graminearum* using single and mixture strain of antagonistic bacteria on resistance and susceptible cultivars. Afr. J. Microbiol. Res. 6 (2), 326–333.

Altomare, C., Norvell, W., Björkman, T., Harman, G., 1999. Solubilization of phosphates and micronutrients by the plant-growth-promoting and biocontrol fungus *Trichoderma harzianum* Rifai 1295-22. Appl. Environ. Microbiol. 65 (7), 2926–2933.

Anand, G., Rajeshkumar, K.C., 2022. Challenges and threats posed by plant pathogenic fungi on agricultural productivity and economy. In: Fungal Diversity, Ecology and Control Management. Springer.

Atehnkeng, J., Ojiambo, P., Cotty, P., Bandyopadhyay, R., 2014. Field efficacy of a mixture of atoxigenic *Aspergillus flavus* Link: Fr vegetative compatibility groups in preventing aflatoxin contamination in maize (*Zea mays* L.). Biol. Control 72, 62–70.

Audenaert, K., Callewaert, E., Höfte, M., De Saeger, S., Haesaert, G., 2010. Hydrogen peroxide induced by the fungicide prothioconazole triggers deoxynivalenol (DON) production by *Fusarium graminearum*. BMC Microbiol. 10 (1), 1–14.

Audenaert, K., Vanheule, A., Höfte, M., Haesaert, G., 2013. Deoxynivalenol: a major player in the multifaceted response of *Fusarium* to its environment. Toxins 6 (1), 1–19.

Avila, C.F., Moreira, G.M., Nicolli, C.P., Gomes, L.B., Abreu, L.M., Pfenning, L.H., Haidukowski, M., Moretti, A., Logrieco, A., Del Ponte, E.M., 2019. *Fusarium incarnatum-equiseti* species complex associated with Brazilian rice: phylogeny, morphology and toxigenic potential. Int. J. Food Microbiol. 306, 108267.

Bacon, C.W., Yates, I.E., Hinton, D.M., Meredith, F., 2001. Biological control of *Fusarium moniliforme* in maize. Environ. Health Perspect. 109 (Suppl. 2), 325–332.

Baffoni, L., Gaggia, F., Dalanaj, N., Prodi, A., Nipoti, P., Pisi, A., Biavati, B., Di Gioia, D., 2015. Microbial inoculants for the biocontrol of *Fusarium* spp. in durum wheat. BMC Microbiol. 15 (1), 1–10.

Balconi, C., Berardo, N., Locatelli, S., Lanzanova, C., Torri, A., Redaelli, R., 2014. Evaluation of ear rot (*Fusarium verticillioides*) resistance and fumonisin accumulation in Italian maize inbred lines. Phytopathol. Mediterr., 14–26.

Bartók, T., Szécsi, Á., Szekeres, A., Mesterházy, Á., Bartók, M., 2006. Detection of new fumonisin mycotoxins and fumonisin-like compounds by reversed-phase high-performance liquid chromatography/electrospray ionization ion trap mass spectrometry. Rapid Communications in Mass Spectrometry: An International Journal Devoted to the Rapid Dissemination of Up-to-the-Minute Research in Mass Spectrometry 20 (16), 2447–2462.

Bartók, T., Tölgyesi, L., Szekeres, A., Varga, M., Bartha, R., Szécsi, Á., Bartók, M., Mesterházy, Á., 2010. Detection and characterization of twenty-eight isomers of fumonisin B1 (FB1) mycotoxin in a solid rice culture infected with *Fusarium verticillioides* by reversed-phase high-performance liquid chromatography/electrospray ionization time-of-flight and ion trap mass spectrometry. Rapid Commun. Mass Spectrom. 24 (1), 35–42.

Benítez, T., Rincón, A.M., Limón, M.C., Codon, A.C., 2004. Biocontrol mechanisms of *Trichoderma* strains. Int. Microbiol. 7 (4), 249–260.

Berthiller, F., Crews, C., Dall'Asta, C., Saeger, S.D., Haesaert, G., Karlovsky, P., Oswald, I.P., Seefelder, W., Speijers, G., Stroka, J., 2013. Masked mycotoxins: a review. Mol. Nutr. Food Res. 57 (1), 165–186.

Beyer, M., Pogoda, F., Pallez, M., Lazic, J., Hoffmann, L., Pasquali, M., 2014. Evidence for a reversible drought induced shift in the species composition of mycotoxin producing *Fusarium* head blight pathogens isolated from symptomatic wheat heads. Int. J. Food Microbiol. 182, 51–56.

Blakeman, J.P., Fokkema, N., 1982. Potential for biological control of plant diseases on the phylloplane. Ann. Rev. Phtopathol. 20 (1), 167–190.

Bolechová, M., Benešová, K., Běláková, S., Čáslavský, J., Pospíchalová, M., Mikulíková, R., 2015. Determination of seventeen mycotoxins in barley and malt in the Czech Republic. Food Control 47, 108–113.

Bottalico, A., 1998. *Fusarium* diseases of cereals: species complex and related mycotoxin profiles, in Europe. J. Plant Pathol., 85–103.

Brazelton, J.N., Pfeufer, E.E., Sweat, T.A., Gardener, B.B.M., Coenen, C., 2008. 2, 4-Diacetylphloroglucinol alters plant root development. Mol. Plant Microbe Inter. 21 (10), 1349–1358.

Brimner, T.A., Boland, G.J., 2003. A review of the non-target effects of fungi used to biologically control plant diseases. Agric. Ecosyst. Environ. 100 (1), 3–16.

Bronzwaer, S., Catchpole, M., de Coen, W., Dingwall, Z., Fabbri, K., Foltz, C., Ganzleben, C., van Gorcom, R., Humphreys, A., Jokelainen, P., 2022. One health collaboration with and among EU Agencies–Bridging research and policy. One Health, 100464.

Calistru, C., McLean, M., Berjak, P., 1997. In vitro studies on the potential for biological control of *Aspergillus flavus* and *Fusarium moniliforme* by *Trichoderma* species. Mycopathologia 137 (2), 115–124.

Caloni, F., Cortinovis, C., 2010. Effects of fusariotoxins in the equine species. Vet. J. 186 (2), 157–161.

Cao, A., Santiago, R., Ramos, A.J., Souto, X.C., Aguín, O., Malvar, R.A., Butrón, A., 2014. Critical environmental and genotypic factors for *Fusarium verticillioides* infection, fungal growth and fumonisin contamination in maize grown in northwestern Spain. Int. J. Food Microbiol. 177, 63–71.

Carnegie, A., Callaghan, S., Laurence, M., Plett, K., Plett, J., Green, P., Wildman, O., Daly, A., Summerell, B., 2022. Response to the detection of an exotic fungal pathogen, *Fusarium commune*, in a *Pinus radiata* production nursery in Australia. Aust. For., 1–12.

Carneiro-Leão, M.P., Tiago, P.V., Medeiros, L.V., da Costa, A.F., de Oliveira, N.T., 2017. *Dactylopius opuntiae*: control by the *Fusarium incarnatum–equiseti* species complex and confirmation of mortality by DNA fingerprinting. J. Pest. Sci. 90 (3), 925–933.

Castlebury, L., Sutherland, J., Tanner, L., Henderson, A., Cerniglia, C., 1999. Use of a bioassay to evaluate the toxicity of beauvericin to bacteria. World J. Microbiol. Biotechnol. 15 (1), 119–121.

Cavaglieri, L., Orlando, J., Etcheverry, M., 2005. In vitro influence of bacterial mixtures on *Fusarium verticillioides* growth and fumonisin B1 production: effect of seeds treatment on maize root colonization. Lett. Appl. Microbiol. 41 (5), 390–396.

Chandra Nayaka, S., Udaya Shankar, A.C., Reddy, M.S., Niranjana, S.R., Prakash, H.S., Shetty, H.S., Mortensen, C.N., 2009. Control of *Fusarium verticillioides*, cause of ear rot of maize, by *Pseudomonas fluorescens*. Pest Manag. Sci.: Formerly Pestic. Sci. 65 (7), 769–775.

Chang, P.-K., Ehrlich, K., 2011. Cyclopiazonic acid biosynthesis by *Aspergillus flavus*. Toxin Rev. 30 (2–3), 79–89.

Chatterjee, S., Kuang, Y., Splivallo, R., Chatterjee, P., Karlovsky, P., 2016. Interactions among filamentous fungi *Aspergillus niger*, *Fusarium verticillioides* and *Clonostachys rosea*: fungal biomass, diversity of secreted metabolites and fumonisin production. BMC Microbiol. 16 (1), 1–13.

Chelkowski, J., Zawadzki, M., Zajkowski, P., Logrieco, A., Bottalico, A., 1990. Moniliformin production by *Fusarium* species. Mycotoxin Res. 6 (1), 41–45.

Chen, L., Yang, J., Wang, H., Yang, X., Zhang, C., Zhao, Z., Wang, J., 2022. NX toxins: new threat posed by *Fusarium graminearum* species complex. Trends Food Sci. Technol. 119, 179–191.

Choudhary, D.K., Prakash, A., Johri, B., 2007. Induced systemic resistance (ISR) in plants: mechanism of action. Indian J. Microbiol. 47 (4), 289–297.

Chukwudi, U., Kutu, F., Mavengahama, S., 2021. Mycotoxins in maize and implications on food security: a review. Agric. Rev. 42 (1).

Comby, M., Gacoin, M., Robineau, M., Rabenoelina, F., Ptas, S., Dupont, J., Profizi, C., Baillieul, F., 2017. Screening of wheat endophytes as biological control agents against *Fusarium* head blight using two different in vitro tests. Microbiol. Res. 202, 11–20.

Corkley, I., Fraaije, B., Hawkins, N., 2022. Fungicide resistance management: maximizing the effective life of plant protection products. Plant Pathol. 71 (1), 150–169.

Cortinovis, C., Caloni, F., Schreiber, N.B., Spicer, L.J., 2014. Effects of fumonisin B1 alone and combined with deoxynivalenol or zearalenone on porcine granulosa cell proliferation and steroid production. Theriogenology 81 (8), 1042–1049.

Cotty, P., 1989. Virulence and cultural characteristics of two *Aspergillus flavus* strains pathogenic on cotton. Phytopathology 79 (7), 808–814.

Cotty, P., 1990. Effect of atoxigenic strains of *Aspergillus flavus* on aflatoxin contamination of developing cottonseed. Plant Dis. (USA) 74 (3), 233–235.

Cotty, P.J., Bhatnagar, D., 1994. Variability among atoxigenic *Aspergillus flavus* strains in ability to prevent aflatoxin contamination and production of aflatoxin biosynthetic pathway enzymes. Appl. Environ. Microbiol. 60 (7), 2248–2251.

Couillerot, O., Prigent-Combaret, C., Caballero-Mellado, J., Moënne-Loccoz, Y., 2009. *Pseudomonas fluorescens* and closely-related fluorescent pseudomonads as biocontrol agents of soil-borne phytopathogens. Lett. Appl. Microbiol. 48 (5), 505–512.

Covarelli, L., Beccari, G., Prodi, A., Generotti, S., Etruschi, F., Juan, C., Ferrer, E., Mañes, J., 2015. *Fusarium* species, chemotype characterisation and trichothecene contamination of durum and soft wheat in an area of central Italy. J. Sci. Food Agric. 95 (3), 540–551.

Crane, J., Gibson, D., Vaughan, R., Bergstrom, G., 2013. Iturin levels on wheat spikes linked to biological control of *Fusarium* head blight by *Bacillus amyloliquefaciens*. Phytopathology 103 (2), 146–155.

Daguerre, Y., Siegel, K., Edel-Hermann, V., Steinberg, C., 2014. Fungal proteins and genes associated with biocontrol mechanisms of soil-borne pathogens: a review. Fungal Biol. Rev. 28 (4), 97–125.

Dal Bello, G., Monaco, C., Simon, M., 2002. Biological control of seedling blight of wheat caused by *Fusarium graminearum* with beneficial rhizosphere microorganisms. World J. Microbiol. Biotechnol. 18 (7), 627–636.

Danielsen, S., Jensen, D.F., 1999. Fungal endophytes from stalks of tropical maize and grasses: isolation, identification, and screening for antagonism against *Fusarium verticillioides* in maize stalks. Bicontrol Sci. Technol. 9 (4), 545–553.

Dawson, W., Jestoi, M., Rizzo, A., Nicholson, P., Bateman, G., 2004. Field evaluation of fungal competitors of *Fusarium culmorum* and *F. graminearum*, causal agents of ear blight of winter wheat, for the control of mycotoxin production in grain. Biocontrol Sci. Technol. 14 (8), 783–799.

de HC Maciel, M., do Amaral, A.C.T., da Silva, T.D., Bezerra, J.D., de Souza-Motta, C.M., da Costa, A.F., Tiago, P.V., de Oliveira, N.T., 2021. Evaluation of mycotoxin production and phytopathogenicity of the entomopathogenic fungi *Fusarium caatingaense* and *F. pernambucanum* from Brazil. Curr. Microbiol. 78 (4), 1218–1226.

De La Fuente, L., Mavrodi, D.V., Landa, B.B., Thomashow, L.S., Weller, D.M., 2006. phlD-based genetic diversity and detection of genotypes of 2, 4-diacetylphloroglucinol-producing *Pseudomonas fluorescens*. FEMS Microbiol. Ecol. 56 (1), 64–78.

De Ruyck, K., De Boevre, M., Huybrechts, I., De Saeger, S., 2015. Dietary mycotoxins, co-exposure, and carcinogenesis in humans: short review. Mutat. Res. Rev. Mutat. Res. 766, 32–41.

De Silva, D.D., Crous, P.W., Ades, P.K., Hyde, K.D., Taylor, P.W.J., 2017. Life styles of *Colletotrichum* species and implications for plant biosecurity. Fungal Biol. Rev. 31 (3), 155–168.

Deshmukh, S., Kogel, K.-H., 2007. *Piriformospora indica* protects barley from root rot caused by *Fusarium graminearum*. J. Plant Dis. Protect. 114 (6), 263–268.

Desjardins, A.E., 2006. Fusarium Mycotoxins: Chemistry, Genetics, and Biology. APS Press, St. Paul, MN.

Diniz, A.G., Cerqueira, L.V.-B.M.P.d., Ribeiro, T.K.d.O., da Costa, A.F., Tiago, P.V., 2020. Pathogenicity of isolates of *Fusarium incarnatum-equiseti* species complex to *Nasutitermes corniger* (Blattodea: Termitidae) and *Spodoptera frugiperda* (Lepidoptera: Noctuidae). Int. J. Pest Manag., 1–10.

Diniz, A.G., Cerqueira, L.V.-B.M.P.d., Ribeiro, T.K.d.O., da Costa, A.F., Tiago, P.V., 2022. Pathogenicity of isolates of *Fusarium incarnatum-equiseti* species complex to *Nasutitermes corniger* (Blattodea: Termitidae) and *Spodoptera frugiperda* (Lepidoptera: Noctuidae). Int. J. Pest Manag. 68 (2), 103–112.

Dorner, J.W., 2009. Biological control of aflatoxin contamination in corn using a nontoxigenic strain of *Aspergillus flavus*. J. Food Prot. 72 (4), 801–804.

Dorner, J.W., 2010. Efficacy of a biopesticide for control of aflatoxins in corn. J. Food Prot. 73 (3), 495–499.

Edel-Hermann, V., Aimé, S., Cordier, C., Olivain, C., Steinberg, C., Alabouvette, C., 2011. Development of a strain-specific real-time PCR assay for the detection and quantification of the biological control agent Fo47 in root tissues. FEMS Microbiol. Lett. 322 (1), 34–40.

Edwards, S., 2011. Zearalenone risk in European wheat. World Mycotoxin J. 4 (4), 433–438.

EFSA Panel on Contaminants in the Food Chain (CONTAM), 2011. Scientific Opinion on the risks for animal and public health related to the presence of T-2 and HT-2 toxin in food and feed. EFSA J. 9 (12), 2481.

EFSA Panel on Contaminants in the Food Chain (CONTAM), 2013. Scientific Opinion on risks for animal and public health related to the presence of nivalenol in food and feed. EFSA J. 11 (6), 3262.

EFSA Panel on Contaminants in the Food Chain (CONTAM), 2014. Scientific Opinion on the risks to human and animal health related to the presence of beauvericin and enniatins in food and feed. EFSA J. 12 (8), 3802.

Ehrlich, K.C., 2014. Non-aflatoxigenic *Aspergillus flavus* to prevent aflatoxin contamination in crops: advantages and limitations. Front. Microbiol. 5, 50.

El-Sayed, R.A., Jebur, A.B., Kang, W., El-Esawi, M.A., El-Demerdash, F.M., 2022. An overview on the major mycotoxins in food products: characteristics, toxicity, and analysis. J. Futr. Foods 2 (2), 91–102.

El-Tarabily, K.A., Sivasithamparam, K., 2006a. Non-streptomycete actinomycetes as biocontrol agents of soil-borne fungal plant pathogens and as plant growth promoters. Soil Biol. Biochem. 38 (7), 1505–1520.

El-Tarabily, K.A., Sivasithamparam, K., 2006b. Potential of yeasts as biocontrol agents of soil-borne fungal plant pathogens and as plant growth promoters. Mycoscience 47 (1), 25–35.

Emran, M.G.I., Ahmed, K.T., Anzum, S.A., Raihan, M.A., Khan, A.-S., Banerjee, S., 2022. Evaluating the agricultural status of some unions of Kalapara Upazila, Bangladesh: preliminary investigation. In: Evaluating the Agricultural Status of Some Unions of Kalapara Upazila, Bangladesh: Preliminary Investigation, p. 119.

Etcheverry, M.G., Scandolara, A., Nesci, A., Vilas Boas Ribeiro, M.S., Pereira, P., Battilani, P., 2009. Biological interactions to select biocontrol agents against toxigenic strains of *Aspergillus flavus* and *Fusarium verticillioides* from maize. Mycopathologia 167 (5), 287–295.

European Food Safety Authority (EFSA), 2004. Opinion of the Scientific Panel on contaminants in the food chain [CONTAM] related to Deoxynivalenol (DON) as undesirable substance in animal feed. EFSA J. 2 (6), 73.

Fandohan, P., Hell, K., Marasas, W., Wingfield, M., 2003. Infection of maize by *Fusarium* species and contamination with fumonisin in Africa. Afr. J. Biotechnol. 2 (12), 570–579.

Ferrigo, D., Raiola, A., Piccolo, E., Scopel, C., Causin, R., 2014. *Trichoderma harzianum* T22 induces in maize systemic resistance against *Fusarium verticillioides*. J. Plant Pathol. 96 (1), 133–142.

Ferrigo, D., Raiola, A., Causin, R., 2016. *Fusarium* toxins in cereals: occurrence, legislation, factors promoting the appearance and their management. Molecules 21 (5), 627.

Figlan, S., Mwadzingeni, L., 2022. Breeding tools for assessing and improving resistance and limiting mycotoxin production by *Fusarium graminearum* in wheat. Plants 11 (15), 1933.

Fisher, M.C., Henk, D., Briggs, C.J., Brownstein, J.S., Madoff, L.C., McCraw, S.L., Gurr, S.J., 2012. Emerging fungal threats to animal, plant and ecosystem health. Nature 484 (7393), 186–194.

Fornelli, F., Minervini, F., Logrieco, A., 2004. Cytotoxicity of fungal metabolites to lepidopteran (*Spodoptera frugiperda*) cell line (SF-9). J. Invertebr. Pathol. 85 (2), 74–79.

Fotso, J., Leslie, J.F., Smith, J.S., 2002. Production of beauvericin, moniliformin, fusaproliferin, and fumonisins B1, B2, and B3 by fifteen ex-type strains of *Fusarium* species. Appl. Environ. Microbiol. 68 (10), 5195–5197.

Fukuda, T., Arai, M., Yamaguchi, Y., Masuma, R., Tomoda, H., Omura, S., 2004. New beauvericins, potentiators of antifungal miconazole activity, produced by *Beauveria* sp. FKI-1366 I. Taxonomy, fermentation, isolation and biological properties. J. Antibiot. 57 (2), 110–116.

Gale, L.R., Harrison, S.A., Ward, T.J., O'Donnell, K., Milus, E.A., Gale, S.W., Kistler, H.C., 2011. Nivalenol-type populations of *Fusarium graminearum* and *F. asiaticum* are prevalent on wheat in southern Louisiana. Phytopathology 101 (1), 124–134.

Ganassi, S., Moretti, A., Pagliai, A.M.B., Logrieco, A., Sabatini, M.A., 2002. Effects of beauvericin on *Schizaphis graminum* (Aphididae). J. Invertebr. Pathol. 80 (2), 90–96.

Garbetta, A., Debellis, L., De Girolamo, A., Schena, R., Visconti, A., Minervini, F., 2015. Dose-dependent lipid peroxidation induction on ex vivo intestine tracts exposed to chyme samples from fumonisins contaminated corn samples. Toxicol. in Vitro 29 (5), 1140–1145.

García-Moraleja, A., Font, G., Mañes, J., Ferrer, E., 2015. Analysis of mycotoxins in coffee and risk assessment in Spanish adolescents and adults. Food Chem. Toxicol. 86, 225–233.

Gil-Serna, J., Mateo, E., González-Jaén, M., Jiménez, M., Vázquez, C., Patiño, B., 2013. Contamination of barley seeds with *Fusarium* species and their toxins in Spain: an integrated approach. Food Addit. Contam. Part A 30 (2), 372–380.

Goettel, M., Glare, T., 2010. 11 Entomopathogenic fungi and their role in regulation of insect populations. In: Insect Control: Biological and Synthetic Agents. Academic Press, pp. 387–431.

Gonçalves Diniz, A., Barbosa, L.F.S., Santos, A.C.d.S., Oliveira, N.T.d., Costa, A.F.d., Carneiro-Leão, M.P., Tiago, P.V., 2020. Bio-insecticide effect of isolates of *Fusarium caatingaense* (Sordariomycetes: Hypocreales) combined to botanical extracts against *Dactylopius opuntiae* (Hemiptera: Dactylopiidae). Biocontrol Sci. Technol. 30 (4), 384–395.
Goswami, R.S., Kistler, H.C., 2004. Heading for disaster: *Fusarium graminearum* on cereal crops. Mol. Plant Pathol. 5 (6), 515–525.
Gromadzka, K., Chelkowski, J., Popiel, D., Kachlicki, P., Kostecki, M., Golinski, P., 2009. Solid substrate bioassay to evaluate the effect of *Trichoderma* and *Clonostachys* on the production of zearalenone by *Fusarium* species. World Mycotoxin J. 2 (1), 45–52.
Grove, J.F., Pople, M., 1980. The insecticidal activity of beauvericin and the enniatin complex. Mycopathologia 70 (2), 103–105.
Gupta, S., Krasnoff, S.B., Underwood, N.L., Renwick, J., Roberts, D.W., 1991. Isolation of beauvericin as an insect toxin from *Fusarium semitectum* and *Fusarium moniliforme* var. *subglutinans*. Mycopathologia 115 (3), 185–189.
Gupta, C., Dubey, R., Kang, S., Maheshwari, D., 2001. Antibiosis-mediated necrotrophic effect of *Pseudomonas* GRC 2 against two fungal plant pathogens. Curr. Sci., 91–94.
Gurikar, C., Shivaprasad, D., Sabillón, L., Gowda, N.N., Siliveru, K., 2022. Impact of mycotoxins and their metabolites associated with food grains. Grain Oil Sci. Technol. 6, 1–9.
Habschied, K., Krstanović, V., Zdunić, Z., Babić, J., Mastanjević, K., Šarić, G.K., 2021. Mycotoxins biocontrol methods for healthier crops and stored products. J. Fungi 7 (5), 348.
Hamill, R., Higgens, C., Boaz, H., Gorman, M., 1969. The structure op beauvericin, a new depsipeptide antibiotic toxic to *Artemia salina*. Tetrahedron Lett. 10 (49), 4255–4258.
Haratian, M., Sharifnabi, B., Alizadeh, A., Safaie, N., 2008. PCR analysis of the Tri13 gene to determine the genetic potential of *Fusarium graminearum* isolates from Iran to produce nivalenol and deoxynivalenol. Mycopathologia 166 (2), 109–116.
Harman, G.E., Howell, C.R., Viterbo, A., Chet, I., Lorito, M., 2004. *Trichoderma* species—opportunistic, avirulent plant symbionts. Nat. Rev. Microbiol. 2 (1), 43–56.
Heil, M., Bostock, R.M., 2002. Induced systemic resistance (ISR) against pathogens in the context of induced plant defences. Ann. Bot. 89 (5), 503–512.
Henkes, G.J., Jousset, A., Bonkowski, M., Thorpe, M.R., Scheu, S., Lanoue, A., Schurr, U., Röse, U.S., 2011. *Pseudomonas fluorescens* CHA0 maintains carbon delivery to *Fusarium graminearum*-infected roots and prevents reduction in biomass of barley shoots through systemic interactions. J. Exp. Bot. 62 (12), 4337–4344.
Hernández-Rodríguez, A., Heydrich-Pérez, M., Acebo-Guerrero, Y., Velazquez-Del Valle, M.G., Hernandez-Lauzardo, A.N., 2008. Antagonistic activity of Cuban native rhizobacteria against *Fusarium verticillioides* (Sacc.) Nirenb. in maize (*Zea mays* L.). Appl. Soil Ecol. 39 (2), 180–186.
Hoque, N., Hasan, C.M., Rana, M., Varsha, A., Sohrab, M., Rahman, K.M., 2018. Fusaproliferin, a fungal mycotoxin, shows cytotoxicity against pancreatic cancer cell lines. Molecules 23 (12), 3288.
Horn, B.W., Dorner, J.W., 2009. Effect of nontoxigenic *Aspergillus flavus* and *A. parasiticus* on aflatoxin contamination of wounded peanut seeds inoculated with agricultural soil containing natural fungal populations. Biocontrol Sci. Technol. 19 (3), 249–262.
Howell, C., 2003. Mechanisms employed by *Trichoderma* species in the biological control of plant diseases: the history and evolution of current concepts. Plant Dis. 87 (1), 4–10.
Hruska, Z., Rajasekaran, K., Yao, H., Kinkaid, R., Darlington, D., Brown, R.L., Bhatnagar, D., Cleveland, T.E., 2014. Co-inoculation of aflatoxigenic and non-aflatoxigenic strains of *Aspergillus flavus* to study fungal invasion, colonization, and competition in maize kernels. Front. Microbiol. 5, 122.
Hsu, D.-Z., Chen, Y.-W., Chu, P.-Y., Periasamy, S., Liu, M.-Y., 2013. Protective effect of 3, 4-methylenedioxyphenol (sesamol) on stress-related mucosal disease in rats. Biomed. Res. Int. 2013.
Hu, W., Gao, Q., Hamada, M.S., Dawood, D.H., Zheng, J., Chen, Y., Ma, Z., 2014. Potential of *Pseudomonas chlororaphis* subsp. *aurantiaca* strain Pcho10 as a biocontrol agent against *Fusarium graminearum*. Phytopathology 104 (12), 1289–1297.
Hue, A., Voldeng, H., Savard, M., Fedak, G., Tian, X., Hsiang, T., 2009. Biological control of fusarium head blight of wheat with *Clonostachys rosea* strain ACM941. Can. J. Plant Pathol. 31 (2), 169–179.
Inch, S., Gilbert, J., 2007. Effect of *Trichoderma harzianum* on perithecial production of *Gibberella zeae* on wheat straw. Biocontrol Sci. Technol. 17 (6), 635–646.

Ivanova, L., Egge-Jacobsen, W., Solhaug, A., Thoen, E., Fæste, C., 2012. Lysosomes as a possible target of enniatin B-induced toxicity in Caco-2 cells. Chem. Res. Toxicol. 25 (8), 1662–1674.

Jensen, B.D., Knorr, K., Nicolaisen, M., 2016. In vitro competition between *Fusarium graminearum* and *Epicoccum nigrum* on media and wheat grains. Eur. J. Plant Pathol. 146 (3), 657–670.

Jeong, J.W., Singhvi, M., Kim, B.S., 2022. Improved extracellular enzyme-mediated production of 7, 10-dihydroxy-8 (E)-octadecenoic acid by *Pseudomonas aeruginosa*. Biotechnol. Bioprocess Eng., 1–8.

Jestoi, M., 2008. Emerging *Fusarium*-mycotoxins fusaproliferin, beauvericin, enniatins, and moniliformin—a review. Crit. Rev. Food Sci. Nutr. 48 (1), 21–49.

Jochum, C., Osborne, L., Yuen, G., 2006. Fusarium head blight biological control with *Lysobacter enzymogenes* strain C3. Biol. Control 39 (3), 336–344.

Johansson, P., Johnsson, L., Gerhardson, B., 2003. Suppression of wheat-seedling diseases caused by *Fusarium culmorum* and *Microdochium nivale* using bacterial seed treatment. Plant Pathol. 52 (2), 219–227.

Juan, C., Mañes, J., Raiola, A., Ritieni, A., 2013. Evaluation of beauvericin and enniatins in Italian cereal products and multicereal food by liquid chromatography coupled to triple quadrupole mass spectrometry. Food Chem. 140 (4), 755–762.

Jung, B., Park, S.-Y., Lee, Y.-W., Lee, J., 2013. Biological efficacy of *Streptomyces* sp. strain BN1 against the cereal head blight pathogen *Fusarium graminearum*. Plant Pathol. J. 29 (1), 52.

Jurjevic, Z., Wilson, D., Wilson, J., Geiser, D.M., Juba, J., Mubatanhema, W., Widstrom, N., Rains, G.C., 2005. *Fusarium* species of the *Gibberella fujikuroi* complex and fumonisin contamination of pearl millet and corn in Georgia, USA. Mycopathologia 159 (3), 401–406.

Karlsson, M., Durling, M.B., Choi, J., Kosawang, C., Lackner, G., Tzelepis, G.D., Nygren, K., Dubey, M.K., Kamou, N., Levasseur, A., 2015. Insights on the evolution of mycoparasitism from the genome of *Clonostachys rosea*. Genome Biol. Evol. 7 (2), 465–480.

Kavitha, T., Nelson, R., 2013. Exploiting the biocontrol activity of *Trichoderma* spp against root rot causing phytopathogens. ARPN J. Agric. Biol. Sci. 8 (7), 571–574.

Kelly, A.C., Clear, R.M., O'Donnell, K., McCormick, S., Turkington, T.K., Tekauz, A., Gilbert, J., Kistler, H.C., Busman, M., Ward, T.J., 2015. Diversity of Fusarium head blight populations and trichothecene toxin types reveals regional differences in pathogen composition and temporal dynamics. Fungal Genet. Biol. 82, 22–31.

Kepenekci, I., Saglam, H., Oksal, E., Yanar, D., Yanar, Y., 2017. Nematicidal activity of *Beauveria bassiana* (Bals.-Criv.) Vuill. against root-knot nematodes on tomato grown under natural conditions. Egypt. J. Biol. Pest Control 27 (1), 117–120.

Keswani, C., Mishra, S., Sarma, B.K., Singh, S.P., Singh, H.B., 2014. Unraveling the efficient applications of secondary metabolites of various *Trichoderma* spp. Appl. Microbiol. Biotechnol. 98 (2), 533–544.

Khan, M.R., Doohan, F.M., 2009. Bacterium-mediated control of Fusarium head blight disease of wheat and barley and associated mycotoxin contamination of grain. Biol. Control 48 (1), 42–47.

Khan, N., Schisler, D., Boehm, M., Lipps, P., Slininger, P., 2004. Field testing of antagonists of Fusarium head blight incited by *Gibberella zeae*. Biol. Control 29 (2), 245–255.

Khan, M.R., Fischer, S., Egan, D., Doohan, F.M., 2006. Biological control of Fusarium seedling blight disease of wheat and barley. Phytopathology 96 (4), 386–394.

Kim, S.H., Vujanovic, V., 2016. Relationship between mycoparasites lifestyles and biocontrol behaviors against *Fusarium* spp. and mycotoxins production. Appl. Microbiol. Biotechnol. 100 (12), 5257–5272.

Koch, P., 2004. State of the art of trichothecenes analysis. Toxicol. Lett. 153 (1), 109–112.

Köhl, J., De Haas, B., Kastelein, P., Burgers, S., Waalwijk, C., 2007. Population dynamics of *Fusarium* spp. and *Microdochium nivale* in crops and crop residues of winter wheat. Phytopathology 97 (8), 971–978.

Kokkonen, M., Ojala, L., Parikka, P., Jestoi, M., 2010. Mycotoxin production of selected Fusarium species at different culture conditions. Int. J. Food Microbiol. 143 (1–2), 17–25.

Kosawang, C., Karlsson, M., Vélëz, H., Rasmussen, P.H., Collinge, D.B., Jensen, B., Jensen, D.F., 2014. Zearalenone detoxification by zearalenone hydrolase is important for the antagonistic ability of *Clonostachys rosea* against mycotoxigenic *Fusarium graminearum*. Fungal Biol. 118 (4), 364–373.

Kumar, P., Mahato, D.K., Gupta, A., Pandey, S., Paul, V., Saurabh, V., Pandey, A.K., Selvakumar, R., Barua, S., Kapri, M., 2022. Nivalenol mycotoxin concerns in foods: an overview on occurrence, impact on human and animal health and its detection and management strategies. Toxins 14 (8), 527.

Lakhesar, D.P.S., Backhouse, D., Kristiansen, P., 2010. Nutritional constraints on displacement of *Fusarium pseudograminearum* from cereal straw by antagonists. Biol. Control 55 (3), 241–247.

Lattanzio, V.M., Ciasca, B., Haidukowski, M., Infantino, A., Visconti, A., Pascale, M., 2013. Mycotoxin profile of *Fusarium langsethiae* isolated from wheat in Italy: production of type-A trichothecenes and relevant glucosyl derivatives. J. Mass Spectrom. 48 (12), 1291–1298.

Leland, J.E., McGuire, M.R., Grace, J.A., Jaronski, S.T., Ulloa, M., Park, Y.-H., Plattner, R.D., 2005. Strain selection of a fungal entomopathogen, *Beauveria bassiana*, for control of plant bugs (*Lygus* spp.)(Heteroptera: Miridae). Biol. Control 35 (2), 104–114.

Li, C., Zuo, C., Deng, G., Kuang, R., Yang, Q., Hu, C., Sheng, O., Zhang, S., Ma, L., Wei, Y., 2013. Contamination of bananas with beauvericin and fusaric acid produced by *Fusarium oxysporum* f. sp. *cubense*. PLoS One 8 (7), e70226.

Lindblad, M., Gidlund, A., Sulyok, M., Börjesson, T., Krska, R., Olsen, M., Fredlund, E., 2013. Deoxynivalenol and other selected *Fusarium* toxins in Swedish wheat—occurrence and correlation to specific *Fusarium* species. Int. J. Food Microbiol. 167 (2), 284–291.

Liu, T., Wang, L., Duan, Y.-X., Wang, X., 2008. Nematicidal activity of culture filtrate of *Beauveria bassiana* against *Meloidogyne hapla*. World J. Microbiol. Biotechnol. 24 (1), 113–118.

Liuzzi, V.C., Mirabelli, V., Cimmarusti, M.T., Haidukowski, M., Leslie, J.F., Logrieco, A.F., Caliandro, R., Fanelli, F., Mulè, G., 2017. Enniatin and beauvericin biosynthesis in *Fusarium* species: production profiles and structural determinant prediction. Toxins 9 (2), 45.

Logrieco, A., Moretti, A., Castella, G., Kostecki, M., Golinski, P., Ritieni, A., Chelkowski, J., 1998. Beauvericin production by *Fusarium* species. Appl. Environ. Microbiol. 64 (8), 3084–3088.

Logrieco, A., Mule, G., Moretti, A., Bottalico, A., 2002. Toxigenic *Fusarium* species and mycotoxins associated with maize ear rot in Europe. In: Mycotoxins in Plant Disease. Springer.

Lugtenberg, B.J., Caradus, J.R., Johnson, L.J., 2016. Fungal endophytes for sustainable crop production. FEMS Microbiol. Ecol. 92 (12).

Magazù, S., Migliardo, F., Telling, M., 2008. Structural and dynamical properties of water in sugar mixtures. Food Chem. 106 (4), 1460–1466.

Majumdar, A., Boetel, M.A., Jaronski, S.T., 2008. Discovery of *Fusarium solani* as a naturally occurring pathogen of sugarbeet root maggot (Diptera: Ulidiidae) pupae: prevalence and baseline susceptibility. J. Invertebr. Pathol. 97 (1), 1–8.

Mallebrera, B., Maietti, A., Tedeschi, P., Font, G., Ruiz, M.-J., Brandolini, V., 2017. Antioxidant capacity of trans-resveratrol dietary supplements alone or combined with the mycotoxin beauvericin. Food Chem. Toxicol. 105, 315–318.

Mallikarjunaiah, N.H., Jayapala, N., Puttaswamy, H., Ramachandrappa, N.S., 2017. Characterization of non-aflatoxigenic strains of *Aspergillus flavus* as potential biocontrol agent for the management of aflatoxin contamination in groundnut. Microb. Pathog. 102, 21–28.

Mamarabadi, M., Jensen, D.F., Lübeck, M., 2009. An N-acetyl-β-d-glucosaminidase gene, cr-nag1, from the biocontrol agent *Clonostachys rosea* is up-regulated in antagonistic interactions with *Fusarium culmorum*. Mycol. Res. 113 (1), 33–43.

Mancini, V., Romanazzi, G., 2014. Seed treatments to control seedborne fungal pathogens of vegetable crops. Pest Manag. Sci. 70 (6), 860–868.

Mantelin, S., Touraine, B., 2004. Plant growth-promoting bacteria and nitrate availability: impacts on root development and nitrate uptake. J. Exp. Bot. 55 (394), 27–34.

Manyes, L., Escrivá, L., Ruiz, M., Juan-García, A., 2018. Beauvericin and enniatin B effects on a human lymphoblastoid Jurkat T-cell model. Food Chem. Toxicol. 115, 127–135.

Marasas, W.F., 1996. Fumonisins: history, world-wide occurrence and impact. Adv. Exp. Med. Biol. 392, 1–17.

Marin, S., Magan, N., Ramos, A.J., Sanchis, V., 2004. Fumonisin-producing strains of Fusarium: a review of their ecophysiology. J. Food Protect. 67 (8), 1792–1805.

Marin, S., Ramos, A., Cano-Sancho, G., Sanchis, V., 2013. Mycotoxins: occurrence, toxicology, and exposure assessment. Food Chem. Toxicol. 60, 218–237.

Matarese, F., Sarrocco, S., Gruber, S., Seidl-Seiboth, V., Vannacci, G., 2012. Biocontrol of *Fusarium* head blight: interactions between *Trichoderma* and mycotoxigenic *Fusarium*. Microbiology 158 (1), 98–106.

Maurhofer, M., Baehler, E., Notz, R., Martinez, V., Keel, C., 2004. Cross talk between 2, 4-diacetylphloroglucinol-producing biocontrol pseudomonads on wheat roots. Appl. Environ. Microbiol. 70 (4), 1990–1998.

Mauro, A., Battilani, P., Cotty, P.J., 2015. Atoxigenic *Aspergillus flavus* endemic to Italy for biocontrol of aflatoxins in maize. BioControl 60 (1), 125–134.

Mavrodi, O.V., McSpadden Gardener, B.B., Mavrodi, D.V., Bonsall, R.F., Weller, D.M., Thomashow, L.S., 2001. Genetic diversity of phlD from 2, 4-diacetylphloroglucinol-producing fluorescent *Pseudomonas* spp. Phytopathology 91 (1), 35–43.

Meca, G., Sospedra, I., Soriano, J., Ritieni, A., Moretti, A., Manes, J., 2010a. Antibacterial effect of the bioactive compound beauvericin produced by *Fusarium proliferatum* on solid medium of wheat. Toxicon 56 (3), 349–354.

Meca, G., Zinedine, A., Blesa, J., Font, G., Mañes, J., 2010b. Further data on the presence of Fusarium emerging mycotoxins enniatins, fusaproliferin and beauvericin in cereals available on the Spanish markets. Food Chem. Toxicol. 48 (5), 1412–1416.

Mei, L., Zhang, L., Dai, R., 2009. An inhibition study of beauvericin on human and rat cytochrome P450 enzymes and its pharmacokinetics in rats. J. Enzym. Inhib. Med. Chem. 24 (3), 753–762.

Mikunthan, G., Manjunatha, M., 2006. Pathogenicity of *Fusarium semitectum* against crop pests and its biosafety to non-target organisms. Commun. Agric. Appl. Biol. Sci. 71 (2 Pt B), 465–473.

Mohammadi, A., Shams-Ghahfarokhi, M., Nazarian-Firouzabadi, F., Kachuei, R., Gholami-Shabani, M., Razzaghi-Abyaneh, M., 2016. *Giberella fujikuroi* species complex isolated from maize and wheat in Iran: distribution, molecular identification and fumonisin B1 in vitro biosynthesis. J. Sci. Food Agric. 96 (4), 1333–1340.

Mohammadipour, M., Mousivand, M., Salehi Jouzani, G., Abbasalizadeh, S., 2009. Molecular and biochemical characterization of Iranian surfactin-producing *Bacillus subtilis* isolates and evaluation of their biocontrol potential against *Aspergillus flavus* and *Colletotrichum gloeosporioides*. Can. J. Microbiol. 55 (4), 395–404.

Mottaleb, K.A., Kruseman, G., Snapp, S., 2022. Potential impacts of Ukraine-Russia armed conflict on global wheat food security: a quantitative exploration. Glob. Food Secur. 35, 100659.

Mousa, W.K., Shearer, C.R., Limay-Rios, V., Zhou, T., Raizada, M.N., 2015. Bacterial endophytes from wild maize suppress *Fusarium graminearum* in modern maize and inhibit mycotoxin accumulation. Front. Plant Sci. 6, 805.

Mousa, W.K., Shearer, C., Limay-Rios, V., Ettinger, C.L., Eisen, J.A., Raizada, M.N., 2016. Root-hair endophyte stacking in finger millet creates a physicochemical barrier to trap the fungal pathogen *Fusarium graminearum*. Nat. Microbiol. 1 (12), 1–12.

Mukherjee, M., Mukherjee, P.K., Horwitz, B.A., Zachow, C., Berg, G., Zeilinger, S., 2012. *Trichoderma*–plant–pathogen interactions: advances in genetics of biological control. Indian J. Microbiol. 52 (4), 522–529.

Müller, T., Behrendt, U., Ruppel, S., von der Waydbrink, G., Müller, M.E., 2016. Fluorescent pseudomonads in the phyllosphere of wheat: potential antagonists against fungal phytopathogens. Curr. Microbiol. 72 (4), 383–389.

Munkvold, G.P., 2003. Epidemiology of *Fusarium* diseases and their mycotoxins in maize ears. Eur. J. Plant Pathol. 109 (7), 705–713.

Munkvold, G.P., Desjardins, A.E., 1997. Fumonisins in maize: can we reduce their occurrence? Plant Dis. 81 (6), 556–565.

Musyimi, S.L., Muthomi, J.W., Narla, R.D., Wagacha, J.M., 2012. Efficacy of biological control and cultivar resistance on Fusarium head blight and T-2 toxin contamination in wheat. Am. J. Plant Sci. 3, 599–607.

Nagórska, K., Bikowski, M., Obuchowski, M., 2007. Multicellular behaviour and production of a wide variety of toxic substances support usage of *Bacillus subtilis* as a powerful biocontrol agent. Acta Biochim. Pol. 54 (3), 495–508.

Navale, V.D., Vamkudoth, K., 2022. Toxicity and preventive approaches of *Fusarium* derived mycotoxins using lactic acid bacteria: state of the art. Biotechnol. Lett., 1–16.

Nganje, W.E., Bangsund, D.A., Leistritz, F.L., Wilson, W.W., Tiapo, N.M., 2004. Regional economic impacts of Fusarium head blight in wheat and barley. Appl. Econ. Perspect. Policy 26 (3), 332–347.

Nilanonta, C., Isaka, M., Kittakoop, P., Palittapongarnpim, P., Kamchonwongpaisan, S., Pittayakhajonwut, D., Tanticharoen, M., Thebtaranonth, Y., 2000. Antimycobacterial and antiplasmodial cyclodepsipeptides from the insect pathogenic fungus *Paecilomyces tenuipes* BCC 1614. Planta Med. 66 (08), 756–758.

Nilanonta, C., Isaka, M., Kittakoop, P., Trakulnaleamsai, S., Tanticharoen, M., Thebtaranonth, Y., 2002. Precursor-directed biosynthesis of beauvericin analogs by the insect pathogenic fungus *Paecilomyces tenuipes* BCC 1614. Tetrahedron 58 (17), 3355–3360.

O'Donnell, K., Humber, R.A., Geiser, D.M., Kang, S., Park, B., Robert, V.A., Crous, P.W., Johnston, P.R., Aoki, T., Rooney, A.P., 2012. Phylogenetic diversity of insecticolous fusaria inferred from multilocus DNA sequence data and their molecular identification via FUSARIUM-ID and *Fusarium* MLST. Mycologia 104 (2), 427–445.

O'Donnell, K., Libeskind-Hadas, R., Hulcr, J., Bateman, C., Kasson, M.T., Ploetz, R.C., Konkol, J.L., Ploetz, J.N., Carrillo, D., Campbell, A., Duncan, R.E., Liyanage, P.N.H., Eskalen, A., Lynch, S.C., Geiser, D.M., Freeman, S., Mendel, Z., Sharon, M., Aoki, T., Cossé, A.A., Rooney, A.P., 2016. Invasive Asian *Fusarium – Euwallacea ambrosia* beetle mutualists pose a serious threat to forests, urban landscapes and the avocado industry. Phytoparasitica 44 (4), 435–442.

O'Donnell, K., McCormick, S.P., Busman, M., Proctor, R.H., Ward, T.J., Doehring, G., Geiser, D.M., Alberts, J.F., Rheeder, J.P., 2018. Marasas et al. 1984 "Toxigenic *Fusarium* species: identity and mycotoxicology" revisited. Mycologia 110 (6), 1058–1080.

Odeya, G.O., Adegbiteb, M.A., Denkyirac, S.A., Alhajd, S.M., IIIe, D.E.L.-P., 2022. Women and food security in Africa: the double burden in addressing gender equality and environmental sustainability. Adv. Food Secur. Sustain. 7, 35.

Ogórek, R., Plaskowska, E., 2011. *Epicoccum nigrum* for biocontrol agents in vitro of plant fungal pathogens. Commun. Agric. Appl. Biol. Sci. 76 (4), 691–697.

Oueslati, S., Meca, G., Mliki, A., Ghorbel, A., Mañes, J., 2011. Determination of *Fusarium* mycotoxins enniatins, beauvericin and fusaproliferin in cereals and derived products from Tunisia. Food Control 22 (8), 1373–1377.

Ouhaibi Ben Abdeljalil, N., Vallance, J., Gerbore, J., Daami-Remadi, M., Rey, P., 2022. Single and combined effects of *Pythium oligandrum* Po37 and a consortium of three rhizobacterial strains on *Sclerotinia* stem rot severity and tomato growth promotion. J. Plant Pathol., 1–15.

Pal, K.K., Gardener, B.M., 2006. Biological control of plant pathogens. Plant Health Instr., 1–25. https://doi.org/10.1094/PHI-A-2006-1117-02.

Palazzini, J.M., Ramirez, M.L., Torres, A.M., Chulze, S.N., 2007. Potential biocontrol agents for Fusarium head blight and deoxynivalenol production in wheat. Crop Protect. 26 (11), 1702–1710.

Palazzini, J.M., Alberione, E., Torres, A., Donat, C., Köhl, J., Chulze, S., 2016. Biological control of *Fusarium graminearum* sensu stricto, causal agent of Fusarium head blight of wheat, using formulated antagonists under field conditions in Argentina. Biol. Control 94, 56–61.

Palmieri, D., Ianiri, G., Del Grosso, C., Barone, G., De Curtis, F., Castoria, R., Lima, G., 2022. Advances and perspectives in the use of biocontrol agents against fungal plant diseases. Horticulturae 8 (7), 577.

Palumbo, J.D., O'keeffe, T.L., Abbas, H.K., 2007. Isolation of maize soil and rhizosphere bacteria with antagonistic activity against *Aspergillus flavus* and *Fusarium verticillioides*. J. Food Protect. 70 (7), 1615–1621.

Parsons, M., Munkvold, G., 2012. Effects of planting date and environmental factors on *Fusarium* ear rot symptoms and fumonisin B1 accumulation in maize grown in six North American locations. Plant Pathol. 61 (6), 1130–1142.

Patel, R., Mehta, K., Prajapati, J., Shukla, A., Parmar, P., Goswami, D., Saraf, M., 2022. An anecdote of mechanics for *Fusarium* biocontrol by plant growth promoting microbes. Biol. Control, 105012.

Peczyńska-Czoch, W., Urbańczyk, M., Bałazy, S., 1991. Formation of beauvericin by selected strains of *Beauveria bassiana*. Arch. Immunol. Ther. Exp. 39 (1–2), 175–179.

Pei, P., Xiong, K., Wang, X., Sun, B., Zhao, Z., Zhang, X., Yu, J., 2022. Predictive growth kinetic parameters and modelled probabilities of deoxynivalenol production by *Fusarium graminearum* on wheat during simulated storing conditions. J. Appl. Microbiol. 133 (2), 349–361.

Peltonen, K., Jestoi, M., Eriksen, G., 2010. Health effects of moniliformin: a poorly understood Fusarium mycotoxin. World Mycotoxin J. 3 (4), 403–414.

Pereira, P., Nesci, A., Etcheverry, M., 2007. Effects of biocontrol agents on *Fusarium verticillioides* count and fumonisin content in the maize agroecosystem: impact on rhizospheric bacterial and fungal groups. Biol. Control 42 (3), 281–287.

Pereira, P., Nesci, A., Castillo, C., Etcheverry, M., 2010. Impact of bacterial biological control agents on fumonisin B1 content and *Fusarium verticillioides* infection of field-grown maize. Biol. Control 53 (3), 258–266.

Pereira, P., Nesci, A., Castillo, C., Etcheverry, M., 2011. Field studies on the relationship between *Fusarium verticillioides* and maize (*Zea mays* L.): effect of biocontrol agents on fungal infection and toxin content of grains at harvest. Int. J. Agron. 2011.

Perez-Nadales, E., Almeida Nogueira, M.F., Baldin, C., Castanheira, S., El Ghalid, M., Grund, E., Lengeler, K., Marchegiani, E., Mehrotra, P.V., Moretti, M., Naik, V., Oses-Ruiz, M., Oskarsson, T., Schäfer, K.,

Wasserstrom, L., Brakhage, A.A., Gow, N.A.R., Kahmann, R., Lebrun, M.-H., Perez-Martin, J., Di Pietro, A., Talbot, N.J., Toquin, V., Walther, A., Wendland, J., 2014. Fungal model systems and the elucidation of pathogenicity determinants. Fungal Genet. Biol. 70, 42–67.

Perrone, G., Ferrara, M., Medina, A., Pascale, M., Magan, N., 2020. Toxigenic fungi and mycotoxins in a climate change scenario: ecology, genomics, distribution, prediction and prevention of the risk. Microorganisms 8 (10), 1496.

Petrisor, C., Paica, A., Constantinescu, F., 2017. Effect of secondary metabolites produced by different *Trichoderma* spp. isolates against *Fusarium oxysporum* f. sp. *radicis-lycopersici* and *Fusarium solani*. Sci. Pap. B Hortic. 61, 407–411.

Pieterse, C.M., Leon-Reyes, A., Van der Ent, S., Van Wees, S.C., 2009. Networking by small-molecule hormones in plant immunity. Nat. Chem. Biol. 5 (5), 308–316.

Pieterse, C.M., Zamioudis, C., Berendsen, R.L., Weller, D.M., Van Wees, S.C., Bakker, P.A., 2014. Induced systemic resistance by beneficial microbes. Annu. Rev. Phytopathol. 52, 347–375.

Pitt, J.I., Miller, J.D., 2017. A concise history of mycotoxin research. J. Agric. Food Chem. 65 (33), 7021–7033.

Pizzo, F., Caloni, F., Schutz, L.F., Totty, M.L., Spicer, L.J., 2015. Individual and combined effects of deoxynivalenol and α-zearalenol on cell proliferation and steroidogenesis of granulosa cells in cattle. Environ. Toxicol. Pharmacol. 40 (3), 722–728.

Placinta, C., D'Mello, J.F., Macdonald, A., 1999. A review of worldwide contamination of cereal grains and animal feed with *Fusarium* mycotoxins. Anim. Feed Sci. Technol. 78 (1–2), 21–37.

Porras-Alfaro, A., Bayman, P., 2011. Hidden fungi, emergent properties: endophytes and microbiomes. Ann. Rev. Phtopathol. 49, 291–315.

Punja, Z.K., Utkhede, R.S., 2003. Using fungi and yeasts to manage vegetable crop diseases. Trends Biotechnol. 21 (9), 400–407.

Rabiey, M., Shaw, M.W., 2016. *Piriformospora indica* reduces fusarium head blight disease severity and mycotoxin DON contamination in wheat under UK weather conditions. Plant Pathol. 65 (6), 940–952.

Rabiey, M., Ullah, I., Shaw, M., 2015. The endophytic fungus *Piriformospora indica* protects wheat from fusarium crown rot disease in simulated UK autumn conditions. Plant Pathol. 64 (5), 1029–1040.

Randazzo, G., Fogliano, V., Ritieni, A., Mannina, L., Rossi, E., Scarallo, A., Segre, A.L., 1993. Proliferin, a new sesterterpene from *Fusarium proliferatum*. Tetrahedron 49 (47), 10883–10896.

Ritieni, A., Fogliano, V., Randazzo, G., Scarallo, A., Logrieco, A., Moretti, A., Manndina, L., Bottalico, A., 1995. Isolation and characterization of fusaproliferin, a new toxic metabolite from *Fusarium proliferatum*. Nat. Toxins 3 (1), 17–20.

Ritieni, A., Monti, S.M., Moretti, A., Logrieco, A., Gallo, M., Ferracane, R., Fogliano, V., 1999. Stability of fusaproliferin, a mycotoxin from *Fusarium* spp. J. Sci. Food Agric. 79 (12), 1676–1680.

Roberti, R., Veronesi, A., Cesari, A., Cascone, A., Di Berardino, I., Bertini, L., Caruso, C., 2008. Induction of PR proteins and resistance by the biocontrol agent *Clonostachys rosea* in wheat plants infected with *Fusarium culmorum*. Plant Sci. 175 (3), 339–347.

Romero, D., de Vicente, A., Rakotoaly, R.H., Dufour, S.E., Veening, J.-W., Arrebola, E., Cazorla, F.M., Kuipers, O.P., Paquot, M., Pérez-García, A., 2007. The iturin and fengycin families of lipopeptides are key factors in antagonism of *Bacillus subtilis* toward *Podosphaera fusca*. Mol. Plant-Microbe Interact. 20 (4), 430–440.

Sahayaraj, K., Namasivayam, S.K.R., Rathi, J.M., 2011. Compatibility of entomopathogenic fungi with extracts of plants and commercial botanicals. Afr. J. Biotechnol. 10 (6), 933–938.

Samsudin, N.I.P., Magan, N., 2016. Efficacy of potential biocontrol agents for control of *Fusarium verticillioides* and fumonisin B1 under different environmental conditions. World Mycotoxin J. 9 (2), 205–213.

Samsudin, N.I.P., Rodriguez, A., Medina, A., Magan, N., 2017. Efficacy of fungal and bacterial antagonists for controlling growth, FUM1 gene expression and fumonisin B1 production by *Fusarium verticillioides* on maize cobs of different ripening stages. Int. J. Food Microbiol. 246, 72–79.

Santini, A., Meca, G., Uhlig, S., Ritieni, A., 2012. Fusaproliferin, beauvericin and enniatins: occurrence in food—a review. World Mycotoxin J. 5 (1), 71–81.

Santos, C.A., Nobre, B., da Silva, T.L., Pinheiro, H., Reis, A., 2014. Dual-mode cultivation of *Chlorella protothecoides* applying inter-reactors gas transfer improves microalgae biodiesel production. J. Biotechnol. 184, 74–83.

Santos, A.d.S., Diniz, A., Tiago, P., Oliveira, N.d., 2020. Entomopathogenic *Fusarium* species: a review of their potential for the biological control of insects, implications and prospects. Fungal Biol. Rev. 34, 41–57.

Sasson, A., 2012. Food security for Africa: an urgent global challenge. Agric. Food Secur. 1 (1), 1–16.

Scarpino, V., Blandino, M., Negre, M., Reyneri, A., Vanara, F., 2013. Moniliformin analysis in maize samples from North-West Italy using multifunctional clean-up columns and the LC-MS/MS detection method. Food Addit. Contam. Part A 30 (5), 876–884.

Schisler, D.A., Core, A.B., Boehm, M.J., Horst, L., Krause, C., Dunlap, C.A., Rooney, A.P., 2014. Population dynamics of the Fusarium head blight biocontrol agent *Cryptococcus flavescens* OH 182.9 on wheat anthers and heads. Biol. Control 70, 17–27.

Schollenberger, M., Müller, H.-M., Ernst, K., Sondermann, S., Liebscher, M., Schlecker, C., Wischer, G., Drochner, W., Hartung, K., Piepho, H.-P., 2012. Occurrence and distribution of 13 trichothecene toxins in naturally contaminated maize plants in Germany. Toxins 4 (10), 778–787.

Schöneberg, A., Musa, T., Voegele, R., Vogelgsang, S., 2015. The potential of antagonistic fungi for control of *Fusarium graminearum* and *Fusarium crookwellense* varies depending on the experimental approach. J. Appl. Microbiol. 118 (5), 1165–1179.

Schuhmacher-Wolz, U., Heine, K., Schneider, K., 2010. Report on toxicity data on trichothecene mycotoxins HT-2 and T-2 toxins. EFSA Support. Publ. 7 (7), 65E.

Sempere, F., Santamarina, M., 2007. In vitro biocontrol analysis of *Alternaria alternata* (Fr.) Keissler under different environmental conditions. Mycopathologia 163 (3), 183–190.

Sempere, F., Santamarina, M.P., 2009. Antagonistic interactions between fungal rice pathogen *Fusarium verticillioides* (Sacc.) Nirenberg and *Trichoderma harzianum* Rifai. Ann. Microbiol. 59 (2), 259–266.

Sewram, V., Mshicileli, N., Shephard, G.S., Vismer, H.F., Rheeder, J.P., Lee, Y.-W., Leslie, J.F., Marasas, W.F., 2005. Production of fumonisin B and C analogues by several *Fusarium* species. J. Agric. Food Chem. 53 (12), 4861–4866.

Shanahan, P., O'Sullivan, D.J., Simpson, P., Glennon, J.D., O'Gara, F., 1992. Isolation of 2, 4-diacetylphloroglucinol from a fluorescent pseudomonad and investigation of physiological parameters influencing its production. Appl. Environ. Microbiol. 58 (1), 353–358.

Shank, R.A., Foroud, N.A., Hazendonk, P., Eudes, F., Blackwell, B.A., 2011. Current and future experimental strategies for structural analysis of trichothecene mycotoxins—a prospectus. Toxins 3 (12), 1518–1553.

Sharma, L., Marques, G., 2018. Fusarium, an entomopathogen—a myth or reality? Pathogens 7 (4), 93.

Sherameti, I., Shahollari, B., Venus, Y., Altschmied, L., Varma, A., Oelmüller, R., 2005. The endophytic fungus *Piriformospora indica* stimulates the expression of nitrate reductase and the starch-degrading enzyme glucan-water dikinase in tobacco and *Arabidopsis* roots through a homeodomain transcription factor that binds to a conserved motif in their promoters. J. Biol. Chem. 280 (28), 26241–26247.

Shimada, A., Fujioka, S., Koshino, H., Kimura, Y., 2010. Nematicidal activity of beauvericin produced by the fungus *Fusarium bulbicola*. Zeitschrift für Naturforschung C 65 (3–4), 207–210.

Shin, C.-G., An, D.-G., Song, H.-H., Lee, C., 2009. Beauvericin and enniatins H, I and MK1688 are new potent inhibitors of human immunodeficiency virus type-1 integrase. J. Antibiot. 62 (12), 687–690.

Shishido, M., Miwa, C., Usami, T., Amemiya, Y., Johnson, K.B., 2005. Biological control efficiency of Fusarium wilt of tomato by nonpathogenic *Fusarium oxysporum* Fo-B2 in different environments. Phytopathology 95 (9), 1072–1080.

Shoresh, M., Harman, G.E., Mastouri, F., 2010. Induced systemic resistance and plant responses to fungal biocontrol agents. Annu. Rev. Phytopathol. 48, 21–43.

Singh, K., Kumari, A., 2022. Emerging mycotoxins and their clinicopathological effects. Springer, Mycotoxins and Mycotoxicoses.

Sørensen, J.L., Nielsen, K.F., Thrane, U., 2007. Analysis of moniliformin in maize plants using hydrophilic interaction chromatography. J. Agric. Food Chem. 55 (24), 9764–9768.

Steiniger, C., Hoffmann, S., Mainz, A., Kaiser, M., Voigt, K., Meyer, V., Süssmuth, R.D., 2017. Harnessing fungal nonribosomal cyclodepsipeptide synthetases for mechanistic insights and tailored engineering. Chem. Sci. 8 (11), 7834–7843.

Steyn, M., Thiel, P.G., Van Schalkwyk, G.C., 1978. Isolation and purification of moniliformin. J. Assoc. Off. Anal. Chem. 61 (3), 578–580.

Streit, E., Naehrer, K., Rodrigues, I., Schatzmayr, G., 2013. Mycotoxin occurrence in feed and feed raw materials worldwide: long-term analysis with special focus on Europe and Asia. J. Sci. Food Agric. 93 (12), 2892–2899.

Stumpf, R., Santos, J.d., Gomes, L.B., Silva, C., Tessmann, D.J., Ferreira, F., Machinski Junior, M., Del Ponte, E.M., 2013. *Fusarium* species and fumonisins associated with maize kernels produced in Rio Grande do Sul State for the 2008/09 and 2009/10 growing seasons. Braz. J. Microbiol. 44 (1), 89–95.

Takahashi-Ando, N., Kimura, M., Kakeya, H., Osada, H., Yamaguchi, I., 2002. A novel lactonohydrolase responsible for the detoxification of zearalenone: enzyme purification and gene cloning. Biochem. J. 365 (1), 1–6.

Takahashi-Ando, N., Ohsato, S., Shibata, T., Hamamoto, H., Yamaguchi, I., Kimura, M., 2004. Metabolism of zearalenone by genetically modified organisms expressing the detoxification gene from *Clonostachys rosea*. Appl. Environ. Microbiol. 70 (6), 3239–3245.

Tao, Y.-w., Lin, Y.-c., She, Z.-g., Lin, M.-t., Chen, P.-x., Zhang, J.-y., 2015. Anticancer activity and mechanism investigation of beauvericin isolated from secondary metabolites of the mangrove endophytic fungi. Anti-Cancer Agents Med. Chem. (Formerly Curr. Med. Chem. Anti-Cancer Agents) 15 (2), 258–266.

Thiel, P.G., 1978. A molecular mechanism for the toxic action of moniliformin, a mycotoxin produced by *Fusarium moniliforme*. Biochem. Pharmacol. 27 (4), 483–486.

Thrane, U., Adler, A., Clasen, P.-E., Galvano, F., Langseth, W., Lew, H., Logrieco, A., Nielsen, K.F., Ritieni, A., 2004. Diversity in metabolite production by *Fusarium langsethiae*, *Fusarium poae*, and *Fusarium sporotrichioides*. Int. J. Food Microbiol. 95 (3), 257–266.

Tian, Y., Tan, Y., Yan, Z., Liao, Y., Chen, J., De Boevre, M., De Saeger, S., Wu, A., 2018. Antagonistic and detoxification potentials of *Trichoderma* isolates for control of Zearalenone (ZEN) producing *Fusarium graminearum*. Front. Microbiol. 8, 2710.

Tijerino, A., Hermosa, R., Cardoza, R.E., Moraga, J., Malmierca, M.G., Aleu, J., Collado, I.G., Monte, E., Gutierrez, S., 2011. Overexpression of the *Trichoderma brevicompactum* tri5 gene: effect on the expression of the trichodermin biosynthetic genes and on tomato seedlings. Toxins 3 (9), 1220–1232.

Tittlemier, S.A., Roscoe, M., Trelka, R., Gaba, D., Chan, J.M., Patrick, S.K., Sulyok, M., Krska, R., McKendry, T., Gräfenhan, T., 2013. *Fusarium* damage in small cereal grains from Western Canada. 2. Occurrence of *Fusarium* toxins and their source organisms in durum wheat harvested in 2010. J. Agric. Food Chem. 61 (23), 5438–5448.

Tolosa, J., Font, G., Mañes, J., Ferrer, E., 2017. Mitigation of enniatins in edible fish tissues by thermal processes and identification of degradation products. Food Chem. Toxicol. 101, 67–74.

Tong, Y., Liu, M., Zhang, Y., Liu, X., Huang, R., Song, F., Dai, H., Ren, B., Sun, N., Pei, G., 2016. Beauvericin counteracted multi-drug resistant *Candida albicans* by blocking ABC transporters. Synth. Syst. Biotechnol. 1 (3), 158–168.

Torbati, M., Arzanlou, M., Sandoval-Denis, M., Crous, P.W., 2018. Multigene phylogeny reveals new fungicolous species in the *Fusarium tricinctum* species complex and novel hosts in the genus *Fusarium* from Iran. Mycol. Prog. 18 (1), 119–133.

Tralamazza, S.M., Bemvenuti, R.H., Zorzete, P., de Souza Garcia, F., Corrêa, B., 2016. Fungal diversity and natural occurrence of deoxynivalenol and zearalenone in freshly harvested wheat grains from Brazil. Food Chem. 196, 445–450.

Tretiakova, P., Voegele, R.T., Soloviev, A., Link, T.I., 2022. Successful silencing of the mycotoxin synthesis gene TRI5 in *Fusarium culmorum* and observation of reduced virulence in VIGS and SIGS experiments. Genes 13 (3), 395.

Uhlig, S., Torp, M., Jarp, J., Parich, A., Gutleb §, A., Krska, R., 2004. Moniliformin in Norwegian grain. Food Addit. Contam. 21 (6), 598–606.

Uka, V., Moore, G.G., Arroyo-Manzanares, N., Nebija, D., De Saeger, S., Diana Di Mavungu, J., 2017. Unravelling the diversity of the cyclopiazonic acid family of mycotoxins in *Aspergillus flavus* by UHPLC triple-TOF HRMS. Toxins 9 (1), 35.

Van Der Fels-Klerx, H., Klemsdal, S., Hietaniemi, V., Lindblad, M., Ioannou-Kakouri, E., Van Asselt, E., 2012. Mycotoxin contamination of cereal grain commodities in relation to climate in North West Europe. Food Addit. Contam. Part A 29 (10), 1581–1592.

van der Lee, T., Zhang, H., van Diepeningen, A., Waalwijk, C., 2015. Biogeography of *Fusarium graminearum* species complex and chemotypes: a review. Food Addit. Contam. Part A 32 (4), 453–460.

van Diepeningen, A.D., de Hoog, G.S., 2016. Challenges in *Fusarium*, a trans-kingdom pathogen. Mycopathologia 181 (3), 161–163.

Van Kan, J.A., Shaw, M.W., Grant-Downton, R.T., 2014. *Botrytis* species: relentless necrotrophic thugs or endophytes gone rogue? Mol. Plant Pathol. 15 (9), 957–961.

Varga, E., Wiesenberger, G., Hametner, C., Ward, T.J., Dong, Y., Schöfbeck, D., McCormick, S., Broz, K., Stückler, R., Schuhmacher, R., 2015. New tricks of an old enemy: isolates of *Fusarium graminearum* produce a type A trichothecene mycotoxin. Environ. Microbiol. 17 (8), 2588–2600.

Varma, A., Verma, S., Sudha, Sahay, N., Bütehorn, B., Franken, P., 1999. *Piriformospora indica*, a cultivable plant-growth-promoting root endophyte. Appl. Environ. Microbiol. 65 (6), 2741–2744.

Varo, A., Raya-Ortega, M., Trapero, A., 2016. Selection and evaluation of micro-organisms for biocontrol of *Verticillium dahliae* in olive. J. Appl. Microbiol. 121 (3), 767–777.

Veras, F.F., Correa, A.P.F., Welke, J.E., Brandelli, A., 2016. Inhibition of mycotoxin-producing fungi by *Bacillus* strains isolated from fish intestines. Int. J. Food Microbiol. 238, 23–32.

Verena, S.-S., Alfredo, H.-E., Enrique, M., Susanne, Z., 2011. *Trichoderma*: the genomics of opportunistic success. Nat. Rev. Microbiol. 9 (10).

Verma, M., Brar, S.K., Tyagi, R., Surampalli, R.n., Valero, J., 2007. Antagonistic fungi, *Trichoderma* spp.: panoply of biological control. Biochem. Eng. J. 37 (1), 1–20.

Vickers, N.J., 2017. Animal communication: when i'm calling you, will you answer too? Curr. Biol. 27 (14), R713–R715.

Villani, A., Moretti, A., De Saeger, S., Han, Z., Di Mavungu, J.D., Soares, C.M., Proctor, R.H., Venâncio, A., Lima, N., Stea, G., 2016. A polyphasic approach for characterization of a collection of cereal isolates of the *Fusarium incarnatum-equiseti* species complex. Int. J. Food Microbiol. 234, 24–35.

Villaverde, J.J., Sevilla-Morán, B., Sandín-España, P., López-Goti, C., Alonso-Prados, J.L., 2014. Biopesticides in the framework of the European Pesticide Regulation (EC) No. 1107/2009. Pest Manag. Sci. 70 (1), 2–5.

Vinale, F., Sivasithamparam, K., Ghisalberti, E.L., Marra, R., Woo, S.L., Lorito, M., 2008. Trichoderma–plant–pathogen interactions. Soil Biol. Biochem. 40 (1), 1–10.

Voss, K., Smith, G., Haschek, W., 2007. Fumonisins: Toxicokinetics, mechanism of action and toxicity. Anim. Feed Sci. Technol. 137 (3–4), 299–325.

Vujanovic, V., Goh, Y.K., 2009. *Sphaerodes mycoparasitica* sp. nov., a new biotrophic mycoparasite on *Fusarium avenaceum*, *F. graminearum* and *F. oxysporum*. Mycol. Res. 113 (10), 1172–1180.

Vujanovic, V., Goh, Y.K., 2012. qPCR quantification of *Sphaerodes mycoparasitica* biotrophic mycoparasite interaction with *Fusarium graminearum*: in vitro and in planta assays. Arch. Microbiol. 194 (8), 707–717.

Waller, F., Achatz, B., Baltruschat, H., Fodor, J., Becker, K., Fischer, M., Heier, T., Hückelhoven, R., Neumann, C., von Wettstein, D., 2005. The endophytic fungus *Piriformospora indica* reprograms barley to salt-stress tolerance, disease resistance, and higher yield. Proc. Natl. Acad. Sci. 102 (38), 13386–13391.

Wang, Q., Xu, L., 2012. Beauvericin, a bioactive compound produced by fungi: a short review. Molecules 17 (3), 2367–2377.

Wang, J., Liu, J., Chen, H., Yao, J., 2007. Characterization of *Fusarium graminearum* inhibitory lipopeptide from *Bacillus subtilis* IB. Appl. Microbiol. Biotechnol. 76 (4), 889–894.

Wang, Y., Liu, S., Zheng, H., He, C., Zhang, H., 2013. T-2 toxin, zearalenone and fumonisin B1 in feedstuffs from China. Food Addit. Contam. Part B 6 (2), 116–122.

Ward, T.J., Clear, R.M., Rooney, A.P., O'Donnell, K., Gaba, D., Patrick, S., Starkey, D.E., Gilbert, J., Geiser, D.M., Nowicki, T.W., 2008. An adaptive evolutionary shift in Fusarium head blight pathogen populations is driving the rapid spread of more toxigenic *Fusarium graminearum* in North America. Fungal Genet. Biol. 45 (4), 473–484.

Weller, D.M., 1988. Biological control of soilborne plant pathogens in the rhizosphere with bacteria. Annu. Rev. Phytopathol. 26 (1), 379–407.

Wenda-Piesik, A., Sun, Z., Grey, W.E., Weaver, D.K., Morrill, W.L., 2009. Mycoses of wheat stem sawfly (Hymenoptera: Cephidae) larvae by *Fusarium* spp. isolates. Environ. Entomol. 38 (2), 387–394.

Wu, F., 2022. Mycotoxin risks are lower in biotech corn. Curr. Opin. Biotechnol. 78, 102792.

Wu, W., Flannery, B.M., Sugita-Konishi, Y., Watanabe, M., Zhang, H., Pestka, J.J., 2012. Comparison of murine anorectic responses to the 8-ketotrichothecenes 3-acetyldeoxynivalenol, 15-acetyldeoxynivalenol, fusarenon X and nivalenol. Food Chem. Toxicol. 50 (6), 2056–2061.

Wudil, A.H., Usman, M., Rosak-Szyrocka, J., Pilař, L., Boye, M., 2022. Reversing years for global food security: a review of the food security situation in Sub-Saharan Africa (SSA). Int. J. Environ. Res. Public Health 19 (22), 14836.

Xue, A., Chen, Y., Sant'anna, S., Voldeng, H., Fedak, G., Savard, M., Längle, T., Zhang, J., Harman, G., 2014. Efficacy of CLO-1 biofungicide in suppressing perithecial production by *Gibberella zeae* on crop residues. Can. J. Plant Pathol. 36 (2), 161–169.

Yang, L., Yu, Z., Hou, J., Deng, Y., Zhou, Z., Zhao, Z., Cui, J., 2016. Toxicity and oxidative stress induced by T-2 toxin and HT-2 toxin in broilers and broiler hepatocytes. Food Chem. Toxicol. 87, 128–137.

Yang, D., Ye, Y., Sun, J., Wang, J.-S., Huang, C., Sun, X., 2022. Occurrence, transformation, and toxicity of fumonisins and their covert products during food processing. Crit. Rev. Food Sci. Nutr., 1–14.

Yates, I., Meredith, F., Smart, W., Bacon, C., Jaworski, A., 1999. *Trichoderma viride* suppresses fumonisin B1 production by *Fusarium moniliforme*. J. Food Prot. 62 (11), 1326–1332.

Yazar, S., Omurtag, G.Z., 2008. Fumonisins, trichothecenes and zearalenone in cereals. Int. J. Mol. Sci. 9 (11), 2062–2090.

Yoo, S., Kim, M.-Y., Cho, J.Y., 2017. Beauvericin, a cyclic peptide, inhibits inflammatory responses in macrophages by inhibiting the NF-κB pathway. Korean J. Physiol. Pharmacol. 21 (4), 449–456.

Zaied, C., Zouaoui, N., Bacha, H., Abid, S., 2012. Natural occurrence of zearalenone in Tunisian wheat grains. Food Control 25 (2), 773–777.

Zain, M.E., 2011. Impact of mycotoxins on humans and animals. J. Saudi Chem. Soc. 15 (2), 129–144.

Zanon, M.A., Chiotta, M., Giaj-Merlera, G., Barros, G., Chulze, S., 2013. Evaluation of potential biocontrol agent for aflatoxin in Argentinean peanuts. Int. J. Food Microbiol. 162 (3), 220–225.

Zanon, M.S.A., Barros, G.G., Chulze, S.N., 2016. Non-aflatoxigenic *Aspergillus flavus* as potential biocontrol agents to reduce aflatoxin contamination in peanuts harvested in Northern Argentina. Int. J. Food Microbiol. 231, 63–68.

Zhang, L., Yan, K., Zhang, Y., Huang, R., Bian, J., Zheng, C., Sun, H., Chen, Z., Sun, N., An, R., 2007. High-throughput synergy screening identifies microbial metabolites as combination agents for the treatment of fungal infections. Proc. Natl. Acad. Sci. 104 (11), 4606–4611.

Zhang, J., Xue, A., Tambong, J., 2009. Evaluation of seed and soil treatments with novel *Bacillus subtilis* strains for control of soybean root rot caused by *Fusarium oxysporum* and *F. graminearum*. Plant Dis. 93 (12), 1317–1323.

Zhao, D., Liu, B., Wang, Y., Zhu, X., Duan, Y., Chen, L., 2013. Screening for nematicidal activities of *Beauveria bassiana* and associated fungus using culture filtrate. Afr. J. Microbiol. Res. 7 (11), 974–978.

Zhao, Y., Selvaraj, J.N., Xing, F., Zhou, L., Wang, Y., Song, H., Tan, X., Sun, L., Sangare, L., Folly, Y.M.E., 2014. Antagonistic action of *Bacillus subtilis* strain SG6 on *Fusarium graminearum*. PLoS One 9 (3), e92486.

Zhou, L., Wei, D.-D., Selvaraj, J.N., Shang, B., Zhang, C.-S., Xing, F.-G., Zhao, Y.-J., Wang, Y., Liu, Y., 2015. A strain of *Aspergillus flavus* from China shows potential as a biocontrol agent for aflatoxin contamination. Biocontrol Sci. Technol. 25 (5), 583–592.

Zinedine, A., Meca, G., Mañes, J., Font, G., 2011. Further data on the occurrence of *Fusarium* emerging mycotoxins enniatins (A, A1, B, B1), fusaproliferin and beauvericin in raw cereals commercialized in Morocco. Food Control 22 (1), 1–5.

Zinedine, A., Fernández-Franzón, M., Mañes, J., Manyes, L., 2017. Multi-mycotoxin contamination of couscous semolina commercialized in Morocco. Food Chem. 214, 440–446.

Žužek, M.C., Grandič, M., Jakovac Strajn, B., Frangež, R., 2016. Beauvericin inhibits neuromuscular transmission and skeletal muscle contractility in mouse hemidiaphragm preparation. Toxicol. Sci. 150 (2), 283–291.

PART 2

General applications, commercialization, and remediation

CHAPTER

12

Applications of nanofungicides in plant diseases control

Rajkuberan Chandrasekaran[a], *P. Rajiv*[b], *Farah K. Ahmed*[c], *and Karungan Selvaraj Vijai Selvaraj*[d]

[a]Department of Biotechnology, Karpagam Academy of Higher Education, Coimbatore, Tamil Nadu, India [b]Department of Biotechnology, PSG College of Arts & Science, Coimbatore, Tamil Nadu, India [c]Biotechnology English Program, Faculty of Agriculture, Cairo University, Giza, Egypt [d]Vegetable Research Station, Tamil Nadu Agricultural University, Palur, Cuddalore, Tamil Nadu, India

1 Introduction

Microbial diseases in plants are a major setback for the growth of the plants and cause huge economic losses to the farmers. Among the microbial diseases, 70%–80% of plant diseases are caused by fungi. Around 8000 species of fungus are known to cause diseases in plants (Bhandari et al., 2021). In the present scenario, the incidence of fungi infections is more severe due to global change, pathological behavior, lower host fitness, and resistance (Greenspan et al., 2017). At the same time, people's awareness is very low regarding the negative effects of synthetic fungicides. So the present need is to understand the molecular dynamics of the infections and to formulate a new drug entity to combat the infections.

The conventional use of synthetic fungicides has proven effective in managing the infestation of diseases. But the side effects associated with the fungicides are lethal to the environment, human beings, mammals, soil, and aquatic organisms also (Pogăcean and Gavrilescu, 2009). Despite the negative hazards, the use of synthetic fungicides is still practiced vigorously. Consequently, the farmers have focused their agriculture practices on the method of organic farming. As a result, botanical pesticides, insecticides, fertilizers, micronutrients, and herbicides were developed and are environmentally friendly. But the demand for botanical-based products is very high on the market. At the same time, it is very difficult to produce ample quantities of botanical products to meet the demand.

Nanohybrid Fungicides
https://doi.org/10.1016/B978-0-443-23950-2.00015-1

Nanotechnology is a new boon to researchers for developing smart nanomaterials for the development of sustainable agriculture (Alghuthaymi et al., 2021b). These nanoparticles' unique optical properties, reactive surface area sites, ease of conjugation with biomolecules and chemicals, and nontoxicity have led to the intrinsic and extrinsic applications of nanoparticles in health hazards and agriculture (Prabukumar et al., 2018). As a result, nanotechnology assisted in the development of nano-based pesticides, insecticides, and fertilizers and was successfully applied in the crop ecosystem.

Developing long-term policies and strategies as an alternative to the traditional use of synthetic fungicides is always welcomed by researchers. The combination of fungicides and biofungicides in nanotechnology will result in a new tool for phytopathogen management. This review will emphasize the new developments made in nanofungicides. Further, the pros and cons of the nanofungicides will be discussed prospectively. As a result, the chapter gives researchers a basic understanding of nanofungicides.

2 Synthetic fungicides

The devastating effects of phytopathogens in plants cause a great economic loss to the farmers and also to the nation. Likewise all around the world, a huge sum of damage is caused by phytopathogenic fungi, which ultimately affect the production. For instance, *Botrytis* causes a major profit reduction of AUS $ 52 million/annum in wine production due to the infection in grapes (Abbey et al., 2019). The use of synthetic fungicides poses severe toxicity to the plants, humans, and soil microbial population. This is because the synthetic pesticides contain salts of toxic metals, organic compounds containing sulfur and mercury, quinones and heterocyclic nitrogenous compounds, and trace metals. Table 1 highlights major fungicides and their target diseases.

3 Biofungicides

Biofungicides are the feasible alternative for the intensive and extensive usage of synthetic fungicides. The so-called biofungicides were derived from plants and microbes that are nontoxic, eco-friendly, and not lethal to humans and the environment (Al-Samarrai et al., 2012). The biofungicides have a large spectrum of activity against diverse phytopathogenic fungi. For instance, *Trichoderma harzianum* was reported to have antagonistic activity against *Fusarium* and *Rhizoctonia*, and *Coniothyrium minitans* are widely used for controlling diseases in lettuce, oilseed, and carrots caused by *Sclerotinia sclerotiorum* (Chitrampalam et al., 2008). Likewise, *Streptomyces cellulosae* Actino 48 can inhibit the phytopathogen *Sclerotium rolfsii*, a causative agent for peanut diseases (Abo-Zaid et al., 2021). The *T. harzianum* isolates T-39, T-161, and T-166 were effective in controlling *Botrytis cinerea* in strawberries (Freeman et al., 2004). Among the bacterial species, *Bacillus subtilis* is effective against *B. cinerea* in the treatment of gray mold infection in strawberries (Hang et al., 2005). *Pseudomonas* spp. is also effective in treating the diseases caused by *B. cinerea*.

TABLE 1 Synthetic fungicides and its target diseases.

S.No.	Fungicide	Diseases
1	Elemental sulfur	Powdery mildews of fruits, vegetables, flowers, and tobacco
2	Zineb (zinc ethylene bisdithiocarbamate)	Early and late blight of potato and tomato, downy mildews, rusts of cereals, and blast of rice
3	Maneb (manganese ethylene bisdithiocarbamate)	Blight of potato and tomato
4	Vapam (SMDC) (sodium methyl dithiocarbamate)	Damping-off disease of papaya
5	Ferbam (ferric dimethyl dithiocarbamate)	Anthracnose of citrus
6	Captafol (*cis*-*N*-1,1,2,2-tetrachlorohexane 1,2-dicarboximide)	Foliage and fruit diseases of potato, tobacco, and coffee
7	Folpet (Folpet) [*N*-(trichloromethyl-thio) phthalimide]	Leaf spots and powdery mildews
8	Quintozene	Botrytis diseases
9	Dichlone (2,3-dichloro1,4-naphthoquinone)	Apple scab and Peach leaf curl
10	Fentin chloride (TPTC—triphenyltin chloride)	Cercospora leaf spot of sugarbeet
11	Carboxin	Loose smut of wheat
12	Fuberidazole	Diseases caused by *Fusarium*
13	Thiophanate	Diseases caused by *Venturia* spp.
14	Pyrazophos	Mildews of cereals, fruits, and vegetables
15	Tricyclazole (5-methyl-1,2,4 triazole(3,4*b*)-benzothiazole)	Paddy blast fungus *P. oryzae*
16	Chlorothalonil (tetra chloroisophthalonitrile)	Groundnut diseases

Other notable biofungicides are plant oils and plant-based compounds that have proved their efficacy in controlling the phytopathogen. Essential oils from peppermint (*Mentha piperita*) and sweet basil (*Ocimum basilicum*) showed effective inhibitory activity against the stored peach (*Prunus persica*) (Sumalan et al., 2020). Likewise, essential oils from *Zingiber officinale*, *Eupatorium cannabinum*, and *Ocimum canum* display antagonistic activity against the *B. cinerea* (Abbey et al., 2019). The biofungicides are one of the best prospects for controlling the plant pathogen, but there are various challenges associated with the production of biofungicides. Though biofungicides proved their efficacy in the laboratory, their transformation into a commercial scale is very difficult and low. The key factors are due to the product development and formulation. It very much needs special formulation techniques to maintain the biocontrol agents alive, intact, and effective. Climatic changes also contribute to the efficacy of biocontrol agents. The viability of microbial culture is also a parameter for the development of biofungicides. In plants, oils is present at very low concentrations, but it is very difficult to extract and purify on a large commercial scale. Due to the synthesis challenge, the researchers turned into semisynthetic derivatives; for example, fludioxonil is a synthetic fungicide developed from the derivatives of pyrrolnitrin produced by *Pseudomonas* spp. (Ligon et al., 2000). Thus, biofungicides are facing various challenges in developing and marketing. Therefore, it needs to look for hyphenated technology to encompass biofungicides for the development of a sustainable agriculture ecosystem that is environmentally friendly.

4 Nanotechnology

The Greek word "nano" is inspiring the research community to develop nanoparticles and nanomaterials (Schwarz et al., 2004). These tailored particles have specific optical, physical, and chemical properties that are unique when compared with their parent materials (Alghuthaymi et al., 2021b). The nanomaterials dominate in solar cells, chemical industries, equipment, automobile industries, LED, LCD, mobiles, dye industries, textiles, drug delivery, drug encapsulation, sensors, diagnosis, and antimicrobial and anticancer agents (Behera, 2022). Henceforth, these nanoparticles are widely gaining recognition for their efficiency and longevity. The use of nanotechnology-based concepts in the agroecosystem is indeed time-limited. In agriculture, the nanoparticles were used as nanoinsecticides, nanopesticides, nanofungicides, growth regulators, nanofertilizers, and gene delivery vesicles. The use of nanoparticles in agricultural practices gives them sustained release, prolonged activity, less toxicity, and maximum efficacy (Duhan et al., 2017).

In the agroecosystem, various types of nanomaterials are used, such as metalloids, metal oxides, carbon nanotubes, quantum dots, liposomes, and dendrimers (Fig. 1). These

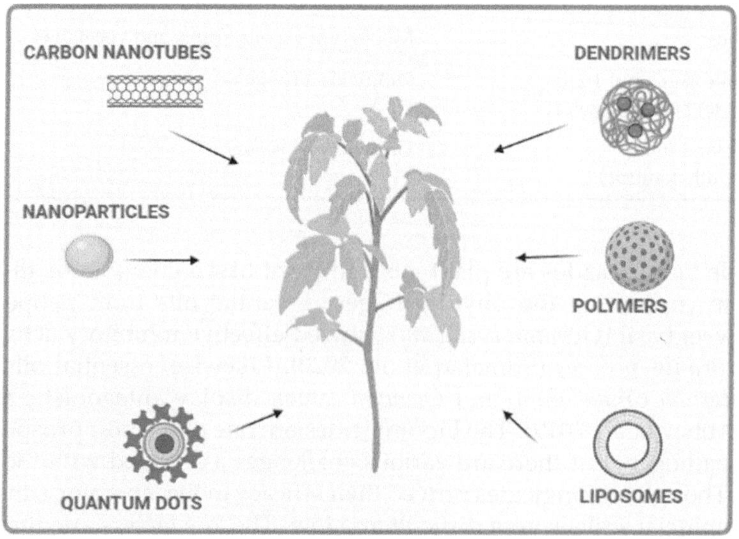

FIG. 1 Some nanomaterials used in plant growth promotion and protection.

nanoparticles were routinely used in plant disease management as fungicides, insecticides, pesticides, micronutrients, and gene transfer agents (Davari et al., 2017). Each of the nanomaterials differed in its properties and also exerted the requisite activity with good efficacy in plants, which increased crop production as well. The advantages of using nanoparticles in phytopathogen management are that the active ingredients entering the plant system are protected from extrinsic factors and significantly reduced; as nanoparticles possess high reactive active surfaces, as a result, the active ingredients can easily translocate into the plant cells and target the site (Pérez-de-Luque, 2017). Further, the nanoparticles can easily interact with bimolecular components and activate the reaction.

4.1 Nanofungicides

The performance of fungal disease management relies on the efficient performance of fungicides and minimizes crop loss. Conventional fungicides are toxic to other target organisms and also lethal to humans (Worrall et al., 2018). However, despite being highly effective and toxic to phytopathogens, it has serious drawbacks, which demand the need for alternative fungicides. The development of nano-based fungicides is still mostly unknown and in its early stages, but they have some promising properties that can be employed in plant protection and production.

Table 2 provides a linguistic review of a broad spectrum of nanofungicides in the agroecosystem.

4.2 Silver nanoparticles

Silver nanoparticles (AgNPs) inherently possess broad-spectrum antimicrobial activity against bacteria and fungi (Alghuthaymi et al., 2023). In the agroecosystem, the use of silver nanoparticles is commercially established due to their unique properties and specificity. *Macrophomina phaseolina* is a soil-borne fungus causing the diseases charcoal rot in faba bean plant. To control the infection, the authors developed green synthesized AgNPs from (*Citrus sinensis* L.) the peel extracts and chemical synthesized AgNPs (Mohamed and Elshahawy, 2022). As a result, the orange AgNPs inhibited the radial development by 100% at 100 ppm concentration, while chemical AgNPs inhibited 60.5% at 100 ppm concentration. At the same time, in greenhouse conditions, biogenic AgNPs stimulated seed germination and inhibited the infection of damping and charcoal rot diseases.

In another study, *Buchanania lanzan* Spreng aqueous leaf extract was utilized as a reducing agent to synthesize and stabilize the silver nanoparticles (Purohit et al., 2022). The aforementioned silver nanoparticles strongly interrogated the phytopathogenic fungi *Rhizoctonia solani* and *Fusarium oxysporum* f. sp. *lycopersici* under in vitro conditions estimated by food poison techniques in potato dextrose agar media (PDA). The biogenic nanoparticles inhibited the mycelial growth 75% at 150 ppm in *R. solani* while in *F. oxysporum* f. sp. *lycopersici* 47% at 150 ppm. *Cymodocea serrulata* (Seagrass) and *Padina australis* (Seaweed) were employed as a template for the synthesis of silver nanoparticles (Kailasam et al., 2022). The nanoparticles *C. serrulata* AgNPs escalate the antifungal efficacy against *Helminthosporium oryzae* (12 ± 1.4 mm), *Alternaria* sp. (10 ± 1.2 mm), and *Pyricularia oryzae* (14 ± 1.5 mm), while seaweed

TABLE 2 Nanofungicides and its target pathogen.

S.No.	Name of the metal nanofungicides	Target fungi	Reference
1	Silver nanoparticles from *A. indica*	*A. alternata, S. sclerotiorum, M. phaseolina, R. solani, B. cinerea,* and *C. lunata*	Krishnaraj et al. (2012)
2	Chemical-mediated synthesis of silver nanoparticles	*R. solani*	Elgorban et al. (2015)
3	Copper nanoparticles from *Persea americana*	*A. flavus, A. fumigates,* and *F. oxysporum*	Rajeshkumar and Rinitha (2018)
4	Chemical synthesis of copper nanoparticles	*Fusarium* sp.	Van Viet et al. (2016)
5	Chitosan—Clove essential oil	*A. niger*	Hasheminejad et al. (2019)
6	Chlorothalonil (M 05) KATHON 930—PVC	Turkey tail (*T. versicolor*), *G. trabeum*	Liu et al. (2002)
7	Pyraclostrobin chitosan-PLA graft copolymer	*C. gossypii* Southw	Xu et al. (2014)
8	Tebuconazole, propineb, fludioxonil-silver nanoparticles	*B. maydis*	Huang et al. (2018)
9	Pyraclostrobin—Chitosan-PLA graft copolymer	*C. gossypii* Southw	Xu et al. (2014)
10	Pyraclostrobin mesoporous silica nanoparticles	*P. asparagi*	Cao et al. (2018)

AgNPs inhibited *R. solani* (16 ± 1.6 mm) and *Xanthomonas oryzae* (15 ± 1.3 mm). Interestingly, yeast *Cryptococcus laurentii* and *Rhodotorula glutinis* were employed for the synthesis of silver nanoparticles (Fernández et al., 2016). The nanoparticles effectively inhibited the growth of phytopathogenic fungi *B. cinerea, Penicillium expansum, Aspergillus niger, Alternaria* sp., and *Rhizopus* sp. while compared with chemical AgNPs. Further, at 3 ppm concentration, the AgNPs nanoparticles exert a similar effect to iprodione synthetic fungicide.

The extracellular culture filtrate *Pseudomonas aeruginosa* MTCC 424 was utilized to synthesize silver nanoparticles (Deshmukh et al., 2012); the synthesized silver nanoparticles produced 40–60 nm spherical-shaped nanoparticles. The nanoparticles exhibited antimycotic activity against plant pathogenic fungi *Fusarium moniliforme* (MTCC 1848) and *Phoma glomerata* (MTCC 2710). Also, the nanoparticles conjugated with the antibiotic clotrimazole escalate the activity twofold higher than AgNPs and antibiotics alone. Zaki et al. (2022) formulated the extracellular culture filtrate of *T. harzianum* for the fine synthesis of AgNPs. The synthesized AgNPs deteriorate the growth activity of *Fusarium fujikuroi* (FF10), *R. solani* (RS9), and *M. phaseolina* isolate (MP4) under in vitro conditions. Further, under greenhouse conditions, the cotton seeds were infected with FF10, RS 9, and MP4 pathogens and assayed. The AgNPs at different dose levels significantly reduce the infection and promote the growth of the plant. Kumar et al. (2022) formulated chemical-mediated AgNPs and applied as foliar spray 50 ppm against leaf spot disease in okra plants for 10–12 weeks. As a result, the disease severity was decreased by 23.2%, and an increase in the crop yield was observed, which was 1.6 kg/plant. Moreover, the biological efficacy of AgNPs in controlling leaf spot diseases in okra plant was 55%.

4.3 Iron oxide nanoparticles

Iron oxide nanoparticles, another variant of metallic nanoparticles, possess excellent catalytic properties and optoelectronic properties (Alghuthaymi et al., 2021b). The influence of iron oxide nanoparticles in controlling phytopathogens is critically investigated by various researchers. Parveen et al. (2018) developed the iron oxide nanoparticles using tannic acid as a reducing agent. The synthesized iron oxide nanoparticles displayed strong antagonistic activity against the vegetable and fruit rot disease-causing fungus *Cladosporium herbarum*, *Trichothecium roseum*, *Alternaria alternata*, *Penicillium chrysogenum*, and *A. niger*. The antimycotic data showed that iron oxide nanoparticles inhibit the *T. roseum* by 87% and *C. herbarum* by 84%. Further, the MIC value ranges between 0.063 and 0.016 mg/mL against the tested pathogens. Ahmad et al. (2017) synthesized polydispersed nanoparticles using the leaves extracts of *Azadirachta indica*. The nanoparticles escalate the growth inhibitory activity against the phytopathogens (*Botryosphaeria dothidea*, *Alternaria mali*, and *Diplodia seriata*). In another study, oriental bananas (*Platanus orientalis*) leaves were used to synthesize iron oxide nanoparticles inhibiting the *Mucor piriformis* and *A. niger* at a concentration of 100 μg/mL (Devi et al., 2019). Microalgae *Chlorella*-K01-mediated synthesis of iron oxide nanoparticles exhibited antifungal activity against the rot-causing fungi *F. moniliforme*, *R. solani*, *F. oxysporum*, *Fusarium tricinctum*, and *Pythium* sp. (Win et al., 2021). The tested pathogens drastically inhibited the growth with ZOI diameters ranging from 10 to 25 mm. Extracellular culture filtrate *T. harzianum* was used as a stabilizing agent for the formation of biogenic nanoparticles. The biogenic iron oxide nanoparticles strongly inhibited the white mold *S. sclerotiorum* at varying concentrations (Bilesky-José et al., 2021). Saqib et al. (2019) developed hybrid nanomaterials (chitosan-iron oxide nanoparticles) developed using the hydrothermal method. The hybrid material was evaluated against the *Rhizopus oryzae* under in vitro and in vivo conditions. Different concentrations (0.25%, 0.50%, 0.75%, and 1%) of the hybrid nanocomposite synergistically inhibited the growth of *R. oryzae*. Likewise, chitosan-coated iron oxide nanoparticles stimulated the inhibitory activity against the *Rhizopus stolonifer* under in vitro and in vivo studies (Saqib et al., 2020).

4.4 Copper oxide nanoparticles

Eichhornia-mediated copper oxide nanoparticles were spherical with an average size of 28 nm. These copper oxide nanoparticles exhibited rapid antifungal activity against the plant pathogens *A. niger* and *Fusarium culmorum* (Vanathi et al., 2015). Neem leaves extract was used as a reducing and stabilizing agent for the synthesis of copper oxide nanoparticles (Ahmad et al., 2020). By using the food poisoning technique, the pathogens *A. mali*, *D. seriata*, and *B. dothidea* were progressively inhibited by the copper oxide nanoparticles. Likewise, *P. chrysogenum* extracellular culture filtrate produced polydispersed copper oxide nanoparticles and inhibited the phytopathogens. The maximum ZOI activity was observed in *F. oxysporum*—37 mm; *Alternaria solani*—28 mm; and *A. niger*—26 mm (El-Batal et al., 2020). Hasanin et al. (2021) synthesized copper oxide nanoparticles using *A. niger* filtrate (CuONPs) and nanocomposite based on starch (St-CuONPs). The nanocomposite showed better antifungal activity than CuONPs against *R. solani*, *F. oxysporum*, and *P. expansum*. Citrus black disease in citrus plants causes severe infection, while synthetic fungicides are not effective in controlling the disease. Meanwhile, copper oxide nanoparticles synthesized from the lemon peel extract showed effective inhibitory activity against *Alternaria citri*. At concentration 10–100 mg/mL, the ZOI ranges from 18 to 50 mm. The minimum inhibitory concentration (MIC) and minimum fungicidal concentration (MFC) were found to be 90 mg/mL and 100 mg/mL. Likewise, the copper oxide nanoparticles produced from the extracellular filtrate of *Streptomyces zaomyceticus* Oc-5 and *Streptomyces pseudogriseolus* Acv-11 showed the dominant antifungal activity with minimal concentration against the plant pathogens *A. niger*, *F. oxysporum*, *Pythium ultimum*, and *A. alternata* (Hassan et al., 2019). Interestingly, copper oxide nanoparticles from the leaves extract of *Cassia fistula* displayed prominent antifungal activity against the *F. oxysporum* f. sp. *lycopersici* at varying concentrations (Ashraf et al., 2021). Additionally, the copper oxide nanoparticles were applied as a foliar spray in infected tomato seedlings, which significantly suppressed the infection and promote plant growth and fruit. El-Abeid et al. (2020) decorated reduced graphene oxide (rGO) in CuONPs synthesized by the precipitation method. The authors produced three sizes (5, 20, and 25 nm) of rGO-CuONPs and evaluated them in *F. oxysporum* strains FORL, FLOC1, and FOC2, respectively. The nanocomposite at 5 nm size effectively inhibited the mycelia growth of the strains with the inhibitory percentage of 94%, 92%, and 87%. Further, the author performed the rGO-CuONPs and their interaction with fungal cells by microscopy analysis (Fig. 2). The positive charge rGO-CuONPs interacts with the negative cell wall of fungi and collapse the internal bound membrane and organelles and cause cell death.

Al-Dhabaan et al. (2017) fabricated the CuNPs (50–100 nm) and assayed them on the tomato fruit infected by *A. alternata*. The CuNPs exert the activity, and the lesion diameter size was reduced in the treatment of CuNPs from one to the third week. Further, the CuNPs degrade the liner DNA by DNA strand scission.

4.5 Zinc oxide nanoparticles

Colletotrichum sp. is the causative agent for anthracnose disease in coffee cultivation. Zinc oxide nanoparticles synthesized by the chemical method effectively inhibited the growth of mycelium. On days 2–6, the growth of the fungus was drastically inhibited in the ZnOPs concentrations (9 mmol L^{-1}, 12 mmol L^{-1}, 15 mmol L^{-1}) with a significant inhibition percentage (Mosquera-Sánchez et al., 2020). Jamdagni et al. (2018) attempt to study the comparative effect of chemical ZnO and biogenic ZnO against phytopathogens. The biogenic ZnO synthesized from the flower extract displayed effective antifungal activity against the pathogens *A. alternata*, *A. niger*, *P. expansum*, *F. oxysporum*, and *B. cinerea*. The ZnO from the chemical method also exerts an activity similar to biogenic ZnO. Neem leaves (*A. indica*) fabricate ZnONPs with a good polydispersity index (PDI) of 0.37 and are spherical (Ali et al., 2021). The nanoparticles showed an effective antifungal activity against the *A. alternata*, *S. rolfsii*,

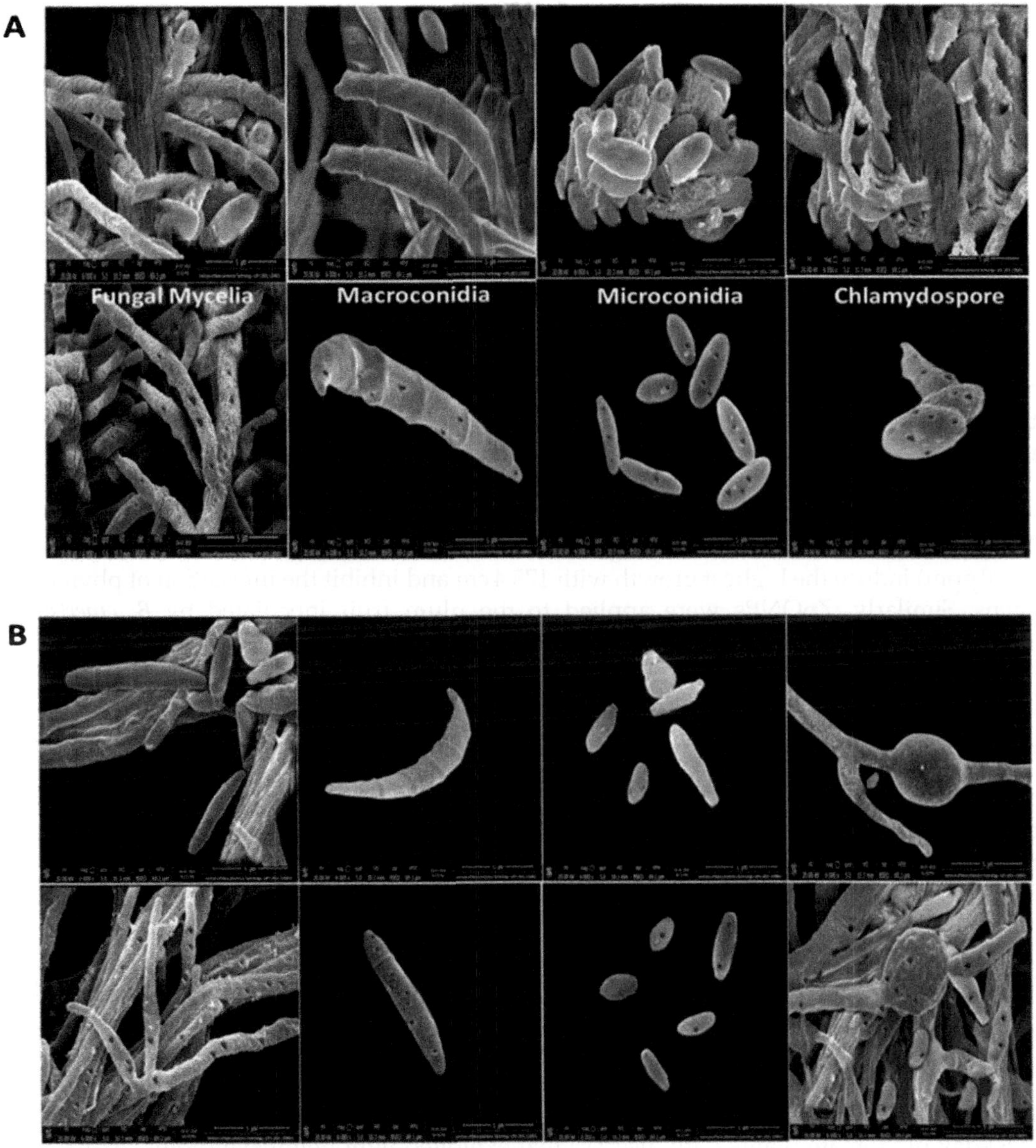

FIG. 2 Effect of rGO-CuONPs against *F. oxysporum* strain. The picture describes the structural alteration of rGO-CuONPs with respect to microconidia, macroconidia, fungal mycelia, and chlamydospores (A, B). *Reproduced with permission from El-Abeid, S.E., Ahmed, Y., Daròs, J.A., Mohamed, M.A., 2020. Reduced graphene oxide nanosheet-decorated copper oxide nanoparticles: a potent antifungal nanocomposite against* Fusarium *root rot and wilt diseases of tomato and pepper plants. Nanomaterials (Basel, Switzerland) 10 (5), 1001. The article is licensed under an open-access Creative Commons CC BY 4.0 license with permission from MDPI.*

and *Stemphylium solani*. As a result, at 40 μg, ZnONPs demonstrated an inhibitory percentage of 45% for *A. alternata*, 38% for *S. rolfsii*, and 42% for *S. solani*. At the same time, the chemical ZnO also possesses antifungal activity to a lesser extent compared with biogenic ZnO. The above-cited studies proved that biogenic ZnO has strong biocidal activity compared with chemical ZnO.

Antimycotic activity of ZnONPs was investigated against dark spore formers *Bipolaris sorokiniana* and *Alternaria brassicicola* (Kriti et al., 2020). At different concentrations (10–100 ppm), the ZnONPs effectively arrested the inhibition of spore formation of *A. brassicicola* at 10 ppm, while ZnONPs at 20 ppm inhibited the spore of *B. sorokiniana*. ZnONPs synthesized from *Melia azedarach* portrayed the growth inhibitory of the seed-borne pathogens of soybean (*F. oxysporum* and *Cladosporium cladosporioides*) (Lakshmeesha et al., 2020). Further, the author demonstrated that by microdilution assay the ZnONPs completely stop the growth of pathogens at 200 and 250 μg/mL concentrations. Additionally, the author proved that ergosterol, the component in the fungal cell wall, is absent in the treatment of ZnONPs followed by intracellular ROS generation and elevating malondialdehyde levels causing apoptosis.

ZnONPs are prepared by a precipitation method possessing a semispherical shape with an average diameter of 10–40 nm. These nanoparticles were applied as a foliar spray in *F. oxysporum*-infected tomato seedlings (González-Merino et al., 2021). As a result, ZnONPs (1500 ppm) induce the highest growth with 175.4 cm and inhibit the infestation of phytopathogens. Similarly, ZnONPs were applied to the plum fruit inoculated by *B. cinerea*. At 100 μg/mL, the ZnONPs exhibited an inhibition rate of 41.17% (Malandrakis et al., 2019).

4.6 Titanium dioxide nanoparticles

Irshad et al. (2020) synthesized highly structured titanium dioxide (TiO_2) nanoparticles by the chemical and green synthesis method. The synthesized TiO_2NPs were evaluated against the wheat rust (*Ustilago tritici*) by the food poisoning technique. The TiO_2NPs was applied at 25 μL, 50 μL, and 75 μL from 0.10 mg/mL concentrations. The nanoparticles effectively inhibited the growth of fungus with 31%, 45%, and 62%.

TiO_2/Ag_3PO_4 nanocomposite exhibits photocatalytic antifungal activity against the *Fusarium graminearum* (Liu et al., 2017). The nanocomposite under visible light illumination inhibits the spore conidia germination by 2%. Henceforth, this nanocomposite can be used as a photocatalytic biocidal agent also. Vegetable crops such as potatoes and tomatoes are often infected with *A. alternata*. To alleviate the infection, El-Gazzar and Ismail (2020) devised TiO_2NPs using *Aspergillus versicolor* KY509550. The synthesized nanoparticles were applied in 60-day-old tomato plants with diseases that exhibited light blight symptoms. At 100 ppm concentration of TiO_2NPs, the nanoparticles effectively reduce the disease severity by 36%, and the efficacy was about 59%. Likewise, rose petals were inoculated with spores of *B. cinerea* and inoculated with TiO_2NPs and kept in a water agar plate for 72 h incubation (Hao et al., 2017). After the period, the nanoparticles reduced the mycelia growth diameter with respect to varying concentrations. Habibi-Yangjeh et al. (2021) studied the nanocomposite TiO_2/AgBr to induce high antifungal activity against the *F. graminearum* than the TiO_2. This is due to the synergistic activity of TiO_2 and AgBr. Within 60 min, the nanocomposite completely inhibits the fungal spores by 98.4%. Likewise, Ag-doped TiO_2 nanoparticles (solid and hollow) were evaluated against the *Fusarium solani* and *Venturia inaequalis* (Boxi et al., 2016). Under dark and day conditions, the assay is performed. As a result, the Ag-doped TiO_2 nanoparticles (hollow) perform superior activity against both pathogens. Additionally, at a very low dose (0.0015 mg/plate), the Ag-doped nanoparticles inhibited the production of naphthoquinone, an important pigment related to fungal pathogenicity. Parveen and Siddiqui (2022) applied TiO_2 at $0.10\,mL\,L^{-1}$ and $0.20\,mL\,L^{-1}$ concentrations in tomato plants inoculated with *F. oxysporum* f. sp. *lycopersici*. Among the concentrations used, $0.20\,mL\,L^{-1}$ concentration of TiO_2 was effective in managing the fungal pathogens (disease severity) and increasing the growth, chlorophyll, carotenoid, proline, SOD, CAT, APX, and PAL defensive enzymes. Scanning microscopy images reveal that TiO_2 adsorbed on the surface of trichomes, stomata, and seed surfaces.

4.7 Other metallic/metallic oxide nanoparticles

Barium ferrite nanoparticles were synthesized using the rhizome extract of *Acorus calamus* (Thakur et al., 2020). The synthesized nanoparticles were evaluated against the *Colletotrichum gloeosporioides*, *Marssonina rosae*, *F. oxysporum*, and *A. alternata*. At different doses of concentrations of 200–600 mg/L, the maximum activity was observed at the concentration of 600 mg/L. The nanoparticles inhibited the mycelia growth of *F. oxysporum* (76.67%), *A. alternata* (75%), *C. gloeosporioides* (72%), and *M. rosae* (70%).

Magnesium oxide nanoparticles are a safe food agent promulgated by US Food and Drug Administration. Abdel-Aziz et al. (2020) synthesized MgO nanoparticles from the endophytic bacterium *Burkholderia rinojensis*. The authors assayed the antifungal effect of MgO nanoparticles at various concentrations against the *F. oxysporum*. At 15 μg/mL, the MgO completely suppressed the growth of mycelia by 95% effectively.

Ismail et al. (2021) synthesized MgO nanoparticles and applied them as a foliar spray in pepper plants infected with powdery mildew disease. Different concentrations of MgO nanoparticles (100–200 mg/L) were applied as a foliar spray on pepper plants infected with powdery mildew disease under greenhouse conditions. As a result, the severity of the disease in response to the treatment of MgONPs was significant, and it is also noted that MgONPs increase the chlorophyll content and polyphenol oxidase activity.

Sidhu et al. (2020) developed a green approach to sepiolite MgO (SE-MgO) nanocomposite and evaluated it against various phytopathogens of rice. At 250 μg/mL, the SE-MgO inhibited the *Fusarium verticillioides* ED_{50} 133, *Bipolaris oryzae* ED_{50} 152, and *F. fujikuroi* ED_{50} 122 with significant ED_{50} values. Sepiolite and MgO are also nontoxic to the living beings, and hence, the present study proved the green approach to controlling the rice diseases.

Zirconium oxide nanoparticles were synthesized through the chemical precipitation method (Derbalah et al., 2019). The synthesized nanoparticles effectively inhibit the *R. solani* with an 85% inhibition percentage. Further, the nanoparticles were applied as a foliar spray in cucumber plants under greenhouse conditions. The disease severity was noticeable in the plant, and the plant growth characteristics and fruit yield were upregulated and downregulated with the treatments.

Palladium-modified nitrogen-doped titanium oxide nanocomposite was activated under visible light illumination to deactivate the growth of *F. graminearum*. Under light illumination, the nanocomposite performs the photocatalytic disinfection by adsorbing into the macroconidium and causing ROS and disruption of cellular membrane and integrity followed by cell death (Zhang et al., 2013).

5 Nanoformualtions

Nanoformualtions are smart delivery systems in which nanoparticles and other inorganic/organic are combined on a nanoscale (Jampílek and Kráľová, 2015). These formulations increase the stability and minimize the usage of fungicides. The rate of release of fungicides is limited by external factors; nanoemulsions improve the solubility of the active compound and promote sustained release to the active sites (Kah and Hofmann, 2014). Application of nanoemulsion before pathogen attacks stimulate the protection effectively and inhibit the growth of invasive pathogens for a long period due to the sustained release of fungicides (Khan et al., 2014). Wang et al. (2007) developed a nanoemulsion to deliver the water-insoluble pesticide β-cypermethrin using the surfactant water/poly(oxyethylene) nonionic surfactant/methyl decanoate. The prepared nanoemulsion was performed with greater stability and release kinetics pattern.

6 Nanogels

Nanogels are smart, soft, deformable, penetrable objects having an internal gel-like structure encompassed by the dispersing solvent. The advantage of nanogels is to respond to external stimuli like pH, ionic strength, pressure, and temperature. These attributes make the nanogel extend its applications in all kinds of biological sectors (Karg et al., 2019). Brunel et al. (2013) formulated a chitosan nanogels adsorbed with copper (II); the adsorption of copper (II) in nanogels was more and stable than chitosan solutions. At pH 5, the maximum adsorption of Cu was achieved in chitosan. Further, the chitosan copper complex hydrolyzes the chitinolytic enzymes present in many phytopathogenic fungi. The chitosan copper complexes effectively inhibit the phytopathogenic fungi *F. graminearum* growth with an MIC value of 17.5 mg/mL.

7 Nanocapsule

Nanocapsule consists of active drugs in the central cavity surrounded by the polymer membrane. The nanocapsule is attractive for researchers due to its unique properties such as sustained release, protective coating, usually pyrophoric, and easy to oxidize (Kothamasu et al., 2012). Carbendazim (MBC) and Tebuconazole (TBZ) are widely used fungicides for the treatment of various diseases. These two fungicides were designed in the form of solid lipid nanoparticles (SLNs) and polymeric nanocapsules (NCs). The fungicides were mixed and delivered in the form of the nanoparticles; the nanoparticles exhibited good stability and release kinetics. The in vitro cytotoxicity of the nanoparticles was evaluated against the normal cell line preosteoblast (MC3T3-E1), fibroblast (Balb/c 3T3), and cancer cell line adenocarcinoma (HeLa). The nanoparticles pose the least toxicity against the normal cell line and possess toxic effects against the HeLa cells. This is due to the uptake of nanoparticles by the cells. Further, the plant emergence rate was estimated for the treatment of nanoparticles and fungicides. The germination rate of *Proteus vulgaris* seeds was about 92% in nanoparticle treatment. The plant growth was also increased due to the sustained release of fungicides from the complex.

8 Polymer-mediated delivery of fungicides

Polymers are another class of smart materials ideally suited for the delivery of various biomolecules such as DNA/RNA, proteins, hormones, and siRNA; antibiotics, drugs, and others. In agriculture, the polymer is intensively used for the delivery of nutraceuticals, fertilizers, pesticides, insecticides, and others. The polymer has special attributes such as targeted delivery, conjugation with organic/inorganic compounds, and stimuli-responsive and biodegradable. Henceforth, the use of polymers is gaining special momentum for researchers. Wang et al. (2021) decisively formulated a star polymer chitosan complex sprayed on late blight disease in potato plants. The antiviral agent eugenol delivered by the star polymer chitosan complex effectively inhibited the disease more than bare eugenol. The chitosan complex completely inhibited the onset of diseases by inducing a defense resistance mechanism. Lignin is a biopolymer present in plants; many wood-decaying fungi are capable of degrading the lignin by secreting the ligases. With this rationale, Machado et al. (2020) prepared a chemically crossed lignin carrier encapsulated with synthetic fungicides namely, boscalid, pyraclostrobin, azoxystrobin, and tebuconazole. The nanocomplex was evaluated against the lignin decaying fungi *Phaeomoniella chlamydospora*, *Neonectria ditissima*, *B. cinerea*, *Neofusicoccum parvum*, *Phytophthora infestans*, and *Magnaporthe oryzae*. The lignin nanocarrier effectively suppressed the growth of fungi with lower MIC values. This is due to the higher availability of the fungicides when released from the nanocarrier. Further, esca-associated fungi *P. chlamydospora* and *Phaeoacremonium minimum* were studied for their inhibition of spores and mycelia growth by nanocarrier under in vitro conditions. After 96h incubation, the lignin nanocarriers showed prominent activity in inhibition growth, and particularly, boscalid nanocarriers exhibited better activity. Further, the boscalid nanocarriers were injected by trunk method for 5 years from 2015 to 2019 in the Portugieser grapevine plants. During these years, the plant did not show any sign or symptom of esca disease. Interestingly, nanocarrier capable of penetrating phloem tissue is developed to treat vascular disease. This is achieved by developing fludioxonil (FLU)-loaded glycine methyl ester-conjugated polysuccinimide nanoparticles (PGA) (Wu et al., 2021). The maximum release of the FLU-PGA nanoparticles in the plant phloem was achieved in alkaline conditions at pH 8. By using the filter paper method, *F. oxysporum* f. sp. *cubense* (Foc) TR4 mycelial growth was inhibited by FLU-PGA nanoparticles at 200 mg/L. Further, the author studied the FLU-PGA nanoparticles translocation and its distribution in potted banana plants. By foliar spray, the FLU-PGA nanoparticles were sprayed and after 2–10 days, the FLU was noticed in all parts of the parts including rhizomes and roots, while bare FLU did not translocate inside

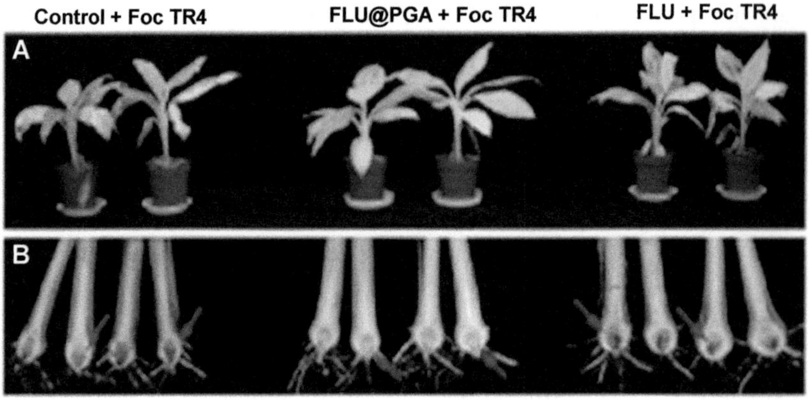

FIG. 3 Effect of FLU-PGA nanoparticles in banana plants infected with Foc TR4. *Reproduced with permission from Wu, H., Hu, P., Xu, Y., Xiao, C., Chen, Z., Liu, X., et al. 2021. Phloem delivery of fludioxonil by plant amino acid transporter-mediated polysuccinimide nanocarriers for controlling fusarium wilt in banana. J. Agric. Food Chem. 69 (9), 2668–2678. The copyright permission needs to be obtained.*

the rhizomes and roots even after 10-day period. In Fig. 3, the effect of FLU-PGA nanoparticles was studied in *Fusarium* wilt-infected banana plants under greenhouse conditions. The nanoparticle-treated banana plants showed less disease severity 33% and no symptoms of *Fusarium* wilt diseases after 50 days. This showed the practical application of polymer-mediated delivery of nanoparticles in field conditions.

Cellulase-segregating fungi *N. ditissima* (syn. *Nectria galligena*) causes severe diseases like apple canker and leads to severe damage in fruit crops where there is no curative treatment available until so far. Machado et al. (2021) developed the cellulose nanocarrier loaded with hydrophobic fungicides captan and pyraclostrobin. The developed nanocarriers were evaluated against the esca and apple canker disease-causing fungus *M. oryzae*, *B. cinerea*, *P. chlamydospora*, *N. ditissima*, *P. infestans*, and *N. parvum*. The MIC values were more significant against the tested pathogens than the parent fungicides. This is due to the high surface area and water dispersible property which make high bioavailability of the fungicides in the medium. For the management of *Sclerotinia* disease, a smart nanocomposite prochloraz (Pro) and pH-jump reagent-loaded zeolitic imidazolate framework-8 (PD@ZIF-8) were conceived and developed. In pot experiments, the nanocomposite showed high efficacy in controlling the infection after 14 days post spraying. The nanocomposite delivers the ZIF 8 in the oil seed rape plants precisely in various parts of the plant as proved in fluorescence tracking and does not impose toxicity on the HepG2 cells (Liang et al., 2021). Similarly, prochloraz was constructed in the mesoporous organosilica nanoparticles (MON) with calcium carbonate (CAC) to deliver the fungicide (Gao et al., 2020). The nanoparticles effectively suppressed the growth of the fungi *S. sclerotiorum* with the EC_{50} 0.142 mg/L; under field conditions, the nanoparticles were sprayed in rapeseed plants. After 1 h—7 days post spraying, the plant disease severity was dramatically decreased compared with bare MON and CAC. Likewise, pectinase nanocarrier was designed to control the blast disease in rice (Abdelrahman et al., 2021). Prochloraz is loaded into the mesoporous silica nanoparticles and coated with pectin. The authors studied the translocation of nanoparticles in rice seedlings. As illustrated in

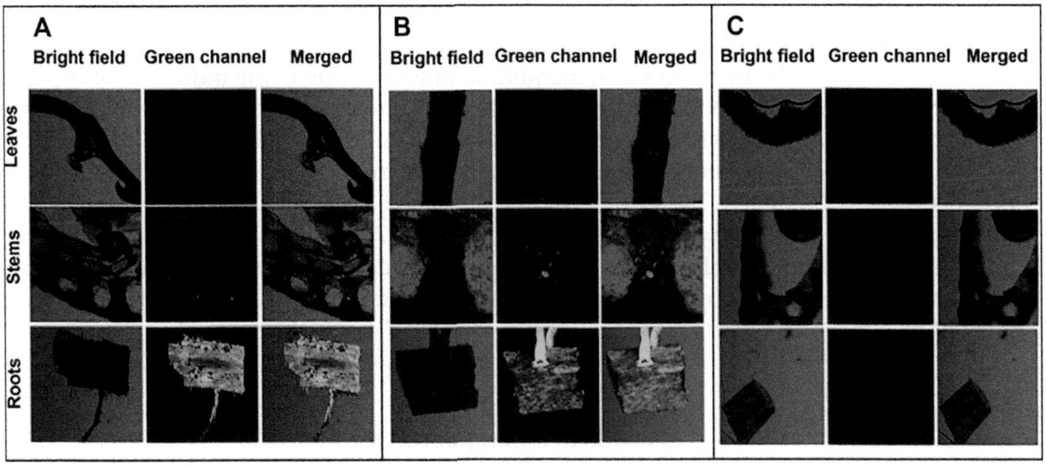

FIG. 4 Confocal microscopy images of the translocation of mesoporous silica nanoparticles in the rice seedlings. *Reproduced with permission from Abdelrahman, T.M., Qin, X., Li, D., Senosy, I.A., Mmby, M., Wan, H., et al., 2021. Pectinase-responsive carriers based on mesoporous silica nanoparticles for improving the translocation and fungicidal activity of prochloraz in rice plants. Chem. Eng. J. 404, 126440 with permission from Elsevier.*

Fig. 4, the mesoporous silica nanoparticles transfer the prochloraz into the stem, root, and leaves of the rice seedlings.

Mancozeb-loaded chitosan carrageen nanoparticles were developed and evaluated against the *A. alternata*, *Septoria lycopersici*, *A. solani*, and *S. sclerotiorum*. At three different concentrations (0.5 ppm, 1 ppm, and 1.5 ppm), the nanoparticles effectively inhibited the growth in *S. sclerotiorum*, *S. lycopersici*, and *A. alternata*, and negligible activity was observed in *A. solani*. Further, the disease control efficacy (DCE%) in controlling early blight and stem rot (or southern blight) in potato (*Solanum tuberosum* L.) and early blight and leaf spot in tomato (*Lycopersicon esculentum* L.) was studied in pot house conditions. As a result, the nanoparticles showed a significant DCE in controlling the disease. The nanoparticles showed a stupendous effect on plant growth and development (Kumar et al., 2021). Thus, with appropriate scientific evidence, it is postulated that polymer-mediated delivery of fungicides precisely deliver the fungicides, and the plant cells easily uptake the polymeric nanoparticles and spur the reaction.

9 Antifungal mechanism of nanofungicides

The important phenomenon to be understood by the researchers is the antifungal activity; it is an important aspect of developing nanofungicides. The antifungal activity of nanoparticles is governed by various factors such as size, shape, pH, crystallinity, surface area, and synthesis methods (Cruz-Luna et al., 2021).

There is the various hypothesis put forward for the antifungal activity of nanoparticles. The metal ions released from the nanoparticles bind to the essential protein and disrupt the function, thereby leading to a collapse in cell permeability (Mikhailova, 2020).

Nanoparticles interact with the electron transport chain and respiratory pathway and lead to the failure of the internal respiratory pathway (Alghuthaymi et al., 2021a). In addition, the nanoparticles disrupt the mitochondrial membrane potential due to the induction of reactive oxygen species (ROS) (Huerta-García et al., 2014). Further, the induced ROS severely damages the biomolecules DNA/RNA/proteins and causes cell death (Zhao et al., 2017). Moreover, the nanoparticles inhibit the germination of the spore/conidia of the mycelium (Rana et al., 2020).

To support the above hypothesis, Hao et al. (2017) demonstrated the engineered nanomaterials and their lethal effect on rose petals infected with *B. cinerea*. After 72h incubation with the engineered nanoparticles, the hyphae were slender, fragile, and lack viability, as presented in Fig. 5.

10 Future perspectives

Globally, the plant diseases caused by fungal infection are still challenging the researchers to develop alternative fungicides. Conventional fungicides were utilized intensively causing heavy pollution and unwarranted toxic effects on the living beings and the environment. In the present scenario, nanotechnology incorporates the delivery of fungicides as an alternative paradigm for synthetic fungicide usage.

Nanofungicide is a new concept for agriculturalists to understand what the material is and how efficient it is. To date, the nanofungicide concept is at an early stage, and limited success is established. The problem in developing the nanofungicides is that the shareholders and public entrepreneurs are not yet familiarized with the concepts and practices methods. To alleviate the problem, government, nongovernment organizations, researchers, and farmers should get aware of these technologies and their progress.

Toxicity is the main concern for developing nanofungicides. Hence, the utilization of safe biodegradable polymers with metal nanoparticles will minimize the toxicity. Before applying the nanofungicides, the nontargeted effects and public health should be considered. Intelligent, automated technologies should be utilized to deliver the nanofungicides so that we can use a minimal amount of nanofungicides. The nanofungicides should be broad spectrum, and phytotoxicity, ecotoxicity, soil health-related problems, and other problems related to nanofungicides should be addressed before bringing this product to market.

11 Conclusion

The development of next-generation nanofungicides is indeed important for sustainable agriculture management. Fungicides developed from synthetic or biological sources must be delivered in the form of nanotechnology. The different nanofungicides proved their efficacy in a variety of diseases and were also nontoxic and safe for nontargeted organisms. Among the metallic nanoparticles, silver nanoparticles are very feasible nanofungicides that can be employed in various monocot and dicot crops. Polymer-mediated delivery of fungicides is to be intensively studied in various diseases due to its potential to become

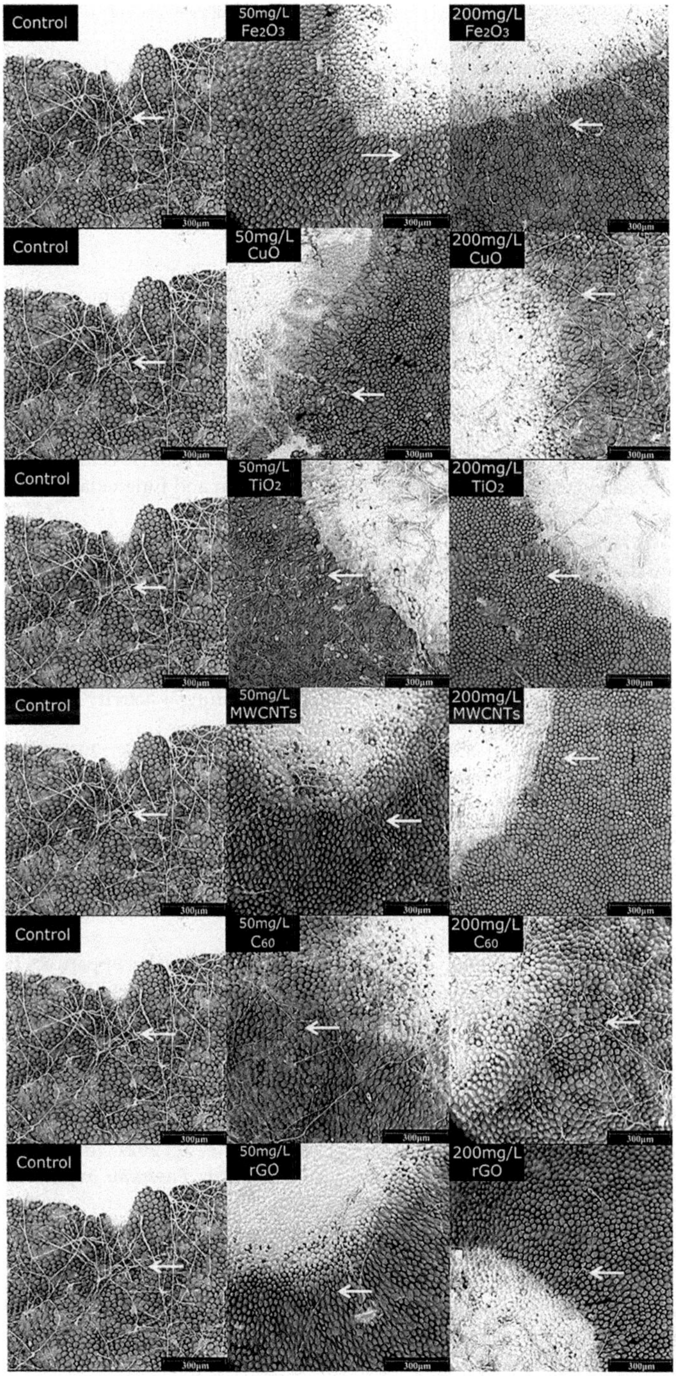

FIG. 5 Effect of engineered nanoparticles on the rose petals infected with *B. cinerea. Reproduced with permission from \Hao, Y., Cao, X., Ma, C., Zhang, Z., Zhao, N., Ali, A., et al., 2017. Potential applications and antifungal activities of engineered nanomaterials against gray mold disease agent* Botrytis cinerea *on rose petals. Front. Plant Sci. 8, 1332. This is an open-access article distributed under the terms of the Creative Commons Attribution License (CC BY).*

commercialized. Thus, with appropriate scientific evidence, the chapter covers the ongoing prospects for nanofungicides. Henceforth, the chapter benefits the researchers, agriculturalists, and industrialists who were interested in developing nanofungicides.

References

Abbey, J.A., Percival, D., Abbey, L., Asiedu, S.K., Prithiviraj, B., Schilder, A., 2019. Biofungicides as an alternative to synthetic fungicide control of grey mould (*Botrytis cinerea*)–prospects and challenges. Biocontrol Sci. Technol. 29 (3), 207–228.

Abdel-Aziz, M.M., Emam, T.M., Elsherbiny, E.A., 2020. Bioactivity of magnesium oxide nanoparticles synthesized from cell filtrate of endobacterium *Burkholderia rinojensis* against *Fusarium oxysporum*. Mater. Sci. Eng. C 109, 110617.

Abdelrahman, T.M., Qin, X., Li, D., Senosy, I.A., Mmby, M., Wan, H., et al., 2021. Pectinase-responsive carriers based on mesoporous silica nanoparticles for improving the translocation and fungicidal activity of prochloraz in rice plants. Chem. Eng. J. 404, 126440.

Abo-Zaid, G., Abdelkhalek, A., Matar, S., Darwish, M., Abdel-Gayed, M., 2021. Application of bio-friendly formulations of chitinase-producing *Streptomyces cellulosae* Actino 48 for controlling peanut soil-borne diseases caused by *Sclerotium rolfsii*. J. Fungi 7 (3), 167.

Ahmad, H., Rajagopal, K., Shah, A.H., Bhat, A.H., Venugopal, K., 2017. Study of bio-fabrication of iron nanoparticles and their fungicidal property against phytopathogens of apple orchards. IET Nanobiotechnol. 11 (3), 230–235.

Ahmad, H., Venugopal, K., Bhat, A.H., Kavitha, K., Ramanan, A., Rajagopal, K., et al., 2020. Enhanced biosynthesis synthesis of copper oxide nanoparticles (CuO-NPs) for their antifungal activity toxicity against major phytopathogens of apple orchards. Pharm. Res. 37 (12), 1–12.

Al-Dhabaan, F.A., Shoala, T., Ali, A.A., Alaa, M., Abd-Elsalam, K., Abd-Elsalam, K., 2017. Chemically produced copper, zinc nanoparticles and chitosan-bimetallic nanocomposites and their antifungal activity against three phytopathogenic fungi. Int. J. Agric. Technol. 13 (5), 753–769.

Al-Samarrai, G., Singh, H., Syarhabil, M., 2012. Evaluating eco-friendly botanicals (natural plant extracts) as alternatives to synthetic fungicides. Ann. Agric. Environ. Med. 19 (4).

Alghuthaymi, M.A., Rajkuberan, C., Rajiv, P., Kalia, A., Bhardwaj, K., Bhardwaj, P., Abd-Elsalam, K.A., Valis, M., Kuca, K., 2021a. Nanohybrid antifungals for control of plant diseases: current status and future perspectives. J. Fungi 7 (1), 48.

Alghuthaymi, M.A., Rajkuberan, C., Santhiya, T., Krejcar, O., Kuča, K., Periakaruppan, R., Prabukumar, S., 2021b. Green synthesis of gold nanoparticles using *Polianthes tuberosa* L. floral extract. Plants 10 (11), 2370.

Alghuthaymi, M.A., Patil, S., Rajkuberan, C., Krishnan, M., Krishnan, U., Abd-Elsalam, K.A., 2023. *Polianthes tuberosa*-mediated silver nanoparticles from flower extract and assessment of their antibacterial and anticancer potential: an in vitro approach. Plants 12 (6), 1261.

Ali, J., Mazumder, J.A., Perwez, M., Sardar, M., 2021. Antimicrobial effect of ZnO nanoparticles synthesized by different methods against food borne pathogens and phytopathogens. Mater. Today Proc. 36, 609–615.

Ashraf, H., Anjum, T., Riaz, S., Ahmad, I.S., Irudayaraj, J., Javed, S., et al., 2021. Inhibition mechanism of green-synthesized copper oxide nanoparticles from *Cassia fistula* towards *Fusarium oxysporum* by boosting growth and defense response in tomatoes. Environ. Sci. Nano 8 (6), 1729–1748.

Behera, A., 2022. Nanomaterials. In: Advanced Materials. Springer, Cham, pp. 77–125.

Bhandari, S., Yadav, P., Sarhan, A., 2021. Botanical fungicides; current status, fungicidal properties and challenges for wide scale adoption: a review. Rev. Food Agric. 2 (2), 63–68.

Bilesky-José, N., Maruyama, C., Germano-Costa, T., Campos, E., Carvalho, L., Grillo, R., et al., 2021. Biogenic α-Fe2O3 nanoparticles enhance the biological activity of *Trichoderma* against the plant pathogen *Sclerotinia sclerotiorum*. ACS Sustain. Chem. Eng. 9 (4), 1669–1683.

Boxi, S.S., Mukherjee, K., Paria, S., 2016. Ag doped hollow TiO2 nanoparticles as an effective green fungicide against *Fusarium solani* and *Venturia inaequalis* phytopathogens. Nanotechnology 27 (8), 085103.

Brunel, F., El Gueddari, N.E., Moerschbacher, B.M., 2013. Complexation of copper (II) with chitosan nanogels: toward control of microbial growth. Carbohydr. Polym. 92 (2), 1348–1356.

Cao, L., Zhang, H., Zhou, Z., Xu, C., Shan, Y., Lin, Y., Huang, Q., 2018. Fluorophore-free luminescent double-shelled hollow mesoporous silica nanoparticles as pesticide delivery vehicles. Nanoscale 10, 20354–20365.

Chitrampalam, P., Figuli, P.J., Matheron, M.E., Subbarao, K.V., Pryor, B.M., 2008. Biocontrol of lettuce drop caused by *Sclerotinia sclerotiorum* and *S. minor* in desert agroecosystems. Plant Dis. 92 (12), 1625–1634.

Cruz-Luna, A.R., Cruz-Martínez, H., Vásquez-López, A., Medina, D.I., 2021. Metal nanoparticles as novel antifungal agents for sustainable agriculture: current advances and future directions. J. Fungi 7 (12), 1033.

Davari, M.R., Bayat Kazazi, S., Akbarzadeh Pivehzhani, O., 2017. Nanomaterials: implications on agroecosystem. In: Nanotechnology. Springer, Singapore, pp. 59–71.

Derbalah, A., Elsharkawy, M.M., Hamza, A., El-Shaer, A., 2019. Resistance induction in cucumber and direct antifungal activity of zirconium oxide nanoparticles against *Rhizoctonia solani*. Pest. Biochem. Phys. 157, 230–236.

Deshmukh, S.D., Deshmukh, S.D., Gade, A.K., Rai, M., 2012. *Pseudomonas aeruginosa* mediated synthesis of silver nanoparticles having significant antimycotic potential against plant pathogenic fungi. J. Bionanoscience 6 (2), 90–94.

Devi, H.S., Boda, M.A., Shah, M.A., Parveen, S., Wani, A.H., 2019. Green synthesis of iron oxide nanoparticles using *Platanus orientalis* leaf extract for antifungal activity. Green Process. Synth. 8 (1), 38–45.

Duhan, J.S., Kumar, R., Kumar, N., Kaur, P., Nehra, K., Duhan, S., 2017. Nanotechnology: the new perspective in precision agriculture. Biotechnol. Rep. 15, 11–23.

El-Abeid, S.E., Ahmed, Y., Daròs, J.A., Mohamed, M.A., 2020. Reduced graphene oxide nanosheet-decorated copper oxide nanoparticles: a potent antifungal nanocomposite against fusarium root rot and wilt diseases of tomato and pepper plants. Nanomaterials 10 (5), 1001.

El-Batal, A.I., El-Sayyad, G.S., Mosallam, F.M., Fathy, R.M., 2020. *Penicillium chrysogenum*-mediated mycogenic synthesis of copper oxide nanoparticles using gamma rays for in vitro antimicrobial activity against some plant pathogens. J. Clust. Sci. 31 (1), 79–90.

El-Gazzar, N., Ismail, A.M., 2020. The potential use of titanium, silver, and selenium nanoparticles in controlling leaf blight of tomato caused by *Alternaria alternata*. Biocatal. Agric. Biotechnol. 27, 101708.

Elgorban, A.M., El-Samawaty, A.E.-R.M., Yassin, M.A., Sayed, S.R., Adil, S.F., Elhindi, K.M., Bakri, M., Khan, M., 2015. Antifungal silver nanoparticles: synthesis, characterization, and biological evaluation. Biotechnol. Biotechnol. Equip. 30, 56–62.

Fernández, J.G., Fernández-Baldo, M.A., Berni, E., Camí, G., Durán, N., Raba, J., Sanz, M.I., 2016. Production of silver nanoparticles using yeasts and evaluation of their antifungal activity against phytopathogenic fungi. Process Biochem. 51 (9), 1306–1313.

Freeman, S., Minz, D., Kolesnik, I., Barbul, O., Zveibil, A., Maymon, M., et al., 2004. Trichoderma biocontrol of *Colletotrichum acutatum* and *Botrytis cinerea* and survival in strawberry. Eur. J. Plant Pathol. 110 (4), 361–370.

Gao, Y., Liang, Y., Dong, H., Niu, J., Tang, J., Yang, J., et al., 2020. A bioresponsive system based on mesoporous organosilica nanoparticles for smart delivery of fungicide in response to pathogen presence. ACS Sustain. Chem. Eng. 8 (14), 5716–5723.

González-Merino, A.M., Hernández-Juárez, A., Betancourt-Galindo, R., Ochoa-Fuentes, Y.M., Valdez-Aguilar, L.A., Limón-Corona, M.L., 2021. Antifungal activity of zinc oxide nanoparticles in *Fusarium oxysporum-Solanum lycopersicum* pathosystem under controlled conditions. J. Phytopathol. 169 (9), 533–544.

Greenspan, S.E., Bower, D.S., Roznik, E.A., Pike, D.A., Marantelli, G., Alford, R.A., et al., 2017. Infection increases vulnerability to climate change via effects on host thermal tolerance. Sci. Rep. 7 (1), 1–10.

Habibi-Yangjeh, A., Davari, M., Manafi-Yeldagermani, R., Alikhah Asl, S., Enaiati, S., Ebadollahi, A., Feizpoor, S., 2021. Antifungal activity of TiO2/AgBr nanocomposites on some phytopathogenic fungi. Food Sci. Nutr. 9 (7), 3815–3823.

Hang, N.T.T., Oh, S.O., Kim, G.H., Hur, J.S., Koh, Y.J., 2005. *Bacillus subtilis* S1-0210 as a biocontrol agent against *Botrytis cinerea* in strawberries. Plant Pathol. J. 21 (1), 59–63.

Hao, Y., Cao, X., Ma, C., Zhang, Z., Zhao, N., Ali, A., et al., 2017. Potential applications and antifungal activities of engineered nanomaterials against gray mold disease agent *Botrytis cinerea* on rose petals. Front. Plant Sci. 8, 1332.

Hasanin, M., Hashem, A.H., Lashin, I., Hassan, S.A., 2021. In vitro improvement and rooting of banana plantlets using antifungal nanocomposite based on myco-synthesized copper oxide nanoparticles and starch. Biomass Convers. Biorefinery, 1–11.

Hasheminejad, N., Khodaiyan, F., Safari, M., 2019. Improving the antifungal activity of clove essential oil encapsulated by chitosan nanoparticles. Food Chem. 275, 113–122.

Hassan, S.E.D., Fouda, A., Radwan, A.A., Salem, S.S., Barghoth, M.G., Awad, M.A., et al., 2019. Endophytic actinomycetes *Streptomyces* spp mediated biosynthesis of copper oxide nanoparticles as a promising tool for biotechnological applications. JBIC, J. Biol. Inorg. Chem. 24 (3), 377–393.

Huang, W., Wang, C., Duan, H., Bi, Y., Wu, D., Du, J., Yu, H., 2018. Synergistic antifungal effect of biosynthesized silver nanoparticles combined with fungicides. Int. J. Agric. Biol. 20, 1225–1229.

Huerta-García, E., Pérez-Arizti, J.A., Márquez-Ramírez, S.G., Delgado-Buenrostro, N.L., Chirino, Y.I., Iglesias, G.G., López-Marure, R., 2014. Titanium dioxide nanoparticles induce strong oxidative stress and mitochondrial damage in glial cells. Free Radic. Biol. Med. 73, 84–94.

Irshad, M.A., Nawaz, R., ur Rehman, M.Z., Imran, M., Ahmad, J., Ahmad, S., et al., 2020. Synthesis and characterization of titanium dioxide nanoparticles by chemical and green methods and their antifungal activities against wheat rust. Chemosphere 258, 127352.

Ismail, A.M., El-Gawad, A., Mona, E., 2021. Antifungal activity of MgO and ZnO nanoparticles against powdery mildew of pepper under greenhouse conditions. Egypt. J. Agric. Res. 99 (4), 421–434.

Jamdagni, P., Rana, J.S., Khatri, P., Nehra, K., 2018. Comparative account of antifungal activity of green and chemically synthesized zinc oxide nanoparticles in combination with agricultural fungicides. Int. J. Nano Dimens. 9 (2), 198–208.

Jampílek, J., Králová, K., 2015. Application of nanotechnology in agriculture and food industry, its prospects and risks. Ecol. Chem. Eng. S 22, 321–361.

Kah, M., Hofmann, T., 2014. Nanopesticide research: current trends and future priorities. Environ. Int. 63, 224–235.

Kailasam, S., Sundaramanickam, A., Tamilvanan, R., Kanth, S.V., 2022. Macrophytic waste optimization by synthesis of silver nanoparticles and exploring their agro-fungicidal activity. Inorg. Nano-Metal Chem., 1–10.

Karg, M., Pich, A., Hellweg, T., Hoare, T., Lyon, L.A., Crassous, J.J., et al., 2019. Nanogels and microgels: from model colloids to applications, recent developments, and future trends. Langmuir 35 (19), 6231–6255.

Khan, M.R., Ashraf, S., Rasool, F., Salati, K.M., Mohiddin, F.A., Haque, Z., 2014. Field performance of Trichoderma species against wilt disease complex of chickpea caused by *Fusarium oxysporum* f. sp. *ciceri* and *Rhizoctonia solani*. Turk. J. Agric. For. 38 (4), 447–454. https://doi.org/10.3906/tar-1209-10.

Kothamasu, P., Kanumur, H., Ravur, N., Maddu, C., Parasuramrajam, R., Thangavel, S., 2012. Nanocapsules: the weapons for novel drug delivery systems. BioImpacts: BI 2 (2), 71.

Krishnaraj, C., Ramachandran, R., Mohan, K., Kalaichelvan, P., 2012. Optimization for rapid synthesis of silver nanoparticles and its effect on phytopathogenic fungi. Spectrochim. Acta A Mol. Biomol. Spectrosc. 93, 95–99.

Kriti, A., Ghatak, A., Mandal, N., 2020. Antimycotic efficacy of zinc nanoparticle on dark spore forming phytopathogenic fungi. J. Pharmacogn. Phytochem. 9 (2), 750–754.

Kumar, R., Najda, A., Duhan, J.S., Kumar, B., Chawla, P., Klepacka, J., et al., 2021. Assessment of antifungal efficacy and release behavior of fungicide-loaded chitosan-carrageenan nanoparticles against phytopathogenic fungi. Polymers 14 (1), 41.

Kumar, I., Bhattacharya, J., Das, B.K., 2022. Efficacy of silver nanoparticles-based foliar spray application to control plant diseases, its effect on productivity, and risk assessment. Arab. J. Geosci. 15 (5), 1–11.

Lakshmeesha, T.R., Murali, M., Ansari, M.A., Udayashankar, A.C., Alzohairy, M.A., Almatroudi, A., et al., 2020. Biofabrication of zinc oxide nanoparticles from *Melia azedarach* and its potential in controlling soybean seed-borne phytopathogenic fungi. Saudi J. Biol. Sci. 27 (8), 1923–1930.

Liang, W., Xie, Z., Cheng, J., Xiao, D., Xiong, Q., Wang, Q., et al., 2021. A light-triggered pH-responsive metal–organic framework for smart delivery of fungicide to control sclerotinia diseases of oilseed rape. ACS Nano 15 (4), 6987–6997.

Ligon, J.M., Hill, D.S., Hammer, P.E., Torkewitz, N.R., Hofmann, D., Kempf, H.J., Pée, K.H.V., 2000. Natural products with antifungal activity from *Pseudomonas* biocontrol bacteria. Pest Manag. Sci. 56 (8), 688–695.

Liu, Y., Laks, P., Heiden, P., 2002. Controlled release of biocides in solid wood. II. Efficacy against *Trametes versicolor* and *Gloeophyllum trabeum* wood decay fungi. J. Appl. Polym. Sci. 86, 608–614.

Liu, B., Xue, Y., Zhang, J., Han, B., Zhang, J., Suo, X., et al., 2017. Visible-light-driven TiO2/Ag3PO4 heterostructures with enhanced antifungal activity against agricultural pathogenic fungi *Fusarium graminearum* and mechanism insight. Environ. Sci. Nano 4 (1), 255–264.

Machado, T.O., Beckers, S.J., Fischer, J., Müller, B., Sayer, C., de Araújo, P.H., et al., 2020. Bio-based lignin nanocarriers loaded with fungicides as a versatile platform for drug delivery in plants. Biomacromolecules 21 (7), 2755–2763.

Machado, T.O., Beckers, S.J., Fischer, J., Sayer, C., de Araújo, P.H., Landfester, K., Wurm, F.R., 2021. Cellulose nanocarriers via miniemulsion allow pathogen-specific agrochemical delivery. J. Colloid Interface Sci. 601, 678–688.

Malandrakis, A.A., Kavroulakis, N., Chrysikopoulos, C.V., 2019. Use of copper, silver, and zinc nanoparticles against foliar and soil-borne plant pathogens. Sci. Total Environ. 670, 292–299.

Mikhailova, E.O., 2020. Silver nanoparticles: mechanism of action and probable bio-application. J. Funct. Biomater. 11, 84.

Mohamed, Y.M.A., Elshahawy, I.E., 2022. Antifungal activity of photo-biosynthesized silver nanoparticles (AgNPs) from organic constituents in orange peel extract against phytopathogenic *Macrophomina phaseolina*. Eur. J. Plant Pathol. 162 (3), 725–738.

Mosquera-Sánchez, L.P., Arciniegas-Grijalba, P.A., Patiño-Portela, M.C., Guerra-Sierra, B.E., Muñoz-Florez, J.E., Rodríguez-Páez, J.E., 2020. Antifungal effect of zinc oxide nanoparticles (ZnO-NPs) on *Colletotrichum* sp., causal agent of anthracnose in coffee crops. Biocatal. Agric. Biotechnol. 25, 101579.

Parveen, A., Siddiqui, Z.A., 2022. Foliar spray and seed priming of titanium dioxide nanoparticles and their impact on the growth of tomato, defense enzymes and some bacterial and fungal diseases. Arch. Phytopathol. Plant Protect. 55 (5), 527–548.

Parveen, S., Wani, A.H., Shah, M.A., Devi, H.S., Bhat, M.Y., Koka, J.A., 2018. Preparation, characterization, and antifungal activity of iron oxide nanoparticles. Microb. Pathog. 115, 287–292.

Pérez-de-Luque, A., 2017. Interaction of nanomaterials with plants: what do we need for real applications in agriculture? Front. Environ. Sci. 5, 12.

Pogăcean, M.O., Gavrilescu, M., 2009. Plant protection products and their sustainable and environmentally friendly use. Environ. Eng. Manag. J. 8 (3), 607–627.

Prabukumar, S., Rajkuberan, C., Sathishkumar, G., Illaiyaraja, M., Sivaramakrishnan, S., 2018. One pot green fabrication of metallic silver nanoscale materials using *Crescentia cujete* L. and assessment of their bactericidal activity. IET Nanobiotechnol. 12 (4), 505–508.

Purohit, A., Sharma, R., Shiv Ramakrishnan, R., Sharma, S., Kumar, A., Jain, D., et al., 2022. Biogenic synthesis of silver nanoparticles (AgNPs) using aqueous leaf extract of *Buchanania lanzan* Spreng and evaluation of their antifungal activity against phytopathogenic fungi. Bioinorg. Chem. Appl. 2022.

Rajeshkumar, S., Rinitha, G., 2018. Nanostructural characterization of antimicrobial and antioxidant copper nanoparticles synthesized using novel *Persea americana* seeds. OpenNano 3, 18–27.

Rana, A., Yadav, K., Jagadevan, S.A., 2020. Comprehensive review on green synthesis of nature-inspired metal nanoparticles: mechanism, application, and toxicity. J. Clean. Prod. 272, 122880.

Saqib, S., Zaman, W., Ullah, F., Majeed, I., Ayaz, A., Hussain Munis, M.F., 2019. Organometallic assembling of chitosan-iron oxide nanoparticles with their antifungal evaluation against *Rhizopus oryzae*. Appl. Organomet. Chem. 33 (11), e5190.

Saqib, S., Zaman, W., Ayaz, A., Habib, S., Bahadur, S., Hussain, S., et al., 2020. Postharvest disease inhibition in fruit by synthesis and characterization of chitosan iron oxide nanoparticles. Biocatal. Agric. Biotechnol. 28, 101729.

Schwarz, J.A., Contescu, C.I., Putyera, K. (Eds.), 2004. Dekker Encyclopedia Nanoscience Nanotechnology. Vol. 5. CRC Press.

Sidhu, A., Bala, A., Singh, H., Ahuja, R., Kumar, A., 2020. Development of MgO-sepoilite nanocomposites against phytopathogenic fungi of rice (*Oryzae sativa*): a green approach. ACS Omega 5 (23), 13557–13565.

Sumalan, R.M., Kuganov, R., Obistioiu, D., Popescu, I., Radulov, I., Alexa, E., et al., 2020. Assessment of mint, basil, and lavender essential oil vapor-phase in antifungal protection and lemon fruit quality. Molecules 25 (8), 1831.

Thakur, A., Sharma, N., Bhatti, M., Sharma, M., Trukhanov, A.V., Trukhanov, S.V., et al., 2020. Synthesis of barium ferrite nanoparticles using rhizome extract of *Acorus calamus*: characterization and its efficacy against different plant phytopathogenic fungi. Nanostruct. Nano-Objects 24, 100599.

Van Viet, P., Nguyen, H.T., Cao, T.M., Van Hieu, L., Pham, V., 2016. *Fusarium* antifungal activities of copper nanoparticles synthesized by a chemical reduction method. J. Nanomater. 2016, 1957612.

Vanathi, P., Rajiv, P., Sivaraj, R., 2015. Eichhornia mediated copper oxide nanoparticles: in vitro analysis of antimicrobial activity. Int. J. Pharm. Pharm. Sci. 7 (11), 422–424.

Wang, L., Li, X., Zhang, G., Dong, J., Eastoe, J., 2007. Oil-in-water nanoemulsions for pesticide formulations. J. Colloid Interface Sci. 314 (1), 230–235.

Wang, X., Zheng, K., Cheng, W., Li, J., Liang, X., Shen, J., et al., 2021. Field application of star polymer-delivered chitosan to amplify plant defense against potato late blight. Chem. Eng. J. 417, 129327.

Win, T.T., Khan, S., Bo, B., Zada, S., Fu, P., 2021. Green synthesis and characterization of Fe3O4 nanoparticles using *Chlorella*-K01 extract for potential enhancement of plant growth stimulating and antifungal activity. Sci. Rep. 11 (1), 1–11.

Worrall, E.A., Hamid, A., Mody, K.T., Mitter, N., Pappu, H.R., 2018. Nanotechnology for plant disease management. Agronomy 8 (12), 285.

Wu, H., Hu, P., Xu, Y., Xiao, C., Chen, Z., Liu, X., et al., 2021. Phloem delivery of fludioxonil by plant amino acid transporter-mediated polysuccinimide nanocarriers for controlling fusarium wilt in banana. J. Agric. Food Chem. 69 (9), 2668–2678.

Xu, L., Cao, L.D., Li, F.M., Wang, X.J., Huang, Q.L., 2014. Utilization of chitosan-lactide copolymer nanoparticles as controlled release pesticide carrier for pyraclostrobin against *Colletotrichum gossypii* Southw. J. Dispers. Sci. Technol. 35, 544–550.

Zaki, S.A., Ouf, S.A., Abd-Elsalam, K.A., Asran, A.A., Hassan, M.M., Kalia, A., Albarakaty, F.M., 2022. Trichogenic silver-based nanoparticles for suppression of fungi involved in damping-off of cotton seedlings. Microorganisms 10 (2), 344.

Zhang, J., Liu, Y., Li, Q., Zhang, X., Shang, J.K., 2013. Antifungal activity and mechanism of palladium-modified nitrogen-doped titanium oxide photocatalyst on agricultural pathogenic fungi *Fusarium graminearum*. ACS Appl. Mater. Interfaces 5 (21), 10953–10959. https://doi.org/10.1021/am4031196.

Zhao, X., Zhou, L., Rajoka, M.S.R., Dongyan, S., Jiang, C., Shao, D., Zhu, J., Shi, J., Huang, Q., Yang, H., et al., 2017. Fungal silver nanoparticles: synthesis, application, and challenges. Crit. Rev. Biotechnol. 38, 817–835.

CHAPTER

13

Nanostructures for fungal disease management in the agri-food industry

R. Britto Hurtado[a], S. Horta-Piñeres[b], J.M. Gutierrez Villarreal[c], M. Cortez-Valadez[d], and M. Flores-Acosta[a]

[a]Department of Physics Research, University of Sonora, Hermosillo, Sonora, Mexico [b]Physics Department, Popular University of Cesar, Valledupar, Colombia [c]Technological University of South Sonora, Obregon, Sonora, Mexico [d]CONACYT—Department of Physics Research, University of Sonora, Hermosillo, Sonora, Mexico

1 Introduction

According to the Food and Agriculture Organization (FAO), millions of tons of food are wasted worldwide in the production and distribution of products. It happens in homes, restaurants, supermarkets, and other food services (Nordin et al., 2020). Food waste can have serious environmental and economic effects. For example, food lost or wasted by humans generates methane, a greenhouse gas. Between 8% and 10% of global greenhouse gas emissions are associated with unconsumed food (Sharma et al., 2020). It is important to consider that food waste also means the loss of other resources such as energy and water that were used for the production and distribution of the product. In addition, its social impact is considerable, especially bearing in mind that many households worldwide undergo famine due to the food lack for their consumption, increasing the mortality rate due to hunger and malnutrition. In recent years, substantial technological advances have been developed in the field of agribusiness to reduce food and plant waste (Kim et al., 2020), as well as in the innovation of practical solutions and new methodologies to improve distribution and meet the demand for food due to the growth of the world's population. In that sense, nanotechnology often differs significantly from their bulk properties through nanomaterials' unique physical and chemical properties. It can provide new tools in the agricultural sector to combat phytopathogenic fungi and improve the ability of plants to absorb nutrients. Nanotechnology can also provide safe fertilizers for plants and soil preservation and nanoparticles as active agents in

https://doi.org/10.1016/B978-0-443-23950-2.00013-8

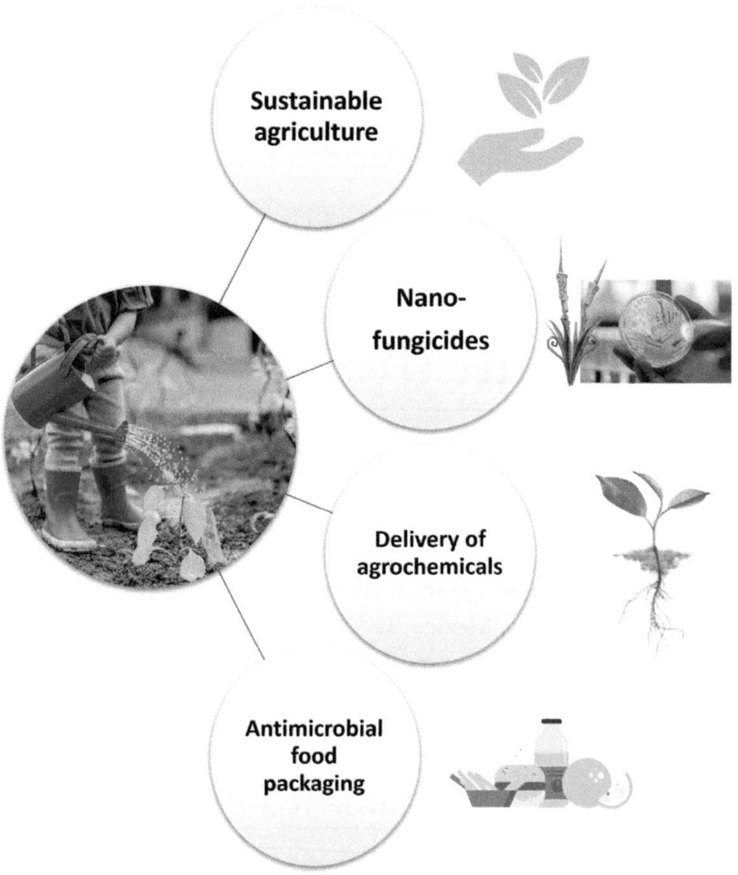

FIG. 1 Some applications of nanotechnology in agriculture.

food packaging materials (Fig. 1). Due to their size (1–100 nm), nanoparticles have interesting antimicrobial properties and have been used as antifungicides to prevent the growth of some types of fungi or cause their death. For example, it was demonstrated that copper nanoparticles (CuNPs) could be used as a source of micronutrients in cases of copper deficiency in the soil and as fungicides(Lopez-Lima et al., 2021). The cited study showed that CuNPs at different concentrations (0.1, 0.25, 0.5, 0.75, and 1.0 mg/mL) decrease the incidence and severity of wilt by *Fusarium oxysporum f.* sp. *Lycopersici* (FOL), because Cu NPs reduce FOL growth in tomato plants by applying the NPs in the root/soil system by establishing direct contact between the NPs and FOL spores; thus, Cu NPs can inhibit the growth of the spores before they enter the roots. A study on titanium dioxide nanoparticles (TiO_2 NPs), silver nanoparticles (AgNPs), and selenium nanoparticles (SeNPs) with potent nanofungicidal properties for the biological control of blight caused by *Alternaria alternata* on tomato leaves, using concentrations for TiO_2 NPs, AgNPs, and SeNPs of 50, 75, and 100 ppm, respectively, were also reported (El-Gazzar and Ismail, 2020). Pariona et al. reported that CuNPs are potential fungicides against *F. solani*, *Neofusicoccum* sp., and *F. oxysporum*, which are pathogenic fungi that invade the vascular tissues of plants. It inhibits water transport through the xylem by inducing vessel plugging and leading to wilting of the foliage. Their study found that concentrations of CuNPs at 0.75 mg/mL almost completely inhibited the growth of the three fungi evaluated based on percentage inhibition of radial growth (IRG) close to 100% (Pariona et al., 2019).

Moreover, nanotechnology contributes to the controlled release of fertilizers, pesticides, and herbicides using agrochemical carriers to develop an intelligent delivery system to improve the efficiency of nutrient and active ingredient use (Panpatte et al., 2016). One of the global goals to protect life on earth is to stop soil degradation. Additionally, nanofertilizers also improve soil health by decreasing the toxic effects of excessive fertilizer use. Saleem et al. reported a study on the coating of potassium ferrite nanoparticles ($KFeO_2$ NPs) with sizes between 7 and 18 nm on diammonium phosphate (DAP) fertilizer for a slow and stable release of nitrogen (N) and phosphorus (P) (Saleem et al., 2021). Also, the development of slow-release fertilizers coated with superhydrophobic biopolymers (SBSF) self-assembled with Fe_3O_4 magnetic nanoparticles was reported to have longer slow-release longevity (more than 100 days) compared to slow-release fertilizers coated with unmodified biopolymers (Xie et al., 2019). Other authors reported the encapsulation of active fungicide agents (hexaconazole) in chitosan nanoparticles, intending to develop a more effective nanodelivery system against the pathogenic fungus *Ganoderma boninense* in oil palm. The in vitro fungicide showed that the nanoparticles are sustainably released with a prolonged release time of 86 h (Maluin et al., 2019). Similarly, other nanomaterials have been reported as essential auxiliaries for delivering nutrients to crops in a controlled release form to enhance plant productivity and minimize environmental pollutants (Fatima et al., 2021; Guo et al., 2018). Silicon nanoparticles with nanosensor applications are the potential for agricultural use as agents releasing proteins, nucleotides, and other chemicals in plants to monitor soil conditions (Rastogi et al., 2019). Therefore, the importance of the properties of nanomaterials applied in agribusiness is presented as a technological tool to improve food quality and minimize crop and product losses while ensuring sustainability and environmental care.

2 Nanostructures in the agri-food sector

Researchers in agriculture are currently facing the problem of food demand generated by the growing world population. In addition to the soil deterioration caused by the excessive synthetic agrochemicals use and the growing resistance of pesticides used due to the mutations of pathogens that damage a large part of the plants and crops. Therefore, it is necessary to consider innovations and the generation of new technologies to help counteract these harmful effects. It is essential to develop technologies that make it possible to harvest and produce safe food and extend the shelf life while preserving quality and freshness to reduce food waste as much as possible. Reduce energy consumption in water care and transportation and, at the same time, protect the health of the soil and the environment (Singh Sekhon, 2014). Increasingly common in the industry, inorganic nanostructures are used throughout the food production chain because of their attractive physical and chemical properties. Modern industrial processes require continuous improvement and effort guided by the need for more authentic, efficient, and environmental-friendly applications (Bhagat et al., 2013). At the nanoscale, these structures manifest extraordinary capacities in surface area, ion exchange, and ion adsorption, among others, which significantly enhance their industrial use. Through their most basic use, such as improving soils to cultivate a wide variety of food plants, these structures can be used in a wide range of industrial applications. The appropriate use of nanomaterials can make nutrient availability more efficient since the transport of nutrients between plants and soil depends on ion exchange processes (Mukhopadhyay, 2014). Therefore, nanotechnology would allow the development of easily transported systems available for use in crops. The capacity of nanomaterials to improve the germination and growth capacity of some plants has been reported. Such as the use of nanopowders based on iron (Fe) and copper (Cu) in low concentrations (Churilov et al., 2018). Additionally, in this same report, the positive accumulation of active organic compounds was studied because of the action of these nanopowders, stimulating the growth of crops such as corn, wheat, and peas, among others. The advance in nanotechnology in agriculture is undeniable, as seen by a quick look at a wide variety of scientific articles. However, it is also crucial to be aware of its use with little planning since there are reports of the risk caused by the accumulation of nanoparticles in soils due to their excessive use in some industrial processes. Innovations in this field have led to the development of a new generation of nanofertilizers that aim to deliver nutrients only where they are needed, minimizing their dispersion in the ecosystem (Singh Sekhon, 2014).

On the other hand, modern pesticides have the critical task of improving crop efficiency while reducing their environmental impact, especially on water sources. In this sense, nanotechnology could play a relevant role in this field by being able to develop nanopesticides that fulfill their traditional role. At the same time, reducing their ability to spread in nature and thus avoid affecting especially insects, which are the most affected by modern agriculture methods (Musters et al., 2021). Consequently, techniques have been developed that allow the combining of different compounds such as surfactants, polymers, and nanoparticles to obtain hydrophobic pesticides that allow controlled delivery of the active agent where it is needed (Singh Sekhon, 2014). Recent studies show zein nanoparticles (zein is a protein found in corn used in food and other industries) to encapsulate and perform controlled delivery of active agents, allowing the reduction of environmental impact (Pascoli et al., 2018). Likewise, nanopesticides have been developed to fulfill the purpose of sustainable agriculture, as is the case of Neem oil, which has been shown to have a wide variety of active biological compounds. It is encapsulated in Zein nanoparticles and allows controlled delivery, reducing the environmental hazards of its massive use (Pascoli et al., 2019). Fig. 2 shows the potential nanotechnological applications of nanoparticles in agriculture.

Another necessary point is the nanotechnology applied in developing light, strong, and resistant plastics with antimicrobial and gas barrier properties to prevent food spoilage and, therefore, prolong its shelf life and reduce food waste (Souza and Fernando, 2016). Silver nanoparticles have antimicrobial, antifungal, antiyeast, and antiviral activities. They can be combined with nondegradable and edible polymers for the active packaging of foods such as meat, fruit, and dairy products (Carbone et al., 2016). The antimicrobial action can be obtained by biocide released directly into the food or the space around the food. Also, the use of ZnO nanoparticles in polymeric matrices has been reported to provide antimicrobial activity to the packaging material and improve the properties of the packaging (Espitia et al., 2012). Starch-based nanocomposite (St) films containing Ag, ZnO, and CuO nanoparticles with antimicrobial properties were also reported for potential use in food packaging

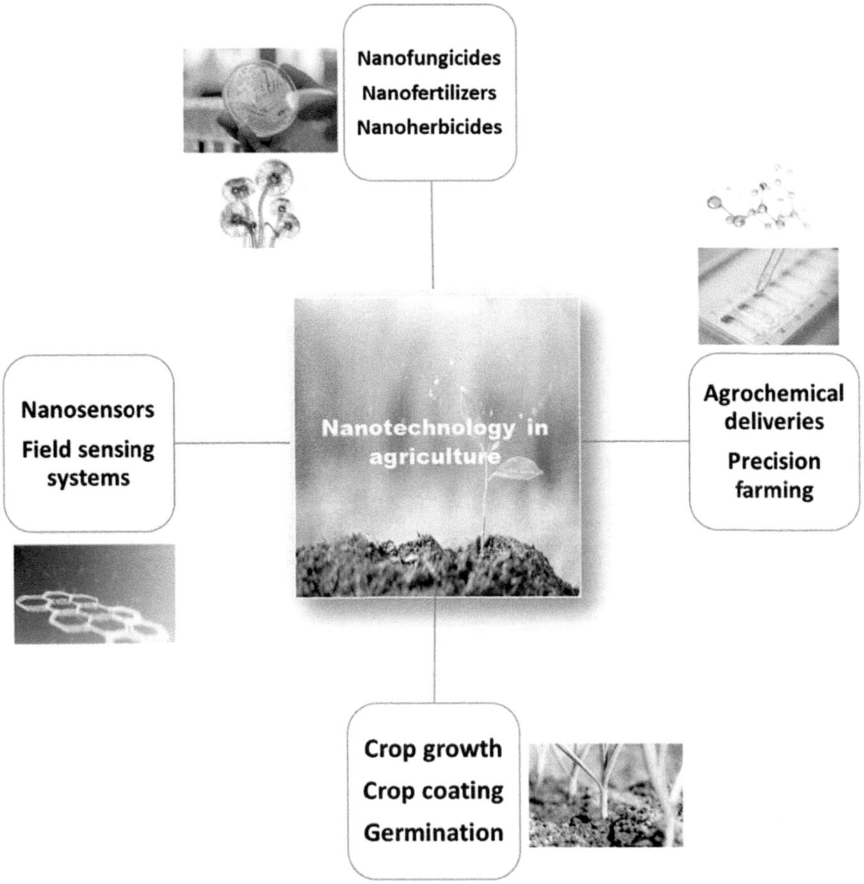

FIG. 2 Potential applications of nanotechnology in the agri-food industry.

applications. The reported microbial tests showed that St-Ag and St-CuO films had the highest antibacterial activity against *E. coli* and *S. aureus*, respectively (Peighambardoust et al., 2019). Physical and chemical methods are generally used to obtain nanoparticles; however, this synthesis route requires many chemicals that can be toxic to people and pollute the environment. Therefore, for applications that require nanomaterials in agroindustry, ecological or green synthesis methods are preferred because the synthesis processes use ecological sources. Such plant or fruit extracts, organic molecules, algae, bacteria, and more, are low toxicity, economic, easy, and effective alternatives. For example, recently, zinc oxide nanoparticles with sizes between 52 and 70 nm with fungicidal capacity against apple orchard pathogenic fungi (*Alternaria mali*, *Botryosphaeria dothidea*, and *Diplodia seriata*) were detected reported (Hilal Ahmad et al., 2020). In addition, it was reported that CuO and ZnO NPs synthesized using lemon peel extract can inhibit the fungal growth of *Alternaria citri*, which causes citrus black rot disease (Sardar et al., 2022). However, further research is needed to address nanoparticles' safety and health risks. Table 1 lists some known plant pathogens and literature reports of nanoparticles that have been used against these microorganisms.

TABLE 1 Nanoparticles used against phytopathogenic microorganisms.

Fungi phytopathogens	Plant pathology	NPs used as antifungal agents (Size; forms)	Fungicidal effect	Reference
Aspergillus flavus	*Aspergillus flavus* (*A. Flavus*) is an opportunistic pathogen that adopts a necrotrophic lifestyle, causing cell death in the host. *A. flavus* infects important cereal crops (Musungu et al., 2020), affecting oil-rich seeds (León Peláez et al., 2012). Cases of poultry infected with *A. flavus* have also been reported (Xue et al., 2022)	Ag (10–100 nm; N/A)	The results suggest that the synthesized AgNPs are capable of inhibiting *Aspergillus flavus* isolated fungi. In this case, the MIC is 7.45 ± 0.18 (μg/mL)	Al-Zubaidi et al. (2019)
		Au (38.5 ± 10.6 nm; spherical)	A marked antifungal activity of AuNPs against *Aspergillus* isolates was observed. Three out of 11 (28%) *A. flavus* isolates strains (AM2, AM11, and AM15) were inhibited. MIC was 1000 ppm for *A. flavus* (AM15)	Almansob et al. (2022)
		CuO (40–60 nm; spherical)	The CuO-NPs presented a marked antifungal activity against *A. Flavus*. The maximum inhibition was a zone of ~10 mm in diameter at a concentration of 75 μL	Rajeshkumar and Rinitha (2018)
		Cu (26–40 nm; spherical)	The maximum reduction of aflatoxins after treatment with different concentrations of CuNPs was 77.2% for 100 μg/mL	Asghar et al. (2018)
		ZnO (700–800 nm; Flower-Shaped)	A notable reduction in mycelial production of *A. flavus* was observed at concentrations of 1.25, 2.5, and 5 mM ZnO, showing reductions up to 78% of the control level	Hernández-Meléndez et al. (2018)
Venturia inaequalis	The ascomycete *Venturia inaequalis* (Cooke) G. Winter is responsible for the most severe disease affecting apples, as well as many ornamental Malus species in many countries, including Poland: apple scab. To date, pathogenic *V. inaequalis* races have been reported on *Malus* spp. (Michalecka and Puławska, 2021)	Ag (10–20 nm; spherical)	The inhibitions percent of *V. inaequalis* was 73.76%–75.65% while the inhibition value for the variant treated with a standard chemical product was 87.39%	Ungureanu et al. (2021)
Alternaria alternata	*Alternaria alternata* causes black spot in many fruits and vegetables around the world (Troncoso-Rojas and Tiznado-Hernández, 2014) Several pathotypes of *A. alternata* attack different host plants and on each, they produce one of several multiple forms of related compounds that are toxic only to the particular host plant of each pathotype (Agrios, 2005)	Ag (86–100 nm; spherical)	A significantly higher mycelial growth inhibition was observed against *A. alternate* 62.12 ± 4.50%of radial growth inhibition	Bernardo-Mazariegos et al. (2019)
		CuO (18.18–43.37 nm; spherical)	The highest zone of inhibition was recorded as 18.5 ± 1–50 ± 0.5 mm at a concentration of 10–100 mg/mL	Sardar et al. (2022)
		ZnO (14.57–49.88 nm; spherical in shape and elongated in shape)	The highest zone of inhibition was recorded as 51.5 ± 0.5 mm at a concentration of 100 mg/mL.	Sardar et al. (2022)

Fusarium spp.	The filamentous fungal genus *Fusarium* has a worldwide distribution and contains at least 300 phylogenetically distinct species/species complexes. This ascomycete genus is among the world's most economically destructive plant pathogens (Dongzhen et al., 2020). The genus *Fusarium* is the most economically important genera of fungi globally. Its species, such as *Fusarium oxysporum*, *F. solani*, *F. equisetti*, *F. chlamydosporum*, and others, cause root and stem rot, vascular wilt, and fruit rot in some crop species (Perveen et al., 2022)	Ag (10–20 nm; spherical)	The results suggest that the Silver Nanoparticles Phytosynthesized is capable of inhibiting *F. Oxysporum*. The MIC is 53.93 ± 0.23 (μg/mL)	Ungureanu et al. (2021)
		CuO (40–60 nm; spherical)	The CuO-NPs presented a marked antifungal activity against *A. Flavus*. The maximum inhibition was a zone of ~12 mm in diameter at a concentration of 25 μL	Rajeshkumar and Rinitha (2018)
		Cu (8 nm; spherical)	The diameters of the zones of complete inhibition including the diameter of the disc were 16.67 ± 1.15 mm	Shende et al. (2021)
		Cu (23–82 nm; polygonal)	The antifungal study showed that CuNPs have potential antifungal activity against *F. oxysporum* where 3.2 μg/mL of nanoparticle solution inhibited the growth	Ammar et al. (2019)
		Ag (729 nm; N/A)	The inhibitions percent of *F. Solani* was 83.05% compared to control sample	Ruiz-Romero et al. (2018)
		Ag (86–100 nm; spherical)	A significantly higher mycelial growth inhibition was observed against *A. alternate* 35.60 ± 3.55% of radial growth inhibition	Bernardo-Mazariegos et al. (2019)
Aspergillus niger	*Aspergillus niger* belongs to *Aspergillus* section *nigri*, a group of black *aspergilli*. As a typical food spoilage fungus, *A. niger* naturally exists in many agricultural products (e.g., different kinds of fruits, onion, nuts, and corn) under a wide range of temperatures (6–47°C) and pH levels (1.5–9.8) (Li et al., 2020)	Ag (10–20 nm; spherical)	The results suggest that the silver nanoparticles synthesized are capable of inhibiting *A. niger*. The MIC is 53.93 ± 0.23 (μg/mL)	Ungureanu et al. (2021)
		Au (12–22 nm; spherical)	For the different concentrations (10, 25, and 50 μg/mL) of synthesized AuNPs, the mean value of the zone of inhibition was 5, 8, and 9 mm	Piruthiviraj et al. (2016)
		Cu (40–60 nm)	In general, inhibition % at 200 ppm was 58.10% and at 300 ppm inhibition % reached 86.89%	Abdel Ghany et al. (2020)
		Cu (8 nm; spherical)	The diameters of the zones of complete inhibition including the diameter of the disc were 27.67 ± 0.58 mm	Shende et al. (2021)

Continued

TABLE 1 Nanoparticles used against phytopathogenic microorganisms.—Cont'd

Fungi phytopathogens	Plant pathology	NPs used as antifungal agents (Size; forms)	Fungicidal effect	Reference
Colletotrichum gloeosporioides	The genus *Colletotrichum* includes several plant pathogens of major importance, causing diseases in a wide variety of woody and herbaceous plants. *Colletotrichum* is an opportunistic plant pathogen able to produce anthracnose disease in a wide range of postharvest fruits, almond, apple, avocado, banana, cacao, citrus, coffee, cranberry, mango, papaya, passion fruit, pear, soursop, strawberry, etc. (Shi et al., 2021)	Ag (22–40 nm; spherical)	The synthesis of AgNPs was confirmed by UV-Vis spectroscopy. The shape of AgNPs was found to be spherical. The Mc-AgNPs from goat, cow, and buffalo urine exhibited 146.15%, 133.33%, and 114.28% more antifungal activity than the fungicides alone respectively	Raghavendra et al. (2020)
Sclerotinia sclerotiorum	*Sclerotinia sclerotiorum* is a phytopathogenic fungus, which causes many diseases in crops and uncultivated plants. It is one of the causes of the disease known as white rot (Bolton et al., 2006)	Cu (10 to 50 nm, and 2 to 12 nm; spherical)	The antifungal activities of NPs (5–100 μg/mL), were compared to those of the recommended chemical fungicide Topsin-M 70 WP at a dose of 1000 μg/mL	Sadek et al. (2022)
Macrophomina phaseolina	*Macrophomina phaseolina* is a soil- and seed-borne pathogenic fungus belonging to the Botryosphaeriaceae family that causes seedling blight, collar rot, stem rots, and root rot, among other diseases in many plant species (Kaur et al., 2012)	Ag (729 nm; N/A)	The inhibitions percent of M. Phaseolin was 67.05% compared to the control sample	Ruiz-Romero et al. (2018)
		Ag (32–47 nm; spherical)	The antifungal activity of the aqueous extract of orange peel, chemo- and bio-AgNPs at doses of 0, 10, 25, 50, and 100 ppm were tested in vitro against the most virulent isolate of *M. phaseolina*	Mohamed and Elshahawy (2022)

3 Use of nanostructures to control plant pathogenic fungi in the seedlings stage

One of the objectives of using nanomaterials in agriculture is to stop the spread of pathogenic microorganisms at different stages of plant growth and harvesting. Nanoparticles could have different functions during the food production process and efficiently take advantage of potent antimicrobial properties. Plant diseases caused by fungi can appear at different stages of growth, so treatment at an early planting stage and during preharvest and postharvest is essential. Seed germination is a process that depends on certain factors such as temperature, humidity, oxygen, and soil conditions, among others, that help to achieve optimal conditions for the seeds to germinate. Due to the different types, varying ranges of optimum temperature, amount of water, and time, some seeds do not sprout in a short period. Many times it can take days or months, or even several years. Nanoparticles applied in this field can help improve water uptake processes, ROS production and act directly against pathogens. For example, metal and metal oxide nanoparticles of Fe, Cu, Co, Ag, and ZnO were used to pretreat cereal seeds to analyze germination rate, plant development, and inhibition effects against pathogenic fungi (Hoang et al., 2022). These results showed a favorable response in increasing seedling length and reducing the number of infected kernels twice for wheat and 3.6 times for barley against the three fungi *Alternaria* sp., *Fusarium* sp., and *Helminthosporium teres*. The highest sensitivity shown to the nano preparation with the greatest inhibition effect when the seeds were grown on a medium with Co nanoparticles and Ag-chitosan was *Helminthosporium teres*. Metallic and metal-oxide nanoparticles are essential for seed germination and seedling growth. For example, Ag nanoparticles with a diameter of ~10 nm were applied at concentrations of 0.25, 1.25, and 2.5 mg/dm^3 as a presowing treatment on seeds for rapid seedling germination of two bean cultivars, "Bali" and "Delfina," under normal and cold temperatures (Prażak et al., 2020). Likewise, iron oxide nanoparticles (Fe-NPs) with a diameter of 19–30 nm were synthesized by green synthesis using onion extract for the priming of watermelon seeds using different concentrations of Fe-NPs (20, 40, 80, and 160 mg/L). It with favorable results for seedling development, and chlorophyll biosynthesis does not represent toxicity for germination (Kasote et al., 2019). Moreover, nonpathogenic fungi have been used as a green synthesis method to obtain nanoparticles with applications in agriculture. In this sense, the synthesis of AgNPs using fungal strains of *T. harzianum* and *A. fumigatus* was recently reported as an economical, safe, and ecological method. The NPs were used in seed germination and seedling growth of *Solanum lycopersicum* (tomato) (Noshad et al., 2019). Iron oxide nanoparticles (Fe_3O_4 NPs) were also reported as a plant growth stimulant for commercialized seeds (rice, corn, mustard, chickpea, and watermelon) and antifungal agent against *Fusarium oxysporum, Fusarium tricinctum, Fusarium maniliforme, Rhizoctonia solani,* and *Phythium* sp. (Win et al., 2021). In this case, the researchers immersed the seeds in the synthesized nanoparticles solution for 1 h and stirred at 100 rpm. Anna Gorczyca et al. studied the germination process and morphology of soft wheat seedlings using AgNPs treatment and found a significant beneficial reduction of *Fusarium culmorum* seedling infection as a result of AgNP treatment. However, AgNP treatment was found to cause significant disintegration of root cell membranes (Gorczyca et al., 2015). Alternatively, a green route was used by Dawoud et al. as an alternative method to evaluate the antifungal activity against *Fusarium* spp. instead of using synthetic chemicals (Dawoud et al., 2021). A study of watermelon seed priming with AgNPs to improve seed germination, growth, and yield with high fruit quality through an environment-friendly synthesis approach using an aqueous extract of onion peel as a reducing agent was also reported (Acharya et al., 2020). Likewise, a study on the effect of copper oxide nanoparticles (CuO NPs) on wheat seedlings' germination capacity, development, and biochemical status was reported. They found that CuO NPs at concentrations of 0.01 and 1 g/L inhibited the mycelial growth of *Alternaria solani*. (Zakharova et al., 2019). Biosynthesized selenium nanoparticles (Se-NPs) using *Bacillus megaterium* were applied as growth promoters and antifungal agents against *Rhizoctonia solani* in vitro and in vivo, the Se-NPs were able to decrease the wilting caused by *R. solani* and minimize the severity of root rot disease (Hashem et al., 2021). Therefore, the physicochemical properties of the nanoparticles are sought to promote growth and inhibit the proliferation of certain phytopathogenic fungi to reduce chemical pesticides that affect the environment. Fig. 3 shows an illustrative image of nanomaterials' effects on plant germination, growth, and care.

Some nanoparticles are applied as foliar sprays to kill pathogens that cause different plant diseases and fruits and increase crop yields. For example, CuZn bimetallic nanoparticles were sprayed on tomato leaves to study ROS formation, phytotoxicity, and antifungal activity against *Saccharomyces cerevisiaese* (Antonoglou et al., 2018). Also, it was demonstrated that the foliar application of chitosan nanoparticles to tomato leaves improved plant growth and flowering (El Amerany et al., 2020). Recently, nanoparticles of silicon dioxide, zinc oxide, selenium, and graphene were reported as foliar sprays on sugarcane leaves to understand the ameliorating effect of NPs against the negative impact of cold stress on photosynthesis and photoprotection (Elsheery et al., 2020). Likewise, the foliar application of silver nanoparticles with an average diameter of 16.7 nm in two varieties (Bronco and Nebraska) of common bean (*Phaseolus vulgaris* L.) significantly increased plant height, root length, and the number of

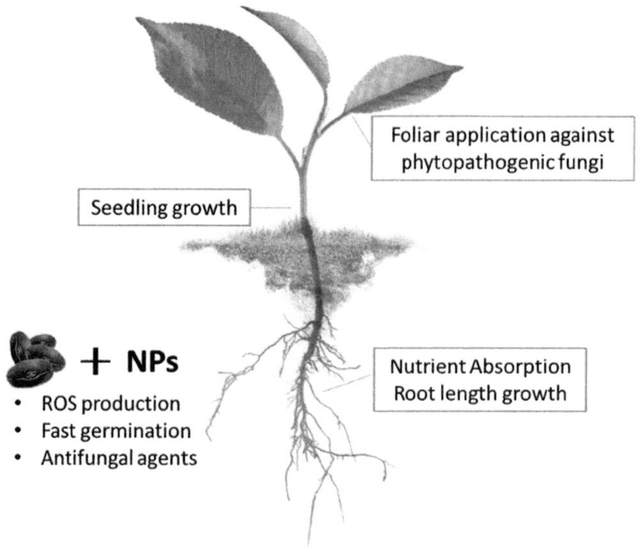

FIG. 3 Applications of nanostructures in early life stages of seeds and seedlings.

leaves (El-Batal et al., 2016). Although the effect of foliar NPs on plants is still widely debated, due to their potential for yield improvement and disease resistance, and disadvantages, such as toxicity and genotoxicity to plants. It is a vast field of research with great potential in agribusiness. As more research is done on the adverse effects of foliar application, it will be of great use to society.

4 Use of nanostructures to control plant pathogenic fungi at preharvest and postharvest

Without control of pathogenic microorganisms affecting plants and food, many populations worldwide will suffer terrible epidemics and outbreaks, causing famine and many deaths. In order to prevent the harmful effects of plant pathogenic fungi, several agrochemicals have been developed and used. Over time, the excessive use of some synthetic agrochemicals has caused soil deterioration and negative environmental effects. It has proved to be toxic to humans, which has led to the controlled application of these substances. Therefore, nanoscience has focused on finding solutions in the agro-industrial field to face this problem. The species of the phytopathogenic fungus *Pythium ultimum*, which is responsible for a variety of diseases in a wide range of crops and ornamental species, are well known (Lévesque et al., 2010). Most species are soil-dwelling. Diseases caused by *Pythium* spp. on plants include wilt, root rot, and a variety of field and postharvest rots. Nanoparticles can act directly in the soil to control pathogenic fungi. A green synthesis method of AuNPs with diameters between 5 and 20 nm with potent antifungal activity against the oomycete *Pythium ultimum* var. *ultimum* (SR1) and *Pythium* sp. (BP1120), at a concentration of 0.8 mg/mL, was recently reported (Khatua et al., 2019). AuNPs are often used in nanomedicine for therapeutic applications due to their high stability and biocompatibility (Liu et al., 2021). However, as mentioned earlier, AuNPs are used in agriculture, although less often due to the precious metal cost. In another literature report, the antifungal activity of gold-chitosan (with an average diameter of 80 nm) and carbon nanoparticles (with a diameter of ~23 nm) was studied against two different strains of *Fusarium oxysporum* (an ascomycete fungus known as a soil-plant pathogen) (Lipsa et al., 2020). Also, paper dipped in gold nanoparticles was used to monitor pesticide migration behaviors in various fruits and vegetables (i.e., apple, cucumber, bell pepper, plum, carrot, and strawberry) (Qin et al., 2021). Similarly, the antifungal activity of AgNPs has been evaluated against sheath blight caused by *Rhizoctonia solani*, which is one of the most widespread and destructive diseases of rice and causes substantial yield losses in endemic areas. AgNPs with spherical shape, polydisperse and average size of 16.5 ± 6.2 nm showed enhanced mycelial inhibition (81.7–96.7%) at 10 μg/mL (Kora et al., 2020).

The benefits of consuming fruits and vegetables help maintain good health and prevent diseases. However, sometimes these food products are affected by the incidence of phytopathogenic microorganisms during planting, development, preharvest, and postharvest. The postharvest spoilage of fruits, vegetables, and grains by phytopathogenic fungi causes significant economic losses and prevents the availability of healthy foods for a more extended period. Some fungi present in food cause allergic reactions and respiratory problems, and others produce mycotoxins that pose a risk to human health, especially in processed foods intended for children (Wang et al., 2022). Aflatoxins are a type of toxin produced by certain fungi (mainly *Aspergillus flavus* and *Aspergillus parasiticus*) in crops such as corn, peanuts, and some nuts. Aflatoxin-producing fungi can contaminate crops in the field during harvest or storage (Pankaj et al., 2018). Exposure to aflatoxins is associated with an increased risk of liver cancer and may cause weight gain or loss, loss of appetite, or infertility in men (Claeys et al., 2020). Therefore, preharvest antifungal treatment can help reduce economic losses, product damage, and potential human health impacts. In this respect, Horky et al. reported on the use of different nanomaterials with antifungal activity in mold inhibition, mycotoxin adsorption, and reduction of the toxic effect (Horky et al., 2018). On the other hand, preharvest spraying of soybean plants with different nanoparticles (CuO, ZnO, and MgO) against aflatoxin-producing fungi *Aspergillus flavus* and *A. parasiticus* (Alghandour et al., 2019). In this case, the treatment inhibited the *Aspergillus flavus* group at the preharvest stage and contributed to the pyrotechnical aspects of the plant, such as germination and seed weight. Another phytopathogenic agent that causes plant diseases is the fungus *Colletotrichum gloeosporioides*, responsible for anthracnose disease in several fruits and crops, such as avocado, papaya, mango, pitaya, tomato, citrus, and almonds, among others (Zakaria, 2021). Anthracnose invades leaves, flowers, and fruits during preharvest and invades fresh produce postharvest (Xue et al., 2019). Against this pathogen, De la Rosa-García et al. used MgO and ZnO nanoparticles and demonstrated enhanced antifungal activity against *C. gloeosporioides*, which causes anthracnose in avocado and papaya (De La Rosa-García et al., 2018). The antifungal effect of zinc oxide nanoparticles was reported against *Colletotrichum* sp. to protect coffee crops. ZnO NPs showed an appreciable percentage inhibition of fungal growth, ~96% for the concentration of 15 mmol/L at 6 days, causing a loss in the continuity of some hyphae (Mosquera-Sánchez et al., 2020). Likewise, the antifungal capacity of ZnO NPs obtained by the green synthesis in a garlic extract was reported against pathogenic coffee fungi, *Mycena citricolor,* and *Colletotrichum* sp., with growth inhibition of 97% and 93%, respectively (Arciniegas-Grijalba et al., 2019). Anthracnose also tends to attack chili peppers both preharvest and postharvest. Therefore, the use of Ag/Cu-chitosan nanoparticles as an antifungal agent has been reported to inhibit the growth of the hyphae of the fungus *Colletotrichum capsici* (Eris et al., 2019). Table 2 shows some basic food products and some reports of nanoparticles that have been used to control plant pathogenic fungi.

5 Management toxigenic fungi

Diseases caused by phytopathogenic fungi and the toxins they produce seriously threaten agricultural production before and during harvest or storage. They affect the quality and quantity of plant products available for consumption. Toxigenic fungi produce a wide variety of toxic compounds, known as mycotoxins, which attack cereals, fruits, and vegetables, among other foods, and can cause diseases in humans and animals (Yang et al., 2020). Mycotoxins are naturally occurring low molecular weight compounds produced by microscopic fungi that generate a toxic response upon ingestion, inhalation, or skin contact. (Omotayo et al., 2019). Among the most common mycotoxin-secreting fungi, genera are *Aspergillus* sp., *Fusarium* sp., *Alternaria* sp., and *Penicillium* sp. (Jampílek and Králová, 2020). Examples

TABLE 2 Nanoparticles against phytopathogenic fungi in agricultural foodstuffs.

Food product	NPs	Antifungal activity	Reference
Apple	CuO	CuO-NPs were evaluated against *Alternaria mali, Diplodia seriata,* and *Botryosphaeria dothidea* to estimate their antifungal activity using a modified food poisoning technique (FPT)	Ahmad et al. (2020)
Tomato	TiO_2, Ag, Se	The antifungal activity of nanoparticles was studied against *Alternaria alternata* to control tomato leaf blight	El-Gazzar and Ismail (2020)
Mango	Ag	Silver nanoparticles conjugated with carbendazim (Cz-AgNP) are an alternative to control anthracnose disease caused by *Colletotrichum gloeosporioides* in mango	Shivamogga Nagaraju et al. (2019)
Grains (barley, wheat, and corn)	Ag	The results demonstrated that Ag-NPs could inhibit the linear growth of *Fusarium* spp. and eliminate the toxin deoxynivalenol (DON)	El-Naggar et al. (2018)
Potato	ZnO	Zinc oxide nanoparticles (ZnO NPs) were used as antifungal agents against *Alternaria solani* (early potato blight)	Singh et al. (2022)
Rice	Ag	The antifungal activity of Ag NP was evaluated against *Rhizoctonia solani* fungus, the causal agent of sheath blight disease in rice	Kora et al. (2020)
Banana	silica-copper and silica-chitosan-copper	The nanoparticles showed excellent results against the fungal pathogen *Musicillium theobromae,* the cause of cigar-end rot, a postharvest disease that affects the banana fruit quality	Youssef et al. (2020)
Orange	Se, and chitosan	Selenium and chitosan nanoparticles were used as effective fungicidal agents to control *P. digitatum* strains to protect against citrus green molds	Salem et al. (2022)

of mycotoxins of significant health importance include aflatoxins, trichothecenes, deoxynivalenol, fumonisins, ochratoxins, zearalenone, patulin, etc. The control of mycotoxins produced by mycotoxigenic fungi is of great interest to many scientists. Through the antimicrobial properties of nanoparticles, nanotechnology attempts to provide solutions to inhibit the growth of toxic fungi and their mechanism of action. The use of Ag NPs with antifungal activity against three toxigenic *Aspergillus* species was recently reported (*Aspergillus flavus, Aspergillus nomius,* and *Aspergillus parasiticus*) as an ideal method to control the growth of fungi during storage (Bocate et al., 2019). Likewise, the antifungal effect of Ag NPs against the toxigenic *Fusarium* spp. was reported and their effect on mycotoxin accumulation was evaluated, obtaining excellent results for applications as an antifungal ingredient in bioactive polymers (paints, films, or coatings) in the agri-food sector (Tarazona et al., 2019). The effectiveness of Ag NPs in controlling the growth of the main aflatoxigenic and ochratoxigenic species was also evaluated, significantly affecting fungal growth (Gómez et al., 2019). Furthermore, mesoporous silica nanoparticles (MSNs) incorporated with zinc oxide (MSNs-ZnO) were reported as a potential antifungal agent against toxigenic fungi strains *Fusarium graminearum* and *Aspergillus flavus*. The nanoparticles caused a morphological alteration in both fungi, showing ruptures and deformations in the fungal hyphae, affecting their growth and toxin production (Savi et al., 2022). The effect of nanoencapsulation of chitosan nanoparticles (CS) was reported as an inhibitor of aflatoxins produced mainly by *Aspergillus* spp. (Mekawey and El-Metwally, 2019). The contribution of nanotechnology is essential in developing new methods to design new antifungal agents to help improve the quality and safety of food, nutrition, and human health.

6 Nanostructures and their antifungal activity

Nanomaterials have been very useful in developing new technologies for different sectors of society. Nanotechnology allows us to have control of materials on a 10–9 m scale and apply their optical, mechanical, electronic, and magnetic properties in some specific regions. Different types of nanoparticles can be obtained, which generally tend to be classified into two groups: organic nanoparticles and inorganic nanoparticles (Klapper et al., 2008). Organic nanoparticles are widely used in medicine for the transport of drugs and the treatment of cancerous diseases. Organic nanoparticles include those made from polymers, micelles, dendrimers, and liposomes. Inorganic nanoparticles are composed of metallic elements, metal oxides, carbon allotropes, semiconductors, and alloys. The latter has the most significant activity in catalysis, electronics, antiviral, and antifungal activities. In recent years, there has been increasing interest in nanoparticles for antimicrobial applications (Dizaj et al., 2014; Wang et al., 2017). Although some microbial agents are harmless and, in some cases, suitable for health and food manufacturing, others are pathogenic. Pathogenic microorganisms cause severe damage to the health of animals and humans and cause the spoilage of a large amount of food. A pathogenic agent can cause damage to the host. In the case of plants, it is called phytopathogenic, from the Greek phyton meaning "plant" and pathogen meaning "that originates or causes a disease." Therefore, a phytopathogenic agent is a microorganism that generates diseases in plants. In that sense, inorganic nanomaterials have been widely used to combat different species of fungi. Due to the small size of nanoparticles, they can adhere to the surface of the cell membrane and even manage to penetrate them, causing significant intracellular lethal damage, generating toxicity, and oxidative stress and inhibiting the capacity of reproduction (growth and division) and, therefore, causing bacterial death (Jian et al., 2021) (Fig. 4). For example, silver nanoparticles (AgNPs) with sizes between 5 and 45 nm and of diversified shapes (spherical, triangular, quadrilateral, and hexagonal) exhibited a highly antifungal effect against *Aspergillus niger* (the fungus that produces a black mold on vegetables such as lettuce and tomato), *Aspergillus flavus* (the fungus that causes corneal infections, is allergenic, occurs frequently on corn and peanuts), and *Fusarium oxysporum* (the fungus that attacks the roots of some banana varieties) (Le et al., 2020). The ecotoxicity of gold nanoparticles (AuNPs) was reported against the fungi *Aspergillus niger, Mucor hiemalis,* and *Penicillium chrysogenumse*, and it was found that larger AuNPs that did not have a spherical geometry had relatively stronger toxicities (Liu et al., 2018a,b). Vivek et al. reported an eco-friendly synthesis method using the red algae extract *Gelidiella acerosa como* to obtain AgNPs with spherical geometry and an average size of 22 nm, with antifungal activity against

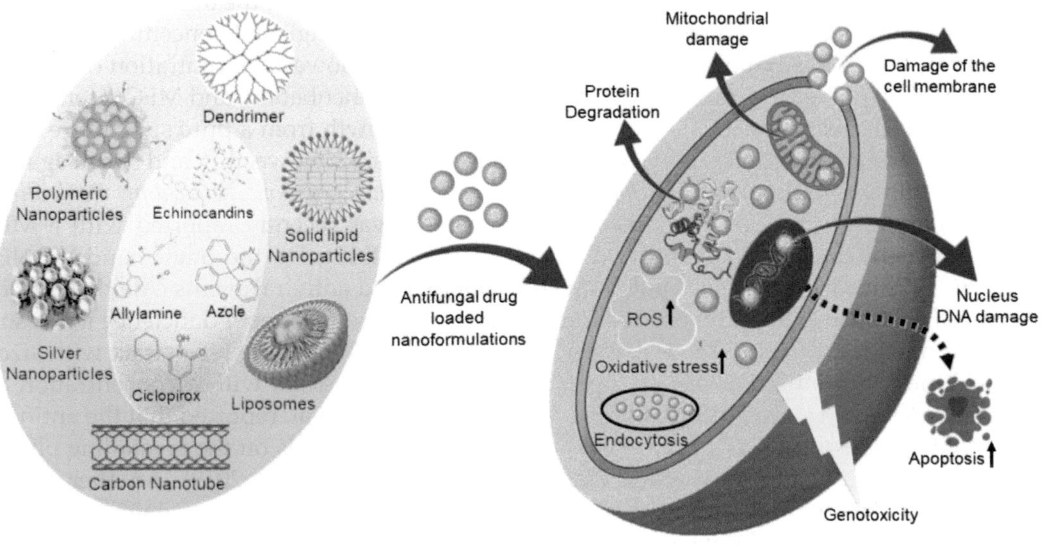

FIG. 4 Schematic mechanisms of cytotoxic activity of antifungal drug-loaded nanocarriers. *Reproduced from Mishra, V., Singh, M., Mishra, Y., Charbe, N., Nayak, P., Sudhakar, K., Aljabali, A. A. A., Shahcheraghi, S. H., Bakshi, H., Serrano-Aroca, Á., Tambuwala, M. M. 2021. Nanoarchitectures in management of fungal diseases: an overview. Appl. Sci. 11(15) 7119. The article is licensed under an open-access Creative Commons CC BY 4.0 license with permission from MDPI.*

Humicola insolens, *Fusarium dimerum*, *Mucor indicus*, and *Trichoderma reesei* (Vivek et al., 2011). Likewise, AgNPs with diameters of 5 to 20 nm were reported against the pathogen *Trichosporon asahii*, inhibiting its growth by penetrating the fungal cell and damaging the cell wall and cellular components (Xia et al., 2016). The possible mechanism of action of metal nanoparticles on phytopathogenic fungi is described by Cruz-Luna et al. in seven steps: (1) ions are released by nanoparticles and bind to certain protein groups, which affect the function of essential membrane proteins and interfere with cell permeability. (2) The nanoparticles inhibit the germination of the conidia and suppress their development. (3) Nanoparticles and released ions disrupt electron transport, protein oxidation, and alter membrane potential. (3) They also interfere with protein oxidative electron transport. (5) They affect the potential of the mitochondrial membrane by increasing the levels of transcription of genes in response to oxidative stress (ROS). (6) ROS induces the generation of reactive oxygen species, triggering oxidation reactions catalyzed by the different metallic nanoparticles, causing severe damage to proteins, membranes, and deoxyribonucleic acid (DNA) and interfering with nutrient absorption. (7) The ions of the nanoparticles have a genotoxic effect that destroys DNA, therefore causing cell death (Cruz-Luna et al., 2021).

On the other hand, zinc oxide nanoparticles (ZnO NPs) and iron oxide nanoparticles (FeO NPs) have been used as important antifungal agents. For example, the antifungal effect of ZnO NPs nanoparticles against *Colletotrichum* species (responsible for causing anthracnose disease in food and fiber crops) was recently reported with a very significant percentage inhibition of fungal growth at the concentration of 15 mmol/L at 6 days (Mosquera-Sánchez et al., 2020). Miri et al. reported the synthesis of hexagonal-shaped ZnO NPs with a size range between 40 and 50 nm using the aqueous extract of *Prosopis farcta* fruit and performed antifungal assays against *Candida albicans,* finding values of 32–64 for the minimum inhibitory concentration (MIC) and 128–512 μg/mL for the minimum fungicidal concentration (MFC) (Miri et al., 2020). It is important to mention that MIC is the lowest concentration of an antimicrobial that inhibits the growth of a microorganism after incubation, and MFC refers to the solution that decreases by 99.9% the colonies or fungal growth from a pure subculture inoculum. The antifungal activity of Fe_2O_3 nanoparticles with sizes between 10 and 30 nm against different fungal pathogens (*Trichothecium roseum, Cladosporium herbarum, Penicillium chrysogenum, Alternaria alternata,* and *Aspergillus niger*) was recently reported with a MIC value between 0.063 and 0.016 mg/mL (Parveen et al., 2018). On the other hand, CuNPs and CuO NPs have been used as potential antimicrobial and antiviral agents (Alavi and Kennedy, 2021). However, they are used as micronutrients in cases of copper deficiency in the soil and as fungicides. Khatami M et al. reported obtaining CuNPs using tea extract with sizes below 80 nm with inhibitory effects up to 90% on *Fusarium solani* with a concentration of 80 μg/mL (Khatami et al., 2019). Likewise, Hermida-Montero et al. reported that the antifungal activity of Cu/Cu_xO NPs against *Fusarium oxysporum* depends on the crystalline phase, the synthesis route, and the particle size (Hermida-Montero et al., 2019). Another report showed that the smallest CuNPs exhibited the best antifungal activities against *Fusarium oxysporum* (30 ppm after 3 days of incubation) and *Phytophthora capsici* (7.5 ppm after 1 day of incubation) because the smaller size can easily penetrate cell membranes through their surface (Pham et al., 2019). Other metal oxides have also been used as fungicides, such as magnesium oxide, which has low toxicity and is not considered hazardous to health. Magnesium oxide is widely used as an antacid to relieve heartburn and as a nutritional supplement. Therefore, nanostructured systems based on this compound are promising for food and biomedical applications. It was recently reported that MgO NPs significantly inhibited the growth of *Candida albicans* (Kong et al., 2020). Sharmila G, et al. reported the synthesis of MgO NPs with sizes between 50 and 100 nm using *Pisonia alba* leaf extract with antifungal activity against *Aspergillus flavus* with the zone of inhibition (ZOI). They use concentrations of 75 and 100 mg/mL of 2 and 4 mm, respectively) and *Fusarium solani* (with a ZOI of 2 and 3 mm at the concentrations of 75 and 100 mg/mL, respectively) (Sharmila et al., 2019). Abdel-Aziz et al. reported the biosynthesis of MgO NPs with an average diameter of 26.7 nm with antifungal activity, completely inhibiting the mycelial growth of the fungus *Fusarium oxysporum* at the concentration of 15.36 μg/mL (Abdel-Aziz et al., 2020). On the other hand, inorganic nanomaterial nanoalloys have also been used as potent fungicides. For example, hollow Ag-doped TiO_2 nanoparticles were more effective than solid ones against the plant pathogens *Venturia inaequalis* and *Fusarium solani* using minimum inhibitory doses of 0.75 and 0.43 mg/plate, respectively (Boxi et al., 2016). Likewise, the synthesis of CuZn bimetallic nanoparticles with the antifungal activity of BNP against *Saccharomyces cerevisiaese* in tomato plants was reported (Antonoglou et al., 2018). Bimetallic Ag-Cu and trimetallic Ag-Cu-Co nanoparticles have also been used as potent fungicides (Kamli et al., 2021; Paszkiewicz et al., 2016).

7 Future trends

With the increase in world population, the decrease in available land for cultivation, the extensive use of fungicides for disease management, the resistance of pathogenic microorganisms, etc., humanity will face a challenge in the agri-food sector. Mainly in meeting the demand for food and reducing the extensive use of fungicides for disease management, which generates contamination and causes damage to the ecosystem. One of the main challenges is to find strategies to inhibit the spread of phytopathogenic agents in seedlings, preharvest, and postharvest. Fungicide chemicals are often toxic to nontarget organisms such as earthworms and microbes, causing imbalances in ecosystems. They are known to be poisonous and, in some cases, fatal, causing contact dermatitis, chronic skin disease, pulmonary edema, and other effects (genotoxicity) in humans. In the process of production and distribution of foodstuffs, minimal loss must be guaranteed, and it is expected that foodstuffs can be available for consumption for long periods. To mitigate damage, the physicochemical properties of nanomaterials will play an essential role in the further development of new technologies. The high demands of food production will automatically lead to nanotechnology implementation to replace fertilizers and pesticides that degrade the soil and pollute the environment. Among the applications that nanomaterials could have in agriculture are mainly nanofertilizers, nanopesticides, and nanosensors. Controlled delivery of micronutrients and natural bioactive molecules to plants (targeted delivery) would prevent the compounds from being dispersed in other areas that are not of interest. The antifungal potential of metal nanoparticles to counteract the effects of phytopathogenic fungi and as candidates to prevent mycotoxin production is undisputed. In the future, we expect to be able to control particle size, minimum inhibitory concentration (MIC), and minimum fungicidal concentration (MFC). As well as to advance toxicological studies on the use of nanomaterials in the agri-food sector. Nanotechnology applied to the industry is a very new topic, the use of nanomaterials must be economically advantageous and socially acceptable to achieve a positive impact. The methods of ecological synthesis of nanoparticles that seek to reduce costs and toxic substances represent an environment-friendly alternative, these methods usually use plant extracts, algae, organic molecules, etc., for the synthesis of organic and inorganic nanoparticles. This is a large area of exploration to advance the application of new, highly efficient, and environment-friendly nanostructured materials. Another important aspect to consider is that the authorities of each country should inform people about issues related to food care. Society should be aware of the agribusiness problems and generate social awareness of food waste and optimal storage conditions. Such as ventilation, temperature, and humidity, to avoid the generation of mycotoxins. With the advance in green and sustainable methodologies, nanoparticles will have a greater impact on food. Regulatory agencies will raise regulations and risk assessments on the use of these technologies soon. There is a huge potential for nanotechnology applications in the food industry sector, but more research is urgently needed to advance the development of this area.

8 Conclusions

Finally, the importance of nanomaterials in the technological development of a sustainable agro-industrial sector has been shown. The physical and chemical properties of inorganic nanoparticles allow them to be used as antimicrobial agents to counteract the damage caused by phytopathogenic fungi in plants and food. Also, nanoparticles are essential for monitoring the condition of food products and can improve quality and preserve food for long periods. In addition to protection against UV rays, they create barrier properties against moisture and gases, to minimize food losses and provide safe food. Nanotechnology also allows the "smart delivery" of agrochemicals directly to the target sites of plants, avoiding the spraying of fertilizers and pesticides on the soil. However, a lot remains to be done to research adverse effects and advance the field of nano-agroindustry. More toxicity-related studies are needed on health risks and environmental impact. The use of pesticides obtained through the green synthesis of nanomaterials could help to reduce environmental and health risks in agriculture.

Acknowledgment

R. Britto Hurtado and J.M. Gutierrez Villarreal acknowledge the Postdoctoral Fellowship by CONAHCYT Mexico. This work was supported by Project A1-S-46242 of the CONAHCYT Basic Science. M. Cortez-Valadez appreciates the support by "Investigadores por México" program. S. Horta-Piñeres is very grateful to Bicentennial Doctoral Excellence Program-Colombia, and to the project "Formation of High Level Human Capital" Universidad Popular del Cesar-Nacional with BPIN code 2019000100010.

References

Abdel Ghany, T.M., Bakri, M.M., Al-Rajhi, A.M.H., Al Abboud, M.A., Alawlaqi, M.M., Shater, A.R.M., 2020. Impact of copper and its nanoparticles on growth, ultrastructure, and laccase production of *Aspergillus niger* using corn cobs wastes. Bioresources 15, 3289–3306.

Abdel-Aziz, M.M., Emam, T.M., Elsherbiny, E.A., 2020. Bioactivity of magnesium oxide nanoparticles synthesized from cell filtrate of endobacterium Burkholderia rinojensis against *Fusarium oxysporum*. Mater. Sci. Eng. C 109, 110617.

Acharya, P., Jayaprakasha, G.K., Crosby, K.M., Jifon, J.L., Patil, B.S., 2020. Nanoparticle-mediated seed priming improves germination, growth, yield, and quality of watermelons (*Citrullus lanatus*) at multi-locations in Texas. Sci. Rep. 10, 1–16.

Agrios, G.N., 2005. How pathogens attack plants. Plant Pathol., 175–205.

Ahmad, H., Venugopal, K., Bhat, A.H., Kavitha, K., Ramanan, A., Rajagopal, K., Srinivasan, R., Manikandan, E., 2020. Enhanced biosynthesis synthesis of copper oxide nanoparticles (CuO-NPs) for their antifungal activity toxicity against major phyto-pathogens of apple orchards. Pharm. Res. 37, 1–12.

Alavi, M., Kennedy, J.F., 2021. Recent advances of fabricated and modified Ag, Cu, CuO and ZnO nanoparticles by herbal secondary metabolites, cellulose and pectin polymers for antimicrobial applications. Cellulose 28, 3297–3310.

Alghandour, S.A., El-Sanhoty, R.M., Magdy Rahhal, M.H., Abdel-Halim, K.Y., Abdel-Mageed, W.S., Roshdy, T.M., 2019. The antifungal efficacy of pre-harvest spraying with organic compounds and nanoparticles on aflatoxins production on stored soybean seeds. Egypt. J. Chem. Environ. Health 5 (1), 1–19.

Almansob, A., Bahkali, A.H., Ameen, F., 2022. Efficacy of gold nanoparticles against drug-resistant nosocomial fungal pathogens and their extracellular enzymes: resistance profiling towards established antifungal agents. Nano 12, 814.

Al-Zubaidi, S., Al-Ayafi, A., Abdelkader, H., 2019. Biosynthesis, characterization and antifungal activity of silver nanoparticles by *Aspergillus niger* isolate. J. Nanotechnol. Res. 1 (1), 23–36.

Ammar, H.A., Rabie, G.H., Mohamed, E., 2019. Novel fabrication of gelatin-encapsulated copper nanoparticles using *Aspergillus versicolor* and their application in controlling of rotting plant pathogens. Bioprocess Biosyst. Eng. 42, 1947–1961.

Antonoglou, O., Moustaka, J., Adamakis, I.D.S., Sperdouli, I., Pantazaki, A.A., Moustakas, M., Dendrinou-Samara, C., 2018. Nanobrass CuZn nanoparticles as foliar spray nonphytotoxic fungicides. ACS Appl. Mater. Interfaces 10, 4450–4461.

Arciniegas-Grijalba, P.A., Patiño-Portela, M.C., Mosquera-Sánchez, L.P., Guerra Sierra, B.E., Muñoz-Florez, J.E., Erazo-Castillo, L.A., Rodríguez-Páez, J.E., 2019. ZnO-based nanofungicides: synthesis, characterization and their effect on the coffee fungi *Mycena citricolor* and *Colletotrichum* sp. Mater. Sci. Eng. C 98, 808–825.

Asghar, M.A., Zahir, E., Shahid, S.M., Khan, M.N., Asghar, M.A., Iqbal, J., Walker, G., 2018. Iron, copper and silver nanoparticles: green synthesis using green and black tea leaves extracts and evaluation of antibacterial, antifungal and aflatoxin B1 adsorption activity. LWT 90, 98–107.

Bernardo-Mazariegos, E., Valdez-Salas, B., González-Mendoza, D., Abdelmoteleb, A., Tzintzun Camacho, O., Baceña Duran, C., Gutiérrez-Miceli, F., 2019. Silver nanoparticles from Justicia spicigera and their antimicrobial potentialities in the biocontrol of foodborne bacteria and phytopathogenic fungi. Rev. Argent. Microbiol. 51, 103–109.

Bhagat, D., Samanta, S.K., Bhattacharya, S., 2013. Efficient management of fruit pests by pheromone nanogels. Sci. Rep. 3, 1–8.

Bocate, K.P., Reis, G.F., de Souza, P.C., Oliveira Junior, A.G., Durán, N., Nakazato, G., Furlaneto, M.C., de Almeida, R.-S., Panagio, L.A., 2019. Antifungal activity of silver nanoparticles and simvastatin against toxigenic species of *Aspergillus*. Int. J. Food Microbiol. 291, 79–86.

Bolton, M.D., Thomma, B.P.H.J., Nelson, B.D., 2006. *Sclerotinia sclerotiorum* (Lib.) de Bary: biology and molecular traits of a cosmopolitan pathogen. Mol. Plant Pathol. 7, 1–16.

Boxi, S.S., Mukherjee, K., Paria, S., 2016. Ag doped hollow TiO_2 nanoparticles as an effective green fungicide against *Fusarium solani* and *Venturia inaequalis* phytopathogens. Nanotechnology 27, 085103.

Carbone, M., Donia, D.T., Sabbatella, G., Antiochia, R., 2016. Silver nanoparticles in polymeric matrices for fresh food packaging. J. King Saud. Univ. Sci. 28, 273–279.

Churilov, G.I., Polischuk, S.D., Kuznetsov, D., Borychev, S.N., Byshov, N.V., Churilov, D.G., 2018. Agro ecological grounding for the application of metal nanopowders in agriculture. Int. J. Nanotechnol. 15, 258–279.

Claeys, L., Romano, C., De Ruyck, K., Wilson, H., Fervers, B., Korenjak, M., Zavadil, J., Gunter, M.J., De Saeger, S., De Boevre, M., Huybrechts, I., 2020. Mycotoxin exposure and human cancer risk: a systematic review of epidemiological studies. Compr. Rev. Food Sci. Food Saf. 19, 1449–1464.

Cruz-Luna, A.R., Cruz-Martínez, H., Vásquez-López, A., Medina, D.I., 2021. Metal nanoparticles as novel antifungal agents for sustainable agriculture: current advances and future directions. J. Fungi 7, 1033.

Dawoud, T.M., Yassin, M.A., El-Samawaty, A.R.M., Elgorban, A.M., 2021. Silver nanoparticles synthesized by *Nigrospora oryzae* showed antifungal activity. Saudi J. Biol. Sci. 28, 1847–1852.

De La Rosa-García, S.C., Martínez-Torres, P., Gómez-Cornelio, S., Corral-Aguado, M.A., Quintana, P., Gómez-Ortíz, N.M., 2018. Antifungal activity of ZnO and MgO nanomaterials and their mixtures against *Colletotrichum gloeosporioides* strains from tropical fruit. J. Nanomater. 3498527, 1–9.

Dizaj, S.M., Lotfipour, F., Barzegar-Jalali, M., Zarrintan, M.H., Adibkia, K., 2014. Antimicrobial activity of the metals and metal oxide nanoparticles. Mater. Sci. Eng. C 44, 278–284.

Dongzhen, F., Xilin, L., Xiaorong, C., Wenwu, Y., Yunlu, H., Yi, C., Jia, C., Zhimin, L., Litao, G., Tuhong, W., Xu, J., Chunsheng, G., 2020. *Fusarium* species and *Fusarium oxysporum* species complex genotypes associated with yam wilt in South-Central China. Front. Microbiol. 11, 1964.

El Amerany, F., Meddich, A., Wahbi, S., Porzel, A., Taourirte, M., Rhazi, M., Hause, B., 2020. Foliar application of chitosan increases tomato growth and influences mycorrhization and expression of endochitinase-encoding genes. Int. J. Mol. Sci. 21, 535.

El-Batal, A.I., Gharib, F.A.E.L., Ghazi, S.M., Hegazi, A.Z., El Hafz, A.G.M.A., 2016. Physiological responses of two varieties of common bean (*Phaseolus Vulgaris* L.) to foliar application of silver nanoparticles. Nanomater. Nanotechnol. 6, 1–16.

El-Gazzar, N., Ismail, A.M., 2020. The potential use of titanium, silver and selenium nanoparticles in controlling leaf blight of tomato caused by *Alternaria alternata*. Biocatal. Agric. Biotechnol. 27, 101708.

El-Naggar, M.A., Alrajhi, A.M., Fouda, M.M., Abdelkareem, E.M., Thabit, T.M., Bouqellah, N.A., 2018. Effect of silver nanoparticles on toxigenic *fusarium* spp. and deoxynivalenol secretion in some grains. J. AOAC Int. 101, 1534–1541.

Elsheery, N.I., Sunoj, V.S.J., Wen, Y., Zhu, J.J., Muralidharan, G., Cao, K.F., 2020. Foliar application of nanoparticles mitigates the chilling effect on photosynthesis and photoprotection in sugarcane. Plant Physiol. Biochem. 149, 50–60.

Eris, D.D., Wahyuni, S., Mismana Putra, S., Yusup, C.A., Sri Mulyatni, A., Siswanto, Krestini, E.H., Winarti, C., 2019. The effect of Ag/Cu-nanochitosan on development of anthracnose disease in chili. J. Ilmu Pertan. Indones. 24, 201–208.

Espitia, P.J.P., de Fátima Ferreira Soares, N., dos Reis Coimbra, J.S., de Andrade, N.J., Medeiros, E.A.A., 2012. Zinc oxide nanoparticles: synthesis, antimicrobial activity and food packaging applications. Food Bioprocess. Technol. 5, 1447–1464.

Fatima, F., Hashim, A., Anees, S., 2021. Efficacy of nanoparticles as nanofertilizer production: a review. Environ. Sci. Pollut. Res. 28, 1292–1303.

Gómez, J.V., Tarazona, A., Mateo, F., Jiménez, M., Mateo, E.M., 2019. Potential impact of engineered silver nanoparticles in the control of aflatoxins, ochratoxin A and the main aflatoxigenic and ochratoxigenic species affecting foods. Food Control 101, 58–68.

Gorczyca, A., Pociecha, E., Kasprowicz, M., Niemiec, M., 2015. Effect of nanosilver in wheat seedlings and *Fusarium culmorum* culture systems. Eur. J. Plant Pathol. 142, 251–261.

Guo, H., White, J.C., Wang, Z., Xing, B., 2018. Nano-enabled fertilizers to control the release and use efficiency of nutrients. Curr. Opin. Environ. Sci. Health 6, 77–83.

Hashem, A.H., Abdelaziz, A.M., Askar, A.A., Fouda, H.M., Khalil, A.M.A., Abd-Elsalam, K.A., Khaleil, M.M., 2021. *Bacillus megaterium*-mediated synthesis of selenium nanoparticles and their antifungal activity against *Rhizoctonia solani* in faba bean plants. J. Fungi 7, 195.

Hermida-Montero, L.A., Pariona, N., Mtz-Enriquez, A.I., Carrión, G., Paraguay-Delgado, F., Rosas-Saito, G., 2019. Aqueous-phase synthesis of nanoparticles of copper/copper oxides and their antifungal effect against *Fusarium oxysporum*. J. Hazard. Mater. 380, 120850.

Hernández-Meléndez, D., Salas-Téllez, E., Zavala-Franco, A., Téllez, G., Méndez-Albores, A., Vázquez-Durán, A., 2018. Inhibitory effect of flower-shaped zinc oxide nanostructures on the growth and aflatoxin production of a highly toxigenic strain of *Aspergillus flavus* link. Materials 11, 1265.

Hoang, A.S., Cong, H.H., Shukanov, V.P., Karytsko, L.A., Poljanskaja, S.N., Melnikava, E.V., Mashkin, I.A., Nguyen, T.H., Pham, D.K., Phan, C.M., 2022. Evaluation of metal nano-particles as growth promoters and fungi inhibitors for cereal crops. Chem. Biol. Technol. Agric. 9, 1–9.

Horky, P., Skalickova, S., Baholet, D., Skladanka, J., 2018. Nanoparticles as a solution for eliminating the risk of mycotoxins. Nano 8, 727.

Jampílek, J., Kráľová, K., 2020. Impact of nanoparticles on toxigenic fungi. In: Nanomycotoxicology—Treating Mycotoxins in the Nano Way, pp. 309–348.

Jian, Y., Chen, X., Ahmed, T., Shang, Q., Zhang, S., Ma, Z., Yin, Y., 2021. Toxicity and action mechanisms of silver nanoparticles against the mycotoxin-producing fungus *Fusarium graminearum*. J. Adv. Res. 38, 1–12.

Kamli, M.R., Srivastava, V., Hajrah, N.H., Sabir, J.S.M., Hakeem, K.R., Ahmad, A., Malik, M.A., 2021. Facile Bio-Fabrication of Ag-Cu-Co Trimetallic Nanoparticles and Its Fungicidal Activity against *Candida auris*. J. Fungi 7, 62.

Kasote, D.M., Lee, J.H., Jayaprakasha, G.K., Patil, B.S., 2019. Seed priming with iron oxide nanoparticles modulate antioxidant potential and defense-linked hormones in watermelon seedlings. ACS Sustain. Chem. Eng. 7, 5142–5151.

Kaur, S., Dhillon, G.S., Brar, S.K., Vallad, G.E., Chand, R., Chauhan, V.B., 2012. Emerging phytopathogen Macrophomina phaseolina: biology, economic importance and current diagnostic trends. Crit. Rev. Microbiol. 38, 136–151.

Khatami, M., Varma, R.S., Heydari, M., Peydayesh, M., Sedighi, A., Agha Askari, H., Rohani, M., Baniasadi, M., Arkia, S., Seyedi, F., Khatami, S., 2019. Copper oxide nanoparticles greener synthesis using tea and its antifungal efficiency on *Fusarium solani*. Geomicrobiol J. 36, 777–781.

Khatua, A., Priyadarshini, E., Rajamani, P., Patel, A., Kumar, J., Naik, A., Saravanan, M., Barabadi, H., Prasad, A., Ghosh, L., Paul, B., Meena, R., 2019. Phytosynthesis, characterization and fungicidal potential of emerging gold nanoparticles using *Pongamia pinnata* leave extract: a novel approach in nanoparticle synthesis. J. Clust. Sci. 31, 125–131.

Kim, J., Rundle-Thiele, S., Knox, K., Burke, K., Bogomolova, S., 2020. Consumer perspectives on household food waste reduction campaigns. J. Clean. Prod. 243, 118608.

Klapper, M., Clark, C.G., Müllen, K., 2008. Application-directed syntheses of surface-functionalized organic and inorganic nanoparticles. Polym. Int. 57, 181–202.

Kong, F., Wang, J., Han, R., Ji, S., Yue, J., Wang, Y., Ma, L., 2020. Antifungal activity of magnesium oxide nanoparticles: effect on the growth and key virulence factors of *Candida albicans*. Mycopathologia 185, 485–494.

Kora, A.J., Mounika, J., Jagadeeshwar, R., 2020. Rice leaf extract synthesized silver nanoparticles: an in vitro fungicidal evaluation against *Rhizoctonia solani*, the causative agent of sheath blight disease in rice. Fungal Biol. 124, 671–681.

Le, N.T.T., Nguyen, D.H., Nguyen, N.H., Ching, Y.C., Nguyen, D.Y.P., Ngo, C.Q., Nhat, H.N.T., Thi, T.T.H., 2020. Silver nanoparticles ecofriendly synthesized by *Achyranthes aspera* and *Scoparia dulcis* leaf broth as an effective fungicide. Appl. Sci. 10, 2505.

León Peláez, A.M., Serna Cataño, C.A., Quintero Yepes, E.A., Gamba Villarroel, R.R., De Antoni, G.L., Giannuzzi, L., 2012. Inhibitory activity of lactic and acetic acid on *Aspergillus flavus* growth for food preservation. Food Control 24, 177–183.

Lévesque, C.A., Brouwer, H., Cano, L., Hamilton, J.P., Holt, C., Huitema, E., Raffaele, S., Robideau, G.P., Thines, M., Win, J., Zerillo, M.M., Beakes, G.W., Boore, J.L., Busam, D., Dumas, B., Ferriera, S., Fuerstenberg, S.I., Gachon, C.-M.M., Gaulin, E., Govers, F., Grenville-Briggs, L., Horner, N., Hostetler, J., Jiang, R.H.Y., Johnson, J., Krajaejun, T., Lin, H., Meijer, H.J.G., Moore, B., Morris, P., Phuntmart, V., Puiu, D., Shetty, J., Stajich, J.E., Tripathy, S., Wawra, S., van West, P., Whitty, B.R., Coutinho, P.M., Henrissat, B., Martin, F., Thomas, P.D., Tyler, B.M., De Vries, R.P., Kamoun, S., Yandell, M., Tisserat, N., Buell, C.R., 2010. Genome sequence of the necrotrophic plant pathogen *Pythium ultimum* reveals original pathogenicity mechanisms and effector repertoire. Genome Biol. 11, 1–22.

Li, C., Zhou, J., Du, G., Chen, J., Takahashi, S., Liu, S., 2020. Developing *Aspergillus niger* as a cell factory for food enzyme production. Biotechnol. Adv. 44, 107630.

Lipsa, F.D., Ursu, E.L., Ursu, C., Ulea, E., Cazacu, A., 2020. Evaluation of the antifungal activity of gold-chitosan and carbon nanoparticles on *Fusarium oxysporum*. Agronomy 10, 1143.

Liu, K., He, Z., Byrne, H.J., Curtin, J.F., Tian, F., 2018b. Investigating the role of gold nanoparticle shape and size in their toxicities to fungi. Int. J. Environ. Res. Public Health 15, 998.

Liu, J., Sui, Y., Wisniewski, M., Xie, Z., Liu, Y., You, Y., Zhang, X., Sun, Z., Li, W., Li, Y., Wang, Q., 2018a. The impact of the postharvest environment on the viability and virulence of decay fungi. Crit. Rev. Food Sci. Nutr. 58, 1681–1687.

Liu, X.Y., Wang, J.Q., Ashby, C.R., Zeng, L., Fan, Y.F., Chen, Z.S., 2021. Gold nanoparticles: synthesis, physiochemical properties and therapeutic applications in cancer. Drug Discov. Today 26, 1284–1292.

Lopez-Lima, D., Mtz-Enriquez, A.I., Carrión, G., Basurto-Cereceda, S., Pariona, N., 2021. The bifunctional role of copper nanoparticles in tomato: effective treatment for *Fusarium* wilt and plant growth promoter. Sci. Hortic. 277, 109810.

Maluin, F.N., Hussein, M.Z., Yusof, N.A., Fakurazi, S., Idris, A.S., Hilmi, N.H.Z., Daim, L.D.J., 2019. Preparation of chitosan–hexaconazole nanoparticles as fungicide nanodelivery system for combating ganoderma disease in oil palm. Molecules 24, 2498.

Mekawey, A.A.I., El-Metwally, M.M., 2019. Impact of nanoencapsulated natural bioactive phenolic metabolites on chitosan nanoparticles as aflatoxins inhibitor. J. Basic Microbiol. 59, 599–608.

Michalecka, M., Puławska, J., 2021. Multilocus sequence analysis of selected housekeeping- and pathogenicity-related genes in *Venturia inaequalis*. PathoGenetics 10, 447.

Miri, A., Khatami, M., Ebrahimy, O., Sarani, M., 2020. Cytotoxic and antifungal studies of biosynthesized zinc oxide nanoparticles using extract of *Prosopis farcta* fruit. Green Chem. Lett. Rev. 13, 27–33.

Mohamed, Y.M.A., Elshahawy, I.E., 2022. Antifungal activity of photo-biosynthesized silver nanoparticles (AgNPs) from organic constituents in orange peel extract against phytopathogenic *Macrophomina phaseolina*. Eur. J. Plant Pathol. 162, 725–738.

Mosquera-Sánchez, L.P., Arciniegas-Grijalba, P.A., Patiño-Portela, M.C., Guerra-Sierra, B.E., Muñoz-Florez, J.E., Rodríguez-Páez, J.E., 2020. Antifungal effect of zinc oxide nanoparticles (ZnO-NPs) on *Colletotrichum* sp., causal agent of anthracnose in coffee crops. Biocatal. Agric. Biotechnol. 25, 101579.

Mukhopadhyay, S.S., 2014. Nanotechnology in agriculture: prospects and constraints. Nanotechnol. Sci. Appl. 7, 63.

Musters, C.J.M., Evans, T., Wiggers, J.M.R., van 't-Zelfde, M., de Snoo, G.R., 2021. Distribution of flying insects across landscapes with intensive agriculture in temperate areas. Ecol. Indic. 129, 107889.

Musungu, B., Bhatnagar, D., Quiniou, S., Brown, R.L., Payne, G.A., O'Brian, G., Fakhoury, A.M., Geisler, M., 2020. Use of dual RNA-seq for systems biology analysis of Zea mays and *Aspergillus flavus* interaction. Front. Microbiol. 11, 853.

Nordin, N.H., Kaida, N., Othman, N.A., Akhir, F.N.M., Hara, H., 2020. Reducing food waste: strategies for household waste management to minimize the impact of climate change and contribute to Malaysia's sustainable development. IOP Conf. Ser. Earth Environ. Sci. 479, 012035.

Noshad, A., Hetherington, C., Iqbal, M., 2019. Impact of AgNPs on seed germination and seedling growth: A focus study on its antibacterial potential against *Clavibacter michiganensis* subsp. *michiganensis* infection in *Solanum lycopersicum*. J. Nanomater. 2019, 1–12.

Omotayo, O.P., Omotayo, A.O., Mwanza, M., Babalola, O.O., 2019. Prevalence of mycotoxins and their consequences on human health. Toxicol. Res. 35, 1–7.

Pankaj, S.K., Shi, H., Keener, K.M., 2018. A review of novel physical and chemical decontamination technologies for aflatoxin in food. Trends Food Sci. Technol. 71, 73–83.

Panpatte, D.G., Jhala, Y.G., Shelat, H.N., Vyas, R.V., 2016. Nanoparticles: the next generation technology for sustainable agriculture. In: Microbial Inoculants in Sustainable Agricultural Productivity. Vol. 2: Functional Applications, pp. 289–300.

Pariona, N., Mtz-Enriquez, A.I., Sánchez-Rangel, D., Carrión, G., Paraguay-Delgado, F., Rosas-Saito, G., 2019. Green-synthesized copper nanoparticles as a potential antifungal against plant pathogens. RSC Adv. 9, 18835–18843.

Parveen, S., Wani, A.H., Shah, M.A., Devi, H.S., Bhat, M.Y., Koka, J.A., 2018. Preparation, characterization and antifungal activity of iron oxide nanoparticles. Microb. Pathog. 115, 287–292.

Pascoli, M., De Lima, R., Fraceto, L.F., 2018. Zein nanoparticles and strategies to improve colloidal stability: a mini-review. Front. Chem. 6, 6.

Pascoli, M., Jacques, M.T., Agarrayua, D.A., Avila, D.S., Lima, R., Fraceto, L.F., 2019. Neem oil based nanopesticide as an environmentally-friendly formulation for applications in sustainable agriculture: an ecotoxicological perspective. Sci. Total Environ. 677, 57–67.

Paszkiewicz, M., Golabiewska, A., Rajski, L., Kowal, E., Sajdak, A., Zaleska-Medynska, A., Zaleska-Medynska, A., 2016. The antibacterial and antifungal textile properties functionalized by bimetallic nanoparticles of Ag/Cu with different structures. J Nanomater 2016, 1–13.

Peighambardoust, S.J., Peighambardoust, S.H., Pournasir, N., Pakdel, P.M., 2019. Properties of active starch-based films incorporating a combination of Ag, ZnO and CuO nanoparticles for potential use in food packaging applications. Food Packag. Shelf Life 22, 100420.

Perveen, K., Bukhari, N.A., Al Masoudi, Alqahtani, A.N., Alruways, M.W., Alkhattaf, F.S., 2022. Antifungal potential, chemical composition of *Chlorella vulgaris* and SEM analysis of morphological changes in *Fusarium oxysporum*. Saudi J. Biol. Sci. 29, 2501–2505.

Pham, N.D., Duong, M.M., Le, M.V., Hoang, H.A., Pham, L.K.O., 2019. Preparation and characterization of antifungal colloidal copper nanoparticles and their antifungal activity against *Fusarium oxysporum* and *Phytophthora capsici*. C. R. Chim. 22, 786–793.

Piruthiviraj, P., Margret, A., Krishnamurthy, P.P., 2016. Gold nanoparticles synthesized by *Brassica oleracea* (Broccoli) acting as antimicrobial agents against human pathogenic bacteria and fungi. Appl. Nanosci. 6, 467–473.

Prażak, R., Święciło, A., Krzepiłko, A., Michałek, S., Arczewska, M., 2020. Impact of ag nanoparticles on seed germination and seedling growth of green beans in normal and chill temperatures. Agriculture 10, 312.

Qin, R., Li, P., Du, M., Ma, L., Huang, Y., Yin, Z., Zhang, Y., Chen, D., Xu, H., Wu, X., 2021. Spatiotemporal visualization of insecticides and fungicides within fruits and vegetables using gold nanoparticle-immersed paper imprinting mass spectrometry imaging. Nanomaterials 11, 1327.

Raghavendra, S.N., Raghu, H.S., Chaithra, C., Rajeshwara, A.N., 2020. Potency of mancozeb conjugated silver nanoparticles synthesized from Goat, Cow and Buffalo urine against *Colletotrichum gloeosporioides* causing anthracnose disease. Nat. Environ. Pollut. Technol. 19, 969–979.

Rajeshkumar, S., Rinitha, G., 2018. Nanostructural characterization of antimicrobial and antioxidant copper nanoparticles synthesized using novel *Persea americana* seeds. OpenNano 3, 18–27.

Rastogi, A., Tripathi, D.K., Yadav, S., Chauhan, D.K., Živčák, M., Ghorbanpour, M., El-Sheery, N.I., Brestic, M., 2019. Application of silicon nanoparticles in agriculture. 3Biotech. 9, 1–11.

Ruiz-Romero, P., Valdez-Salas, B., González-Mendoza, D., Mendez-Trujillo, V., 2018. Antifungal effects of silver phytonanoparticles from Yucca shilerifera against strawberry soil-borne pathogens: *Fusarium solani* and *Macrophomina phaseolina*. Mycobiology 46, 47–51.

Sadek, M.E., Shabana, Y.M., Sayed-Ahmed, K., Tabl, A., Sadek, M.E., Shabana, Y.M., Sayed-Ahmed, K., Tabl, A.H.A., Eg, A.H.A.T., 2022. Antifungal activities of sulfur and copper nanoparticles against cucumber postharvest diseases caused by *Botrytis cinerea* and *Sclerotinia sclerotiorum*. J. Fungi 8, 412.

Saleem, I., Maqsood, M.A., Rehman, M.Z., Aziz, T., Bhatti, I.A., Ali, S., 2021. Potassium ferrite nanoparticles on DAP to formulate slow release fertilizer with auxiliary nutrients. Ecotoxicol. Environ. Saf. 215, 112148.

Salem, M.F., Abd-Elraoof, W.A., Tayel, A.A., Alzuaibr, F.M., Abonama, O.M., 2022. Antifungal application of biosynthesized selenium nanoparticles with pomegranate peels and nanochitosan as edible coatings for citrus green mold protection. J. Nanobiotechnol. 20, 1–12.

Sardar, M., Ahmed, W., Al Ayoubi, S., Nisa, S., Bibi, Y., Sabir, M., Khan, M.M., Ahmed, W., Qayyum, A., 2022. Fungicidal synergistic effect of biogenically synthesized zinc oxide and copper oxide nanoparticles against *Alternaria citri* causing citrus black rot disease. Saudi J. Biol. Sci. 29, 88–95.

Savi, G.D., Zanoni, E.T., Furtado, B.G., de Souza, H.M., Scussel, R., Machado-de-Ávila, R.A., Angioletto, E., 2022. Mesoporous silica nanoparticles incorporated with zinc oxide as a novel antifungal agent against toxigenic fungi strains. J. Environ. Sci. Health B 57, 176–183.

Sharma, P., Gaur, V.K., Kim, S.H., Pandey, A., 2020. Microbial strategies for bio-transforming food waste into resources. Bioresour. Technol. 299, 122580.

Sharmila, G., Muthukumaran, C., Sangeetha, E., Saraswathi, H., Soundarya, S., Kumar, N.M., 2019. Green fabrication, characterization of *Pisonia alba* leaf extract derived MgO nanoparticles and its biological applications. Nano-Struct. Nano-Objects 20, 100380.

Shende, S., Bhagat, R., Raut, R., Rai, M., Gade, A., 2021. Myco-fabrication of copper nanoparticles and its effect on crop pathogenic fungi. IEEE Trans. Nanobiosci. 20, 146–153.

Shi, X.C., Wang, S.Y., Duan, X.C., Wang, Y.Z., Liu, F.Q., Laborda, P., 2021. Biocontrol strategies for the management of Colletotrichum species in postharvest fruits. Crop Prot. 141, 105454.

Shivamogga Nagaraju, R., Holalkere Sriram, R., Achur, R., 2019. Antifungal activity of carbendazim-conjugated silver nanoparticles against anthracnose disease caused by *Colletotrichum gloeosporioides* in mango. J. Plant Pathol. 102, 39–46.

Singh, A., Gaurav, S.S., Shukla, G., Rani, P., 2022. Assessment of mycogenic zinc nano-fungicides against pathogenic early blight (*Alternaria solani*) of potato (*Solanum tuberosum* L.). Mater. Today Proc. 49, 3528–3537.

Singh Sekhon, B., 2014. Nanotechnology in agri-food production: an overview. Nanotechnol. Sci. Appl. 7, 31.

Souza, V.G.L., Fernando, A.L., 2016. Nanoparticles in food packaging: biodegradability and potential migration to food—a review. Food Packag. Shelf Life 8, 63–70.

Tarazona, A., Gómez, J.V., Mateo, E.M., Jiménez, M., Mateo, F., 2019. Antifungal effect of engineered silver nanoparticles on phytopathogenic and toxigenic *Fusarium* spp. and their impact on mycotoxin accumulation. Int. J. Food Microbiol. 306, 108259.

Troncoso-Rojas, R., Tiznado-Hernández, M.E., 2014. Alternaria alternata (Black Rot, Black Spot). In: Postharvest Decay. Academic Press, pp. 147–187.

Ungureanu, C., Fierascu, I., Fierascu, R.C., Costea, T., Avramescu, S.M., Călinescu, M.F., Somoghi, R., Pirvu, C., 2021. In vitro and in vivo evaluation of silver nanoparticles phytosynthesized using *Raphanus sativus* L. waste extracts. Materials 14, 1845.

Vivek, M., Kumar, P.S., Steffi, S., Sudha, S., 2011. Biogenic silver nanoparticles by *Gelidiella acerosa* extract and their antifungal effects. Avicenna J. Med. Biotechnol. 3, 143.

Wang, L., Hu, C., Shao, L., 2017. The antimicrobial activity of nanoparticles: present situation and prospects for the future. Int. J. Nanomedicine 12, 1227.

Wang, Z., Sui, Y., Li, J., Tian, X., Wang, Q., 2022. Biological control of postharvest fungal decays in citrus: a review. Crit. Rev. Food Sci. Nutr. 62, 861–870.

Win, T.T., Khan, S., Bo, B., Zada, S., Fu, P.C., 2021. Green synthesis and characterization of Fe3O4 nanoparticles using Chlorella-K01 extract for potential enhancement of plant growth stimulating and antifungal activity. Sci. Rep. 11, 1–11.

Xia, Z.K., Ma, Q.H., Li, S.Y., Zhang, D.Q., Cong, L., Tian, Y.L., Yang, R.Y., 2016. The antifungal effect of silver nanoparticles on *Trichosporon asahii*. J. Microbiol. Immunol. Infect. 49, 182–188.

Xie, J., Yang, Y., Gao, B., Wan, Y., Li, Y.C., Cheng, D., Xiao, T., Li, K., Fu, Y., Xu, J., Zhao, Q., Zhang, Y., Tang, Y., Yao, Y., Wang, Z., Liu, L., 2019. Magnetic-sensitive nanoparticle self-assembled superhydrophobic biopolymer-coated slow-release fertilizer: fabrication, enhanced performance, and mechanism. ACS Nano 13, 3320–3333.

Xue, W., Li, Y., Zhao, Q., Liang, T., Wang, M., Sun, P., Zhu, A., Wu, X., Chen, L., Zhang, T., Huo, S., Li, Y., 2022. Research note: study on the antibacterial activity of Chinese herbal medicine against *Aspergillus flavus* and *Aspergillus fumigatus* of duck origin in laying hens. Poult. Sci. 101, 101756.

Xue, Y., Zhou, S., Fan, C., Du, Q., Jin, P., 2019. Enhanced antifungal activities of eugenol-entrapped casein nanoparticles against anthracnose in postharvest fruits. Nano 9, 1777.

Yang, Y., Li, G., Wu, D., Liu, J., Li, X., Luo, P., Hu, N., Wang, H., Wu, Y., 2020. Recent advances on toxicity and determination methods of mycotoxins in foodstuffs. Trends Food Sci. Technol. 96, 233–252.

Youssef, K., Mustafa, Z.M.M., Kamel, M.A.M., Mounir, G.A., 2020. Cigar end rot of banana caused by *Musicillium theobromae* and its control in Egypt. Arch. Phytopathol. Plant Protect. 53, 162–177.

Zakaria, L., 2021. Diversity of *Colletotrichum* species associated with anthracnose disease in tropical fruit crops—a review. Agriculture 11, 297.

Zakharova, O., Kolesnikov, E., Shatrova, N., Gusev, A., 2019. The effects of CuO nanoparticles on wheat seeds and seedlings and Alternaria solani fungi: in vitro study. IOP Conf. Ser. Earth Environ. Sci. 226, 012036.

CHAPTER

14

Nano-biofungicides for the reduction of mycotoxin contamination in food and feed

Mohamed Amine Gacem[a], *Badreddine Boudjemaa*[a], *Valeria Terzi*[b], *Aminata Ould El Hadj-Khelil*[c], *and Kamel A. Abd-Elsalam*[d]

[a]Department of Biology, Faculty of Science, University of Amar Tlidji, Laghouat, Algeria [b]Council for Research in Agriculture and Economics, Research Centre for Genomics and Bioinformatics (CREA-GB), Fiorenzuola d'Arda, Italy [c]Laboratory of Ecosystems Protection in Arid and Semi-Arid Area, University of Kasdi Merbah, Ouargla, Algeria [d]Plant Pathology Research Institute, Agricultural Research Center, Giza, Egypt

1 Introduction

A country's economic growth and poverty reduction are based on powerful and sustainable food security, which is based on sustainable agriculture with high production and covering the country's food needs (Webb and Block, 2012). This backbone of a country can only be ensured through rigorous agricultural practices. Indeed, agricultural sustainability suggests both a better understanding of the benefits of agronomic management and improved genotypes through the modern biological approaches being developed (Pretty, 2008). Currently, the management of crops relies heavily on disease control as a priority. The use of agrochemicals, especially pesticides (insecticides, fungicides, herbicides, rodenticides, molluscicides, and nematicides) and plant growth regulators contributes to the control of plant pathogens, protects plants against significant economic losses, and therefore increases crop yields (Aktar et al., 2009). However, the health services are concerned about the intensive use of chemical pesticides and are beginning to point out the risk. Moreover, several scientists alert the population to the risks and harm of pesticides to humans and the environment (Nicolopoulou-Stamati et al., 2016). The extension of pesticides to nontarget living organisms is also among

Nanohybrid Fungicides
https://doi.org/10.1016/B978-0-443-23950-2.00011-4

the negative impacts of pesticides like tebuconazole; this fungicide mainly accumulates in the liver and induces histopathological damage. Tebuconazole significantly deregulates enzymes and genes related to acid metabolism and causes liver enlargement and steatosis (Ku et al., 2021). In addition, estimates have indicated that approximately 80%–90% of sprayed fungicides are lost in different ecosystems after or during their applications (Kutawa et al., 2021). It is observed that the effectiveness of pesticides is threatened by the evolution of resistant pathogens; indeed, target species are often able to develop resistance soon after the introduction of a new chemical compound (Hawkins et al., 2018). A recent study showed that about 70%–80% of yield losses are due to fungal pathogens, while only 15%–18% of agricultural losses are due to animal pests (Kutawa et al., 2021). To curb the fatal manifestations of phytopathogenic fungi and the negative effects of pesticides on human health and the environment, dynamic and rigorous approaches must be implemented. In other words, effective and sustainable crop protection relies on new innovative strategies and technologies that combine smart materials and organic ingredients.

Recently, nanoscale-engineering approaches have sparked the advancement of innovative ideas with enormous potential to solve problems in the fields of pharmacology and medicine (Pelaz et al., 2017; Soares et al., 2018). Several published studies highlight the potential of nanomaterials in water control (Jain et al., 2021), gene transfer by nano-vectors (Riley and Vermerris, 2017), the controlled release of agrochemicals (nanopesticide and nanofertilizer) (Raliya et al., 2018; Chaud et al., 2021), easy diagnosis of diseases by nano-sensors (Li et al., 2020), and seed germination (Szőllősi et al., 2020). Despite the growing interest in future applications of nanotechnology, this field has not developed as much in agricultural applications.

Many researchers have succeeded in designing nanoparticles and nanostructures with perfectly defined characteristics, such as shape, pore size, and surface properties, so that they can be used as active fungicides to control phytopathogenic fungi (Chen et al., 2020). The most widely used nanoparticles (NPs) in plant disease control are based on carbon, silver, copper, silica, and other metals. Silica nanoparticles control the suppression of the Fusarium wilt disease caused by *Fusarium oxysporum f.* sp. *niveum* in watermelon (*Citrullus lanatus*) (Kang et al., 2021), Chen et al. demonstrated the fungicidal activity of metallic MgO-NPs against soilborne fungal phytopathogens (*Phytophthora nicotianae*) (Chen et al., 2020). ZnO-NPs control the growth and mycotoxins synthesis of *Fusarium graminearum* (Lakshmeesha et al., 2019). Multiwalled carbon nanotubes, fullerene, reduced graphene oxide, CuO-NPs, FeO-NPs, and TiO_2-NPs inhibited the growth of *Botrytis cinerea* (Hao et al., 2017). Bio-based NPs like chitosan NPs also showed significant inhibition of *F. graminearum* growth (Kheiri et al., 2016), Spadola and his group have developed a nano-formulation based on thiosemicarbazone for the control of *Aspergillus flavus* (Spadola et al., 2020). Advances in nanotechnology have created a footprint that enables the efficient development of multiple forms and increased efficiency of NPs through the development of nanohybrids and nanocomposites. Nano-hybrids can have various origins, spanning biological-inorganic materials as well as organic-inorganic or even natural-synthetic materials. These hybrid nanostructures can have an additive effect by increasing the effects of the antagonistic properties of individual NPs.

This chapter focuses on the factors that promote the biosynthesis of mycotoxins as well as their toxicities, the different types of bio-nanoparticles and hybrid nanostructures, and finally, the various antifungal applications of nano-biofungicides and their mode of action in the fungal cell and the inhibition of mycotoxin biosynthesis.

2 Occurrence, toxicities, and factors affecting mycotoxins biosynthesis

Mycotoxins are fungal secondary metabolites; they are produced by several fungal genera and are present in a wide variety of foods and water. Ochratoxin, aflatoxin, fumonisin, zearalenone, and other mycotoxins are present in cereals such as wheat, maize, and barley. They are also present in cereal by-products. Fruit and vegetables can also be contaminated by mycotoxins according to their composition, which is rich in vitamins and carbohydrates. In other food categories, the biotransformation of mycotoxins in the animal body makes them present in cow's milk, meat, and even eggs. Fermented food made from milk or cereal products is also contaminated by mycotoxins. Mycotoxins have high toxicity for humans, animals, and even plants. They can cause nephrotoxicity, genotoxicity, teratogenicity, neurotoxicity, hepatotoxicity, immunotoxicity, gastrointestinal toxicity, cardiotoxicity, and pulmonary toxicity, as well as cancer in the case of more advanced toxicities. Their mechanisms of toxicity are controlled by reactive oxygen species (ROS) (see Fig. 1).

For their biosynthesis, mycotoxins require multiple factors and precursors, including abiotic factors such as nutritional factors (nitrogen and carbon sources), pH, temperature, water activity, and osmotic pressure. Understanding these factors is essential to applying effective techniques capable of inhibiting the synthesis of mycotoxins. Carbon is abundant in carbohydrates (glucose, sucrose, maltose, and fructose); however, it can be present in complex forms in cereals. However, nitrogen is abundant in amino acids and proteins. The use of the two compounds by fungal cells requires the presence of a gene cluster able to degrade and use carbohydrates and proteins. In addition, the persistence of expression is also recommended. In contrast, certain compounds, such as nitrates, can suppress the biosynthesis of aflatoxins despite the presence of nitrogen in their structure. Biosynthesis of mycotoxins occurs when water activity is high; however, biosynthesis can also occur when water activity is low. The temperature has a direct influence on mycotoxin biosynthesis. Indeed, the optimal production of mycotoxins is localized between 25°C and 30°C. The influence of the temperature on its rise or fall is explained by the disruption of genes that regulate biosynthesis. For the pH, the biosynthesis of mycotoxins is optimal in an acid medium, whereas in an alkaline medium, the biosynthesis is inhibited for most mycotoxins. Oxidative stress is another key factor that induces the biosynthesis of mycotoxins. The presence of an undamaged gene cluster is a crucial point for signal transduction and mycotoxin biosynthesis (Gacem et al., 2020a,b,c). The factors affecting the biosynthesis of mycotoxins during the different phases of cereal production are illustrated in Fig. 2.

3 Types of bio-nanoparticles used in the fight against phytopathogenic fungi

3.1 Green nanoparticles as protectors against fungal plant pathogens and mycotoxin synthesis

NPs ranging in size from 10 to 100 nm can be used alone on plant foliage, roots, or seeds to protect crops against various pathogens, such as fungi, insects, viruses, and bacteria. Several types of metallic NPs are studied for their antifungal properties, in particular Ag-NPs synthesized by green processes. The synthesis of Ag-NPs by aqueous *Melia azedarach* leaves

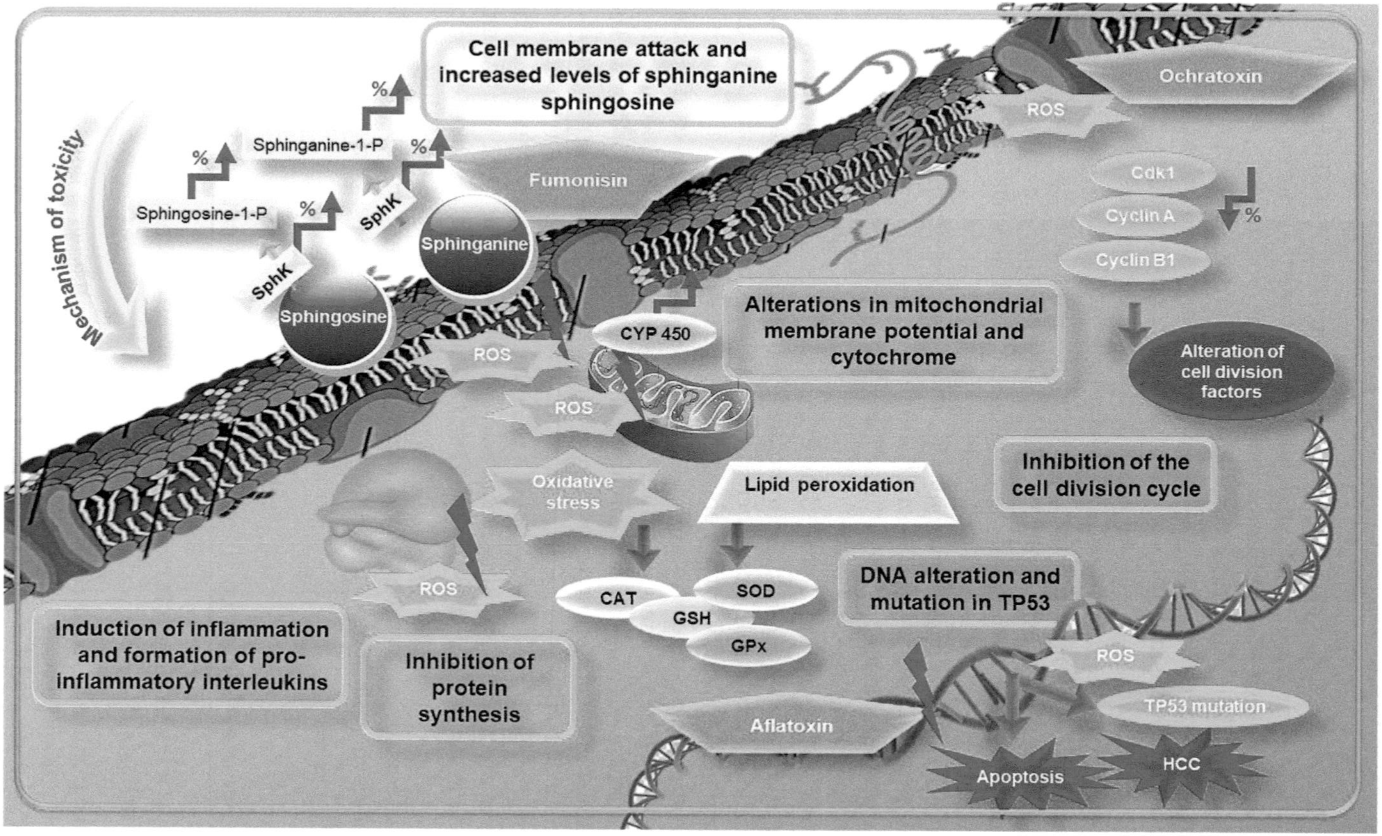

FIG. 1 Mechanisms of mycotoxin toxicity via ROS.

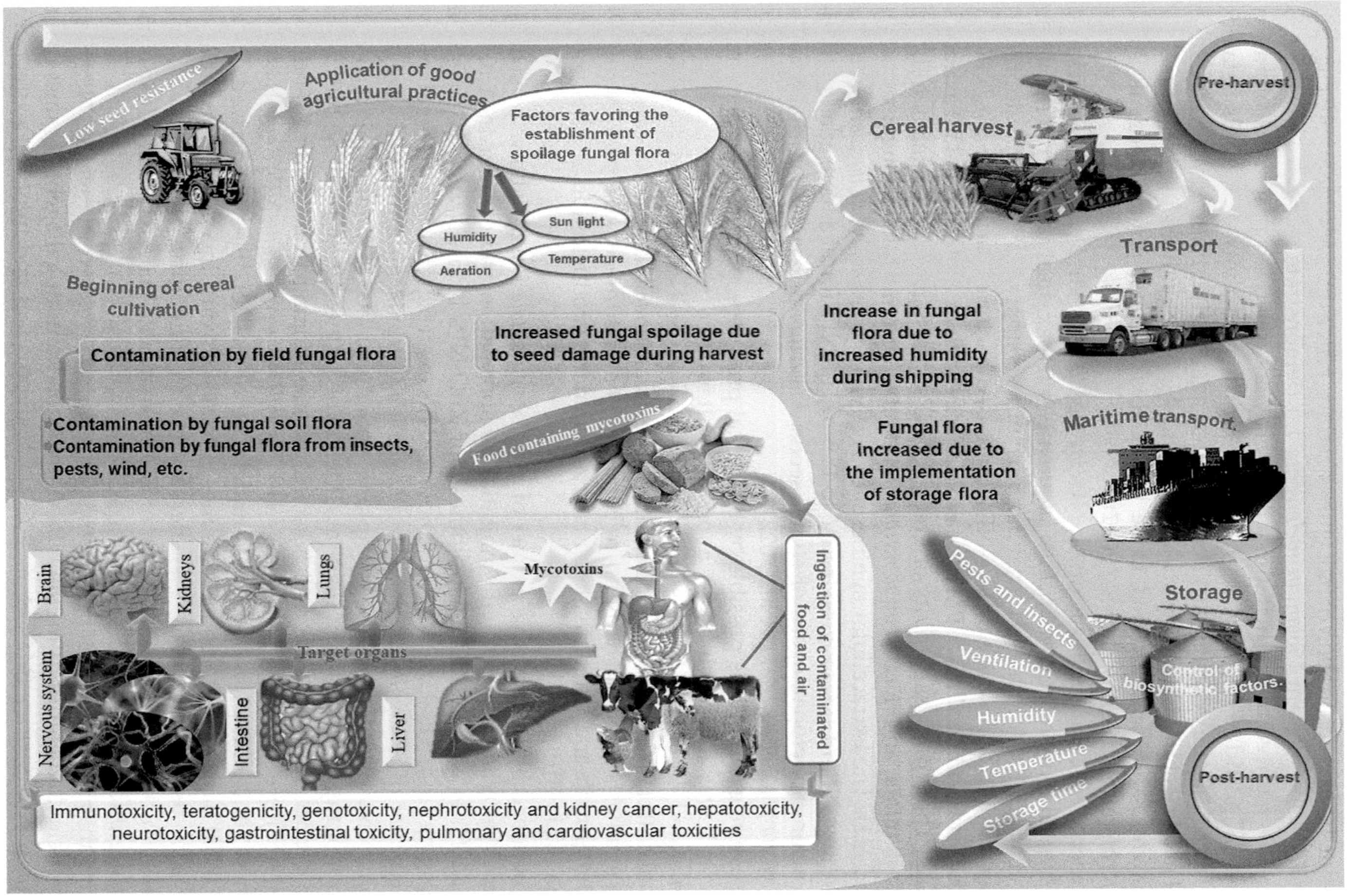

FIG. 2 Toxicity of mycotoxins and fungal spoilage of cereals during the different production phases.

extracts as a reducing and stabilizing agent made it possible to obtain NPs of sizes ranging from 18 to 30 nm. In vivo application of biosynthesized Ag-NPs as an antifungal agent against *Verticillium dahliae* in eggplant (*Solanum melongena* L.) demonstrated that the presence of 20 ppm of Ag-NPs reduced the verticillium wilt severity and relative vascular discoloration, respectively, by 87% and 97% (Jebril et al., 2020). Antifungal activity of Ag-NPs obtained from *Ginkgo biloba* L. leaf extracts is also demonstrated against *Setosphaeria turcica*, the causative agent of maize leaf blight (Huang et al., 2017), and Ag-NPs synthesis using aqueous leaf extracts of *Tagetes patula* L was found to be highly toxic against phytopathogenic fungi *Colletotrichum chlorophyti* (Sukhwal et al., 2017).

In other experiments, the synthesis of Ag-NPs by *Bacillus* sp. strain AW1-2 significantly inhibited the hyphal growth of *Colletotrichum falcatum* Went, the causal agent of red rot of sugarcane (Ajaz et al., 2021). Another extracellular biosynthesis approach has been adopted by another research group to produce bio Ag-NPs, they used bacteria (*Bacillus siamensis* PM13, *B. xiamenensis* PM14, and *Bacillus* sp. PM15), a mold (*F. oxysporum*), and supernatant extract of sugarcane husk. The antifungal activity of green Ag-NPs investigated for sugarcane fungal pathogens *C. falcatum* and *Fusarium moniliforme* exhibit prominent antifungal activities (Amna et al., 2021). The extracellular synthesis of Ag-NPs by *Trichoderma harzianum* made it possible to obtain NPs of size ranging from 6 to 15 nm. The NPs formed have good antifungal potential against phytopathogens of cotton seeds: *Fusarium fujikuroi, Rhizoctonia solani,* and *Macrophomina phaseolina* (Zaki et al., 2022).

The manufacture of Zn-NPs is possible by biological processes, phytochemicals found in *Salvia officinalis* leaves extract, used as a possible alternative to conventional chemical methods, generated ZnO-NPs with significant antifungal and photocatalytic activities (Abomuti et al., 2021). Flower bud extracts from *Syzygium aromaticum* are a good choice for the green synthesis of NPs. The NPs are between 30 and 40 nm in size, and these NPs can reduce the growth and deoxynivalenol and zearalenone production in *F. graminearum* (Lakshmeesha et al., 2019).

ZnO-NPs obtained by the aqueous extract of *Zingiber officinale* Roscoe showed maximal activity against *Aspergillus awamori* with inhibition of sporulation and ochratoxin A (OTA) production (Raafat et al., 2021). The biosynthesis of ZnO-NPs by *T. harzianum* generated NPs of size ringing from 8 to 23 nm, trichogenic ZnO-NPs exert significant antifungal potential against *Fusarium* sp., *Rhophitulus solani*, and *Modiolula phaseolina*. In addition, a considerable reduction in cotton seedling disease symptoms under greenhouse conditions is recorded in the presence of mycogenic NPs (Zaki et al., 2021). The phytofabrication of Se-NPs from *Carica papaya* extract produced spherical NPs with a size ranging from 101 to 137 nm. At 40 μg/mL in the culture of *Aspergillus ochraceus* and *Penicillium verrucosum*, NPs demonstrated reduced growth and OTA production (Vundela et al., 2022). Se-NPs derived from *T. harzianum* JF309 have good potential to reduce *Alternaria* toxins. Indeed, NPs recorded a reduction of 83% tenuazonic acid, 79% alternariol, 63% fumonisin B_1 (FB_1), and 76% deoxynivalenol. Furthermore, the expression of key biosynthetic genes of FUM1, PA, TRI5, and TRI6 was significantly reduced (Hu et al., 2019).

The green synthesis of Cu-NPs using *Celastrus paniculatus* leaves extract revealed NPs with an average particle diameter of 5 nm, the NPs exhibit good antifungal agents against plant pathogenic fungi *F. oxysporum* (Mali et al., 2020). Eucalyptus leaf and mint leaf extract are good choices for Cu-NPs synthesis, and they allow the synthesis of NPs of sizes ranging from

10 to 130 nm and from 23 to 39 nm, respectively. The antifungal activity of the NPs is tested against *Colletotrichum capsici*, a pathogen of chili crop, which showed a better activity for the NPs obtained with the extract of mint leaves (Iliger et al., 2020).

The extracellular and intracellular synthesis of gold nanoparticles (Au-NPs) from the green algae *Chlorella sorokiniana* generated NPs of different shapes. They are 5 to 15 nm and 20 to 40 nm for extracellular and intracellular synthesis, respectively. Au-NPs showed significant antifungal activity against isolates of *Candida tropicalis, C. glabrata,* and *C. albicans* (Gürsoy et al., 2021). The use of an aqueous extract of *Moringa oleifera* leaf for the reduction of titanium dioxide salt leads to the biogenesis TiO_2-NPs in the size range of 10–100 nm. The application of the biogenic NPs exogenously on wheat plants infected with *Bipolaris sorokiniana* reduced disease severity (Satti et al., 2021). The treatment of wheat rust caused by *Ustilago tritici* by TiO_2-NPs prepared by *Chenopodium quinoa* demonstrated that green TiO_2-NPs have been shown to have the best antifungal activity in comparison to those prepared by the sol-gel method (Irshad et al., 2020).

Chitosan is beneficial for the elimination of mycotoxins; this requires good optimization of mycotoxin elimination conditions. On palm kernel cake, chitosan allowed maximum elimination for aflatoxin B_1 (AFB_1), AFB_2, AFG_1, AFG_2, OTA, zearalenone (ZEA), FB_1 and FB_2 under the optimized conditions (pH4, 8h and 35°C), the removal rate were 94.35%, 45.90%, 82.11%, 84.29%, 90.03%, 51.30%, 90.53% and 90.18%, respectively (Abbasi Pirouz et al., 2020). The elimination of *Fusarium solani* and *Aspergillus niger* when studying the antifungal activity of Chitosan-NPs is influenced by the particle size and the zeta potential of the applied NPs (Ing et al., 2012). On strawberry leaves, 25 mg/mL of green chitosan-NPs (obtained by the bioconversion of chitosan to chitosan-NPs using *Pelargonium graveolens* leaves extract) recorded significant antifungal activity against the phytopathogen *B. cinerea.* Chitosan-NPs reduced gray mold severity symptoms by up to 3% (El-Naggar et al., 2022).

3.2 Green nanoparticles as nanocarriers to provide antifungal activities against plant pathogens

Green nanoparticles are used in the design of nanosupports, and the nanocomposite thus formed serves to encapsulate, trap, and fix active substances. It should be noted that formed nanocomposites are more powerful formulations than simple NPs. Silica NPs play a very effective role in the preparation of green NCs; however, guaranteed delivery of the silica-coated substance requires control of shape, structure, and size. The potential of this type of NC has already been demonstrated in reducing water pollution. Hybrid tannic acid-silica-based porous nanoparticles (TA-SiO_2-NPs) have been synthesized under mild conditions in the presence of a green and renewable tannic acid biopolymer, and the NCs formed showed good results in the elimination of Cu^{2+} ions in aqueous solutions (Tescione et al., 2022). The design of silica-based nanocomposites advocates the presence of pores on the surface of NCs, such as microporous silica nanoparticles. (diameter $<$ 2), mesoporous (2 $<$ diameter $<$ 50), and macroporous (Diameter $>$ 50) (Prabha et al., 2020), these nanostructures guarantee and prevent the degradation of the active particles present in the nanoparticles, they also prevent the active particles from escaping due to their volatile nature (essential oil), and, therefore, they ensure an accurate distribution (Kutawa et al., 2021). Safavinia and his group generated silica-based NCs and ZnO-NPs using methanol leaf extract of *Daphne oleoides*, the ZnO-NPs are loaded on silica gel matrix (ZnO/SG nanocomposite), the surface of the NCs has been increased compared to the ZnO-NPs. Moreover, the antimicrobial activity of the formed NCs was better than that of the bare ZnO-NPs. (Safavinia et al., 2021). Good antifungal activity is recorded against *Phytophthora capsici* by the application of copper- and silica-based NCs coated with carboxymethyl cellulose. The initial SiO_2-NPs were prepared from rice husk ash and were coated by a copper ultrathin film using hydrazine. The core/shell SiO_2@Cu-NPs with an average size of ~19 nm were surrounded by carboxyméthyl cellulose (Hai et al., 2021).

Solid lipid nanoparticles (SLNs) are nanostructures with very relevant functionalities, they serve as nano-systems capable of coating bioactive substances or NPs, they also avoid losses during applications, and also the modification of coated substances or particles. In the medical field, the application of this combinatorial drug delivery system represents a promising strategy to overcome drug resistance, this system is capable of simultaneously transporting two or more substances to the targeted site in a human body. In agriculture, these nano-systems offer a healthy alternative for the management of phytopathogenic diseases of fungal origin with more fungicidal potential (Carbone et al., 2020). Evaluation of solid lipid nanoparticles (SLNs) containing *Zataria multiflora* essential oil (ZEO) with a size around 255.5 ± 3 nm demonstrated that the antifungal efficacy of ZEO-SLNs was significantly more than ZEO. The minimum inhibitory concentration (MIC) under in vitro conditions for the ZEO-SLNs on the fungal pathogens of *A. ochraceus, A. niger, A. flavus, R. solani, Alternaria solani,* and *Rhizopus stolonifer* was 200, 200, 200, 100, 50, and 50 ppm, respectively, for ZEO, MIC was 300, 200, 300, 200, 200, and 200 ppm, respectively (Nasseri et al., 2016). the encapsulation of carbendazim and tebuconazole in SLNs allowed good management of the release of bioactive compounds and their transfer to the target site, a decrease in losses by leaching or degradation, and a decrease in toxicity in the environment. The encapsulation of the two fungicides in combination in SLNs makes it possible to obtain particles of an average size of 191.6 nm, these NCs show low toxicity to normal cells (3T3 and MC3T3) compared to commercial fungicides. Indeed, only 30% of the carbendazim is released after 6 days (Campos et al., 2015). Food protection by NCs-based food packaging films has been demonstrated against post-harvest fungi: *Ranunculus stolonifer, Alternaria* spp., and *A. niger*, pullulan-based food packaging films are modified by doping essential oil active compounds (cinnamaldehyde, eugenol, and thymol) into liquid and solid lipid nanodroplets. Encapsulation of essential oil compounds dramatically enhances their antifungal activity by improving their accessibility and ensuring even distribution of the antimicrobial compound. SLNs showed higher antifungal activity due to increased concentration of bioactive substances (McDaniel et al., 2019).

Chitosan NCs is an excellent biopolymer with promising antimicrobial potential against fungal phytopathogens. Nowadays, engineered nanostructures have gained much attention due to their potential use in plant disease management. The chitosan-magnesium nanocomposite (CS-Mg-NCs) having particles size ranging from 29 to 60 nm are obtained by green synthesis. *Bacillus* sp. strain RNT3 is the bacterial species used in the manufacture of NCs. CS-Mg-NCs have demonstrated impressive antifungal activity against *R. solani* at 100 μg/mL (Ahmed et al., 2021). Another hybrid bionanomaterial (oligochitosan) was synthesized from low-cost waste products, and it is produced from chitosan of crab shells, nanosilica [$nSiO_2$] obtained from rice husk, and carboxymethyl cellulose. The produced nanostructured oligochitosan-silica/carboxymethyl cellulose has a particle size ranging between 3 and 8 nm. A concentration of 800 mg/L was the lowest concentration where the material completely inhibited *Phytophthora infestans* growth (Nguyen et al., 2019). Previously, the same group demonstrated that nanostructured oligochitosan-silica exhibits a significative antifungal activity in chili plants against *Colletotrichum* sp. (Nguyen et al., 2017). Chitosan-Ag-NPs NCs with ranging sizes from 495 to 616 nm exhibited significantly higher antifungal activity against *Colletotrichum gloeosporioides*. In vivo assay, anthracnose was significantly inhibited in mango fruit by the use of chitosan-Ag-NPs NCs (Chowdappa et al., 2012). Similarly, coatings designed by synergistic chitosan and *Cymbopogon citratus* essential oil combinations decreased anthracnose lesion development in guava, mango, and papaya inducted by *Colleotrichum* species during storage (Lima Oliveira et al., 2018). Soybean protein isolate/cinnamaldehyde/ZnO-NPs bio-NCs used to overcome and ensure delayed ripening, maintain fruit firmness, and reduce fruit weight loss demonstrated good activities in improving the postharvest quality of bananas (Li et al., 2019). Table 1 shows the different types of bio-nanocomposites providing antifungal activities against plant pathogens.

TABLE 1 Different types of antimicrobial bio-nanocomposites.

Types of nanocarrier	Food product	Antimicrobial effect	References
Sodium alginate-gelatin microcapsules enriched/carbon nanotubes/SiO_2 NPs encapsulate *Bacillus velezensis*	Pistachio gummosis	Synergistic control of *Phytophthora drechsleri*	Moradi Pour et al. (2022)
Chitosan NPS	Strawberry leaves	Excellent antimicrobial activities against *B. cinerea*	El-Naggar et al. (2022)
Polysaccharide bionanocomposites		Excellent antimicrobial activities and can be used for food packing	Pal et al. (2021)
Gelatin/cellulose nanofibers/ZnONPs/SeNPs)		Used in the development of active food packaging products	Ahmadi et al. (2021)
Cellulose/glycine/zinc nano-biocomplex		Excellent antifungal activity of toward *Aspergillus niger, A. terreus, A. flavus,* and *A. fumigatus*	Dacrory et al. (2021)
Corn starch/chitosan nanoparticles/thymol NCs	Cherry tomatoes	The NCs extend the shelf life of cherry tomatoes, and retard mold growth up to day 6 of storage	Othman et al. (2021)
Pectin/nanochitosan NCs	Mango fruit	The pectin/nanochitosan coatings could maintain the number of total microbials, below the acceptable limit of 10^6 CFU/g fruit, even after 24 days of storage	Ngo et al. (2021)
Corn starch/AgNPs/ guava leaves extract NCs films.	Can be utilized as a reinforcing and UV-blocking agent in the nanocomposite film.	Excellent antimicrobial activities	Kumar et al. (2020)

Continued

TABLE 1 Different types of antimicrobial bio-nanocomposites—cont'd

Types of nanocarrier	Food product	Antimicrobial effect	References
Whey protein isolate/polydextrose-based nanocomposite film/cellulose nanofiber/*Lactobacillus plantarum*:		The nanocomposite film can be applied to bioactive food packaging system	Karimi et al. (2020)
tarch films/ChitosanNPs NCs	Cherry tomatoes	Used as films for food packaging against microbes	Shapi'I et al. (2020)
Carboxymethyl cellulose/cellulose nanofiber/inulin/probiotic nanocomposite film.	Chicken fillet	Used for improving the shelf life of food products and providing health benefits for consumers	Zabihollahi et al. (2020)
Pectin/nanochitosan films NCs	Used as active films to extend the shelf life of food	The NCs film inhibits the growth of *C. gloeosporioides, Saccharomyces cerevisiae*, and *Aspergillus niger*	Ngo et al. (2020)
Pectin/Ags NCs		The NCs exhibit potent antifungal activities against *Aspergillus japonicus*	Su et al. (2019)
Starch/pectin/titanium oxide nanoparticles ((TiO_2)-NPs) films		The films can be used as food-grade UV screening biodegradable packaging material	Dash et al. (2019)
Poly(3-hydroxybutyrate)-thermoplastic starch/organically modified montmorillonite (OMMT)/eugenol	Used as food packaging materials	The bio-NCs exhibited antioxidant and antifungal activity against *Botritys cinerea*	Garrido-Miranda et al. (2018)
Cellulose/Cu NCs	Pineapple and melon juice	A strong antimicrobial activity against *S. cerevisiae*	Llorens et al. (2012)
Cellulose/Ag NCs	Melon	Good antimicrobial activity against spoilage-related microorganisms	Fernández et al. (2010)

4 Application of nanomaterials and nanohybrid biomaterials against fungal phytopathogens

Plant pathogen management using nanohybrid materials focuses on the fabrication of nanostructures composed of multiple components. The preparation of sepiolite-MgO (SE-MgO) nanocomposite formed of nontoxic sepiolite with metabolizable MgO nanoforms (nMgO) intended for antifungal evaluation against various phytopathogenic fungi of rice (*Fusarium verticillioides, Bipolaris oryzae*, and *F. fujikuroi*) demonstrated that SE-MgO was more potent than MgO-NPs in an aqua dispersed form (aqMgO-NPs). Microscopic results reveal antifungal activity due to distortion of hyphae and collapse of spores (Sidhu et al., 2020). Magnetically separable Fe_3O_4/ZnO/AgBr nanocomposites prepared by microwave-assisted method exhibit good antifungal activity against *F. graminearum* and *F. oxysporum*. The microwave irradiation time and a weight ratio of Fe_3O_4 to ZnO/AgBr have considerable influence on the antifungal activity (Hoseinzadeh et al., 2016). Silver-titanate nanotubes (AgTNTs) functionalized with silver nanoparticles exhibit unique surface and antifungal properties for the photoinactivation of *B. cinerea* under visible light. Oxidative stress caused by reactive oxygen and silver cytotoxicity species such as Ag and Ag^+ expands the conidia to induce cell death (Rodríguez-González et al., 2016).

Multiwalled carbon nanotubes (MWCNT) have significant antifungal activity; the addition of surface functional groups (OH, COOH, and NH2) to MWCNT enhanced its antifungal activity against the plant pathogen *F. graminearum*. All the modified MWCNTs showed inhibition of spore growth and germination compared to the blank MWCNTs. The length of spores decreased by almost a half from 54.5 μm to 28.3, 27.4, and 29.5 μm, after being treated with 500 μg/mL MWCNTs-COOH, MWCNTs-OH, and MWCNTs-NH_2 separately (Wang et al., 2017a,b). The surface of carboxylated MWCNTs grafted with some pyrazole derivatives (pyrazole and pyrazolone moieties) has significant antimicrobial activity (Metwally et al., 2019). Hybrid nanomaterials (f-MWCNTs-CdS and f-MWCNTs-$Ag^{(2)}$S) developed by covalent grafting of cationic hyperbranched dendritic polyamidoamine (PAMAM) onto MWCNTs and successive deposition of CdS and $Ag^{(2)}$S quantum dots (QDs) have similarly good antimicrobial effects. The germicidal action of MWCNTs was improved by the grafting of PAMAM and the immobilization of CdS and $Ag^{(2)}$S QDs (Neelgund et al., 2012). The effect of graphene oxide-silver nanocomposite (GO-Ag-NPs) against *F. graminearum* showed good antifungal activity, in addition, inhibition efficiency increases three- and sevenfold over pure Ag-NPs and GO suspension, respectively. In the detached leaf experiment, the GO-Ag-NPs showed a significant effect in controlling the leaf spot disease infected by *F. graminearum* (Chen et al., 2016). GO-Fe_3O_4 as nanocomposite formed of graphene oxide and iron oxide is tested against *Plasmopara viticola*, a causal agent of grapevine downy, which is a destructive disease in field-grown grapevines. The results of the study showed that the treatment of grapevine leaves in the field with 250 μg/mL GO-Fe_3O_4 decrease the severity of downy mildew with no significant toxic effects on grapevine plants. The control efficiency was 65.08%. In addition, the protective, fungicidal, and curative activities of GO-Fe_3O_4 could be regulated in a time-dependent manner. This effect may be attributed to the strong inhibition of spore germination. Surface-adsorbed NMs play an important role in the inhibition of sporangium germination by the blockage of the water channel of the sporangia (Wang et al., 2017a,b).

Green-synthesized iron-oxide NPs were prepared using spinach (*Spinacia oleracea* L) as a starting material and black coffee extract as a stabilizer, iron-oxide NPs caused a significant reduction in mycelial growth and spore germination of *F. oxysporum*. Malformed mycelia, DNA fragmentation, and ROS production indicated the antimicrobial potential of iron-oxide NPs. Treatment of tomatoes reduced disease severity by an average of 47.8% after exposure to 12.5 μg/mL iron-oxide NPs. In addition, the content of flavonoids, proteins, and vitamin C was significantly improved in the processed tomatoes (Ashraf et al., 2022). The synthesis of copper oxide/carbon (CuO/C) NCs was possible through the green method using the leaf extract of *Adhatoda vasica*, the leaf extract serves as a source of carbon, a capping, and a reducing agent. The CuO/C nanoflakes obtained had an average thickness of 7–11 nm. The NCs showed significant antifungal activity against *A. niger* (Bhavyasree and Xavier, 2020). Similarly, CuO/C NCs were obtained using sucrose as a capping agent, and the antifungal activities of CuO/C NCs showed 70% restraint on *A. flavus* and 90% hindrance on *A. niger*

(Roopan et al., 2019). In another study, Cu-NPs and starch were prepared to synthesize Cu-NPs-based NCs, the NCs exhibited outstanding antifungal activity toward *A. niger*, *A. terreus*, and *A. fumigatus* (Hasanin et al., 2021). From these published reports, it is remarkable that metals, carbon, inorganic materials, and green plant compounds possess inherent properties when combined as nanocomposites. The increased antifungal activity is attributed to the synergistic effects of the components of the nanocomposites compared to the metals alone. However, the antifungal activity strongly depends on the contact time, the components of the nanocomposites, the method of synthesis, and the fungal species.

Nanochitosan-based NCs show excellent antimicrobial potential against various plant pathogens, e.g., clay/chitosan nanocomposite exhibits a complete inhibition of *Penicillium digitatum* at 20 μg/mL for clay/chitosan (1:0.5), (1:1) and (1:2). A reduction of 70% of rot was reported for clay/chitosan (1:2) in vivo onto orange (Youssef and Hashim, 2020). Chitosan-NPs, silica-NPs and chitosan-silica NCs were tested in vitro and in vivo against *B. cinerea*. In vitro tests showed that chitosan-NPs, silica-NPs, and chitosan-silica reduced fungal growth by 72%, 76%, and 100%, respectively. Under natural infection, chitosan-silica NCs were the most effective treatment and reduced the development of gray mold by 59% and 83%, for "Italia" and "Benitaka" grapes, respectively (Youssef et al., 2019). During another investigation, chitosan zinc oxide and copper NCs (CS-Zn-Cu-NCs) showed a significant antifungal potential against three phytopathogenic fungi including *A. alternata*, *Rhophitulus solani*, and *B. cinerea*. At 90 μg/mL, CS-Zn-Cu-NCs present the highest antifungal activity compared to Zn- and Cu-NPs (Al-Dhabaan et al., 2017), In another study, Cu-chitosan NCs showed good antifungal efficacy against *R. solani*. The NCs cause DNA damage in the absence of DNA amplification (Abd-Elsalam et al., 2018). In tomato (*Solanum lycopersicum* Mill), Cu-chitosan NPs induce growth promotory effect on tomato seed germination and seedling length. Cu-chitosan caused 70.5% and 73.5% inhibition of mycelia growth and 61.5% and 83.0% inhibition of spore germination in *A. solani* and *F. oxysporum*, respectively (Saharan et al., 2015). Silver/chitosan NCs portrayed an incremental growth-inhibitory effect against *Aspergillus* sp., *Rhizoctonia* sp., and *Alternaria* sp. isolated from chickpea seeds with rate inhibition of 94%, 67%, and 78%, respectively (Kaur et al., 2012). Silver nanoparticle-encapsulated chitosan functionalized with 4-((E)-2-(3-hydroxynaphthalen-2-yl)diazen-1-yl) benzoic acid demonstrated improved effectiveness against *A. flavus* and *A. niger* (Mathew and Kuriakose, 2013), chitosan-silver nanocomposites have significant antifungal activity against *P. expansum* (Alghuthaymi et al., 2020).

Biopolymeric chitosan-Carrageenan NCs containing the chemical fungicide mancozeb showed that mancozeb-loaded nanoformulations exhibited significant inhibition of mancozeb. The NCs were tested for antifungal activity against *Alternaria alternata*, *Septoria lycopersici*, *A. solani*, and *Sclerotinia sclerotiorum* at two concentrations (1.0 and 1.5 ppm), and the results showed that the NCs exhibited 100% radial growth inhibition for *S. lycopersici* and *S. sclerotiorum* at 1.0 and 1.5 ppm, equivalent to commercial mancozeb at the same concentrations. Mancozeb-loaded NPs at 0.5 and 1.0 ppm also exhibited a mere inhibition of $67.7 \pm 1.4\%$ and $67.7 \pm 0\%$, respectively, against *A. solani*. However, at 1.5 ppm, good inhibition of $83.1 \pm 0\%$ was exhibited, comparable to commercial mancozeb ($84.6 \pm 0\%$) at this concentration. The carrying capacity of mancozeb is a function of the ionic interaction between the fungicides and the nanoparticles. Therefore, fungicide release is prolonged for NCs containing low fungicide concentration and NCs containing higher doses may not be properly entrapped by the nanosystem. In vivo tests carried out on tomatoes and potatoes have shown that NCs improve the effectiveness of the control of phytopathogens, except for *A. alternata* (Kumar et al., 2021). Carbendazim-loaded chitosan-pectin NCs demonstrated sustained release of nanoformulated carbendazim compared to unentrapped carbendazim release rates, this is explained by the trapping of the active principle in the polymer. In contrast, the rapid release rate of the nanostructured fungicide at acidic (4) and basic (10) pH may be related to the solubility and stability of carbendazim at high and low pH. In addition, the NCs exhibited higher inhibition than pure carbendazim and commercial formulations. At 0.5 and 1.0 ppm, the NCs showed 100% inhibition of radial growth for *F. oxysporum* and *A. parasiticus*, respectively (Sandhya et al., 2017). Similarly, carbendazim and tebuconazole incorporated in polymeric and solid lipid nanoparticles were used with enhanced antifungal activity (Campos et al., 2015).

5 Antifungal mechanism of nano-biofungicide and nanohybrid biofungicide

Multiple mechanisms are involved in the fungicidal activity exerted by nano-biofungicides. Generalized antifungal activity is shown in Fig. 3. According to Fig. 3, the antifungal activity of bio-NCs is accomplished by several events that first start on the outer components of the fungal cell and then move to the intracellular components. According to Kalia and her group, the internalization of nanomaterials occurs through three mechanisms: (1) direct internalization of NMs into the cell wall, (2) mediated adsorption by specific

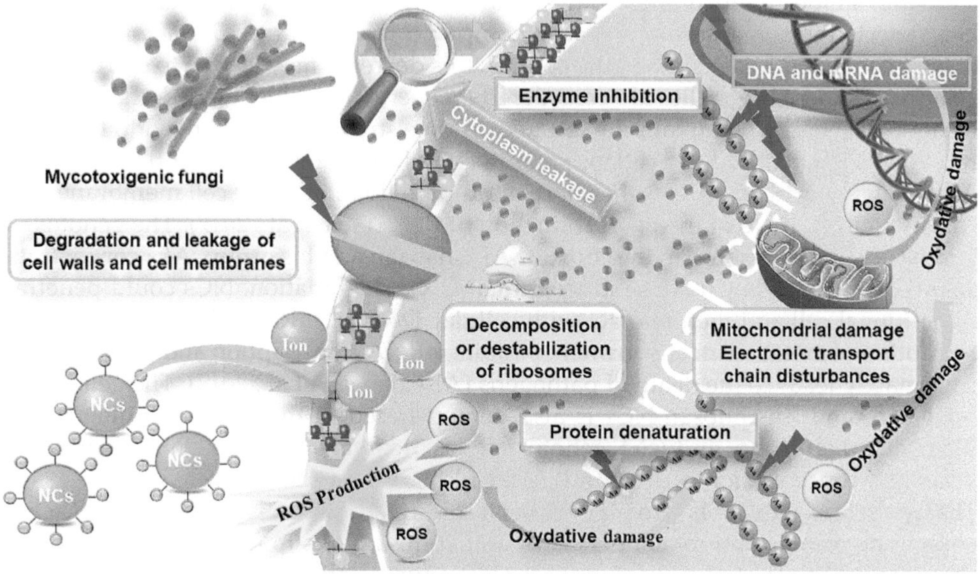

FIG. 3 Mechanisms governing the antifungal and antimycotoxigenic potential of nano-biofungicide and nanohybrid biofungicide.

receptors followed by internalization, and (3) internalization of NMs by ion transport proteins (Kalia et al., 2020).

Microscopic observation of *Rhophitulus solani* hyphae exposed to Cu-chitosan and Zn-chitosan nanocomposites showed severe damage presented by the separation of wall layers and collapse of fungal hyphae (Abd-Elsalam et al., 2018). The fungal wall interacts with the external environment by controlling cell permeability, and therefore, it interferes with the NCs present in the extracellular environment. This vital part of the fungal cell is composed mainly of polysaccharides such as chitin, glucan, mannan or galactomannan, lipid, and β-1,3-D-glucan and β-1.6-D-glucan macroproteins, NMs inhibit *N*-acetylglucosamine (*N*-acetyl-D-glucose-2-amine) or on β-1,3-D-glucan synthase by the action of ion and/or ROS. *N*-acetylglucosamine is involved in the biosynthesis of the chitin while the β-1,3-D-glucan synthase participates in the synthesis of β-1.3-D-glucan (important components of the cell wall in fungi) (Arciniegas-Grijalba et al., 2017). The interaction between NMs and fungal bio-membranes can lead to destabilization, denaturation, and inactivation of proton pumps. This leads to leakage of intracellular contents. NCs can also increase membrane permeability and ultimately lead to dysregulation of cellular bio-membrane functions and the death of the pathogen (Alghuthaymi et al., 2020).

At the intracellular level, NMs interact with various biomolecules and thus cause the deformation of biomolecules and the inactivation of catalytic proteins. NPs can bind to sulfurized proteins by denaturing them; they interact with thiol groups of proteins, which are important for microbial respiration; and they can also damage mitochondria and ribosomes. NPs can also cross the fungal nuclear membrane and bind to its DNA, which can inhibit mRNA synthesis (Alghuthaymi et al., 2020).

ROS plays an essential role in the mechanism of antifungal activity of bio-NCs. During the treatment of grapes with chitosan-silica-NCs to fight against *B. cinerea*, an accumulation of ROS is detected in the spores of *B. cinerea* compared to the control. *B. cinerea* spores treated with NCs contained less ATP than the control, this decrease may be due to disruption of mitochondria, as these organelles are considered an energy machine. Indeed, the disruption of mitochondrial energy potential is linked to oxidative stress, which is linked to high levels of intracellular ROS that lead to a collapse of mitochondrial membrane potential and a decrease in ATP (Youssef et al., 2019). Furthermore, lipid peroxidation of the cell membrane is one of the major consequences of the overproduction of ROS (Hua et al., 2017). The genotoxicity of hybrid NCs is demonstrated in the study of Al-Dhabaan and his group. At a concentration of 60 μg/mL, Chitosan-Zn-Cu-NCs induce fungal DNA degradation. NCs could penetrate inside the fungal cell and then interact with cellular DNA or mRNA, this subsequently leads to disruption of metabolism and growth, inhibits mRNA transcription and protein synthesis, and ultimately leads to the death of pathogens (Al-Dhabaan et al., 2017). The genotoxicity of Cu-chitosan- and Zn-chitosan-NCs is also demonstrated in the study conducted by Abd-Elsalam and his group. Induced genotoxicity is demonstrated by DNA degradation of *R. solani* at a concentration of 40 μg/mL (Abd-Elsalam et al., 2018).

The ions released by NMs also play a crucial role in the mechanism of antifungal activity of bio-NCs. The combination of copper, zinc, and chitosan NPs to form CS-Zn-Cu NCs effectively reduces the rot of strawberries through the fungistatic effect of chitosan and the fungicidal activities of Cu and ZnO. Indeed, the biocidal activity can be attributed to the copper ions released by the Cu-NPs. NPs could be strongly adsorbed on the surface of fungal cells, this interaction results in disruption of cell permeability, denaturation of functional biomolecules, and oxidative damage. The released ions (Cu^{2+}) can cross the fungal membrane and move inside the cell; hence, they lead to cell apoptosis via cell membrane disruption and protein denaturation (Al-Dhabaan et al., 2017). Besides, during the infection of plants, phytopathogenic fungi tend to produce acids, the acidic pH induces the protonation of the amino groups of chitosan, which then leads to the release of Cu ions from the chitosan NCs. The increased antifungal activity of Cu^{2+} is linked to the production of highly reactive hydroxyl radicals that can then damage biomolecules (Saharan et al., 2015). Other mechanisms may be involved in the antifungal activity of NCs. Indeed, NCs can also target the host plant and improve its resistance to pathogens.

6 Bio-nanofungicide for mycotoxins degradations

The advantages of NMs are not limited only to plant breeding and the fight against phytopathogens, and they contain enormous potential in the management of mycotoxins. NMs have mycotoxin detection and detoxification capabilities, and they have more advantages than NPs, due to the constituents that synergistically interact with each other and accelerate the kinetics of the detoxification reaction (Alghuthaymi et al., 2021). Several studies have demonstrated the antimycotoxinogenic potential of NCs synthesized by green pathways containing components of biological origin. Hamza et al. demonstrated that the design of NCs based on glucan mannan lipid particles (GMLPs) capable of adsorbing aflatoxins is possible. The idea is supported by the strong ability of *Saccharomyces cerevisiae* cell wall GMLPs to adsorb AFB_1 in gastric fluid and simulated intestinal fluid. The new system formed can encapsulate humic acid NPs (HA-NPs) by complexation with ferric chloride. Encapsulation of HA-NPs in GMLPs increased the stability of HA-NPs in gastric fluid and simulated intestinal fluid, furthermore, the GMLP-HA-NP hybrid formulation synergistically enhanced AFB_1 binding. In simulated gastric fluid, the AFB_1 adsorption capacity of free HA was 8.5 μg AFB_1/mg HA, and for free HA-Fe complexed NPs was 7.3 μg AFB_1/mg HA. The efficient trapping of HA by ferric chloride inside GMLPs increased the adsorbed AFB_1 mass from 10.8 μg AFB_1/mg HA for GMLP HA to 13.5 μg AFB_1/mg HA for GMLP HA-Fe NPs (Hamza et al., 2019), the in vivo study in male Sprague-Dawley rats confirmed that AFB_1 detoxification is effective. Indeed, the co-administration of GMLP/HA-Fe NPs with AFB_1 has a protective effect against hepato-nephrotoxicity, oxidative stress, and histological alterations of the liver and kidneys induced by AFB_1 (Hamza et al., 2022).

The design of NCs based on poly-(ε-caprolactone) and then loaded with benzophenone or valerophenone thiosemicarbazone showed that aflatoxin production in the aflatoxigenic species *A. flavus* was inhibited (Spadola et al., 2020). Similarly, the determination of the antifungal activity of Ag-chitosan-NCs (particle sizes ranging from 4 to 10 nm) at various concentrations (0.30, 0.60, and 0.100 mg/mL) is effective. NCs induce overexpression against dairy cattle *Penicillium expansum*-producing patulin and citrinin (Alghuthaymi et al., 2020), and the antifungal activity of chitosan-NPs has been demonstrated previously against two aflatoxin producers such as *A. flavus* and *A. parasiticus* (Mekawey, 2018).

Selenium nanoparticles derived from *T. harzianum* JF309 cover good antimycotoxinogenic potential. Various metabolites are present in the formed NPs including organic acids and their derivates, such as psoromic acid, β-decyloxybenzoic acid, and glucaric acid lactone. In *Fusarium verticillium* BJ6 and *F. graminearum* PH1 mainly produce FB_1 and DON, NPs cause a decrease in FUM1 expression, which resulted in a 70% decrease in FB1 production. For DON production, expressions of structural genes, TRI5 and TRI6, showed a great reduction, while the expression of a regulated gene of TRI10 was not changed. The incubation of maize kernel with the spores of BJ6 and NPs showed good protector against rot (Hu et al., 2019). In another study, control of wheat growth and productivity by green selenium NPs (BioSe-NPs) synthesized by *Lactobacillus acidophilus* ML14 could be recommended. The NPs were spherical with a size of 46nm. Under greenhouse conditions, the wheat treated with 100μg/mL BioSe-NPs was significantly improved, and a reduction in 75% in the incidence of root rot diseases caused by *Fusarium culmorum* and *F. graminearum* is recorded (El-Saadony et al., 2021).

Ag-NPs could be a good strategy in the management of the main aflatoxigenic and ocratoxigenic species contaminating food. Ag-NPs with an average diameter size of 30nm and at doses of 0–45μg/mL were active in the inhibition and reduction of *A. flavus* (AFB_1 and AFB_2), *A. parasiticus* (AFB_1, AFB_2, AFG_1, and AFG_2), *A. carbonarius, A. niger, A. ochraceus, A. steynii, A. westerdijkiae, and P. verrucosum* (OTA). Ag-NPs dose, contact time, and their interactions significantly affect fungal development and mycotoxins accumulation (Gómez et al., 2019). *Fusarium chlamydosporum* NG30 and *Penicillium chrysogenum* NG85 have good potential in the synthesis of Ag-NPs. Transmission electron microscopy revealed their spherical shape and a size range between 6 and 26nm *for F. chlamydosporum* Ag-NPs (FAg-NPs) and from 9 to 17.5nm for *P. chrysogenum* Ag-NPs (Pag-NPs). The antifungal activity of the NPs against *A. flavus* recorded a MIC of 48 and 45μg/mL for Fag-NPs and Pag-NPs, respectively. When testing *A. ochraceus,* MIC values were 51 and 47μg/mL for FAg-NPs and PAg-NPs, respectively. In addition, the statistical MIC values to inhibit completely the total aflatoxin and OTA production by *A. flavus* and *A. ochraceus* were 5.9 and 5.6μg/mL, and 6.3 and 6.1μg/mL for FAgNPs and PAgNPs, respectively (Khalil et al., 2019). The biosynthesis of Ag-NPs by using the fungi *Aspergillus terreus* HA1N and *P. expansum* HA2N revealed Ag-NPs size ranging between 14 and 25nm in the case of *P. expansum* and 10–18nm in the case of *A. terreus.* At the concentration 220 μg per100 mL media, Ag-NPs gave the highest reduction of OTA produced by *A. ochraceus,* the percentages of reduction were 58.87% and 52.18% for Ag-NPs produced by *P. expansum* and *A. terreus,* respectively (Ammar and El-Desouky, 2016). The biosynthesis of Ag-NPs from *A. flavus* PNU05 isolated from date palm (*Phoenix dactylifera* L.) rhizosphere soil revealed Ag-NPs with a high antimycotoxigenic effect, the percentage of reduction of AFB_1 was 56.45% (Zaban et al., 2019). Ag-NPS exhibit good antifungal and mycotoxin accumulation effects against toxigenic *F. graminearum* and *F. culmorum* (DON, 3-acetyldeoxynivalenol and (ZEA), *F. sporotrichioides* and *F. langsethiae* (T-2 and HT-2 toxins), *F. poae* (nivalenol), *F. verticillioides,* and *F. proliferatum* (FB_1 and FB_2) (Tarazona et al., 2019).

The green synthesis of ZnO-NPs from *Syzygium aromaticum* (SaZnO NPs) flower bud extract revealed the development of triangular and hexagonal-shaped NPs with a size ranging from 30 to 40nm. The synthesized NPs reduced the growth and biosynthesis of DON and ZEA of *F. graminearum*. The complete elimination of fungal growth and mycotoxins was noticed at 140μg/mL of SaZnO NPs (Lakshmeesha et al., 2019). Chitosan-encapsulated *Cymbopogon martinii* essential oil nanoparticles (Ce-CMEO-NPs) have presented potent fungicidal and antimycotoxin activities in maize grains. Under laboratory conditions over a storage period of 28 days, the reductions of *F. graminearum* growth and mycotoxins in maize grains were dependent on the concentration of Ce-CMEO-NPs. The complete reductions of fungal growth and mycotoxins were noticed at 700 ppm of Ce-CMEO-NPs in comparison with CMEO noticed at 900 ppm, and this suggests that essential oils captured in chitosan are less prone to decomposition, perseverance bio-functional characters, and gradually release the bio-active compounds in a controlled way (Kalagatur et al., 2018). Chitosan-based *Litsea cubeba* essential oil emulsion had good antifungal effects and minimal activity on the germinability of barley, a germination rate of 87.7%, and a DON concentration of 690 μg/kg, which was 20.9% lower than that of control was recorded (Peng et al., 2021). The biosynthesis of ZnO-NPs using *Zingiber officinale* aqueous extract had good inhibition potential of OTA production by *A. awamori* in a concentration-dependent manner. Reduction percentages were 45.6%, 84.7%, and 95.6% at 10, 15, and 20 μg of ZnO-NPs/mL, respectively (Raafat et al., 2021). In living organisms, CNs may provide other benefits, The encapsulation of cinnamon essential oil in whey protein has made it possible to obtain emulsion droplets with an average diameter of 235 ± 1.4 nm. The treatment of male Sprague-Dawley rats by these NCs does not induce toxic effects; on the other hand, in rats treated with fumonisin and/or AFB_1, the NCs improve the expression of the apoptotic enzyme. NCs also improve the percentage of DNA fragmentation in the kidneys and liver in rats treated with mycotoxins, they participate in the increase of antioxidant enzymes and normalize the levels of creatinine and uric acid (Abdel-Wahhab et al., 2018). Curcumin nanoparticles loaded with hydrogels (Cur-NPs-Hgs) are excellent candidates against AFB_1-induced genotoxicity in rat liver. In vivo tests showed that Cur-NPs-Hgs induce significant protection against AFB_1. Co-treatment with AFB_1 and Cur-NPs-Hgs succeeded to reduce the elevation in the total number of chromosomal aberrations in bone marrow, liver, and spleen in a dose-dependent manner (Abdel-Wahhab et al., 2016).

7 Conclusion

However, nanomaterial applications to the management of mycotoxin contamination are still in the early stages, and the unknown health outcomes of nanomaterials may limit their widespread use in food security. The use of nanotechnology in agriculture has enormous promise for lowering mycotoxin generation. Metal nanoparticles in nanocomposites used to make food packaging have suppressed the growth of fungus and other diseases, enhancing food safety and quality for extended shelf life. In this chapter, we concentrated on the applications of developing nanomaterials to toxicogenic fungus contamination, mycotoxin inhibition, mycotoxin adsorption, and mycotoxin removal. The nanoencapsulation techniques were developed in appropriate polymeric matrices, allowing them to be recommended as innovative green preservatives against foodborne mold and mycotoxin-induced deterioration of stored food items. Thus, new nanomaterials with superior biocompatibility and nontoxicity must be developed further for efficient and safe mycotoxin contamination reduction. In-depth research into the mechanisms of interaction between nano-encapsulated bioactive chemicals and food components, as well as their impacts on human and animal health, should be conducted.

Acknowledgment

Part of the work was funded by the Italian Ministry of Agriculture and Forestry in the DiBio-BIOPRIME project (Prot. 76381, MiPAAF PQAI I).

References

Abbasi Pirouz, A., Selamat, J., Zafar Iqbal, S., Iskandar Putra Samsudin, N., 2020. Efficient and simultaneous chitosan-mediated removal of 11 mycotoxins from palm kernel cake. Toxins 12, 115.

Abd-Elsalam, K.A., Vasilkov, A.Y., Said-Galiev, E.E., Rubina, M.S., Khokhlov, A.R., Naumkin, A.V., Shtykova, E.V., Alghuthaymi, M.A., 2018. Bimetallic blends and chitosan nanocomposites: novel antifungal agents against cotton seedling damping-off. Eur. J. Plant Pathol. 151, 57–72.

Abdel-Wahhab, M.A., El-Nekeety, A.A., Hassan, N.S., Gibriel, A.A.Y., Abdel-Wahhab, K.G., 2018. Encapsulation of cinnamon essential oil in whey protein enhances the protective effect against single or combined sub-chronic toxicity of fumonisin B1 and/or aflatoxin B1 in rats. Environ. Sci. Pollut. Res. Int. 25, 29144–29161.

Abdel-Wahhab, M.A., Salman, A.S., Ibrahim, M.I., El-Kady, A.A., Abdel-Aziem, S.H., Hassan, N.S., Waly, A.I., 2016. Curcumin nanoparticles loaded hydrogels protects against aflatoxin B1-induced genotoxicity in rat liver. Food Chem. Toxicol. 94, 159–171.

Abomuti, M.A., Danish, E.Y., Firoz, A., Hasan, N., Malik, M.A., 2021. Green synthesis of zinc oxide nanoparticles using *Salvia officinalis* leaf extract and their photocatalytic and antifungal activities. Biology 10, 1075.

Ahmadi, A., Ahmadi, P., Sani, M.A., Ehsani, A., Ghanbarzadeh, B., 2021. Functional biocompatible nanocomposite films consisting of selenium and zinc oxide nanoparticles embedded in gelatin/cellulose nanofiber matrices. Int. J. Biol. Macromol. 175, 87–97.

Ahmed, T., Noman, M., Luo, J., Muhammad, S., Shahid, M., Ali, M.A., Zhang, M., Li, B., 2021. Bioengineered chitosan-magnesium nanocomposite: a novel agricultural antimicrobial agent against *Acidovorax oryzae* and *Rhizoctonia solani* for sustainable rice production. Int. J. Biol. Macromol. 168, 834–845.

Ajaz, S., Ahmed, T., Shahid, M., Noman, M., Shah, A.A., Mehmood, M.A., Abbas, A., Cheema, A.I., Iqbal, M.Z., Li, B., 2021. Bioinspired green synthesis of silver nanoparticles by using a native *Bacillus sp.* strain AW1-2: characterization and antifungal activity against *Colletotrichum falcatum* Went. Enzym. Microb. Technol. 144, 109745.

Aktar, M.W., Sengupta, D., Chowdhury, A., 2009. Impact of pesticides use in agriculture: their benefits and hazards. Interdiscip. Toxicol. 2, 1–12.

Al-Dhabaan, F.A., Shoala, T., Ali, A.A.M., Alaa, M., Abd-Elsalam, K., 2017. Chemically-produced copper, zinc nanoparticles and chitosan-bimetallic nanocomposites and their antifungal activity against three phytopathogenic fungi. Int. J. Agric. Technol. 13, 753–769.

Alghuthaymi, M.A., Abd-Elsalam, K.A., Shami, A., Said-Galive, E., Shtykova, E.V., Naumkin, A.V., 2020. Silver/chitosan nanocomposites: preparation and characterization and their fungicidal activity against dairy cattle toxicosis *Penicillium expansum*. J. Fungi 6, 51.

Alghuthaymi, M.A., Rajkuberan, C., Rajiv, P., Kalia, A., Bhardwaj, K., Bhardwaj, P., Abd-Elsalam, K.A., Valis, M., Kuca, K., 2021. Nanohybrid antifungals for control of plant diseases: current status and future perspectives. J. Fungi 7, 48.

Ammar, H.A., El-Desouky, T.A., 2016. Green synthesis of nanosilver particles by *Aspergillus terreus* HA1N and *Penicillium expansum* HA2N and its antifungal activity against mycotoxigenic fungi. J. Appl. Microbiol. 121, 89–100.

Amna, Mahmood, T., Khan, U.N., Amin, B., Javed, M.T., Mehmood, S., Farooq, M.A., Sultan, T., Munis, M.F.H., Chaudhary, H.J., 2021. Characterization of bio-fabricated silver nanoparticles for distinct anti-fungal activity against sugarcane phytopathogens. Microsc. Res. Tech. 84, 1522–1530.

Arciniegas-Grijalba, P.A., Patiño-Portela, M.C., Mosquera-Sánchez, L.P., Guerrero-Vargas, J.A., Rodríguez-Páez, J.E., 2017. ZnO nanoparticles (ZnO-NPs) and their antifungal activity against coffee fungus *Erythricium salmonicolor*. Appl. Nanosci. 7, 225–241.

Ashraf, H., Anjum, T., Riaz, S., Batool, T., Naseem, S., Li, G., 2022. Sustainable synthesis of microwave-assisted IONPs using *Spinacia oleracea* L. for control of fungal wilt by modulating the defense system in tomato plants. J. Nanobiotechnol. 20, 8.

Bhavyasree, P.G., Xavier, T.S., 2020. Green synthesis of copper oxide/carbon nanocomposites using the leaf extract of *Adhatoda vasica* nees, their characterization and antimicrobial activity. Heliyon 6, e03323.

Campos, E.V., de Oliveira, J.L., da Silva, C.M., Pascoli, M., Pasquoto, T., Lima, R., Abhilash, P.C., Fraceto, L.F., 2015. Polymeric and solid lipid nanoparticles for sustained release of carbendazim and tebuconazole in agricultural applications. Sci. Rep. 5, 13809.

Carbone, C., Fuochi, V., Zielińska, A., Musumeci, T., Souto, E.B., Bonaccorso, A., Puglia, C., Petronio Petronio, G., Furneri, P.M., 2020. Dual-drugs delivery in solid lipid nanoparticles for the treatment of *Candida albicans* mycosis. Colloids Surf. B: Biointerfaces 186, 110705.

Chaud, M., Souto, E.B., Zielinska, A., Severino, P., Batain, F., Oliveira-Junior, J., Alves, T., 2021. Nanopesticides in agriculture: benefits and challenge in agricultural productivity, toxicological risks to human health and environment. Toxics 9, 131.

Chen, J., Sun, L., Cheng, Y., Lu, Z., Shao, K., Li, T., Hu, C., Han, H., 2016. Graphene oxide-silver nanocomposite: novel agricultural antifungal agent against *Fusarium graminearum* for crop disease prevention. ACS Appl. Mater. Interfaces 8, 24057–24070.

Chen, J., Wu, L., Lu, M., Lu, S., Li, Z., Ding, W., 2020. Comparative study on the fungicidal activity of metallic MgO nanoparticles and macroscale MgO against soilborne fungal phytopathogens. Front. Microbiol. 11, 365.

Chowdappa, P., Shivakumar, G., Chettana, C.S., Madhura, S., 2012. Antifungal activity of chitosan-silver nanoparticle composite against *Colletotrichum gloeosporioides* associated with mango anthracnose. Afr. J. Microbiol. Res. 8, 1803–1812.

Dacrory, S., Hashem, A.H., Hasanin, M., 2021. Synthesis of cellulose-based amino acid functionalized nano-biocomplex: characterization, antifungal activity, molecular docking and hemocompatibility. Environ. Nanotechnol. Monitor. Manag. 15, 100453.

Dash, K.K., Ali, N.A., Das, D., Mohanta, D., 2019. Thorough evaluation of sweet potato starch and lemon-waste pectin based-edible films with nano-titania inclusions for food packaging applications. Int. J. Biol. Macromol. 139, 449–458.

El-Naggar, N.E., Saber, W.I.A., Zweil, A.M., Bashir, S.I., 2022. An innovative green synthesis approach of chitosan nanoparticles and their inhibitory activity against phytopathogenic *Botrytis cinerea* on strawberry leaves. Sci. Rep. 12, 3515.

El-Saadony, M.T., Saad, A.M., Najjar, A.A., Alzahrani, S.O., Alkhatib, F.M., Shafi, M.E., Selem, E., Desoky, E.M., Fouda, S.E.E., El-Tahan, A.M., Hassan, M.A.A., 2021. The use of biological selenium nanoparticles to suppress *Triticum aestivum* L. crown and root rot diseases induced by Fusarium species and improve yield under drought and heat stress. Saudi J. Biol. Sci. 28, 4461–4471.

Fernández, A., Picouet, P., Lloret, E., 2010. Cellulose-silver nanoparticle hybrid materials to control spoilage-related microflora in absorbent pads located in trays of fresh-cut melon. Int. J. Food Microbiol. 142, 222–228.

Gacem, M.A., Gacem, H., Telli, A., Ould-El-Hadj-Khelil, A., 2020a. Mycotoxins: decontamination and nanocontrol methods. In: Nanomycotoxicology: 1st Edition Treating Mycotoxins in the Nano Way. Academic Press Books—Elsevier.

Gacem, M.A., Gacem, H., Telli, A., Ould-El-Hadj-Khelil, A., 2020b. Mycotoxin-induced toxicities and diseases. In: Nanomycotoxicology: 1st Edition Treating Mycotoxins in the Nano Way. Academic Press Books—Elsevier.

Gacem, M.A., Ould-El-Hadj-Khelil, A., Boudjemaa, B., Gacem, H., 2020c. Mycotoxins occurrence, toxicity and detection methods. In: Lichtfouse, E. (Ed.), Sustainable Agriculture Reviews. vol. 40. Springer.

Garrido-Miranda, K.A., Rivas, B.L., Pérez-Rivera, M.A., Sanfuentes, E.A., Peña-Farfal, C., 2018. Antioxidant and antifungal effects of eugenol incorporated in bionanocomposites of poly(3-hydroxybutyrate)-thermoplastic starch. LWT 98, 260–267.

Gómez, J.V., Tarazona, A., Mateo, F., Jiménez, M., Mateo, E.M., 2019. Potential impact of engineered silver nanoparticles in the control of aflatoxins, ochratoxin A and the main aflatoxigenic and ochratoxigenic species affecting foods. Food Control 101, 58–68.

Gürsoy, N., Yilmaz Öztürk, B., Dağ, İ., 2021. Synthesis of intracellular and extracellular gold nanoparticles with a green machine and its antifungal activity. Turk. J. Biol. 45, 196–213.

Hai, N.T.T., Cuong, N.D., Quyen, N.T., Hien, N.Q., Hien, T.T.D., Phung, N.T.T., Toan, D.K., Huong, N.T.T., Phu, D.V., Hoa, T.T., 2021. Facile synthesis of carboxymethyl cellulose coated core/shell SiO2@Cu nanoparticles and their antifungal activity against *Phytophthora capsici*. Polymers Basel 13, 888.

Hamza, Z., El-Hashash, M., Aly, S., Hathout, A., Soto, E., Sabry, B., Ostroff, G., 2019. Preparation and characterization of yeast cell wall beta-glucan encapsulated humic acid nanoparticles as an enhanced aflatoxin B1 binder. Carbohydr. Polym. 203, 185–192.

Hamza, Z.K., Hathout, A.S., Ostroff, G., Soto, E., Sabry, B.A., El-Hashash, M.A., Hassan, N.S., Aly, S.E., 2022. Assessment of the protective effect of yeast cell wall β-glucan encapsulating humic acid nanoparticles as an aflatoxin B1 adsorbent in vivo. J. Biochem. Mol. Toxicol. 36, e22941.

Hao, Y., Cao, X., Ma, C., Zhang, Z., Zhao, N., Ali, A., Hou, T., Xiang, Z., Zhuang, J., Wu, S., Xing, B., Zhang, Z., Rui, Y., 2017. Potential applications and antifungal activities of engineered nanomaterials against gray mold disease agent *Botrytis cinerea* on rose petals. Front. Plant Sci. 8, 1332.

Hasanin, M., Al Abboud, M.A., Alawlaqi, M.M., Abdelghany, T.M., Hashem, A.H., 2021. Ecofriendly synthesis of biosynthesized copper nanoparticles with starch-based nanocomposite: antimicrobial, antioxidant, and anticancer activities. Biol. Trace Elem. Res. 200, 2099–2112.

Hawkins, N.J., Bass, C., Dixon, A., Neve, P., 2018. The evolutionary origins of pesticide resistance. Biol. Rev. Camb. Philos. Soc. 94, 135–155.

Hoseinzadeh, A., Habibi-Yangjeh, A., Davari, M., 2016. Antifungal activity of magnetically separable Fe3O4/ZnO/AgBr nanocomposites prepared by a facile microwave-assisted method. Prog. Nat. Sci. Mat. Int. 26, 334–340.

Hu, D., Yu, S., Yu, D., Liu, N., Tang, Y., Fan, Y., Wang, C., Wu, A., 2019. Biogenic *Trichoderma harzianum*-derived selenium nanoparticles with control functionalities originating from diverse recognition metabolites against phytopathogens and mycotoxins. Food Control 106, 106748.

Hua, X., Chi, W., Su, L., Li, J., Zhang, Z., Yuan, X., 2017. ROS-induced oxidative injury involved in pathogenesis of fungal keratitis via p38 MAPK activation. Sci. Rep. 7, 10421.

Huang, W., Bao, Y., Duan, H., Bi, Y., Yu, H., 2017. Antifungal effect of green synthesised silver nanoparticles against *Setosphaeria turcica*. IET Nanobiotechnol. 11, 803–808.

Iliger, K.S., Sofi, T.A., Bhat, N.A., Ahanger, F.A., Sekhar, J.C., Elhendi, A.Z., Al-Huqail, A.A., Khan, F., 2020. Copper nanoparticles: green synthesis and managing fruit rot disease of chilli caused by *Colletotrichum capsici*. Saudi J. Biol. Sci. 28, 1477–1486.

Ing, L.Y., Zin, N.M., Sarwar, A., Katas, H., 2012. Antifungal activity of chitosan nanoparticles and correlation with their physical properties. Int. J. Biomater. 2012, 632698.

Irshad, M.A., Nawaz, R., Zia Ur Rehman, M., Imran, M., Ahmad, J., Ahmad, S., Inam, A., Razzaq, A., Rizwan, M., Ali, S., 2020. Synthesis and characterization of titanium dioxide nanoparticles by chemical and green methods and their antifungal activities against wheat rust. Chemosphere 258, 127352.

Jain, K., Patel, A.S., Pardhi, V.P., Flora, S.J.S., 2021. Nanotechnology in wastewater management: a new paradigm towards wastewater treatment. Molecules 26, 1797.

Jebril, S., Ben Jenana, R.K., Dridi, C., 2020. Green synthesis of silver nanoparticles using *Melia azedarach* leaf extract and their antifungal activities: in vitro and in vivo. Mater. Chem. Phys. 248, 122898.

Kalagatur, N.K., Nirmal Ghosh, O.S., Sundararaj, N., Mudili, V., 2018. Antifungal activity of chitosan nanoparticles encapsulated with *Cymbopogon martinii* essential oil on plant pathogenic fungi *Fusarium graminearum*. Front. Pharmacol. 9, 610.

Kalia, A., Abd-Elsalam, K.A., Kuca, K., 2020. Zinc-based nanomaterials for diagnosis and management of plant diseases: ecological safety and future prospects. J. Fungi 6, 222.

Kang, H., Elmer, W., Shen, Y., Zuverza-Mena, N., Ma, C., Botella, P., White, J.C., Haynes, C.L., 2021. Silica nanoparticle dissolution rate controls the suppression of fusarium wilt of watermelon (*Citrullus lanatus*). Environ. Sci. Technol. 55, 13513–13522.

Karimi, N., Alizadeh, A., Almasi, H., Hanifian, S., 2020. Preparation and characterization of whey protein isolate/polydextrose-based nanocomposite film incorporated with cellulose nanofiber and *L. plantarum*: a new probiotic active packaging system. LWT 121, 108978.

Kaur, P., Thakur, R., Choudhary, A., 2012. An in vitro study of the antifungal activity of silver/chitosan nanoformulations against important seed borne pathogens. Int. J. Sci. Technol. Res. 1, 83–86.

Khalil, N.M., Abd El-Ghany, M.N., Rodríguez-Couto, S., 2019. Antifungal and anti-mycotoxin efficacy of biogenic silver nanoparticles produced by *Fusarium chlamydosporum* and *Penicillium chrysogenum* at non-cytotoxic doses. Chemosphere 218, 477–486.

Kheiri, A., Moosawi Jorf, S.A., Malihipour, A., Saremi, H., Nikkhah, M., 2016. Application of chitosan and chitosan nanoparticles for the control of Fusarium head blight of wheat (*Fusarium graminearum*) in vitro and greenhouse. Int. J. Biol. Macromol. 93, 1261–1272.

Ku, T., Zhou, M., Hou, Y., Xie, Y., Li, G., Sang, N., 2021. Tebuconazole induces liver injury coupled with ROS-mediated hepatic metabolism disorder. Ecotoxicol. Environ. Saf. 220, 112309.

Kumar, R., Ghoshal, G., Goyal, M., 2020. Development and characterization of corn starch based nanocomposite film with AgNPs and plant extract. Mater. Sci. Energy Technol. 3, 672–678.

Kumar, R., Najda, A., Duhan, J.S., Kumar, B., Chawla, P., Klepacka, J., Malawski, S., Kumar Sadh, P., Poonia, A.K., 2021. Assessment of antifungal efficacy and release behavior of fungicide-loaded chitosan-carrageenan nanoparticles against phytopathogenic fungi. Polymers 14, 41.

Kutawa, A.B., Ahmad, K., Ali, A., Hussein, M.Z., Abdul Wahab, M.A., Adamu, A., Ismaila, A.A., Gunasena, M.T., Rahman, M.Z., Hossain, M.I., 2021. Trends in nanotechnology and its potentialities to control plant pathogenic fungi: a review. Biology Basel 10, 881.

Lakshmeesha, T.R., Kalagatur, N.K., Mudili, V., Mohan, C.D., Rangappa, S., Prasad, B.D., Ashwini, B.S., Hashem, A., Alqarawi, A.A., Malik, J.A., Abd Allah, E.F., Gupta, V.K., Siddaiah, C.N., Niranjana, S.R., 2019. Biofabrication of zinc oxide nanoparticles with *Syzygium aromaticum* flower buds extract and finding its novel application in controlling the growth and mycotoxins of *Fusarium graminearum*. Front. Microbiol. 10, 1244.

Li, J., Sun, Q., Sun, Y., Chen, B., Wu, X., Le, T., 2019. Improvement of banana postharvest quality using a novel soybean protein isolate/cinnamaldehyde/zinc oxide bionanocomposite coating strategy. Sci. Hortic. 258, 108786.

Li, Z., Yu, T., Paul, R., Fan, J., Yang, Y., Wei, Q., 2020. Agricultural nanodiagnostics for plant diseases: recent advances and challenges. Nanoscale Adv. 2, 3083–3094.

Lima Oliveira, P.D., de Oliveira, K.Á.R., Vieira, W.A.D.S., Câmara, M.P.S., de Souza, E.L., 2018. Control of anthracnose caused by *Colletotrichum* species in guava, mango and papaya using synergistic combinations of chitosan and *Cymbopogon citratus* (D.C. ex Nees) Stapf. essential oil. Int. J. Food Microbiol. 266, 87–94.

Llorens, A., Lloret, E., Picouet, P., Fernandez, A., 2012. Study of the antifungal potential of novel cellulose/copper composites as absorbent materials for fruit juices. Int. J. Food Microbiol. 158, 113–119.

Mali, S.C., Dhaka, A., Githala, C.K., Trivedi, R., 2020. Green synthesis of copper nanoparticles using *Celastrus paniculatus* Willd. leaf extract and their photocatalytic and antifungal properties. Biotechnol. Rep. 27, e00518.

Mathew, T.V., Kuriakose, S., 2013. Photochemical and antimicrobial properties of silver nanoparticle-encapsulated chitosan functionalized with photoactive groups. Mater. Sci. Eng. C Mater. Biol. Appl. 33, 4409–4415.

McDaniel, A., Tonyali, B., Yucel, U., Trinetta, V., 2019. Formulation and development of lipid nanoparticle antifungal packaging films to control postharvest disease. J. Agric. Food Res. 1, 100013.

Mekawey, A.I.I., 2018. Effects of chitosan nanoparticles as antimicrobial activity and on mycotoxin production. Acad. J. Agric. Res. 6, 101–106.

Metwally, N.H., Saad, G.R., Abd El-Wahab, E.A., 2019. Grafting of multiwalled carbon nanotubes with pyrazole derivatives: characterization, antimicrobial activity and molecular docking study. Int. J. Nanomedicine 14, 6645–6659.

Moradi Pour, M., Saberi Riseh, R., Skorik, Y.A., 2022. Sodium alginate-gelatin nanoformulations for encapsulation of bacillus velezensis and their use for biological control of *Pistachio gummosis*. Materials. 15, 2114.

Nasseri, M., Golmohammadzadeh, S., Arouiee, H., Jaafari, M.R., Neamati, H., 2016. Antifungal activity of *Zataria multiflora* essential oil-loaded solid lipid nanoparticles in-vitro condition. Iran. J. Basic Med. Sci. 19, 1231–1237.

Neelgund, G.M., Oki, A., Luo, Z., 2012. Antimicrobial activity of CdS and Ag2S quantum dots immobilized on poly(amidoamine) grafted carbon nanotubes. Colloids Surf. B: Biointerfaces 100, 215–221.

Ngo, T.M.P., Nguyen, T.H., Dang, T.M.Q., Do, T.V.T., Reungsang, A., Chaiwong, N., Rachtanapun, P., 2021. Effect of pectin/nanochitosan-based coatings and storage temperature on shelf-life extension of "elephant" mango (*Mangifera indica* L.) fruit. Polymers 13, 3430.

Ngo, T.M.P., Nguyen, T.H., Dang, T.M.Q., Tran, T.X., Rachtanapun, P., 2020. Characteristics and antimicrobial properties of active edible films based on pectin and nanochitosan. Int. J. Mol. Sci. 21, 2224.

Nguyen, T.N., Huynh, T.N., Hoang, D., Nguyen, D.H., Nguyen, Q.H., Tran, T.H., 2019. Functional nanostructured oligochitosan–silica/carboxymethyl cellulose hybrid materials: synthesis and investigation of their antifungal abilities. Polymers 11, 628.

Nguyen, N.T., Nguyen, D.H., Pham, D.D., Nguyen, Q.H., Hoang, D.Q., 2017. New oligochitosan-nanosilica hybrid materials: preparation and application on chili plants for resistance to anthracnose disease and growth enhancement. Polym. J. 49, 861–869.

Nicolopoulou-Stamati, P., Maipas, S., Kotampasi, C., Stamatis, P., Hens, L., 2016. Chemical pesticides and human health: the urgent need for a new concept in agriculture. Front. Public Health 4, 148.

Othman, S.H., Othman, N.F.L., Shapi'I, R.A., Ariffin, S.H., Yunos, K.F.M., 2021. Corn starch/chitosan nanoparticles/thymol bio-nanocomposite films for potential food packaging applications. Polymers 13, 390.

Pal, K., Sarkar, P., Anis, A., Wiszumirska, K., Jarzębski, M., 2021. Polysaccharide-based nanocomposites for food packaging applications. Materials. 14, 5549.

Pelaz, B., Alexiou, C., Alvarez-Puebla, R.A., Alves, F., Andrews, A.M., Ashraf, S., Balogh, L.P., Ballerini, L., Bestetti, A., Brendel, C., Bosi, S., Carril, M., Chan, W.C., Chen, C., Chen, X., Chen, X., Cheng, Z., Cui, D., Du, J., Dullin, C., Escudero, A., Feliu, N., Gao, M., George, M., Gogotsi, Y., Grünweller, A., Gu, Z., Halas, N.J., Hampp, N., Hartmann, R.K., Hersam, M.C., Hunziker, P., Jian, J., Jiang, X., Jungebluth, P., Kadhiresan, P., Kataoka, K., Khademhosseini, A., Kopeček, J., Kotov, N.A., Krug, H.F., Lee, D.S., Lehr, C.M., Leong, K.W., Liang, X.J., Ling, Lim, M., Liz-Marzán, L.M., Ma, X., Macchiarini, P., Meng, H., Möhwald, H., Mulvaney, P., Nel, A.E., Nie, S., Nordlander, P., Okano, T., Oliveira, J., Park, T.H., Penner, R.M., Prato, M., Puntes, V., Rotello, V.-M., Samarakoon, A., Schaak, R.E., Shen, Y., Sjöqvist, S., Skirtach, A.G., Soliman, M.G., Stevens, M.M., Sung, H.W., Tang, B.Z., Tietze, R., Udugama, B.N., VanEpps, J.S., Weil, T., Weiss, P.S., Willner, I., Wu, Y., Yang, L., Yue, Z., Zhang, Q., Zhang, Q., Zhang, X.E., Zhao, Y., Zhou, X., Parak, W.J., 2017. Diverse applications of nanomedicine. ACS Nano 11, 2313–2381.

Peng, Z., Feng, W., Cai, G., Wu, D., Lu, J., 2021. Enhancement effect of chitosan coating on inhibition of deoxynivalenol accumulation by *Litsea cubeba* essential oil emulsion during malting. Foods 10, 3051.

Prabha, S., Durgalakshmi, D., Rajendran, S., Lichtfouse, E., 2020. Plant-derived silica nanoparticles and composites for biosensors, bioimaging, drug delivery and supercapacitors: a review. Environ. Chem. Lett. 2020, 1–25.

Pretty, J., 2008. Agricultural sustainability: concepts, principles and evidence. Philos. Trans. R. Soc. Lond. Ser. B Biol. Sci. 363 (1491), 447–465.

Raafat, M., El-Sayed, A.S.A., El-Sayed, M.T., 2021. Biosynthesis and anti-mycotoxigenic activity of *Zingiber officinale* roscoe-derived metal nanoparticles. Molecules 26, 2290.

Raliya, R., Saharan, V., Dimkpa, C., Biswas, P., 2018. Nanofertilizer for precision and sustainable agriculture: current state and future perspectives. J. Agric. Food Chem. 66, 6487–6503.

Riley, M.K., Vermerris, W., 2017. Recent advances in nanomaterials for gene delivery—a review. Nanomaterials 7, 94.

Rodríguez-González, V., Domínguez-Espíndola, R.B., Casas-Flores, S., Patrón-Soberano, O.A., Camposeco-Solis, R., Lee, S.W., 2016. Antifungal nanocomposites inspired by titanate nanotubes for complete inactivation of *Botrytis cinerea* isolated from tomato infection. ACS Appl. Mater. Interfaces 8, 31625–31637.

Roopan, S.M., Devi Priya, D., Shanavas, S., Acevedo, R., Al-Dhabi, N.A., Arasu, M.V., 2019. CuO/C nanocomposite: synthesis and optimization using sucrose as carbon source and its antifungal activity. Mater. Sci. Eng. C Mater. Biol. Appl. 101, 404–414.

Safavinia, L., Akhgar, M.R., Tahamipour, B., Ahmadi, S.A., 2021. Green synthesis of highly dispersed zinc oxide nanoparticles supported on silica gel matrix by *Daphne oleoides* extract and their antibacterial activity. Iran. J. Biotechnol. 19, e2598.

Saharan, V., Sharma, G., Yadav, M., Choudhary, M.K., Sharma, S.S., Pal, A., Raliya, R., Biswas, P., 2015. Synthesis and in vitro antifungal efficacy of Cu–chitosan nanoparticles against pathogenic fungi of tomato. Int. J. Biol. Macromol. 75, 346–353.

Sandhya, Kumar, S., Kumar, D., Dilbaghi, N., 2017. Preparation, characterization, and bio-efficacy evaluation of controlled release carbendazim-loaded polymeric nanoparticles. Environ. Sci. Pollut. Res. Int. 24, 926–937.

Satti, S.H., Raja, N.I., Javed, B., Akram, A., Mashwani, Z.U., Ahmad, M.S., Ikram, M., 2021. Titanium dioxide nanoparticles elicited agro-morphological and physicochemical modifications in wheat plants to control *Bipolaris sorokiniana*. PLoS One 16, e0246880.

Shapi'I, R.A., Othman, S.H., Nordin, N., Kadir Basha, R., Nazli Naim, M., 2020. Antimicrobial properties of starch films incorporated with chitosan nanoparticles: in vitro and in vivo evaluation. Carbohydr. Polym. 230, 115602.

Sidhu, A., Bala, A., Singh, H., Ahuja, R., Kumar, A., 2020. Development of MgO-sepoilite nanocomposites against phytopathogenic fungi of rice (*Oryzae sativa*): a green approach. ACS. Omega 5, 13557–13565.

Soares, S., Sousa, J., Pais, A., Vitorino, C., 2018. Nanomedicine: principles, properties, and regulatory issues. Front. Chem. 6, 360.

Spadola, G., Sanna, V., Bartoli, J., Carcelli, M., Pelosi, G., Bisceglie, F., Restivo, F.M., Degola, F., Rogolino, D., 2020. Thiosemicarbazone nano-formulation for the control of *Aspergillus flavus*. Environ. Sci. Pollut. Res. Int. 27, 20125–20135.

Su, D.-l., Li, P.-j., Ning, M., Li, G.-y., Shan, Y., 2019. Microwave assisted green synthesis of pectin based silver nanoparticles and their antibacterial and antifungal activities. Mater. Lett. 244, 35–38.

Sukhwal, A., Jain, D., Joshi, A., Rawal, P., Kushwaha, H.S., 2017. Biosynthesised silver nanoparticles using aqueous leaf extract of *Tagetes patula* L. and evaluation of their antifungal activity against phytopathogenic fungi. IET Nanobiotechnol. 11, 531–537.

Szőllősi, R., Molnár, Á., Kondak, S., Kolbert, Z., 2020. Dual effect of nanomaterials on germination and seedling growth: stimulation vs. phytotoxicity. Plan. Theory 9, 1745.

Tarazona, A., Gómez, J.V., Mateo, E.M., Jiménez, M., Mateo, F., 2019. Antifungal effect of engineered silver nanoparticles on phytopathogenic and toxigenic *Fusarium spp.* and their impact on mycotoxin accumulation. Int. J. Food Microbiol. 306, 108259.

Tescione, F., Tammaro, O., Bifulco, A., Del Monaco, G., Esposito, S., Pansini, M., Silvestri, B., Costantini, A., 2022. Silica meets tannic acid: designing green nanoplatforms for environment preservation. Molecules 27, 1944.

Vundela, S.R., Kalagatur, N.K., Nagaraj, A., Kadirvelu, K., Chandranayaka, S., Kondapalli, K., Hashem, A., Abd Allah, E.F., Poda, S., 2022. Multi-biofunctional properties of phytofabricated selenium nanoparticles from *Carica papaya* fruit extract: antioxidant, antimicrobial, antimycotoxin, anticancer, and biocompatibility. Front. Microbiol. 12, 769891.

Wang, X., Cai, A., Wen, X., Jing, D., Qi, H., Yuan, H., 2017a. Graphene oxide-Fe3O4 nanocomposites as high-performance antifungal agents against Plasmopara viticola. Sci. China Mater. 60, 258–268.

Wang, X., Zhou, Z., Chen, F., 2017b. Surface modification of carbon nanotubes with an enhanced antifungal activity for the control of plant fungal pathogen. Materials. 10, 1375.

Webb, P., Block, S., 2012. Support for agriculture during economic transformation: impacts on poverty and undernutrition. Proc. Natl. Acad. Sci. U. S. A. 109, 12309–12314.

Youssef, K., de Oliveira, A.G., Tischer, C.A., Hussain, I., Roberto, S.R., 2019. Synergistic effect of a novel chitosan/silica nanocomposites-based formulation against gray mold of table grapes and its possible mode of action. Int. J. Biol. Macromol. 141, 247–258.

Youssef, K., Hashim, A.F., 2020. Inhibitory effect of clay/chitosan nanocomposite against *Penicillium digitatum* on citrus and its possible mode of action. Jordan J. Biol. Sci. 13, 349–355.

Zaban, M.I., Azim, N.S.A., Abd El-Aziz, A.R.M., 2019. Antifungal and anti-aflatoxin efficacy of biogenic silver nanoparticles produced by *Aspergillus* species: molecular study. Pak. J. Pharm. Sci. 32, 2509–2526.

Zabihollahi, N., Alizadeh, A., Almasi, H., Hanifian, S., Hamishekar, H., 2020. Development and characterization of carboxymethyl cellulose based probiotic nanocomposite film containing cellulose nanofiber and inulin for chicken fillet shelf life extension. Int. J. Biol. Macromol. 160, 409–417.

Zaki, S.A., Ouf, S.A., Abd-Elsalam, K.A., Asran, A.A., Hassan, M.M., Kalia, A., Albarakaty, F.M., 2022. Trichogenic silver-based nanoparticles for suppression of fungi involved in damping-off of cotton seedlings. Microorganisms 10, 344.

Zaki, S.A., Ouf, S.A., Albarakaty, F.M., Habeb, M.M., Aly, A.A., Abd-Elsalam, K.A., 2021. *Trichoderma harzianum*-mediated ZnO nanoparticles: a green tool for controlling soil-borne pathogens in cotton. J. Fungi 7, 952.

CHAPTER

15

Nanobiopesticides: Significance, preparation technologies, safety aspects, and commercialization for sustainable agriculture

P. Karthik[a], *A. Saravanaraj*[b], *V. Vijayalakshmi*[c], *K.V. Ragavan*[d,e], *and Vinoth Kumar Vaidyanathan*[f]

[a]Centre for Food Nanotechnology, Department of Food Technology, Faculty of Engineering, Karpagam Academy of Higher Education (Deemed to be University), Coimbatore, India [b]Department of Chemical Engineering, Vel Tech High Tech Dr. Rangarajan Dr. Sakunthala Engineering College, Chennai, India [c]Department of Biotechnology, Vel Tech High Tech Dr. Rangarajan Dr. Sakunthala Engineering College, Chennai, India [d]Agro-Processing and Technology Division, CSIR-National Institute for Interdisciplinary Science and Technology, Thiruvananthapuram, India [e]Academy of Scientific and Innovative Research (AcSIR), Ghaziabad, India [f]Integrated Bioprocessing Laboratory, Department of Biotechnology, School of Bioengineering, Faculty of Engineering and Technology, SRM Institute of Science and Technology (SRM IST), Kattankulathur, Chengalpattu, Tamil Nadu, India

1 Introduction

Nanotechnology involves the study of matter at the nanoscale (1–100 nm), where physical, chemical, and biological properties are enhanced (Ezhilarasi et al., 2013; Sharma et al., 2012). Nanotechnology has found its application in a plethora of sectors such as agriculture, healthcare, pharmaceuticals, textile, and energy with significant outcomes. In agriculture, nanotechnology was introduced in various fields such as precision farming (nanosensors to monitor and detect environmental conditions, soil nutrients, etc.), crop protection (nanopesticides, nanoherbicides, etc.), and crop improvement (nanofertilizers, nanoparticles

https://doi.org/10.1016/B978-0-443-23950-2.00009-6

as gene delivery systems). Moreover, recent research on nanotechnology has emphasized the ecofriendly application of nanomaterials for slow and sustained nutrient delivery from nanofertilizers and active ingredients from nanopesticides and nanofilms to increase the shelf life of perishables (Jampílek and Králová, 2019; Acharya and Pal, 2020; Bapat et al., 2022).

Since its inception in the 1960s, "Green Revolution" technology has resulted in remarkable developments in agricultural productivity with the use of high-yielding varieties, chemical fertilizers, and pesticides (Pingali, 2012). Therefore, around 60% of the world population is dependent on agricultural activities for their livelihood (FAO, 2013). Currently, diverse factors threaten the agricultural sector on the local and global scales such as climate change, low productivity due to declining organic matter, micro- and macronutrients in the soil, decreased water availability and cultivable lands, environment safety and health issues related to pesticide residues, and plant disease management strategies (Acharya and Pal, 2020; Bapat et al., 2022). These constraints are further aggravated by the food demand imposed by the steadily growing global population (an estimated 9 billion by 2050) at the rate of 4% (Chen and Yada, 2011). Thus, the agriculture sector must be equipped with sustainable and innovative techniques to produce more food and livelihood opportunities from less land and water per capita.

In general, the global population is drastically increasing, which results in a scarcity of food. Hence, there is a need for new innovative technologies to improve soil quality and enhance crop production. Nanotechnology and its recent advancements have shown promising potential for sustainable agriculture. Nanotechnology provides various applications in the agricultural sector such as promoting crop growth, improving crop yield, ensuring crop safety, managing stress and drought conditions, enhancing soil fertility, monitoring and controlling pollution, etc., to promote sustainable agriculture. The application of nanotechnology for the advancement in agriculture is shown in Fig. 1. Also, nanotechnology has a great impact on the reduction and reuse of agricultural waste (Koul, 2019).

The worldwide use of pesticides in crops has increased significantly during the past couple of decades. Chemical pesticides in various forms are available in local and online markets that are accessible to all farmers. Excessive use of chemical pesticides helps to increase productivity but may cause major health-related side effects such as cytotoxicity, endocrine disruption, breast cancer, and reproductive toxicity (Nicolopoulou-Stamati et al., 2016). During the last few decades, biopesticides have gained attention as an alternative to synthetic pesticides (Koul, 2019; Sahu et al., 2022). To minimize the risk factors, nano-based biopesticides are developed as nanobiopesticides (NBPs) with enhanced properties. The efficiency of fertilizers and pesticides has been improved using nano-based carriers, which reduce the amount to be supplied without affecting productivity (Fraceto et al., 2016). NBPs are naturally occurring bioactive substances or organisms used to protect crops from pests. Comparatively, NBPs are safer than synthetic pesticides. This chapter mainly focuses on NBP uses for sustainable agriculture, the fundamentals and development of various types of NBPs, technologies for the preparation of NBPs, mode of action, benefits, bioavailability, safety aspects, regulatory measures, future perspectives, and commercialization.

2 NBPs for sustainable agriculture

Sustainable agriculture employs the thoughtful and constant use of significant resources for today's needs without affecting the needs of future generations by adopting the concept

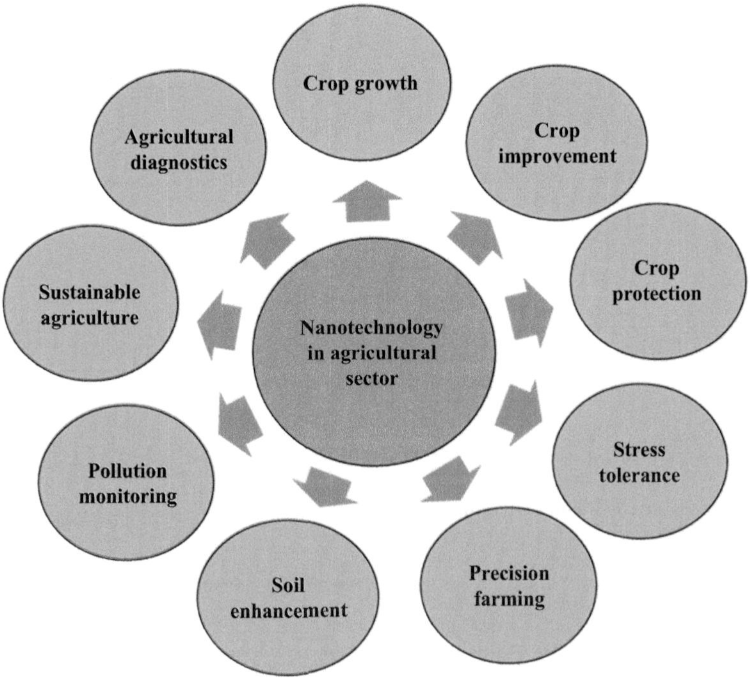

FIG. 1 Applications of nanotechnology in the agricultural sector.

of utilization of natural resources (Burton, 1987). Sustainable agriculture focuses on the triad of achieving economic viability, environmental objectives, and social acceptability (Fenibo et al., 2022), as depicted in Fig. 2. NBP application creates a balance among these three-dimensional concepts as the keystone to sustainable agriculture. Globally, annual food loss caused by pests is 40% and about 67,000 species are identified as agricultural crop pests, including insects, rodents, weeds, pathogenic microbes (fungi, bacteria, and viruses), and nematodes (Ross and Lembi, 1985; IPPC Secretariat, 2021). Chemical pesticides are efficient agents for controlling pests; however, excessive use has led to some undesirable effects on environmental safety and human health. Biopesticides, which are derived from sources such as plants, animals, and microorganisms, are considered an ecofriendly and safer alternative. Additionally, the use of nanotechnology in biopesticide formulation in the form of nanoemulsions, nanoencapsulation, nanogels, nanovesicles, nanofibers, etc., has enhanced the scope of biopesticides. Nano-based systems are generally more efficient than their bulk counterparts due to increased stability, bioavailability, controlled release, and overall activity of the active ingredient (Jampílek and Kráľová, 2019). Thus, the reliance on chemical pesticides can be further reduced by the use of NBPs.

2.1 Biopesticides as alternatives to chemical pesticides

Chemical pesticides have had a major role in pest management that has significantly contributed to increasing crop yield in many developed and developing countries during the last

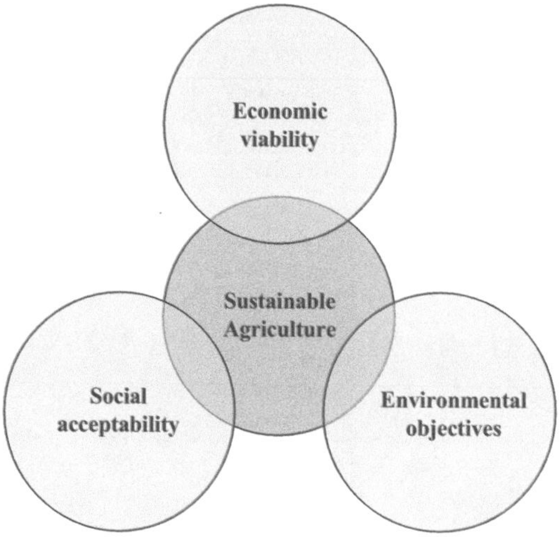

FIG. 2 Triad to connect sustainable agriculture.

few decades (Pingali, 2012). Chemical pesticides are produced by synthetic methods and act as a repellent to kill or stop the growth of pests (EPA, 2018). According to the Environmental Protection Agency (EPA), pesticides should not harm human beings, plants, animals, or the environment. However, their unsystematic application in agricultural lands has resulted in environmental pollution reducing soil microflora, diminishing nitrogen fixation, and affecting normal flora and fauna (Ombódi and Saigusa, 2000; Trenkel, 1997; Tilman et al., 2002). Pesticides can affect any living tissue and traverse from organism to organism through the food chain on land, air, and water (Aktar et al., 2009). Pesticide manufacturers, dealers, and farmers are among the highly exposed groups. In humans, hormone disruption, cancer, immune suppression, reduced intelligence, and reproductive defects are caused by prolonged and low-dose exposure to pesticides (Street et al., 2018). Some pesticides belonging to carbamates, organophosphates, organochlorines, and pyrethroids are linked to neurological disorders (Alzheimer's and Parkinson's) (Sánchez-Santed et al., 2016). Furthermore, the continued use of pesticides has led to the development of pesticide resistance in pests and weeds. More than 500 insecticide-resistant arthropod pest species and more than 260 herbicide-resistant weed species have emerged worldwide (Hajek and Eilelnberg, 2004; Heap, 2022). This scenario makes it necessary for other alternatives to chemical pesticides, including biopesticides as a replacement.

Biopesticides are pesticides that are derived from plants, animals, microorganisms (bacteria, fungi, viruses, protozoa, etc.,) and certain minerals. Because biopesticides are made of naturally occurring compounds, they are generally safer than chemical pesticides. A biopesticide is a kind of formulation that controls pests in an ecofriendly manner by adopting nontoxic mechanisms. Hence, biopesticides are attractive worldwide because of their potential benefits. They may be derived from plants (chrysanthemum, azadirachta), animals, (e.g., nematodes), and microorganisms (e.g., *Bacillus thuringiensis, Trichoderma*). Also, their products (phytochemicals, microbial products) and byproducts (semiochemicals) can be used to manage pests (Mazid et al., 2011). Various common biopesticides used in the agricultural field are reviewed and shown in Table 1. Biopesticides are strongly recommended by the National Farmer Policy 2007 to increase the production of crops while also sustaining the

TABLE 1 Various common biopesticides as well as their sources, types, and targets based on various research reports.

Source	Type	Organism	Targets	References
Bacteria	Insecticide	*Bacillus thuringiensis var kurstaki*	Caterpillars and fungi (*Botrytis*)	Koul (2011), Bravo et al. (2007)
		B. thuringiensis var tenebrionis	Elm leaf beetle and alfalfa weevil	Saberi et al. (2020)
		B. thuringiensis subsp. *israelensis*	Lepidopteran pests	Kabaluk et al. (2010)
	Fungicide	*Bacillus subtilis*	*Botrytis* spp.	Koul (2011), Bravo et al. (2007)
		Pseudomonas fluorescens	Plant soil-borne diseases	Kabaluk et al. (2010)
Fungi	Insecticide	*Beauveria bassiana*	Whitefly	McGuire and Northfield (2020)
		Metarhizium anisopliae	Coleoptera, Lepidoptera, termites, mosquitoes, leafhoppers, beetles, and grubs	Kabaluk et al. (2010)
		Paecilomyces fumosoroseus	Whitefly	Kabaluk et al. (2010)
		Paecilomyces lilacinus	Whitefly	Kabaluk et al. (2010)
		Verticillium lecanii	Whitefly, coffee green bug, and homopteran pests	Kabaluk et al. (2010)
	Fungicide	*Ampelomyces quisqualis*	Powdery mildew	Kabaluk et al. (2010)
		Coniothyrium minitans	*Sclerotinia* spp. *S. sclerotiorum*	Gams et al. (2004)
		Trichoderma harzianum	*Sclerotinia* spp. *S sclerotiorum*	Dolatabadi et al. (2011)
		Trichoderma viride	Soil-borne pathogens	Kabaluk et al. (2010)
	Herbicide	*Chondrostereum purpureum*	Cut stumps of hardwood trees and shrubs	Bailey (2014)
	Nematicide	*Paecilomyces lilacinus*	Plant-parasitic nematodes in soil	Moreno-Gavíra et al. (2020)
		Verticillium chlamydosporium	Nematodes	Kabaluk et al. (2010)

Continued

TABLE 1 Various common biopesticides as well as their sources, types, and targets based on various research reports—Cont'd

Source	Type	Organism	Targets	References
Virus	Insecticide	*Cydia pomonella granulovirus*	Codling moth	Kabaluk et al. (2010)
		Helicoverpa armigera nucleopolyhedrosis virus	*Helicoverpa armigera*	Kabaluk et al. (2010)
		Spodoptera litura nucleopolyhedrosis virus	*S. litura*	Kabaluk et al. (2010)
Plant extracts	Fungicide	*Reynoutria sachalinensis (giant knotweed) extract*	Powdery mildew, downy mildew, botrytis, late blight, and citrus canker	Marrone (2002)
	Nematicide	*Quillaja saponaria*	Plant-parasitic nematodes	Isman (2020)

health of farmers and the environment. This policy also stated that biopesticides would be considered equal to chemical pesticides in terms of promotion and support. Biopesticides, in contrast to chemical pesticides, are less toxic to humans, leave no harmful residues (biodegradable) so they do not pollute the environment, and are more specific to target pests and closely related organisms, thereby reducing the impact on nontarget organisms (Kaya and Lacey, 2007). Biopesticides do not lead to bioaccumulation; they are also effective at a very low concentration and hence can reduce the use of chemical pesticides. Widespread concerns over risks associated with chemical pesticides and recent efforts by researchers and policymakers have led to many bioactive compounds being registered as biopesticides in the agro market (Kookana et al., 2014; Damalas and Koutroubas, 2018; Kumar et al., 2021). Advances in delivery and application options are other factors contributing to their increased use (Glare et al., 2012).

The current research focuses on the improvement of biopesticides in their mode of action, including new mechanisms that will replace the use of chemical pesticides in pest management (Nawaz et al., 2016). The main challenges of new biopesticides in their growth and use are how to market or commercialize them and how to improve the mode of action and stability (Tripathi et al., 2020; Damalas and Koutroubas, 2018). All these developments will lead to understanding the benefit of biopesticides as a suitable green alternative to chemical pesticides.

2.2 Importance of nanotechnology in biopesticide formulations

Materials at the nanoscale show unique properties due to their increased surface area-to-volume ratio. Most biomolecules function at the nanoscale, so nanomaterials can play a significant role in biological processes as well (Bapat et al., 2022). When biopolymers are used in NBP formulation, characteristics such as increased spreadability, quicker degradability in soil, slower degradability in host plants, and permissible residue levels under regulatory criteria in foodstuffs can be achieved (Bergeson, 2010). One of the drawbacks of biopesticides is rapid degradation in the field due to chemical conversions such as oxidation, polymerization, and isomerization (Chandler et al., 2011; Turek and Stintzing, 2013). Nanotechnology is an emergent research area that can be used for crop protection by adopting NBP. It will be used to design novel bioactive molecules at the nanoscale with different structures. NBPs are more attractive than chemical pesticides because of their properties such as the high surface area-to-volume ratio, small size, stability, better solubility, enhanced efficiency, mobility, and decreased toxicity.

The importance of nanotechnology for biopesticide formulation is shown in Fig. 3. NBPs possess certain key characteristics that overcome the limitations caused by synthetic or chemical pesticides by promoting stability and improving toxicity toward pests but avoiding secondary effects (De Oliveira et al., 2014). Chemical pesticides directly applied to plants possess toxins that are released into the air and mixed in the food chain, which causes environmental issues. As natural source-based alternatives, biopesticides with different nanoformulations

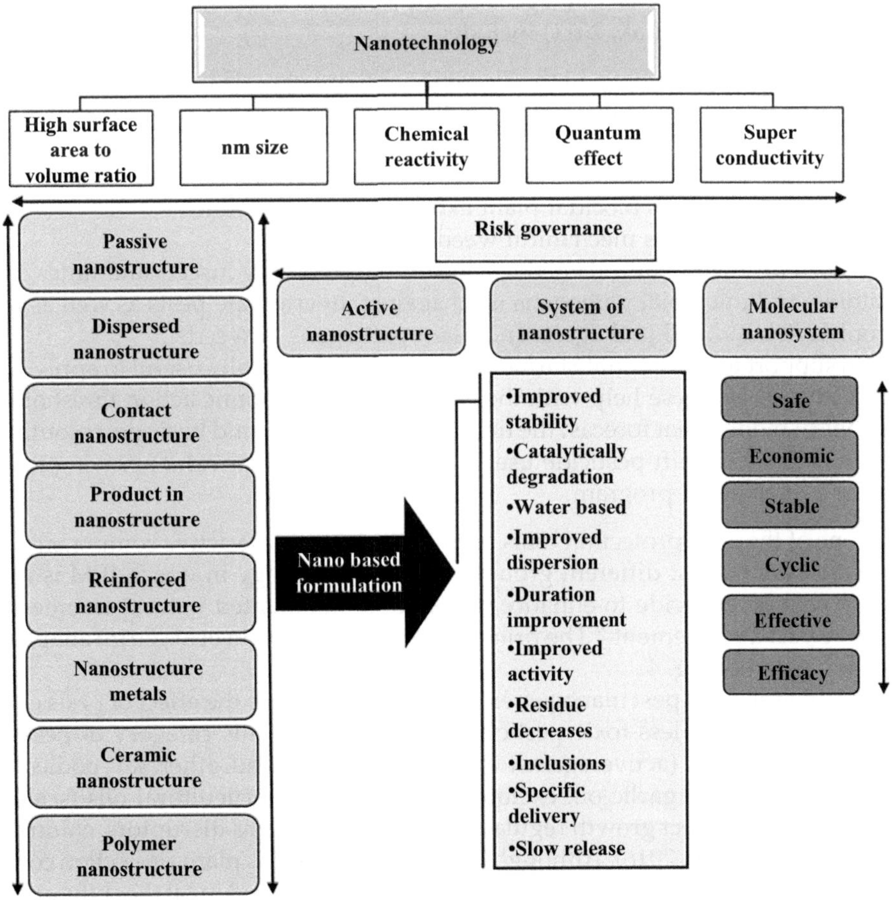

FIG. 3 Importance of nanotechnology in biopesticide formulations (Lade et al., 2019).

such as nanoparticles, nanocomposites, micelles, etc., were designed to reduce the chances of both health and environmental problems. The majority of registered chemical pesticides are neurotoxins that directly affect the insect nervous system; this kind of pesticide may cause risks to mammals and nontarget insects. Therefore, NBPs should be synthesized with a different mode of action to improve pest management and benefit sustainable agriculture (Palermo et al., 2021).

2.3 Role of NBPs in integrated pest management

New alternative techniques are required to improve crop protection with greater sustainability. Integrated pest management (IPM) is one of the best ways to promote sustainable crop protection (Bajwa and Kogan, 2002). IPM is a systematic approach in which different crop protection practices are combined to improve the monitoring and controlling of pests (Flint, 2012). The main idea behind IPM implementation is that different practices combined can overcome the shortcomings of individual practices. Though we aim is to eradicate or manage pests, preventing the economic damage caused by pests is the prime importance. The main IPM strategies (Chandler et al., 2011) include:

- Chemical pesticides that have high selectivity and are classed by regulators as low-risk compounds, such as synthetic insect growth regulators.
- Crop cultivars bred with partial or total pest resistance.
- Cultivation practices such as intercropping or undersowing, and crop rotation.
- Natural products such as biocidal plant extracts or semiochemicals.
- Physical methods such as mechanical weeders.
- Biological control with natural enemies, including predatory insects and mites, parasites, parasitoids, and microbial pathogens used against invertebrate pests as well as microbial antagonists of microbial pathogens and plant pathogens of weeds.
- Decision support tools to update farmers when it is economically useful to apply pesticides and other controls. These help with the calculation of economic action thresholds, phenological models that forecast the timing of pest activity, and basic pest scouting. These tools can be used to shift pesticide use away from routine calendar spraying to a supervised or targeted program.

NBP is one of the crop protection tools used in IPM. NBPs are active components of living organisms that act against different groups of pests. A strategy in which IPM is integrated with a microbial biopesticide to enhance the effect of augmented natural enemies is called "biointensive pest management." The principles of IPM (Barzman et al., 2015) adapted to protect crops are given in Fig. 4.

However, biointensive pest management is not able to reduce the effect of pests or the population density. Further, less-toxic pesticides may be used. This category of pesticides includes insecticidal soaps (active against aphids, whiteflies, and other soft-bodied insects), botanicals (neem oil and garlic oil, excluding pyrethrum), horticultural oils (scales, mites, smother aphids), and insect growth regulators (molting synthesis disruptors, chitin synthesis inhibitors) (Rahman et al., 2016). Although biopesticides used as plant protection compounds are intensified in IPM principles 4–7, their efficacy is also reflected in all the different stages of the IPM pyramid shown in Fig. 5.

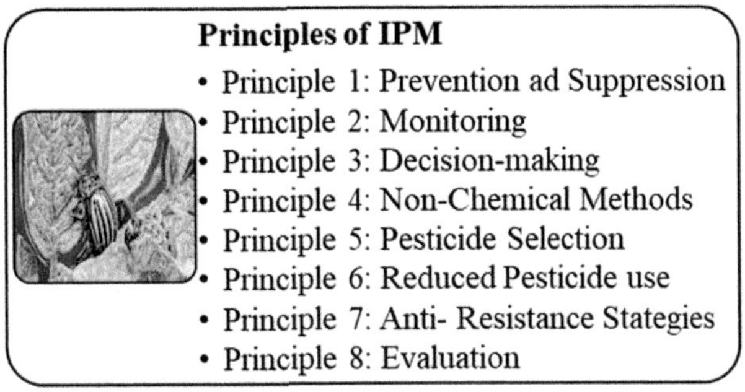

FIG. 4 Principles of integrated pest management.

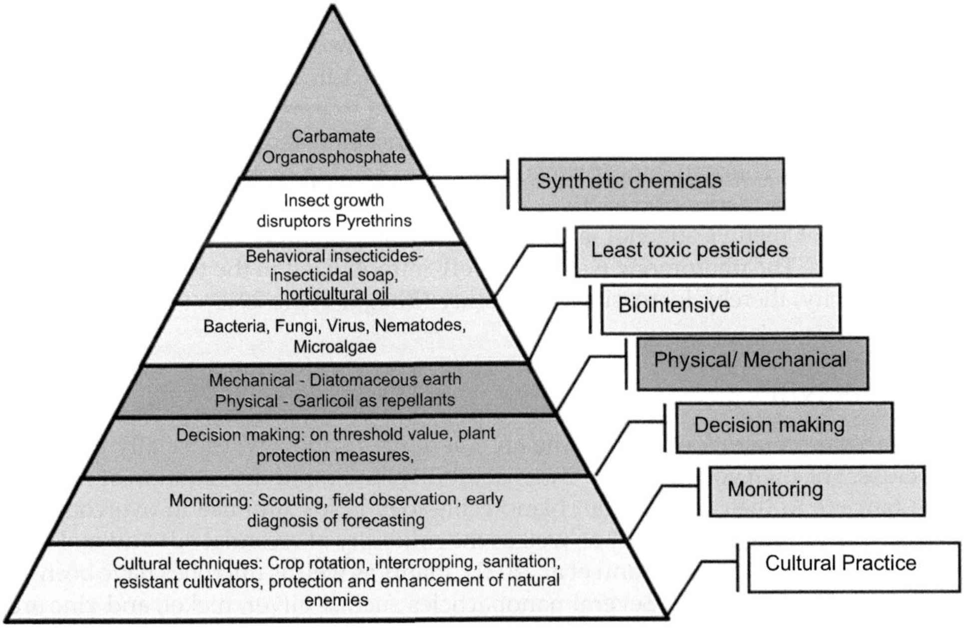

FIG. 5 Biopesticide-driven integrated pest management pyramid (Fenibo et al., 2022).

3 Fundamentals and development of NBPs

3.1 Concept of NBPs

When complexed with nano-sized bioactives or any other nanomaterials intended to prevent, repel,or kill pests, biopesticides are referred to as NBPs. NBPs have enhanced properties over their biopesticide counterparts in terms of increased bioavailability, slow and controlled release of bioactives, improved stability, and higher efficiency (Ragaei and Sabry, 2014). Unique properties observed in NBPs are attributed to nano-sized bioactives or bioactives loaded in nano-sized materials as carriers (Lade et al., 2019).

Nanoparticles containing biopesticides can be used very effectively against a range of agricultural pests. For biopesticides that are poorly soluble in water, the nanoemulsion method is most suitable for increasing their bioavailability and efficacy, where the bioactives in the oil phase are dispersed as nano-sized (20–200 nm) droplets in the water phase by the addition of surfactants (Pavoni et al., 2019). In the case of highly volatile biopesticides, nanoencapsulation and nanogel methods aid in the slow release of bioactives by using coating/carrier materials that provide protection from environmental interaction, which in turn increases their stability (Perlatti et al., 2013; Bhagat et al., 2013). Furthermore, nanoformulations improve the penetration and cellular uptake of bioactives (Zhao et al., 2018), thus reducing the amount of biopesticide applied. For instance, the efficacy of the *Geranium maculatum* essential oil nanoemulsion against *Culex pipiens* and *Plodia interpunctella* larvae was increased two-fold over its bulk counterpart. The particle size of the nanoemulsion was thought to increase the bioavailability of essential oil components that helped in higher penetration through insect cuticles, causing neurological disruption in target pests (Jesser et al., 2020). *Carum copticum* essential oil and *Peganum harmala* extract encapsulated in chitosan was found to be nearly four times more effective than the bulk extract in controlling *Alternaria alternata*, a plant pathogenic fungus causing early blight disease in tomatoes. The suggested mechanism behind the increased antifungal activity of the encapsulated formulation was attributed to the attachment of chitosan nanoparticles (used as a coating material) to the fungal surface and/or chitosan as a carrier of bioactives directly into fungal cells (Izadi et al., 2021). An effective nanogel of methyl eugenol, a pheromone, was prepared to control the guava pest *Bactrocera dorsalis*. The pheromone was effectively immobilized in the nanogel for prolonged retention capacity, thereby increasing its stability (Bhagat et al., 2013).

3.2 Types of NBPs

Nanomaterial preparation has become an emerging technology, especially in the field of biopesticides. The high volatility, low water solubility, and rapid degradation of essential oils stand as barriers to their application. Nanotechnology holds promise in overcoming these problems by increasing the stability as well as the solubility of essential oils and making them more suitable for application (Giunti et al., 2021). Plants as well as microbes have been used in the synthesis of nanoparticles. Several nanoparticles such as silver, nickel, and zinc are synthesized from plants, and are used for pest management (Devakumar and Dureja, 2002).

3.2.1 *Nanofungicides*

Fungal diseases are a top reason for crop destruction across the globe. Biofungicides can be defined as a class of fungicides derived from biological systems, including fungi and bacteria, that are effective against plant pathogens. The term nanofungicide can be used to describe any fungicide on a nanometer scale with all the properties associated with a small length. The use of biofungicides could be an ecofriendly way of controlling pests. The fungal system has become effective for the synthesis of nanoparticles because of the different characteristics of fungi such as the high wall binding capacity as well as the ease of culturing and biomass handling. Nanofungicides have vast physiochemical as well as functionalization properties. Several species of plant pathogenic fungi can be managed by nano-based biofungicides

(Yadav et al., 2015). Nanofungicides can be isolated from different bacterial strains such as *Bacillus amyloliquefaciens* from which silver nanoparticles are synthesized and are used as nanofungicides under greenhouse conditions (Abd El Aty and Zohair, 2020).

3.2.2 *Nanobioinsecticides*

Biopesticides based on plant-based essential oils have become an effective alternative to synthetic pesticides, even in integrated pest management strategies. The drawbacks of essential oils, including their volatile nature, can be solved by the nanoformulation of essential oils. It has been reported that the nanoformulation of pesticides shows less toxicity against non-target organisms (Abasse, 2018). Tia et al. (2021) reported the effect of A nanoemulsion of *Lippia multiflora* essential oil with chitosan as a natural surfactant in the management of the cabbage pest *Brassica oleracea* L. Sabbour and Solieman (2022) reported the effectiveness of nanochitosan against harmful sugarcane pests. Moreover, the advantages of plant-mediated synthesis of metal nanoparticles are greater compared to microorganism-mediated synthesis (Ibrahim, 2015).

3.2.3 *Other NBPs*

The eradication of weeds is important in agriculture. Nanotechnology is used for the development of nanoherbicides, which are found to be ecofriendly and will not leave any toxic residues in the soil or environment. Nanoparticles have been referred to as "magic bullets" that have been loaded with particles, including herbicides and chemicals, that target specific plants (Pérez-de-Luque and Rubiales, 2009). It is possible to reduce the quantity of the herbicide with a proper delivery system (Ali et al., 2014). Abbassy et al. (2017) reported the nematocidal activity of silver nanoparticles of different botanical products against the root-knot nematode *Meloidogyne incognita*.

4 Technologies for preparation of NBP

There are two major categories for NBP synthesis: The high-energy technique and the low-energy technique. In the high-energy technique, high energy is transferred to break the dispersed phase with the dispersion medium. This energy will be supplied for a short duration to obtain nano-sized particles. There are different types of high-energy methods such as high-pressure homogenization, high sheer stirring, ultrasonic emulsification, and microfluidization. In the low-energy method, a water-in-oil macroemulsion will be converted into an oil-in-water emulsion by changing the composition of the sample and temperature. There are different categories of low-energy methods such as phase inversion temperature (PIT), phase inversion composition, and spontaneous nanoemulsion (Karthik et al., 2017; Pan et al., 2023).

The plant-mediated synthesis of highly stable nanoparticles is very fast as well as economical on a large scale (Iravani, 2011). The industrial-scale production of metallic nanoparticles also can be done through this approach by the tissue culture method (Jha et al., 2009). The phytochemical-based synthesis of nanoparticles (Ahmed et al., 2014) and the plant-mediated synthesis of bimetallic nanoparticles (Lu et al., 2014) are also more advantageous. The synthesis involves nanoparticles and pesticides that have been trapped in a protein or carbohydrate polymer such as chitosan and polyethylene glycol (PEG), which can hold the nanoparticles (Sayed and Behle, 2017). There are various techniques reported for NBP synthesis including electro-spinning, grinding, electro-spraying, ball-milling, etc.

5 Mode of action of NBPs

Although many studies focusing on the efficacy of NBPs on plant pathogens/pests, very few studies focus on understanding their mode of action. Nevertheless, the mode of action of NBPs derived from known biopesticides would be the same as their bulk counterpart. However, the exact mode of action of novel NBPs has yet to be identified.

5.1 Insect pests

The NBP, depending on the concentration and type applied, can affect insect physiology and morphology by entering insects through the outer cuticle, spiracles, mouth openings, setae, abdominal prolegs, and anal prolegs. After entry, the NBP could be absorbed in the blood, affect the internal organs directly, and manifest in other body parts (Gilligan and Passoa, 2014; Lade and Gogle, 2019). The modes of action of the NBP against various insect life stages such as eggs, larvae, pupae, and adults are not completely understood; however, studies on the mode of action of mineral-based NBPs have indicated damaging effects on the head, thorax, abdomen, and siphon of insects (Benelli, 2018). Furthermore, silica and alumina nanoparticles were reported to be physisorbed by the insect cuticular lipids, causing severe dehydration in insects such as *Plutella xylostella* (diamondback moth) and *Sitophilus oryzae* (rice weevil), which ultimately led to their death (Stadler et al., 2017; Shoaib et al., 2018). Likewise, silver nanoparticles were reported to have reduced the activities of the digestive enzymes amylase, protease, lipase, and invertase while also significantly affecting the normal functioning of the gut microflora in *Spodoptera litura* (tobacco caterpillar) (Bharani and Namasivayam, 2017). At higher concentrations, silver nanoparticles induced oxidative stress, thereby enhancing antioxidant enzyme levels in the gut of *S. litura* larvae (Yasur and Usha-Rani, 2015). In addition to these effects, nanoparticles may bind to sulfur in proteins and phosphorous in nucleic acids, causing mutations that lead to cell death (Benelli, 2018).

5.2 Fungal pathogens

Numerous studies have illustrated various mechanisms by which metallic NBPs control/mitigate phytopathogens. In vitro and in planta assays of silver nanoparticles have shown the ability to affect the viability of spores, thereby inhibiting the colony formation and disease progression of fungal pathogens. Silica nanoparticles are reported to stimulate plant physiological mechanisms and enhance their resistance against diseases. The fungicidal activity of copper and zinc nanoparticles is attributed to their ability to form highly reactive hydroxyl and superoxide radicals, which, upon interaction with cellular materials (such as DNA, proteins, lipids, and other biomolecules), cause damage leading to cell death. These nanoparticles interfere with the electron transfer chain and thereby disrupt the normal functioning of biological processes. Further, zinc nanoparticles deform vegetative hyphae while also inhibiting sporulating bodies (conidiophores) and spore development, ultimately causing fungal hyphae death. Iron nanoparticles interrupt adenosine triphosphate forming and arrest the cell cycle. Additionally, iron nanoparticles induce oxidative stress and bind to fungal cell membranes forming pits to increase their permeability and eventually leading to cell death (Ul Haq and Ijaz, 2019).

5.3 Other pests and vectors

The plant bacterial pathogens *Pectobacterium betavasculorum, Xanthomonas campestris* pv. *Beticola,* and *Pseudomonas syringae* were inhibited by metal oxide nanoparticles (ZnO and TiO2). ZnO nanoparticles along with hydrogen peroxide produced by ZnO and hydrogen ions penetrate and change the bacterial cell membrane permeability, leading to cell lysis. Furthermore, ZnO and TiO2 nanoparticles were reported to increase chlorophyll content, improve plant growth, and enhance enzymes involved in oxidative stress induced by pathogens, thereby indirectly controlling bacterial disease in the host plant (Siddiqui et al., 2019). The silver nanoparticle was successful in mitigating gall formation in Bermuda grass caused by *Meloidogyne* (root-knot nematode). The nematocidal activity of the silver nanoparticle is not identified but associated with cell membrane permeability disruption, inhibition of ATP synthesis, and inducing oxidative stress in both eukaryotic and prokaryotic cells as the silver nanoparticle is a broad-spectrum pesticide (Cromwell et al., 2014). The antiviral activity of various nanoparticles has been studied against some major plant viruses such as tobacco mosaic virus (TMV), turnip mosaic virus (TuMV), tomato mosaic virus (ToMV), barley yellow mosaic virus (BaYMV), and so on. In vitro studies suggested that metal and metal oxide nanoparticles (Ag, ZnO, SiO2, and Fe2O3) inhibited the replication of mosaic viruses (TMV, ToMV, TuMV) by binding to the viral proteins and disrupting the capsid structure, leading to the aggregation and deactivation of viral particles (Farooq et al., 2021).

6 Benefits of NBPs over traditional pest control strategies

Traditional agricultural practices are more focused on avoiding pests rather than controlling them, such that the time and season of plowing, tilling, and cultivating crops the type are chosen carefully to avoid pest incidents (Morales, 2002). Traditional pest control strategies are practiced by small farmers. For example, in some parts of India and Mexico, predators (birds, ladybugs, spiders, grasshoppers, sunflower beetle, frog, etc.) and insecticidal plant parts (neem leaves, mahua flowers, chili, garlic, turmeric, basil, etc.) are used for pest control (Morales and Perfecto, 2000; Patel et al., 2020). Moreover, cow dung, cow urine, and cow dung ash have prominent roles in storage and field pest management in some Indian villages (Samal and Dhyani, 2007). Traditional pest strategies are environmentally sustainable; however, they are labor-intensive and low-yielding (Patel et al., 2020). Thus, the advantage of NBPs can be realized in improving the performance of biopesticides being used in traditional pest control strategies. For example, silica, a major constituent of ash, when made into a nanoformulation has been reported to have enhanced insecticidal activity and has also been shown to improve the host plant immune systems (Stadler et al., 2017; Ul Haq and Ijaz, 2019).

7 Bioavailability of NBP

The rate at which the NBP reaches the target site and the quantity available for targeted action are the two major concerns that decide NBP bioavailability. All three categories of biopesticides such as botanical pesticides, microbial pesticides, and plant-incorporated protectants can be formulated in the nanoscale to obtain NBPs. Bioavailability is affected by pest species, formulation composition, and particle size (Bo Cui et al., 2019). The sustained release of active ingredients of NBP will be regulated by suitable carriers and additives. NBP enhances the growth of mutualistic microbes, which in turn regulate plant growth (Ghormade et al., 2011).

Imidacloprid, a controlled-release NBP encapsulated by polymers, was reported to be more effective than conventionally formulated imidacloprid in controlling the steam and white fly pests of soybean crops. Interestingly, 50% by weight of the nano-based pesticide was found to be more efficient than the conventional pesticide. The yellow mosaic virus (YMV) infection caused by Imidacloprid white fly was recorded by the YMV infestation rating. It was found to be 7 compared to the control treated with Imidacloprid commercial formulation, with a YMV infestation value of 6.33 after 60 days of sowing soybean seeds. Advantageously, residues in the crops and soil were found to be negligible while using sustained-release NBP for insect management (Adak et al., 2012). An NBP of carbofuran formulated by a two-step milling process was tested against one of the most destructive insect pests, the diamondback moth. Microsuspensions of pesticide produced by a conventional milling process were scaled down using 0.2 mm beads of zirconium oxide to produce nanosuspensions of pesticide. The formulated NBP was found to be stable for around 2 years. Around a 13% improvement in efficacy compared to the carbofuran microsuspensions was observed, with a lethal dosage value of around 11 mg/L. The reduced absorption rate was the significant reason for the improved efficiency, and a minimal dose of NBP compared to a microsuspension was sufficient to achieve the targeted efficacy. The spontaneity of dispersion was also found to be 100%, which clearly shows the maximum bioavailability of the formulated carbofuran NBP (Chin et al., 2011).

The solubility of 227 nm zoxamide NBP was around 40 mg/L, which was almost 1.5 times higher compared to conventional pesticides used for tomato plants. Crystals of the zoxamide NBP were produced by a media milling process with polysorbate as stabilizers. The retention of the formulated NBP in the leaves was 3.9 mg/cm^2 vs. 2.4 mg/cm^2 for the conventional formulation, clearly indicating the higher extent of bioavailability of the zoxamide NBP. It was also suggested that the retention of pesticides in the leaves and berries of tomato plants must use janus particles (nanoformulations with distinctive physical properties) for improved interaction between the NBP and the plant parts (Corrias et al., 2021). Oil-based biopesticides in the form of gels and cold aerosols tested against *Tribolium confusum* (confused flour beetle) show a significant level of toxicity against the pest. There is higher relative efficacy in the cold aerosol than the gels because of the fumigant and contact nature of the insecticide (Palermo et al., 2021). Also, it has been demonstrated that nano-sized aqueous dispersion formulations improve the bioavailability of pesticides. When *Artemisia arborescence* essential oil was added to solid nanoparticles of about 250 nm, it slowed down how quickly it evaporated compared to when it was in an emulsion form (Manchikanti, 2019).

Diyarex Gold is a new generation of biofungicide used to control powdery mildews, downy mildews, rust, and early and late blight diseases in vegetables, herbs, grape vines, and orchards. It was developed and processed using molecular nanotechnology and is an effective bactericide and fungicide, which is attributed to the product's natural components. According to certification from Biocert India, the product can be used as an environmentally friendly biofungicide that doesn't leave any residue in plants or their tissues. This ensures both environmental and human safety, especially because bees and other economically significant insect communities aren't harmed (Bhattacharyya et al., 2016). Citronella essential oil and neem oil mixture-based nanoemulsions demonstrated potent antifungal activity against phytopathogenic fungi when compared to individual citronella and neem oils. This discovery has paved the way for the use of citronella oil in conjunction with neem oil or vice versa as biopesticides to control plant diseases caused by *Rhizoctonia solani* and *Sclerotium rolfsii* (Ali et al., 2017). The biological activity of an abamectin-based nanosuspension was compared to the biological activity of a conventional emulsion. The nanosuspension was found to be 2.8 times more toxic against diamondback moths (*Plutella xylostella* L.) than the emulsion. Because of the small size and high surface-to-volume ratio effects of nanoparticles, it has been reported that the bioavailability of a nanoformulation can be improved when compared to its corresponding conventional formulation. The improved biological activity can be attributed to improvements in formulation performance in retention and photostability caused by size reduction and composition. The retention of the nanosuspension-based pesticide was measured as $36\,mg/cm^2$ compared to the conventional form, having around $25–36\,mg/cm^2$ on the cucumber. Enhanced retention and adhesion are conducive to enhancing pesticide efficacy through improved action concentration and contact probability (Cui et al., 2019).

The National Science Foundation says that materials such as nanoclays and layered double hydroxides are biocompatible, have low toxicity, and have the potential to encapsulate agrochemicals to form controlled-release biopesticides. Layered double hydroxides with a high affinity for anionic species are acidic soluble. In the case of hydrophobic chemicals, the arrangement increased solubility and thus bioavailability by preventing recrystallization (Ghormade et al., 2011). Chlorantraniliprole has a low Kow value ($\log K_{ow} = 2.77$); it is assumed that the CAP uptake from soils to earthworms occurs primarily via a passive diffusion pathway, determining CAP bioavailability to earthworms (Kow-octanol/water partition coefficient) (Wang et al., 2012). The CSNP system encapsulated Harpin*Pss* with 90% efficiency. The sustained release of Harpin*Pss* (Harpin*Pss*, an elicitor from *P. syringae* pv. *Syringae)* from HCSNPs after foliar spray significantly increased Harpin*Pss* bioavailability, induced plant defense enzymes, and reduced *R. solani* infection (p0.05). Confocal analysis revealed NP entry into the plant cell and Harpin*Pss* localization in chloroplasts (Nadendla et al., 2018).

Nanobiopesticides formulated to act against a plethora of pathogens and their significant action upon them are listed in Table 2. For instance, cinnamon oil as an active component of a nanobiopesticide to act upon *Botrytis cinerea*, a necrotrophic fungal pathogen, was found to prevent the decay of fruits (Yousef et al., 2019). Cyhalothrin was efficient in enhancing water

TABLE 2 NBPs formulated with various nanocarriers against several pathogens.

Sl. no.	Active component	Additives	Pathogen	Targeted activity	References
1.	Cinnamon essential oil	Tween 80, Span 80, and lecithin	*Botrytis cinerea*	Fruit decay was delayed	Yousef et al. (2019)
2.	Cyhalothrin	Alkyd resin	Insects	Improvement in water stability	Qin et al. (2017)
3.	*Satureja hortensis* oil	Tween 80	*Chenopodium album*	Bioherbicidal activity was improved	Hazrati et al. (2017)
4.	*Eucalyptus globules* oil	*Pongamia glabra*	*Tribolium castaneum*	Improved toxicity	Pant et al. (2014)
5.	Glyphosate Isopropylamine	Fatty acid methyl esters	*Eleusine indica*	Improved leaf wettability and toxicity	Jiang et al. (2012)
6.	Pyridalyl	Sodium alginate	*Helicoverpa armigera*	Improved toxicity	Saini et al. (2014)
7.	Novaluron	Isopropanol	*Spodoptera littoralis*	High efficiency	Elek et al. (2010)
8.	Avermectin	Polydopamine	Weeds	Sustained release	Jia et al. (2014)
9.	Atrazine	Poly(ε-caprolactone)	*Zabrotes Subfasciatus*	Lesser ultraviolet degradation	Da Costa et al. (2014)
10.	Tebuconazole	Methyl methacrylate	*Gloeophyllum Trabeum*	Less leaching loss	Salma et al. (2010)

stability against insects (Qin et al., 2017). *Satureja hortensis* oil was active in improving the bioherbicidal activity of *Chenopodium album*, a weedy annual plant (Hazrati et al., 2017). Likewise, *Eucalyptus globules* oil against *Tribolium castaneum* (Pant et al., 2014), glyphosate isopropylamine against *Eleusine indica* (Jiang et al., 2012), pyridalyl against *Helicoverpa armigera* (Saini et al., 2014), novaluron against *Spodoptera littoralis* (Elek et al., 2010), atrazine against *Zabrotes subfasciatus* (Da Costa et al., 2014), and tebuconazole against *Gloeophyllum trabeum* (Salma et al., 2010) were other active ingredients of nanobiopesticides that are reported to have activity with the fullest bioavailability upon a variety of targeted plant pathogens.

8 Safety aspects of NBPs

In addition to the many advantages of NBPs, there are many unanswered questions related to their biomolecular interactions. Important aspects such as a simple method of preparation, multiple pest toxicity, and safety to nontarget organisms and the environment are to be considered before selecting nanopesticides (Manchikanti, 2019).

8.1 Environmental risk and safety assessment

Biopesticides are made of naturally occurring compounds (plants, animals, microorganisms, and certain minerals), so are generally safer than chemical pesticides. Properties such as biodegradability, biocompatibility, availability of natural resources, controlled release, and ecological safety support the use of NBPs over chemical pesticides (Manchikanti, 2019). Bioactives used in NBP preparation are biodegradable, and biopolymers used for the slow release of bioactives dissolve in water and soil. These biopolymers have fewer side effects in humans and plants and are considered safe to some extent (Lade and Gogle, 2019). Therefore, NBPs are safe; however, precautions should be taken by operators to avoid absorption during NBP application (Lade and Gogle, 2019).

Concerns over NBP safety persist as they can be phytotoxic and indirectly affect animals and the environment (Monica and Cremonini, 2009). Several studies indicate the negative effects of metal nanopesticides such as slowed germination, reduced root growth, cell division obstruction, and biomass reduction in plants as well as toxicity due to bioaccumulation in soil and water organisms (Makarenko and Makarenko, 2019). The unpredictability, phytotoxicity at higher concentrations, and reduced photosynthesis must be addressed to assess the reactivity of nanomaterials in plant systems (Nair et al., 2010). In the environment, nanomaterials undergo physical and chemical changes that are subjected to factors such as pH, temperature, organic matter, and interaction with biological systems, which as a consequence leads to an altered ecological role and nanomaterial toxicity (He et al., 2015). Nanomaterials could migrate into the soil surface, groundwater, or open water systems by wind and rain (Ray et al., 2009). In the soil, a portion of nanomaterials is controlled by the sorption process of organic substances while destruction is influenced by solar radiation. The environmental degradation process of nanomaterials has still not been completely examined (Turco et al., 2011). Moreover, studies on NBP ecotoxicity as such are lacking.

To differentiate natural nanoparticles from synthesized nanoparticles, changes in particle polydispersity would be useful in tracking synthesized nanoparticles in living systems. NBP detection in living systems is difficult, as some are coated in polymer- or organism-derived materials (protein, surfactants, ligands) that may undergo changes upon interacting with plant and animal systems. Their transport (bound or unbound), fate, and ecotoxicological aspects have still not been established. Due to the high reactivity of nanoparticles, they are expected to be transformed in the biological environment based on their form (individual, aggregate, or encapsulated). Besides biotransformation, the extent of reactivity, interaction, and persistence of NBPs in the pest with the environment needs to be studied. Studies on the durability, bioavailability, toxicity, and behavior of NBPs with reproducible outcomes are needed. Moreover, the high durability of some nanoparticles is also a major issue in terms of environmental safety. Further, the ability to detect and quantify NBPs in organisms and various parts of the environment such as soil, water, and air would be a significant challenge (Manchikanti, 2019; Ragaei and Sabry, 2014).

Techniques used for assessing the risk of general materials would be insufficient for nanoparticles. Therefore, the development of specific approaches such as toxicokinetics, toxicity mechanisms, early nanotoxicity bioindicators, nanotoxicity prediction models, and nanomaterial exposure limit detection that could affect human health has been suggested (Pietroiusti et al., 2018). Registration guidelines for botanical-, microbial-, and semiochemicalbased pest control agents for plant protection and public health use are provided by the World Health Organization International Code of Conduct on pesticide management (World Health Organization, & Food and Agriculture Organization of the United Nations, 2016). These guidelines, however, do not include aspects related to NBPs (Manchikanti, 2019). Thus, a concerted effort is required between industries and private and government departments to develop protocols for risk assessment of NBP as well as standardized models (Ragaei and Sabry, 2014). Additionally, scientists are obligated to have a public debate regarding research and concerns over nanoproducts as well as acknowledging the gap between our understanding of the few functions at the nanometric scale and the complexity of life (Larrouturou, 2005; Ragaei and Sabry, 2014).

8.2 Human health impacts

One of the desirable qualities of NBPs is their ability to cross the membrane barrier in pest organisms to effect toxicity; the same may be detrimental in humans. The possibility of nanoparticles affecting humans directly or indirectly is a growing concern. Nanoparticles show dermal absorption (due to their nanosize); they pass through the cellular barrier and cause DNA damage in human cells, further promoting oxidative stress and inflammatory response (Bhabra et al., 2009). Nanoparticles absorbed via inhalation (due to their light weight) can cause respiratory problems and may reach the brain by crossing the blood–brain barrier, inducing neurotoxicity (Hu and Gao, 2010). Evidence also shows that nanoparticles (multiwalled carbon nanotubes) can deposit in lung tissues, indicating their affinity to blood plasma lipids and proteins (Gasser et al., 2010). To understand and determine the safety of NBPs in relation to human health, extensive research is required.

8.3 Impacts on nontarget organisms

Studies of the impact of NBPs on nontarget organisms are scarce in the literature. However, biopesticides are generally believed to have less effect on nontarget organisms due to their specificity to target pests and closely related organisms (Kaya and Lacey, 2007). Studies on the effect of biopesticides on nontarget organisms are limited. One study on the side effects of Bt toxin on nontarget insect pests revealed that a diet based on Bt maize delayed larval development as well as caused 45% mortality in *Acarus siro* (grain mite) (Dabrowski et al., 2006). An *Artemisia absinthium*-based biopesticide was evaluated for its ecotoxicity on nontarget aquatic organisms. The study revealed acute toxicity of the biopesticide toward *Daphnia magna* (water flea), *Vibrio fisheri* (aquatic Gram-negative bacteria), and *Chlamydomonas reinhardtii* (green algae) at the lowest concentration of 0.2% while the natural microbial community was not affected (Pino-Otín et al., 2019a). Another study evaluated the ecotoxicity of *Lavandula luisieri*-based biopesticide on nontarget soil organisms, namely *Eisenia fetida* (earthworm), *Allium cepa* (onion), and the natural soil microbial community. Acute toxicity was reported for the earthworm and onion plants at LC50 values of >0.4mL/g and 2.2%, respectively. While the growth of the natural microbial community was significantly reduced, their carbohydrate-degrading capacity was increased (Pino-Otín et al., 2019b). These studies highlight the need for a detailed study on risk assessment and ecotoxicological studies on biopesticides. It can be further deduced that biopesticides with ecotoxicity toward nontarget organisms could have elevated adverse effects in nanoformulations, as the bioactives are more stable and active.

9 Regulatory measures of NBPs

Each country has various departments to regulate the launch of nano-based products in the market by determining the quality, safety, and toxicity aspects. The Soil Association (United Kingdom) and the Biological Farmers of Australia, the leading global organic food certifiers, consider nanomaterial-based agri-food on organic standards (Scrinis and Lyons, 2010). The International Federation of Organic Agriculture Movements (IFOAM, Germany), however, refuse to consider food products grown with artificial nanomaterials as organic (Naturland, 2011; IFOAM, 2011). In the United States, the Environmental Protection Agency's Pesticide Program Dialogue Committee (PPDC) is concerned with policy for the approval of nanoformulations (including NBP) from the lab to the field (Ragaei and Sabry, 2014). So far, there are limited reports available from food safety authorities on the safety and standards of nano-based agri-products; hence more research studies are required to support nanoproducts. To give the nano-based agri-product sector a better chance in the market, intense research including organic chemistry, green chemistry, biology, and material science is needed, along with successful field trials (Lade and Gogle, 2019).

10 Future perspectives and global demand for NBPs

The need to increase food production for the burgeoning population requires a complete overhaul of the current agricultural practices and food production systems (United Nations, 2021). The greatest challenge is to produce almost twice the amount of food by 2050 with the existing agricultural resources and inputs (land, water, and agrochemicals) while confronting challenges such as degradation of soil quality, irregular weather patterns, etc. (Food and Agriculture Organization of the United Nations, 2017). The quantum of synthetic pesticides applied for the cultivation of food and cash crops required for human life is a crucial factor in helping food production. However, this devastates the environment, crops, and human health (Mesnage and Séralini, 2018; Nicolopoulou-Stamati et al., 2016). Among the alternatives envisaged for a sustainable food production system, biopesticides in nanoformulations with superlative properties are projected to grab a huge share of the pesticide market in the coming years. This is expected to reduce the impacts of synthetic pesticides on the land, water, environment, crops, and human health (Fenibo et al., 2021). Another considerable challenge facing farmers at a large scale is the constant rise in pesticide resistance over the years, mainly driven by the smart population genetics of pests. This is evident in the burgeoning crop losses despite of applying higher concentrations of pesticides to grow crops (Gould et al., 2018; Hawkins et al., 2019; Pinho, 2021). All these events require a holistic solution to combat the challenges and threats faced by the food production systems. Biopesticides in nanoformulation are the only viable option that can be scaled to the levels to replace the synthetic pesticides in use.

The biopesticides market is projected to have an annual growth rate of 13%–15% and reach a market value of around $10 billion by 2027 (Biopesticides Market, 2020; Mordor Intelligence, 2021). The biopesticides market is mainly driven by the adoption of organic farming practices, pest resistance to synthetic pesticides, and high research and development costs associated with pesticide production. Major challenges include the lack of technological expertise to improve the stability and shelf life of biopesticides, a limited range of products to combat pests, and the preference for synthetic pesticides by farmers in developing countries. Europe is expected to be a major market for biopesticides in the coming years. Novozymes, BASF, BAYER, Syngenta, and Marrone Bio Innovations are projected to be the key manufacturers of biopesticides (Biopesticides Market, 2020).

A new set of regulations is required to answer some critical questions regarding the composition and size of NBPs and their classification (Roduner, 2006). For example, will it be classified as synthetic if a nanoformulation of the biopesticide contains a synthetic encapsulating compound/nanomaterial as a major ingredient? What is the threshold limit for the synthetic components to be a part of the nanoformulations to still classify as a bioproduct? Life cycle assessment and the toxicity of the components of NBP formulation are necessary to ascertain the ecological footprint of its synthetic counterparts (Furxhi, 2022; Klaper, 2020). Absolute clarity is required in the above aspects for the successful commercialization of these NBPs as an alternative to pesticides. It is essential to evaluate the role of the encapsulating material of the biopesticide nanoformulation on the crops, the environment, and human health as well as its ecological footprint through an interdisciplinary research approach. As the information on the safety of nanocomponents is that it is benign, it is expected to take significant time and effort to ascertain the facts, gather critical information, and ensure the safety and ecological aspects of NBPs (Klaper, 2020). It would be prudent for regulatory agencies such as the Insecticide Resistance Action Committee and the EPA to develop a set of good agricultural practices for applying nanoformulations of biopesticides to combat or slow pests developing resistance toward NBPs as a part of insect resistance management and IPM (Gallagher, 2021).

11 Commercialization of NBP

The commercialization of NBP depends on a multitude of aspects, including production strategies, marketing, promotion, efficiency, cost, and the impact of encapsulation on the crop and soil, etc. At present, biopesticides command a premium price over their synthetic counterparts. It is of utmost importance for NBPs to match the price and efficiency of synthetic pesticides currently applied for specific crops. This is a herculean task, as it requires collective efforts from various sectors but is not limited to arranging resources such as raw materials, processing, extraction, and formulation. We must develop the necessary infrastructure such as plant sources, bioprocess/fermentation units, and expertise in nanoformulation techniques to meet the forecasted demand for biopesticides (De la Cruz Quiroz et al., 2015). At the same time, it is necessary to create sufficient awareness among farmers to readily switch to NBPs as an appropriate alternative to pesticides. Global agencies, in coordination with local governments, must come up with incentives and schemes to develop the necessary infrastructure to realize the desired change in the commercialization and adoption of NBPs in farming. It is advised to exercise caution not to hasten the process without proper technology or infrastructure, as this might lead to food scarcity, food insecurity, and skyrocketing food prices. Similar undesired events were observed in certain Asian countries that had a complete changeover of the entire farming system to organic practices without sufficient trials, resources, and infrastructure (Bhattacharya and Singh, 2021; Nordhaus and Shah, 2021). NBPs are certainly a viable alternative to unsustainable synthetic pesticides. However, improper implementation might jeopardize the entire society and lead to food insecurity, poverty, and the collapse of food production systems.

The market share of nano-based agri-products in the agricultural sector remains marginal despite its many proven potential advantages (Parisi et al., 2015). Many are involved in the chain of development and commercialization of pest control products such as scientists, regulators, stakeholders, marketers, and end users. Sometimes, unresolved issues and disagreements between the participants of this chain, such as marketers with different opinions from regulators and scientists, often lead to end-user confusion regarding the weakness of the final product (Isman, 2015). Quicker procedures and enforcement of time limits are important if launching new products and generating income are considered. Moreover, the high registration cost is another limiting aspect of the commercialization of new products (Pavela, 2014). The trend of patent applications, especially from agrochemical companies, continues to grow. However, there are no new nano-based products in the agricultural sector. This indicates that active patenting by the applicants is an assurance to exploit the broad claim on patents in future commercial product development from the patent. Generally, high production costs and legislative uncertainties result in low acceptance of such innovative products (Parisi et al., 2015). Additionally, support from the regulatory systems to small- and medium-sized firms in terms of reliable tools to produce economically feasible products would further help the commercialization of biopesticides (Damalas and Koutroubas, 2018).

12 Conclusions

Increasing food demand to meet the growing population, controlling pests, and protecting the harvest/food while not further degrading the environment are worldwide challenges in the current scenario. Although chemical pesticides are an easy and quick solution to pest control, their detrimental effects on the environment and humans cannot be disregarded. As a result, biopesticides are sought as a safer alternative with many products already available on the market. However, instability and multiple applications in the field are setbacks experienced by biopesticides. Nevertheless, nanotechnology application in agriculture is becoming prevalent. Nano-based agri-products have novel properties that lead to improved stability, enhanced efficacy, and controlled release of biopesticides. These factors further allow the use of NBPs at lower doses. Successful large-scale field trials as well as adequate understanding of NBPs concerning the environment and human health are still needed to promote NBPs beyond the lab and patents.

Acknowledgments

The authors are grateful to Dr. S. Ezhil Vendan, Scientist, CSIR–Central Food Technological Research Institute, Mysore, for the valuable comments and discussion to improve this work. The author (PK) thanks the Karpagam Academy of Higher Education for the support and help.

References

Abasse, A.A., 2018. Nano bioinsecticides based on essential oils against *Phenacoccus solenopsis*. Egypt. Acad. J. Biolog. Sci. (A. Entomology) 11 (5), 1–12.

Abbassy, M.A., Abdel-Rasoul, M.A., Nassar, A.M., Soliman, B.S., 2017. Nematicidal activity of silver nanoparticles of botanical products against root-knot nematode, *Meloidogyne incognita*. Arch. Phytopathol. Plant Prot. 50 (17–18), 909–926.

Abd El Aty, A.A., Zohair, M.M., 2020. Green-synthesis and optimization of an eco-friendly nanobiofungicide from *Bacillus amyloliquefaciens* MH046937 with antimicrobial potential against phytopathogens. Environ. Nanotechnol. Monit. Manag. 14, 100309.

Acharya, A., Pal, P.K., 2020. Agriculture nanotechnology: translating research outcome to field applications by influencing environmental sustainability. NanoImpact 19, 100232.

Adak, T., Kumar, J., Dey, D., Shakil, N.A., Walia, S., 2012. Residue and bio-efficacy evaluation of controlled release formulations of imidacloprid against pests in soybean (*Glycine max*). J. Environ. Sci. Heal. - part B Pestic. Food Contam. Agric. Wastes 47, 226–231.

Ahmed, K.B.A., Subramaniam, S., Veerappan, G., Hari, N., Sivasubramanian, A., Veerappan, A., 2014. β-Sitosterol-d-glucopyranoside isolated from *Desmostachya bipinnata* mediates photoinduced rapid green synthesis of silver nanoparticles. RSC Adv. 4 (103), 59130–59136.

Aktar, W., Sengupta, D., Chowdhury, A., 2009. Impact of pesticides use in agriculture: their benefits and hazards. Interdiscip. Toxicol. 2, 1–12.

Ali, M.A., Rehman, I., Iqbal, A., Din, S., Rao, A.Q., Latif, A., Samiullah, T.R., Azam, S., Husnain, T., 2014. Nanotechnology, a new frontier in agriculture. Adv. Life Sci. 1 (3), 129–138.

Ali, E.O.M., Shakil, N.A., Rana, V.S., Sarkar, D.J., Majumder, S., Kaushik, P., Singh, B.B., Kumar, J., 2017. Antifungal activity of nano emulsions of neem and citronella oils against phytopathogenic fungi, *Rhizoctonia solani* and *Sclerotium rolfsii*. Ind. Crop Prod. 108, 379–387.

Bailey, K.L., 2014. The bioherbicide approach to weed control using plant pathogens. In: Abrol, D.P. (Ed.), Integrated Pest Management: Current Concepts and Ecological Perspective. Academic Press, pp. 245–266.

Bajwa, W.I., Kogan, M., 2002. Compendium of IPM definitions (CID)—What is IPM and how is it defined in the worldwide literature. IPPC Publ. 998 (998), 1–14.

Bapat, M.S., Singh, H., Shukla, S.K., Singh, P.P., Vo, D.V.N., Yadav, A., Goyal, A., Sharma, A., Kumar, D., 2022. Evaluating green silver nanoparticles as prospective biopesticides: an environmental standpoint. Chemosphere 286, 131761.

Barzman, M., Bàrberi, P., Birch, A.N.E., Boonekamp, P., Dachbrodt-Saaydeh, S., Graf, B., et al., 2015. Eight principles of integrated pest management. Agron. Sustain. Dev. 35 (4), 1199–1215.

Benelli, G., 2018. Mode of action of nanoparticles against insects. Environ. Sci. Pollut. Res. 25 (13), 12329–12341.

Bergeson, L.L., 2010. Nanosilver: US EPA's pesticide office considers how best to proceed. Environmental Quality Management. 19 Wiley Periodicals, pp. 79–85.

Bhabra, G., Sood, A., Fisher, B., Cartwright, L., Saunders, M., Evans, W.H., Case, C.P., 2009. Nanoparticles can cause DNA damage across a cellular barrier. Nat. Nanotechnol. 4 (12), 876–883.

Bhagat, D., Samanta, S.K., Bhattacharya, S., 2013. Efficient management of fruit pests by pheromone nanogels. Sci. Rep. 3, 1294.

Bharani, R.A., Namasivayam, S.K.R., 2017. Biogenic silver nanoparticles mediated stress on developmental period and gut physiology of major lepidopteran pest *Spodoptera litura* (Fab.) (Lepidoptera: Noctuidae)—an eco-friendly approach of insect pest control. J. Environ. Chem. Eng. 5 (1), 453–467.

Bhattacharya, S., Singh, R., 2021. Multifaceted challenges led to economic emergency of Sri Lanka in 2021. Int. J. Res. Eng. Sci. Manag. 4 (10), 162–164.

Bhattacharyya, A., Duraisamy, P., Govindarajan, M., Buhroo, A.A., Prasad, R., 2016. Nano-biofungicides: Emerging trend in insect pest control. In: Prasad, R. (Ed.), Advances and Applications through Fungal Nanobiotechnology. Fungal biology, Springer, Cham, pp. 307–319.

Biopesticides Market, 2020, September. Biopesticides Market Size, Share | Industry Analysis Report [2020–2027]. Available at: https://www.fortunebusinessinsights.com/industry-reports/biopesticides-market-100073.

Bravo, A., Gill, S.S., Soberon, M., 2007. Mode of action of *Bacillus thuringiensis* Cry and Cyt toxins and their potential for insect control. Toxicon 49 (4), 423–435.

Burton, I., 1987. Report on reports: our common future. Environ. Sci. Policy Sustain. Dev. 29, 25–29.

Chandler, D., Bailey, A.S., Tatchell, G.M., Davidson, G., Greaves, J., Grant, W.P., 2011. The development, regulation and use of biopesticides for integrated pest management. Philos. Trans. R. Soc. B Biol. Sci. 366, 1987–1998.

Chen, H., Yada, R., 2011. Nanotechnologies in agriculture: new tools for sustainable development. Trends Food Sci. Technol. 22, 585–594.

Chin, C.P., Wu, H.S., Wang, S.S., 2011. New approach to pesticide delivery using nanosuspensions: research and applications. Ind. Eng. Chem. Res. 50 (12), 7637–7643.

Corrias, F., Melis, A., Atzei, A., Marceddu, S., Dedola, F., Sirigu, A., Pireddu, R., Lai, F., Angioni, A., 2021. Zoxamide accumulation and retention evaluation after nanosuspension technology application in tomato plant. Pest Manag. Sci. 77 (7), 3508–3518.

Cromwell, W.A., Yang, J., Starr, J.L., Jo, Y.K., 2014. Nematicidal effects of silver nanoparticles on root-knot nematode in bermudagrass. J. Nemato. 46 (3), 261–266.

Cui, B., Lv, Y., Gao, F., Wang, C., Zeng, Z., Wang, Y., Sun, C., Zhao, X., Shen, Y., Liu, G., Cui, H., 2019. Improving abamectin bioavailability via nanosuspension constructed by wet milling technique. Pest Manag. Sci. 75 (10), 2756–2764.

Da Costa, J.T., Forim, M.R., Costa, E.S., De Souza, J.R., Mondego, J.M., Junior, A.L.B., 2014. Effects of different formulations of neem oil-based products on control *Zabrotes subfasciatus* (Boheman, 1833) (Coleoptera: Bruchidae) on beans. J. Stored Prod. Res. 56, 49–53.

Dabrowski, Z.T., Czajkowska, B., Bocinska, B., 2006. Ecological impact of genetically modified organisms. In: Romeis, J., Meissle, M. (Eds.), Proceedings of the Meeting of the Ecological Impact of Genetically Modified Organisms. International Organization for Biological and Integrated Control of Noxious Animals and Plants, West Palearctic Regional Section, Spain, pp. 39–42.

Damalas, C.A., Koutroubas, S.D., 2018. Current status and recent developments in biopesticide use. Agriculture 8, 13.

De la Cruz Quiroz, R., Roussos, S., Hernández, D., Rodríguez, R., Castillo, F., Aguilar, C.N., 2015. Challenges and opportunities of the bio-pesticides production by solid-state fermentation: filamentous fungi as a model. Crit. Rev. Biotechnol. 35 (3), 326–333.

De Oliveira, J.L., Campos, E.V.R., Bakshi, M., Abhilash, P.C., Fraceto, L.F., 2014. Application of nanotechnology for the encapsulation of botanical insecticides for sustainable agriculture: prospects and promises. Biotechnol. Adv. 32 (8), 1550–1561.

Devakumar, C., Dureja, P., 2002. Global News on Pesticides (2002). Pestic. Res. J. 14 (2), 365–370.

Dolatabadi, K.H., Goltapeh, E.M., Varma, A., Rohani, N., 2011. In vitro evaluation of arbuscular mycorrhizal-like fungi and *Trichoderma* species against soil borne pathogens. J. Agric. Technol. 7 (1), 73–84.

Elek, N., Hoffman, R., Raviv, U., Resh, R., Ishaaya, I., Magdassi, S., 2010. Novaluron nanoparticles: formation and potential use in controlling agricultural insect pests. Colloids Surf. A Physicochem. Eng. Asp. 372 (1–3), 66–72.

EPA, 2018. United States Environmental Protection Agency. Available at: https://www.epa.gov/home/forms/contact-epa. (Accessed 25 August 2018).

Ezhilarasi, P.N., Karthik, P., Chhanwal, N., Anandharamakrishnan, C., 2013. Nanoencapsulation techniques for food bioactive components: a review. Food Bioproc. Tech. 6 (3), 628–647.

FAO, 2013. Statistical Year Book. World Food and Agriculture, FAO, Rome.

Farooq, T., Adeel, M., He, Z., Umar, M., Shakoor, N., da Silva, W., Elmer, W., White, J.C., Rui, Y., 2021. Nanotechnology and plant viruses: an emerging disease management approach for resistant pathogens. ACS Nano 15 (4), 6030–6037.

Fenibo, E.O., Ijoma, G.N., Matambo, T., 2021. Biopesticides in sustainable agriculture: a critical sustainable development driver governed by green chemistry principles. Front. Sustain. Food Syst. 141.

Fenibo, E.O., Ijoma, G.N., Matambo, T., 2022. Biopesticides in sustainable agriculture: Current status and future prospects. In: Mandal, S.D., Ramkumar, G., Karthi, S., Jin, F. (Eds.), New and Future Development in Biopesticide Research: Biotechnological Exploration. Springer, Singapore, pp. 1–53.

Flint, M.L., 2012. IPM in Practice: Principles and Methods of Integrated Pest Management. vol. 3418 University of California Agriculture and Natural Resources.

Food and Agriculture Organization of the United Nations (Ed.), 2017. The Future of Food and Agriculture: Trends and Challenges. Food and Agriculture Organization of the United Nations.

Fraceto, L.F., Grillo, R., De Medeiros, G.A., Scognamiglio, V., Rea, G., Bartolucci, C., 2016. Nanotechnology in agriculture: which innovation potential does it have? Front. Environ. Sci. 4, 20.

Furxhi, I., 2022. Health and environmental safety of nanomaterials: O data, where art thou? NanoImpact 25, 100378.

Gallagher, N., 2021. Building a common-sense solution with IPM and IRM. Syngenta. https://www.syngentapmp.com/articles/newsarticle.aspx?type=tech&paid=221190. (Accessed 29 October 2021).

Gams, W., Diederich, P., Poldmaa, K., 2004. Fungicolous fungi. In: Mueller, G.M., Bills, G.F., Foster, M.S. (Eds.), Biodiversity of Fungi. Academic Press, Cambridge, MA, pp. 343–392.

Gasser, M., Rothen-Rutishauser, B., Krug, H.F., Gehr, P., Nelle, M., Yan, B., Wick, P., 2010. The adsorption of biomolecules to multi-walled carbon nanotubes is influenced by both pulmonary surfactant lipids and surface chemistry. J. Nanobiotechnol. 8, 1–9.

Ghormade, V., Deshpande, M.V., Paknikar, K.M., 2011. Perspectives for nano-biotechnology enabled protection and nutrition of plants. Biotechnol. Adv. 29 (6), 792–803.

Gilligan, T.M., Passoa, S.C., 2014. LepIntercept—an identification resource for intercepted Lepidoptera larvae. In: Identification Technology Program (ITP). Fort Collins, CO. Available at: http://idtools.org/id/leps/lepintercept/morphology.html. (Accessed 8 August 2022).

Giunti, G., Campolo, O., Laudani, F., Zappalà, L., Palmeri, V., 2021. Bioactivity of essential oil-based NBPs toward *Rhyzopertha dominica* (Coleoptera: Bostrichidae). Ind. Crop Prod. 162, 113257.

Glare, T., Caradus, J., Gelernter, W., Jackson, T., Keyhani, N., Köhl, J., Marrone, P., Morin, L., Stewart, A., 2012. Have biopesticides come of age? Trends Biotechnol. 30, 250–258.

Gould, F., Brown, Z.S., Kuzma, J., 2018. Wicked evolution: can we address the sociobiological dilemma of pesticide resistance? Science 360 (6390), 728–732.

Hajek, A., Eilelnberg, J., 2004. Natural Enemies: An Introduction to Biological Control, second ed. Cambridge University Press, Cambridge, UK.

Hawkins, N.J., Bass, C., Dixon, A., Neve, P., 2019. The evolutionary origins of pesticide resistance. Biol. Rev. 94 (1), 135–155.

Hazrati, H., Saharkhiz, M.J., Niakousari, M., Moein, M., 2017. Natural herbicide activity of *Satureja hortensis* L. essential oil nanoemulsion on the seed germination and morphophysiological features of two important weed species. Ecotoxicol. Environ. Saf. 142, 423–430.

He, X., Aker, W.G., Fu, P.P., Hwang, H.M., 2015. Toxicity of engineered metal oxide nanomaterials mediated by nano-bio-eco-interactions: a review and perspective. Environ. Sci. Nano 2, 564–582.

Heap, I., 2022. The international herbicide-resistant weed database. Available at: www.weedscience.org. (Accessed 22 April 2022).

Naturland, 2011. Naturland standards for organic aquaculture. Available at: http://www.nocaagro.com/standards/Naturland/Naturland%20Standards%20Aquaculture.pdf. (Accessed 2 August 2022).

Hu, Y.L., Gao, J.Q., 2010. Potential neurotoxicity of nanoparticles. Int. J. Pharm. 394 (1–2), 115–121.

Ibrahim, H.M., 2015. Green synthesis and characterization of silver nanoparticles using banana peel extract and their antimicrobial activity against representative microorganisms. J. Radiat. Res. Appl. Sci. 8 (3), 265–275.

IFOAM, 2011. IFOAM position paper on "The use of nanotechnologies and nanomaterials in organic agriculture". Available at: https://www.ifoam.bio/use-nanotechnologies-and-nanomaterials-organic-agriculture. (Accessed 8 August 2022).

IPPC Secretariat, 2021. Scientific Review of the Impact of Climate Change on Plant Pests – A Global Challenge to Prevent and Mitigate Plant Pest Risks in Agriculture, Forestry and Ecosystems. FAO on behalf of the IPPC Secretariat, Rome.

Iravani, S., 2011. Green synthesis of metal nanoparticles using plants. Green Chem. 13 (10), 2638–2650.

Isman, M.B., 2015. A renaissance for botanical insecticides? Pest Manag. Sci. 71, 1587–1590.

Isman, M.B., 2020. Bioinsecticides based on plant essential oils: a short overview. Z. Naturforsch. C 75, 179–182.

Izadi, M., Jorf, S.A.M., Nikkhah, M., Moradi, S., 2021. Antifungal activity of hydrocolloid nano encapsulated *Carum copticum* essential oil and *Peganum harmala* extract on the pathogenic fungi *Alternaria alternata*. Physiol. Mol. Plant Pathol. 116, 101714.

Jampílek, J., Králová, K., 2019. NBP in agriculture: State of the art and future opportunities. In: Koul, O. (Ed.), NBPs Today and Future Perspectives. Academic Press, pp. 397–447.

Jesser, E., Lorenzetti, A.S., Yeguerman, C., Murray, A.P., Domini, C., Werdin-González, J.O., 2020. Ultrasound assisted formation of essential oil nanoemulsions: emerging alternative for *Culex pipiens pipiens* Say (Diptera: Culicidae) and *Plodia interpunctella* Hübner (Lepidoptera: Pyralidae) management. Ultrason. Sonochem. 61, 104832.

Jha, A.K., Prasad, K., Prasad, K., Kulkarni, A.R., 2009. Plant system: nature's nanofactory. Colloids Surf. B Biointerfaces 73 (2), 219–223.

Jia, X., Sheng, W.B., Li, W., Tong, Y.B., Liu, Z.Y., Zhou, F., 2014. Adhesive polydopamine coated avermectin microcapsules for prolonging foliar pesticide retention. ACS Appl. Mater. Interfaces 6, 19552–19558.

Jiang, L.C., Basri, M., Omar, D., Rahman, M.B.A., Salleh, A.B., Rahman, R.N.Z.R.A., Selamat, A., 2012. Green nanoemulsion intervention for water-soluble glyphosate isopropylamine (IPA) formulations in controlling *Eleusine indica* (*E. indica*). Pestic. Biochem. Phys. 102 (1), 19–29.

Kabaluk, J.T., Svircev, A.M., Goettel, M.S., Woo, S.G., 2010. The use and Regulation of Microbial Pesticides in Representative Jurisdictions Worldwide. International Organization for Biological Control of Noxious Animals and Plants (IOBC). pp. 99.

Karthik, P., Ezhilarasi, P.N., Anandharamakrishnan, C., 2017. Challenges associated in stability of food grade nanoemulsions. Crit. Rev. Food Sci. Nutr. 57 (7), 1435–1450.

Kaya, H.K., Lacey, L.A., 2007. Introduction to microbial control. In: Lacey, L.A., Kaya, H.K. (Eds.), Field Manual of Techniques in Invertebrate Pathology. Springer, Dordrecht, pp. 3–7.

Klaper, R.D., 2020. The known and unknown about the environmental safety of nanomaterials in commerce. Small 16 (36), 2000690.

Kookana, R.S., Boxall, A.B.A., Reeves, P.T., Ashauer, R., Beulke, S., Chaudhry, Q., Cornelis, G., Fernandes, T.F., Gan, J., Kah, M., Lynch, I., Ranville, J., Sinclair, C., Spurgeon, D., Tiede, K., Van Den Brink, P.J., 2014. Nanopesticides: guiding principles for regulatory evaluation of environmental risks. J. Agric. Food Chem. 62, 4227–4240.

Koul, O., 2011. Microbial biopesticides: opportunities and challenges. CAB Rev. 6, 1–26.

Koul, O. (Ed.), 2019. NBPs Today and Future Perspectives. Academic Press.

Kumar, J., Ramlal, A., Mallick, D., Mishra, V., 2021. An overview of some biopesticides and their importance in plant protection for commercial acceptance. Plan. Theory 10 (6), 1–15.

Lade, B.D., Gogle, D.P., 2019. NBPs: synthesis and applications in plant safety. In: Abd-Elsalam, K.A., Prasad, R. (Eds.), Nanobiotechnology Applications in Plant Protection, Nanotechnology in the Life Sciences. Springer, Switzerland, pp. 169–189.

Lade, B.D., Gogle, D.P., Lade, D.B., Moon, G.M., Nandeshwar, S.B., Kumbhare, S.D., 2019. NBP formulations: Application strategies today and future perspectives. In: Koul, O. (Ed.), NBPs Today and Future Perspectives. Academic Press, pp. 179–206.

Larrouturou, L., 2005. The nanosciences. Centre national De La Recherché. Science 42.

Lu, F., Sun, D., Huang, J., Du, M., Yang, F., Chen, H., Li, Q., 2014. Plant-mediated synthesis of Ag–Pd alloy nanoparticles and their application as catalyst toward selective hydrogenation. ACS Sustain. Chem. Eng. 2 (5), 1212–1218.

Makarenko, N.A., Makarenko, V.V., 2019. Nanotechnologies in crop cultivation: ecotoxicological aspects. Biosyst. Divers. 27 (2), 148–155.

Manchikanti, P., 2019. Bioavailability and environmental safety of NBP. In: Koul, O. (Ed.), NBPs Today and Future Perspectives. Academic Press, pp. 207–222.

Marrone, P.G., 2002. An effective biofungicide with novel modes of action. Pesticide Outlook 13 (5), 193–194.

Mazid, S., Kalita, J.C., Rajkhowa, R.C., 2011. A review on the use of biopesticides in insect pest management. Int. J. Sci. Adv. Technol. 1 (7), 169–178.

McGuire, A.V., Northfield, T.D., 2020. Tropical occurrence and agricultural importance of *Beauveria bassiana* and *Metarhizium anisopliae*. Front. Sustain. Food Syst. 4, 6.

Mesnage, R., Séralini, G.-E., 2018. Editorial: toxicity of pesticides on health and environment. Front. Public Health 6.

Monica, R.C., Cremonini, R., 2009. Nanoparticles and higher plants. Caryologia 62, 161–165.

Morales, H., 2002. Pest management in traditional tropical agroecosystems: lessons for pest prevention research and extension. Integr. Pest Manag. Rev. 7 (3), 145–163.

Morales, H., Perfecto, I., 2000. Traditional knowledge and pest management in the Guatemalan highlands. Agric. Human Values 17 (1), 49–63.

Mordor Intelligence, 2021. Biopesticides Market Share, Trends, Analysis (2022–27). Available at: https://www.mordorintelligence.com/industry-reports/global-biopesticides-market-industry.

Moreno-Gavíra, A., Huertas, V., Diánez, F., Sánchez-Montesinos, B., Santos, M., 2020. Paecilomyces and its importance in the biological control of agricultural pests and diseases. Plan. Theory 9 (12), 1746.

Nadendla, S.R., Rani, T.S., Vaikuntapu, P.R., Maddu, R.R., Podile, A.R., 2018. HarpinPss encapsulation in chitosan nanoparticles for improved bioavailability and disease resistance in tomato. Carbohydr. Polym. 199, 11–19.

Nair, R., Varghese, S.H., Nair, B.G., Maekawa, T., Yoshida, Y., Kumar, D.S., 2010. Nanoparticulate material delivery to plants. Plant Sci. 179, 154–163.

Nawaz, M., Mabubu, J.I., Hua, H., 2016. Current status and advancement of biopesticides: microbial and botanical pesticides. J. Entomol. Zool. Stud. 4 (2), 241–246.

Nicolopoulou-Stamati, P., Maipas, S., Kotampasi, C., Stamatis, P., Hens, L., 2016. Chemical pesticides and human health: the urgent need for a new concept in agriculture. Front. Public Health 4, 148.

Nordhaus, T., Shah, S., 2021. In Sri Lanka, organic farming went catastrophically wrong. In: Foreign Policy. Available at: https://foreignpolicy.com/2022/03/05/sri-lanka-organic-farming-crisis/. (Accessed 3 August 2021).

Ombódi, A., Saigusa, M., 2000. Broadcast application versus band application of polyolefin-coated fertilizer on green peppers grown on andisol. J. Plant Nutr. 23 (10), 1485–1493.

Palermo, D., Giunti, G., Laudani, F., Palmeri, V., Campolo, O., 2021. Essential oil-based NBPs: formulation and bioactivity against the confused flour beetle *Tribolium confusum*. Sustainability 13 (17), 9746.

Pan, X., Guo, X., Zhai, T., Zhang, D., Rao, W., Cao, F., Guan, X., 2023. Nanobiopesticides in sustainable agriculture: developments, challenges, and perspectives. Environ. Sci. Nano.

Pant, M., Dubey, S., Patanjali, P.K., Naik, S.N., Sharma, S., 2014. Insecticidal activity of eucalyptus oil nanoemulsion with karanja and jatropha aqueous filtrates. Int. Biodeter. Biodegr. 91, 119–127.

Parisi, C., Vigani, M., Rodríguez-Cerezo, E., 2015. Agricultural nanotechnologies: what are the current possibilities? Nano Today 10, 124–127.

Patel, S.K., Sharma, A., Singh, G.S., 2020. Traditional agricultural practices in India: an approach for environmental sustainability and food security. Energy Ecol. Environ. 5 (4), 253–271.

Pavela, R., 2014. Limitation of plant biopesticides. In: Singh, D. (Ed.), Advances in Plant Biopesticides. Springer Publishing, New Delhi, India, pp. 347–359.

Pavoni, L., Benelli, G., Maggi, F., Bonacucina, G., 2019. Green nanoemulsion interventions for biopesticide formulations. In: Koul, O. (Ed.), NBPs Today and Future Perspectives. Academic Press, pp. 133–160.

Pérez-de-Luque, A., Rubiales, D., 2009. Nanotechnology for parasitic plant control. Pest Manag. Sci. 65 (5), 540–545.

Perlatti, B., de Bergo, P.L.S., da Silva, M.F.G.F., Fernandes, J.B., Forim, M.R., 2013. Polymeric nanoparticle-based insecticides: a controlled release purpose for agrochemicals. In: Trdan, S. (Ed.), Insecticides—Development of Safer and More Effective Technologies. IntechOpen, pp. 523–550.

Pietroiusti, A., Stockmann-Juvala, H., Lucaroni, F., Savolainen, K., 2018. Nanomaterial exposure, toxicity, and impact on human health. Wiley Interdiscip. Rev. Nanomed. Nanobiotechnol. 10, 1–21.

Pingali, P.L., 2012. Green revolution: impacts, limits, and the path ahead. Proc. Natl. Acad. Sci. U. S. A. 109, 12302–12308.

Pinho, B., 2021. The growing problem of pesticide resistance. In: Chemistry World. Available at: https://www.chemistryworld.com/features/the-growing-problem-of-pesticide-resistance/4013465.article. (Accessed 6 April 2021).

Pino-Otín, M.R., Ballestero, D., Navarro, E., González-Coloma, A., Val, J., Mainar, A.M., 2019a. Ecotoxicity of a novel biopesticide from *Artemisia absinthium* on non-target aquatic organisms. Chemosphere 216, 131–146.

Pino-Otín, M.R., Val, J., Ballestero, D., Navarro, E., Sánchez, E., González-Coloma, A., Mainar, A.M., 2019b. Ecotoxicity of a new biopesticide produced by *Lavandula luisieri* on non-target soil organisms from different trophic levels. Sci. Total Environ. 671, 83–93.

Qin, H., Zhang, H., Li, L., Zhou, X., Li, J., Kan, C., 2017. Preparation and properties of lambda-cyhalothrin/polyurethane drug-loaded nanoemulsions. RSC Adv. 7, 52684–52693.

Ragaei, M., Sabry, A.H., 2014. Nanotechnology for insect pest. Int. J. Sci. Environ. 3, 528–545.

Rahman, S., Biswas, S.K., Barman, N.C., Ferdous, T., 2016. Plant extract as selective pesticide for integrated pest management. Biotechnol. Res. 2 (1), 6–10.

Ray, P.C., Yu, H., Fu, P.P., 2009. Toxicity and environmental risks of nanomaterials: challenges and future needs. J. Environ. Sci. Heal. - Part C Environ. Carcinog. Ecotoxicol. Rev. 27, 1–35.

Roduner, E., 2006. Size matters: why nanomaterials are different. Chem. Soc. Rev. 35 (7), 583–592.

Ross, M., Lembi, C., 1985. Applied Weed Science, Simon & Schuster Books for Young Readers. Burgess Pub Co, Minneapolis.

Sabbour, M.M., Solieman, N.Y., 2022. Effect of biological insecticides on three harmful sugarcane pests. J. Posit. School Psychol. 6 (2), 6241–6249.

Saberi, F., Marzban, R., Ardjmand, M., Shariati, F.P., Tavakoli, O., 2020. Optimization of culture media to enhance the ability of local *bacillus thuringiensis var. tenebrionis*. J. Saudi Soc. Agric. Sci. 19 (7), 468–475.

Sahu, U., Malik, T., Ibrahim, S.S., Vendan, S.E., Karthik, P., 2022. Pest management with green nanoemulsions. In: Bio-Based Nanoemulsions for Agri-Food Applications. Elsevier, pp. 177–195.

Saini, P., Gopal, M., Kumar, R., Srivastava, C., 2014. Development of pyridalyl nanocapsule suspension for efficient management of tomato fruit and shoot borer (*Helicoverpa armigera*). J. Environ. Sci. Health B 49, 344–351.

Salma, U., Chen, N., Richter, D.L., Filson, P.B., Dawson-Andoh, B., Matuana, L., Heiden, P., 2010. Amphiphilic core/shell nanoparticles to reduce biocide leaching from treated wood, 1—Leaching and biological efficacy. Macromol. Mater. Eng. 295, 442–450.

Samal, P.K., Dhyani, P.P., 2007. Indigenous soil fertility maintenance and pest control practices in traditional agriculture in the Indian Central Himalaya: empirical evidence and issues. Outlook Agric. 36 (1), 49–56.

Sánchez-Santed, F., Colomina, M.T., Herrero Hernández, E., 2016. Organophosphate pesticide exposure and neurodegeneration. Cortex 74, 417–426.

Sayed, A.M., Behle, R.W., 2017. Comparing formulations for a mixed-microbial biopesticide with *Bacillus thuringiensis* var. kurstaki and *Beauveria bassiana* blastospores. Arch. Phytopathol. Plant Prot. 50 (15–16), 745–760.

Scrinis, G., Lyons, K., 2010. Nanotechnology and the techno-corporate Agri-food paradigm. In: Lawrence, G., Lyons, K., Wallington, T. (Eds.), Food Security Nutrition and Sustainability. Earthscan, London, pp. 252–270.

Sharma, V., Kumar, A., Dhawan, A., 2012. Nanomaterials: exposure, effects and toxicity assessment. Proc. Natl. Acad. Sci. India Sect. B Biol. Sci. 82, 3–11.

Shoaib, A., Elabasy, A., Waqas, M., Lin, L., Cheng, X., Zhang, Q., Shi, Z.H., 2018. Entomotoxic effect of silicon dioxide nanoparticles on *Plutella xylostella* (L.) (Lepidoptera: Plutellidae) under laboratory conditions. Toxicol. Environ. Chem. 100 (1), 80–91.

Siddiqui, Z.A., Khan, M.R., Abd-Allah, E.F., Parveen, A., 2019. Titanium dioxide and zinc oxide nanoparticles affect some bacterial diseases, and growth and physiological changes of beetroot. Int. J. Veg. Sci. 25 (5), 409–430.

Stadler, T., Lopez-Garcia, G.P., Gitto, J.G., Buteler, M., 2017. Nanostructured alumina: biocidal properties and mechanism of action of a novel insecticide powder. Bull. Insectology 70 (1), 17–25.

Street, M.E., Angelini, S., Bernasconi, S., Burgio, E., Cassio, A., Catellani, C., Cirillo, F., Deodati, A., Fabbrizi, E., Fanos, V., Gargano, G., Grossi, E., Iughetti, L., Lazzeroni, P., Mantovani, A., Migliore, L., Palanza, P., Panzica, G., Papini, A.M., Parmigiani, S., Predieri, B., Sartori, C., Tridenti, G., Amarri, S., 2018. Current knowledge on endocrine disrupting chemicals (EDCs) from animal biology to humans, from pregnancy to adulthood: Highlights from a national Italian meeting. Int. J. Mol. Sci. 19 (6), 1647.

Tia, V.E., Gueu, S., Cissé, M., Tuo, Y., Gnago, A.J., Konan, E., 2021. Bio-insecticidal effects of essential oil nanoemulsion of *Lippia multiflora* Mold. on major cabbage pests. J. Plant Prot. Res. 61 (1), 103–109.

Tilman, D., Knops, J., Wedin, D., Reich, P., 2002. Plant diversity and composition: Effects on productivity and nutrient dynamics of experimental grasslands. In: Loreau, M., Naeem, S., Inchausti, P. (Eds.), Biodiversity and Ecosystem Functioning: Synthesis and Perspectives. Oxford University Press, UK, p. 21.

Trenkel, M.E., 1997. Controlled-Release and Stabilized Fertilizers in Agriculture: Improving Fertilizer Use Efficiency. IFA.

Tripathi, Y.N., Divyanshu, K., Kumar, S., Jaiswal, L.K., Khan, A., Birla, H., Gupta, A., Singh, S.P., Upadhyay, R.S., 2020. Biopesticides: Current status and future prospects in India. In: Keswani, C. (Ed.), Bioeconomy for Sustainable Development. Springer, Singapore, pp. 79–109.

Turco, R.F., Bischoff, M., Tong, Z.H., Nies, L., 2011. Environmental implications of nanomaterials: Are we studying the right thing? Curr. Opin. Biotechnol. 22, 527–532.

Turek, C., Stintzing, F.C., 2013. Stability of essential oils: a review. Compr. Rev. Food Sci. Food Saf. 12, 40–53.

Ul Haq, I., Ijaz, S., 2019. Use of metallic nanoparticles and nanoformulations as nanofungicides for sustainable disease management in plants. In: Prasad, R., Kumar, V., Kumar, M., Choudhary, D. (Eds.), Nanobiotechnology in Bioformulations. Springer, Cham, pp. 289–316.

United Nations, 2021. Population, Food Security, Nutrition and Sustainable Development (UN Department of Economic and Social Affairs (DESA) Policy Briefs Policy Brief no 102; UN Department of Economic and Social Affairs (DESA) Policy Briefs). Available at:, https://doi.org/10.18356/27081990-102.

Wang, T.T., Cheng, J., Liu, X.J., Jiang, W., Zhang, C.L., Yu, X.Y., 2012. Effect of biochar amendment on the bioavailability of pesticide chlorantraniliprole in soil to earthworm. Ecotoxicol. Environ. Saf. 83, 96–101.

World Health Organization, & Food and Agriculture Organization of the United Nations, 2016. The International Code of Conduct on Pesticide Management: Guidelines on Highly Hazardous Pesticides. Food and Agriculture Organization of the United Nations.

Yadav, A., Kon, K., Kratosova, G., Duran, N., Ingle, A.P., Rai, M., 2015. Fungi as an efficient mycosystem for the synthesis of metal nanoparticles: progress and key aspects of research. Biotechnol. Lett. 37 (11), 2099–2120.

Yasur, J., Usha-Rani, P., 2015. Lepidopteran insect susceptibility to silver nanoparticles and measurement of changes in their growth, development and physiology. Chemosphere 124, 92–102.

Yousef, N., Niloufar, M., Elena, P., 2019. Antipathogenic effects of emulsion and nanoemulsion of cinnamon essential oil against Rhizopus rot and grey mold on strawberry fruits. Foods Raw Mater. 7, 210–216.

Zhao, X., Cui, H., Wang, Y., Sun, C., Cui, B., Zeng, Z., 2018. Development strategies and prospects of nano-based smart pesticide formulation. J. Agric. Food Chem. 66, 6504–6512.

CHAPTER

16

Nanoagrochemicals start-up for sustainable agriculture

Bipin D. Lade[a], *Avinash P. Ingle*[b], *Mangesh Moharil*[b], *and Bhimanagouda S. Patil*[a]

[a]Vegetable and Fruit Improvement Center, USDA National Center of Excellence Department of Horticultural Sciences, Texas A&M University, College Station, TX, United States
[b]Biotechnology Centre, Department of Agricultural Botany, Dr. Panjabrao Deshmukh Agricultural University, Akola, Maharashtra, India

1 Introduction

Agriculture is a farming trade that includes livestock and crop production, and numerous several countries rely on this trade (Mendhe et al., 2008; Dhawan, 2017). Primitive agriculture practices that assembled food for a particular area have transformed into improved farming, considering large fields and resources integrated with high-level technologies. Agriculture produces food and feed enterprises for the worldwide population of almost 7.5 billion, which is projected to reach 9 billion by 2050. This will require 50%–70% more production (Hossain et al., 2020; Singh et al., 2021). Agriculture and allied agribusiness consulting are facing many challenges, such as decreased crop yield, loss of soil fertility, soil degradation, climate change, excess use of agrochemicals, pests, and insect attacks. Various abiotic (drought, flooding, heat, hail, salinity, heavy metals, mineral deficiency) and biotic (living entities such as viruses, bacteria, insects, fungi, weeds, and nematodes) stresses are usually responsible for crop yield losses (Seleiman et al., 2021). However, among these, biotic stress, particularly plant fungi, is the most important factor contributing to major crop losses. Some of the common fungal species causing plant diseases are listed in Table 1. Among the fungi, the *Fusarium* species including *Fusarium oxysporum*, *Fusarium solani*, *Fusariumfujikuroi*, *Fusarium graminearum*, *Fusarium culmorum*, and *Fusarium avenaceum* are reported to attack and destroy wheat, barley, and other small grain cereals. *Fusarium* is mainly responsible for the formation of crown rot (FCR) disease and Fusarium head blight (FHB). The *Fusarium* species in FHB produce asexual conidia,

Nanohybrid Fungicides
https://doi.org/10.1016/B978-0-443-23950-2.00018-7

TABLE 1 The fungal species that affect various crops (Gea et al., 2021; Perrone et al., 2020; Logrieco et al., 2003).

S. no.	Fungal species	Crops affected
1	*F. graminearum*	Maize, wheat
2	*F. culmorum*	
3	*F. crookwellense*	
4	*F. verticillioides*	Maize
5	*F. proliferatum*	
6	*A. flavus*	Nuts, peanuts, spices, maize, dried fruit
7	*A. parasiticus*	
8	*F. sporotrichioides*	Barley, rye, oats
9	*F. langesthiae*	
10	*F. poae*	
11	*Aspergillus niger*	Grapes, apricot
12	*Trichorderma viride*	Maize
13	*Rhizopus oryzae*	Vegetables and fruits
14	*Rhizopus stolonifer*	Tomato, cherry
15	*Albugo candida*	Brassicas
16	*Sclerotiorum rolfii*	Beans, beets, potato, tomato
17	*S. cepivorum*	Onions

but some have sexual stages producing ascospores such as *F. graminearum* (teleomorph *Gibberella zeae*) and *F. anaceum* (teleomorph *G. avenaceave*) (Karlsson et al., 2017; Logrieco et al., 2003; Bernhoft et al., 2012). In Japan, *F. oxysporum* f. sp. *batatas* have emerged that effects sweet potato wilt disease. The *F. oxysporum* f. sp. *pisi* has extended to North America, Europe, Australia, New Zealand, and Japan (Arie, 2019).

Pathogen problems are solved using agrochemicals (pesticides and fertilizers), which are usually formulated using synthetic chemicals. Overall, approximately 60% of fertilizers and 2.5 million tons of pesticides are used every year for total world food production (War et al., 2020; Abd-Elsalam and Alghuthaymi, 2015). Agrochemical application not only affects the quality of the crop stock, but it also causes serious damage to the environment and other living organisms (Bhandari, 2014). Excessive use of agrochemicals may hinder sustainable agriculture, and repeated use has led to resistance developing in pathogens. Recently, agrochemicals have been formulated using organic compounds and plant extract-based solutions to avoid the synthesis of toxic byproducts. These organic agrochemicals are ecofriendly, cost-effective, maintain higher crop productivity, and improve soil fertility (Yadav et al., 2013). To overcome the disadvantages of conventional methods, nanotechnology has been applied in agriculture to ensure precision bioactive delivery in agriculture and protect crops against pathogen attacks and various environmental stresses.

Therefore, the aim of this chapter is to focus on the role of nanotechnology in general and nano-based start-ups in particular in sustainable agriculture. Various nano-based start-ups and the products they developed, in addition to other nano-based agro-products, are also discussed. Apart from these, other important topics such as the current status of agribusiness, the economic importance of nanoagrochemicals, the requirement to set up nano-based start-ups, etc., are critically explained.

2 Nanotechnology and agriculture

Nanotechnology has numerous applications as far as the agriculture field is concerned, which mainly include the development of mechanical components with nanocoating and the use of nanobiosensors for the control of weeds and pests. Furthermore, a wide range of smart machines useful in agriculture has been developed using different nanomaterials such as metal oxides, magnetic materials, ceramics, quantum dots, lipids, polymers, and dendrimers. Nanotechnological innovation has assisted in the transformation of farming, food production, and harvest yields while at the same time protecting the natural environmental equilibrium (Singh et al., 2021). The most successful nanotechnological interventions in farming frameworks mainly include water and soil remediation, improvements in soil attributes, and reducing plant pathogen infections. The unique and extraordinary properties of nanomaterials make them suitable for various agricultural applications. These mainly include: (a) higher charge density and higher reactivity of nanoparticles due to their small size, (b) nanoparticles have higher catalytic activity when they are present in a tetrahedral structure, followed by cubic and spherical structures, recognized for the improvement of chemical reactivity at the sharp edges and corners of the former, (c) high surface area and controlled size, shape, and structure as a result of having a high antimicrobial effect, and (d) increased pest target activity. Nanoparticles that control plant diseases are nanoforms of carbon, silver, silica, alumina-silicates, zinc oxide, titanium oxide, platinum, and gold (Duhana et al., 2017). To date, several nano-based agrochemicals, such as nanopesticides, nanofungicides, nanoweedicides, etc., have been developed and used to control pre- and postharvest diseases. Moreover, there has been success in the development of nanosensors and diagnostic devices for monitoring agroecosystems for crop protection (Elizabath et al., 2019). Some of the important nano-based agro-products are discussed here.

2.1 Nanoagrochemicals

The global demand for agrochemicals increased from $90,480 million in 2021 and forecasted to generate revenue of $328.0 billion (Zobir et al., 2021) during 2018–2026 (Global newswire, 2021). Nanoagrochemicals are considered a new generation of chemicals with enormous advantages and a few drawbacks. The organic agrochemicals integrated with nanotechnology are called "smart agrochemicals," also known as "nanoagrochemicals," for the development of effective, precise, and sustainable agriculture. Agrochemicals, such as pesticides, herbicides, insecticides, fertilizers, fungicides, hormones, growth agents, acidifying agents, etc., in nanoform or complexed with other nanomaterials are used to maintain the agriculture ecosystem. Agrochemical efficacy has been improved with the use of nanotechnology, and this is now called "nanoagrochemical" (Prasad and Mahawer, 2020). Certain nanoagrochemicals possess potential antimicrobial properties; hence, they can effectively manage the microbial pathogens responsible for plant diseases. In addition to nanofungicides and nanopesticides, these agents can significantly promote plant growth and protect plants from attack by insects and pests. These are used for improving crop productivity, food quality, quantity, and nutrients in foods. Nanofertilizers used for efficient nutrient management, nanopesticides inhibits the growth of pests, nanoherbicides assist selective weed control, and nanomaterials play critical role in efficient delivery of agents for eco-friendly agrochemicals (Qazi and Dar, 2020). Nanotechnology allows the manipulation and control of matter at a dimension between 1 and 100nm when complexed with organic, biopolymeric, and bioactive molecules for the formulation of nanoagrochemicals that reduce toxicity risk to plants, animals, human health, and the environment. These synthesized nanofertilizers and nanopesticides improve plant growth and control pathogens. For example, silica nanoparticles (SiO_2 NPs) showed increased tolerance of the hawthorn (*Crataegus* sp.) seedling against drought stress. Similarly, SiO_2 NPs assist with problems of salinity. The use of SiO_2 NPs nanoproducts decreased the Na^+ ion toxicity and improved the growth of plants under salt stress (Seleiman et al., 2021). The utilization of phosphorus-enriched hydroxyapatite nanoparticle (HA-NPs) fertilizer was found to remarkably improve the plant height, shoot growth, and high grain yield. For example, in soybeans (*Glycine* max. L.), zeolites and chitosan were found to improve the intake and uptake efficiency by 18% (Seleiman et al., 2021). The different types of nanoagrochemicals are discussed in the following.

2.1.1 *Nanofertilizers (NF)*

Nanofertilizers (NFs) are smart fertilizers that play a crucial role in agriculture to improve crop productivity, soil fertility, and the quality of agricultural products. Nano-enabled fertilizers help plants absorb nutrients and are used to enhance seed germination, for example. NFs are of different types and are fabricated using organic and synthetic chemicals to achieve controlled and targeted release. Nanofertilizers are divided into two types: (a) nano-supported fertilizers and (b) nano-sized fertilizers, which are fertilizers manufactured at the nanoscale (Shao et al., 2022).

(a) Nano-supported fertilizers

Nano-supported fertilizers are synthesized via the application of nanostructured materials as additives for the controlled release of fertilizers. The fertilizers are entrapped in liposomes via solvent injection and extrusion techniques (Farshchi et al., 2021). In addition, nanofibers are used for the entrapment of fertilizers using electrospinning techniques. As active ingredients, polyvinyl alcohol (PVA) was functional in the core and polylactic acid (PLA) formed the outer covering shell (Nooeaid et al., 2021). Porous block copolymers have been employed to entrap Fe in the palygorskite nanoparticles for temperature-based release to the crop plant (Chi et al., 2018). In another study, urea was loaded into the porous halloysite nanotubes coated with chitosan, which showed release upon glutathione produced by crops by disturbing the sulfur bonds of chitosan (Wang et al., 2020).

(b) Nano-sized fertilizers

Bottom-up approach techniques can convert insoluble nutrients such as minerals to nanoscale materials. Nano-enabled fertilizers are used for the delivery of nutrients using different nanomaterials such as hydroxyapatite, mesoporous silica, chitosan, calcite, carbon nanotubes, meta-oxide nanoparticles (zinc, copper), zeolites, magnetite, and nanoclays. HA nanoparticles have the ability to increase the phosphorous uptake efficiency in crop plants (Xiong et al., 2018). The Fe nano-sized fertilizers showed a decrease in the Fe deficiency symptoms of soybean plants in iron-deficient calcareous soil. Fe is applied to plants in both foliar and spray forms for effective transfer (Farshchi et al., 2021). Manganese zinc ferrite nanoparticles were synthesized using a microwave-assisted hydrothermal technique, and thus synthesized nanoparticles were reported to improve the growth and yield of squash (*Cucurbita pepo* L.) by 52.9% in comparison to control samples (Shebl et al., 2020; Shao et al., 2022). Nano-zinc oxide (NP-ZnO) enhances the germination rate of peanut seeds; likewise, nano-silicon dioxide (NP-SiO_2) was found to increase germination in soybean seeds (Singh et al., 2021). Zinc NFs in the form of ZnO are used regularly because they are more efficient and cost-effective than synthetic Zn fertilizers and may be used for soil mixing and foliar spray. Zinc nanofertilizers have a tendency to increase short growth, dry height, leaf area, and protein content in the sunflower (*Helianthus annus* L.), rice, maize, potato, and sugarcane (*Saccharum officinarum* L.) (Seleiman et al., 2021). Zn particles improve cell division, auxin production, and mineral absorption. The bimetallic manganese oxide/iron oxide (MnO/FeO) NPs synthesized from bacteria showed the best outcome on plant development, particularly in germination rates, root development, and new weight in maize plantlets; this demonstrates that these can be utilized as micronutrient nanofertilizers (Bratovcic et al., 2021). The application of $10\,mg\,L^{-1}$ of Se-NPs under extreme temperature stress during the booting stage of sorghum (*Sorghum bicolor* L. Moench) increased pollen germination and enhancee antioxidant defense (Seleiman et al., 2021). Cu is a constituent of regulatory proteins that participates in photosynthesis and plant respiration while also being a cofactor of antioxidants such as ascorbate oxidase. In crops, copper deficiency leads to various disorders such as a low number of seeds, stunted growth, a low number of fruits, necrosis, and low productivity of crops. CuO NP nanofertilizers improved the germination and root growth of soybeans and chickpeas (*Cicer arietinum* L.). Silicon NPs (Si-NPs) were successfully used in agriculture for maintaining soil moisture, increasing water retention of soil, etc. Therefore, Si-NPs are widely used as a fertilizer for the growth and improvement of the plant. Another important nanoagrochemical nanopesticide is discussed next.

2.1.2 Nanopesticides

Nanopesticides are any nanoformulation that involves either a very small quantity of a pesticide active ingredient or any small, engineered structures with useful pesticidal properties (Lade et al., 2017). Thus, nanopesticide formulations use low dosages and reduce exposure to humans (Prasad and Mahawer, 2020). Nanopesticides (NPc) are of different types and are fabricated using organic and synthetic chemicals. There are two types of NPc: (a) nano-entrapped pesticides and (b) nano-sized pesticides.

(a) Nano-entrapped pesticides

The pesticide of interest is entrapped in a matrix using biopolymers such as polysaccharides and proteins, which are commonly used as nanocarrier matrices (Sinha et al., 2019). The polymers, such as cyclodextrin, have the advantage of being hydrophobic inside and hydrophilic outside. The inner hydrophobic site serves as a binding site for pesticide entrapment and assists in slow release. Insect pesticide avermectin-loaded microcapsules with starch acetate as the carrier matrix have been created (Li et al., 2016). The 90% active loading was achieved on abamectin nanoparticles by using hypromellose acetate succinate (HPMCAS) with lecithin (Chun and Feng, 2021). In addition, 64% loading efficiency was reached for the λ-cyhalothrin-entrapped nanopesticides with benzoyl lignin as a nanoprecipitation method (Zhou et al., 2018). The biodegradable polylactic acid (PLA) and poly(lactic-*co*-glycolic acid) (PLGA) provided the entrapment site for the pesticide (Liu et al., 2017). Biodegradable castor oil-based polyurethane nanoparticles were prepared using a prepolymer dispersion method (Zhang et al., 2018). Mesoporous silica NPs have been used in agriculture to encapsulate and deliver pesticides that show greater activity against the biotic stresses of plants (War et al., 2020). Therefore, several biodegradable and nontoxic polymers are used for the preparation and entrapment of pesticides for slow release and long-term action.

(b) Nano-sized pesticides

Nanopesticide formulations can increase water solubility as well as the control of pathogens, insects, and weeds in crops (Chaud et al., 2021). Inorganic nanomaterials are more suited for processing with surface adsorption. For example, Cu-NPs as well as cobalt and nickel ferrite were used as a nanofungicide against *Fusarium solani*, *Neofusicoccum* sp., *Fusarium oxysporum*, and *Dematophora necatrix*. Nanostructured alumina manifested a strong insecticidal property to the families *Coleoptera* (*Sitophilus oryzae*, *Oryzaephilus surinamensis*) and *Dipteran* (*Ceratitis capitate*) (Zobir et al., 2021). For instance, inorganic nanoparticles such as Cu and Al have shown insecticidal and antibacterial activity in crop plants. Engineered and encapsulated NMs such as magnetite NPs, TiO_2 NPs, C60 MWCNTS, chitosan NMS, ZnO NPs, and SiO_2 NPs provides resistance against viral infections during plant growth. Furthermore, for greater application of nanopesticides, it is important to verify risk and gain knowledge of the nanopesticide action at the cellular level. Metal organic frameworks (MOFs) are a porous inorganic–organic hybrid material, with typical frameworks consisting of inorganic metal centers and organic pesticidal ligands. MOFs after their action get degrade and absorbed by soil. There, zirconium-, aluminum-, and iron-based MOFs that are synthesized via hydrothermal or microwave heating method are effective in nature (Meng et al., 2020; Gao et al., 2021). Nanopesticides including nanofungicides, nanobactericides, nanoweedicides, and nanoinsecticides are used for their efficient delivery system, low cost, easier application, and promotion of sustainable development of agriculture (Lade et al., 2017; Liu et al., 2021; Singh et al., 2021).

(c) Nanofungicides

Nanofungicides are as fungicides developed using nanomaterials. Nanofungicides are found to be most effective against a wide range of fungi, including resistant species, due to the strong fungicidal activity of certain metal nanoparticles such as silver, copper, sulfur, etc. Fungicides, also known as "antimycotic," are any toxic substance used to kill or inhibit the growth of fungi. Various plants and crops get infected by plant pathogens at any stage. The fungal pathogens such as *Fusarium graminearum*, *Puccinia* spp., *Bottrytis cinerea*, *Magnaporthe oryzae*, *Rhizoctonia solani*, and various fusarium species are able to attack plants and produce symptoms such as wilting, chlorosis, necrosis, stunting, yellowing of leaves, and plant death. The conventional use of fungicides to control plant pathogens can be damaging to the environment, farmers, and human health. About 2.5 million tons of pesticides are used on crops each year. Thus, there is a need to produce ecosafe antifungals such as bio-based nanomaterials (Abd-Elsalam and Alghuthaymi, 2015).

2.1.3 Nanoagrochemical benefits

Nanoagrochemicals are gaining popularity due to their high potential when compared to synthetic agrochemicals. They have aided farmers financially by increasing crop yields in both qualitative and quantitative terms. The various nanoagrochemicals such as nanofertilizers increase soil fertility and improve grain yield over conventional fertilizers, and nanopesticides may reduce environmental pollution by using biological nanoemulsions. Likewise, nano-based devices and materials have a considerable role in agro-industry. On the other hand, nanoemulsions remarkably reduce organic solvents and surfactants compared to conventional agrochemicals. Nano-encapsulated fertilizers and pesticides are useful for the protection of nontarget species. A recent study showed that nano-based stretch films synthesized with polyethelene have been used for hay packing systems to protect plants against *Escherichia coli* spoilage (Rajput et al., 2021). Also, carbon nanotubes are applicable in improving the germination of tomato seeds (Elizabath et al., 2019). Nanosensors (dust and gas) have helped in determining the amount of pollutants and dust in the environment. Thus, nanotechnology has a greater role in crop production due to its environmental and economical safety. Moreover, nanomaterial-based biomolecules such as DNA and RNA transfer are becoming promising tools in research fields involving genome editing and gene silencing in plants (Liu et al., 2021).

2.1.4 Nanoagrochemical toxicity

The application of nanoagrochemicals benefits the agriculture sector, but at the same time, it has a negative impact on the environment. Some researchers have shown that engineered nanoparticles have chronic and acute toxicological effects. Certain nanomaterials, including metal nanoparticles of heavy metals such as silver, have been reported to have a toxic effect. Therefore, the development of agrochemicals using such nanomaterials may cause toxicity to beneficial soil microflora, aquatic organisms, and plants if not applied in the desired concentration. For example, silver NPs may cause mitochondrial disinfection due to a change in cell permeability for K^+ and Na^+ ions. Al_2O_3, TiO_2, ZnO, and SiO_2 nanoparticles show toxicity among *Bacillus subtilis* and *Pseudomonas fluorscenes* (Prasad and Mahawer, 2020). Apart from this, the application of nanoagrochemicals can damage human health by reducing sperm quality and count, leading to sterility in men (Bhandari, 2014). Moreover, a recent study showed that the use of ZnO in earthworms leads to in vivo toxicity and causes oxidative damage to their coelomocytes. The earthworm coelomocytes were directly exposed to DNA lipids, polysaccharides, and proteins. Hence, the intracellular reactive oxygen species (ROS) caused potential damage to the earthworm (Li et al., 2019). Although most copper-based fungicides deplete the population of nitrogen-fixing bacteria, fungicidal residue such as amest, captan, etc., remains in the soil and affects N-fixing bacteria (Rhizobium) (Meena et al., 2020). The NPs, including Ag NPs, copper oxide NPs, and zinc oxide NPs, may lead to effects such as disruption of the stages of cell division, cell disintegration, chromosomal breaks, inhibition of seed germination, DNA damage, and elongation of roots in plants. Nowadays, nanoagrochemicals are emerging products, and their application in farming has led to a lot of difficulties in the market. Hence, nanoagrochemicals may be regarded as critical in terms of environmental impact.

2.1.5 Nanoscale products

Nanoscale products are those containing any ingredient in nanoform at any dimension. There are more than 232 nano-enabled products manufactured by 75 companies in 26 countries all over the world, and they are used in plant protection, soil improvement, animal husbandry, and plant breeding. As we discussed above, nano-enabled products such as nanofertilizers, nanopesticides, nanofungicides, and nanoherbicides can be used to destroy any pests, microbes, and unwanted herbs that harm agriculture. Table 2 shows various nano-enabled products and their respective applications.

TABLE 2 Some nano-enabled products available on the market and their respective applications.

S. no.	Nanoproducts	Applications	References
1	Nano-sized nutrients (ZnO and TiO_2 nanoparticles)	Boost in antioxidants in tomatoes	Duhana et al. (2017)
2	Biodegradable thermoplastic Starch (TPS)	Good tensile strength and lowered water permeability	Duhana et al. (2017)
3	Primo MAXX	Grass growth regulatory and nanopesticide	Duhana et al. (2017)
4	Nanocides	Pesticides encapsulated in nanoparticles for controlled release of nanoemulsions for greater efficiency	
5	Nanosensors	Detection of pathogens in contaminated packaged food.	Mukhopadhyay (2014)
6	Silver NPs	Control of phytopathogens	Mishra et al. (2014)
7	Chitosan polymer NPs	Used for controlled release of NPK	Mishra et al. (2014)
8	Nano-encapsulated herbicides	Control of parasitic weeds	Mishra et al. (2014)
9	Nanofertilizer	Increase the efficiency of plant nutrients and reduce soil toxicity	Rajput et al. (2021)
10	Lithocal	An essential element for crops, promoting root system, enhancing nutrient uptake, and increasing tillering sprouts and vegetative vigor	Rajput et al. (2021)

TABLE 2 Some nano-enabled products available on the market and their respective applications—cont'd

S. no.	Nanoproducts	Applications	References
11	Cu NPs	Nanofungicide in Fusarium solani and Fusarium oxysporum in various crops	Singh et al. (2021)
12	Cobalt and nickel ferrite nanoparticles	Act as nanofungicide against Fusarium oxysporum in woody trees	Singh et al. (2021)
13	AuNPs, AgNPs, QDs, CNts, and graphene	Detection of heavy metals (Pb^{2+}, Cd^{2})	Singh et al. (2021)
14	Quantum dots, carbon nanomaterials, polymers	Detection of pesticide residue	Singh et al. (2021)
15	Nanosilver (20–70nm)	Fresh box nano silver food container	Singh et al. (2021)
16	Nanosilver	Antibacterial	
17	PEG-400 (acephate)	Reduced toxicity of acephate and improved stability	Prasad and Mahawer (2020)
18	Subdue MAXX	For the control of pythium and phytophthera blight	Bhan et al. (2018)
19	Cruiser MAXX	Application in protection of seeds and seedlings	Bhan et al. (2018)
20	Nano Green	Increase rice yield by 25% and also acts as nanopesticide	Bhan et al. (2018)
21	Nanoemulsion	Nanoscale droplets in the range of 10–400nm in insect pest control	Bhan et al. (2018)
22	Nano silica	Act as nanoinsecticide in *Sitophilus oryzae* in *Zea mays*	Singh et al. (2021)

2.1.6 Nanoagrochemical start-up

Start-up

A start-up is usually defined as a company in the first stage of its operations, and it is often financed by its entrepreneurial founders during the initial starting period. Start-ups are generally founded by one or more entrepreneurs who want to develop a product or service. Such companies generally start with high costs and limited revenue, which is why they look for capital from a variety of sources, such as venture capitalists (https://www.investopedia.com/terms/s/startup.asp). As far as the start-ups in nanotechnology are concerned, there are several examples around the globe. Many new start-ups have been founded by entrepreneurs for a variety of nano-based products. Table 3 shows some globally important nano-based start-ups.

TABLE 3 Some important nanotechnology-based start-ups.

S. no.	Name of start-up company	Company profile	Head office/ Headquarters
1	Digid-Digital Diagnostics AG	Developing a new kind of diagnostics healthcare platform solution to enable real-time identification of different biomarkers	Mainz, Rheinland-Pfalz, Germany
2	PredaSAR	An emerging nanosatellite data provider that develops SAR satellite constellation	Boca Raton, Florida, United States
3	It's Nanoed	Engaged in manufacture of sanitization and barrier products.	Las Vegas, Nevada, United States.
4	LIGC Application	Maker of laser-induced graphene filters.	Caesarea, Hefa, Israel Caesarea, Hefa, Israel
5	FLEEP Technologies	Designs and develops printed electronics solutions for the biomedical and smart packaging sectors.	Via Giovanni Pascoli, Milano MI, Italy
6	QustomDot	An advanced materials start-up that provides customized solutions for next-generation applications.	Technologiepark-Zwijnaarde, Gent, Belgium
7	ELDICO Scientific	Develops, manufactures and sells novel instruments for electron diffraction in crystallography, enabling nanocrystalline solid state analysis in the submicrometer range	Villigen, Switzerland
8	AegiQ	Engaged in making high-performance quantum photonic systems with on-demand single photons	Sheffield, Sheffield, United Kingdom
9	FootPrint Coalition	Goal is to scale the adoption of technologies to restore our environment	Venice, California, United States
10	EpinovaTech	Focuses on a new generation of power conversion technology using gallium nitride; it offers fast charging solutions	Skane Lan, Sweden
11	Nanofy	An Entrepreneur First portfolio company developing and producing self-disinfection solutions via nanotechnology innovation	Singapore
12	TruSpin Nanomaterial Innovation, Inc.	Engaged in the development of advanced materials and nanomaterials	Birmingham, Alabama, United States
13	Axorus	A developer of neuroelectronic interfaces for therapeutic applications	Loos, Nord-Pas-de-Calais, France
14	Laska	Engaged in waste upcycling applications, and also produces high-value, in-demand products	Kağıthane, İstanbul
15	Anavo medical	Uses nanotechnology to boost wound healing and keep infections in check	Bern, Switzerland
16	XRnanotech	Offers innovative nanostructured x-ray optics with outstanding precision for high-tech applications	Villigen-PSI, Switzerland

TABLE 3 Some important nanotechnology-based start-ups—cont'd

S. no.	Name of start-up company	Company profile	Head office/ Headquarters
17	Matrix Meats	Develops and manufactures nanofiber scaffolds to support the production of clean, healthy, tasty, and environmentally friendly cellular meat	Dublin, Ohio, United States
18	Nano Diamond Battery (NDB)	Nanotechnology-based start-up engaged in developing a green, safe, and lifelong battery made by recycling waste	San Francisco, California, United States
19	Infrascreen	Uses nanotechnology to design the next generation of greenhouse screens, light filters using nanotechnologies that enable drastic improvements in climate management within greenhouses	Neuchâtel, Switzerland
20	Gisens Biotech	Nanotechnology-powered bioelectronic sensors for real-time and portable health monitoring	Buenos Aires, Argentina
21	Nanoseen	Engaged in obtaining innovative nanomaterials such as nanotubes, nanofibers, nanoneedles, nanospheres, graphene, etc. Also creates unique solutions for the desalination of both sea and brackish water using its next-generation nanomaterials obtained on recycled metal wafers	Sopot, Pomorskie, Poland
22	Nanolyse Technologies	Engaged in creating ground-breaking innovations in analytical science	Didcot, United Kingdom
23	Sicora Technologies	Turns smart technology and science into a quality, high-performing solution	Mumbai, Maharashtra
24	Sonus Microsystems	Engaged in creating a special type of ultrasound transducer fabricated with polymer materials	Vancouver, British Columbia, Canada
25	Sygnis Bio Technologies	A leading provider of scientific equipment on the Polish market, with a strong focus on high-resolution live cell imaging and tissue engineering	Warszawa, Poland
26	SiTration	Improves ultradurable filtration membranes intended for efficient separations in the harshest environments	Cambridge, Massachusetts, United States
27	KPM-Accelerate	Mandate is to support chem-tech start-ups to accelerate their development on the road to commercialization	Kingston, Canada
28	BikeHive	Provides ultrafast charging technology for electric scooters and e-bikes	Ramat Gan, Tel Aviv, Israel
29	UNIBIO	Produces bionanocarriers from shrimp waste to be mixed in agrochemical formulations to make them more powerful and sustainable	Roskilde, Denmark
30	Ike Scientific	Designed to help small, mid-size, and large businesses obtain the science solutions they need	Salt Lake City, Utah, United States

Continued

TABLE 3 Some important nanotechnology-based start-ups—cont'd

S. no.	Name of start-up company	Company profile	Head office/ Headquarters
31	Ekidna Sensing	Provides rapid, simple to operate, and on demand cannabinoid testing for the growing cannabis industry	Ottawa, Canada
32	NS Nanotech	Patented nitride semiconductors and nano light emitting diode technology to provide light in the invisible ultraviolet spectrum	Ann Arbor, Michigan, United States
33	Hummink	Uses a patented technology that combines a nanometric pen with an oscillating microresonator to perform a capillary deposition of various liquids, thereby providing semiconductor molecular diagnostic industries with the tools to build large objects with the same resolution and precision on millimetric to centimetric substrates	Paris, France
34	CENmat	Nanomaterial company that offers consultation and contractual research to nanomaterial synthesis, upscale, and integration	Stuttgart, Germany
35	AtomicAI	Builds data science solutions for nanotechnology companies	Philadelphia, Pennsylvania, United States
36	PGM NanoSensing	Based on medical lab on a chip	Rotterdam, Netherlands
37	Nanorbital	Chemical and nanomaterial company	Ahmedabad Gujrat, India
38	Kimialys	Develops and sells proprietary and patented surface chemistry for biosensors	Paris, France

3 Commercialization of nanotechnology

3.1 Origins

The majority of attempts at commercialization begin with the submission of patent applications to safeguard intellectual property. A significant number of nanotechnology firms get at least some of their early intellectual property from universities or government facilities. Another method to form nanotech start-up enterprises is the "spinning out" of a business unit from an existing parent company. The third method by which independent entrepreneurs might create nanotechnology businesses (Waitz and Bokhari, 2003) is via the generation of intellectual property (IP). This is an unusual occurrence, and we have not come across any other successful businesses that got their start in the same way, at least not in the beginning. Maintaining a strong intellectual property position during this period is one of the most important success factors. A business strategy that is crystal clear, succinct, well thought out, and captivating is another component that contributes to success.

3.2 Financing

There is a variety of financing available. The following types of potential investors are often explored for a nanotechnology start-up: friends and family, angel investors, venture capitalists, government agencies, and corporate partners. The process of fundraising involves a lot of different moving parts. One important step in obtaining financing from the government is creating a solid application. For venture capitalists, having a solid team that has "done it before" and a substantial potential in the market are smart places to start. However, the idea of having high-profile academics who have actually produced some of the intellectual property that the firm is founded on is one element that stands out as being relatively distinctive to the field of nanotechnology (Waitz and Bokhari, 2003).

3.3 Expansion

When it comes to expanding their businesses, leaders of nanotech start-ups often use a few distinct growth tactics, which we have witnessed. One option is to form a partnership with a bigger, existing company. Through strategic collaboration, a firm may have access to production and sales channels, both of which are costly for a start-up to build on their own. A further method for achieving expansion is to derive several technologies from a single underlying technology. According to Waitz and Bokhari (2003), the presence of a capable management team that has market expertise is the single most essential element in determining success in this phase.

4 Overview of nano-based product in market

The global market size associated with nanomaterials was estimated to be about $8 billion in 2020, and this is expected to expand at a compound annual growth rate of 14.1% from 2021 to 2028. The unique physicochemical properties and growing applications of nanomaterials in different fields such as cosmetics, textiles, medicine, electronics, healthcare, the environment, the automotive industry, food, agriculture, sports, home appliances, aerospace, etc., are expected to drive the market in the future. Due to the COVID-19 pandemic, the market for nanomaterials plummeted, leading to a decline in the purchasing power of the population. However, in 2021 overall demand, including demand for nano-based products, has increased due to the ease of restrictions and financial packages announced by governments to jumpstart economies (https://www.grandviewresearch.com/industry-analysis/nanotechnology-and-nanomaterials-market).

However, the COVID-19 pandemic has initiated event to explore the potential of nanomaterials in healthcare in the diagnosis, prevention, and treatment of COVID-19 and other viral diseases. Recently, it has been reported that a variety of nanomaterials are being used in biomedical sectors for various applications such as imaging, implants, photothermal therapy, and drug delivery as well as for their antimicrobial, antioxidant, and anticancer properties. In addition to the biomedical sector, nanomaterials are being used in the development of a number of products for different industries (Fig. 1).

As far as nano-based products in agriculture are concerned, according to information from the nanotechnology products database (https://product.statnano.com), to date about 230 different

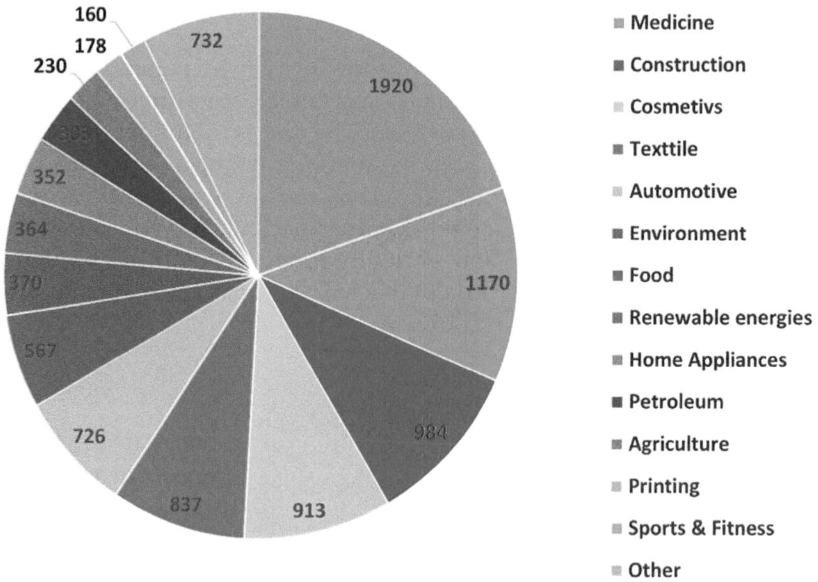

FIG. 1 Different industry-wise number of nano-based products currently developed. *Figure prepared using data available at https://product.statnano.com/.*

nano-based products have been developed for use in various sectors in agriculture, mainly including animal husbandry, fertilizers, plant breeding, plant protection, and soil improvement. Fig. 2 shows a number of nano-based products developed for use in different sectors of agriculture. Table 4 shows some important approved and commercially available nanofertilizers.

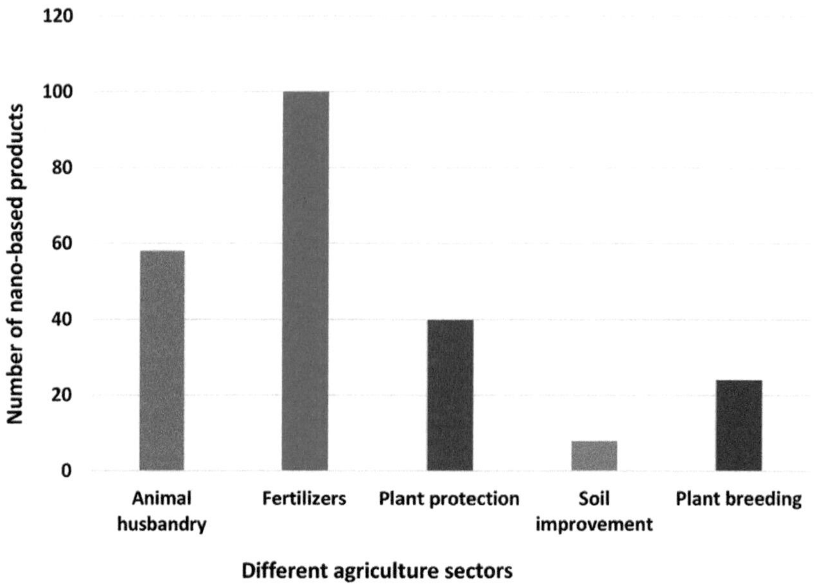

FIG. 2 Number of nano-based products developed for use in different sectors of agriculture.

TABLE 4 Some important approved and commercially available nanofertilizers.

S. no.	Nanofertilizers	Constituents	Name of manufacturer
1	Nano Ultra-Fertilizer (500) g	Organic matter, 5.5%; nitrogen, 10%; P_2O_5, 9%; K_2O, 14%; P_2O_5, 8%; K_2O, 14%; MgO, 3%	SMTET Eco-technologies Co., Ltd., Taiwan
2	Nano Calcium (Magic Green) (1) kg	$CaCO_3$, 77.9%; $MgCO_3$, 7.4%; SiO_2, 7.47%; K, 0.2%; Na, 0.03%; P., 0.02%; Fe 7.4 ppm; Al_2O_3, 6.3 ppm; Sr, 804 ppm; sulfate, 278 ppm; Ba, 174 ppm; Mn, 172 ppm; Zn, 10 ppm	AC International Network Co., Ltd., Germany
3	Nano Capsule	N, 0.5%; P_2O_5, 0.7%; K_2O, 3.9%; Ca, 2.0%; Mg, 0.2%; S, 0.8%; Fe, 2.0%; Mn, 0.004%; Cu, 0.007%; Zn, 0.004%	The Best International Network Co., Ltd., Thailand
4	Nano Micro Nutrient (EcoStar) (500) g	Zn, 6%; B, 2%; Cu, 1%; Fe, 6%+; EDTA Mo, 0.05%; Mn, 5%+; AMINOS, 5%	Shan Maw Myae Trading Co., Ltd., India
5	PPC Nano (120) mL	M protein, 19.6%; Na2O, 0.3%; K2O, 2.1%; (NH4) 2SO4, 1.7%; diluent, 76%	WAI International Development Co., Ltd., Malaysia
6	Nano Max NPK Fertilizer	Multiple organic acids chelated with major nutrients, amino acids, organic carbon, organic micronutrients/trace elements, vitamins, and probiotics	JU Agri Sciences Pvt. Ltd., Janakpuri, New Delhi, India
7	TAG NANO (NPK, PhoS, zinc, cal, etc.) fertilizers	Proteino-lacto-gluconate chelated with micronutrients, vitamins, probiotics, seaweed extracts, and humic acid	Tropical Agrosystem India (P) Ltd., India
8	Nano Green	Extracts of corn, grain, soybeans, potatoes, coconut, and palm	Nano Green Sciences, Inc., India
9	Biozar Nanofertilizer	Combination of organic materials, micronutrients, and macromolecules	Fanavar Nano-Pazhoohesh Markazi Company, Iran
10	Nano urea liquid	30nm urea particles (4.0% total nitrogen (*w/v*))	Indian Farmers Fertilizer Cooperative Ltd., India
11	Plant Nutrition Powder (Green Nano)	N, 0.5%; P_2O_5, 0.7%; K_2O, 3.9%; Ca, 2.0%; Mg, 0.2%; S, 0.8%; Fe, 1.0%; Mn, 49 ppm; Cu, 17 ppm; Zn, 12 ppm	Green Organic World Co., Ltd., Thailand
12	Hero Super Nano	N, 0.7%; P_2O5, 2.3%; K_2O, 8.9%; Ca, 0.5%; Mg, 0.2%; S, 0.4%	World Connet Plus Myanmar Co., Ltd., Thailand
13	Supplementary Powder (The Best Nano)	N, 0.5%; P_2O_5, 0.7%; K_2O, 3.9%; Ca, 2.0%; Mg, 0.2%; S, 0.75%; Fe, 0.03%; Mn, 0.004%; Cu, 0.007%; Zn, 0.004%	The Best International Network Co., Ltd., Thailand
14	Zinc oxide [ZnO]–universal additive agent. 1–50nm	Zinc oxide 99.9%	Land Green & Technology Co., Ltd., Taiwan

Continued

TABLE 4 Some important approved and commercially available nanofertilizers—cont'd

S. no.	Nanofertilizers	Constituents	Name of manufacturer
15	Titanium dioxide [TiO_2]–universal pigment. [20nm]	Titanium dioxide 99%	Land Green & Technology Co., Ltd., Taiwan
16	Silicon dioxide [SiO_2]–universal stabilizer agent [20–60nm]	Silicon dioxide 99%	
17	Manganese dioxide [MnO_2]–universal purifier [1–50nm]	Manganese dioxide 99.9%	
18	Selenium colloid [Se]–universal antioxidant [1–20nm]	Selenium colloid 99.9%	
19	NanoCS of NanoShield products. 1–100nm	NPK, zinc	Aqua-Yield®, USA
20	NanoGro 1–100nm	NPK	
21	NanoN+	Nitrogen	
22	NanoK	Potassium	
23	NanoPhos	Phosphorus	
24	NanoZn	Zinc	
25	NanoPack	Sulfur, copper, iron, manganese, and zinc	
26	NanoCalSi	Calcium and silicate molecules	
27	NanoFe	Iron	
28	Nano-Ag Answer	NPK = 1.0–0.1 - 5.5 Total nitrogen 1.0%; Available phosphate 0.1%; Soluble potash 5.5%; Other ingredients 93.4%	Urth Agriculture, USA
29	Hibong biological fulvic acid	Nanofertilizer, humic acid. Chitosan Oligosacchairides $\geq$30g/L, $N \geq$46g/L, $P_2O_5 \geq$21g/L, $K_2O \geq$62g/L, organic matter: 130g/L	Qingdao Hibong Fertilizer Co., Ltd., China
30	Humic acid granular fertilizer	Humic acid: 55%, organic matter: 70%	
31	Seaweed Nano Organic carbon Fertilizer	NPK: 2–3-3, seaweed extract $\geq$5%, organic matter: 35%, humic acid $\geq$5%, amino acid $\geq$5%	

Reproduced from Avila-Quezada, G.D., Ingle, A.P., Golinska, P., Rai, M., 2022. Strategic applications of nano-fertilizers for sustainable agriculture: benefits and bottlenecks. Nanotechnol. Rev. 11, 1–18. https://doi.org/10.1515/ntrev-2022-0126; open access article.

5 Requirement for approval of nano product company

Nanotechnology is an emerging science of the 21st century due to its widespread applications in numerous industries. Due to their excellent physicochemical properties, several nanomaterials are currently being used in the development of a variety of products in different sectors. However, the highly reactive nature of nanomaterials and their minute size may lead to undesirable effects, such as cytotoxicity. Therefore, it is important to validate different extrinsic properties of nanomaterials, such as biological interactions, physiological effects, biokinetics, their uptake and distribution, and also biological effects in different scenarios of exposure (Avila-Quezada et al., 2022). To date, several expert meetings have been organized to define concepts for nanomaterial hazards and their assessment in different regulatory frameworks as well as understand the application and extrapolate potential nanomaterial regulatory hazards. Based on such discussion, different countries have developed their own legislative frameworks for the application of nanomaterials for different sectors or purposes.

As a public health agency, the US Food and Drug Administration (FDA) makes regulatory decisions for a wide range of products, ranging from cosmetics to chemotherapy agents to food packaging, using available scientific information. In this context, the FDA has long encountered the combination of promise, risk, and uncertainty that accompanies emerging technologies, and nanotechnology is not an exception in this regard. It is well known that materials can exhibit new or altered physicochemical properties at nanoscale dimensions, which can enable the development of novel products. The unique changes in the biological, chemical, and other properties of nanomaterials make them suitable for different applications (https://www.fda.gov/science-research/nanotechnology-programs-fda/fdas-approach-regulation-nanotechnology-products).

The products developed using nanomaterials may differ from those of conventionally manufactured products. Therefore, evaluations of the safety or effectiveness are necessary, leading to the FDA performing such assessments. However, the FDA does not categorically judge all products containing nanomaterials or otherwise involving the application of nanotechnology as intrinsically benign or harmful. The FDA usually regulates nanotechnology products under existing statutory authorities, in accordance with the specific legal standards applicable to each type of product under its jurisdiction. Consistent with Executive Order 13563 on improving regulation as well as with White House policy statements on regulating emerging technologies and applications of nanotechnology, the FDA supports innovation and the safe use of nanotechnology in FDA-regulated products under appropriate and balanced regulatory oversight. Considering the safety aspects of nano-based products, the FDA's nanotechnology regulatory science research plan will provide all the required necessary guidelines for research in nanotechnology (https://www.fda.gov/science-research/nanotechnology-programs-fda/fdas-approach-regulation-nanotechnology-products).

In this context, the FDA has issued several guidance documents for nanotechnology applications in FDA-regulated products. These documents are being issued as part of the FDA's ongoing implementation of recommendations from its 2007 Nanotechnology Task Force Report. However, guidance documents do not create or confer any rights for or on any person and do not operate to bind the FDA or the public; they represent the FDA's current thinking on a topic. All such guidance documents that discuss the use of nanotechnology or nanomaterials in FDA-regulated products are available at: https://www.fda.gov/science-research/nanotechnology-programs-fda/nanotechnology-guidance-documents. Implementation of these guidelines is essentially required to develop any nano-based product in any company or start-up.

6 Nanoagrochemicals: Marketing and sales strategies

It is anticipated that agribusiness will have a worldwide investment of $2.9 trillion by 2030 (World Bank, 2013). Within the next 10 years, nanotechnology will almost certainly have a significant influence on all aspects of the agriculture industry. The possibilities of growing both traditional and stranded agriculture resources might be increased with the use of nanotechnology. The nanoinventions has enormous potential for the use of nanotechnology in agricultural settings. Nanotechnology has already began to applied in several sectors such as agricultural, food production, food safety, biomedicine, environmental engineering, security, and water and energy resources. Regular implementation of nanoscale innovative technologies is an important factor determining growth in the global wealth and market value. Nanotechnology innovation in the agri-food industry has enhanced the agricultural production, food security, and industrial economic growth by at least 30% (on average, $0.9 trillion).

6.1 Agribusiness nanotechnology market size

Agribusiness market around the world was worth between $20.7 billion and $0.98 trillion in 2010 which is anticipated to climb value of the market to $3.4 trillion by 2020. Bakucs et al. (2014) foreseen the applications of nanotechnology in agriculture as a new industrial revolution to improve the market size worldwide. The nanotechnology-based research is under expansion in developed and developing countries are investing in nanotechnology to secure a sustainable market share. United States leads with investments totalling $3.7 billion over the course of 4 years as part of the National Nanotechnology Initiative. Japan is the second-largest contributor, followed by the European Union, which has pledged massive sums of $1.2 billion and $750 million per year (including individual country payments). According to Lopez-Salazar et al. (2014), nanotechnology has been applied in the process related to plant products and postharvest technologies to enhance the production with low cost investment with an emphasis on the commercial sector.

6.2 Nanotechnology adoption strategies in agriculture

Ranjan et al. (2014) suggested that adoption strategy often involves the implementation of installations of the new, less expensive technology to the consumer. Nanoscale and technologies require significant expenditures, which will be reduced in the future and available at cost effective way. The technical expertized, repair contracts are utilized by the producing companies to make replacement costs or servicing easier (Scoones et al., 2014). In addition, the collaborative alliance across industry, academia, and government is having important role for improving awareness of nanotechnology and their influence on the development of the agriculture industries. This cooperative working are reflected in the collaborative projects among the university and industry. Wilson et al. (2014) reported that there is cooperation and coordination among agribusiness industries, the government, and academic institutions to advance nanotech-driven fields. The investment in the form of funding in areas of nanoscience, technology, agriculture, and agribusiness are critical to maintaining sustainable global economy. A report by Ali et al. (2014) indicated that regional innovation clusters in agribusiness markets composed of large companies, start-up, public and private research and development centers, universities, specialized suppliers, investors, and regional and government agencies. The partnership within this cluster to pursue a shared strategy is responsible for the growth of nanotechnology in agriculture markets. This strategy was developed to establish synergies in a particular nanotech application area (Ortega, 2013). As the technologies advance, the price discrimination in agricultural markets is expected (Raynolds et al., 2014). Alishahi et al. (2014) stated that agricultural enterprises located in same region may charge various prices of technology to customers. Yusuf (2014) agribusiness-based businesses with nanobased innovative product design, development, and production has possibility to corner the market. The constant development of new nanotechnologies has shifting preferences and requirements of consumers, and the intense level of competition in agribusiness markets, to enhance the features, benefits, and performance of outcomes (Kafi and Ghomi, 2014).

6.3 Nanotechnology in agribusiness: A competitive market

The global demand for the agricultural improvement is dynamics, where raw material and energy import based on efficient materials. The nanotechnology has impacted the agriculture developing AI-based and nanosensor-based technologies. Ali et al. (2014) suggested that the importance of the nanoscale materials has impacted the global market. The use of nanotechnology in the fertilizer business influences on the sector's competitiveness in increasing which will be reflected in the future (Yusuf, 2014). In the packaging and pesticide sectors is developing and criteria such as technology, raw materials, the readiness of human resources, infrastructure, economic effect, have critical value in preserving the competitiveness. Start-up companies in the agricultural industry have chosen nanotechnologies to achieve market value (Hirsch et al., 2014). Companies in the food and agriculture industries have created strategies aimed at researching new nanobased products, materials, and technologies to attract financial institutions for investment. The outreach programs focusing on the impact and advantages of nanotechnology to the farmers and other agribusiness entrepreneurs assisted to spread knowledge. The programs at national and international levels are in urgent need.

6.4 Economic importance of nano agrochemicals

Nanoagrochemicals have shown significant benefits in agriculture by helping farmers economically with the increasing yield of vegetation, both qualitatively and quantitatively, by substituting for synthetic agrochemicals. The Business Communication Enterprise study of 2016 indicated that the global nanotechnology market will reach $95 billion by 2021, up from $39.2 billion in 2016, at an annual growth rate of 18.2%. The nanotechnology market is growing at a rate of 30% per year. Nowadays, farmers are using new formulations, technologies, and improvements that assist in increasing the economic benefits of farming. Biological nanoagrochemicals demonstrated their favorable impact on agricultural sectors by increasing manufacturing yield. Improved nanodevices might truly cause a revolution in the agroindustry, including nanotablets for agrochemical delivery; nanosensors/biosensors for detecting pathogens; nanozeolites for slow release of water and fertilizers; vitamins and capsules for cattle; and antimicrobial nanoemulsions for decontamination of food, improving plants, and animal breeding. The use of nanoagrochemicals has the potential to significantly impact and benefit sustainable agriculture (Lade et al., 2019; Qazi and Dar, 2020).

7 Future perspective

Nanotechnology research at the nanoscale is still in its initial stages, but it is developing quickly. Natural resources, rather than artificial and standard media, should be used in the future with the interaction of nanoagrochemicals. To achieve precision techniques in the agrochemical sector, researchers want to take a severe look at the hazards to create more comfortable surroundings. The controlled release of agrochemicals by nanotechnology will become necessary in the future to protect the environment and human health. There are numerous future opportunities for targeted drug delivery as well as the delivery of essential nutrients, genetic materials, and nanoagrochemicals using nanoemulsions, which will provide new opportunities for the agricultural revolution. Some nanoparticles, including metals and others, might be dangerous to plant life as well as human health. Therefore, these materials must be incorporated carefully. On the other hand, researchers must find new renewable methods to fabricate nanoscale nanomaterials that can be employed in agriculture (Abdollahdokht et al., 2022). The organic synthesis of nanoagrochemicals may improve the cost, time, and energy requirements, which will help decrease the amount of synthetic chemicals used for the production of nanocomposites through physical and chemical techniques. The use of biodegradable polymers can be encouraged in future research due to their ecofriendly characteristics. There are tremendous applications of green nanoparticles that can be identified and incorporated in agriculture, including the utilization of agricultural wastes such as banana and orange peels, wheat whiskers, straw, cotton, coconut, rice husks, etc. Novel nanocomposites have two crucial factors, size and stability, that will be key to their success in antifungal management. But, before applying nanocomposites, researchers should check their toxicity because several nanomaterials are made from toxic materials; for example, $TiO2$ is reported to cause cancer.

8 Conclusion

With the progress and establishment of start-ups in the nanoagrochemicals sector, sustainable crops and the environment seem achievable and reliable. The nanoagrochemicals, including nanofertilizers, nanopesticides, nanofungicides, nanosensors, and nanoclays, are efficient in managing plant diseases. The nanofertilizers and nanopesticides are promising to assist plants in taking nutrients and help reduce biotic and abiotic stress, respectively. Nanoclays and nanosensors play important roles in detecting disease and drug delivery systems. We have cited CuNPs, AgNPs, MgO, SiO2, CTEO, and chitosan nanoparticles as effective nanofungicides to constrain the Fusarium species. The new nanoscale products are commercialized and emerging in the market with persistence in crop protection, minimum environmental damage, and elevated economic growth in sustainable agriculture. In the future, a physiobiochemical study of nanoparticles and plant interactions will be important to explore NP translocation and transformation in plants for toxicology studies using TEM and SEM analytical instrumentation.

Author contribution

BDL, API, MM, and BSP contributed equally to this work.

Acknowledgments

This research was supported by the USDA-NIFA-SCRI-2017-51181-26834 through the National Center of Excellence for Melon at the Vegetable and Fruit Improvement Center of Texas A&M University. Also, BP acknowledges the Institute for Advancing Health Through Agriculture for providing financial support.

Competing interest

The author declares no competing financial interest.

References

Abd-Elsalam, K.A., Alghuthaymi, M.A., 2015. Nanobiofungicides: is it the next-generation of fungicides? J. NanoTech Mater. Sci. 2 (2), 1–3.

Abdollahdokht, D., Gao, Y., Faramarz, S., Poustforoosh, A., Abbasi, M., Asadikaram, G., Nematollahi, M.H., 2022. Conventional agrochemicals towards nano-biopesticides: an overview on recent advances. Chemical and biological technologies in agriculture. Chem. Biol. Technol. Agric. 9, 13.

Ali, M.A., Rehman, I., Iqbal, A., ud Din, S., Rao, A.Q., Latif, A., Husnain, T., 2014. Nanotechnology: a new frontier in agriculture. Adv. Life Sci. 1 (3), 129–138.

Alishahi, A., Proulx, J., Aider, M., 2014. Chitosan as bio-based nanocomposite in seafood industry and aquaculture. Seafood science. Adv. Chem., Technol. Appl. 211. https://doi.org/10.1201/b17402-12.

Arie T. 2019. Fusarium diseases of cultivated plants, control, diagnosis, and molecular and genetic studies. J. Pestic. Sci. 44(4): 275–281. https://doi.org/10.1584/jpestics.J19-03. PMID: 31777447; PMCID: PMC6861427.

Avila-Quezada, G.D., Ingle, A.P., Golinska, P., Rai, M., 2022. Strategic applications of nano-fertilizers for sustainable agriculture: benefits and bottlenecks. Nanotechnol. Rev. 2022 (11), 1–18. https://doi.org/10.1515/ntrev-2022-0126.

Bakucs, Z., Fałkowski, J., Fertő I. 2014. Does market structure influence price transmission in the agro-food sector? A meta-analysis perspective. J. Agric. Econ., 65 (1), 1–25. https://doi.org/10.1111/1477-9552.12042.

Bernhoft, A., Torpa, M., Clasena, P.E., Loesband, A.K., Kristoffersena, A.B., 2012. Influence of agronomic and climatic factors on fusarium infestation and mycotoxin contamination of cereals in Norway. Food Addit. Contam. 29 (7), 1129–1140.

Bhan, S., Mohan, L., Srivastava, C., 2018. Nanopesticides: a recent novel ecofriendly approach in insect pest management. J. Entomol. Res. 42 (2), 263–270.

Bhandari, G., 2014. An overview of agrochemicals and their effects on environment in Nepal. Appl. Ecol. Environ. Sci. 2 (2), 66–73.

Bratovcic, A., Hikal, W.M., Said-Al, A.H.A.H., Tkachenko, K.G., Baeshen, R.S., Sabra, A.S., Sany, H., 2021. Nanopesticides and nanofertilizers and agricultural development: scopes, advances and applications. Open J. Ecol. 11, 301–316.

Chaud, M., Souto, E.B., Zielinska, A., Severino, P., Batain, F., Oliveira-Junior, J., Alves., 2021. Nanopesticides in agriculture: benefits and challenge in agricultural productivity, toxicological risks to human health and environment. Toxics 9, 131.

Chi, Y., Zhang, G., Xiang, Y., Cai, D., Wu, Z., 2018. Fabrication of reusable temperature-controlled-released fertilizer using a palygorskite-based magnetic nanocomposite. Appl. Clay Sci. 161, 194.

Chun, S., Feng, J., 2021. Preparation of abamectin nanoparticles by flash nanoprecipitation for extended photostability and sustained pesticide release. ACS Appl. Nano Mater. 4 (2), 1228–1234.

Dhawan, V., 2017. Water and agriculture in India. In: Background Paper for the South Asia Expert Panel during the Global Forum for Food and Agriculture (GFFA).

Duhana, J.S., Kumar, R., Kumar, N., Kaur, P., Nehrab, K., Duhan, S., 2017. Nanotechnology: the new perspective in precision agriculture. Biotechnol. Rep. 15, 11–23.

Elizabath, A., Babychan, M., Mathew, A.M., Syriae, G.M., 2019. Application of nanotechnology in agriculture. Int. J. Pure Appl. Biosci. 7 (2), 131–139.

Farshchi, H.K., Azizi, M., Teymouri, M., Nikpoor, A.R., Jaafari, M.R., 2021. Synthesis and characterization of nanoliposome containing Fe2+ element: A superior nano-fertilizer for ferrous iron delivery to sweet basil. Sci. Hortic. 283, 110110.

Gao, Y., Liang, Y., Zhou, Z., Yang, J., Tian, Y., Niu, J., Tang, G., Tang, J., Chen, X., Li, Y., Cao, Y., 2021. Metal-organic framework Nanohybrid carrier for precise pesticide delivery and Pest management. Chem. Eng. J. 422 (2), 130143.

Gea, F.J., Navarro, M.J., Diánez, S.M., Carrasco, J., 2021. Control of fungal diseases in mushroom crops while dealing with fungicide resistance: a review. Microorganisms 9, 585.

Global newswire, 2021. New York, USA, September 13, 2021. https://www.globenewswire.com/news-release/2021/09/13/2295845/0/en/Global-Agrochemicals-Market-to-Reach-328-0-Billion-at-a-CAGR-of-4-1-from-2018-to-2026-Exclusive-COVID-19-Impact-Analysis-210-pages-Report-by-Research-Dive.html.

Hirsch, S., Schiefer, J., Gschwandtner, A., Hartmann, M., 2014. The determinants of firm profitability differences in EU food processing. J. Agric. Econ. https://doi.org/10.1111/1477-9552.12061.

Hossain, K., Abbas, S.Z., Ahmad, A., Rafatullah, M., Ismail, N., Pant, G., Avasn, M., 2020. Nanotechnology: A boost for the urgently needed second green revolution in Indian agriculture. Nanobiotechnology in Agriculture, Nanotechnology in the Life Sciences. https://doi.org/10.1007/978-3-030-39978-8_2.

Kafi, F., Ghomi, F.S., 2014. A game-theoretic model to analyze value creation with simultaneous cooperation and competition of supply chain partners. Math. Probl. Eng. 2014.

Karlsson, I., Friberg, H., Kolseth, A.-K., Steinberg, C., Persson, P., 2017. Agricultural factors affecting fusarium communities in wheat kernels. Int. J. Food Microbiol. 252, 53–60. https://doi.org/10.1016/j.ijfoodmicro.2017.04.011.

Lade, B.D., Gogle, D.P., Nandeshwar, S.B., 2017. Nano bio pesticide to constraint plant destructive pests. J. Nanomed. Res. 6 (3), 00158.

Lade, B.D., Gogle, D.P., Lade, D.B., Moon, G.M., Nandeshwar, S.B., Kumbhare, S.D., 2019. Nanobiopesticide formulations: application strategies today and future perspectives. In: Nano-Biopesticides Today and Future Perspectives, pp. 179–206.

Li, D., Liu, B., Yang, F., Wang, X., Shen, H., Wu, D., 2016. Carbohydr. Polym. 136, 341.

Li, M., Yang, Y., Xied, J., Xua, G., Yua, Y., 2019. In-vivo and in-vitro tests to assess toxic mechanisms of nano ZnO to earthworms. Sci. Total Environ. 687, 71–76.

Liu, B., Wang, Y., Yang, F., Cui, H., Wu, D., 2017. Development of a chlorantraniliprole microcapsule formulation with a high loading content and controlled-release property. J. Agric. Food Chem. 66 (26), 6561–6568.

Liu, C., Zhou, H., Zhou, J., 2021. The applications of nanotechnology in crop production. Molecules 26, 7070.

Logrieco, A., Bottalico, A., Mule, G., Moretti, A., Perrone, G., 2003. Eur. J. Plant Pathol. 109, 645–667.
Lopez-Salazar, A., López-Mateo, C., Molina-Sánchez, R., 2014. What determines the technological capabilities of the agribusiness sector in Mexico? Int. Bus. Res. 7 (10), 47.
Meena, R.S., Kumar, S., Datta, R., Lal, R., Vijayakumar, V., et al., 2020. Impact of agrochemicals on soil microbiota and management: a review. Land 9, 34. https://doi.org/10.3390/land9020034.
Mendhe, S.N., Pawar, W.S., Ghadekar, S.R., 2008. Principles of Agronomy. Agromet Publishers, Nagpur, India.
Meng, W., Tian, Z., Yao, P., Fang, X., Wu, T., Cheng, J., Zou, A., 2020. Preparation of a novel sustained-release system for pyrethroids by using metal-organic frameworks (MOFs) nanoparticle. Colloids Surf. A Physicochem. Eng. Asp. 604, 125266.
Mishra, S., Singh, A., Keswani, C., Singh, H.B., 2014. Nanotechnology: exploring potential application in agriculture and its opportunities and constraints. Biotech Today 4 (1), 9–14.
Mukhopadhyay, S.S., 2014. Nanotechnology in agriculture: prospects and constraints. Nanotechnol. Sci. Appl. 7, 63–71.
Nooeaid, P., Chuysinuan, P., Pitakdantham, W., Aryuwananon, D., Techasakul, S., Dechtrirat, D., 2021. Eco-friendly polyvinyl alcohol/polylactic acid core/shell structured fibers as controlled-release fertilizers for sustainable agriculture. J. Polym. Environ. 29 (2), 552.
Ortega, D.L., 2013. Modernizing agrifood chains in China: implications for rural development. AME J. Agric. Econ. 95 (4), 1046–1048.
Perrone, G., Ferrara, M., Medina, A., Pascale, M., Magan, N., 2020. Toxigenic fungi and mycotoxins in a climate change scenario: ecology, genomics, distribution, prediction and prevention of the risk. Microorganisms 8, 1496.
Prasad, M., Mahawer, S.K., 2020. Nano-agrochemicals: risk assessment and management strategies. J. Plant Health Issues 1 (2), 49–54.
Qazi, G., Dar, F.A., 2020. Nano-agrochemicals: Economic Potential and Future Trends. Springer, pp. 185–193.
Rajput, V.D., Singh, A., Minkina, T., Rawat, S., Mandzhieva, S., Sushkova, S., Shuvaeva, V., Nazarenko, O., Rajput, P., Komariah, Verma, K.K., et al., 2021. Nano-enabled products: challenges and opportunities for sustainable agriculture. Plan. Theory 10, 2727.
Ranjan, S., Dasgupta, N., Chakraborty, A., Samuel, S., Ramalingam, C., Shanker, R., Kumar, A., 2014. Nanoscience and nanotechnologies in food industries: opportunities and research trends. J. Nanopart. Res. 16, 2464–2487. https://doi.org/10.1007/s11051-014-2464-5.
Raynolds, L.T., Long, M.A., Murray, D.L., 2014. Regulating corporate responsibility in the American market: A comparative analysis of voluntary certifications. Competition & Change 18 (2), 91–110.
Scoones, I., Smalley, R., Hall, R., Tsikata, D., 2014. Narratives of Scarcity: Understanding the 'Global Resource Grab'.
Seleiman, M.F., Almutairi, K.F., Alotaibi, M., Shami, A., Alhammad, B.A., Battaglia, M.L., 2021. Nano-fertilization as an emerging fertilization technique: why can modern agriculture benefit from its use? Plants 10, 2.
Shao, C., Zhao, H., Wang, P., 2022. Recent development in functional nanomaterials for sustainable and smart agricultural chemical technologies. Nano Convergence 9 (1), 1–17.
Shebl, A., Hassan, A.A., Salama, D.M., Abd El-Aziz, M.E., Abd Elwahed, M.S.A., 2020. Template-free microwave-assisted hydrothermal synthesis of manganese zinc ferrite as a nanofertilizer for squash plant (*Cucurbita pepo* L.). Heliyon 6 (3), e03596.
Singh, H., Sharma, A., Bhardwaj, S.K., Arya, S.K., Bhardwaj, N., Khatri, M., 2021. Recent advances in the applications of nano-agrochemical for sustainable agriculture development. Environ. Sci.: Processes Impacts 23, 213–239.
Sinha, T., Bhagwatwar, P., Krishnamoorthy, C., Chidambaram, R., 2019. In: Gutiérrez, T.J. (Ed.), Polymers for Agri-Food Applications. Springer International Publishing, Cham. p. 5.
Waitz, A., Bokhari, W., 2003. Nanotechnology Commercialization Best Practices.
Wang, C., He, Z., Liu, Y., Zhou, C., Jiao, J., Li, P., Sun, D., Lin, L., Yang, Z., 2020. Appl. Clay Sci. 198, 105802.
War, J.M., Fazili, M.A., Mushtaq, W., Wani, A.H., Bhat, M.Y., 2020. Role of nanotechnology in crop improvement. Nanobiotechnology in Agriculture, Nanotechnology in the Life Sciences. https://doi.org/10.1007/978-3-030-39978-8_4.
Wilson, W.W., Shakya, S., Dahl, B., 2014. Dynamic changes in spatial competition for fertilizer., https://doi.org/10.1016/j.agsy.2014.11.006.
World Bank, 2013. Growing Africa: Unlocking the Potential of Agribusiness. AFTFP/AFTAI. World Bank, Washington DC.
Xiong, L., Wang, P., Hunter, M.N., Kopittke, P.M., 2018. Bioavailability and movement of hydroxyapatite nanoparticles (HA-NPs) applied as a phosphorus fertiliser in soils. Environ. Sci. Nano 5 (12), 2888.

Yadav, S.K., Babu, S., Yadav, M.K., Singh, K., Yadav, G.S., Pal, S., 2013. A review of organic farming for sustainable agriculture in Northern India. Int. J. Agron. Volume 2013, Article ID718145.
Yusuf, S.A., 2014. The Analysis of Export Performance of Newly Industrialized Countries (NICs): The Lesson for African Countries.
Zhang, H., Qin, L., Li, X., Zhou, W., Wang, C., Kan, J., 2018. Preparation and characterization of controlled-release Avermectin/Castor oil-based polyurethane nanoemulsions. Agric. Food Chem. 66 (26), 6552.
Zhou, M., Xiong, Z., Yang, D., Pang, Y., Wang, D., Qiu, X., 2018. Preparation of slow release nanopesticide microspheres from benzoyl lignin. Holzforschung 72 (7), 599–607.
Zobir, S.A.M., Ali, A., Adzmi, F., Sulaiman, M.R., Ahmad, K., 2021. A review on nanopesticides for plant protection synthesized using the supramolecular chemistry of layered hydroxide hosts. Biology 10, 1077.

CHAPTER

17

Patent landscape in biofungicides, nanofungicides, and nano-biofungicides

Prabuddha Ganguli

Vision-IPR, Mumbai, India and Adjunct Faculty, Indian Institute of Technology, Jodhpur, Rajasthan, India

1 Introduction

The global fungicide market has been forecasted to touch 3.4 bn US$ by 2025 at a CAGR of 16.1%. The market dynamics will be strongly influenced by the type of fungicides (Microbial species, Botanical), Mode of Application (Soil treatment, Foliar application, Seed treatment), Species (Bacillus, Trichoderma, Streptomyces, Pseudomonas), Crop Type, Formulation, and Region (Marketsandmarkets, 2020). The transdisciplinary field of biofungicides, nanofungicides, and nanobiofungicides is a convergence of knowledge and practices from plant pathology, plant genetics, entomology, agriculture, cultural practices, materials science, and device technologies. The literature on biopesticides and biofungicides is replete with research papers, conference proceedings, and other scholarly publications. However, in comparison, the number of publications related to nanofungicides and nano-biofungicides is relatively smaller as compared to the general field of biofungicides. Developments in nanotechnology in fields such as medicine, structural materials, thin-films, entrapping and encapsulating agents, and smart materials with environmentally responsive controlled release properties are finding new applications in agriculture (Huang et al., 2018; Elmer and White, 2018; Alghuthaymi et al., 2021). Parallel to scholarly works, selective findings with market potential have been protected through patent filings in various jurisdictions. This chapter provides an overall patent landscape of biofungicides, nanofungicides, and nano-biofungicides from 2001 to 2022.

https://doi.org/10.1016/B978-0-443-23950-2.00010-2

2 Patentscape of biofungicides

Biofungicides are a part of an integrated disease management system, wherein biological control is achieved by biocides containing naturally occurring substances. The modes of action of biocides are generally achieved by:

- inhibiting/suppressing plant pathogens using organism(s) capable of attacking or competing with a pathogen or pest
- improving plant immunity by priming, inducing, or strengthening plant defense responses using pesticidal substances that plants produce from genetic material that has been added to the plant thereby stimulating "systemic acquired resistance" (SAR)
- modifying the environment through the effects of beneficial microorganisms, compounds
- adopting healthy cropping systems

(see He et al., 2021; Thambugala et al., 2020).

The last two decades have seen steady progress in the development of scientific and technological knowledge related to fungicides. Biopesticides as narrow-spectrum agents with low toxicity that decompose quickly, and potentially having low negative impact on the environment have been investigated.

Several products in preharvest and postharvest operations have been commercialized. Inventions have been patented primarily by commercial entities to retain competitiveness in the marketplace.

Fig. 1 illustrates the steady rise in the number of BioFungicide patent applications filed and published from 2001 to 2022. The fall in the publication numbers post-2019 is an aberration due to the COVID-19 pandemic. The top patent applicants in biofungicides in the 20-year period from 2001 to 2022 are presented in Fig. 1 and Table 1.

Analysis of the patents filed by the top 10 applicants in various jurisdictions is given in Table 2. Patent applications by the top 10 applicants/assignees are concentrated in the USA, European Union, Canada, China, Brazil, Japan, Korea Russia, and Australia.

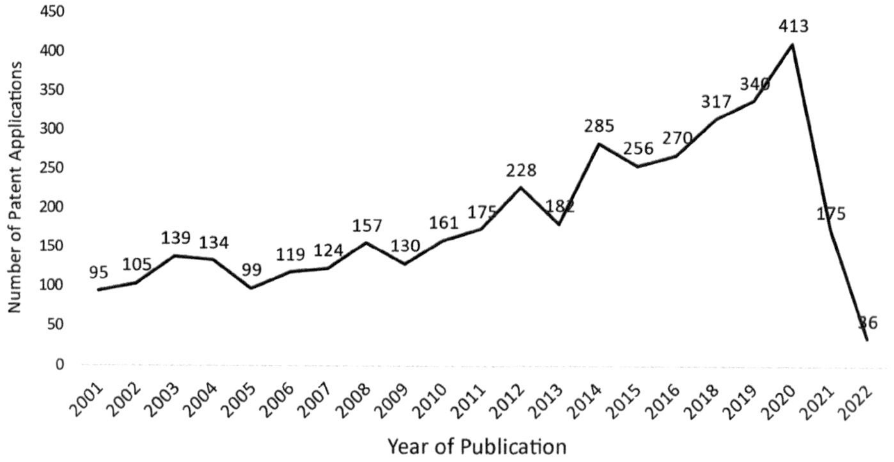

FIG. 1 Patent applications in biofungicides published from 2001 to March 2022.

TABLE 1 Top 10 biofungicide patent applicants in the period 2001–22.

Top patent applicants	# Applications
BAYER	420
BASF	182
NOVO NORDISK	162
DUPONT	81
POLYONE	75
CHEMCHINA	72
DOW	67
CORTEVA	50
FMC	46
US AGRICULTURE	45

TABLE 2 Biofungicides patent application by top 10 applicants/assignees in various jurisdictions.

	Jurisdictions									
Patent applicants	**US**	**WO**	**EP**	**CA**	**CN**	**BR**	**JP**	**KR**	**RU**	**AU**
BAYER	70	49	48	44	34	30	15	22	18	15
BASF	27	29	14	19	15	8	12	8	16	6
NOVO NORDISK	52	26	17	18	9	6	6	3	2	7
DUPONT	12	13	3	5	11	3	9	4	5	2
POLYONE	19	31	14	2	0	0	0	2	0	0
CHEMCHINA	9	14	6	6	5	6	5	1	3	6
DOW	12	5	9	3	4	3	8	5	2	2
CORTEVA	16	8	4	1	5	3	1	3	0	4
FMC	11	9	3	2	2	1	2	3	2	4
US AGRICULTURE	5	9	7	6	1	2	4	0	1	3

Only seven patent applications related to biofungicides have been filed in India as listed in Table 3.

Indian Patent No. IN 227691 (Application 3456/DEL/2005) titled "Biofertiliser cum biofungicide/biobactericide composition B5" was granted to The Indian Council of Agricultural Research in 2009.

Surprisingly India has not been a preferred destination for patent applications related to biofungicides.

TABLE 3 Biofungicides patent applications filed in India.

Application number	Title	Application date	Status
202,117,028,738	USE OF GUAR GUM IN *BIOFUNGICIDE* COMPOSITIONS	25/06/2021	Published
202,117,028,187	USE OF GUAR DERIVATIVES IN *BIOFUNGICIDE* COMPOSITIONS	23/06/2021	Published
201,841,044,650	TECHNIQUES TO PREPARE FORMULATION OF MYCOPARASITIC *BIOFUNGICIDE* FOR USE AGAINST POWDERY MILDEW DISEASE OF PLANTS	27/11/2018	Published
201,811,008,387	"AQUATIC WEED (DAL-WEED) SUCROSE" MEDIUM FOR COMMERCIAL MASS PRODUCTION OF TRICHODERMA *BIOFUNGICIDE*	07/03/2018	Published
201,721,018,512	*BIOFUNGICIDE.*	26/05/2017	Published
3456/DEL/2005	BIOFERTILIZER CUM *BIOFUNGICIDE*/BIOBACTERICIDE COMPOSITION B5	23/12/2005	Published
1419/MAS/1997	A PROCESS FOR THE PREPARATION OF A *BIOFUNGICIDE*	27/06/1997	Published

The published patent application related to biofungicides was analyzed to identify the sub-technologies and the CPC Codes. The results are given in Tables 4 and 5.

Patents filed/granted cover biocides having microorganisms, viruses, microbial fungi, enzymes, fermentates, materials from algae, lichens, bryophyta, multicellular fungi, plants, or extracts thereof, and formulations comprising heterocyclic compounds, carboxylic acid, carbonic acid and diverse arthropodicides, including pest repellants or attractants, or plant growth regulators.

TABLE 4 Sub-technologies in biofungicides patent applications.

Sub-technologies	Applications
Biocides having micro-organisms, viruses, microbial fungi, enzymes, fermentates	2646
Biocides having heterocyclic compounds	1041
Micro-organisms, Processes of treating, culturing	1012
Fungicides	878
Biocides forms, ingredients, and application	774
Processes using micro-organisms	611
Biocides having carboxylic acids	545
Biocides having carbonic acid	389
Biocides having material from algae, lichens, bryophyta, multicellular fungi, plants	347
Arthropodicides	316

TABLE 5 Top CPC codes of biofungicide patent applications.

CPC code	Description	Applications
A01N 63/04	Biocides having microorganisms, viruses, microbial fungi, enzymes, fermentates≫Microbial fungi or extracts thereof	558
A01N 63/00	Biocides having micro-organisms, viruses, microbial fungi, enzymes, fermentates≫Biocides, pest repellants or attractants, or plant growth regulators containing microorganisms, viruses, microbial fungi, animals or substances produced by, or obtained from microorganisms, viruses, microbial fungi or animals, e.g., enzymes or fermentates	211
A01N 63/02	Biocides having micro-organisms, viruses, microbial fungi, enzymes, fermentates≫Fermentates or substances produced by, or extracted from, microorganisms or animal material	141
C12R 1/645	Processes using micro-organisms≫using fungi	103
C12N 1/14	Microorganisms, Processes of treating, culturing≫Fungi; Culture media Therefore	73
A01N 43/16	Biocides having heterocyclic compounds≫with oxygen as the ring hetero atom	61
A01N 65/00	Biocides having material from algae, lichens, bryophyta, multi-cellular fungi, plants≫Biocides, pest repellants or attractants, or plant growth regulators containing material from algae, lichens, bryophyta, multi-cellular fungi or plants, or extracts thereof	60
A01N 43/40	Biocides having heterocyclic compounds≫six-membered rings	48
A01N 43/653	Biocides having heterocyclic compounds≫1,2,4-Triazoles; Hydrogenated 1,2,4-triazoles	44
C05F 11/08	Other organic fertilizers≫Organic fertilizers containing added bacterial cultures, mycelia, or the like	44

Abstracts of a few selected patents/patent applications illustrate the subject matter of the inventions in the category of biopesticides/biofungicides as follows:

WO2017049355A1 discloses the use of *Metarhiziuim* spp. as a fungal pesticide to control pests that affect cotton crops.

WO2015011615A1 discloses compositions comprising active components of *Trichoderma harzianum* strain SK-55, or a cell-free extract thereof or at least one metabolite having pesticidal activity. This patent application also discloses compositions comprising a mutant of *Trichoderma harzianum* SK-55 having pesticidal activity and producing at least one pesticidal metabolite, or a pesticidal metabolite or extract of the mutant, and a pesticide.

US2016000091A1 discloses compositions comprising least one biological control agent selected from the group consisting of *Paecilomyces lilacinus* strain 251 (AGAL No. 89/030550), *Trichoderma atroviride* SC1 (CBS No. 122089), and *Coniothyrium minitans* CON/M/91–08 (DSM 9660) and/or a mutant of these strains having all the identifying characteristics of the respective strain, and/or a metabolite produced by the respective strain that exhibits activity against nematodes, insects and/or phytopathogens, and at least one insecticide.

US2018064110A1 discloses compositions of fungicidally active yeast and fungicides and methods for reducing overall damage of plants and plant parts as well as losses in harvested fruits or vegetables caused by plant pathogenic fungi or other unwanted microorganisms. These compositions are useful during the growth phase, around harvesting, and after harvesting.

TW201740810A discloses microbial compositions and methods of use for benefiting plant growth and treating plant disease.

AR106309A1 discloses compositions comprising an entomopathogenic fungal strain selected from *Metarhizium robertsii* and *Metarhizium anisopliae* to control or eradicate insect pests in plants.

US10433557B2 discloses insecticidal cry toxins including methods for using DNA segments as diagnostic probes and templates for protein production, and the use of proteins, fusion protein carriers, and peptides for insect control and in various immunological and diagnostic applications. This patent also disclosed methods of making and using nucleic acid segments in the development of plant-incorporated protectants in transgenic plant cells containing the DNA segments.

Commercially available products are based on biological control formulations of *Trichoderma harzianum, A. pullulans, Bacillus subtilis, Streptomyces griseoviridis,* and *Gliocladium* virens for application against different plant fungal pathogens. The challenge has been to provide shelf-stable formulated products that maintain biocontrol activity similar to that of fresh cells. In addition, a majority of the fungal infections happen in the soil and root zone area of the plant, which delays plant growth and eventually leads to plant death. Therefore, farmers adopt soil treatment solutions to ensure a healthy crop yield. These limitations have led to the exploitation of nanotechnology in the field of fungicides.

3 Nanofungicides

The term nanofungicide is used to describe any fungicidal formulation that intentionally includes entities in the nanometric range (<100nm). Biohybrid nanocide materials include environment-friendly antimicrobial agents against different pathogenic fungal organisms in plants.

Nanoparticles of Ag, Cu, Se, Ni, Mg, and Fe have been investigated for their potential usefulness as suppressive control agents against phytopathogenic fungi. Further, silver, copper, and zinc have been investigated for their antifungal activities. Solid nanomaterials and liquid state nanophases of various types are being investigated as effective carriers and/or controlled delivery systems/stabilizers for active ingredients.

The present section presents an analysis of 862 nanofungicides published patent applications, during the period 2001 to 2021. Fig. 2 captures the filing trends in the last two decades. The peaking for nanofungicides around 2019 coincides with the peak for patent applications for biofungicides. The decline during 2020 and 2021 can be attributed to the general slowdown in R&D during COVID-19 pandemic.

Nanofungicides draw upon knowledge from both fungicide and materials technologies. It involves the preparation of nanostructures of diverse materials both as biocides and carrier

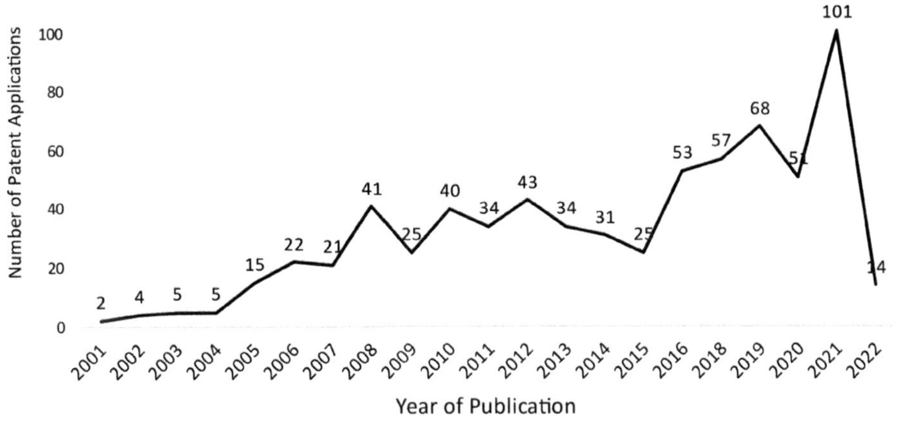

FIG. 2 Patent applications in nanofungicides published from 2001 to March 2022.

substrates, encapsulation techniques, coating & thin film materials and processes, creation of dispersions, emulsions, and their like for carrying and delivering the actives, etc. Polymeric drug delivery systems are being adopted for systematic and localized administration of fungicides reducing doses with a prolonged delivery. Recent trends have been in the synthesis of polymeric composites, grafted polymers, or the modification of polymeric fungicides, by the coupling of cationic moieties on polymers, using metallic salts and nanoparticles, and by the loading of fungicides in micelles and metallic nanoparticles (Velazco-Medel et al., 2021). This has opened the avenues for the use of naturally occurring polymers such as chitosan.

Analysis of the filed patent applications has been grouped into the relevant subtechnologies and CPC Classifications as presented in Tables 6 and 7.

Tables 6 and 7 show that the patent applications/granted patents involve metals/metal oxides particles in the nanorange, biocide compositions in the form of pastes, dispersions,

TABLE 6 Subtechnologies in nanofungicides patent applications.

Subtechnologies	Applications
Biocides forms, ingredients, and application	150
Medicinal preparations characterized by special physical form	123
Biocides having elements or inorganic compounds	113
Coating compositions characterized by the physical nature or effects produced, Filling Pastes	101
Fungicides	91
Medicinal preparations containing organic active ingredients	79
Biocides having heterocyclic compounds	71
Antibiotics, antiseptics, chemotherapeutics	70
Disinfectants, antimicrobials	70
Medicinal preparations characterized by nonactive ingredients used	69

TABLE 7 The top CPC codes of nanofungicides patent applications.

CPC code	Description	Applications
C09D 5/14	Coating compositions characterized by the physical nature or effects produced, Filling pastes≫Paints containing biocides, e.g., fungicides, insecticides or pesticides	22
A01N 59/16	Biocides having elements or inorganic compounds≫Heavy metals; Compounds thereof	21
A01N 25/04	Biocides forms, ingredients, and application≫Dispersions, emulsions, suspoemulsions, suspension concentrates	15
C08K 2201/011	Specific properties of additives used as compounding ingredients≫Nanostructured additives	13
A01N 25/28	Biocides forms, ingredients, and application≫Microcapsules; or nanocapsules	9
A01N 59/20	Biocides having elements or inorganic compounds≫Copper	9
B82Y 30/00	Nanotechnology for materials or surface science≫Nanotechnology for materials or surface science, e.g., nanocomposites	9
B82Y 5/00	Nano-biotechnology or nano-medicine≫Nanobiotechnology or nanomedicine, e.g., protein engineering or drug delivery	8
C08L 2205/035	Polymer mixtures characterized by other features≫Containing four or more polymers in a blend	8
C09D 175/04	Coating compositions based on polyureas or polyurethanes≫Polyurethanes	8

emulsions, suspoemulsions, suspension concentrates, microcapsules, nanostructured additives, nanocapsules, nanocomposites, polyurethanes, specialty coatings, specialty polymers and their blends, materials, and processes used in nanobiotechnology and nanomedicine.

Abstracts of a few selected patent applications/patents are presented to illustrate the type of inventions being protected as patents in the category of nanofungicides:

US 11,248,119 discloses an emulsion that comprises a nonaqueous phase comprising a solid silicone resin and a siloxane carrier vehicle having an average of at least one silicon-bonded functional group per molecule and capable of carrying the solid silicone resin, an aqueous phase comprising water, and a surfactant, wherein the emulsion is substantially free from organic solvents. This emulsion can serve as an efficient carrier of various biocides such as fungicides, herbicides, pesticides, antimicrobial agents, or a combination thereof. Alternatively, the biocide may comprise a boron-containing material, e.g., boric anhydride, borax, or disodium octaborate tetrahydrate, which may function as a pesticide, fungicide,

CN105123753A discloses nano-cuprous oxide fungicide composite prepared by in situ reduction of lignin.

EP3668312 discloses "liquid fertilizer-dispersible compositions" using a fibril or microfibril or nanofibril structuring agent. It also discloses methods of producing them. The compositions comprising: (a) an agricultural bioactive material, a fungicide, insecticide, herbicide, and/or plant growth regulator; (b) a fibril or microfibril or nanofibril structuring agent; and (c) optionally, an auxiliary surface-active agent.

US7476698 discloses an antimicrobial adhesive and coating material that contains as the antimicrobial component metallic silver particles with <5ppm of silver, sodium, and potassium ions. The silver particles are formed of highly porous aggregates having an average grain size of between 1 μm and 20 μm. The primary particles are connected by sinter necks.

Indian Patent Application IN1564/DEL/2010 discloses the synthesis of surface-modified monoclinic nanosulfur for use as a potent fungicide.

WO2021084549 discloses a method of manufacturing nanocopper fungicide for slow release, and enhanced utilization by the plants.

Indian Patent IN371380 discloses a method of manufacturing nanozinc fungicide for slow release enhanced utilization by plants.

WO2021130763 discloses a method of manufacturing solid, liquid, or aerosol nanonitrogen using urea for slow release, enhanced utilization by plants. This product may be used as a fertilizer, pesticide, herbicide, fungicide, or antimicrobial by foliar application, root drenching, or by amending with soil.

CN105457663 discloses a BI_2WO_6/AG_3PO_4 composite photocatalytic fungicide and methods of preparation.

CN1337163 discloses an eco-friendly, high-efficiency, nontoxic, low-cost fungicide composite of nanoinorganic SIO_2/TIO_2, wherein nano-TiO_2 of 10nm is dispersed in the nanocolloid SiO_2, and synthesized into Si/Ti composite colloidal sol.

US11,172,675 discloses a stable nanolipid emulsion delivery system in the form of a nanoconcentrate for use as a carrier for industrial, medical, animal, horticultural, and agricultural chemicals.

The top 10 patent applicants in the field of nanofungicides as indicated in Tables 8 and 9 demonstrate that a lot of work going on in academic institutions and the corporate sector.

As an example of corporate strategy, Vive Crop Protection has protected it's Allosperse Delivery System that uses patented polymer "shuttles" to carry an active ingredient such

TABLE 8 Top 10 patent applicants in nanofungicides.

Top patent applicants	# Applications
BASF	23
INMOLECULE INTERNATIONAL	10
BLUEBERRY THERAPEUTICS	9
VIVE CROP PROTECTION	9
NANOBIO CORP	8
UNIVERSITY OF CENTRAL FLORIDA RESEARCH	8
NSF USA	7
CHINESE ACADEMY OF SCIENCES	6
HEBEI CHENYANG IND & TRADE GROUP	6
SIB LAB	6

TABLE 9 Patent application in nanofungicides by top 10 applicants/assignees in various jurisdictions.

Patent applicants	Jurisdictions									
	CN	US	EP	WO	JP	CA	DE	KR	MX	AU
BASF	0	3	5	3	5	2	3	2	0	0
INMOLECULE INTERNATIONAL	1	1	1	0	1	1	0	0	0	1
BLUEBERRY THERAPEUTICS	1	1	4	1	1	0	0	0	0	0
VIVE CROP PROTECTION	1	5	0	0	1	1	0	0	1	0
NANOBIO CORP	1	1	1	2	1	1	0	0	0	1
UNIVERSITY OF CENTRAL FLORIDA RESEARCH	0	1	1	1	0	0	0	0	1	0
NSF USA	0	4	0	1	0	0	0	0	0	0
CHINESE ACADEMY OF SCIENCES	6	0	0	0	0	0	0	0	0	0
HEBEI CHENYANGIND & TRADE GROUP	6	0	0	0	0	0	0	0	0	0
SIB LAB	1	0	1	1	0	1	0	1	0	0

as Azoxystrobin, *Reynoutria sachalinensis* extract and mixtures thereof, or Bifenthrin for targeted delivery. They are covered by the following patent portfolio:

Patent numbers: US10966422, US 10455830, US 9686979, US 20140187424 A1 (Publication) disclose Pyrethroid formulations comprising nanoparticles of polymer-associated pyrethroid compounds along with various formulating agents.

Patent Numbers US10206391 and US 20140364310 disclose Strobilurin formulations including a nanoparticle including a polymer-associated strobilurin compound with an average diameter of between about 1 nm and about 500 nm; wherein the polymer is a polyelectrolyte, and a dispersant or a wetting agent.

US 10070650 discloses methods to produce polymer nanoparticles and formulations of active ingredients including a polymer nanoparticle and at least one agricultural active compound incorporated with the nanoparticle, wherein the nanoparticles are less than 100 nm in diameter, and the polymer includes a polyelectrolyte.

US 9961901, US 9392786, and US 20150366186 A1(publication) disclose mectin and milbemycin nanoparticle formulations formulation including a nanoparticle including a polymer-associated mectin and/or milbemycin compounds with an average diameter of between about 1 nm and about 500 nm, wherein the polymer is a polyelectrolyte and a dispersant or a wetting agent.

US 9961900 discloses herbicide formulations comprising a nanoparticle including a polymer-associated herbicide, such as fenoxaprop or pyroxsulam with an average diameter of between about 1 nm and about 500 nm, wherein the polymer is a polyelectrolyte and a dispersant or a wetting agent.

US 9363994, US 9648871, 20,140,287,010 A1, US 8741808, and US 20130130904 A1 disclose methods to produce polymer nanoparticles and formulations of active ingredients including a polymer nanoparticle and at least one agricultural active compound incorporated with the nanoparticle, wherein the nanoparticle are less than 100 nm in diameter, and the polymer includes a polyelectrolyte.

US 20150359221 A1 discloses a formulation including a nanoparticle including a polymer-associated triazole compound with an average diameter of between about 1 nm and about 500 nm, wherein the polymer is a polyelectrolyte and a dispersant or a wetting agent.

US 20130034650 A1 discloses a composition including a collapsed, polymer nanoparticle and at least one organic, neutral compound associated with the nanoparticle, wherein the nanoparticle is less than 100 nm in diameter, and the polymer comprises a water-soluble polyelectrolyte, has a molecular weight of at least about 100,000 Da and is cross-linked.

US 8283036 and US 8,084,397 discloses a composite nanoparticle comprising a nanoparticle confined within a cross-linked collapsed polyelectrolyte polymer, wherein the nanoparticle comprises a charged organic ion.

US RE45848 discloses methods for producing a nanoparticle within a cross-linked, collapsed polymeric material. It also discloses the production of a nonconfined nanoparticle by complete pyrolysis of the confined nanoparticle and a carbon-coated nanoparticle by incomplete pyrolysis of the confined nanoparticle.

US 8257785 discloses a method for producing a composite nanoparticle. The method of preparation involves changing the conformation of a dissolved polyelectrolyte polymer from a first extended conformation to a more compact conformation by changing a solution condition so that at least a portion of the polyelectrolyte polymer is associated with a precursor moiety to form a composite precursor moiety with a mean diameter in the range between about 1 nm and about 100 nm; and cross-linking the polyelectrolyte polymer of the composite precursor moiety to form a composite nanoparticle.

This comprehensive sample patent portfolio of Vive Crop Protection illustrates how a company creates a protective fence around its key inventions that forms the basis of their respective commercial products.

Similarly, BASF and other corporations are fencing their technologies with strong patent portfolios.

4 Nanobiofungicides

Developments of new nanostructures in the form of dispersions, quantum dots, metal oxide nanoparticles, nanocarrier substrates comprising silica, lipids, or polymers for controlled release of actives in combination with biocides are continually being explored for applications in phytopathology. These nanomaterials with appropriate combinations with actives, nutrients, and growth regulators are emerging as promising candidates for targeted sustained delivery of fertilizers, pesticides, herbicides, nematocides, fungicides, and insecticides.

Nature-derivatized polymers as encapsulators/carriers of active ingredients are also being investigated as effective agents for plant disease management. Nano-based materials are capable of penetrating the leaves for effective translocation throughout the plant. Such nanocarriers/media can effectively be used for seed treatment, soil, and/or foliar application.

Efforts in the last two decades have been directed to the creation of efficacious nanobiofungicides that can be cost-effectively prepared and capable of being formulated with a variety of biocides to provide new types of environment-friendly biohybrid-nano-antimicrobial agents. Such materials are being tested for different pathogenic fungal organisms in plants.

The thrust areas of have been the development of nano-based eco-friendly media, substrates, and bio-actives that can be used for targeted and sustained delivery of the natural actives to the plants. Crops treated with safe nanobiofungicides are expected to be free of chemical residues and pathogens that affect human health. Nanobiofungicides are also expected to be used in combination with other plant growth agents/fertilizers in a variety of cropping patterns to enhance agricultural productivity. (Mittal et al., 2020).

The present section presents the results of a study on 426 published patent applications addressing the technologies and applications related to nanobiofungicides.

As illustrated in Fig. 3, the number of patent filings has steadily grown in the last two decades though the patenting scenario in nanobiofungicides is still in its infancy. The lower numbers in 2019 can be attributed to the slowdown during the COVID-19 pandemic.

The field of nanobiofungicides is a convergence of four technologies namely nanomaterials, bioactives, biochemistry of fungi, and devices for delivery to plants/crops.

Cross-fertilization of knowledge has been the hallmark of the development of nanobiofungicides. For example, teachings derived from the exploitation of nanotechnology in the pharmaceutical industry have effectively been adapted to nanobiofungicides.

Interestingly, the facile adaptation of knowledge from allied and nonallied fields has created a challenge in establishing inventive steps in the patentability of nanobiofungicide-related inventions. Novelty and industrial applicability (utility) have rarely been problematic in the prosecution of patents in the field of nano-biofungicides.

Tables 10 and 11 listing the sub-technologies and CPC Codes demonstrate the transdisciplinary nature of inventions for which patents are filed in the field of nanobiofungicides. These encompass teachings from diverse elements of medicinal preparations, biocides having microorganisms, viruses, microbial fungi, enzymes, fermentates, coating compositions,

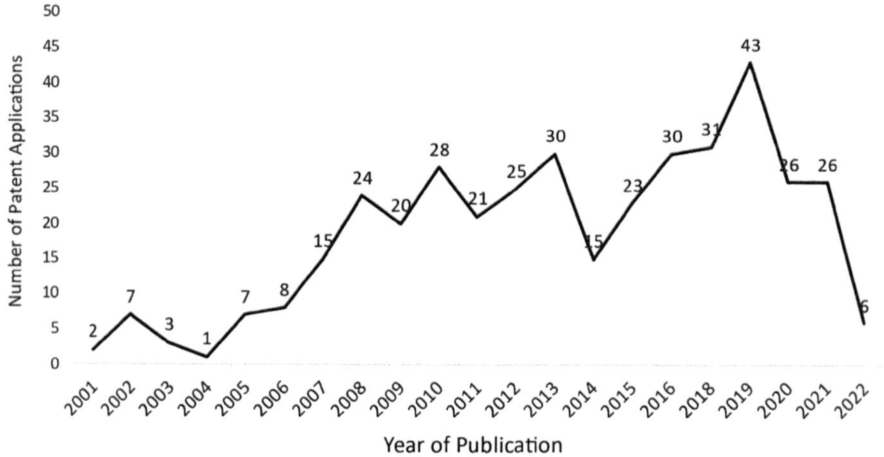

FIG. 3 Patent applications in nanobiofungicides published from 2001 to March 2022.

TABLE 10 Subtechnologies in nano-biofungicides patent applications.

Sub-technologies	Applications
Medicinal preparations characterized by special physical form	76
Biocides forms, ingredients, and application	70
Biocides having elements or inorganic compounds	39
Medicinal preparations containing organic active ingredients	39
Fungicides	37
Biocides having microorganisms, viruses, microbial fungi, enzymes, fermentates	35
Medicinal preparations characterized by nonactive ingredients used	33
Coating compositions characterized by the physical nature or effects produced, filling pastes	32
Use of inorganic substances as compounding ingredients	29
Antibiotics, antiseptics, chemotherapeutics	25

TABLE 11 Top CPC codes of nano-biofungicide patent applications.

CPC code	Description	Applications
A01N 63/04	Biocides having microorganisms, viruses, microbial fungi, enzymes, fermentates≫Microbial fungi or extracts thereof	9
B82Y 30/00	Nanotechnology for materials or surface science≫Nanotechnology for materials or surface science, e.g., nanocomposites	9
C08K 9/04	Use of pretreated substances as compounding ingredients≫Ingredients treated with organic substances	9
A01N 25/28	Biocides forms, ingredients, and application≫Microcapsules; or nanocapsules	7
A01N 59/16	Biocides having elements or inorganic compounds≫Heavy metals; Compounds thereof	6
A01N 25/00	Biocides forms, ingredients and application≫Biocides, pest repellants or attractants, or plant growth regulators, characterized by their forms, or by their nonactive ingredients, or by their methods of application, e.g., seed treatment or sequential application; substances for reducing the noxious effect of the active ingredients to organisms other than pests	5
A01N 25/04	Biocides forms, ingredients, and application≫Dispersions, emulsions, suspoemulsions, suspension concentrates	5
A01N 25/10	Biocides forms, ingredients, and application≫Macromolecular compounds	5
A01N 25/26	Biocides forms, ingredients, and application≫in coated particulate form	5
A01N 25/34	Biocides forms, ingredients, and application≫Shaped forms, e.g., sheets, not provided for in any other subgroup of this main group	5

fungicides, active ingredients, various techniques, materials, and states of matter utilized in nanotechnology, macromolecular chemistry, biocides, pest repellants or attractants, or plant growth regulators, characterized by their forms, or by their nonactive ingredients or by their methods of application; e.g., seed treatment or sequential application, substances for reducing the noxious effect of the active ingredients to organisms other than pests.

The top 10 patent applicants/patent assignees (Tables 12 and 13) in nanobiofungicides are those organizations that have been able to draw upon multidisciplinary knowledge, skills, and expertise from diverse sources. PolyOne that tops the list has been discussed in some detail under the section titled Strategic Mergers, Acquisitions, Joint Ventures, and Licensing

TABLE 12 Top 10 patent applicants in nano-biofungicides.

Top patent applicants	# Applications
POLYONE	22
INSTILLO	12
INMOLECULE INTERNATIONAL	10
SICPA HOLDING	9
BASF	8
CELANESE	8
NANOBIO CORP	8
UNIVERSITY OF CENTRAL FLORIDA RESEARCH	7
SIB LAB	6
ACETATE INTERNATIONAL	5

TABLE 13 Patent application in nano-biofungicides by top 10 applicants/assignees in various jurisdictions.

Patent applicants	Jurisdictions									
	WO	US	EP	CA	JP	CN	AU	IN	KR	EA
POLYONE	11	10	0	0	0	0	0	1	0	0
INSTILLO	0	1	1	1	1	1	1	1	1	1
INMOLECULE INTERNATIONAL	0	1	1	1	1	1	1	1	0	1
SICPA HOLDING	1	1	1	1	1	0	1	0	0	0
BASF	2	0	3	1	1	0	0	0	1	0
CELANESE	3	2	1	1	1	0	0	0	0	0
NANOBIO CORP	2	1	1	1	1	1	1	0	0	0
UNIVERSITY OF CENTRAL FLORIDA RESEARCH	1	1	1	0	0	0	0	1	0	0
SIB LAB	1	0	1	1	0	1	0	0	1	0
ACETATE INTERNATIONAL	1	1	1	1	1	0	0	0	0	0

Arrangement later in this chapter. The United States is the preferred jurisdiction of the patenting activities in this field.

Abstracts of a few selective patents/patent applications given below illustrate the type of inventions that have been protected in the emerging field of nanobiofungicides.

US10,258,048 discloses a Bacillus strain, Bacillus sp. isolate F727, that produces metabolites with pesticidal activities. This patent further provides bioactive compositions and metabolites derived from cultures of Bacillus sp. isolate F727 capable of controlling pests. It also discloses methods of use of the strain and its metabolites for controlling pests. The compositions can be formulated as emulsifiable concentrates (EC), wettable powders (WP), soluble liquids (SL), aerosols, ultra-low volume concentrate solutions (ULV), soluble powders (SP), microencapsulates, water-dispersed granules, flowables (FL), microemulsions (ME), nano-emulsions (NE), etc.

US10,058,093 discloses nanoparticle compositions and methods for targeted delivery of a bioactive agent to a plant. The nanoparticle composition includes a coronatine-coated nanoparticle formulated to deliver one or more bioactive agents through plant stomata. A variety of bioactive agents may be included in the nanoparticles, including one or more bactericides, fungicides, insecticides, acaricides, miticides, nematicides, molluscicides, herbicides, plant nutrients, fertilizers, plant growth regulators, or combinations thereof.

US9,918,479 provides compounds and compositions derived from Pseudomonas sp., particularly, *Pseudomonas fluorescens* or *Pseudomonas protegens* and more particularly strain having the identifying characteristics of Pseudomonas ATCC 55799 having antimicrobial properties and particularly, antibacterial properties. This also provides a method of inhibiting one or more phytopathogenic microorganisms in a targeted location such as in soil. The patent also mentions that the composition may contain an amount of a cell suspension, supernatant, filtrate, cell fraction, or whole-cell broth derived from a Pseudomonas ATCC55799. For the purpose of this invention, the phytopathogenic microorganisms are selected from the group consisting of *Bacillus subtillus, Bacillus cereus, Xanthomonas campestris, Xanthamonas arboricola, Xanthamonas vesicatoria, Streptomyces scabie, Botrytis cinerea, Erwinia carotovora,* and *Sphaerotheca fulginea.*

US11,124,460 discloses a composition on a matrix or support comprising a fertilizer and a free enzyme or expansion protein for stimulating plant growth and/or promoting plant health. The enzyme is selected from a phospholipase, a lipase, a xylosidase, a lactonase, a mannanase, a pectinase, a chitosanase, a protease, a glucanase, an ACC deaminase, and combinations of any thereof. The fertilizer in the composition comprises nitrogen, phosphate, potassium, zinc, iron, selenium, boron, copper, or a combination of any thereof. The matrix or support comprises charcoal, biochar, nanocarbon, agarose, alginate, cellulose, a cellulose derivative, silica, plastic, stainless steel, glass, polystyrene, ceramic, dolomite, clay, diatomaceous earth, talc, a polymer, a gum, a water-dispersible material, or a combination of any thereof.

US 8,741,808 discloses a composition comprising a polymer nanoparticle and at least one organic, neutral, nonionic agricultural compound associated with the polymer nanoparticle, wherein the polymer nanoparticle is less than 100 nm in diameter and crosslinked; wherein the polymer nanoparticle comprises a collapsed water-soluble polyelectrolyte; wherein the polyelectrolyte has a molecular weight of at least about 100,000 Da prior to collapse and cross-linking; and wherein the agricultural compound is a fungicide.

US9,363,994 discloses a composition comprising a polymer nanoparticle and a combination of at least two organic, neutral, nonionic agricultural compounds associated with the polymer nanoparticle, wherein the polymer nanoparticle is less than 100 nm in diameter and crosslinked; wherein the polymer nanoparticle comprises a collapsed water-soluble polyelectrolyte; wherein the polymer has a molecular weight of at least about 100,000 Da prior to collapse and cross-linking; and wherein the combination of at least two organic, neutral, nonionic agricultural compounds comprises a combination of an insecticide and a fungicide.

US10,070,650 discloses a composition comprising a polymer nanoparticle, and at least one organic, neutral, nonionic agricultural active compound associated with the polymer nanoparticle; wherein the polymer nanoparticle is less than 100 nm in diameter and crosslinked; the polymer nanoparticle comprises a collapsed, water-soluble polyelectrolyte having a molecular weight of at least about 100,000 Da; and the polymer nanoparticle is swollen due to the presence of a hydrophobic solvent, as compared to the polymer nanoparticle without the hydrophobic solvent.

US11,168,314 discloses a recombinant, modified extracellular chitinase having an amino acid sequence outlined in SEQ ID NO.4 or SEQ ID NO.5. The modified chitinase exhibits both exochitinase and endochitinase activity. The recombinant, modified enzyme is thermoactive, has thermostability, functions in a broad pH range (pH 3.0–11.0), and also exhibits improved solubility and enhanced efficacy for control of insects and phytopathogenic fungi.

US10,470,466 discloses formulations containing anthraquinone derivatives derived from root of *R. sachalinensis* with increased effectiveness as pesticides. Water-based formulations include suspension concentration (SC), microemulsion (ME), nanoemulsion (NE), soluble liquid (SL), emulsion in water (EW), ready-to-use (RTU), and microencapsulate or nanoencapsulate formulation. Powder formulations include but are not limited to water-soluble powder (WSP), water-dispersible granules (WDG), and water-dispersible tablet (WGT).

US10,159,250 discloses a species of *Burkholderia* sp. with no known pathogenicity to vertebrates but with pesticidal activity (e.g., plants, insects, fungi, weeds, and nematodes). The composition may further comprise a biofungicide such as an extract of *R. sachalinensis* (Regalia) or a fungicide.

US 9,586,871 discloses a pesticide composition that includes at least one biologically inert carrier; and at least one ferrite or at least one doped component including at least one fixed copper compound doped with at least one compound selected from the group consisting of iron compounds, zinc compounds, magnesium compounds, calcium compounds, and combinations and/or mixtures thereof. It also discloses a method of controlling fungal and other diseases in banana plants, by applying the aforementioned composition to their growth habitat in banana plant groves.

US11,021,408 discloses a nanoparticulate fertilizer that comprises water-soluble nanoparticles having a contact area to total surface area ratio greater than 1:4. The nanoparticles may have an overall positive surface charge in water. The fertilizer may be suitable for application as a seed coat. Also provided is a method of applying a seed coat, and a method of fertilizing a plant. The fertilizer is drawn into the plant by penetration through either or both of the stomatal openings and cuticle, into the leaf epidermis.

US10,433,543 discloses a multilayer bioactive and biodegradable film that includes one or more bioactive compounds or microorganisms for promoting the growth and health of a plant. The bioactive compounds or microorganisms are contained between layers of the film, wherein each one of the layers comprises about 60% to about 75% (m/m) polyhydroxyalkanoate. The bioactive compounds or microorganisms may include any one of or a combination of a metabolite, an antimicrobial compound, an enzyme, a live microorganism, a fertilizer, a plant growth hormone, a preservative, a pesticide, or an herbicide. The release of one or more bioactive compounds may be achieved in a timed and controlled manner.

US8,992,960 discloses compositions, gels, methods for synthesizing multifunctional silica-based nanoparticle gel, method of treating, preventing, or both treating and preventing, a disease in a plant species. The patent further discloses a method for simultaneously treating citrus plants for citrus canker and preventing the invasion of an Asian Citrus Psyllid (ACP) vector that carries the pathogen and spreads the citrus greening disease in citrus plants, and the like.

US10,405,550 discloses a method for preventing, controlling, or treating fungal infection on a plant organ includes applying to the plant organ a nonfungicidal amount or a potentiating amount of a composition including a potentiating agent of a plant defense molecule, in association with a phytopharmaceutical vehicle.

US 20090292055 discloses polypropylene-based compatibilizer with a mixture of organoclay and a polyethylene resin matrix. Nonlimiting examples of optional additives include adhesion promoters; FDA-compliant biocides, if any, (antibacterials, fungicides, and mildewcides).

5 Regulatory requirements

The key challenge has been to create stable formulations to meet regulatory requirements of efficacy and environmental safety as mandated by various regulatory agencies. Under the Federal Insecticide, Fungicide, and Rodenticide Act (FIFRA), the US Environmental Protection Agency (US EPA) requires a manufacturer to register a product as a pesticide if that product incorporates a substance intended to destroy pests, including microbes (US FIFRA 1996). The Department of Biotechnology, Government of India issued its guidelines for the evaluation of nano-based agri-input and food products in India in 2020 (DBT, 2020). In Europe, Nanomaterials fall under the existing "Registration, Evaluation, and Authorization of Chemical Substances – EC1907/2006" (REACH) and "Classification, Labeling and Packaging" (CLP) definition of a substance, and provisions set by both regulations apply. As of January 1, 2020, explicit legal requirements under REACH apply for companies that manufacture or import nanoforms.

6 Strategic mergers, acquisitions, joint ventures, and licensing arrangement

In view of the transdisciplinary nature of this field, knowledge resources and infrastructure converging from various sectors have been consolidated by way of strategic collaborations, alliances, mergers, joint ventures, acquisitions, and structured licensing arrangements.

PolyOne Corporation the leading global provider of specialized polymer materials, services, and sustainable solutions, purchased the color masterbatch businesses of Clariant and Clariant Chemicals India Ltd. PolyOne also announced that it has changed its name to Avient. Further, PolyOne has created strategic distributor alliances with Dow Corning and has access to the specialty silcone elastomers patent portfolio of Dow. PolyOne tops the list of the top 10 patent holders in the field of nano-biofungicides. Polyone has become a global leader in biompolymers industry (Polyone, 2008).

In 2001, Dow Chemical Company's agricultural business acquired the agrochemical unit of Rohm and Haas's fungicide lines, insecticides, herbicides, and other product lines, as well as trademarks and licenses to all agricultural uses of its biotechnology assets.

In 2013, FMC Corp. acquired exclusive rights to a patented, broad-spectrum crop protection product from Bayer Crop Science to develop and market the same. In 2013, FMC acquired the rights to develop and sell patented fungicide in key global Markets. FMC Corp entered into an exclusive licensing agreement with Belchim Crop Protection (Belchim) to develop, register, manufacture, and sell the patented fungicide valifenalate mixtures in North America, Latin America, and some select countries outside of the Americas. In 2015, FMC Corp acquired all global rights to a novel, proprietary herbicide from Kumiai Chemical Industry Co., Ltd., and Ihara Chemical Industry Co., Ltd. The new herbicide was highly effective in controlling broadleaf weeds and was initially developed for use in corn, cereals, soybeans, and sugarcane in key countries around the world. In 2017, FMC acquired DuPont's Cereal Broadleaf Herbicides and Chewing Insecticides portfolios, including Rynaxypyr, Cyazypyr and Indoxacarb. In addition, FMC acquired DuPont Crop Protection R&D pipeline and organization, excluding seed treatment, nematicides, and late-stage R&D programs, which DuPont would continue to develop and bring to market. The deal excluded the personnel needed to support marketed products and R&D programs that were retained with DuPont.

In August 2014, Japan's Mitsui acquired the global business operations for DuPont's copper fungicide Kocide, a key fungal and bacterial disease management tool for farmers, in particular for organically grown fruits and vegetables such as grapes and citrus fruits. Mitsui's US subsidiary established a local company to buy the business trademarks for Kocide, as well as product registrations, registration data, manufacturing know-how, and certain third-party contracts, along with the US group's copper fungicide production facility in Houston, Texas. Under the terms of the deal, DuPont continued to sell Kocide in the Asia-Pacific region for up to five years. Mitsui supplied Kocide and copper fungicide mixtures to DuPont. Copper fungicides are.

In December 2021, FMC Corporation signed multiyear agreements with Corteva Agriscience a leader in developing and supplying commercial seed combining advanced germplasm and novel traits to provide FMC's Rynaxypyr and Cyazypyr actives for seed treatment. These agreements are examples of FMC's exploitation of its patent-protected products. FMC collaborates with several crop protection companies to provide growers around the world with its leading insecticide technology. Products containing Rynaxypyr and Cyazypyr actives are registered in over 120 countries for use on hundreds of crops.

In April 2017 ChemChina acquired Syngenta. In August 2017, DuPont and Dow merged to form DowDuPont. In 2019, DowDuPont split to form DuPont and Corteva. BASF signed agreements in October 2017 and April 2018 to acquire the businesses and assets Bayer offered to divest in the context of its acquisition of Monsanto. BASF renamed the division from Crop Protection to Agricultural Solutions. In June 2018, Bayer successfully completed the acquisition of Monsanto.

All the mergers and acquisitions including alliances have resulted in re-adjustments in the IPR Portfolios through assignments, licensing, cross-licensing, and other forms of IPR Pooling, etc.

"Easyconnect" is a closed transfer system (CTS) for liquid crop protection products that consists of two components: a unique cap—prefitted on the containers—and a coupler. It makes filling the sprayer faster, easier and safer than conventional methods. BASF developed "Easyconnect" in collaboration with third-party equipment manufacturers. "Easyconnect" is an open technology, accessible to all interested parties in the market. The "Easyconnect" Working Group consisting of interested companies have piloted it in selected countries. As of June 2021, the working group includes ADAMA, BASF, Belchim Crop Protection, Certis Europe, Corteva Agriscience, FMC Corporation, Nufarm, Rovensa Group, and Syngenta.

Recent reports on the Antimicrobial Additives market include lead companies such as The Dow Chemical Company, BASF SE, Biocote Limited, Milliken Chemical Company, AkzoNobel N.V., Clariant AG, Bayer Material Sciences, Ticona and Victres, Nanobiomatters Industries S.L., Sanitized AG, A. Schulman, Inc., Polyone Corporation.

7 Litigations

IPR litigations in the field of agribusinesses have not been as frequent as in the pharmaceuticals and healthcare sector. However, a few litigations in the last few years in various jurisdictions indicate the firm intentions of companies to enforce their respective IPRs.

(1) In July 2021, the Delhi High Court in India granted FMC an interim injunction against NATCO Pharma Limited ("Natco") with respect to the granted Indian patents IN201307 and 213,332. The interim injunction restrained Natco from manufacturing, using, distributing, advertising, exporting, offering to sell, and/or selling any product, which contains chlorantraniliprole. FMC acquired these patents through a deed of assignment from E.I duPont de Nemours in 201. Coragen and Ferterra insecticides are approved for controlling pests on rice, sugarcane, vegetables, maize, and other important crops (Delhi High Court, 2021).

(2) On Sept. 13, 2021, in a patent infringement suit between FMC Corporation and Shandong Weifang Rainbow Chemical Co. Ltd. ("Rainbow"), the Qingdao Intermediate Court in China ruled in favor of FMC Corporation. A permanent injunction was ordered directing Rainbow to immediately stop manufacturing, selling, offering to sell, and using chlorantraniliprole. The Court found Rainbow infringed on FMC's composition of matter patent for the insecticidal active ingredient chlorantraniliprole and a key intermediate to manufacture chlorantraniliprole. The court also ordered the China-based crop protection manufacturer to compensate FMC for related damages. FMC markets its products that contain chlorantraniliprole under several brand names around the world, including Rynaxypyr active, Coragen insect control, Altacor insect control, Prevathon insect control, Premio insect control, and Ferterra insect control (FMC, 2021).

(3) On March 27, 2015, Syngenta sued Willowood in the United States District Court for the Middle District of North Carolina, for infringement of four patents related to the fungicide azoxystrobin [U.S. Patent Nos. 5,602,076; 5,633,256; 5,847,138; and 8,124,761]. Two of the asserted patents claimed the azoxystrobin compound that expired in 2014, while the other two were directed to processes for manufacturing azoxystrobin. Syngenta also asserted copyright infringement claims against Willowood about the labels accompanying the defendants' generic azoxystrobin products. Before trial, the district court dismissed the copyright infringement claims, determining them to be precluded by the Federal Insecticide Fungicide and Rodenticide Act. In 2017, the court granted summary judgment in favor of Syngenta, finding three of the four patents-in-suit valid and finding infringement with respect to two of the four patents. The court also agreed with Syngenta that, with respect to one of the process patents, the standards for shifting the burden of proof on infringement under 35 U.S.C. § 295 had been met. In September 2017, a jury found that Willowood infringed that process patent and awarded Syngenta nearly $1 million in damages. In November 2017, the court granted Syngenta's motion for a permanent injunction against Willowood and denied Willowood's motion to sell certain products it had imported but not yet sold prior to the jury verdict.

After a jury trial, the district court entered judgment in favor of Willowood Limited on all patent infringement claims; in favor of all defendants on infringement of one patent at issue; and against Willowood, LLC, and Willowood USA, LLC, on infringement of the remaining three patents. The district court denied Syngenta Crop Protection, LLC's motions for judgment as a matter of law. Syngenta Crop Protection, LLC, appealed the district court's denials of its motions for judgment as a matter of law and its final judgment. Defendants conditionally cross-appealed the district court's partial denial of their motion to exclude expert testimony on damages. The court affirmed-in-part, reversed-in-part, vacated-in-part, and remanded for further proceedings consistent with the opinion. (Syngenta Corp Protection, LLC v Willowood LLC and others, 2019).

Willowood brought the matter before the Supreme Court on two issues:

(i) Whether, by requiring the EPA to grant expedited review and approval of labels for generic pesticides that are "identical or substantially similar" to the previously approved labels for the same product, Congress intended to preclude claims of copyright infringement with respect to generic pesticide labels.

(ii) Whether liability for patent infringement under 35 U.S.C. § 271(g) requires that all steps of a patented process must be practiced by, or at least attributable to, a single entity, a requirement that the Supreme Court previously recognized is a prerequisite for infringement under 35 U.S.C. §§ 271(a) and (b) in Limelight Networks Inc. v. Akamai Technologies Inc.

Willowood's petition was denied as per the Supreme Court decision on October 5, 2020 (Willowood, LLC v. Syngenta Crop Protection, LLC, 2020).

(4) On February 1, 2018, in the US District Court of the District of Colorado the Plaintiff BASF ("BASF") filed a lawsuit against Defendants Willowood, LLC ("W-LLC"), Willowood USA, LLC ("W-USA"), Greenfields Marketing, Limited ("Greenfields"), RightLine, LLC ("RightLine"), and Willowood Limited ("W-Limited") for patent infringement. Relating to the manufacture and use of its fungicide pyraclostrobin. On 14 January 2019, the Court granted the Defendant's Motion to Dismiss and thereby the Defendant Willowood Limited was DISMISSED from this civil action. It was further ordered that the caption on all subsequent filings shall reflect the removal of Defendant Willowood Limited as a Defendant in this case (BASF Corp. v. Willowood, LLC, 2019).

(5) On November 12, 2019, Syngenta Crop Protection, LLC, Plaintiff (Syngenta) sued ATTICUS, LLC, Defendant. for patent infringement of U.S. Patent No. 5,602,076 ("the '076 patent") and US Patent No. 8,552,185 ("the '185 Patent") in the U.S. District Court for the Eastern District of North Carolina, alleging, among other things, that Atticus' Acadia 2 SC, Acadia ESQ, Aquila XL, Artavia 2 SC, and Artavia Xcel products each infringe certain Syngenta patents relating to the manufacture of azoxystrobin fungicide. On Sep. 21, 2020, the court granted Atticus motion to seal (i.e. the defendant need not disclose the details of their processes). On June 29, 2021, Atticus filed an amended answer and counterclaims. Atticus raised counterclaims alleging noninfringement and the invalidity of Syngenta's patents, violations of the Sherman Act, 15 U.S.C. §§ 1, et seq., violations of the North Carolina Unfair and Deceptive Trade Practices Act ("UDTPA"), N.C. Gen. Stat. §§ 75–1.1, et seq., defamation, and a claim for attorney's fees. See [D.E. 272] 256–96. On July 13, 2021, Syngenta moved to dismiss Atticus's antitrust, defamation, and UDTPA claims under Federal Rule of Civil Procedure 12(b)(6) and filed a memorandum in support. On July 20, 2021, Syngenta moved to stay discovery concerning Atticus's counterclaims while Syngenta's motion to dismiss was pending On August 3, 2021, Atticus responded in opposition to Syngenta's motion to dismiss and in opposition to Syngenta's motion to stay. On August 17, 2021, Syngenta replied to Atticus's response in opposition to Syngenta's motion to dismiss. The court on March 21, 2022 granted Syngenta's motion to dismiss and denied as moot Syngenta's motion to stay (Syngenta Corp Protection LLC v Atticus LLC, 2022.)

8 Conclusion

In conclusion, the field of nano-hybrid antifungals, comprising inorganic or organic polymers (as supports and/or carriers), metals or metal oxides (as actives), and organic molecules (as actives and/or structures in emulsions and other fluid states), holds promise as efficacious plant disease management agents. The development of packaging materials, devices, and delivery equipment are related areas of significance. Translational research and development activities are gradually resulting in proprietary technologies that are increasingly being protected as intellectual property rights by various organizations. There is also a flurry of mergers and acquisitions, as well as strategic restructuring of IPR portfolios to stay ahead of market competition. The transdisciplinary nature of this field, coupled with multiple transactions involving patent and trademark portfolios, including mergers and acquisitions, has made the mapping of patent claims and proprietorship very complex. It is therefore critical to conduct detailed patent due diligence by navigating the patent landscape to ensure "freedom to operate" in the commercial interest territory to avoid infringement of existing patent rights, even if no product has previously been made and sold commercially in those territories. With proprietary technologies are also consolidating their positions through tactical IPR litigation. Further, all mergers, acquisitions, collaborations, JVs, etc., will be constantly under the radar of national competition authorities. It is to be appreciated that many of the developments in nano-biofungicides may be novel; they may have been claimed in granted patents in various jurisdictions. In the years ahead, IPR will have to be managed as an integral component of agricultural operations. Various countries are now in the process of formulating stringent regulatory guidelines for the use of nanobiocides. The key challenge will continue to be creating stable formulations to meet regulatory requirements for efficacy and environmental safety as mandated by various regulatory agencies.

Acknowledgments

The author is grateful to Dr. Ginish George and Dr. George Koomullil of Relecura Technologies, Bangalore, India, for their invaluable contribution by way of the exhaustive patent searches, data, and figures included in this chapter. The author is also thankful to them for their permission to include all the tables and figures in this chapter.

References

Alghuthaymi, M.A., Kalia, A., Bhardwaj, K., Bhardwaj, P., Abd-Elsalam, K.A., Valis, M., Kuca, K., 2021. Nanohybrid antifungals for control of plant diseases: current status and future perspectives. J. Fungi 7 (1), 48.

BASF Corp. v. Willowood, LLC, 2019.

DBT, 2020. Guidelines for Evaluation of Nano-based Agri-input and Food Products in India. Department of Biotechnology, Government of India, New Delhi 110003, March 2020.

Delhi High Court 2021, Fmc Corporation & Anr. vs Natco Pharma Limited on 7 July, 2021, https://indiankanoon.org/doc/83320267/.

Elmer, W., White, J.C., 2018. The future of nanotechnology in plant pathology. Annu. Rev. Phytopathol. 56, 111–133.

FMC, 2021. Court Rules in FMC Corporation's favour in patent infringement case against Shandong Weifang Rainbow Chemical Co. Ltd. https://investors.fmc.com/news/news-details/2021/Court-rules-in-FMC-Corporations-favor-in-patent-infringement-case-against-Shandong-Weifang-Rainbow-Chemical-Co.-Ltd/default.aspx.

He, D.C., He, M.H., Amalin, D.M., Liu, W., Alvindia, D.G., Zhan, J., 2021. Biological control of plant diseases: an evolutionary and eco-economic consideration. Pathogens 10 (10), 1311. https://doi.org/10.3390/pathogens10101311.

Huang, B., Chen, F., Shen, Y., Qian, K., Wang, Y., Sun, C., Zhao, X., Cui, B., Gao, F., Zeng, Z., Cui, H., 2018. Advances in targeted pesticides with environmentally responsive controlled release by nanotechnology. Nano 8 (2), 102.

Marketsandmarkets, 2020. Biofungicides Market Report Code: AGI 7638. https://www.marketsandmarkets.com/PressReleases/biofungicide.asp.

Mittal, D., Kaur, G., Singh, P., Yadav, K., Ali, S.A., 2020. Nanoparticle-based sustainable agriculture and food science: recent advances and future outlook. Front. Nanotechnol. 2, 10.

Polyone, (2008) Biomaterials Development in the Polymer Industry—Technical Bulletin.

Syngenta Corp Protection LLC v Atticus LLC (2022).

Syngenta Corp Protection, LLC v. Willowood, LLC and others, (2019).

Thambugala, K.M., Daranagama, D.A., Phillips, A.J., Kannangara, S.D., Promputtha, I., 2020. Fungi vs. fungi in biocontrol: an overview of fungal antagonists applied against fungal plant pathogens. Front. Cell. Infect. Microbiol. 10, 604923. https://doi.org/10.3389/fcimb.2020.604923.

Velazco-Medel, M.A., Camacho-Cruz, L.A., Lugo-González, J.C., Bucio, E., 2021. Antifungal polymers for medical applications. Med. Dev. Sensors 4 (1), e10134.

Willowood, LLC v. Syngenta Crop Protection, LLC (2020).

CHAPTER

18

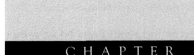

Microbial bioremediation of fungicides

Abdelmageed M. Othman[a] *and Alshaimaa M. Elsayed*[b]

[a]Microbial Chemistry Department, Biotechnology Research Institute, National Research Centre, Giza, Egypt [b]Molecular Biology Department, Biotechnology Research Institute, National Research Centre, Giza, Egypt

1 Introduction

As a consequence of various human activities, chemical leakage from industrialization and associated practices has generated several environmental hazards. As a result of urbanization and industrialization, a massive amount of sewage and radioactive waste are poured into the environment, damaging the ground, water, and atmosphere (Ferronato and Torretta, 2019; Tak et al., 2022). Contaminated ground, as well as poisoned soil and open water, is a major problem worldwide, and the consequences of these dangerous elements on the ecosystem are long-lasting. Hydrocarbon deposition in soil harms the surrounding ecology, and accumulation in living things can lead to genetic abnormalities (Ali et al., 2019). Contaminants and pollutants may be handled improperly using traditional cleanup methods. The bioremediation process would be a much more effective, environmentally benign, and nonexpensive procedure to remove dangerous substances from the ecosystem. The process of converting dangerous substances into a state that is not toxic to life forms and may be reused is known as bioremediation (Bhatt et al., 2020).

Microorganisms can be located in their normal locations in ecosystems and might be a moderate substitute for traditional approaches. Microbes like bacteria and fungi are eco-friendly in nature and have the ability to read environmental combinations because of their hopeful genetic materials. Pesticides, sewage waste, heavy metals, petroleum pollutants, hydrocarbons, dyes, and oils may be converted into nonhazardous chemicals by microorganisms, which are able to oxidize, degrade, or immobilize them (Bhatt et al., 2021). Bioremediation is mostly dependent on microbe enzymatic processes that are controlled by ecologic conditions, for instance, oxygen content, temperature, and soil pH. Microscopic organisms may act in and outside their environments, and yet, these organisms can cleanse severe environmental pollutants (Ojuederie and Babalola, 2017). Microbes with a genetically

Nanohybrid Fungicides
https://doi.org/10.1016/B978-0-443-23950-2.00007-2

modified genome, either explicitly or indirectly, can boost remedial bioactivity and have a stronger capacity to breakdown dangerous chemicals. Bioremediation is an inexpensive, environmentally friendly approach that can be applied to control large regions of ground water and soil (Sales da Silva et al., 2020; Tak et al., 2022). Bioremediation is frequently broken down into substrategies, such as bioattenuation (basically letting the polluted site's native organisms do the cleaning with no or negligible assistance), biostimulation (adding restricting nutrients or electron acceptors to promote the beginnings and activity of preexisting degraders), bioaugmentation (culturing of microbial species to boost the level of deterioration of the target pollutant), or bioadsorption (captures the pollutant in a physiologically harmless form on the microbial biomass) (Steffan, 2019). To synthetically study the microbial community, synthetic biology can assist in the collection of genomes via digital libraries and comprehend the link connecting catabolic and metabolic actions (Tak et al., 2022). The main subjects of this chapter include the role of microbes in fungicides remediation, their procedures, methodologies, and factors that influence microbial-aided bioremediation.

2 Fungicides

Fungicides are chemicals that are applied to plants or seeds to avoid or remove fungal infections. They are sprayed directly on cereals, field crops, trees, ornamental plants, and landscaping grasses in agriculture to preserve vegetables, fruits, and tubers while in storage (Gupta, 2011; Raffa and Chiampo, 2021). Fungicides disrupt a number of metabolic pathways in the fungus' cytoplasm and mitochondria. They disrupt a variety of enzymes and proteins involved in fungal respiration, lipid metabolism, and adenosine triphosphate (ATP) synthesis, among other things (Raffa and Chiampo, 2021; Thind and Hollomon, 2018). Table 1 classifies some common fungicides and their modes of action.

Fungicides have indeed been categorized according to their chemical compositions or agriculturally and horticulturally as shown by their method of impact (Ballantyne, 2004). Fungicides are classified as foliar, soil, or dressing fungicides based on their manner of treatment. Foliar fungicides are sprayed onto the apical vegetative sections of plants as solutions or dust, creating a defensive shield on the cuticular surface as well as systemic toxicity in the growing fungus (Gupta and Aggarwal, 2012). Soil fungicides are sprayed as solutions, dry powders, or pellets and work via their gaseous form or systemic characteristics. Dressing fungicides are sprayed on the postharvest crop as solutions or dehydrated powders to avoid fungal infection, especially if the product is kept under less-than-ideal temperature and moisture conditions.

Efficient fungicides would have to be defensive, reparative, or destructive and must have the subsequent characteristics: (1) lesser side effects to plants/animals but potential toxicity to the specific fungus; (2) the capacity to adapt to nontoxic precursors (through fungal or plant enzymes); (3) the aptitude to reach pathogens or rising mycelium to attain the spot of the act; (4) limited ecotoxicity; and (5) the capacity to establish a defensive, persevering film (Phillips, 2001). With some exceptions, the majority of newly discovered compounds have minimal mammalian toxicity. The positive mutagenicity tests achieved with numerous fungicides have sparked public alarm, as has the likelihood of both teratogenic and malignant

TABLE 1 The categorization of the most widespread fungicides and their modes of action.

Fungicide name	Chemical structure	Mode of action
Captan	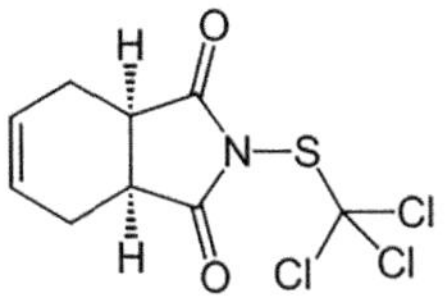	Enzyme activity reduction
Chlorothalonil	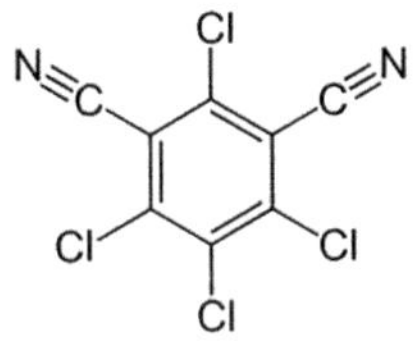	Reduction and inhibition of glutathione
Mancozeb	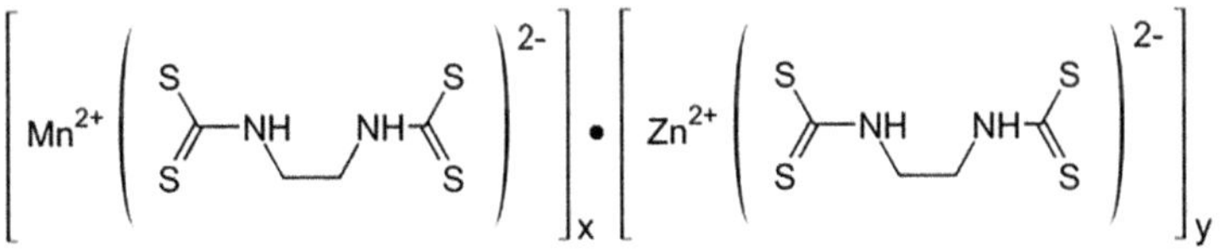	Fungal enzyme activity is inhibited by building a combination with the metalloenzymes
Maneb	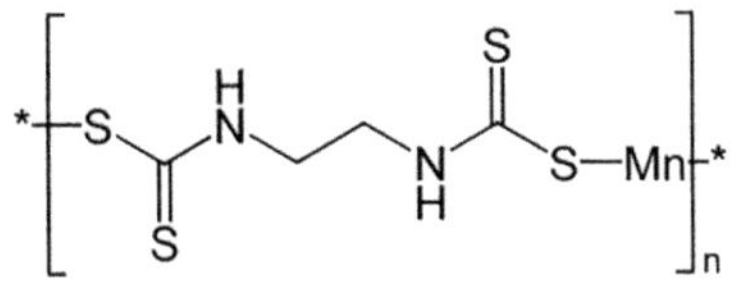	Inactivation of amino acid and enzyme sulfhydryl groups, resulting in disruptions in lipid metabolism, respiration, and adenosine triphosphate synthesis
Tebuconazole	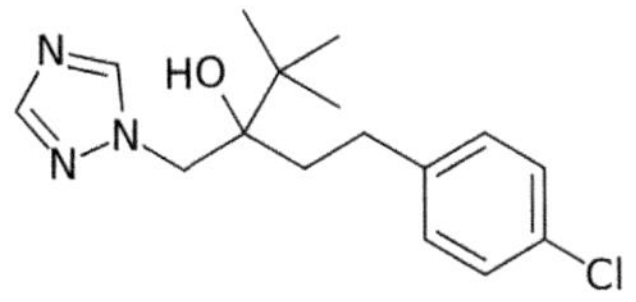	Sterol 14α-demethylase inhibition

Continued

TABLE 1 The categorization of the most widespread fungicides and their modes of action—Cont'd

Fungicide name	Chemical structure	Mode of action
Ziram	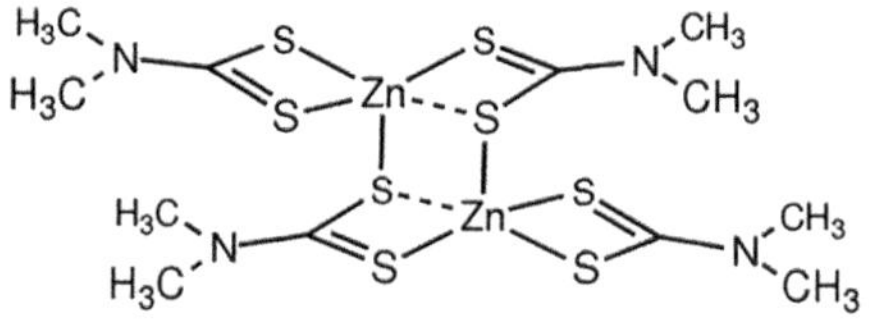	The establishment of a chemical barrier between the plant and a fungus

possibilities. Among 1964 and 1997, the amount of fungicides included in main crops is estimated to have grown 2.3-fold. Since the 1960s, the use of inorganics (mostly copper compounds) and dithiocarbamates has decreased, while captan, chlorothalonils, and other organic compounds currently make up over 90% of all fungicides used. Fungicides from emerging families, such as conazoles, dicarboximides, benzimidazoles, and metal-organic compounds, make up around 10% of overall use (Gupta and Aggarwal, 2012; Osteen and Fernandez-Cornejo, 2013).

As fungicides, a wide range of compounds with highly disparate chemical ingredients are employed. With some exclusion, the majority of newly created compounds have a low order of toxic effects on mammals. Because of their influence on endocrine systems, public attention has centered on their potential to induce developmental and reproductive damage in mammals (Gupta, 2011). Table 2 shows the classification of fungicides based on their chemical composition.

TABLE 2 Fungicides classification according to their chemical composition.

Chemical class	Examples
Amides	Fenhexamid, benalaxyl, metalaxyl, flutolanil, tolyfluanid, dichlofluanid
Anilinopyrimidines	Mepanipyrim, pyrimethanil, cyprodinil
Benzimidazoles	Benomyl, thiophanate-methyl, carbendazim, fuberidazole
Carbamic acid derivatives	Metiram, ferbam, propamocarb, mancozeb, zineb, nabam
Chloroalkylthiodicarboximides	Captan, captafol, folpet
Conazoles	Cyproconazole, etridiazole, penconazole, azaconazole, propiconazole, imazalil
Halogenated substituted monocyclic aromatics	Chlorothalonil, pentachlorophenol, quintozene, dicloran, dichlorophen, chloroneb
Morpholines	Dodemorph (liquid), fenpropimorph (oil), tridemorph
Others	Thiabendazole, cycloheximide, fludioxonil, dimethomorph, trifloxystrobin, fenpyroximate

The first fungicides were inorganic substances such as mercury, copper, lime, and sulfur. Elements of sulfur were advised as fungicides as early as 1803. From the 1940s to the 1950s, hexachlorobenzene (HCB) was widely employed as a fungicidal treatment administered to seed grains. Turkey experienced a poisoning outbreak between 1955 and 1959. HCB is a very poisonous chemical that can cause severe skin reactions as well as hypersensitivity. Many chemicals have since been created and used to combat fungal infections in seeds, plants, and products. For example, carbamic acid derivatives such as mancozeb, zineb, and nabam were created in the 1940s. For nearly 50 years, captafol, folpet, and captan have been used. Chlorothalonil, a halogenated benzonitrile fungicide, was initially approved for use as an agrochemical in the United States in 1966. For more than 35 years, the benzimidazole fungicides carbendazim and benomyl have been used. Anilinopyrimidines, a novel family of fungicides (pyrimethanil, mepanipyrim, and cyprodinil), were developed in 1993 to be applied to cereal grains (Gupta, 2011; Gupta and Aggarwal, 2012).

3 Classification and toxicity of fungicides

Fungicides have a wide and diverse chemical structure range that includes both inorganic and organic chemicals, resulting in cattle toxicity. Some of them are thought to be outdated or no longer in use. So far, WHO has studied numerous fungicides for their hazardous potential, and the Joint Meeting on Pesticide Residues (JMPR) has approved an acceptable daily consumption. A lot of cattle toxicity incidents from fungicides occur as a consequence of treated grains or potatoes, for example. The majority of toxicology information comes from experimental animals; information on farm animals and pets is scarce. Generally, the toxicity of newer groups of fungicides is low to moderate (Gupta and Aggarwal, 2012). The presence of other components (e.g., surfactants and emulsifiers) in the composition may contribute to fungicide toxicity. The specifics of the rules for danger evaluation and categorization have already been presented (Gupta, 2006). However, there is growing scientific and anecdotal evidence that fungicide contact impacts at least some set of conditions and/or fertility for one or many animal species. Human birth abnormalities have been linked to several fungicides. Hexachlorobenzene, captan, benomyl, methylmercury, and zineb, for example, are proven to cause fertility illnesses in humans, whether used alone or in combination with other pesticides. Fungicides, in general, are thought to have a greater overall incidence of causing developmental defects and oncogenesis than other pesticides. Over 80% of all oncogenic hazards from pesticide usage are attributed to a few fungicides; nevertheless, only a tiny number of pesticide-related deaths attributed to fungicides have been documented (Costa, 1997).

3.1 Inorganic fungicides

Titrated or sublimed sulfur, potassium thiocyanate, and potassium azide are all examples of fungicides in this category. Throughout the 19th and early 20th centuries, sulfur was widely used. Fungicides that are routinely employed include elemental sulfur and crude lime sulfur (barium polysulfide and calcium polysulfide). The ability of sulfur to spontaneously oxidize is its most prominent chemical feature. Sulfur's effects on the respiratory tract, skin, and eyes are attributed to this characteristic. Sulfur combustion has safety risks, which can be alleviated by restricting its usage throughout times of heightened ambient temperature. Sulfur poisoning can only be caused by micronized sulfur. Sulfur poisoning can cause respiratory, neurological, and gastrointestinal problems. Postmortem examinations revealed hemorrhagic effusions, intestinal and stomach congestion, and petechiae along the gastrointestinal tract and rarely on the bladder surface (Gammon et al., 2010; Low et al., 1996). Humans have been documented to experience negative effects, particularly on the respiratory tract, eyes, and skin (Gammon et al., 2010). Due to the availability of organic fungicides, their use has currently diminished. After reacting with stomach acid, another fungicide, barium polysulfide, produces barium chloride, which is a strong purgative (Sandhu and Brar, 2009).

3.2 Metallic fungicides

Agriculture was the first to employ inorganic metallic fungicides. They are both defensive and preventative in nature. Because of their toxicity, mercurous and mercuric chemicals have been phased out. Seed treatments for cereals and fodder beets include phenylmercury acetate, phenylmercury chloride, 2-methoxyethoxyethylmercury chloride, and ethylmercury phosphate (Lorgue et al., 1996). Organic metallic fungicides are aliphatic as well as aromatic. Several have a medium to severe human hazard, and some are neurotoxic and immunotoxic. Mercurials can cause skin and central nervous system abnormalities in livestock. They are irritants to the eyes, skin, and mucosae in general, and some are immunotoxic and hepatorenotoxic. They have been demonstrated to reduce T-lymphocyte formation, reduce lymphopoiesis, and enhance infection susceptibility (Gupta and Aggarwal, 2012).

3.3 Halogenated substituted monocyclic aromatics

Chloroneb, tecnazene, dinocap, dichlorophen, quintozene, dicloran, and chlorothalonil are all examples of compounds in this family. Chlorothalonil is a halogenated benzonitrile fungicide that is nontoxic. Dermal irritation was detected after repeated exposure to chlorothalonil, showing that it has the potential to induce skin irritation. Chlorothalonil produces permanent and severe eye injuries. Toxicity symptoms include forestomach hyperplasia, proximal tubular epithelial vacuolar degeneration, increased absolute kidney weight, and decreased body weight and hematological parameters. When tested to dosages that induce severe maternal poison and maternal mortality, chlorothalonil is not a reproductive, developmental, or genotoxic toxicant; nonetheless, a considerable increased rate of postimplantation death owing to premature embryogenesis mortality has been noted (Parsons, 2010). Mostly in rabbit eyes, tecnazene is a moderate allergen. Although the chemical has the ability to cause pulmonary adenoma in mice, it is not teratogenic or embryotoxic (Ballantyne, 2004). Dicloran, quintozene, and chloroneb have the least toxicity. However, dinocap is a mild ocular allergen and has the ability to sensitize mammalian skin (Gupta and Aggarwal, 2012).

3.4 Chloroalkylthiodicarboximides (phthalimides)

This family of compounds includes broad-spectrum fungicides (captafol, folpet, captan, and so on) that are employed as crop exterior preservatives on a variety of crops. Mammalian toxicity is relatively low. Only captan, which is structurally different from a cyclohexene ring, is utilized in this family of compounds; captafol and folpet, which are real phthalimides, have been permanently banned. Because of its high reactivity, the chemical is a strong eye irritant. Folpet causes decreased body weight, decreased food intake, salivation, vomiting, and diarrhea. Cattle are the most affected among ruminants, and captan causes toxicity with laborious breathing, gastroenteritis, ascites, anorexia, depression, and hydrothorax (Sandhu and Brar, 2009). Captafol is distinguished from captan and folpet in several aspects involving structure and chemical action. They both seem to have modest acute toxicity. They do not cause teratogenicity, mutagenesis, or cancer. They are not either reproductive poisons or selective developmental. They irritate mucous membranes, particularly the skin, following reexposure (Gordon, 2010).

3.5 Anilinopyrimidines

Pyrimethanil, mepanipyrim, and cyprodinil are among the anilinopyrimidine fungicides. The chemicals are low in toxicity and are liable to pose severe harm in ordinary usage. Cyprodinil causes elevated thyroid masses, hepatocellular hypertrophy, as well as follicular cell hypertrophy, and hyprochromasia. Mepanipyrim promotes lipofuscin accumulation in hepatocytes and hepatic fatty vacuolation (Terada et al., 1998). Pyrimethanil increases hepatic thyroid hormone metabolism and causes thyroid follicular cell tumors, which may contribute to thyroid carcinogenesis (Hurley, 1998). Cyrodinil and mepanipyrim have opposing impacts on blood and liver lipid markers. Anilinopyrimidines have no negative impacts on development cytotoxicity overall. They really are not genotoxic and do not have carcinogenic potential (Waechter et al., 2010).

3.6 Carbamic acid derivatives

Dithiocarbamates (propamocarb, ziram, thiram, ferbam, etc.) and ethylenebisdithiocarbamates (EDBCs; metiram, nabam, zineb, mancozeb, maneb, etc.) are fungicides of the carbamic acid class. Except for nabam, carbamic acid derivatives have a low to moderate acute hazard via the pulmonary cutaneous and oral routes (Gupta and Aggarwal, 2012). Toxic symptoms comprise flatulence, diarrhea, and anorexia, which are accompanied by neurological consequences such as prostration, muscle spasms, and ataxia. There is indeed a danger of antithyroid effects with repeated consumption, particularly with maneb. The thyroid is the primary target organ after repeated exposure to EBDCs. These fungicides affect thyroid hormone levels and weight. Their developmental toxicity includes embryo-fetotoxic consequences and deformities at maternally toxic dosage levels of EBDCs (Hurt et al., 2010).

3.7 Benzimidazoles

Fuberidazole, carbendazim, and benomyl are the most common benzimidazole fungicides. The toxic effects of carbendazim and benomyl are low, but fuberidazole has considerable toxicity. At high oral dosages, both carbendazim and benomyl cause reproductive and developmental toxicity in experimental animals. Carbendazim is a teratogen and a developmental toxin, and a high carbendazim dosage increases the likelihood of diffuse proliferation of thyroid parafollicular cells (Błaszczak-Świątkiewicz et al., 2016; Gupta and Aggarwal, 2012).

3.8 Conazoles

Diniconazole, triadimefon, triadimenol, propiconazole, imazalil, and cyproconazole are fungicides of the conazole class that have relatively low toxic effects. Triadimenol is a triazole, and triadimefon is chemically similar to triadimenol, with increased toxicity as the isomer A ratio increases (isomer B is less toxic). Triadimenol is not an irritant, but technical grade triadimefon is. Other toxicity signs include liver toxicity, alternating stages of enhanced and decreased motility, abusive behavior, and restlessness (Gupta and Aggarwal, 2012). Durable exposure caused developmental toxicity in the form of lower pup weight at parentally hazardous doses, liver enlargement and tumors, uterine lumen dilatation, and skeletal abnormalities in experimental animals (Baird and DeLorenzo, 2010; Škulcová et al., 2020).

3.9 Morpholines

Tridemorph, fenpropimorph, and dodemorph are examples of morpholine fungicides. Fenpropimorph and dodemorph are unlikely to be dangerous in the short term, but tridemorph is somewhat dangerous. Dodemorph acetate causes some irritation to rabbit skin and is extremely irritating to rabbit eyes. Fenpropimorph is an irritant to rabbit skin, although tridemorph is not. Fenpropimorph and tridemorph cause developmental toxicity, resulting in an increase in the overall number of abnormalities (Gupta and Aggarwal, 2012).

3.10 Amides

Fenhexamid, dichlofluanid, tolylfluanid, flutolanil, metalaxyl, and benalaxyl are examples of amide fungicides that are often used. Except for metalaxyl, which is somewhat poisonous, the other compounds have low toxicity. On the trichloromethylthio moiety of captan and folpet, a fluorine atom replaces one of the three chlorine atoms in dichlofluanid and tolylfluanid. Tolylfluanid is a skin sensitizer that can cause dyspnea, agitation, reduced motility, and drowsiness. Following intraperitoneal administration, signs of increased liver weight, local irritation, changed liver enzyme activity, and histological alterations suggestive of liver toxicity have been seen. These amides are not genotoxic and do not cause cancer in animals (Gupta and Aggarwal, 2012; Nagel et al., 2014).

4 Toxicokinetics of fungicides

Toxicokinetic studies give critical information on the quantity of toxicant administered to a receptor along with species-specific metabolic activity. Fungicides are either ingested or absorbed into the bodies or breathing systems of animals. Their absorption, transportation, processing, and outflow are all influenced by several variables. In general, the liver is the principal site of biotransformation, which includes both detoxifying and activating events (Gupta and Aggarwal, 2012). Some fungicides are not metabolized and instead attach to additional available active sites. The aryl organomercurials methyl- and ethylmercury chloride are weakly eliminated and prefer to collect in brain, muscle, and other tissues, whereas phenylmercury is much more easily expelled through the kidney and is less liable to collect in muscles and brain.

Likewise, hexachlorobenzene (HCB) has all of the qualities of chemical stability, slow degradation and biotransformation, ecological permanence, and bioaccumulation in adipose major organs with a significant lipid membrane concentration (Costa, 2008). The recently added fungicides are quickly absorbed, processed, and expelled and do not concentrate in tissues; nevertheless, some of them were marginally assimilated from the digestive system tract (Parsons, 2010). Before reaching the duodenum, captan is quickly metabolized in the stomach to 1,2,3,6-tetrahydrophthalimide (THPI) and thiophosgene (through thiocarbonyl chloride). THPI has a half-life of 1–4 s and is quickly eliminated after detoxification with glutathione or cysteine. There is no captan found in the urine or the blood. As a result, it's doubtful that these molecules, or even thiophosgene, will survive long enough to reach systemic targets like testes, uterus, or the liver. Eggs, milk, and meat from livestock or poultry would be devoid of the parent components as a result of the quick removal. Captan appears to be metabolized similarly by humans and other species (Gordon, 2010).

Cyprodinil, a fungicide of the anilinopyrimidine family, is rapidly absorbed from the gastrointestinal tract into the systemic circulation. Cyprodinil is metabolized almost entirely. Urine has no intact parent molecule, but feces contain trace levels of unmodified cyprodinil. The majority of the cyprodinil supplied is metabolized by phenyl and pyrimidine ring sequential oxidation. The metabolite profiles of phenyl and pyrimidyl-labeled cyprodinil in urine and feces are identical (Gupta and Aggarwal, 2012). Carbamic acid derivative fungicides like ethylenebisdithiocarbamates (EBDCs) are only slightly absorbed and then quickly metabolized and eliminated, with no indication of long-term bioaccumulation. Oral dosages are rapidly absorbed, and they are expelled within 24 h, with around half eliminated in the urine and half in the feces. Ethylenthiourea is their most frequent metabolite. Tissues, notably the thyroid, have only small levels of residues. Propamocarb, another molecule in this family, is rapidly and practically fully absorbed and distributed, reaching peak concentrations in 1 h. Propamocarb is highly digested, with just trace amounts remaining unaltered in urine. Aliphatic oxidation of the propyl chain (to generate hydroxyl propamocarb) and *N*-demethylation and N-oxidation of the tertiary amine result in propamocarb N-oxide and mono demethyl propamocarb, correspondingly, during metabolism.

In mammals, prior to further metabolism, benomyl is transformed into carbendazim by the elimination of the n-butylcarbamyl side chain. Carbendazim is an aryl hydroxylation-oxidized at the 5 and 6 positions of the benzimidazole ring, preceded by sulfate or glucuronide conjugation before elimination (Oruc, 2010). Likewise, amide fungicides are readily taken up and removed. Metalaxyl-M and metalaxyl may stimulate cytochrome P450 and other drug-metabolizing enzymes in the liver and kidney. Within 48 h, tolylfluanid is swiftly and widely absorbed, followed by fast metabolism and virtually total elimination, primarily in the urine and, to a lesser extent, in the bile. The liver and kidney showed high tissue concentrations immediately after the dose, with lower quantities in the thyroid gonads, brain, and perirenal fat. Fluoride concentrations in bone and teeth rise in most animals in a dose-dependent way. Conazole fungicides like triadimefon and triadimenol are rapidly absorbed and broadly dispersed in the liver and kidney after oral dosage. Discharge and metabolic activity are fast and widespread, mostly via *t*-butyl methyl group oxidation. Propiconazole has a quick and comprehensive absorption (80% of the supplied dosage) and is extensively spread, with the greatest concentrations in the kidney and liver. Fludioxonil, on the other hand, is promptly and widely (80%) absorbed, broadly distributed, substantially digested, and quickly eliminated, predominantly in feces (80%), with a small quantity excreted in urine (20%). Within 1 h of ingestion, the maximal blood concentration is attained. The molecule is extensively metabolized, with oxidation, chain cleavage, shortening, hydroxylation, conjugation, and *O*-demethylation occurring between the trifluoromethyl and glyoxylphenyl moieties (Gupta and Aggarwal, 2012).

5 Bioremediation

Bioremediation is a biological approach that utilizes microorganisms or enzymes to change, transform, cleanse, or negate hazardous substances into safe molecules (Ndeddy Aka and Babalola, 2016; Okoduwa et al., 2017). Several microbe-assisted methods or pathways for environmental pollutant cleanup have been found and characterized. Under regulated settings, bioremediation is an exceptional green technique that facilitates the removal and degradation of harmful pollutants into nontoxic or less dangerous chemicals or substances. It is essentially a microbe-mediated, slow, and natural process. This is beneficial to both microbes and the environment because chemical poisons or radioactive substances are destroyed, supplying resources for microbial metabolism (Hakeem et al., 2020). Aerobic or anaerobic bioremediation involves the total or partial elimination of harmful substances into water, gases, and other inorganic compounds that are harmless to the environment and living organisms. Depending on the kinds of contaminants and their quantities, bioremediation can be performed in situ or ex situ (Sales da Silva et al., 2020). By selecting the appropriate bioremediation approach, pollutant concentrations will be successfully limited.

5.1 Mechanism of microbial bioremediation

Pesticides are primarily used to manage and eliminate agricultural infections and diseases, but pesticide residues have also been linked to adverse effects on public life and the ecosystem. They are, nonetheless, crucial in the field of worldwide ecological rehabilitation research and technology to examine the microbiological breakdown of pesticides. Chlorinated pesticides, such as polycyclic aromatic hydrocarbons, organophosphorus, and polychlorinated diphenyl, have been detoxified by microbes such as bacteria, fungi, and actinomycete. *Acinetobacter*, *Paracoccus*, *Aerobacter*, and other major bacteria are commonly used to aid pesticide clearance. *Aspergillus niger* and *Fusarium oxysporum* are examples of fungi that may degrade pesticides. Pesticides may be removed by enzymes such as laccase, hydrolase, and peroxidase. Pesticides are removed by biosorption, which includes biodegradation (Bhatt and Shrivastav, 2022).

The primary purpose of bioremediation is to encourage microbes with aeration and nutrients to help them destroy contaminants. Microorganisms such as bacteria, fungi, algae, and yeast live in their natural surroundings and remove metals, colors, oils, chemicals, and aromatic hydrocarbons from the environment (Das and Chandran, 2010; Giri et al., 2017a,b, 2021). Microscopic organisms seek an energy resource from their surroundings, making them an ideal agent for bioremediation. Several types of pollutants are found in polluted sites based on their chemical makeup, and polluted site microorganisms decompose or cleanse the pollutants (Juwarkar et al., 2010).

Various microorganisms release enzymes that are able to degrade organic hazardous substances in the environment. Numerous pollutants are detoxified by utilizing oxidoreductase enzymes that transform harmful molecules into inert substances. Oxidoreductase enzymes are typically employed to eliminate hazardous metals and phenolic compounds from the biosphere (Karigar and Rao, 2011). Filamentous fungi may readily access soil contaminants and, with the aid of fungal mycelium, release laccases, and peroxidase enzymes, which neutralize the pollutants (Khatoon et al., 2021). Some bacterial strains in the soil possess oxygenase enzymes that breakdown aromatic contaminants in the environment. To treat organophosphates and oil spills, bacteria release hydrolytic enzymes, including proteases, amylase, and lipases, which hydrolyze toxicants' bonds and cleanse them (Xu et al., 2018).

Pesticides such as linuron, metribuzin, simazine, and trifluralin, as well as chlorpyrifos, are degraded by white rot fungi such as *Phanerochaete chrysosporium, Pycnosporus coccineus, Peniophora gigantea*, and *Trametes versicolor* via producing extracellular enzymes (Gouma et al., 2014). A number of bacteria absorb and recirculate contaminants, incorporate them, biosorb them to membranes, immobilize them, create complexes, precipitate them, and release them. The purple nonsulfur bacteria *Rhodobacter sphaeroides* and *Rhodobium marinum* may remediate heavy metals such as copper, cadmium, and zinc (Li et al., 2016). Genetically engineered *Pseudomonas putida* uses several mechanisms, for instance, changed membrane characteristics and metabolism to breakdown pollutants (Samin et al., 2014).

Heavy metals are incorporated into the cellular walls of soil bacteria or metal-associated proteins via binding sites such as phosphate, hydroxyl, or carbonyl groups, causing metabolic activities to be changed to hinder the uptake of heavy metals and toxic metals from being converted into nontoxic products via numerous enzymatic reactions (Kapahi and Sachdeva, 2019). Yeast employs an ion-exchange mechanism to bioremediate the toxic water. Microalgae such as *Phaeodactylum tricornutum, Cladophora, Chlorella*, and *Spirulina* spp. may function as "hyperaccumulators" and "hyperadsorbents" to cleanse heavy metals from contaminated groundwater (Bwapwa et al., 2017). Aerobic bacteria use microalgae to give oxygen, enzymes, and nutrients for metabolism, and these bacterial strains breakdown oil and manufacturing waste. Joined microalgae-bacteria consortiums breakdown ibuprofen and caffeine quicker than single bacterium ensembles (Sutherland and Ralph, 2019).

Fungi can breakdown pesticides and herbicides by esterification, hydroxylation, and dehydrogenation reactions. Fungi have their own complicated enzyme system for detoxifying environmental contaminants, which includes the cytochrome P450 complex, glutathione transferase, and monooxygenases. For nonactivated hydrocarbons, P450 complexes present in the mitochondria and cytosol of fungi may operate as stereospecific catalysts. The white-rot fungus *Penicillium chrysosporium* possesses CYP63A2 P450 monooxygenase, which oxidizes aliphatic carbon present in oil and polyaromatic hydrocarbons (Lah et al., 2011). *Lentinus squarrosulus* degrades several industrial wastes by activating lignocellulosic modifying enzymes such as hydrogen peroxide-generating enzymes, peroxidases, and laccase (Chukwuma et al., 2020). Polychlorinated biphenyls are degraded by *Fusarium solani, Penicillium chrysogenum*, and *Penicillium digitatum*, whereas *Rhizopus oryzae* detoxifies pentachlorophenol by methylation and dechlorination processes (Aranciaga et al., 2012).

Saccharomyces cerevisiae, for example, decreases ecological pollutants by lowering metal binding capability, permitting radioactive metals to be trapped in biomass, removed, reclaimed, and delivered into ecologically benign chemicals. Through electrostatic interactions, the negatively charged cell wall of yeast adheres to heavy metals, whereas exopolysaccharides boost the cell wall's biosorption capacity (Machado et al., 2009). In addition, *Candida digboiensis* bioaugmented with oily sewage-polluted soil had a greater ability to absorb petroleum hydrocarbons in the laboratory or the field (Sood et al., 2010). *S. cerevisiae, Candida* spp., and *Kluveromyces marxianus* have a stronger function in heavy metal detoxification due to the exterior cell wall mannan-protein layer and the inside cell wall glucan-chitin layer. At appropriate pH values, these yeast species, with their metal binding capabilities and specialized enzymes, could be able to remove 73%–90% of copper (Garcia-Rubio et al., 2020). Fig. 1 illustrates different mechanisms of pollutant bioremediation by microorganisms.

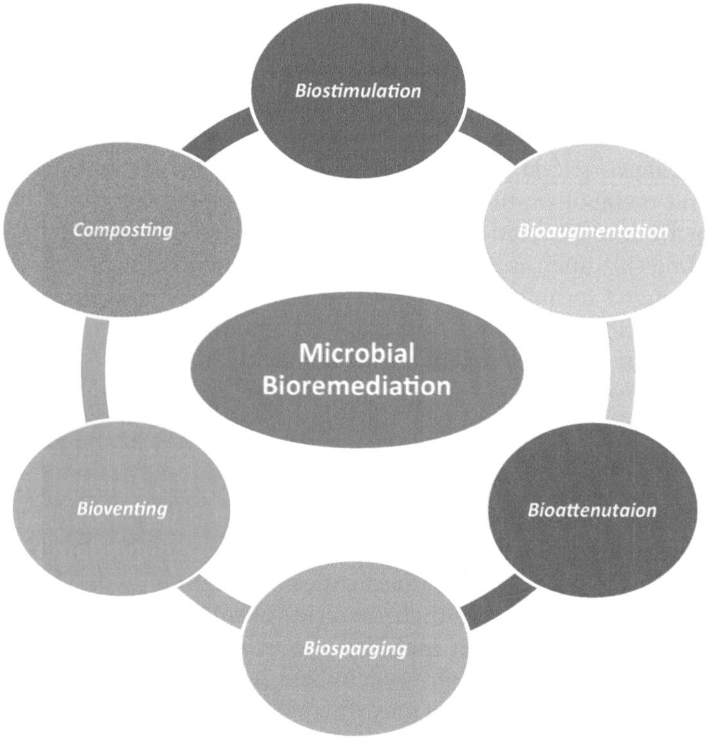

FIG. 1 Mechanisms of pollutants bioremediation by microorganisms.

5.2 Bioremediation methods

5.2.1 *Biostimulation*

Biostimulation is a method of enhancing the activity of indigenous bacteria by infusing certain electron acceptors or donors into soil or ground water (Fig. 2). Increase the inherent or natural ecosystem microbial population and the breakdown of pollutants in this technique by accumulating nutrients or other constraining variables (Kanissery and Sims, 2011). Fungi metabolic growth is boosted by giving nutrients, growth-promoting additives, minerals, oxygen, and temperature to indigenous microorganisms such as soil bacteria. Nutrients like nitrogen, phosphorus, and potassium are essential for the survival of bacteria in order for them to gain energy and cellular biomass (Jacoby et al., 2017). The biostimulation system is fundamentally utilized to eliminate petroleum pollutants from soil; nevertheless, it is also influenced by pH, concentration, oxygen availability, and other soil conditions (Tyagi et al., 2011). By stimulating natural microbes with organic additives and nutrients, the oxidation of 1,2-dichlorobenzene and chlorobenzene in the soil is enhanced (Kurt and Spain, 2013). The most important factor to consider is supplement distribution in such a way that the nutrients are easily accessible to subsurface microbes.

5.2.2 *Bioaugmentation*

Bioaugmentation is the incorporation of endogenous or exogenous microbes able to digest pollutants by increasing their enzyme content at the gene level (Fig. 3). This approach is

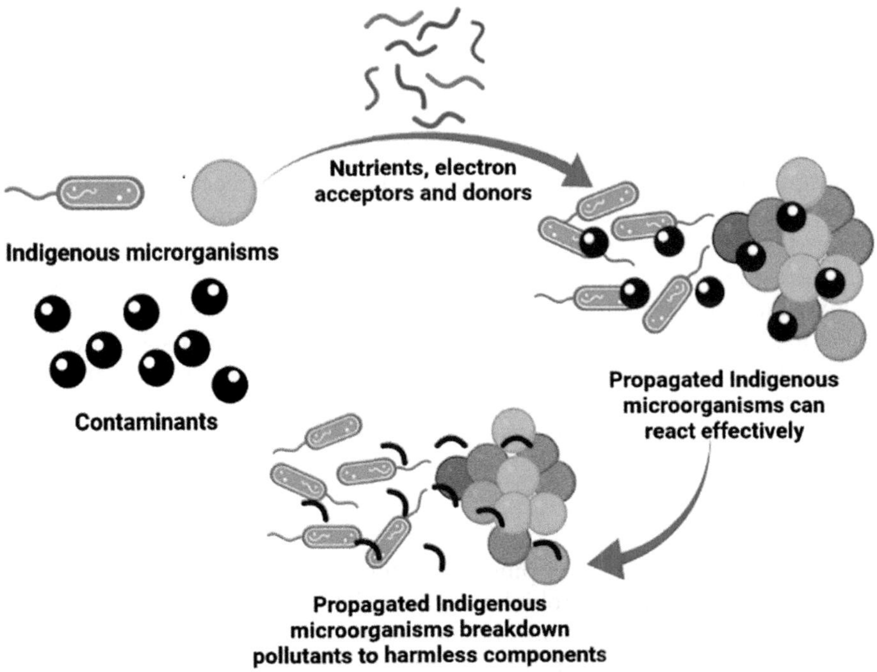

FIG. 2 Biostimulation of indigenous microorganisms to degrade contaminants.

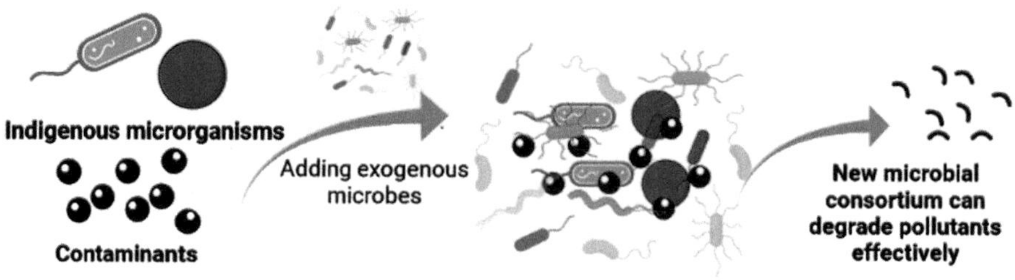

FIG. 3 Bioaugmentation mechanism for pollutants bioremediation.

mostly utilized for oil contaminants, in which a particular strain or consortia, or any genetically engineered organism, improves the breakdown capability of a contaminated pollutant (Nzila et al., 2016). Due to particular DNA alteration and a diverse metabolic pattern, genetically modified microorganisms digest toxins faster and more effectively than native microorganisms. Microorganisms are taken from the diseased spot, isolated, genetically altered, and reintroduced into the same environment (Das and Chandran, 2010). On a laboratory scale, bioaugmentation using bacteria alliances will decompose 2,4-dichlorophenol. When bioaugmented using cocultures of *Ethanoigenens harbinense* B49 and *Clostridium acetobutylicum* X9, the efficiency of cellulose breakdown and hydrogen generation from carboxymethyl cellulose digestion can be increased (Ren et al., 2008). *Stenotrophomonas maltophilia* JR62 and *A. xylosoxidans* 2BC8 bioaugment pyrene and fluorine polluted soil by increasing soil mineralization (Villaverde et al., 2019).

5.2.3 Bioattenutaion

Bioattenutaion is a conventional process in which microbes digest toxic pollutants and substances through diverse biological mechanisms. Biochemical mechanisms are used by microbes to reduce the bulk, duration, and hazard of contaminants in the environment. Toxins can be degraded by transformation, ion exchange, solubilization, sorption, advection, depression, diffusion, volatilization, and other chemical processes. Microbes in groundwater and soil may completely dissolve poisons, transforming them into water and benign gases (Azubuike et al., 2016). In groundwater backup and retrieval systems, bioattenuation using indigenous soil microorganisms may be an optimizing alternative to an organic carbon management long-term plan. Organic degraders could include denitrifying bacteria like *Noviherbaspirillum denitrificans* and iron-reducing bacteria like *Geobacter* spp. (Nguyen et al., 2021).

5.2.4 Biosparging

Biosparging is the practice of pumping air under pressure into a specific zone to supply oxygen and nutrients to local microbial growth while also destroying organic substances. It is often utilized in combination with the bioventing method, which includes blowing air into the zone of saturation to facilitate the aerobic breakdown and mineralization (Parween et al., 2018). It is primarily utilized to breakdown dispersed petroleum components in groundwater sources and capillary fringe. The biosparging process increases redox potential, nitrate, dissolved oxygen, sulfate, and cultivable heterotrophs while decreasing methanogens, cultivable anaerobes, sulfide, and soluble ferrous iron. Pumping air below the polluted location at close well spacing promotes aquifer oxygenation to enhance aerobic biodegradation (Kao et al., 2008).

5.2.5 *Bioventing*

It is a method of bioremediation in situ whereby microorganisms and consortiums increase hydrocarbon bioremediation in soil by enhancing oxygen diffusion in an unsaturated environment. It mostly reduces soil adsorption of leftover fuel or contaminants. Bioventing is a technology that employs low ventilation rates to deliver the oxygen required for biodegradation while minimizing pollutant volatilization and emissions into the atmosphere. Bioventing is particularly effective in removing organic pollutants, nonchlorinated solvents, hazardous pesticides, and petroleum hydrocarbons from soils. Low soil moisture and temperatures might hinder the bioventing approach (Azubuike et al., 2016). The use of bioventing and biosparging together can help remove hydrocarbons and hazardous vapors from the ground surface more quickly (Ahmadnezhad et al., 2021). In terms of aeration, bioventing may be accomplished in either an active or passive mode: in the former, air is blown into the soil by a blower; while in the latter, gas exchange through the vent wells happens only through the action of atmospheric pressure (Raffa and Chiampo, 2021).

5.2.6 *Composting*

Composting is one method of pesticide bioremediation. It entails combining contaminated soil with nonhazardous organic amendments in order to foster the growth of bacterial and fungal populations capable of degrading pesticides via a cometabolic pathway. When pesticide concentrations are low, this strategy is especially appropriate. Microbial bioaccessibility to pollution is critical in composting. As a result, it is critical to keep track of the water content, soil composition, and applied amendment qualities (Raffa and Chiampo, 2021). Biochar can be added to contaminated soils as a supplement to speed up the decomposition process. Biochar is a kind of black carbon produced by thermal transformations of biomass in the presence (gasification) or absence of oxygen (pyrolysis). It has high porosity and large surface area, which helps insecticides adsorb. Furthermore, biochar is a carbon source that increases microbial activity and hence promotes biodegradation. Biochar treatment has been shown to promote soil aeration and increase soil water retention (Varjani et al., 2019). Sun et al. (2020) investigated its potential use in tebuconazole biodegradation. This allowed the bacterial strain involved in the breakdown process, *Alcaligenes faecalis* WZ-2, to be immobilized.

6 Agricultural toxic substances and microbial bioremediation

Agricultural chemicals are employed to increase crop output by minimizing pathogen, disease, and weed damage. Fertilizers, fungicides, herbicides, and pesticides are only a few of the numerous agrochemicals employed in contemporary agriculture that are currently polluting the ecosystem (Liu et al., 2019). Field pollution with agrochemicals is a global issue that

TABLE 3 Microbes having remediation potential for agrochemicals.

Agrochemicals	Microorganisms	References
Decis 2.5 EC, Fitoraz WP 76, Ridomil MZ 68 MG	*Pseudomonas putida, Acinetobacter, Arthrobacter*	Pérez et al. (2016)
Chlorpyrifos and methyl parathion	*Acenetobactor, Pseudomonas, Enterobacter, Photobacterium, Moraxella, Yersinia*	Ravi et al. (2015)
Endosulfan, malation, Chlorpyrifos, Diazinon	*Serratia marcescens, Acinetobacter radioresistens, Bacillus pumilis*	Hussaini et al. (2013)
Cypermethrin	*Pseudomonas, Micrococcus*	Niti et al. (2013)
Tetrachlorvinphos	*Serratia, Yersinia, Vibrio, Proteus*	Niti et al. (2013)
Chlorpyrifos	*Enterobacter, Synechocystis, Brucella, Stenotrophomonas*	Niti et al. (2013)
Oxyfluorfen	*Bacillus, Pseudomonas, Arthrobacter, Aspergillus, Mycobacterium, Micrococcus, Streptomyces*	Mohamed et al. (2011)
Glyphosate	*Pseudomonas pseudomallei*	Peñaloza-Vazquez et al. (1995)

must be addressed immediately. Microorganisms are increasingly being used in agrochemical waste bioremediation. The bioremediation of agrochemical contamination by microorganisms is currently gaining popularity. Numerous microorganisms collected and evaluated from numerous resources have demonstrated their ability to convert pollutants into nonhazardous compounds in the biosphere (Table 3).

7 Microbial bioremediation of fungicides

Given the possibility of resistance and adverse consequences on human health and the environment, the use of systemic fungicides has grown in recent years (Sharma et al., 2019). Fungicides are a diverse group of chemicals with varied modes of action that belong to numerous chemical groups and are used in a total of 400 thousand tons across the world. Many studies have shown that organic fungicides are hazardous to aquatic and terrestrial organisms in the short and long term (Gikas et al., 2022).

The production of new fungicides is a quick process. As a result, new opportunities and challenges have developed in terms of their environmental destiny and removal from contaminated systems using treatment approaches like created wetlands (CWs). According to a variety of studies, CWs may successfully regulate pesticide contamination, including fungicides (Gikas et al., 2018; Papaevangelou et al., 2017; Rajmohan et al., 2020). In comparison to herbicides and insecticides, the environmental destiny of fungicides and ecological risk evaluations for nontarget species have been examined. As a result, studies understate the efficacy of fungicide treatment procedures (Sharma et al., 2019). The stability and maintenance of CW treatment effectiveness are mainly determined by the microbial population. Both the filter bed material (porous media) and the plant roots may include fungicide-degrading bacteria. Also, wetland plants can be colonized by epiphytic or endophytic microbes. The rhizosphere is where the bulk of pesticide biotransformation occurs, which is aided by the emission of oxygen and organic secretions from plants (Sánchez, 2017).

Propiconazole belongs to the triazole family of fungicides and inhibits demethylation. As a foliar spray, this fungicide will spread and enter the soil during treatment. Triazole fungicides are toxic and last a long time in the soil, influencing the fertility of the soil and microflora. The reduction of fungicide toxic effects seems to have been a significant research focus, and common ways of doing so have a number of detrimental environmental consequences. As a result, environmentally beneficial and practicable methods like microbial biodegradation are gaining significance (Satapute and Kaliwal, 2016). Microorganisms are the most preferred biological tools due to their capability to overcome a variety of pesticides and their metabolic ability to break down toxic substances into nontoxic forms. Pesticide-degrading microorganisms have been produced by a variety of bacteria, fungi, algae, and yeast species (Satapute and Kaliwal, 2016). Due to the flexibility of their metabolic pathways and their ability to thrive on complex carbon substrates, bioremediation studies of fungicides utilizing bacteria have been more effective. Several genes involved in the metabolism of hazardous substances have also been discovered. Moreover, cytochrome P450 monooxygenase, which constitutes a large family of protein haem thiolates able to decompose a broad range of toxic substances, has been extensively studied in bacteria (Degtyarenko, 1995). A *Pseudomonas aeruginosa* PS-4 strain was obtained from polluted paddy soil and utilized to investigate propiconazole decomposition in vitro. (Satapute and Kaliwal, 2016).

For many years, strobilurin fungicides have been widely utilized in agriculture. Such pesticides were developed to combat fungal diseases, but their wide-ranging activity might result in unforeseen consequences. As a result, strobilurins removal from the environment has received a lot of attention. Many remediation methods have been established to remove pesticide residues from soil and water ecosystems, including photodecomposition, ozonation, adsorption, incineration, and biodegradation. Bioremediation is a cost-effective and environmentally friendly technique for removing pesticide residues compared to traditional procedures (Feng et al., 2020). Pesticide residues are readily used as carbon and nitrogen sources by a number of microorganisms and microbial communities that breakdown strobilurin. The decomposition mechanisms of strobilurins and the destinations of various metabolites have been established. The most important method for removing strobilurin is microbial remediation (Baćmaga et al., 2015; Chen et al., 2018; Feng et al., 2020). Bacteria and fungi are considered among the strobilurin-degrading microbes, but bacteria play the most important role. Long-standing use of strobilurins has an impact on microbial numbers and biodiversity in environments. *Bacillus, Aphanoascus, Cupriavidus, Rhodanobacter, Arthrobacter, Stenotrophomonas, Klebsiella*, and *Pseudomonas* are among the strobilurin-degrading microbes that have been isolated (Chen et al., 2018).

Strobilurins may directly influence fungal biomass due to their unique mechanism of action by reducing mitochondrial respiration, perhaps leading to a change in soil activity from fungal to bacterial dominance (Baćmaga et al., 2015). The degradation of pyraclostrobin and the triazole fungicide epoxiconazole was demonstrated by a *Klebsiella* strain isolated from the soil (Lopes et al., 2010). Despite the fact that strobilurins have many reaction sites on their molecular structures, the mechanisms underlying the degradability of strobilurins are identical (Chen et al., 2018). Another example indicated that the ability of two freshwater microalgae, *Scenedesmus obliquus* and *Scenedesmus quadricauda*, to remove two fungicides (dimethomorph and pyrimethanil) and one herbicide (isoproturon) from their medium was investigated (Dosnon-Olette et al., 2010).

8 Microbial nanotechnology for bioremediation of fungicides

Nanoparticles are currently widely used in a wide range of fields, including health, cosmetics, agriculture, and food science (Bahrulolum et al., 2021; Kingsley et al., 2013). The use of microorganisms and plants to synthesize metal nanoparticles (MtNPs) has recently been demonstrated as a cost-effective and environmentally benign method of utilizing microorganisms as nanofactories (Bahrulolum et al., 2021; Singh et al., 2016b). The challenges that the global community is facing, particularly in terms of climate change and population growth, demonstrate that nanotechnology has the potential to improve agricultural product quality and boost production and food security while reducing the harmful impacts of agrochemicals on human health and the environment (Bahrulolum et al., 2021). Nanotechnology can enhance agricultural activities like soil characteristics and crop quality by using nanoparticle-based fertilizers or stimulating plant growth. Moreover, by utilizing nanoparticle-based carriers and compounds, the use of fertilizers and pesticides is decreased without affecting productivity (Duhan et al., 2017). Nanotechnology can also reduce waste by developing more efficient products. The goal of green nanotechnology, which uses biological mechanisms to synthesize nanomaterials, is to reduce the amount of hazardous compounds produced.

8.1 Green synthesis of metal nanoparticles by microorganisms

MtNP has been synthesized using a variety of methods, including biological, chemical, and physical approaches. Chemical and physical techniques require the use of expensive equipment, large heat production, and high energy consumption; additionally, low production yields have been recorded (Gahlawat and Choudhury, 2019; Soni et al., 2018). The use of toxic substances, which create a number of environmental problems, is the principal disadvantage of these procedures (Pal et al., 2019). This has enabled the development of an environmentally friendly MtNP synthesis method, with the current focus being on green MtNP synthesis via biological pathways such as microbial enzymes, plants, and microorganisms (Roychoudhury, 2020). As a consequence, green synthesis methods are more beneficial than traditional physical and chemical procedures since they are simple, cost-effective, and free of dangerous and ecologically unfriendly ingredients, and they have advanced greatly in recent years (Pal et al., 2019).

Microorganisms are in charge of green synthesis and have carved out a niche among the different biological sources for MtNP green synthesis because of their rapid growth rate, simplicity of cultivation, and capability to grow under a broad range of temperatures, pH, and pressure conditions (Ali et al., 2020). When compared to traditional physicochemical methods, green MtNP synthesis using microorganisms offers a number of advantages, including being a quick, cost-effective, clean, nontoxic, and environmentally friendly approach for producing MtNPs of diverse sizes, shapes, compositions, and physicochemical qualities (Ovais et al., 2018; Shah et al., 2015).

Several microorganisms could be used as biofactories for the environmentally friendly and low-cost synthesis of different MtNPs, including metals like nickel, palladium, titanium, zinc, copper, gold, and silver (Bahrulolum et al., 2021). This can be accomplished by creating MtNPs with a certain shape, size, structure, and dispersion (Bahrulolum et al., 2021; Gahlawat and Choudhury, 2019; Kato and Suzuki, 2020). Capturing particular metal ions from the ambient and reducing them to an elemental state utilizing an enzymatic reduction process can be used to biosynthesize MtNPs in microorganisms (Bahrulolum et al., 2021). Since MtNPs are created by metabolic activities and cellular enzymes that would not be present in all microorganisms, they are not formed by all of them (Bahrulolum et al., 2021; Mohd Yusof et al., 2019). Also, the synthesis of MtNPs is reliant on the ability of microorganisms to tolerate heavy metals.

High metal stress can influence a variety of microbial functions, and some microorganisms have the ability to reduce metal ions under stress conditions. (Mohd Yusof et al., 2019). For the reason that metals are absorbed and chelated by intracellular and extracellular proteins, microorganisms that live in metal-rich environments are often exceedingly resistant to them. As a consequence, this strategy promotes spontaneous biomineralization and might be an effective way for MtNP production (Mohd Yusof et al., 2019). The internal and extracellular pathways of MtNP production are shown schematically in Fig. 4. (A) Intracellular biosynthesis in

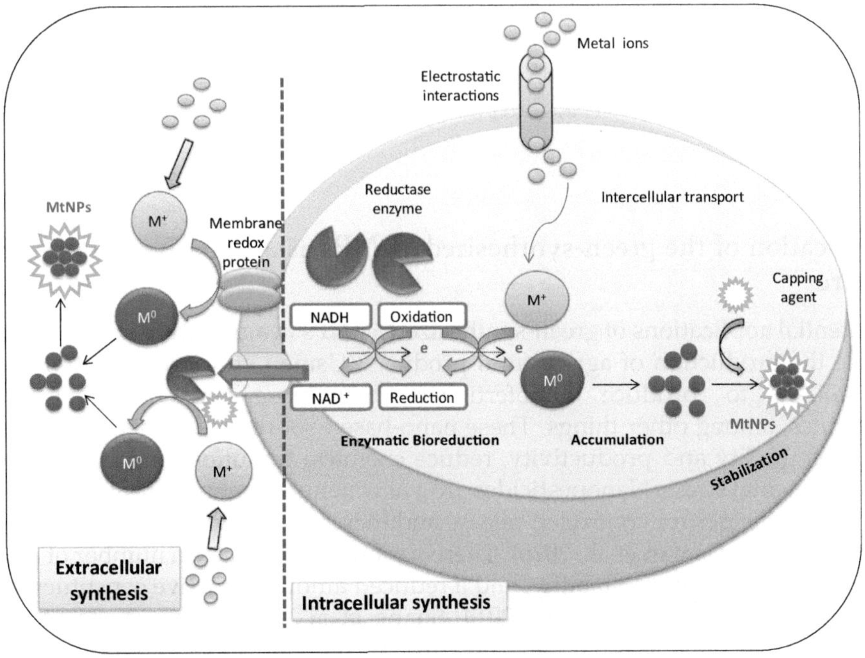

FIG. 4 MtNP biosynthesis internal and extracellular pathways.

microorganisms requires unique transport mechanisms in which the negative charge of the cell wall plays a vital role: electrostatic interactions are used to implant positively charged metal ions into negatively charged cell walls (Tiquia-Arashiro and Rodrigues, 2016). The ions are then reduced by metabolic processes mediated by enzymes such as nitrate reductase in the microorganism's cells, resulting in MtNPs. The MtNPs can then be transferred through the cell wall after aggregating in the periplasmic region (Hulkoti and Taranath, 2014; Tiquia-Arashiro and Rodrigues, 2016). (B) The extracellular biosynthesis of MtNPs also includes a nitrate reductase-mediated synthesis, in which MtNPs are generated by reductase enzymes found in the cell wall or released from the cell into the growth medium. Then the nitrate reductase reduces metal ions to metallic forms during this process (Hulkoti and Taranath, 2014; Mohd Yusof et al., 2019).

The occurrence of several components in microorganisms, such as enzymes, proteins, and other living molecules, also plays a significant role in the reduction of MtNPs (Mohd Yusof et al., 2019). According to some studies, MtNP synthesis is carried out by NADH-dependent enzymes. The reduction mechanisms appear to start with NADH-dependent reductases transporting an electron from NADH as the electron carrier (Jain et al., 2011). Furthermore, proteins produced by microorganisms can serve as a stabilizing agent, maintaining colloidal stability and avoiding MtNP aggregation (Mohd Yusof et al., 2019). Microorganisms are essential nanofactories that can collect and remove heavy metals because of the existence of a variety of reductase enzymes that have the ability to reduce metal salts to MtNP (Singh et al., 2016b). In new studies, bacteria such as *Pseudomonas deptenis* (Jo et al., 2016), *Visella oriza* (Singh et al., 2016a), *Bacillus methylotrophicus* (Wang et al., 2016), *Bhargavaea indica,* and *Brevibacterium frigoritolerans* have the ability to produce gold (Au) and silver (Ag) nanoparticles. MtNPs have also been found in different microbial genera like *Rhodopseudomonas, Pyrobaculum, Plectonemaboryanum, Shewanella, Sargassum, Desulfovibrio, Trichoderma, Brevibacterium, Rhodococcus, Rhodobacter, Weissella, Corynebacterium, Aeromonas, Escherichia, Enterobacter Klebsiella, Streptomyces, Pseudomonas, Bacillus*, and *Lactobacillus* (Singh et al., 2016b).

8.2 Application of the green-synthesized MtNPs as alternative nanofungicides in agriculture

The potential applications of green-synthesized MtNPs in agriculture have become several to increase the production of agricultural products (Usman et al., 2020). MtNPs are extensively utilized to produce nanofertilizers, nanobiosensors, nanofungicides, and nanopesticides, among other things. These nano-based solutions can help improve agricultural product quality and productivity, reduce chemical pollution, and even protect crops from environmental stress. Nanopesticides are a new nanobiotechnological development that encapsulates pesticides for controlled release and increases pesticide selectivity and stability (Paramo et al., 2020; Usman et al., 2020). These nanopesticides have a number of advantages, including improved efficacy, stability, and a reduced amount of active constituents essential in their formulation (Kookana et al., 2014; Shang et al., 2019). The use of MtNPs in the nanoformulation of pesticides has been shown to be more effective in opposition to agricultural pests, insects, phytopathogens, and other. Fungi are the most common plant diseases, accounting for about 70% of all crop losses (Mittal et al., 2020). Common fungicides are now being used to mitigate the damage, but their widespread use for long-term disease management pollutes the environment and has negative consequences for the ecology. Nanofungicides are an efficient technique for controlling fungal diseases. The most widely used application of MtNPs is in the formulation of nanofungicides. Due to their increased solubility and permeability, lower doses, lesser dose-dependent toxicity, and controlled release, these nanofungicides enable targeted targeting and higher bioavailability (Haq and Ijaz, 2019).

Fungi are good candidates for generating intracellular and extracellular MtNPs among the many microorganisms employed to make nanoparticles. Fungi-derived nanoparticles show higher spreading and stability properties (Bahrulolum et al., 2021). The use of fungi in the synthesis of nanomaterials is appealing due to the presence of large numbers of enzyme systems in these microbial cells, the capacity to work with them in the lab, adaptability, and the economically successful development of fungi on an industrial level, all of which make myconanotechnology an environmentally friendly and cost-effective opportunity (Bahrulolum et al., 2021; Gade et al., 2010). Despite the fact that there are several techniques for manufacturing MtNPs from fungi, little is known about their possible drawbacks and limitations. Filamentous fungi may create MtNPs such as iron oxide, silver, gold, and even bimetallic nanoparticles (Bahrulolum et al., 2021; Molnár et al., 2018).

Some studies indicate that numerous different species of fungi, including *Penicillium, Phaenerochaete chrysosporium, Hormoconis resinae, Trichoderma* sp., *Aspergillus niger, Fusarium oxysporum, Humicola* sp., and *Pestalotiopsis* sp., can be employed in the green synthesis of MtNPs with the desired size, surface charge, and shape, as well as other beneficial properties (Bahrulolum et al., 2021). Because of their great efficiency, ease of operation, and minimal residual toxicity, fungi have been proposed as reducing and stabilizing agents for the production of AgNPs. Although the methods of synthesis are unclear, factors such as fungal culture period, pH, temperature, biomass, and silver salt concentration may all be changed to optimize synthesis. More research into the application of MtNPs as antimicrobials in farming is now possible because of these findings. More research into the application of MtNPs as antimicrobials in agriculture is now possible because of these findings. AgNPs stand out among the other types of MtNPs researched so far because of their broad antimicrobial activities (Gupta et al., 2017; Prabhu and Poulose, 2012). These MtNPs attach to the microorganisms' cell walls and membranes, and they may even penetrate the cell. They cause cellular damage, produce reactive oxygen species (ROS), and change signal transduction processes (Bahrulolum et al., 2021; Dakal et al., 2016). Fig. 5 demonstrates how different MtNPs may protect plants from pests and the general mechanism of action of MtNPs as nanofungicides.

9 Enzymatic biodegradation of fungicides

Microbial enzymes from a variety of tiny species, including bacteria and fungus, have been linked to the biodegradation of deadly natural hazardous chemicals. These molecules are the most effective bioremediation agents, and the technique is a cost-effective and environmentally beneficial biotechnology. Processes in this sector have recently evolved, and an

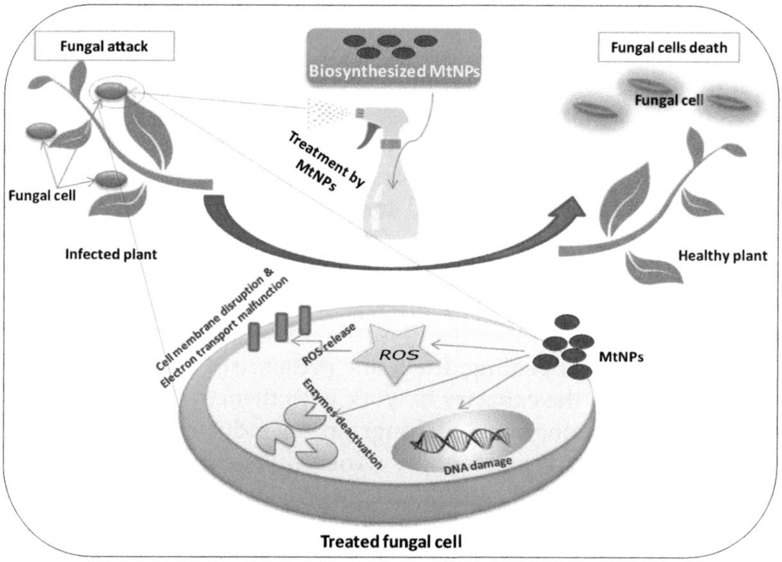

FIG. 5 Application of the green synthesized MtNPs as alternative nanofungicides in agriculture.

expansion in study topics is urgently required. Microbial enzymes have an important role in the reduction of hazardous contaminants as well as the acquisition of novel useful compounds. These microorganisms absorb energy from the bonds and produce harmless molecules as a result of a biological process. The roles of laccases and hydrolases, two microbial enzymes involved in bioremediation, have been widely investigated (Dave and Das, 2021).

In spite of microbial degradation is a cost-effective and environmentally beneficial solution for pollution environmental remediation. However, it has been suggested that microbial enzymatic extracts, instead of microbes, be used to avoid some of the limitations, such as the long adaptation period, the suppressive impact of pollutants on microbe growth and metabolism, or biomass buildup. Furthermore, using microbial-extracted enzymes or raw enzymatic preparations may contribute to the effective and regulated operation (Okino-Delgado et al., 2019; Sosa-Martínez et al., 2022). The kind of enzyme required for bioremediation will be determined by the contaminant. Oxidoreductases, for example, can neutralize contaminants high in free-radicals, whereas hydrolases can aid in the breakdown of organic molecules. In this sense, hydrolases, oxidoreductases, and certain lyases are the most commonly used enzymes in bioremediation. Versatile peroxidase, laccase, lignin peroxidase, and manganese peroxidase are enzymes derived from microbial origin that may breakdown a wide set of advanced harmful and resistant substances, including personal care items, medicines, fungicides, herbicides, pesticides, dyes, and phenol. These so-called green catalysts are environmentally beneficial since they limit the usage of oxidizing chemical compositions, saving energy and time in the biotechnological process (Bhandari et al., 2021).

Despite the fact that enzymes can accelerate the decomposition of the majority of contaminated chemicals, their implementation in bioremediation still faces significant obstacles; the principal problem is the high cost of manufacture and purification. As a result, operations strategy is important in both enzyme synthesis and bioremediation, ensuring that resources are used effectively to address gaps that prevent enzymes from being used on a large scale. There has been a lot of interest in obtaining current information on potential enzymes for biodegradation purposes, technological advancement, enzyme-based enzymatic degradation processes, and the challenges that still need to be solved. Table 4 highlights the most current research on this issue, as well as the chosen process optimization approach, pollutant kinds, enzymes used, and variables considered. Unlike other bioremediation technologies, enzyme-assisted procedures for pollutant removal and biodegradation are more regulated and have fewer variables impacting their efficacy. However, other factors, including those related to the biochemical characteristics and mechanism of action of the enzyme, might potentially influence the process (Sosa-Martínez et al., 2022).

Temperature, pH, enzyme concentration, contaminants, mediators, and reaction rate are all highly researched parameters. Although guaranteeing enzyme stability under process conditions is equally important, the most typical goal response is pollutant degradation/removal percentage. In enzyme-assisted bioremediation procedures, temperature is one of the most important elements to consider. Even within the same enzyme isoform, optimum temperature values varied. Bioremediation with thermostable enzymes is suitable, particularly in industrial environments. Laccase has been shown to degrade quickly at temperatures ranging from 50°C to 60°C (Dauda and Erkurt, 2020; Zhang et al., 2016). Furthermore, records of the enzyme-producing microorganisms' habitat are recommended in order to pick an optimal temperature optimization (Menale et al., 2012). *Trametes orientalis*, for instance, produced

TABLE 4 An overview of the optimization of enzymatic biodegradation.

Pollutants	Enzymes	Optimized variables	Statistical method	References
2- and 4-chlorophenol	Laccase	Time of reaction, quantity of enzyme, rate of flow, pH, temp, and pollutant concentration	OFAT	Menale et al. (2012)
Acetaminophen	Laccase	pH	OFAT	Wang et al. (2018)
Acid Black 10, Direct Black 38, Acid Blue 113	Chloroperoxidase	The quantity of enzyme, the pH, the temperature, and the dye concentration	OFAT	Jin et al. (2018)
Catechol, 2-hydroxybiphenyl (2-HBP), and benzoic acid	Crude extracts (Catechol-2,3-dioxygenase, 2-HBP-3-monooxygenase, benzoate dioxygenase)	Pollutants concentration, enzyme amount, pH, temperature	OFAT	Younis et al. (2020)
Congo Red, Acid Green 27, Cibacron D-Blue, Methylene blue, Maxilon Blue, Acid dye Lanapel Red BM 143-PL	Laccase	Dyes concentrations, enzyme concentration, temperature, HBT-laccase mediator	OFAT	Elshafei et al. (2017)

Continued

TABLE 4 An overview of the optimization of enzymatic biodegradation—Cont'd

Pollutants	Enzymes	Optimized variables	Statistical method	References
Cibacron Blue 3G-A	Laccase	Laccase activity, dye conc., reaction time, mediator concentration	RSM-CCD	Othman et al. (2018)
Dichlorophenol indophenol D5110, Brilliant Green C.I. 42,040, Lanasol Red 6G, Acid Blue C.I. 220, Foron Yellow Brown S2RFLI	Laccase	Dyes concentrations, enzyme concentration, temperature, HBT-laccase mediator	OFAT	Elshafei et al. (2015)
Hydroquinone	Laccase	pH, laccase concentration, reaction time, pollutant concentration	OFAT	Othman et al. (2021)
Lincomycin	Chloroperoxidase	pH, enzyme amount, H_2O_2 concentration, reaction time	RSM	Zhu et al. (2020)
Malachite green	Laccase	pH, temperature	ANN	Rashtbari and Dehghan (2021)
Malachite green	Laccase	Enzyme amount, pH, reaction time, dye concentration	RSM-CCD	Shanmugam et al. (2017)
Polyaromatic hydrocarbons (PAHs), dyes	MnP, laccase	Metal ions concentration, organic solvents concentration	OFAT	Zhang et al. (2016)
Reactive black 5	Laccase	pH, temperature, HBT-laccase mediator	OFAT	Othman et al. (2016)
Reactive blue 19	Laccase	Temperature, reaction time, pH, dye concentration	OFAT	Dauda and Erkurt (2020)
Sulfamethoxazole (SMX)	Peroxidase soybean	SMX concentration, enzyme amount, pH, H_2O_2 concentration,	OFAT	Al-Maqdi et al. (2018)
Synthetic dyes	MnP, LiP	H_2O_2 concentration, temperature, reaction time	RSM-CCD	Sosa-Martínez et al. (2020)
Textile dyes	MnP	H_2O_2 concentration, pH, temperature	OFAT	Rekik et al. (2019)

OFAT, *one factor at a time;* RSM, *response surface methodology;* CCD, *central composite design.*

laccase with a maximum activity of 80°C; this fungus was collected from a very warm temperature zone (Zheng et al., 2017). Nonetheless, laccase loses catalytic capability at temperatures exceeding 60°C, compromising stability and structure (Menale et al., 2012). The ideal degradation temperature is also affected by the kind of pollutant. Using *Phanerochaete chrysosporium* crude extract revealed that different dyes degradation had varied optimum decomposition temperatures (Sosa-Martínez et al., 2020). Peroxidases, in general, have wider temperature limits for optimal decomposition. At an ideal temperature of 70°C, Manganese Peroxidase (MnP) degraded polycyclic aromatic hydrocarbons and dyes (Zhang et al., 2016). *Caldariomyces fumago* chloroperoxidase was utilized to breakdown antibiotics, and it was shown to work best at 60°C (Zhu et al., 2020). Furthermore, if the temperature has a strong effect on the enzymes, it results in decomposition and reduces the effectiveness of pollutant destruction. The stability of enzymes can be increased by the inclusion of redox mediators or cofactors, which, in addition to permitting oxidation-reduction processes, may also offer structural stability to the enzyme. The optimal temperature for MnP degradation of textile dyes in the absence of $MnSO_4$ was 40°C. However with $MnSO_4$ incorporation, the highest breakdown was obtained at 50°C (Rekik et al., 2019).

In terms of pH effects, mainly enzyme-assisted bioremediation activities take place in moderately neutral or acidic settings (Younis et al., 2020). Peroxidases are sustainable in pH values ranging from pH 3.0 to 8.0; although ideal values for MnP decomposition have been shown to be in the pH 4.0–5.0 range (Rekik et al., 2019) and pH 4.0 for chloroperoxidase (Jin et al., 2018). Laccase degrades a wide spectrum of pollutants best at pH levels ranging from pH 4.0 to 5.0 (Dauda and Erkurt, 2020). A research on the adjustment of pH settings indicated the significance of this factor for laccase owing to its structure with four Cu ions in an active center; the acidic pH permits sufficient electron transport across Cu ions (Wang et al., 2018). The appropriate pH value adds to the active site's maintenance in general. Using response surface techniques, it was also shown that there were significant interrelations between pH and enzyme concentration on malachite green degradation by laccase (Shanmugam et al., 2017). Acidic conditions, according to multiple publications, were useful in preserving the catalytic site's conformational stability and therefore boosting bioremediation efficacy. Nevertheless, the kind of pollutant had no effect on the ideal pH level in the *Corynebacterium variabilis* extract, with benzoic acid, catechol, and 2-hydroxybiphenyl (2-HBP), all degrading at comparable rates (Younis et al., 2020). The foregoing suggests that, regardless of the target contaminant, pH is a critical factor that has a direct impact on enzyme effectiveness.

Several scientists have shown that adding redox mediators to enzymatic reactions increases their activity in bioremediation (Othman et al., 2016). It was stated that 1-hydroxybenzotriazole (HBT), a redox-mediator, was utilized in the full decomposition of sulfamethoxazole (Al-Maqdi et al., 2018). Because the concentration of H_2O_2 is crucial for peroxidase enzymes, optimizing it is critical for efficient remediation. Limited H_2O_2 levels result in low enzymatic performance, whereas excessive quantities can induce the oxidation of heme and amino acid functional groups in the enzyme, rendering it inactive. The optimal H_2O_2 concentration level for decolorization by LiP and MnP ranged from 0.01 to 2.0 mM (Rekik et al., 2019; Sosa-Martínez et al., 2020).

Definitely, the concentration of the enzyme and the pollutant are significant elements to manage in terms of operational costs that become much more critical when a pure enzyme is used. Raising the concentration of enzymes may boost contaminant biodegradation;

nevertheless, determining the lowest activity of enzymes to accomplish the required pollutants degradation is critical (Othman et al., 2021; Zhu et al., 2020). For malachite green degradability, a remarkably low concentration of the enzyme was shown to be sufficient to produce a greater decomposition percentage (>90%) (Shanmugam et al., 2017). Nevertheless, dye decomposition at moderate ranges was becoming >90% (Constant coefficient) in this same response optimization investigation. Furthermore, the examined parameters (factor levels) are critical to drawing confident judgments.

Based on the information presented here, one may deduce that the main parameters often examined were reaction time, pH, temperature, and pollutant and enzyme concentration. The fact that fewer factors influence the operation may clarify why most research favors the one-factor-at-a-time technique for analyzing their influence (Table 4). Even though it wasn't the greatest preferred approach (due to possible inter-correlations), the one factor at a time (OFAT) styling, as mentioned previously, allows for a more detailed scan of the impact of only one criterion (more investigated concentrations) while other factors are held constant, especially if those factors necessitate a restricted operating level because of the enzyme's consistency (Sosa-Martínez et al., 2022).

Obviously, understanding the mechanism of action and features of the enzyme is helpful in selecting the components and their ratios for process-optimizing investigations. One plausible strategy would be to test the enzymes under conditions that impact their performance regardless of the contamination, such as temperature and pH. Once the processing parameters that assure enzyme durability have been determined, it would be beneficial to deal with different enzyme/pollutant dosages and define operation kinetics to enhance the effectiveness of both degradation yields and process economics.

10 Conclusion

There are several chemicals in the environment, including herbicides and fungicides. Some of them clearly have the capacity to be hazardous to development and reproduction. Despite their widespread usage, little consideration has been devoted to their impact on animal and human health. Morphological, functional, and physiological changes take place quickly in temporal and spatial patterns throughout the dynamic process of mammalian development, which may impact the potential consequences of toxic substances in a variety of ways.

Fungicides differ greatly in their chemical composition and, hence, in their harm to mammals. Fungicides pose the greatest risk to organisms when used as treatments to preserve deposited grains, potatoes, and other crops. The consumption of treated grains has resulted in numerous incidences of animal toxicity. The majority of current toxicity data comes from animal experiments; limited info is accessible for agricultural animals and pets. Fungicides, generally, have medium to low toxicity. However, over 80% of all oncogenic threats from pesticide usage are derived from a few fungicides; just a few pesticide-related fatalities from fungicides have been observed. Certain fungicides have been shown to alter the endocrine system, potentially resulting in reproductive and developmental problems. Numerous fungicides have been deregistered or banned in several countries, including cycloheximide, due to teratogenicity, yet they are still used in other less controlled parts of the globe. Numerous fungicides are being re-evaluated to adapt to shifting regulatory guidelines; potential toxicity, notably mutagenic, carcinogenic, or reproductively toxic; and an inadequate or out-of-date toxicity database.

Bioremediation is a process that uses indigenous or exogenous microbial communities to remove contaminants from the environment. To repair contaminants, microbe-assisted bioremediation is a simple, less expensive, and far more environmentally acceptable method. The development or construction of genetically engineered microorganisms to digest pollution is of primary interest; this is an environmentally acceptable technique to decompose pollutants. It has a long-lasting influence on marine and terrestrial ecosystems, making it a good technique for mitigating environmental toxicants. With our understanding of microbe metabolic profiles, enzyme construction, and breakdown pathways, we will be able to develop novel genetically engineered microorganisms in the future. Greater knowledge of the microbial genetic structure and metabolic responses might aid bioremediation plans. A comprehensive computerized database should be available to aid in the detoxification of contaminants from a specific location. Scientists can look forward to discovering new genetically modified microbial strains or species that have a greater capacity to digest contaminants in native habitats. At a given pH and temperature, microbes identify and destroy pollutants by enzymatic processes such as elimination hydrolysis, isomerization, reduction, and oxidation. As a result, it is possible to infer that microbial-aided biodegradation of hazardous and toxic pollutants is safe, environmentally benign, and capable of restoring the physicochemical capabilities of groundwater and soil. This clearly demonstrates that additional study into bioremediation technologies and microorganisms can aid in the management of land and marine garbage. Appropriate ecological variables, such as temperature, pH, oxygen, and nutrition, can improve microorganism decomposition capability. To achieve major advancements and establish innovative bioremediation methods, extensive efforts in enzyme characterization and microbial consortium enhancement are required.

References

Ahmadnezhad, Z., Vaezihir, A., Schüth, C., Zarrini, G., 2021. Combination of zeolite barrier and bio sparging techniques to enhance efficiency of organic hydrocarbon remediation in a model of shallow groundwater. Chemosphere 273, 128555. https://doi.org/10.1016/j.chemosphere.2020.128555.

Ali, M.A., Ahmed, T., Wu, W., Hossain, A., Hafeez, R., Islam Masum, M.M., Wang, Y., An, Q., Sun, G., Li, B., 2020. Advancements in plant and microbe-based synthesis of metallic nanoparticles and their antimicrobial activity against plant pathogens. Nanomaterials 10, 1146. https://doi.org/10.3390/nano10061146.

Ali, H., Khan, E., Ilahi, I., 2019. Environmental chemistry and ecotoxicology of hazardous heavy metals: environmental persistence, toxicity, and bioaccumulation. J. Chem. 2019, e6730305. https://doi.org/10.1155/2019/6730305.

Al-Maqdi, K.A., Hisaindee, S., Rauf, M.A., Ashraf, S.S., 2018. Detoxification and degradation of sulfamethoxazole by soybean peroxidase and UV+H_2O_2 remediation approaches. Chem. Eng. J. 352, 450–458. https://doi.org/10.1016/j.cej.2018.07.036.

Aranciaga, N., Durruty, I., González, J.F., Wolski, E.A., 2012. Aerobic biotransformation of 2, 4, 6–trichlorophenol by *Penicillium chrysogenum* in aqueous batch culture: degradation and residual phytotoxicity. Water SA 38, 683–688. https://doi.org/10.4314/wsa.v38i5.5.

Azubuike, C.C., Chikere, C.B., Okpokwasili, G.C., 2016. Bioremediation techniques—classification based on site of application: principles, advantages, limitations and prospects. World J. Microbiol. Biotechnol. 32, 180. https://doi.org/10.1007/s11274-016-2137-x.

Baćmaga, M., Kucharski, J., Wyszkowska, J., 2015. Microbial and enzymatic activity of soil contaminated with azoxystrobin. Environ. Monit. Assess. 187, 615. https://doi.org/10.1007/s10661-015-4827-5.

Bahrulolum, H., Nooraei, S., Javanshir, N., Tarrahimofrad, H., Mirbagheri, V.S., Easton, A.J., Ahmadian, G., 2021. Green synthesis of metal nanoparticles using microorganisms and their application in the agrifood sector. J. Nanobiotechnol. 19, 86. https://doi.org/10.1186/s12951-021-00834-3.

Baird, T.D., DeLorenzo, M.E., 2010. Descriptive and mechanistic toxicity of conazole fungicides using the model test alga *Dunaliella tertiolecta* (*Chlorophyceae*). Environ. Toxicol. 25, 213–220. https://doi.org/10.1002/tox.20493.

Ballantyne, B., 2004. Toxicology of fungicides. In: Marrs, T.C., Ballantyne, B. (Eds.), Pesticide Toxicology and International Regulation. Wiley.com, pp. 194–303.

Bhandari, S., Poudel, D.K., Marahatha, R., Dawadi, S., Khadayat, K., Phuyal, S., Shrestha, S., Gaire, S., Basnet, K., Khadka, U., Parajuli, N., 2021. Microbial enzymes used in bioremediation. J. Chem. 2021, 1–17. https://doi.org/10.1155/2021/8849512.

Bhatt, P., Bhatt, K., Sharma, A., Zhang, W., Mishra, S., Chen, S., 2021. Biotechnological basis of microbial consortia for the removal of pesticides from the environment. Crit. Rev. Biotechnol. 41, 317–338. https://doi.org/10.1080/07388551.2020.1853032.

Bhatt, P., Rene, E.R., Kumar, A.J., Zhang, W., Chen, S., 2020. Binding interaction of allethrin with esterase: bioremediation potential and mechanism. Bioresour. Technol. 315, 123845. https://doi.org/10.1016/j.biortech.2020.123845.

Bhatt, P., Shrivastav, A., 2022. Chapter 17. Removal of pesticides from water and waste water by microbes. In: Shah, M.P., Rodriguez-Couto, S., Kapoor, R.T. (Eds.), Development in Wastewater Treatment Research and Processes. Elsevier, pp. 371–399, https://doi.org/10.1016/B978-0-323-85657-7.00016-X.

Błaszczak-Świątkiewicz, K., Sikora, J., Szymański, J., Danilewicz, M., Mikiciuk-Olasik, E., 2016. Biological evaluation of the toxicity and the cell cycle interruption by some benzimidazole derivatives. Tumour Biol. 37, 11135–11145. https://doi.org/10.1007/s13277-016-4828-1.

Bwapwa, J.K., Jaiyeola, A.T., Chetty, R., 2017. Bioremediation of acid mine drainage using algae strains: a review. S. Afr. J. Chem. Eng. 24, 62–70. https://doi.org/10.1016/j.sajce.2017.06.005.

Chen, X., He, S., Liang, Z., Li, Q.X., Yan, H., Hu, J., Liu, X., 2018. Biodegradation of pyraclostrobin by two microbial communities from Hawaiian soils and metabolic mechanism. J. Hazard. Mater. 354, 225–230. https://doi.org/10.1016/j.jhazmat.2018.04.067.

Chukwuma, O.B., Rafatullah, M., Tajarudin, H.A., Ismail, N., 2020. Lignocellulolytic enzymes in biotechnological and industrial processes: a review. Sustainability 12, 7282. https://doi.org/10.3390/su12187282.

Costa, L.G., 1997. Basic toxicology of pesticides. Occup. Med. 12, 251–268.

Costa, L., 2008. Toxic effects of pesticides. In: Klaassen, C. (Ed.), Casarett & Doull's Toxicology: The Basic Science of Poisons. McGraw Hill Medical, New York, pp. 883–933.

Dakal, T.C., Kumar, A., Majumdar, R.S., Yadav, V., 2016. Mechanistic basis of antimicrobial actions of silver nanoparticles. Front. Microbiol. 7, 1831. https://doi.org/10.3389/fmicb.2016.01831.

Das, N., Chandran, P., 2010. Microbial degradation of petroleum hydrocarbon contaminants: an overview. Biotechnol. Res. Int. 2011, e941810. https://doi.org/10.4061/2011/941810.

Dauda, M.Y., Erkurt, E.A., 2020. Investigation of reactive Blue 19 biodegradation and byproducts toxicity assessment using crude laccase extract from *Trametes versicolor*. J. Hazard. Mater. 393, 121555. https://doi.org/10.1016/j.jhazmat.2019.121555.

Dave, S., Das, J., 2021. Chapter 13. Role of microbial enzymes for biodegradation and bioremediation of environmental pollutants: challenges and future prospects. In: Saxena, G., Kumar, V., Shah, M.P. (Eds.), Bioremediation for Environmental Sustainability. Elsevier, pp. 325–346, https://doi.org/10.1016/B978-0-12-820524-2.00013-4.

Degtyarenko, K.N., 1995. Structural domains of P450-containing monooxygenase systems. Protein Eng. Des. Sel. 8, 737–747. https://doi.org/10.1093/protein/8.8.737.

Dosnon-Olette, R., Trotel-Aziz, P., Couderchet, M., Eullaffroy, P., 2010. Fungicides and herbicide removal in Scenedesmus cell suspensions. Chemosphere 79, 117–123. https://doi.org/10.1016/j.chemosphere.2010.02.005.

Duhan, J.S., Kumar, R., Kumar, N., Kaur, P., Nehra, K., Duhan, S., 2017. Nanotechnology: the new perspective in precision agriculture. Biotechnol. Rep. 15, 11–23. https://doi.org/10.1016/j.btre.2017.03.002.

Elshafei, A., Elsayed, M., Hassan, M., Haroun, B., Othman, A., Farrag, A., 2017. Biodecolorization of six synthetic dyes by *Pleurotus ostreatus* ARC280 laccase in presence and absence of hydroxybenzotriazole (HBT). Annu. Res. Rev. Biol. 15, 1–16. https://doi.org/10.9734/ARRB/2017/35644.

Elshafei, A., Othman, A., Hassan, M., Haroun, B., Elsayed, M., Farrag, A., 2015. Catalyzed mediator-based decolorization of five synthetic dyes by *Pleurotus ostreatus* ARC280 laccase. Br. Biotechnol. J. 9, 1–15. https://doi.org/10.9734/BBJ/2015/19505.

Feng, Y., Huang, Y., Zhan, H., Bhatt, P., Chen, S., 2020. An overview of strobilurin fungicide degradation: current status and future perspective. Front. Microbiol. 11, 389. https://doi.org/10.3389/fmicb.2020.00389.

Ferronato, N., Torretta, V., 2019. Waste mismanagement in developing countries: a review of global issues. Int. J. Environ. Res. Public Health 16, 1060. https://doi.org/10.3390/ijerph16061060.

Gade, A., Ingle, A., Whiteley, C., Rai, M., 2010. Mycogenic metal nanoparticles: progress and applications. Biotechnol. Lett. 32, 593–600. https://doi.org/10.1007/s10529-009-0197-9.

Gahlawat, G., Choudhury, A.R., 2019. A review on the biosynthesis of metal and metal salt nanoparticles by microbes. RSC Adv. 9, 12944–12967. https://doi.org/10.1039/C8RA10483B.

Gammon, D.W., Moore, T.B., O'Malley, M.A., 2010. Chapter 88. A toxicological assessment of sulfur as a pesticide. In: Krieger, R. (Ed.), Hayes' Handbook of Pesticide Toxicology, third ed. Academic Press, New York, pp. 1889–1901, https://doi.org/10.1016/B978-0-12-374367-1.00088-4.

Garcia-Rubio, R., de Oliveira, H.C., Rivera, J., Trevijano-Contador, N., 2020. The fungal cell wall: *Candida, Cryptococcus,* and *Aspergillus* species. Front. Microbiol. 10, 2993. https://doi.org/10.3389/fmicb.2019.02993.

Gikas, G.D., Parlakidis, P., Mavropoulos, T., Vryzas, Z., 2022. Particularities of fungicides and factors affecting their fate and removal efficacy: a review. Sustainability 14, 4056. https://doi.org/10.3390/su14074056.

Gikas, G.D., Pérez-Villanueva, M., Tsioras, M., Alexoudis, C., Pérez-Rojas, G., Masís-Mora, M., Lizano-Fallas, V., Rodríguez-Rodríguez, C.E., Vryzas, Z., Tsihrintzis, V.A., 2018. Low-cost approaches for the removal of terbuthylazine from agricultural wastewater: constructed wetlands and biopurification system. Chem. Eng. J. 335, 647–656. https://doi.org/10.1016/j.cej.2017.11.031.

Giri, B.S., Geed, S., Vikrant, K., Lee, S.S., Kim, K.-H., Kailasa, S.K., Vithanage, M., Chaturvedi, P., Rai, B.N., Singh, R.S., 2021. Progress in bioremediation of pesticide residues in the environment. Environ. Eng. Res. 26. https://doi.org/10.4491/eer.2020.446.

Giri, K., Rai, J.P.N., Pandey, S., Mishra, G., Kumar, R., Suyal, D.C., 2017a. Performance evaluation of isoproturon-degrading indigenous bacterial isolates in soil microcosm. Chem. Ecol. 33, 817–825. https://doi.org/10.1080/02757540.2017.1393535.

Giri, K., Suyal, D.C., Mishra, G., Pandey, S., Kumar, R., Meena, D.K., Rai, J.P.N., 2017b. Biodegradation of isoproturon by *Bacillus pumilus* K1 isolated from foothill agroecosystem of North West Himalaya. Proc. Natl. Acad. Sci. India Sect. B Biol. Sci. 87, 839–848. https://doi.org/10.1007/s40011-015-0667-x.

Gordon, E.B., 2010. Captan and folpet. In: Krieger, R. (Ed.), Hayes' Handbook of Pesticide Toxicology. Elsevier, New York, pp. 1915–1949, https://doi.org/10.1016/B978-0-12-374367-1.00090-2.

Gouma, S., Fragoeiro, S., Bastos, A.C., Magan, N., 2014. 13. Bacterial and fungal bioremediation strategies. In: Das, S. (Ed.), Microbial Biodegradation and Bioremediation. Elsevier, Oxford, pp. 301–323, https://doi.org/10.1016/B978-0-12-800021-2.00013-3.

Gupta, P., 2006. WHO/FAO guidelines for cholinesterase-inhibiting pesticide residues in food. In: Toxicology of Organophosphate & Carbamate Compounds. Elsevier Inc, pp. 643–654.

Gupta, P.K., 2011. Herbicides and fungicides. In: Reproductive and Developmental Toxicology. Elsevier, pp. 503–521, https://doi.org/10.1016/B978-0-12-382032-7.10039-6.

Gupta, P.K., Aggarwal, M., 2012. Toxicity of fungicides. In: Veterinary Toxicology. Elsevier, pp. 653–670, https://doi.org/10.1016/B978-0-12-385926-6.00066-1.

Gupta, R., Kumar, V., Gundampati, R., Malviya, M., Hasan, S.H., Jagannadham, M., 2017. Biosynthesis of silver nanoparticles from the novel strain of *streptomyces Sp.* BHUMBU-80 with highly efficient electroanalytical detection of hydrogen peroxide and antibacterial activity. J Environ. Chem. Eng. 5, 5624–5635. https://doi.org/10.1016/J.JECE.2017.09.029.

Hakeem, K.R., Bhat, R.A., Qadri, H. (Eds.), 2020. Bioremediation and Biotechnology: Sustainable Approaches to Pollution Degradation. Springer International Publishing, Cham, Switzerland, https://doi.org/10.1007/978-3-030-35691-0.

Haq, I.U., Ijaz, S., 2019. Use of metallic nanoparticles and nanoformulations as nanofungicides for sustainable disease management in plants. In: Nanobiotechnology in Bioformulations. Springer International Publishing, Cham, Switzerland, https://doi.org/10.1007/978-3-030-17061-5_12.

Hulkoti, N.I., Taranath, T.C., 2014. Biosynthesis of nanoparticles using microbes—a review. Colloids Surf. B: Biointerfaces 121, 474–483. https://doi.org/10.1016/j.colsurfb.2014.05.027.

Hurley, P.M., 1998. Mode of carcinogenic action of pesticides inducing thyroid follicular cell tumors in rodents. Environ. Health Perspect. 106, 437–445. https://doi.org/10.1289/ehp.98106437.

Hurt, S., Ollinger, J., Arce, G., Bui, Q., Tobia, A.J., van Ravenswaay, B., 2010. Chapter 78. Dialkyldithiocarbamates (EBDCs). In: Krieger, R. (Ed.), Hayes' Handbook of Pesticide Toxicology, third ed. Academic Press, New York, pp. 1689–1710, https://doi.org/10.1016/B978-0-12-374367-1.00078-1.

Hussaini, S.Z., Shaker, M., Iqbal, M.A., 2013. Isolation of bacterial for degradation of selected pesticides. Adv. Bioresourc. 4 (3), 82–85.

Jacoby, R., Peukert, M., Succurro, A., Koprivova, A., Kopriva, S., 2017. The role of soil microorganisms in plant mineral nutrition—current knowledge and future directions. Front. Plant Sci. 1617. https://doi.org/10.3389/fpls.2017.01617.

Jain, N., Bhargava, A., Majumdar, S., Tarafdar, J.C., Panwar, J., 2011. Extracellular biosynthesis and characterization of silver nanoparticles using *Aspergillus flavus* NJP08: a mechanism perspective. Nanoscale 3, 635–641. https://doi.org/10.1039/c0nr00656d.

Jin, X., Li, S., Long, N., Zhang, R., 2018. Improved biodegradation of synthetic azo dye by anionic cross-linking of chloroperoxidase on ZnO/SiO2 nanocomposite support. Appl. Biochem. Biotechnol. 184, 1009–1023. https://doi.org/10.1007/s12010-017-2607-0.

Jo, J.H., Singh, P., Kim, Y.J., Wang, C., Mathiyalagan, R., Jin, C.-G., Yang, D.C., 2016. *Pseudomonas deceptionensis* DC5-mediated synthesis of extracellular silver nanoparticles. Artif. Cells Nanomed. Biotechnol. 44, 1576–1581. https://doi.org/10.3109/21691401.2015.1068792.

Juwarkar, A.A., Singh, S.K., Mudhoo, A., 2010. A comprehensive overview of elements in bioremediation. Rev. Environ. Sci. Biotechnol. 9, 215–288. https://doi.org/10.1007/s11157-010-9215-6.

Kanissery, R.G., Sims, G.K., 2011. Biostimulation for the enhanced degradation of herbicides in soil. Appl. Environ. Soil Sci. 2011, e843450. https://doi.org/10.1155/2011/843450.

Kao, C.M., Chen, C.Y., Chen, S.C., Chien, H.Y., Chen, Y.L., 2008. Application of in situ biosparging to remediate a petroleum-hydrocarbon spill site: field and microbial evaluation. Chemosphere 70, 1492–1499. https://doi.org/10.1016/j.chemosphere.2007.08.029.

Kapahi, M., Sachdeva, S., 2019. Bioremediation options for heavy metal pollution. J. Health Pollut. 9, 191203. https://doi.org/10.5696/2156-9614-9.24.191203.

Karigar, C.S., Rao, S.S., 2011. Role of microbial enzymes in the bioremediation of pollutants: a review. Enzyme Res. 2011, e805187. https://doi.org/10.4061/2011/805187.

Kato, Y., Suzuki, M., 2020. Synthesis of metal nanoparticles by microorganisms. Crystals 10, 589. https://doi.org/10.3390/cryst10070589.

Khatoon, H., Rai, J.P.N., Jillani, A., 2021. Chapter 7. Role of fungi in bioremediation of contaminated soil. In: Sharma, V.K., Shah, M.P., Parmar, S., Kumar, A. (Eds.), Fungi Bio-Prospects in Sustainable Agriculture, Environment and Nano-Technology. Academic Press, pp. 121–156, https://doi.org/10.1016/B978-0-12-821925-6.00007-1.

Kingsley, J.D., Ranjan, S., Dasgupta, N., Saha, P., 2013. Nanotechnology for tissue engineering: need, techniques and applications. J. Pharm. Res. 7, 200–204. https://doi.org/10.1016/j.jopr.2013.02.021.

Kookana, R.S., Boxall, A.B.A., Reeves, P.T., Ashauer, R., Beulke, S., Chaudhry, Q., Cornelis, G., Fernandes, T.F., Gan, J., Kah, M., Lynch, I., Ranville, J., Sinclair, C., Spurgeon, D., Tiede, K., Van den Brink, P.J., 2014. Nanopesticides: guiding principles for regulatory evaluation of environmental risks. J. Agric. Food Chem. 62, 4227–4240. https://doi.org/10.1021/jf500232f.

Kurt, Z., Spain, J.C., 2013. Biodegradation of chlorobenzene, 1,2-dichlorobenzene, and 1,4-dichlorobenzene in the vadose zone. Environ. Sci. Technol. 47, 6846–6854. https://doi.org/10.1021/es3049465.

Lah, L., Podobnik, B., Novak, M., Korošec, B., Berne, S., Vogelsang, M., Kraševec, N., Zupanec, N., Stojan, J., Bohlmann, J., Komel, R., 2011. The versatility of the fungal cytochrome P450 monooxygenase system is instrumental in xenobiotic detoxification. Mol. Microbiol. 81, 1374–1389. https://doi.org/10.1111/j.1365-2958.2011.07772.x.

Li, X., Peng, W., Jia, Y., Lu, L., Fan, W., 2016. Removal of cadmium and zinc from contaminated wastewater using *Rhodobacter sphaeroides*. Water Sci. Technol. 75, 2489–2498. https://doi.org/10.2166/wst.2016.608.

Liu, L., Bilal, M., Duan, X., Iqbal, H.M.N., 2019. Mitigation of environmental pollution by genetically engineered bacteria—current challenges and future perspectives. Sci. Total Environ. 667, 444–454. https://doi.org/10.1016/j.scitotenv.2019.02.390.

Lopes, F.M., Batista, K.A., Batista, G.L.A., Mitidieri, S., Bataus, L.A.M., Fernandes, K.F., 2010. Biodegradation of epoxyconazole and piraclostrobin fungicides by *Klebsiella sp.* from soil. World J. Microbiol. Biotechnol. 26, 1155–1161. https://doi.org/10.1007/s11274-009-0283-0.

Lorgue, G., Lechenet, J., Rivière, A., 1996. In: Chapman, M.J. (Ed.), Clinical Veterinary Toxicology. Blackwell Science, Oxford. English version by.

Low, J.C., Scott, P.R., Howie, F., Lewis, M., FitzSimons, J., Spence, J.A., 1996. Sulphur-induced polioencephalomalacia in lambs. Vet. Rec. 138, 327–329. https://doi.org/10.1136/vr.138.14.327.

Machado, M., Janssens, S., Soares, H., Soares, E.V., 2009. Removal of heavy metals using a brewer's yeast strain of *Saccharomyces cerevisiae*: advantages of using dead biomass. J. Appl. Microbiol. 106, 1792–1804. https://doi.org/10.1111/j.1365-2672.2009.04170.x.

Menale, C., Nicolucci, C., Catapane, M., Rossi, S., Bencivenga, U., Mita, D.G., Diano, N., 2012. Optimization of operational conditions for biodegradation of chlorophenols by laccase-polyacrilonitrile beads system. J. Mol. Catal. B Enzym. 78, 38–44. https://doi.org/10.1016/j.molcatb.2012.01.021.

Mittal, D., Kaur, G., Singh, P., Yadav, K., Ali, S.A., 2020. Nanoparticle-based sustainable agriculture and food science: recent advances and future outlook. Front. Nanotechnol. https://doi.org/10.3389/fnano.2020.579954.

Mohamed, A.T., Hussein, A.A.E., Siddig, M.A.E., Osman, A.G., 2011. Degradation of oxyfluorfen herbicide by soil microorganisms biodegradation of herbicides. Biotechnology 10, 274–279. https://doi.org/10.3923/biotech.2011.274.279.

Mohd Yusof, H., Mohamad, R., Zaidan, U.H., Abdul Rahman, N.A., 2019. Microbial synthesis of zinc oxide nanoparticles and their potential application as an antimicrobial agent and a feed supplement in animal industry: a review. J. Anim. Sci. Biotechnol. 10, 57. https://doi.org/10.1186/s40104-019-0368-z.

Molnár, Z., Bódai, V., Szakacs, G., Erdélyi, B., Fogarassy, Z., Sáfrán, G., Varga, T., Kónya, Z., Tóth-Szeles, E., Szűcs, R., Lagzi, I., 2018. Green synthesis of gold nanoparticles by thermophilic filamentous fungi. Sci. Rep. 8, 3943. https://doi.org/10.1038/s41598-018-22112-3.

Nagel, D.A., Hill, E.J., O'Neil, J., Mireur, A., Coleman, M.D., 2014. The effects of the fungicides fenhexamid and myclobutanil on SH-SY5Y and U-251 MG human cell lines. Environ. Toxicol. Pharmacol. 38, 968–976. https://doi.org/10.1016/j.etap.2014.09.005.

Ndeddy Aka, R.J., Babalola, O.O., 2016. Effect of bacterial inoculation of strains of *Pseudomonas aeruginosa*, *Alcaligenes feacalis* and *Bacillus subtilis* on germination, growth and heavy metal (Cd, Cr, and Ni) uptake of *Brassica juncea*. Int. J. Phytorem. 18, 200–209. https://doi.org/10.1080/15226514.2015.1073671.

Nguyen, H.T., Kim, Y., Choi, J.-W., Jeong, S., Cho, K., 2021. Soil microbial communities-mediated bioattenuation in simulated aquifer storage and recovery (ASR) condition: long-term study. Environ. Res. 197, 111069. https://doi.org/10.1016/j.envres.2021.111069.

Niti, C., Sunita, S., Kamlesh, K., Rakesh, K., 2013. Bioremediation: an emerging technology for remediation of pesticides. Res. J. Chem. Environ. 17, 88–105.

Nzila, A., Razzak, S.A., Zhu, J., 2016. Bioaugmentation: an emerging strategy of industrial wastewater treatment for reuse and discharge. Int. J. Environ. Res. Public Health 13, 846. https://doi.org/10.3390/ijerph13090846.

Ojuederie, O.B., Babalola, O.O., 2017. Microbial and plant-assisted bioremediation of heavy metal polluted environments: a review. Int. J. Environ. Res. Public Health 14, 1504. https://doi.org/10.3390/ijerph14121504.

Okino-Delgado, C.H., Zanutto-Elgui, M.R., do Prado, D.Z., Pereira, M.S., Fleuri, L.F., 2019. Enzymatic bioremediation: current status, challenges of obtaining process, and applications. In: Arora, P.K. (Ed.), Microbial Metabolism of Xenobiotic Compounds, Microorganisms for Sustainability. Springer, Singapore, pp. 79–101, https://doi.org/10.1007/978-981-13-7462-3_4.

Okoduwa, S.I.R., Igiri, B., Udeh, C.B., Edenta, C., Gauje, B., 2017. Tannery effluent treatment by yeast species isolates from watermelon. Toxics 5, E6. https://doi.org/10.3390/toxics5010006.

Oruc, H.H., 2010. Fungicides and their effects on animals. In: Carisse, O. (Ed.), Fungicides. InTech, p. 16.

Osteen, C.D., Fernandez-Cornejo, J., 2013. Economic and policy issues of U.S. agricultural pesticide use trends. Pest Manag. Sci. 69, 1001–1025. https://doi.org/10.1002/ps.3529.

Othman, A., Elshafei, A., Elsayed, M., Hassan, M., 2018. Decolorization of Cibacron Blue 3G-A dye by *Agaricus bisporus* CU13 laccase-mediator system: a statistical study for optimization via response surface methodology. Annu. Res. Rev. Biol. 25, 1–13. https://doi.org/10.9734/ARRB/2018/40772.

Othman, A.M., González-Domínguez, E., Sanromán, Á., Correa-Duarte, M., Moldes, D., 2016. Immobilization of laccase on functionalized multiwalled carbon nanotube membranes and application for dye decolorization. RSC Adv. 6, 114690–114697. https://doi.org/10.1039/C6RA18283F.

Othman, A.M., Mahmoud, M., Abdelraof, M., Abdel Karim, G.S.A., Elsayed, A.M., 2021. Enhancement of laccase production from a newly isolated *Trichoderma harzianum* S7113 using submerged fermentation: optimization of production medium via central composite design and its application for hydroquinone degradation. Int. J. Biol. Macromol. 192, 219–231. https://doi.org/10.1016/j.ijbiomac.2021.09.207.

Ovais, M., Khalil, A.T., Ayaz, M., Ahmad, I., Nethi, S.K., Mukherjee, S., 2018. Biosynthesis of metal nanoparticles via microbial enzymes: a mechanistic approach. Int. J. Mol. Sci. 19, 4100. https://doi.org/10.3390/ijms19124100.

Pal, G., Rai, P., Pandey, A., 2019. Chapter 1. Green synthesis of nanoparticles: a greener approach for a cleaner future. In: Shukla, A.K., Iravani, S. (Eds.), Green Synthesis, Characterization and Applications of Nanoparticles, Micro and Nano Technologies. Elsevier, pp. 1–26, https://doi.org/10.1016/B978-0-08-102579-6.00001-0.

Papaevangelou, V.A., Gikas, G.D., Vryzas, Z., Tsihrintzis, V.A., 2017. Treatment of agricultural equipment rinsing water containing a fungicide in pilot-scale horizontal subsurface flow constructed wetlands. Ecol. Eng. 101, 193–200. https://doi.org/10.1016/j.ecoleng.2017.01.045.

Paramo, L.A., Feregrino-Pérez, A.A., Guevara, R., Mendoza, S., Esquivel, K., 2020. Nanoparticles in agroindustry: applications, toxicity, challenges, and trends. Nanomaterials 10, 1654. https://doi.org/10.3390/nano10091654.

Parsons, P.P., 2010. Chapter 91. Mammalian toxicokinetics and toxicity of chlorothalonil. In: Krieger, R. (Ed.), Hayes' Handbook of Pesticide Toxicology, third ed. Academic Press, New York, pp. 1951–1966, https://doi.org/10.1016/B978-0-12-374367-1.00091-4.

Parween, T., Bhandari, P., Sharma, R., Jan, S., Siddiqui, Z.H., Patanjali, P.K., 2018. Bioremediation: a sustainable tool to prevent pesticide pollution. In: Oves, M., Zain Khan, M., Ismail, M.I., I. (Eds.), Modern Age Environmental Problems and their Remediation. Springer International Publishing, Cham, pp. 215–227, https://doi.org/10.1007/978-3-319-64501-8_12.

Peñaloza-Vazquez, A., Mena, G.L., Herrera-Estrella, L., Bailey, A.M., 1995. Cloning and sequencing of the genes involved in glyphosate utilization by *Pseudomonas pseudomallei*. Appl. Environ. Microbiol. 61, 538–543. https://doi.org/10.1128/aem.61.2.538-543.1995.

Pérez, M., Rueda, O.D., Bangeppagari, M., Johana, J.Z., Ríos, D., Rueda, B.B., Sikandar, I.M., Naga, R.M., 2016. Evaluation of various pesticides-degrading pure bacterial cultures isolated from pesticide-contaminated soils in Ecuador. Afr. J. Biotechnol. 15, 2224–2233. https://doi.org/10.5897/AJB2016.15418.

Phillips, S., 2001. Fungicides and biocides. In: Sullivan, J.B., Krieger, G.R. (Eds.), Clinical Environmental Health and Toxic Exposures. Lippincott Williams & Wilkins, Philadelphia, pp. 1109–1125.

Prabhu, S., Poulose, E.K., 2012. Silver nanoparticles: mechanism of antimicrobial action, synthesis, medical applications, and toxicity effects. Int. Nano Lett. 2, 32. https://doi.org/10.1186/2228-5326-2-32.

Raffa, C.M., Chiampo, F., 2021. Bioremediation of agricultural soils polluted with pesticides: a review. Bioengineering 8, 92. https://doi.org/10.3390/bioengineering8070092.

Rajmohan, K.S., Chandrasekaran, R., Varjani, S., 2020. A review on occurrence of pesticides in environment and current technologies for their remediation and management. Indian J. Microbiol. 60, 125–138. https://doi.org/10.1007/s12088-019-00841-x.

Rashtbari, S., Dehghan, G., 2021. Biodegradation of malachite green by a novel laccase-mimicking multicopper BSA-Cu complex: performance optimization, intermediates identification and artificial neural network modeling. J. Hazard. Mater. 406, 124340. https://doi.org/10.1016/j.jhazmat.2020.124340.

Ravi, R.K., Pathak, B., Fulekar, M.H., 2015. Bioremediation of persistent pesticides in rice field soil environment using surface soil treatment reactor. Int. J. Curr. Microbiol. App. Sci. 4, 359–369.

Rekik, H., Zaraî Jaouadi, N., Bouacem, K., Zenati, B., Kourdali, S., Badis, A., Annane, R., Bouanane-Darenfed, A., Bejar, S., Jaouadi, B., 2019. Physical and enzymatic properties of a new manganese peroxidase from the white-rot fungus *Trametes pubescens* strain i8 for lignin biodegradation and textile-dyes biodecolorization. Int. J. Biol. Macromol. 125, 514–525. https://doi.org/10.1016/j.ijbiomac.2018.12.053.

Ren, N., Wang, A., Gao, L., Xin, L., Lee, D.-J., Su, A., 2008. Bioaugmented hydrogen production from carboxymethyl cellulose and partially delignified corn stalks using isolated cultures. Int. J. Hydrogen Energy 33, 5250–5255. https://doi.org/10.1016/j.ijhydene.2008.05.020.

Roychoudhury, A., 2020. Yeast-mediated green synthesis of nanoparticles for biological applications. Indian J. Pharm. Biol. Res. 8, 26–31. https://doi.org/10.30750/ijpbr.8.3.4.

Sales da Silva, I.G., Gomes de Almeida, F.C., Padilha da Rocha e Silva, N.M., Casazza, A.A., Converti, A., Asfora Sarubbo, L., 2020. Soil bioremediation: overview of technologies and trends. Energies 13, 4664. https://doi.org/10.3390/en13184664.

Samin, G., Pavlova, M., Arif, M.I., Postema, C.P., Damborsky, J., Janssen, D.B., 2014. A *Pseudomonas putida* strain genetically engineered for 1,2,3-trichloropropane bioremediation. Appl. Environ. Microbiol. 80, 5467–5476. https://doi.org/10.1128/AEM.01620-14.

Sánchez, O., 2017. Constructed wetlands revisited: microbial diversity in the -omics era. Microb. Ecol. 73, 722–733. https://doi.org/10.1007/s00248-016-0881-y.

Sandhu, H., Brar, R., 2009. Textbook of Veterinary Toxicology, second ed. Kalyani Publishers, India.

Satapute, P., Kaliwal, B., 2016. Biodegradation of the fungicide propiconazole by *Pseudomonas aeruginosa* PS-4 strain isolated from a paddy soil. Ann. Microbiol. 66, 1355–1365. https://doi.org/10.1007/s13213-016-1222-6.

Shah, M., Fawcett, D., Sharma, S., Tripathy, S.K., Poinern, G.E.J., 2015. Green synthesis of metallic nanoparticles via biological entities. Materials (Basel) 8, 7278–7308. https://doi.org/10.3390/ma8115377.

Shang, Y., Hasan, M.K., Ahammed, G.J., Li, M., Yin, H., Zhou, J., 2019. Applications of nanotechnology in plant growth and crop protection: a review. Molecules 24, E2558. https://doi.org/10.3390/molecules24142558.

Shanmugam, S., Ulaganathan, P., Swaminathan, K., Sadhasivam, S., Wu, Y.-R., 2017. Enhanced biodegradation and detoxification of malachite green by *Trichoderma asperellum* laccase: degradation pathway and product analysis. Int. Biodeterior. Biodegrad. 125, 258–268. https://doi.org/10.1016/j.ibiod.2017.08.001.

Sharma, A., Kumar, V., Shahzad, B., Tanveer, M., Sidhu, G.P.S., Handa, N., Kohli, S.K., Yadav, P., Bali, A.S., Parihar, R.D., Dar, O.I., Singh, K., Jasrotia, S., Bakshi, P., Ramakrishnan, M., Kumar, S., Bhardwaj, R., Thukral, A.K., 2019. Worldwide pesticide usage and its impacts on ecosystem. SN Appl. Sci. 1, 1446. https://doi.org/10.1007/s42452-019-1485-1.

Singh, B., Garg, T., Goyal, A.K., Rath, G., 2016a. Development, optimization, and characterization of polymeric electrospun nanofiber: a new attempt in sublingual delivery of nicorandil for the management of angina pectoris. Artif. Cells Nanomed. Biotechnol. 44, 1498–1507. https://doi.org/10.3109/21691401.2015.1052472.

Singh, P., Kim, Y.-J., Zhang, D., Yang, D.-C., 2016b. Biological synthesis of nanoparticles from plants and microorganisms. Trends Biotechnol. 34, 588–599. https://doi.org/10.1016/j.tibtech.2016.02.006.

Škulcová, L., Chandran, N.N., Bielská, L., 2020. Chiral conazole fungicides–(enantioselective) terrestrial bioaccumulation and aquatic toxicity. Sci. Total Environ. 743, 140821. https://doi.org/10.1016/j.scitotenv.2020.140821.

Soni, M., Mehta, P., Soni, A., Goswami, G.K., 2018. Green nanoparticles: synthesis and applications. IOSR J. Biotechnol. Biochem. 4, 78–83.

Sood, N., Patle, S., Lal, B., 2010. Bioremediation of acidic oily sludge-contaminated soil by the novel yeast strain *Candida digboiensis* TERI ASN6. Environ. Sci. Pollut. Res. 17, 603–610. https://doi.org/10.1007/s11356-009-0239-9.

Sosa-Martínez, J., Balagurusamy, N., Gadi, S.K., Montañez, J., Benavente-Valdés, J.R., Morales-Oyervides, L., 2022. Critical process parameters and their optimization strategies for enhanced bioremediation. In: Suyal, D.C., Soni, R. (Eds.), Bioremediation of Environmental Pollutants: Emerging Trends and Strategies. Springer International Publishing, Cham, Switzerland, pp. 75–110, https://doi.org/10.1007/978-3-030-86169-8.

Sosa-Martínez, J.D., Balagurusamy, N., Montañez, J., Peralta, R.A., de Moreira, R.F.P.M., Bracht, A., Peralta, R.M., Morales-Oyervides, L., 2020. Synthetic dyes biodegradation by fungal ligninolytic enzymes: process optimization, metabolites evaluation and toxicity assessment. J. Hazard. Mater. 400, 123254. https://doi.org/10.1016/j.jhazmat.2020.123254.

Steffan, R.J. (Ed.), 2019. Consequences of Microbial Interactions With Hydrocarbons, Oils, and Lipids: Biodegradation and Bioremediation. Springer International Publishing, Cham, https://doi.org/10.1007/978-3-319-50433-9.

Sun, T., Miao, J., Saleem, M., Zhang, H., Yang, Y., Zhang, Q., 2020. Bacterial compatibility and immobilization with biochar improved tebuconazole degradation, soil microbiome composition and functioning. J. Hazard. Mater. 398, 122941. https://doi.org/10.1016/j.jhazmat.2020.122941.

Sutherland, D.L., Ralph, P.J., 2019. Microalgal bioremediation of emerging contaminants—opportunities and challenges. Water Res. 164, 114921. https://doi.org/10.1016/j.watres.2019.114921.

Tak, Y., Kaur, M., Tilgam, J., Kaur, H., Kumar, R., Gautam, C., 2022. Microbes assisted bioremediation: a green technology to remediate pollutants. In: Suyal, D.C., Soni, R. (Eds.), Bioremediation of Environmental Pollutants: Emerging Trends and Strategies. Springer International Publishing, Cham, Switzerland, pp. 25–52, https://doi.org/10.1007/978-3-030-86169-8.

Terada, M., Mizuhashi, F., Tomita, T., Inoue, H., Murata, K., 1998. Mepanipyrim induces fatty liver in rats but not in mice and dogs. J. Toxicol. Sci. 23, 223–234. https://doi.org/10.2131/jts.23.3_223.

Thind, T.S., Hollomon, D.W., 2018. Thiocarbamate fungicides: reliable tools in resistance management and future outlook. Pest Manag. Sci. 74, 1547–1551. https://doi.org/10.1002/ps.4844.

Tiquia-Arashiro, S., Rodrigues, D., 2016. Nanoparticles Synthesized by Microorganisms. pp. 1–51, https://doi.org/10.1007/978-3-319-45215-9_1.

Tyagi, M., da Fonseca, M.M.R., de Carvalho, C.C.C.R., 2011. Bioaugmentation and biostimulation strategies to improve the effectiveness of bioremediation processes. Biodegradation 22, 231–241. https://doi.org/10.1007/s10532-010-9394-4.

Usman, M., Farooq, M., Wakeel, A., Nawaz, A., Cheema, S.A., Rehman, H.U., Ashraf, I., Sanaullah, M., 2020. Nanotechnology in agriculture: current status, challenges and future opportunities. Sci. Total Environ. 721, 137778. https://doi.org/10.1016/j.scitotenv.2020.137778.

Varjani, S., Kumar, G., Rene, E.R., 2019. Developments in biochar application for pesticide remediation: current knowledge and future research directions. J. Environ. Manag. 232, 505–513. https://doi.org/10.1016/j.jenvman.2018.11.043.

Villaverde, J., Láiz, L., Lara-Moreno, A., González-Pimentel, J.L., Morillo, E., 2019. Bioaugmentation of PAH-contaminated soils with novel specific degrader strains isolated from a contaminated industrial site. Effect of hydroxypropyl-β-cyclodextrin as PAH bioavailability enhancer. Front. Microbiol. https://doi.org/10.3389/fmicb.2019.02588.

Waechter, F., Weber, E., Herner, T., May-Hertl, U., 2010. Cyprodinil: a fungicide of the anilinopyrimidine class. In: Krieger, R. (Ed.), Hayes' Handbook of Pesticide Toxicology, third ed. Elsevier, New York, pp. 1903–1913.

Wang, K., Huang, K., Jiang, G., 2018. Enhanced removal of aqueous acetaminophen by a laccase-catalyzed oxidative coupling reaction under a dual-pH optimization strategy. Sci. Total Environ. 616–617, 1270–1278. https://doi.org/10.1016/j.scitotenv.2017.10.191.

Wang, C., Kim, Y.J., Singh, P., Mathiyalagan, R., Jin, Y., Yang, D.C., 2016. Green synthesis of silver nanoparticles by *Bacillus methylotrophicus*, and their antimicrobial activity. Artif. Cells Nanomed. Biotechnol. 44, 1127–1132. https://doi.org/10.3109/21691401.2015.1011805.

Xu, X., Liu, W., Tian, S., Wang, W., Qi, Q., Jiang, P., Gao, X., Li, F., Li, H., Yu, H., 2018. Petroleum hydrocarbon-degrading bacteria for the remediation of oil pollution under aerobic conditions: a perspective analysis. Front. Microbiol. https://doi.org/10.3389/fmicb.2018.02885.

Younis, S., El-Gendy, N., Nassar, H.N., 2020. Biokinetic aspects for biocatalytic remediation of xenobiotics polluted seawater. J. Appl. Microbiol. 129, 319–334. https://doi.org/10.1111/jam.14626.

Zhang, H., Zhang, S., He, F., Qin, X., Zhang, X., Yang, Y., 2016. Characterization of a manganese peroxidase from white-rot fungus *Trametes sp.*48424 with strong ability of degrading different types of dyes and polycyclic aromatic hydrocarbons. J. Hazard. Mater. 320, 265–277. https://doi.org/10.1016/j.jhazmat.2016.07.065.

Zheng, F., An, Q., Meng, G., Wu, X.-J., Dai, Y.-C., Si, J., Cui, B.-K., 2017. A novel laccase from white rot fungus *Trametes orientalis*: purification, characterization, and application. Int. J. Biol. Macromol. 102, 758–770. https://doi.org/10.1016/j.ijbiomac.2017.04.089.

Zhu, X., Wang, X., Wang, L., Fan, X., Li, X., Jiang, Y., 2020. Biodegradation of lincomycin in wastewater by two-level bio-treatment using chloroperoxidase and activated sludge: degradation route and eco-toxicity evaluation. Environ. Technol. Innov. 20, 101114. https://doi.org/10.1016/j.eti.2020.101114.

Index

Note: Page numbers followed by *f* indicate figures and *t* indicate tables.

D

E

G

N

Q

R